parat

Dictionary of Plastics Technology

English/German

Wörterbuch Kunststofftechnologie

Englisch/Deutsch

herausgegeben von Hans-Dieter Junge

Dr. Hans-Dieter Junge
Cavaillonstr. 78/I
D-6940 Weinheim

Erarbeitet auf der Basis von
Welling, M.S.: Wörterbuch Kunststofftechnologie
Deutsch–Englisch, VCH Verlagsgesellschaft mbH, D-6940 Weinheim
und
Welling, M.S.: German–English Dictionary of Plastics Technologie
Pentech Press Ltd., Plymouth, Devon (England)

Herstellerische Betreuung: Max Denk

CIP-Kurztitelaufnahme der Deutschen Bibliothek:

Junge, Hans-Dieter:
Dictionary of plastics technology: Engl./German
= Wörterbuch Kunststofftechnologie / Hrsg. von
Hans-Dieter Junge. – Weinheim; New York: VCH, 1987.
   ISBN 3-527-26432-9 (Weinheim)
   ISBN 0-89573-525-3 (New York)
   ISSN 0930-6862

© VCH Verlagsgesellschaft mbH, D-6940 Weinheim (Federal Republic of Germany), 1987

Alle Rechte, insbesondere die der Übersetzung in andere Sprachen, vorbehalten. Kein Teil dieses Buches darf ohne schriftliche Genehmigung des Verlages in irgendeiner Form – durch Photokopie, Mikroverfilmung oder irgendein anderes Verfahren – reproduziert oder in eine von Maschinen, insbesondere von Datenverarbeitungsmaschinen, verwendbare Sprache übertragen oder übersetzt werden.
Die Wiedergabe von Warenbezeichnungen, Handelsnamen oder sonstigen Kennzeichen in diesem Buch berechtigt nicht zu der Annahme, daß diese von jedermann frei benutzt werden dürfen. Vielmehr kann es sich auch dann um eingetragene Warenzeichen oder sonstige gesetzlich geschützte Kennzeichen handeln, wenn sie nicht eigens als solche markiert sind.
All rights reserved (including those of translation into other languages). No part of this book may be reproduced in any form – by photoprint, microfilm, or any other means – nor transmitted or translated into a machine language without written permission from the publishers.

Einbandgestaltung: TWI, Herbert J. Weisbrod, D-6943 Birkenau
Texterfassung und EDV-Bearbeitung: Fa. Hellinger, D-6901 Heiligkreuzsteinach
Druck: betz-druck gmbh, D-6100 Darmstadt
Bindung: G. Kränkl, D-6148 Heppenheim
Printed in the Federal Republic of Germany

parat

Dictionary of Plastics Technology
English/German

Wörterbuch Kunststofftechnologie
Englisch/Deutsch

© VCH Verlagsgesellschaft mbH, D-6940 Weinheim (Federal Republic of Germany), 1987

Vertrieb:
VCH Verlagsgesellschaft, Postfach 1260/1280, D-6940 Weinheim (Federal Republic of Germany)
USA und Canada: VCH Publishers, Suite 909, 220 East 23rd Street, New York NY 10010-4606 (USA)

ISSN 0930-6862
ISBN 3-527-26432-9 (VCH Verlagsgesellschaft)
ISBN 0-89573-525-3 (VCH Publisher)

Vorwort

Dieses Wörterbuch wurde auf der Basis des deutsch-englischen parat-Wörterbuches Kunststofftechnologie zusammengestellt. Es enthält mehr als 20000 Einträge aus der Chemie und Physik der Hochpolymeren. Sie bezeichnen Eigenschaften, Verfahren der Prüfung, Herstellung und Umwandlung in Fertigprodukte (z.B. Strangpressen, Spritzgießen, Granulieren, Warmverformen) sowie die wichtigsten Anwendungen der Fertigprodukte.

Hans-Dieter Junge

above average überdurchschnittlich
abrasion Abrieb m
   abrasion characteristics Abriebverhalten n
   abrasion resistance Abrasionswiderstand m, Abrasionsfestigkeit f, Abriebbeständigkeit f, Abriebwiderstand m, Abriebfestigkeit f
   abrasion resistant abrasionsfest, abriebfest
   abrasion test Abriebprüfung f, Verschleißprüfung f
   abrasion tester Abriebprüfmaschine f, Reibtester m
   depth of abrasion Abtragtiefe f
   PTFE improves abrasion resistance PTFE verbessert das Abriebverhalten
abrasive abrasiv, abreibend, abscheuernd
   abrasive belt Schleifband n
   abrasive belt grinder Bandschleifmaschine f
   abrasive belt polisher Bandschleifpoliermaschine f
   abrasive cloth Schleifgewebe n
   abrasive effect Abrasionswirkung f, Abrasivwirkung f, Abriebwirkung f
   abrasive material Schleifmittel n, Abrasivstoff m, Schleifkorn n, Schleifmaterial n
   abrasive wear Abriebverschleiß m
   abrasive wheel Schleifscheibe f
abrasiveness Abrasivität f
absence Nichtvorhandensein n, Fehlen n
   absence of colo(u)r Farblosigkeit f
   absence of pinch-off welds Nichtvorhandensein n von Abquetschmarkierungen
absolute absolut
   absolute determination Absolutmessung f
   absolute pressure Absolutdruck m
absorb/to absorbieren, aufnehmen
absorbency Saugfähigkeit f
absorbent saugend, saugfähig
   energy absorbent energieabsorbierend
   sound absorbent ceiling tile Schallschluckdeckenplatte f
absorber Absorber m
   energy absorber Energieabsorber m
   shock absorber Stoßdämpfer m
   UV absorber UV-Absorber m
absorbing capacity Saugleistung f
absorption Absorption f
   absorption band Absorptionsband n
   absorption coefficient Absorptionskoeffizient m
   absorption of liquid Flüssigkeitsaufnahme f
   absorption spectroscopy Absorptionsspektroskopie f
   absorption spectrum Absorptionsspektrum n
   heat absorption Wärmeabsorption f
   light absorption Lichtabsorption f
   moisture absorption Feuchtigkeitsaufnahme f
   oil absorption value Ölzahl f
   plasticizer absorption Weichmacherabsorption f, Weichmacheraufnahme f
   sound absorption Schallabsorption f, Schallschluckung f
   UV absorption UV-Absorption f
   water absorption Wasseraufnahme f
absorptive absorptiv, saugfähig; dämpfend, schluckend (Schall)
   impact energy absorptive capacity Stoßenergieaufnahmevermögen n
accessible zugänglich
accelerated beschleunigt, zeitraffend
   accelerated ag(e)ing Kurzzeitalterung f [durch Lagerung]
   accelerated bursting test Kurzzeit-Berstversuch m
   accelerated cure Schnellhärtung f
   accelerated tensile test Kurzzeit-Zugversuch m
   accelerated test Kurzprüfung f, Schnelltest m, Zeitraffertest m, Zeitrafferversuch m, Kurzzeitversuch m, Kurzzeitprüfung f
   accelerated weathering Kurzbewitterung f
   accelerated weathering instrument Kurzbewitterungsgerät n, Schnellbewitterungsgerät n
   accelerated weathering resistance Kurzzeitbewitterungsverhalten n
   accelerated weathering test Kurzzeitbewitterungsversuch m, Schnellbewitterung f
   cobalt accelerated Co-beschleunigt, kobaltbeschleunigt
acceleration Beschleunigung f
   acceleration transducer Beschleunigungsaufnehmer m
accelerator Beschleuniger m, Härtungsbeschleuniger m [the accelarator speeds crosslinking of rubber by sulfur]
   accelerator system Beschleunigungssystem n
   amine accelerator Aminbeschleuniger m
   amount of accelerator Beschleunigermenge f
   cobalt accelerator Co-Beschleuniger m, Kobaltbeschleuniger m
   sulphenamide accelerator Sulfenamidbeschleuniger m
   vanadium accelerator Vanadinbeschleuniger m
   vulcanization accelerator Vulkanisationsbeschleuniger m
acceptable tragbar [z.B. Kosten]
acceptor Akzeptor m
   hydrogen acceptor Wasserstoff-Akzeptor m
access Zugang m
   access of air Luftzutritt m
   access time Zugriffszeit f [z.B. zu einem Datenspeicher]
   ease of access Zugänglichkeit f
   easy access bequemer Zugang m
accessibility Zugänglichkeit f
accessible zugänglich
   easily accessible griffgünstig, leicht zugänglich
accessories Zubehör(einrichtungen) n(fpl), Zubehörteile npl
   extruder accessories Extruderzubehör n
   range of accessories Zubehörprogramm n
accident Unfall m
   accident prevention Unfallverhütung f,

Unfallschutz m
accident prevention regulations
Unfallverhütungsvorschriften fpl, UVV fpl
accumulation Anhäufung f, Ansammlung f, Kumulation f
  heat accumulation Wärmestau m
  material accumulation Masseanhäufung f, Werkstoffanhäufung f
  melt accumulation Schmelzedepot n
  pigment accumulation Pigmentnest n
  solid accumulation Feststoffansammlung f
accumulator Akkumulator m, Speicher m
  accumulator chamber Speicherraum m, Sammelraum m
  accumulator cylinder Tauchkammerzylinder m, Speicherzylinder m, Tauchkammer f, Akkumulierzylinder m
  accumulator effect Akku-Wirkung f
  accumulator head Staukopf m, Speicherkopf m, Akku-Kopf m
  accumulator head blow mo(u)lding (process) Staukopfverfahren n
  accumulator head plasticizing unit Staukopfplastifiziergerät n
  hydraulic accumulator Druckmittelakkumulator m, Druckmittelspeicher m, Druckölspeicher m, Systemspeicher m
  melt accumulator Kopfspeicher m, Massespeicher m; Plastikakkumulator m; Schmelzebehälter m, Schmelzespeicher m; Stauzylinder m, Zylinderspeicher m
  melt accumulator machine Kopfspeichermaschine f
  melt accumulator system Kopfspeichersystem n, Massespeichersystem n
  pressure accumulator Druckspeicher m
  ram accumulator Kolbenakkumulator m, Kolbenspeicher m
  reciprocating barrel accumulator Zylinderschubspeicher m
  reciprocating-screw accumulator Schneckenschubspeicher m, Schneckenkolbenspeicher m
  tubular ram accumulator Ringkolbenspeicher m
  tubular ram accumulator head Ringkolbenspeicherkopf m
  type of accumulator Speicherbauart f
accuracy Genauigkeit f, Präzision f
  control accuracy Regelgüte f
  cutting accuracy Schnittpräzision f
  degree of accuracy Genauigkeitsgrad m
  dimensional accuracy Formtreue f
  metering accuracy Dosiergenauigkeit f
  setting accuracy Einstellfeinheit f
accurate genau
  accurate positioning genaue Positionierung f, Positioniergenauigkeit f
  accurate reproduction of detail genaue Abbildung f, Abbildegenauigkeit f
  accurate speed genaue Drehzahl f
  dimensionally accurate maßgenau
accurately adjustable feineinstellbar
accurately bored (drilled) hole Paßbohrung f
acetal Acetal n
  acetal copolymer Acetal-Copolymerisat n
  acetal resin Acetalharz n
  polyvinyl acetal Polyvinylacetal n
acetate Acetat n
  acetate copolymer Copolymerisat n
  acetate group Acetatgruppe f
  alkyl acetate Alkylacetat n
  butyl acetate Butylacetat n, Essigsäurebutylester m
  butyl diglycol acetate Butyldiglykolacetat n
  cellulose acetate Celluloseacetat n
  cellulose acetate butyrate Celluloseacetobutyrat n
  ethyl acetate Ethylacetat n, Essigsäureethylester m
  ethyl acetate test Essigestertest m
  ethyl glycol acetate Ethylglykolacetat n
  ethylene-vinyl acetate [EVA] Ethylen-Vinylacetat n [EVAC]
  ethylene-vinyl acetate rubber Ethylen-Vinylacetat-Kautschuk m
  glycol ether acetate Glykoletheracetat n
  polyvinyl acetate Polyvinylacetat n
  vinyl acetate Vinylacetat n, Essigsäurevinylester m
  vinyl acetate copolymer Polyvinylchloridacetat n, Vinylchlorid-Vinylacetat-Copolymerisat n
acetic acid Essigsäure f
acetone Aceton n
  acetone-soluble acetonlöslich
acetophenone Acetophenon n
acetyl cellulose Acetylcellulose f
acetyl group Acetylgruppe f
acetylacetonate Acetylacetonat n
  alumin(i)um acetylacetonate Aluminiumacetylacetonat n
  zirconium acetylacetonate Zirkonacetylacetonat n
acetylacetone peroxide Acetylacetonperoxid n
acetylated acetyliert
acetylene Acetylen n, Ethin n
acid 1. Säure f; 2. sauer
  acid amide group Säureamidgruppe f
  acid anhydride Säureanhydrid n
  acid anhydride catalyst Säureanhydridhärter m
  acid-catalysed säurekatalysiert
  acid catalyst Säurekatalysator m
  acid chloride Säurechlorid n
  acid content Säuregehalt m
  acid curing säurehärtend
  acid derivative Säurederivat n
  acid-insoluble säureunlöslich
  acid radical Säureradikal n, Säurerest m
  acid resistance Säurebeständigkeit f
  acid resistant säurebeständig
  acid-soluble säurelöslich
  acid value Säurezahl f

acrylic acid Acrylsäure f
acrylic acid ester Acrylsäureester m
adipic acid Adipinsäure f
adipic acid ester Adipinsäureester m
alkanesulphonic acid Alkansulfonsäure f
alkyl carboxylic acid Alkylcarbonsäure f
alkyl sulphonic acid ester Alkylsulfonsäureester m
aminocaproic acid Aminocapronsäure f
aminoundecanoic acid Aminoundecansäure f
aryl carboxylic acid Arylcarbonsäure f
azelaic acid Azelainsäure f
azelaic acid ester Azelainsäureester m
benzoic acid Benzoesäure f
boric acid Borsäure f
butyric acid Buttersäure f
carbamic acid Carbamidsäure f
carboxylic acid Carboxylsäure f, Carbonsäure f
carboxylic acid anhydride Carbonsäureanhydrid n
citric acid Zitronensäure f
coconut oil fatty acid Kokosölfettsäure f
cottonseed oil fatty acid Baumwollsamenölfettsäure f
dehydrated castor oil Ricinensäure f
dicarbolic acid Dicarbonsäure f, Bicarbonsäure f
dicarboxylic acid anhydride Dicarbonsäureanhydrid n
dicarboxylic acid ester Dicarbonsäureester m
dienol fatty acid Dienolfettsäure f
dodecanoic/lauric acid Dodecansäure f
ethylene dicarboxylic acid Ethylendicarbonsäure f
exposure to acids Säurebelastung f
fatty acid Fettsäure f
fatty acid amide Fettsäureamid n
fatty acid derivative Fettsäurederivat n
fatty acid ester Fettsäureester m
fatty acid mono-ester Fettsäuremonoester m
fatty acid soap Fettsäureseife f
formic acid Ameisensäure f
fumaric acid Fumarsäure f
fuming sulphuric acid Oleum n
glutaric acid Glutarsäure f
HET (hexachloroendomethylene tetrahydrophthalic acid) Hetsäure f
hydrochloric acid Chlorwasserstoff m, Salzsäure f
lauric acid Laurinsäure f
Lewis acid Lewissäure f
linseed oil fatty acid Leinölfettsäure f
maleic acid Maleinsäure f
maleic acid ester Maleinsäureester m
monocarboxylic acid Monocarbonsäure f
montanic acid Montansäure f
montanic acid ester Montansäureester m
nitric acid Salpetersäure f
oleic acid ester Ölsäureester m
o(rtho)-phthalic acid Orthophthalsäure f
p-toluenesulphonic acid Paratoluolsulfonsäure f
palmitic acid Palmitinsäure f
peracetic acid Peressigsäure f
peradipic acid Peradipinsäure f
perbenzoic acid Perbenzoesäure f
perbutyric acid Perbuttersäure f
performic acid Perameisensäure f
perpropionic acid Perpropionsäure f
persuccinic acid Perbernsteinsäure f
phosphoric acid Phosphorsäure f
phosphoric acid ester Phosphorsäureester m
phosphorous acid ester Phosphorigsäureester m
phthalic acid Phthalsäure f
phthalic acid ester Phthalsäureester m
polyacrylic acid Polyacrylsäure f
polycarboxylic acid Polycarbonsäure f
polycarboxylic acid anhydride Polycarbonsäureanhydrid n
polyglycol fatty acid ester Polyglykolfettsäureester m
polyphosphoric acid Polyphosphorsäure f
propionic acid Propionsäure f
pyromellitic acid Pyromellitsäure f
residual acid content Restsäuregehalt m
salicylic acid Salicylsäure f
salicylic acid ester Salicylsäureester m
sebacic acid Sebacinsäure f
sebacic acid ester Sebacinsäureester m
soya bean oil fatty acid Sojaölfettsäure f
stearic acid Stearinsäure f
succinic acid Bernsteinsäure f
succinic acid ester Bernsteinsäureester m
sulphonic acid Sulfonsäure f
sulphonic acid ester Sulfonsäureester m
sulphuric acid Schwefelsäure f
tall oil fatty acid Tallölfettsäure f
terephthalic acid Terephthalsäure f
tetracarboxylic acid Tetracarbonsäure f
tetrachlorophthalic acid Tetrachlorphthalsäure f
tetrahydrophthalic acid Tetrahydrophthalsäure f
toluene sulphonic acid Toluolsulfonsäure f
tricarboxylic acid Tricarbonsäure f
trimellitic acid Trimellitsäure f
trimellitic acid ester Trimellitsäureester m
acidity Acidität f
acoustic (e.g. a signal) akustisch
   acoustic emission Schallemission f
   acoustic emission analysis Schallemissionsanalyse f
   acoustic emission curve Schallemissionskurve f
acquisition device/unit/equipment Datenerfassungsgerät n
   data acquisition unit Datenerfassungsanlage f, Datenerfassungsstation f
   production data collecting (or: acquisition) unit Produktionsdaten-Erfassungsgerät n
acrylamide Acrylamid n
acrylate Acryl(säure)ester m
   acrylate-modified acrylatmodifiziert
   alkyl acrylate Acrylsäurealkylester m
   butyl acrylate Acrylsäurebutylester m, Butylacrylat n
   ethyl acrylate Acrylsäureethylester m, Ethylacrylat n
   methyl acrylate Acrylsäuremethylester m

octyl acrylate Octylacrylat n
acrylic acid Acrylsäure f
  acrylic acid ester Acrylsäureester m
acrylic polymer Acrylpolymer n
acrylic resin Acryl(säure)harz n
acrylic sheet Acrylplatte f
acrylic glazing sheet Acrylglas n
acrylonitrile Acrylnitril n
  acrylonitrile-butadiene rubber Acrylnitril-Butadien-Kautschuk m
  acrylonitrile-butadiene-styrene rubber Acrylnitril-Butadien-Styrol-Kautschuk m
  acrylonitrile copolymer Acrylnitril-Copolymerisat n
  acrylonitrile group Acrylnitril-Gruppe f
  acrylonitrile-methacrylate copolymer Acrylnitril-Methacrylat-Copolymerisat n
  original acrylonitrile content Ausgangsacrylnitrilgehalt m
  residual acrylonitrile content Acrylnitrilrestgehalt m
acting 1. Handeln n; 2. handelnd
  acting as a barrier against plasticizer migration weichmachersperrend
  quick acting clamp Schnellverschluß m
  plasticizer action Weichmacherwirkung f
action Handlung f
  action of heat Wärmeeinwirkung f
  action of light Lichteinwirkung f [Sie sind gegen Lichteinwirkung stabilisiert/they are light stabilized]
  action mechanism Wirkungsmechanismus m, Wirkungsweise f
  quick action screen changer Sieb-Schnellwechseleinrichtung f
activated aktiviert
activation Aktivierung f
  activation energy Aktivierungsenergie f
  solvent activation Lösemittelaktivierung f
activator Aktivator m
active aktiv
  active carbon black Aktivruß m
  active oxygen Aktivsauerstoff m
  active oxygen content Aktivsauerstoffgehalt m
  active substance Aktivsubstanz f, Wirkstoff m
  active substance concentration Wirkstoffkonzentration f
  active substance content Wirkstoffgehalt m
  surface active grenzflächenaktiv
actual tatsächlich, wirksam, Ist-
  actual pressure Druckistwert m, Istdruck m
  actual value Istgröße f, Istwert m
  actual value curve Ist(wert)kurve f
  actual value display Istwertanzeige f
  actual value transducer Istwertgeber m
  actual weight Istgewicht n
actuated angesteuert, betätigt
  actuating cam Betätigungsnocken fpl
  actuating lever Betätigungshebel m
  spring actuated federbetätigt

actuation Ansteuerung f
acylation Acylierung f
adaptability Anpaßbarkeit f, Anpassungsfähigkeit f
adaptive control adaptive Regelung f, AC f
adaptor Adapter m, Paßstück n, Übergangsstück n, Zwischenstück n
  adaptor module Anpaßbaustein m
  adaptor ring Zwischenring m
  adaptor system Adaptersystem n
  LED adaptor Leuchtdioden-Adapter m
add/to hinzufügen, einspeisen, beimengen, zudosieren, beimischen
  amount to be added Zugabemenge f
addition Beigabe f, Zudosierung f, Addition f, Zusatz m
  addition crosslinkage/cure Additionsvernetzung f
  addition crosslinking/curing additionsvernetzend
  addition of filler Füllstoffzugabe f
  addition of reclaim/regrind Regeneratzugabe f, Regeneratzusatz m
  addition of resin Harzzusatz m
  addition of styrene Styrolzusatz m
  addition polymer Additionspolymer n
  addition reaction Additionsreaktion f
  capable of addition additionsfähig
  direct addition Direktdosierung f
additional zusätzlich
  additional data Zusatzdaten pl
  additional equipment Zusatzeinrichtung f
  additional information Zusatzinformation f
  additional logic element Zusatzlogik f
  additional metering unit Zusatz-Dosieraggregat n
  additional plasticizer Zusatzweichmacher m
  additional tests Zusatzprüfungen fpl
  additional unit Zusatzgerät n
additive 1. Additiv n, Begleitstoff m, Hilfsstoff m, Zuschlag m, Zuschlagstoff m; 2. additiv
  additive-free zusatzfrei
  additive system Additivsystem n
  liquid additive Flüssigadditiv n
  paint additive Lackadditiv n
  plastics additive Kunststoffadditiv n
  separate additives Einzeladditive npl
address Adresse f
  address bus Adreßbus m
  memory address Speicheradresse f
addressing Adressierung f
adduct Adduct n
  amine adduct Aminaddukt n
adherence Haften n, Haftung f
  wall adherence Wandhaften n
adherent surface Haftgrund m, Klebfläche f
adhesion Adhäsion f, Haftfähigkeit f
  adhesion mechanism Haftmechanismus m
  adhesion problems Haftschwierigkeiten fpl, Haftungsprobleme npl
  adhesion-reducing haftungsmindernd
  cross-hatch adhesion test Gitterschnittest m

having good adhesion haftfest
improved adhesion Haftungssteigerung f
initial adhesion Anfangshaftung f
inter-layer adhesion Zwischenlagenhaftung f, Zwischenschichthaftung f
interlaminar adhesion Lagenbindung f
loss of adhesion Haftungseinbuße f
poor/faulty adhesion Haftungsmängel mpl, Haftungsschäden mpl
adhesive 1. adhäsiv; 2. Kleber m, Klebstoff m
adhesive cartridge Klebstoffpatrone f
adhesive effect Haftwirkung f, Klebwirkung f
adhesive film thickness Klebschichtstärke f
adhesive film/layer Kleb(stoff)schicht f
adhesive force Bindekraft f, Adhäsionskraft f, Haftkraft f
adhesive formulation Klebstoff-Formulierung f
adhesive fracture Adhäsivbruch m, Adhäsionsversagen n, Adhäsionsbruch m
adhesive mix Klebstoffmischung f [Je größer der Klebstoffansatz, um so kürzer ist die Verarbeitungszeit / the greater the amount of adhesive mixed, the shorter will be the pot life]
adhesive power Klebekraft f, Klebevermögen n, Klebkraft f
adhesive properties Hafteigenschaften fpl, Klebeigenschaften fpl
adhesive resin Klebeharz n, Klebstoffharz n, Leimharz n
adhesive strength Adhäsivfestigkeit f, Adhäsionsfestigkeit f, Haftfestigkeit f
adhesive system Kleb(stoff)system n
adhesive tape Klebeband n, Klebestreifen m
adhesives industry Klebstoffindustrie f
application of the adhesive Klebstoffauftrag m
contact adhesive Haftklebstoff m, Kontaktkleber m
cyanacrylate adhesive Cyanacrylatklebstoff m
dispersion adhesive Dispersionskleber m
dry adhesive Trockenklebstoff m
epoxy adhesive Epoxidharzklebstoff m
fast-setting adhesive Schnellkleber m
film adhesive Klebefilm m
hot melt adhesive Schmelzkleber m, Schmelzklebstoff m
hot melt adhesive applicator Schmelzklebstoff-Auftragsgerät n
hot melt contact adhesive Schmelzhaftklebstoff m
one-pack adhesive Einkomponentenkleber m, Einkomponentenklebstoff m
polyester adhesive UP-Harzkleber m
range of adhesives Klebstoffsortiment n
solvent-based adhesive Lösemittelkleber m, Lösemittelklebstoff m
structural adhesive Baukleber m
thick-bed adhesive Dickbettkleber m
thin-bed adhesive Dünnbettkleber m
tile adhesive Fliesenkleber m

two-pack adhesive Reaktionskleber m, Reaktionsklebstoff m, Zweikomponentenklebstoff m
wallpaper adhesive Tapetenkleister m
adiabatic adiabatisch
adipate Adipat n, Adipinsäureester m
butanediol adipate Butandioladipat n
butyl benzyl adipate Butylbenzyladipat n
di-isodecyl adipate Diisodecyladipat n
dibenzyl adipate Dibenzyladipat n
dibutyl adipate Dibutyladipat n
diethylene glycol adipate Diethylenglykoladipat n
dihexyl adipate Dihexyladipat n
dinonyl adipate (DNA) Dinonyladipat n
dioctyl adipate (DOA) Dioctyladipat n
adipic acid Adipinsäure f
adipic acid ester Adipinsäureester m
adjacent benachbart
adjacent molecules Nachbarmoleküle npl
adjust/to justieren, korrigieren
adjust/to (by operating controls) einregeln
adjust/to [e.g. calendar rolls] anstellen
adjustability Einstellbarkeit f, Justierbarkeit f
adjustable anstellbar, justierbar, regelbar, regulierbar, verstellbar
adjustable in height höhenverstellbar
adjustable lip biegsame Lippe f, einstellbare Düsenlippe f, flexible Oberlippe f, einstellbare Lippe f
adjustable roller Stellwalze f
adjustable variable Stellgröße f, Stellwert m
accurately adjustable feineinstellbar
digitally adjustable digital einstellbar
longitudinally adjustable längsverschiebbar
vertically adjustable höhenverstellbar
adjusting Einstell-, Stell-, Verstell-
adjusting mechanism Einstelleinrichtung f
adjusting screw Einstellschraube f, Stellschraube f, Verstellschraube f
back pressure adjusting mechanism, back pressure adjustment Staudruckeinstellung f, Verstelleinrichtung f
die adjusting mechanism, die adjustment Düsenjustiereinrichtung f, Düseneinstellung f
die lip adjusting mechanism Lippenverstellmöglichkeit f
mo(u)ld height adjusting mechanism Werkzeughöhenverstelleinrichtung f
nip adjusting mechanism/device Spaltverstellung f
pressure adjusting device/mechanism Druckeinstellorgan n
pressure adjusting valve Druckeinstellventil f
remote control adjusting mechanism Ferneinstellung f
restrictor bar adjusting mechanism Staubalkenverstellung f
roll adjusting gear Walzenanstellung f
roll adjusting mechanism Walzenverstellung f
screw stroke adjusting mechanism Schnecken-

hubeinstellung f, Schneckenwegeinstellung f
torque adjusting mechanism Drehmomenteinstellung f, Hubeinstellung f
vertical adjusting mechanism Vertikalverstellung f
adjustment Verstellung f, Einstellung f
  adjustment mechanism Verstellmöglichkeit f
  automatic adjustment Selbstregulierung f
  axis/roll crossing adjustment Walzenschrägeinstellung f, Walzenschrägverstellung f, Walzenschrägstellung f
  back pressure adjustment Staudruckeinstellung f
  die adjustment Düseneinstellung f
  die gap adjustment Düsenspaltverstellung f
  die lip adjustment Lippenverstellung f
  extremely fine adjustment Feinstregulierung f
  fine adjustment Feineinstellung f, Feinjustierung f, Feinregulierung f, Feinverstellung f
  mo(u)ld height adjustment Werkzeughöhenverstellung f
  remote control adjustment/setting Ferneinstellung f
  restrictor bar adjustment Staubalkenverstellung f
  screw stroke adjustment Schneckenhubeinstellung f, Schneckenwegeinstellung f
  setpoint adjustment Sollwerteinstellung f
  speed of adjustment Anstellgeschwindigkeit f
  temperature adjustment Temperaturausgleich m
  torque adjustment Drehmomenteinstellung f
  vertical adjustment Vertikalverstellung f
  zero adjustment/setting Nullabgleich m, Nullpunkteinstellung f
administrative costs Verwaltungskosten pl
adsorbent Sorptionsmittel n [used in chromatography]
adsorption Adsorption f
  adsorption process Adsorptionsvorgang m
  adsorption pump Adsorptionspumpe f
advance Vor-, Vorlauf m, Vorschub m
  advance information Vor(ab)information f
  mo(u)ld advance speed Formfahrgeschwindigkeit f
  nozzle advance cylinder Düsenanpreßzylinder m
  nozzle advance speed Düsenvorlaufgeschwindigkeit f
  plunger advance speed Kolbenvorlaufgeschwindigkeit f
  screw advance Schneckenvorschub m, Schneckenvorlauf m [Wichtig ist ein gleichmäßiger Schneckenvorlauf beim Einspritzen / it is important that the screw advances evenly during injection]
  screw advance speed Schneckenvorlaufgeschwindigkeit f
  speed/rate of advance Vorlaufgeschwindigkeit f, Vorschubgeschwindigkeit f
advantage Vorteil m
  cost advantage Kostenvorteil m, Preisvorteil m
  technical/processing advantages verfahrenstechnische Vorteile mpl, Verarbeitungsvorteile mpl
  volume price advantage Volumenpreisvorteil m
aeration Begasen n [Schaumherstellung]
aerator Begasungsanlage f [z.B. zur Begasung von PVC-Paste]
aerodynamic forces Luftkräfte f
aerosol Aerosol n
  aerosol container Aerosolbehälter m
  aerosol valve Aerosolventil n
aerospace Raumfahrt f
  aerospace engineering Raumfahrttechnik f
  aerospace industry Raumfahrtindustrie f
affected empfindlich
  affected by heat hitzeempfindlich, temperaturempfindlich, wärmeempfindlich
  affected by moisture feuchtigkeitsempfindlich
  affected by solvents lösungsmittelempfindlich
  affected by water wasserempfindlich
affinity Affinität f
after-sales service Kundendienst m
afterburner Nachbrenner m
afterburning Nachverbrennung f
ageing [USA: aging] Alterung f
  ag(e)ing characteristics Alterungsverhalten n
  ag(e)ing conditions Alterungsbedingungen fpl
  ag(e)ing period Alterungszeit f, Lagerungsdauer f
  ag(e)ing phenomena Alterungserscheinungen fpl
  ag(e)ing process Alterungsvorgang m, Alterungsprozeß m
  ag(e)ing properties/characteristics Alterungseigenschaften fpl, Alterungswerte mpl, Alterungskennwerte mpl
  ag(e)ing resistance Alterungsbeständigkeit f
  ag(e)ing temperature Alterungstemperatur f
  ag(e)ing test Alterungsprüfung f, Alterungsuntersuchung f, Lagerungsversuch m
  ag(e)ing under standard climatic conditions Klimaraumlagerung f
  ag(e)ing under warm, humid conditions Feucht-Warm-Lagerung f
  accelarated/short-term ag(e)ing Kurzzeitlagerung f
  causes of ag(e)ing Alterungsursachen fpl
  effect of ag(e)ing Alterungseinfluß m, Alterungsauswirkung f
  heat ag(e)ing thermische Alterung f, Wärmealterung f
  heat ag(e)ing behavio(u)r Wärmealterungsverhalten n
  heat ag(e)ing period Warmlagerungszeit f
  heat ag(e)ing properties Wärmealterungswerte mpl
  heat ag(e)ing resistance Wärmealterungsbeständigkeit f
  heat ag(e)ing temperature Warmlagerungstemperatur f
  heat ag(e)ing test Wärmealterungsversuch m, Warmlagerungsversuch m
  hot air ag(e)ing Heißluftalterung f
  improved ag(e)ing resistance verbesserte Alterungswerte mpl
  light ag(e)ing Lichtalterung f
  long-term ag(e)ing Dauerlagerung f
  long-term heat ag(e)ing Dauerwärmelagerung f
  room temperature ag(e)ing

Raumtemperaturlagerung f
susceptibility to ag(e)ing Alterungsanfälligkeit f
susceptible to ag(e)ing alterungsempfindlich
agent Agens n, Mittel n
  amount of blowing agent Treibmittelmenge f, Treibmittelanteil m
  amount of curing agent Vernetzermenge f, Vernetzungsmitteldosierung f
  anti-blocking agent Antiblockmittel n
  anti-corrosive agent Korrosionsschutzmittel n
  anti-floating agent Ausschwimmverhütungsmittel n
  anti-settling agent Antiabsetzmittel n, Absetzverhinderungsmittel n, Schwebemittel n
  anti-skinning agent Antihautmittel n, Hautverhütungsmittel n
  anti-slip agent Antislipmittel n
  antistatic agent Antistatikmittel n
  application of release agent Trennmittelbehandlung f
  blowing agent 1. Treibmittel n; 2. Schäummittel n, Schaumbildner m
  blowing agent concentrate Treibmittelkonzentrat n
  blowing agent content Treibmittelgehalt m
  chelating agent Komplexbildner m
  colo(u)rant agent [which can be a pigment or a dye] Farbmittel n
  coupling agent Haftverbesserer m, Haftvermittler m
  curing/crosslinking agent Vernetzer m, Vernetzungshilfe f, Vernetzungskomponente f, Vernetzungsmittel n
  curing agent concentration Vernetzerkonzentration f
  curing agent decomposition products Vernetzerspaltprodukte npl
  curing agent paste Vernetzerpaste f
  defoaming agent Entschäumer m
  dispersing agent Dispergier(hilfs)mittel n, Dispersionshilfsstoff m, Dispersionsmittel n
  drying agent Trockenmittel n
  emulsifying agent Emulsionshilfsmittel n, Emulgator m
  flow control agent Verlaufmittel n
  finishing agent Appreturmittel n
  free from blowing agent treibmittelfrei
  toughening agent Schlagfestmacher m, Schlagzähigkeitsverbesserer m, Schlagzähkomponente f, Schlagzähmacher m, Schlagzähmodifier m, Schlagzähmodifikator m, Schlagzähmodifizierharz n, Schlagzähmodifiziermittel n
  impregnating agent Tränkmittel n, Imprägniermittel n
  liquifying agent Verflüssigungsmittel n
  modifying agent Modifier n, Modifikator m, Modifizier(ungs)mittel n
  mo(u)ld release agent Entformungs(hilfs)mittel n, Trennmittel n, Formtrennmittel n
  neutralizing agent Neutralisationsmittel n
  nucleating agent Nukleierungs(hilfs)mittel n
  oxidizing agent Oxidationsmittel n
  precipitating agent Fäll(ungs)mittel n
  reducing agent Reduktionsmittel n
  reinforcing agent Verstärkungsmittel n
  release agent residues Trennmittelreste mpl
  silicone release agent Silicontrennmittel n
  slip agent Slipmittel n
  suspending agent Suspensionshilfsmittel n
  swelling agent Quellmittel n
  tackifying agent Klebrigmacher m
  thickening agent Eindick(ungs)mittel n, Verdickungsmittel n
  thixotropic agent Thixotropiermittel n
  vulcanizing/curing agent Vulkanisiermittel n, Vulkanisationsmittel n
  wet strength agent Naßfestmittel n
  wetting agent Benetzungsmittel n, Netzmittel n
  wetting agent concentration Netzmittelkonzentration f
  wetting agent solution Netzmittellösung f
agglomerate Agglomerat n
  pigment agglomerate Pigmentagglomerat n, Farbnest n
agglomerated agglomeriert
agglomeration Agglomerierung f, Agglomeration f
  degree of agglomeration Agglomerationsgrad m
aggressive aggressiv
aggressiveness Aggressivität f
agressivity Aggressivität f
agricultural film/sheeting Agrarfolie f, Landwirtschaftsfolie f
agricultural machinery Landmaschinen fpl
agriculture Landwirtschaft f
aid/to helfen
aid Hilfe f
  decision aid Entscheidungshilfe f
  demo(u)lding aid Entformungshilfe f
  design aids Auslegungshilfen fpl
  feeding aid Einzugshilfe f
  film-forming aid Verfilmungshilfsmittel n
  filtration aid Filterhilfsmittel n
  gelling aid Gelierhilfe f
  plasticizing/plasticating aid Plastifizierhilfe f
  polymerization aid Polymerisationshilfsmittel n
  processing aid Verarbeitungshilfsmittel n
  solution aid Lösungsvermittler m
  spotting aid Tuschierlehre f
  thermoforming aid Formungshilfe f
aided gestützt, unterstützt
  computer aided computerunterstützt, rechnergestützt
  computer aided design (CAD) CAD-Rechentechnik f, rechnergestützte Konstruktion f
  computer aided manufacture (CAM) CAM-Rechentechnik f, rechnergestützte Fertigung f
air Luft f
  air between the particles (e.g. of resin) Zwischenkornluft f
  air blower Luftgebläse n
  air brush Luftbürste f

air bubble Luftblase f
air circulating oven Luftumwälzofen m
air circulating system Luftumwälzungssystem n
air circulation Luftdurchwirbelung f, Luftzirkulation f
air circulation rate Luftzirkulationsrate f
air circulation unit Luftumwälzvorrichtung f
air compressor Luftkompressor m
air conditioning Klimatisierung f
air-conditioning system Klimaanlage f
air-conditioning unit Klimagerät n
air content (e.g. in foam) Luftbeladung f
air cooled luftgekühlt
air cooling device Luftdusche f
air cooling system Kühlluftsystem n
air cooling unit Luftkühlgerät n, Luftkühlaggregat n
air cushion Luftkissen n
air distributing elements Luftverteilungselemente npl
air drying 1.lufttrocknend; 2. Lufttrocknung f
air duct Luft(führungs)kanal m
air filter Luftfilter n,m
air flow conditions Luftströmungsverhältnisse npl
air gap Luftspalte f
air inlet Lufteintritt m, Lufteinlaß m
air inside the bubble Blaseninnenluft f
air jet Luftstrahl m
air knife Luftmesser n, Luftrakel m
air outlet Luftaustritt m, Luftauslaß m
air stream Luftstrom m, Luftströmung f
air supply Luftzufuhr f
air supply elements Luftzuführungselemente npl
air supply line Luftzuführungsleitung f
air throughput Luftdurchsatz m
air travel Luftfahrt f
air-borne sound Luftschall m
air-conditioned klimatisiert
aircraft construction Flugzeugbau m
aircraft hangar Flugzeughalle f
aircraft industry Luftfahrtindustrie f
airless luftlos
airless spraygun Airless-Gerät n, luftlose Spritzpistole f
airless spraying Airless-Spritzen n, Airless-Spritzverfahren n, luftloses Sprühverfahren n
airtight luftdicht
access of air Luftzutritt m
amount of cooling air Kühlluftmenge f
blowing air Aufblasluft f, Blasluft f
blowing air pressure Blasluftdruck m
blowing air supply Blasluftzufuhr f
calibrating air holes Stützluftbohrungen fpl
circulating air (drying) oven Umluftofen m, Umlufttrockenschrank m, Umluftwärmeschrank m
compressed air Druckluft f
compressed air calibrating sleeve Druckluftkalibrierhülse f
compressed air calibration sizing unit Druckluftkalibrierung f
compressed air cylinder Druckluftzylinder m, Druckluftflasche f
compressed air forming Druckluftformung f
compressed air forming machine Druckluftformmaschine f
compressed air jet Preßluftstrahl m, Preßluftstrom m
compressed air supply Druckluftversorgung f
cooling air Kühlluft f
cooling air blower Kühlluftgebläse n
cooling air delivery rate Kühlluftaustrittsgeschwindigkeit f, Kühlluftaustrittsmenge f
cooling air impingement angle Kühlluftanblaswinkel m, Anblaswinkel m
cooling air stream/flow Kühlluftströmung f
dry(ing) air Trockenluft f, Trocknungsluft f
entrapped air Lufteinschlüsse mpl
exclusion of air Luftausschluß m
exhaust air Fortluft f, Abluft f
external air cooling system Außenluftkühlung f
external cooling air stream Außenkühlluftstrom m
forced air circulation Zwangsluftzirkulation f
free from air luftfrei
fresh air supply Frischluftzufuhr f
high speed hot air welding Warmgasschnellschweißen n
hot air blower Warmluftgebläse n
hot air drier Warmlufttrockner m
incoming air Zuluft f
internal air cooling (system) Innenluftkühlung f
internal air pressure calibrating/sizing unit Überdruckkalibrator m
internal cooling air stream Innenkühlluftstrom m
outside air Außenluft f
purging air Spülluft f
recycled air Regenerierluft f
returning air Rückluft f
waste air Fortluft f, Abluft f
(waste air) exhaust fan Abluftventilator m
waste air scrubber Abluftwäscher m
alarm Alarm m
  alarm input Alarmeingang m
  alarm memory Alarmgedächtnis n
  alarm signal Alarmmeldung f, Alarmanzeige f, Alarmgeber m
  malfunction alarm Störmeldung f, Störmeldeinrichtung f
alcohol Alkohol m
  alcohol chain Alkoholkette f
  alcohol soluble alkohollöslich
  butyl alcohol Butylalkohol m
  diacetone alcohol Diacetonalkohol m
  ester alcohol radical Esteralkoholrest m
  ether alcohol Etheralkohol m
  ethyl alcohol Ethylalkohol m
  ethylene-vinyl alcohol copolymer Ethylen-Vinylalkohol-Copolymerisat n
  furfuryl alcohol Furfurylalkohol m

polyvinyl alcohol Polyvinylalkohol m
vinyl alcohol Vinylalkohol m
aldehyde Aldehyd n
   aldehyde group Aldehydgruppe f
algae Algen fpl
algorithm Algorithmus m
   control algorithm Regelalgorithmus m
align/to ausrichten, einstellen, abgleichen, anstellen
   aligned fluchtend, (aus)gerichtet
aliphatic aliphatisch
   aliphatic compound Aliphat n
   aliphatic hydrocarbon Benzinkohlenwasserstoff m, Paraffinkohlenwasserstoff m
alkali Alkali
   alkali alcoholate Alkalialkoholat n
   alkali hydroxide Alkalihydroxid n
   alkali metal Alkalimetall n
   alkali resistance Alkalibeständigkeit f, Laugenbeständigkeit f
   alkali resistant alkalibeständig
   alkali solution Lauge f
   alkali sulphite Alkalisulphit n
   high alkali glass A-Glas n
alkaline alkalisch, basisch
   alkaline earth oxide Erdalkalioxid n
   alkaline earth sulphate Erdalkalisulfat n
alkalinity Alkalität f
alkanesulphonic acid Alkansulfonsäure f
alkenyl group Alkenylgruppe f
alkoxy-functional alkoxyfunktionell
alkoxy radical Alkoxyradikal n
alkoxymethylene group Alkoxymethylengruppe f
alkoxysilane Alkoxysilan n
alkyd Alkyd-
   alkyd finish Alkydharzlackierung f
   alkyd paint Alkydharzlack m
   alkyd resin Alkydharz n
   long-oil alkyd (resin) fettes Alkydharz n, Langölalkydharz n
   medium-oil alkyd (resin) mittelfettes Alkydharz n, Mittelölalkydharz n
   short-oil alkyd (resin) Kurzölalkydharz n
   stoving alkyd Einbrennalkyd n
   wood oil alkyd (resin) Holzölalkydharz n
alkyl Alkyl-
   alkyl acetate Alkylacetat n
   alkyl acrylate Acrylsäurealkylester m
   alkyl benzene Alkylbenzol n
   alkyl carboxylic acid Alkylcarbonsäure f
   alkyl component Alkylkomponente f
   alkyl group Alkylgruppe f
   alkyl hydroperoxide Alkylhydroperoxid n
   alkyl per-ester Alkylperester n
   alkyl phenol Alkylphenol n
   alkyl phosphite Alkylphosphit n
   alkyl polysulphide Alkylpolysulfid n
   alkyl radical Alkylrest m
   alkyl silicone resin Alkylsilikonharz n
   alkyl sulphide Alkylsulfid n
   alkyl sulphonate Alkylsulfonat n
   alkyl sulphonic acid ester Alkylsulfonsäureester m
   alkyl tin carboxylate Alkylzinncarboxylat n
   alkyl tin compound Alkylzinn-Verbindung f
   aryl alkyl phosphate Arylalkylphosphat n
   aryl alkyl phthalate Arylalkylphthalat n
alkylation Alkylierung f
alkylol compound Alkylolverbindung f
alkylperoxy radical Alkylperoxidradikal n
all-foam upholstery Vollschaumstoffpolster n
all-round stabilizer tolerance universelle Stabilisierbarkeit f
all-steel knife Ganzstahlmesser n
allocation data Zuordnungsdaten npl
allophanate Allophanat n
   allophanate linkage Allophanatbindung f
   allophanate linkage content Allophanatbindungsanteil m
   allophanate structure Allophanatstruktur f
allowable erlaubt, zulässig
   maximum allowable concentration (MAC) maximale Arbeitsplatzkonzentration f [von Schadstoffen] (MAK)
alloy Legierung f
   beryllium-copper alloy Berylliumkupfer n
   centrifugal casting alloy Schleuderlegierung f
   magnesium alloy Magnesiumlegierung f
   non-ferrous metal alloy Buntmetall-Legierung f
   polymer alloy Kunststoff-Legierung f
   zinc casting alloy Zink-Gießlegierung f
alloying Legieren n
allyl Allyl-
   allyl chloride Allylchlorid n
   allyl ether Allylether m
   allyl resin Allylharz n
allylidene group Allylidengruppe f
almost trouble-free störungsarm
alphanumeric alphanumerisch
   alphanumeric display Alphanumerik-Bildschirm m
alternating wechselnd
   alternating current [a.c.] Wechselstrom m
   alternating temperature test Temperaturwechselprüfung f
   alternating torsion Wechseltorsion f
   exposure to alternating temperatures Temperaturwechselbeanspruchung f
alumina Tonerde f, Aluminiumoxid n
aluminium Aluminium n (Al)
   aluminium acetylacetonate Aluminiumacetylacetonat n
   aluminium construction Aluminiumkonstruktion f
   aluminium foil Alu-Folie f, Aluminiumfolie f
   aluminium granulex Aluminiumgrieß m
   aluminium hydroxide Aluminiumhydroxid n, Tonerdehydrat n
   aluminium pellets Aluminiumgrieß m

aluminium sheet Aluminiumblech n
aluminium silicate Aluminiumsilikat n
aluminium structure Aluminiumkonstruktion f
bonding of aluminium Aluminiumklebung f, Aluminiumverklebung f
cast aluminium Alu-Guß m, Aluminiumguß m
cast aluminium housing Alu-Gußgehäuse n
cast aluminium mo(u)ld Aluminiumguß-Werkzeug n
aluminum s. aluminium
amber(-colo(u)red) bernsteinfarben
ambient Umgebungs-
  ambient conditions Umgebungsbedingungen fpl, Umfeldbedingungen fpl
  ambient temperature Umgebungstemperatur f
amide Amid n
  amide wax Amidwachs n
  acid amide group Säureamidgruppe f
  fatty acid amide Fettsäureamid n
  polyether amide Polyetheramid n
amine Amin n
  amine accelerator Aminbeschleuniger m
  amine adduct Aminaddukt n
  amine catalyst Aminkatalysator m
  amine cure Aminhärtung f
  amine-cured amingehärtet
  amine group Amingruppe f
  amine hardener Aminhärter m
  amine-type aminisch
  amine-type hardener aminischer Härter m
  amine value Aminzahl f
  ether amine Etheramin n
amino Amino-
  amino-functional aminofunktionell
  amino group Aminogruppe f
  amino mo(u)lding compound Aminoplast-Preßmasse f
  amino radical Aminorest m
  amino resin Amino(plast)harz n
  aminoacid Aminosäure f
  aminoalcohol Aminoalkohol m
  aminoamide Aminoamid n
  aminocaproic acid Aminocapronsäure f
  aminoplastic Aminoplast n
  aminosilane Aminosilan n
  aminoundecanoic acid Aminoundecansäure f
ammeter Amperemeter n
ammonia Ammoniak n
amorphous amorph, nichtkristallin
amortization Amortisierung f
  amortization period Amortisierungszeit f
amount Menge f
  amount added Zugabemenge f, Zusatzmenge f (Alle Dosiermengen können getrennt für beide Komponenten eingestellt werden / all the required amounts can be set separately for each component)
  amount applied [e.g. paint, adhesive, etc.] Auftragsmenge f
  amount conveyed by the pump Pumpenfördermenge f
  amount of accelerator Beschleunigermenge f
  amount of air Luftstrom m, Luftmenge f
  amount of air inside [e.g. a film bubble] Innenluftstrom m
  amount of blowing agent Treibmittelanteil m, Treibmittelmenge m
  amount of cleaning Reinigungsaufwand m
  amount of cooling air Kühlluftmenge f
  amount of curing agent Vernetzermenge f, Vernetzungsmitteldosierung f
  amount of damage Schädigungsgrad m
  amount of dirt [e.g. in an oil filter or a mo(u)lding compound] Verschmutzungsgrad m
  amount of emulsifier Emulgatormenge f
  amount of energy Energiestrom m
  amount of energy required Energieaufwand m
  amount of erosion Erosionsgrad m
  amount of filler (added) Füllstoffmenge f
  amount of heat Wärmemenge f
  amount of hydraulic oil Hydraulikölstrom m
  amount of lubricant content Gleitmittelanteil m
  amount of material [e.g. inside a mixer, a mo(u)ld, in the screw flights, etc.] Füllgrad m
  amount of material in the mo(u)ld Werkzeugfüllungsgrad m
  amount of material in the srew flight(s) Schneckengangfüllung f
  amount of material injected Einspritzmenge f
  amount of melt passing through the die Düsenfluß m
  amount of oil Ölstrom m
  amount of oil needed/required Ölmengenbedarf m
  amount of oxygen available Sauerstoffangebot n
  amount of peroxide Peroxidmenge f, Peroxiddosierung f
  amount of planning (necessary) Planungsaufwand m
  amount of raw material required for a day's production Rohstofftagesmenge f
  amount of resin applied Harzauftragsmenge f
  amount of stabilizer Stabilisatormenge f
  amount of stress applied Beanspruchungshöhe f
  amount to be added Zusatzmenge f
  amount weighed out (material or sample) Einwaage f, Dosiermenge f
  amount used Einsatzmenge f
  approximate amount to be added Richtdosierung f
  extremely small amount Kleinstmenge f
  fixed amount Festmenge f
  incorporation of small amounts anteilige Mitverwendung f
  metered amount Dosiermenge f
  minimizing the amount of flash produced Butzenminimierung f
  required amount of plasticizer Weichmacheranteil m, Weichmachermenge f, Weichmacherbedarf m

residual amount Restmenge f
  residual moisture amount Restmengen fpl an
    Feuchtigkeit
  standard amount Standarddosierung f
  total amount of plasticizer Gesamtweich-
    machermenge f
amplifier Verstärker m
  charge amplifier Ladungsverstärker m
  control amplifier Regelverstärker m
  d.c. amplifier Gleichstromverstärker m
  torque amplifier Drehmomentverstärker m
amplitude Amplitude f
  deforming amplitude Verformungsamplitude f
  force amplitude Kraftamplitude f
  stress amplitude Belastungsamplitude f,
    Lastamplitude f
  stress intensity amplitude Spannungs-
    intensitätsamplitude f
  vibration amplitude Schwingungsamplitude f
anaerobic anaerob
analog(ue) analog, Analog-
  analog(ue) data Analogwerte mpl
  analog(ue)-digital converter
    Analog-Digital-Wandler m, A/D-Wandler m
  analog(ue) input Analogeingabe f, Analogeingang m
  analog(ue) module Analogbaugruppe f
  analog(ue) output Analogausgabe f,
    Analogausgang m
  analog(ue) signal Analogsignal n
  analog(ue) valve Analogventil n
analysis Analyse f
  acoustic emission analysis Schallemissionsanalyse f
  cost analysis Kostenanalyse f
  fault analysis Fehleranalyse f
  fracture analysis Schadensanalyse f
  method of analysis Analysenverfahren n,
    Analysenmethode f
  process analysis Prozeßanalyse f, Verfahrensanalyse f
  sieve analysis Siebanalyse f
  stress analysis Spannungsanalyse f
  thermogravimetric analysis (TGA) thermo-
    gravimetrische Analyse f
  thermomechanical analysis (TMA) thermo-
    mechanische Analyse f
  trace analysis Spurenanalyse f
  X-ray analysis Röntgenanalyse f
analytical balance Analysenwaage f
analyzer Analysegerät n
  data analyzer Datenanalysator m, Datenanalysegerät n
  differential thermal analyzer Differential-
    thermoanlaysengerät n
  laser scanning analyzer Laser-Abtastgerät n
anatase Anatas n
anchor coating Kupplungsschicht f
  anchor coating paste (for PVC paste) Grundie-
    rungspaste f, Grundstrich m, Haftstrich m
anchorage Verankerung f

ancillary Zusatz-, Hilfs-
  ancillary component Zusatzbauteil n
  ancillary equipment Hilfseinrichtungen fpl,
    Hilfsvorrichtungen fpl, Nebeneinrichtungen fpl,
    Zubehör(teile) n(pl), Peripheriegerät n,
    Zusatzausrüstung f
  ancillary extruder Beispritzextruder m,
    Beistellextruder m, Sekundärextruder m,
    Nebenextruder m, Seitenextruder m
  ancillary function Nebenfunktion f
  ancillary screw Nebenschnecke f, Seitenschnecke f,
    Hilfsschnecke f
  ancillary unit Hilfsaggregat n, Nebenaggregat n
angle Winkel m
  angle of contact Berührungswinkel m
  angle of curvature Krümmungswinkel m
  angle of impact Schlagwinkel m
  angle of rotation Drehwinkel m
  angle of slope Neigungswinkel m
  angle of torsion Torsionswinkel m
  bending angle Biegewinkel m
  bubble collapsing angle Flachlegewinkel m
  clearance angle Freiwinkel m
  cooling air impingement angle
    Kühlluftanblaswinkel m
  cutting angle Schnittwinkel m
  deflection angle Umlenkwinkel m
  helix angle Drallwinkel m, Steigungswinkel m,
    Gangsteigungswinkel m
  impingement angle Anblaswinkel m
  lip angle Keilwinkel m
  loss angle Verlustwinkel m
  phase angle Phasenwinkel m
  point angle Spitzenwinkel m
  rake angle Spanwinkel m
  tilting angle Schwenkwinkel m
  wedge angle Keilwinkel m
  winding angle Wickelwinkel m
angled bolt Schrägbolzen m
angled extrusion die head Schräg(spritz)kopf m
angular kantig, winkelig, gewinkelt, eckig
  angular velocity Winkelgeschwindigkeit f
anhydride Anhydrid n
  anhydride cure Anhydridhärtung f,
    Anhydridvernetzung f
  anhydride-cured anhydridgehärtet
  anhydride group Anhydridgruppe f
  anhydride hardener Anhydridhärter m
  acid anhydride Säureanhydrid n
  acid anhydride catalyst Säureanhydridhärter m
  butyric anhydride Buttersäureanhydrid n
  carboxylic acid anhydride Carbonsäureanhydrid n
  dicarboxylic acid anhydride Dicarbonsäureanhydrid n
  dodecanoic anhydride Dodecansäureanhydrid n
  dodecyl succinic anhydride Dodecyl-
    bernsteinsäureanhydrid n
  glutaric anhydride Glutarsäureanhydrid n

maleic anhydride Maleinsäureanhydrid n
phthalic anhydride Phthalsäureanhydrid n
polycarboxylic acid anhydride Polycarbonsäureanhydrid n
pyromellitic anhydride Pyromellitsäureanhydrid n
sebacic anhydride Sebacinsäureanhydrid n
succinic anhydride Bernsteinsäureanhydrid n
tetrachlorophthalic anhydride Tetrachlorphthalsäureanhydrid n
anhydrous wasserfrei
aniline Anilin n
  diethyl aniline Diethylanilin n
  dimethyl aniline Dimethylanilin n
anion Anion n
anionic anionaktiv, anionisch
anisotropic anisotrop
  anisotropic shrinkage Schwindungsanisotropie f
anisotropy Anisotropie f
anneal/to tempern, ausglühen, härten
  bright-annealed blankgeglüht
annealing Glühen n, Ausglühen n, Tempern n
  annealing furnace Temperofen m
  annealing period Temperzeit f
  annealing temperature Temper-Temperatur f
annual jährlich, Jahres-
  annual accounts Jahresabschluß m
  annual capacity Jahreskapazität f
  annual production Jahresproduktion f
  annual profit Jahresgewinn m
  annual report Geschäftsbericht m
  annual surplus Jahresüberschuß m
  annual world capacity Weltjahreskapazität f
  annual world production Weltjahresproduktion f
annular ringförmig
  annular die Ringspaltwerkzeug n, Ring(spalt)düse f, Ringschlitzdüsenwerkzeug n, Ringschlitzdüse f
  annular groove Ringkanal m
  annular piston pump Kreiskolbenpumpe f
  annular slit Ringspalt(e) m(f), Ringspaltöffnung f
  annular space Ringraum m
anodized eloxiert
ante-chamber Vorkammer f, Vorkammerbohrung f, Vorkammerraum m, Angußvorkammer f
  ante-chamber bush Vorkammerbuchse f, Vorkammerkegel m
  ante-chamber direct feed injection Vorkammerdurchspritzverfahren n
  ante-chamber feed system Vorkammer-Angießtechnik f, Vorkammeranguß m
  ante-chamber nozzle Vorkammerdüse f
  ante-chamber pin gate Vorkammerpunktanguß m
  ante-chamber sprue bush Vorkammerangußbuchse f
  antechamber-type pin gate Punktanguß m mit Vorkammer f, Stangenpunktanguß m
anthracite Anthrazit m
  anthracite colo(u)red anthrazitfarbig, anthrazitfarben
anti-blocking Antiblock-

anti-blocking agent Antiblockmittel n, Antiblocksystem n
anti-blocking effect Antiblockwirkung f, Antiblockeffekt m
anti-blocking properties Antiblockeigenschaften fpl
anti-corrosive korrosionsschützend
  anti-corrosive agent Korrosionsschutzmittel n
  anti-corrosive coat(ing) Korrosionsschutzanstrich m, Rostschutzanstrich m
  anti-corrosive effect Korrosionsschutz m, Rostschutzwirkung f
  anti-corrosive paint Korrosionsschutzfarbe f, Korrosionsschutzlack m
  anti-corrosive pigment Korrosionsschutzpigment n, Rostschutzpigment n
  anti-corrosive primer Korrosionsschutzgrundfarbe f, Korrosionsschutzgrundierung f, Rostschutzgrundierung f
  anti-corrosive properties Korrosionsschutzeigenschaften fpl
anti-drift control mechanism Antidrift-Regelung f
anti-drumming coating Antidröhnbeschichtung f
anti-floating agent Ausschwimmverhütungsmittel n
anti-foam schauminhibierend, entschäumungsaktiv, schaumverhindernd
antifoam agent Entschäumer m, Antischaummittel n
anti-fouling coating Antifouling-Anstrich m
anti-fouling paint Antifoulingfarbe f
anti-friction bearing Wälzlager n
anti-friction bearing grease Wälzlagerfett n
anti-friction properties Gleitfähigkeit f, Gleitfähigkeitsverhalten n, Gleiteigenschaften fpl
antimony Antimon n
  antimony compound Antimonverbindung f
  antimony trioxide Antimontrioxid n
antioxidant Oxidationsschutzmittel n, Oxidationsstabilisator m, Alterungsschutzmittel n, Antioxidans n [the antioxidant protects cured rubber against oxidative degradation]
  antioxidant blend Antioxidationssystem n
anti-oxidative antioxidativ
anti(-)ozonant Antiozonans n, Ozonschutzmittel n [antiozonants, e.g. waxes, protect cured rubber products against ozon]
anti-settling agent Antiabsetzmittel n
anti-skinning agent Antihautmittel n, Hautverhütungsmittel n
anti-slip agent Antislipmittel n
antistatic antistatisch, antielektrostatisch
  antistatic agent Antistatikum n, Antistatiksystem n
  antistatic concentrate Antistatikumkonzentrat n
  permanently antistatic dauerantistatisch
apart einzeln, gesondert, getrennt; mit Abstand
  capable of being taken apart zerlegbar
  take apart/to auseinandernehmen, zerlegen
aperture Öffnung f
  nozzle aperture Düsenbohrung f, Düsenöffnung f

apparatus Apparat m, Gerät n
  breathing apparatus Atemschutzgerät n
  Charpy apparatus Charpygerät n
  distillation apparatus Destillationsgerät n
  Dynstat apparatus Dynstatgerät n
  test apparatus Prüfmaschine f, Prüfeinrichtung f
  torsion pendulum apparatus Schwingungsgerät n
  univeral test apparatus Universalprüfmaschine f
apparent scheinbar
  apparent density Fülldichte f, Schüttdichte f, Schüttgewicht n
  apparent modulus of rigidity Torsionssteifheit f
  apparent viscosity scheinbare Viskosität f
  compacted apparent density Stopfdichte f
appearance Aussehen n, Erscheinungsbild n
  film appearance Filmaussehen n
appliance Gerät n
  domestic appliances Haushaltsgeräte npl
  electrical appliance Elektrogerät n
  electrical appliance industry Elektrogeräteindustrie f
  household appliance paint Haushaltsgerätelack m
  instrument appliance, device, unit, equipment, piece of equipment Gerät n
application Anwendung f, Verwendungszweck m, Auftrag m [z.B. Farbe]
  application examples Anwendungsbeispiele npl
  application guidelines Anwendungsrichtlinien fpl
  application of primer Primerlackierung f, Grundierung f
  application of release agent Trennmittelbehandlung f
  application of stress Lastaufbringung f
  application of the adhesive Klebstoffauftrag m
  application possibility Einsatzmöglichkeit f, Anwendungsmöglichkeit f
  application-related anwendungsbezogen
  application software Anwendersoftware f, Benutzersoftware f
  application program Benutzerprogramm n, Anwendungsprogramm n, Anwenderprogramm n
  applications of load(s) Kraftangriff m, Krafteinwirkung f
  consumer goods for food contact applications Lebensmittelbedarfsgegenstände mpl
  electrical applications Elektroanwendungen fpl
  end use application Endanwendung f, Einsatzzweck m
  fields of application Anwendungsgebiete npl
  high temperature applications Hochtemperaturanwendungen fpl
  hot-melt roller application (of an adhesive) Walzenschmelzverfahren n
  important applications Anwendungsschwerpunkte mpl
  knife application Rakel(messer)auftrag m, Rakeln n
  low temperature applications Tieftemperaturanwendungen fpl
  main applications Anwendungsschwerpunkte mpl
  method of application Applikationsmethode f, Auftragsmethode f
  possible applications Anwendungsmöglichkeiten fpl
  principal applications Anwendungsschwerpunkte mpl
  range of applications Anwendungsbreite f, Anwendungsbereich m, Anwendungsspektrum n
  rate of application Auftragsmenge f
  rate of application [e.g. of paint or adhesive, usually expressed in g/sqm] Auftragsgewicht n
  roller application [e.g. of an adhesive or surface coating] Walzenauftrag m
  stress application device Belastungseinrichtung f
  suitable for food contact applications lebensmittelecht
  trowel application Spachteln n
  typical applications Anwendungsbeispiele npl
applicational anwendungstechnisch
applicator [e.g. of adhesive] Auftrag(s)maschine f, Auftrag(s)aggregat n
  applicator roll Auftrag(s)walze f
  glue applicator Leimauftragungsgerät n
  glue applicator roll Beleimungswalze f
  hot-melt adhesive applicator Schmelzklebstoff-Auftragsgerät n
applied angewandt, aufgebracht
  applied load Prüfkraft f
  amount applied [e.g. paint, adhesive, etc.] Auftrag(s)menge f
  amount of resin applied Harzauftrag(s)menge f
  amount of stress applied Beanspruchungshöhe f
apply/to anwenden, auftragen
  easy to apply leicht applizierbar
apprentice Lehrling m, Auszubildender m, Azubi m
approximate/to (sich) nähern
approximate annähernd, ungefähr
  approximate amount to be added Richtdosierung f
  approximate figure Richtgröße f, Anhaltswert m
  approximate processing conditions Verarbeitungsrichtwerte mpl
  approximate temperature Temperaturrichtwert m
  approximate value Anhaltswert m, Richtgröße f
  approximate weight Cirka-Gewicht n
apron Schurz m, Schürze f
  bumper apron Stoßfängerschürze f
  front apron Frontschürze f
  rear apron Heckschürze f
aqueous wäßrig, wasserhaltig
  aqueous extract wäßriger Auszug m
aramid fiber [GB: fibre] Aramidfaser f
arc Lichtbogen m
  arc resistance Bogenwiderstand m, Lichtbogenfestigkeit f
  arc-suppressing lichtbogenlöschend
  xenon arc Xenonbogen m
area Areal n, Fläche f
  area of contact Kontaktfläche f
  areas subject to wear Verschleißstellen fpl
  area terminal Bereichsterminal n

    bonded area Verklebungsfläche f
    catchment area Einzugsbereich m
    cross-sectional area Querschnitt(sfläche) m(f)
    damaged area Schadstelle f
    effective screen area freie Siebfläche f
    floor area Bodenfläche f
    forming area Formfläche f
    gate area Anschnittpartie f, Anschnittbereich m
    having the same surface area flächengleich
    heating area Heizfläche f
    pinch-off area Quetschzone f
    platen area Formaufspannfläche f, Aufspannfläche f, Werkzeugaufspannfläche f
    pressure per unit area Flächenpressung f
    production area (of a factory) Produktionsbereich m
    projected area projizierte Fläche f
    projected mo(u)lding area projizierte Formteilfläche f, Formteilprojektionsfläche f, Spritz(lings)fläche f
    projected runner surface area projizierte Verteilerkanalfläche f
    screen area Filterfläche f
    screw surface area Schneckenoberfläche f
    surface area-weight ratio Oberflächengewichtsverhältnis n
    unit area Flächeneinheit f
    weight per unit area Flächengewicht n
arithmetic arithmetisch
    arithmetic mean arithmetischer Mittelwert m
    arithmetic unit Rechenwerk n
arm Arm m, Hebel m
    cantilever arm Kragarm m
    compensating arm sensor Tänzerarmfühler m
    rocker arm Schwinge f
armature 1. Anker m [z.B. Magnet]; 2. Armatur f, Zubehör n
    magnetic armature Magnetanker m
armrest Armlehne f
aromatic aromatisch
    aromatic compound Aromat m
    aromatic content Aromatengehalt m
    aromatic hydrocarbon aromatischer Kohlenwasserstoff m
    free from aromatic compounds aromatenfrei
arranged angeordnet, geschaltet
    arranged in parallel parallelgeschaltet
    arranged in series in Reihenanordnung f, hintereinandergeschaltet
    runners arranged side by side Reihenverteiler m
arrangement Anordnung f, Vorrichtung f
    eight-runner arrangement Achtfachverteilerkanal m
    electrode arrangement Elektrodenanordnung f
    five-runner arrangement Fünffachverteilerkanal m
    gripper arrangement Greifvorrichtung f
    parallel arrangement Parallelschaltung f
    rotary-table arrangement Rundtischanordnung f
    series arrangement Reihenschaltung f, Serienschaltung f
    single-screw arrangement Einschneckenanordnung f
    six-runner arrangement Sechsfachverteilerkanal m
    star-shaped arrangement Sternanordnung f
    tandem arrangement Tandemanordnung f
    tilting arrangement Kippvorrichtung f
article Teil n, Artikel m
    disposable article Einwegartikel m, Wegwerfartikel m
    injection mo(u)lded rubber article Gummi-Spritzgußteil n
    finished article Fertigartikel m, Fertigerzeugnis n, Fertigprodukt n
    mo(u)lded article Formartikel m
    reject articles Ausschußware f
artificial künstlich
aryl Aryl n
    aryl alkyl phosphate Arylalkylphosphat n
    aryl alkyl phthalate Arylalkylphthalat n
    aryl carboxylic acid Arylcarbonsäure f
    aryl group Arylgruppe f
    aryl phosphite Arylphosphit n
    aryl radical Arylrest m
asbestos Asbest n
    asbestos cement Asbestzement m
    asbestos cloth Asbestgewebe n
    asbestos fiber [GB: fibre] Asbestfaser f
    asbestos sheet Asbestplatte f
ash Asche f, Glührückstand m
    ash content Aschegehalt m
    sulphated ash Sulfatasche f
    sulphated ash content Sulfataschegehalt m
aspect ratio Kantenverhältnis n
assemble/to montieren
assembler Assembler m
assembly Montage f
    assembly belt Montageband n
    assembly drawing Zusammenstellungszeichnung f, Montagezeichnung f
    assembly glue Montageleim m
    assembly instruction 1. Assemblerbefehl m [EDV]; 2. Montageanleitung f
    assembly language Assemblersprache f
    assembly line Fließband n
    assembly line production Fließbandfertigung f
    barrel assembly (of a compounding unit) Knetergehäuse n
    calibrating plate assembly Scheibenpaket n
    candle filter assembly Filterkerzenpaket n
    die assembly Düsensatz m
    draw plate assembly Blendenpaket n, Kalibrierblendenpaket n
    ejector plate assembly Auswerferteller m, Auswerferplatte f
    extrusion die assembly Extruder-Düseneinheit f
    kneading block assembly Knetblockanordnung f
    modular screw assembly Baukastenschneckensatz m
    mo(u)ld assembly Formeinheit f

mo(u)ld plate assembly Plattenpaket n
screw assembly Schneckensatz m, Schneckenbaukasten m, Schneckenausrüstung f, Schneckeneinheit f
screw assembly components Schneckensatzelemente npl
seal assembly Dichtungssatz m
sizing plate assembly Blendenpaket n, Kalibrierblendenpaket n
snap-on assembly Steckmontage f
spring assembly Federpaket n
thrust bearing assembly Rückdrucklagerung f
twin screw assembly Schneckenpaar n
twin vacuum hopper assembly Vakuum-Doppeltrichteranlage f
window frame assembly Rahmenkonfektionierung f

assessment Beurteilung f, Bewertung f
assessment of performance Leistungsbewertung f
quality assessment Qualitätsbeurteilung f

asset Aktivposten m, Gut n, Vermögen n
capital assets Anlagevermögen n
current assets Umlaufvermögen n
disposal of fixed assets Anlagenabgänge mpl
fixed assets Anlagevermögen n
liquid assets flüssige Mittel npl

assistance Beistand m, Hilfe f, Unterstützung f
technical assistance verfahrenstechnische Hilfestellung f

assurance Versicherung f, Sicherung f
quality assurance Gütesicherung f, Qualitätssicherung f

asymmetric unsymmetrisch, asymmetrisch
asymptotic asymptotisch
atactic ataktisch
atmosphere Atmosphäre f
atmosphere of nitrogen Stickstoffatmosphäre f
condensed moisture atmosphere Schwitzwasserklima n
constant test atmosphere Konstantklima n
moisture content of the atmosphere Luftfeuchtegehalt n
standard conditioning atmosphere [This is usually followed by figures such as 23/50, which means 23°C and 50% relative humidity] Normklima n
surrounding atmosphere Umgebungsatmosphäre f, Umgebungsluft f
test atmosphere Prüfklima n

atmospheric atmosphärisch
atmospheric humidity Luftfeuchte f, Luftfeuchtigkeit f
atmospheric oxygen Luftsauerstoff m
atmospheric pollution Luftverschmutzung f
atmospheric pressure Atmosphärendruck m, Luftdruck m
relative atmospheric humidity relative Luftfeuchtigkeit f
sensitivity to atmospheric humidity Luftfeuchteempfindlichkeit f

atom Atom n
carbon atom Kohlenstoffatom n
chlorine atom Chloratom n
group of atoms Atomgruppe f
hydrogen atom Wasserstoffatom n
oxygen atom Sauerstoffatom n

atomic bond Atombindung f
atomic weight Atomgewicht n
attached angebaut, angelagert
attachment 1. Befestigung f, Bindung f; 2. Zusatzeinrichtung f
mo(u)ld attachment Formenanschluß m, Werkzeugaufspannung f, Werkzeugbefestigung f
mo(u)ld attachment holes Aufspannbohrungen fpl

attack/to angreifen, anlösen
attack Angriff m
chemical attack Chemikalienangriff m, Chemikalieneinwirkung f
exposure to chemical attack Chemikalienbeanspruchung f
mo(u)ld attack Schimmelbefall m, Pilzbefall m
radical attack Radikalangriff m
resistance to chemical attack Widerstandsfähigkeit f gegen Chemikalieneinwirkung
surface attack Oberflächenangriff m

attacking medium/agent Angriffsflüssigkeit f, Angriffsmittel n
attenuation constant Abschwächungsbeiwert m
attraction Anziehung f
dipole-dipole attraction Dipol-Dipol-Anziehung f
forces of attraction Anziehungskräfte fpl
audible [e.g. signal] akustisch
audible warning signal akustische Störanzeige f
austenitic austenitisch
autocatalytic autokatalytisch
autoclave Autoklav m, Druckkessel m
automate Automat m
automate/to automatisieren
easy to automate automatisierungsfreundlich
automated automatisiert
fully automated vollautomatisiert
partly automated teilautomatisiert
automatic automatisch, selbsttätig
automatic adjustment Selbstregulierung f
automatic bag-making machine Beutel(herstellungs)automat m
automatic bag-welding machine Beutelschweißautomat m
automatic blow mo(u)lding-filling machine Hohlkörper-Blas- und Füllautomat m
automatic blow mo(u)lding machine Blas(form)automat m, Hohlkörperblasautomat m
automatic bottle blowing machine Flaschenblasautomat m
automatic changing mechanism Wechselautomatik f
automatic circuit breaker Sicherungsautomat m

automatic compression mo(u)lding machine Preß-automat m
automatic controls Steuerautomatik f
automatic carousel-type injection mo(u)lding machine Revolverspritzgießautomat m
automatic cut-out Sicherungsautomat m
automatic deflashing unit Entgratautomat m
automatic device for cutting plastics profiles into lengths Kunststoffprofil-Ablängeautomat m
automatic downstroke press Oberkolbenpreßautomat m
automatic downstroke transfer mo(u)lding press Oberkolbenspritzpreßautomat m
automatic equipment Automat m
automatic fast cycling injection mo(u)lding machine Schnellspritzgießautmoat m
automatic feed unit Vorschubautomat m, Zuführungsautomat m
automatic feeder Dosierautomat m, Zuführungsautomat m, Beschickungsautomat m
automatic feeding equipment Beschickungsautomat m
automatic filling machine Abfüllautomat m, Füllautomat m
automatic film slitter Folienschneidautomat m
automatic filter unit Filterautomatik f
automatic foam mo(u)lding unit Formteilschäumautomat m
automatic gel time tester Gelierzeitautomat m
automatic heat impulse welding instrument Wärmeimpulsschweißautomat m
automatic heat sealer Heißsiegelautomat m
automatic high capacity/performance blow mo(u)lding machine Hochleistungsblasformautomat m
automatic hot gas welding unit Warmgas-Schweißautomat m
automatic injection mo(u)lder Spritzblasautomat m
automatic injection mo(u)lding machine Spritz(gieß)automat m
automatic locking mechanism Zwangsverriegelung f
automatic machine Automat m
automatic metering unit Dosierautomat m
automatic mo(u)ld safety device Automatikwerkzeugschutz m
automatic mo(u)lding machine Formteilautomat m
automatic operating mechanism Bedienungsautomatik f
automatic operation Automatikbetrieb m
automatic oriented PP stretch blow mo(u)lding machine OPP-Streckblasautomat m
automatic pipe cutter Rohrtrennautomat m
automatic processing equipment Verarbeitungsautomat m
automatic profile welding unit Profilschweißautomat m
automatic reciprocating-screw injection mo(u)lding machine Schneckenspritzgießautomat m
automatic rotary injection mo(u)lding machine Rotations-Spritzgußautomat m
automatic rotary-table injection mo(u)lding machine Revolverspritzgießautomat m, Rundläuferautomat m
automatic screw-plunger injection mo(u)lding machine Schneckenspritzgießautomat m
automatic speed control system Drehzahl-Steuerautomatik f
automatic stacker Stapelautomat m
automatic starting/start-up mechanism Anfahrautomatik f
automatic switch-off mechanism Abschaltautomatik f
automatic switch-on mechanism Einschaltautomatik f
automatic thermoforming machine Thermoform-Automat m, Warmformautomat m, Tiefziehautomat m
automatic transfer mo(u)lding machine Spritzautomat m
automatic transfer mo(u)lding press Spritzpreßautomat m
automatic two-colo(u)r injection mo(u)lding machine Zweifarben-Spritzgußautomat m
automatic vacuum forming machine Vakuumformautomat m
automatic winder Wickelautomat m
automatic winding system Wickelautomatik f
automatic workpiece robot Werkstück-Handhabungsautomat m
fully automatic(ally) vollautomatisch
fully automatic machine Vollautomat m
fully automatic system Vollautomatik f
range of automatic blow mo(u)lding machines/equipment Blasformautomaten-Baureihe f
single-station automatic blow mo(u)lding machine Einstationenblasformautomat m
automation Automatisierung f
  degree of automation Automatisierungsgrad m, Automatisierungsaufwand m (Ein höherer Automatisierungsaufwand kann vorgesehen werden / a greater degree of automation can be provided)
  process automation Prozeßautomatisierung f
automotive Kraftfahrzeug-, KFZ-
  automotive application KFZ-Anwendung f
  automotive finish Auto(deck)lack m, Fahrzeuglack m
  automotive paint Fahrzeuglack m, Auto(deck)lack m
autoxidation Autoxidation f
auxiliary 1. helfend, Hilfs-; 2. Hilfsmittel n, Zubehör n
  auxiliary core Hilfskern m
  auxiliary cylinder Hilfszylinder m
  auxiliary drive Hilfsantrieb m
  auxiliary motor Hilfsmotor m
  auxiliary solvent Hilfslösemittel n
  auxiliary take-off (unit) Hilfsabziehwerk n
  auxiliary twin screw extruder Doppelschnecken-Seitenextruder m
  paint auxiliary Lackhilfsmittel n

availability Verfügbarkeit f, Angebot n
  radical availability Radikalangebot n
  raw material availability Rohstoffverfügbarkeit f
available benutzbar, verfügbar
  available dies Werkzeugpark m
  available machines/equipment Maschinenpark m
  available mo(u)lds Werkzeugpark m
  amount of oxygen available Sauerstoffangebot n
average Durchschnitt m, Mittelwert m
  average molecular weight mittleres Molekulargewicht n, Molekulargewicht-Mittelwert m
  average output Durchschnittsleistung f
  average viscosity Viskositätsmittel n
  above average überdurchschnittlich
  number average Zahlenmittel n
  weight average Gewichtsmittel n
awning Markise f
axial axial, Axial-
  axial clearance Axialspiel n
  axial direction Achsrichtung f
  axial displacement axiale Verschiebung f
  axial flow Axialströmung f
  axial loads Axialkräfte fpl
  axial melt stream Axialstrom m
  axial mixing Axialvermischung f
  axial piston motor Axialkolbenmotor m
  axial piston pump Axialkolbenpumpe f
axially movable verschiebbar
axially parallel achsparallel
axially symmetrical rotationssymmetrisch
axis Achse f
  axis crossing Walzenschrägeinstellung f, Walzenschrägstellung f, Walzenschrägverstellung f, Schrägverstellung f
  axis of rotation Rotationsachse f
  cross-axis roll adjustment Walzenschrägeinstellung f, Walzenschrägstellung f, Walzenschrägverstellung f, Schrägverstellung f
  longitudinal machine axis Maschinenlängsachse f
  longitudinal mo(u)ld axis Werkzeuglängsachse f
  screw axis Schneckenachse f, Wellenachse f
azelaic acid Azelainsäure f
azelaic acid ester Azelat n, Azelainsäureester m
azelate Azelat n, Azelainsäureester m
  dioctyl azelate Dioctylazelat n
  di-2-ethylhexyl azelate Di-2-ethylhexylazelat n
azeotropic azeotrop
azimuthal azimutal
azo Azo-
  azo compound Azoverbindung f
  azo initiator Azoinitiator m
  azo pigment Azopigment n
azoisobutyronitrile Azoisobutyronitril n
azodicarbonamide Azodicarbonamid n
back 1. Rücken m; 2. zurück, hinter
  back flushing Rückspülung f
  back of the machine Maschinenrückseite f
  back of the unit/instrument Geräterückseite f
  back plate Rückplatte f
  back pressure Staudruck m, Extrusionsstaudruck m, Rückstaudruck m, Rückstau m, Rückdruck m, Gegendruck m
  back pressure adjustment Staudruckeinstellung f
  back pressure controller Staudruckregler m
  back pressure forces Rückdruckkräfte fpl
  back pressure program(me) Staudruckprogrammverlauf m
  back pressure reduction Staudruckabbau m
  back pressure relief Staudruckentlastung f
  back titration Rücktitration f
  back venting (system) Rückwärtsentgasung f
  cavity back pressure Werkzeugdruck m
  die back pressure Werkzeugrückdruck m, Werkzeuggegendruck m
  melt back-flow Schmelzerückfluß m
  melt back flushing Schmelzenrückspülung f
  screw back pressure Schneckenrückdruck(kraft) m(f), Schneckenstaudruck m, Schneckengegendruck m
backfilling Hinterfüttern n
backflow Rückströmung f, Rückfluß m (bei Rückfluß der Schmelze in den Zylinder / when the melt flows back into the cylinder)
backing Unterstützung f, Verstärkung f, Trägerschicht f, Rückseite f
  backing fabric Trägergewebe n
  backing mix Hinterfütterungsmasse f
  backing paper Trägerpapier n
  backing plate Stützplatte f
  backing roll Gegendruckwalze f, Presseurwalze f
  carpet backing Teppichrückenbeschichtung f, Teppichrückseitenbeschichtung f
bactericidal bakterizid
bad [when describing properties, performance, etc.] schlecht
badly designed falsch ausgelegt
baffle Stauelement n, Prallplatte f, Schikane f, Umlenkblech n
bag Beutel m
  automatic bag-making machine Beutelautomat m, Beutelherstellungsautomat m, Beutelmaschine f
  automatic bag-welding machine Beutelschweißautomat m
  bottom-weld bag Bodennahtbeutel m
  carrier bag Tragetasche f
  polyethylene bag PE-Beutel m
  refuse bag Müllsack m
  vacuum bag mo(u)lding Vakuumfolienverfahren n
bagging unit Absackanlage f
baked eingebrannt
bakery goods Backwaren fpl
baking Einbrennen n, Wärmebehandeln n, Härten n, Trocknen n
balance 1. Waage f; 2. Gleichgewicht n, Balance f; Bilanz f

balance sheet Bilanz f
balance sheet total Bilanzsumme f
analytical balance Analysenwaage f
energy balance Energiebilanz f, Energiehaushalt m
heat balance Wärmebilanz f, Wärmehaushalt m, Wärmestrombilanz f
precision balance Feinwaage f, Genauigkeitswaage f
balanced ausgeglichen, ausgewuchtet, ausgewogen [z.B. Eigenschaften]
balcony Balkon m
ball Ball m, Kugel f
  ball bearing Kugellager n
  ball bearing cage Kugellagerkäfig m
  ball indentation hardness Kugel(ein)druckhärte f
  ball indentation test Kugeldruckprüfung f, Kugeleindruckverfahren n
  ball mill Kugelmühle f
  ball valve Kugelventil n
  falling ball test Kugelfallversuch m
band Band n
  band heater Heizband n, Temperierband n
  band saw Bandsäge f
  absorption band Absorptionsband n
  brake band Bremsband n
  ceramic band heater Keramikheizkörper m
  extra heater band Zusatzheizband n
  heater band Heizband n, Temperierband n
  mica-insulated heater bands Glimmerheizbänder npl
  nozzle heater band Düsenheizband n
  resistance band heater Widerstandsheizband n
bank 1. Damm m, Ufer n; 2. Bank f; 3. Knet m, Masseknet m
  bank loan Bankkredit m
  bank marks Oberflächenmarkierungen fpl
  data bank Datenbank f, Datenspeicher m
bar Stange f, Stab m, Riegel m, Balken m, Steg m, Holm m
  bar code Balkencode m, Barcode m, Strichcode m
  bar display Balkenanzeige f
  bar graph Säulendiagramm n
  bar shaped stabförmig
  choke bar Staubalken m, Stauleiste f
  distance between tie bars lichte Weite f zwischen den Holmen mpl, Holmabstand m
  guide bars Leitstangen fpl
  (heat) sealing bar Siegelbacke f
  pinch-off bars Schneidkanten fpl, Schweißkanten fpl
  restrictor bar Staubalken m, Stauleiste f
  restrictor bar adjusting mechanism Staubalkenverstellung f
  stop bars Anschlagleisten fpl
  tie bar Säule f
  tie bar pre-stressing Säulenvorspannung f
  torsion bar Drehstab m
  two-tie bar clamp unit Zweiholmenschließeinheit f
  two-tie bar design Zweiholmenausführung f
  welding bars Schweißbacken fpl
  with two tie bars zweiholmig
  without tie bars holmenlos
barium Barium n, (Ba)
  barium-cadmium stabilizer Barium-Cadmium-Stabilisator m
  barium metaborate Bariummetaborat n
  barium octoate Bariumoctoat n
  barium soap Bariumseife f
  barium sulphate Bariumsulfat n
  precipitated barium sulphate Blanc fixe n
barometric pressure Barometerdruck m
barrel Trommel f, Walze f, Faß n, Zylinder m, Förderzylinder m, Gehäuse n
  barrel assembly (of a compounding unit) Knetergehäuse n
  barrel bushing Zylinderbuchse f
  barrel cooling sections Zylinderkühlzonen fpl
  barrel cooling unit Zylinderkühlaggregat n
  barrel dimensions Zylinderabmessungen fpl
  barrel head Zylinderkopf m
  barrel heater Zylinderheizelement n, Zylinderheizung f, Zylinderbeheizung f
  barrel heating capacity Zylinderheizleistung f
  barrel heating circuit Zylinderheizkreis m
  barrel heating zones Zylinderheizzonen fpl
  barrel length Zylinderlänge f
  barrel liner Zylinderinnenfläche f, Zylinderinnenwand f, Zylinderauskleidung f, Gehäuseinnenwandung f
  barrel-screw combination Zylinder-Schneckensystem n
  barrel section Gehäuseabschnitt m, Gehäuseschuß m, Zylinderelement n, Gehäuseteile npl
  barrel segment Zylindersegment n
  barrel temperature control Zylindertemperierung f
  barrel temperature Zylindertemperatur f
  barrel vented through longitudinal slits Längsschlitz-Entgasungsgehäuse n
  barrel wall Zylinderwand f, Gehäusewand f
  barrel wall temperature Zylinderwandtemperatur f
  barrel wear Zylinderverschleiß m
  bimetallic barrel Bimetallzylinder m
  extruder barrel Extrudergehäuse n, Extruderzylinder m, Extrusionszylinder m, Schnecken(führungs)zylinder m, Schneckengehäuse n, Schneckenrohr n, Zylindergehäuse n, Zylinderrohr n
  grooved barrel Nutenzylinder m
  nitrided steel barrel Nitrierstahlgehäuse n
  outer barrel surface Zylinderaußenfläche f
  pin-lined barrel Stiftzylinder m
  plasticating barrel Plastifizierzylinder m
  plasticizing barrel Plastifizierzylinder m
  reciprocating barrel accumulator Zylinderschubspeicher m
  single-screw barrel Einschneckenzylinder m
  twin barrel Doppelzylinder m
  twin screw barrel Doppelschneckenzylinder m

vented barrel Entgasungszylinder m
vented barrel extruder Zylinderentgasungsextruder m
barrier Barriere f, Sperre f
  barrier coat(ing) Sperrschicht f, Barriereschicht f
  barrier cream Hautschutzcreme f, Schutzcreme f
  barrier effect Sperrwirkung f, Barriereschicht f
  barrier film Sperrschicht f, Barriereschicht f
  barrier flight Sperrsteg m
  barrier layer Sperrschicht f, Barriereschicht f
  barrier plastic Barriere-Kunststoff m
  barrier properties Barrierewerte mpl, Barriere-eigenschaften fpl, Sperr(schicht)eigenschaften fpl
  barrier sheet(ing) Sperrfolie f
  acting as a barrier against plasticizer migration weichmachersperrend
  heat barrier Wärmesperre f
  sound barrier Schallverkleidung f
  vapo(u)r barrier (sheeting) Dampfsperrbahn f, Dampfsperre f
baryte(s) Schwerspat m
base 1. Basis f, Boden m, Grundfläche f; 2. Base f
  base deflashing device Bodenentgrateinrichtung f
  base flash Bodenabfall m
  base frame Grundrahmen m
  base material Basismaterial n, Grundmaterial n, Grundwerkstoff m
  base plate Grundplatte f
  base resin Basisharz n, Grundharz n
  base tail Bodenbutzen m
  machine base Maschinenbett n
  Mannich base Mannichbase f
  neck and base flash Abquetschlinge f
basic 1. Grund-, Basis-; 2. basisch, alkalisch
  basic components Grundbauteile npl
  basic concept Grundkonzept n
  basic condition Grundvoraussetzung f
  basic data Grunddaten npl
  basic design Grundkonstruktion f, Grundkonzeption f
  basic development Grundentwicklung f
  basic equipment Basisausrüstung f, Grundausstattung f
  basic features Grundmerkmale npl
  basic formulation Basisrezeptur f, Grundrezeptur f
  basic framework Grundgerüst n
  basic grade (of material) Grundtyp m
  basic idea Grundkonzept n
  basic machine Basismaschine f, Grundmaschine f
  basic model Basisausführung f
  basic movements Grundbewegungen fpl
  basic principle Grundprinzip n
  basic properties Grundeigenschaften fpl
  basic range (of machines, equipment etc.) Grundreihe f
  basic screw Ausgangsschnecke f, Grundschnecke f
  basic system Grundsystem n
  basic unit Grundbaustein m, Grundgerät n
basket Korb m

  calibrating basket Führungskorb m, Rollenkorb m
  sizing basket Führungskorb m, Rollenkorb m
batch Charge f, Ansatz m
  batch control Chargenprüfung f, Partiekontrolle f, Partieprüfung f
  batch differences Chargenunterschiede mpl, Chargenschwankungen fpl
  batch number Chargennummer f
  batch processing Stapelverarbeitung f
  batch size [e.g. of a two-pack adhesive after mixing] Chargenmenge f, Ansatzgröße f
  batch-type plant Batchanlage f
  batch variations Chargenunterschiede mpl, Chargenschwankungen fpl
  production batch Fertigungscharge f
batchwise chargenweise
bath Bad n
  cleaning bath Reinigungsbad n
  dip coating bath Tauchbad n
  dipping bath Tauchbad n
  impregnating bath Tränkbad n, Tränkwanne f
  oil bath Ölbad n
  pickling bath Beize f
  quench bath (for cooling extruded pipe or film) Kühlbad n
  solder bath Lötbad n
  solder bath immersion Lötbadlagerung f
  solder bath resistance Lötbadfestigkeit f
bathroom Sanitärzelle f, Badezimmer n
battery Batterie f
  battery-buffered batteriegepuffert
  battery charger Batterieladegerät n
  battery-powered mit Batteriestromversorgung f
  battery separator Batterieseparator m
bayonet coupling joint Bajonettverschluß m
bead 1. Schweißwulst m; 2. Perle f, Kügelchen n
  bead polymerization Perlpolymerisation f
  glass beads Glaskugeln fpl
  glazing bead Glasleiste f
  pre-expanded beads Vorschaum m, Vorschaumperlen fpl
beaker Becherglas n
beam Strahl m
  electron beam Elektronenstrahl m
  electron beam curing EB-Verfahren n, Elektronenstrahlhärtung f
  electron beam oscillograph Elektronenstrahloszillograph m
  laser beam Laserlicht n, Laserstrahl m
  laser beam cutting Laserstrahlschneiden n
  laser beam cutting process Laserschneidverfahren n
  light beam Lichtstrahl m
  light beam guard Lichtschranke f, Lichtvorhang m
  light beam oscillograph Lichtstrahloszillograph m
beaming unit Bäumanlage f
bean Bohne f
  soya bean oil Soja(bohnen)öl n

soya bean oil fatty acid Sojaölfettsäure f
bearing Lager n
   bearing bush Lagerbuchse f, Lagerschale f
   bearing friction Lagerreibung f
   bearing housing Lagergehäuse n
   bearing pressure Lagerkraft f
   bearing seat Lagerstelle f
   anti-friction bearing Wälzlager n
   anti-friction bearing grease Wälzlagerfett n
   ball bearing Kugellager n
   ball bearing cage Kugellagerkäfig m
   bridge bearing pad Brückenlager n
   pin bearing Nadellager n
   rol(ler) bearing Rollenlager n, Zylinderrollenlager n, Walzenlager n, Walzenlagerung f
   rotor bearing Rotorlagerung f
   self-aligning roller bearing Pendelrollenlager n
   sliding bearing Gleitlager n
   thrust bearing Axiallagerung f, Drucklager n, Drucklagerung f
   thrust bearing assembly Rückdrucklagerung f
   thrust bearing unit Axiallagergruppe f
become/to werden
   become brittle verspröden, brüchig werden
   become worn verschleißen
   become yellow vergilben
bed Bett n, Unterlage f
   bed knife Statormesser n, Festmesser n, Gegenmesser n
   bed knife block Statormesserbalken m, Statormesserblock m
   bed knife cutting circle Statormesserkreis m
   electrostatic fluidized bed coating elektrostatisches Wirbelsintern n
   fluidized bed Wirbelbett n
   fluidized bed coater Wirbelsintergerät n [für Überzüge]
   fluidized bed coating Wirbelsintern n [für Überzüge]
   high voltage test bed Hochspannungsprüffeld n
   test bed Prüffeld n, Prüfstand m, Testgelände n, Versuchsfeld n
   test bed trial Prüfstandversuch m
bedding-down Tuschieren n
   bedding-down press Tuschierpresse f
behavior [GB: behaviour] Verhalten n
   behavio(u)r on impact Schlagverhalten n
   blocking behavio(u)r Blockverhalten n
   creep behavio(u)r Kriechverhalten n, Zeitdehnverhalten n, Zeitstandverhalten n
   compressive behavio(u)r Druckverhalten n
   curing behavio(u)r Vernetzungsverhalten n, Härtungsverhalten n, Aushärteverhalten n
   damping behavio(u)r Dämpfungsverhalten n
   decomposition behavio(u)r Zersetzungsverhalten n
   deformation behavio(u)r Verformungsverhalten n
   diffusion behavio(u)r Permeationsverhalten n
   extraction behavio(u)r Extraktionsverhalten n
   fire behavio(u)r Brandverhalten n, Brennverhalten n
   flexural behavio(u)r Biegeverhalten n
   flexural impact behavio(u)r Schlagbiegeverhalten n
   fracture behavio(u)r Bruchverhalten n
   frictional behavio(u)r Reib(ungs)verhalten n
   gelling behavio(u)r Gelierverhalten n
   heat ag(e)ing behavio(u)r Wärmealterungsverhalten n
   high temperature behavio(u)r Hochtemperaturverhalten n
   impact behavio(u)r Schlagverhalten n
   migration behavio(u)r Wanderungsverhalten n
   permeability behavio(u)r Permeabilitätsverhalten n
   power law behavio(u)r (general term) Potenzgesetzverhalten n
   recovery behavio(u)r Erholungsverhalten n
   relaxation behavio(u)r Relaxationsverhalten n
   retardation behavio(u)r Retardationsverhalten n
   short-term behavio(u)r Kurzzeitverhalten n
   slip behavio(u)r Gleitverhalten n
   start-up behavio(u)r Anfahrverhalten n
   stress cracking behavio(u)r Spannungskorrosionsverhalten n
   stress-strain behavio(u)r Spannungs-Dehnungs-Verhalten n, Spannungs-Verformungsverhalten n
   swelling behavio(u)r Quellverhalten n
   tensile behavio(u)r Zugverhalten n
   viscosity behavio(u)r Viskositätsverhalten n
   weathering behavio(u)r Witterungsverhalten n, Bewitterungsverhalten n
bellows Faltenbalg m
belt Gürtel m, Riemen m, Band n
   belt pulley Riemenscheibe f
   abrasive belt Schleifband n
   abrasive belt grinder Bandschleifmaschine f
   abrasive belt polisher Bandschleifpoliermaschine f
   assembly belt Montageband n
   conveyor belt Förderband n, Fördergurt m, Transportband n, Fließband n
bend/to abkanten, falten
bend Biegung f
   pipe bend Rohrbogen m, Rohrkrümmer m
   stiffness in bend Biegesteifigkeit f
bending Biegen n, Biegeumformen n
   bending angle Biegewinkel m
   bending force Biegekraft f
   bending jig Biegeschablone f
   bending moment Biegemoment n
   bending stress Biegespannung f
   bending stress at break Bruchbiegespannung f
   four-point bending Vierpunkt-Biegung f
   four-point bending test Vierpunkt-Biegeversuch m
   hot bending Warmbiegen n
   long-term torsional bending stress Dauertorsionsbiegebeanspruchung f
   modulus of elasticity in bending Biege-E-Modul m, Biegeelastizitätsmodul m
   roll bending Gegenbiegen n, Walzengegenbiegung f

roll bending mechanism
Walzengegenbiegeeinrichtung f,
Walzen(durch)biegevorrichtung f
tensile strength in bending Biegezugfestigkeit f
three-point bending Dreipunktbiegung f
three-point bending test Dreipunkt-Biegeversuch m
under bending stress biegebelastet, biegebeansprucht
bentonite Bentonit n
benzene Benzol n
  benzene extract Benzolextrakt m
  benzene ring Benzolring m, Benzolkern m
  alkyl benzene Alkylbenzol n
  dodecyl benzene Dodecylbenzol n
benzine Benzin n
benzoat Benzoat n
  sodium benzoate Natriumbenzoat n
benzoguanamine resin Benzoguanaminharz n
benzoic acid Benzoesäure f
benzophenone Benzophenon n
benzoquinone Benzochinon n
benzoyl peroxide Benzoylperoxid n
  benzoyl peroxide paste Benzoylperoxid-Paste f
benztriazole Benztriazol n
benzyl- Benzyl-
  benzyl butyl phthalate (BBP) Benzylbutylphthalat n
  benzyl cellulose Benzylcellulose f
  benzyl dimethylamine Benzyldimethylamin n
  butyl benzyl adipate Butylbenzyladipat n
  butyl benzyl phthalate (BBP) Butylbenzylphthalat n
benzylamine Benzylamin n
  dimethyl benzylamine Dimethylbenzylamin n
beryllium-copper alloy Berylliumkupfer n
beverage Getränk n
  beverage bottle Getränkeflasche f
biaxial zweiachsig
  biaxial orienting (of film) biaxiales Recken n
  biaxial stretching (of film) biaxiales Recken n
  biaxial stretching unit/equipment Biaxial-Reckanlage f
biaxially oriented biaxial gestreckt, biaxial orientiert, biaxial verstreckt
bifunctional bifunktionell, difunktionell
bill (of exchange) Wechsel m
bimetallic bimetallisch, Bimetall-
  bimetallic barrel Bimetallzylinder m
  bimetallic thermostat Bimetallthermostat m
binary binär
  binary code Binär-Code m
  binary-coded binär kodiert
  binary number Binärzahl f
binder Bindemittel n
  binder combination Bindemittelkombination f
  binder concentration Bindemittelgehalt m
  binder content Bindemittelgehalt m
  binder twine Bindegarn n
  printing ink binder Druckfarbenbindemittel n
binding Bindung f, Binde-

  binding power Bindekraft f, Bindevermögen n
  pigment binding power Pigmentbindevermögen n
bipolarity Bipolarität f
birefringence Doppelbrechung f
bisphenol Bisphenol n
  bisphenol-A-diglycidyl ether Bisphenol-A-Diglycidylether m
  bisphenol resin Bisphenolharz n
bit Bit n
  bit processor Bitprozessor m
bitumen Bitumen n
  bitumen resistance Bitumenbeständigkeit f
  bitumen resistant bitumenbeständig
bituminous bituminös
biuret Biuret n
  biuret linkage Biuretbindung f
  biuret linkage content Biuretbindungsanteil m
  biuret structure Biuretstruktur f
bivalent zweiwertig
black 1. schwarz; 2. Schwarz n, Schwärze f, Ruß m
  active carbon black Aktivruß m
  carbon black Farbruß m, Kohlenstoffpigment n, Rußpigment n, Ruß m
  carbon black concentrate Rußpaste f
  filled with carbon black rußgefüllt
  furnace black Ofenruß m
  jet black tiefschwarz
  lamp black Flammruß m
blade (of a mixer) Flügel m, Blatt n, Schaufel f
  blade clearance (This is the gap between the bed knife and rotor knife of a granulator) Schneidspalt m
  doctor knife blade Rakelblatt n
  fan blade Lüfterflügel m
  kneader blade Knetarm m
  mixing blade Schaufel f
  rotor blade Dispersionsflügel m
  saw blade Sägeblatt n
  spring steel blade Federstahlmesser n
blanc fixe Blanc fixe n
blank blank, leer, Blind-
  blank test Blindversuch m
  blanks [of sheet stock, prepregs ect.] Zuschnitte mpl
  cut-to-size blank Plattenzuschnitt m
  gear wheel blank Zahnradrohling m
  preform blank Rohling m
blasting Sprengen n, Putzstrahlen n
  grit blasting Schrotstrahlen n
  grit blasting unit Strahlanlage f
bleeding (of pigments or dyes) Ausbluten n, Kontaktbluten n
blend Abmischung f (Abmischungen mit Naturharzen sind ebenfalls möglich / blends with natural resins are also possible)
  additive blend Additivsystem n
  antioxidant blend Antioxidationssystem n
  dry blend Dry-Blend-Mischung f, pulverförmige

Formmasse f, Pulvercompound m, Pulvermischung f
dry blend extruder Pulverextruder m
dry blend processing Pulververarbeitung f
dry blend screw Pulverschnecke f
elastomer blend Elastomerblend n
emulsifier blend Emulgatorgemisch n
epoxy blend Epoxyverschnitt m
liquid resin blend Flüssigharzkombination f
lubricant blend Gleitmittelgemisch n, Gleit(mittel)system n, Kombinationsgleitmittel n
plasticizer blend Weichmachermischung f
polymer blend Kunststoff-Legierung f, Polymermischung f, Polymer(isat)gemisch n, Polymerlegierung f
solid resin blend Festharzkombination f
stabilizer blend Stabilisatorkombination f, Stabilisatormischung f
stabilizer-lubricant blend Stabilisator-Gleitmittel-Kombination f
blended verschnitten
  blended with im Verschnitt mit ...
blending Legieren n, Verschneiden n
  blending resin Verschnittharz n
blind blind
  blind hole Sackloch n, Sacklochbohrung f
  roller blind profile Rolladenprofil n
  roller blind slats Rolladenstäbe mpl
  roller blinds Rolladen m
blister Blase f
  blister pack Blisterpackung f
  blister packaging machine Blisterformmaschine f
blistering Blasenbildung f
block Block m, Rolle f
  block copolymer Blockcopolymer n, Blockpolymerisat n, Segmentpolymerisat n
  block copolymerization Block-Copolymerisation f
  block diagram Blockdarstellung f, Blockdiagramm n
  bed knife block Statormesserblock m, Statormesserbalken m
  circuit block diagram Blockschaltbild n
  distance block Distanzblock m
  hot runner manifold block Heißkanalverteilerbalken m, Heißkanalverteilerblock m, Heizblock m, Querverteiler m
  kneading block Knetblock m
  kneading block assembly Knetblockanordnung f
  manifold block Verteilerstück n, Verteilerbalken m, Verteilerblock m
  nozzle block Düsenblock m
  rotor knife block Rotormesserbalken m
  tendency to block Blockneigung f
blocking Blocken n
  blocking behavio(u)r Blockverhalten n
  blocking force Blockkraft f
  blocking resistance Blockfestigkeit f
  blocking temperature Blockpunkt m
  blocking tendency Blockneigung f

bloom Ausblühung f
blooming [e.g. unwanted migration of sulfur or other substances to rubber surfaces to produce discoloration] Ausblühen n
blow Blasen n
  blow-fill-seal packaging line Blasform-, Füll- und Verschließanlage f
  blow film line Folienblasanlage f
  blow mo(u)ld cavity Blasgesenk n
  blow mo(u)ldable blasbar
  blow mo(u)lded blown geblasen
  blow mo(u)lded part geblasener/blasgeformter Hohlkörper m, Blasteil n, Blaskörper m, Blasformteil n
  blow mo(u)lder Hohlkörperhersteller m
  blow mo(u)lding Blasformen n, Blasformung f, Blasformverfahren n, Hohlkörperblasprozeß m, Blasprozeß m, blasgeformter Hohlkörper m, Blasformteil n, Blasteil n, Blaskörper m
  blow mo(u)lding compound Blasmasse f, Blasformmasse f, Hohlkörpercompound m,n Hohlkörper-Thermoplast m
  blow mo(u)lding grade (of mo(u)lding compound) Blasmarke f
  blow mo(u)lding line Blaslinie f, Blasanlage f, Hohlkörperblasanlage f, Blasformanlage f, Hohlkörperproduktionsanlage f
  blow mo(u)lding machine Blasformmaschine f, Blasmaschine f
  blow mo(u)lding plant Blaslinie f, Blasanlage f, Hohlkörperblasanlage f, Blasformanlage f, Hohlkörperproduktionsanlage f
  blow mo(u)lding process Hohlkörperherstellung f, Hohlkörperfertigung f, Hohlkörperblasen n, Blasverfahren n
  blow mo(u)lding station Blasformstation f
  blow mo(u)lding system Blasformsystem n
  blow mo(u)lding technology Blasformtechnik f
  blow mo(u)lding trials Blasversuche mpl
  blow mo(u)lding unit Blasaggregat n
  blow mo(u)ldings blasgeformte Teile npl
  accumulator head blow mo(u)lding (process) Staukopfverfahren n
  automatic blow mo(u)lding machine Blasformautomat m, Hohlkörperblasmaschine f, Blasautomat m
  automatic high capacity/performance blow mo(u)lding machine Hochleistungsblasformautomat m
  automatic injection blow mo(u)lder/mo(u)lding machine Spritzblasautomat m
  automatic oriented PP stretch blow mo(u)lding machine OPP-Streckblasautomat m
  coextrusion blow mo(u)lding line/plant Coextrusions-Blasanlage f
  coextrusion blow mo(u)lding machine Coextrusionsblasformmaschine f
  coextrusion blow mo(u)lding (process) Coextrusions-Blasverfahren n, Koextrusionsblasen n
  compression blow mo(u)lding Kompressionsblasen n,

Kompressionsformblasen n
compression blow mo(u)lding process Kompressionsblasformverfahren n
dip blow mo(u)lder/mo(u)lding machine Tauchblasmaschine f
dip blow mo(u)lding Tauchblasen n, Tauchblasformen n
dip blow mo(u)lding (process) Tauchblasverfahren n
extrusion blow mo(u)lded extrusionsgeblasen
extrusion blow mo(u)lder Extrusionsblasmaschine f, Extrusionsblasformmaschine f
extrusion blow mo(u)lding (process) Extrusionsblasformprozeß m, Extrusionsblas(form)verfahren n
extrusion blow mo(u)lding line/plant Blasextrusionsanlage f
extrusion blow mo(u)lding machine Blasextrusionsmaschine f, Extrusionsblasmaschine f
extrusion stretch blow mo(u)lding Extrusions-Streckblasformen n, Extrusions-Streckblasen n
foam blow mo(u)lding (process) Schaumblasverfahren n
high-capacity blow mo(u)lding line/plant Großblasformanlage f, Hochleistungsblasformanlage f
injection blow mo(u)ld Spritzblaswerkzeug n
injection blow mo(u)lded spritzgeblasen
injection blow mo(u)lding Spritzblasen n, Spritzblasformen n, Spritzgieß-Blasformen n
injection blow mo(u)lding process Spritzblasverfahren n
injection blow mo(u)lding machine Spritzblasmaschine f
injection stretch blow mo(u)lding Spritzgieß-Streckblasen n, Spritzstreckblasen n
injection stretch blow mo(u)lding machine Spritzstreckblasmaschine f, Spritzgieß-Streckblasmaschine f
machine for blow mo(u)lding drums Faßblasmaschine f
machine for blow mo(u)lding fuel oil storage tanks Heizöltank-Blasmaschine f
oriented polypropylene blow mo(u)lding machine OPP-Maschine f
removal of blow mo(u)ldings Blaskörperentnahme f
single-station automatic blow mo(u)lding machine Einstationenblasformautomat m, Einstationenblas(form)maschine f
single-station extrusion blow mo(u)lding machine Einstationenextrusionsblasformanlage f
small blow mo(u)lding Kleinhohlkörper m
stretch blow mo(u)lder Streckblasmaschine f, Streckblasformmaschine f
stretch blow mo(u)lding Streckblasformen n, Streckblasen n
stretch blow mo(u)lding process Streckblasverfahren n
three-layer blow mo(u)lding Dreischichthohlkörper m
two-cavity injection blow mo(u)ld Zweifachspritzblaswerkzeug n

two-layer blow mo(u)lding Zweischichthohlkörper m
two-stage blow mo(u)lding process Zweistufenblasverfahren n
two-stage extrusion blow mo(u)lding Zweistufen-Extrusionsblasformen n
two-stage injection blow mo(u)lding Zweistufenspritzblasen n
two-stage injection stretch blow mo(u)lding process Zweistufen-Spritzstreckblasverfahren n
two-station blow mo(u)lding machine Zweistationenblasmaschine f
blow up/to aufblasen
blow-up conditions Aufblasbedingungen fpl
blow-up ratio Aufblasverhältnis n
blow-up temperature Aufblastemperatur f
blower Gebläse n
cooling air blower Kühlluftgebläse n
high pressure blower Hochdruckgebläse n
hot air blower Warmluftgebläse n
material transport blower Fördergebläse n
suction blower Sauggebläse n
blowing Blasverfahren n, Blasverarbeitung f, Blasvorgang m
blowing agent Treibsystem n; Schaumbildner m, Schäummittel n
blowing agent concentrate Treibmittelkonzentrat n
blowing agent content Treibmittelgehalt m, Treibmittelanteil m
blowing air Blasluft f, Aufblasluft f
blowing air pressure Blasluftdruck m
blowing air supply Blasluftzufuhr f
blowing device Blasvorrichtung f
blowing gas Treibgas n
blowing mandrel Blasdorn m, Spritzdorn m, Spritzblasdorn m, Blasspindel f, Aufblasdorn m
blowing mandrel support Blasdornträger m
blowing medium Blasmedium n
blowing needle Hohlnadel f
blowing pin Blasnadel f, Blasstift m, Injektionsblasnadel f
blowing position Blasposition f
blowing pressure Blasdruck m
blowing spigot Blasdorn m, Spritzdorn m, Spritzblasdorn m, Blasspindel f, Aufblasdorn m
blowing station Blasstation f
blowing unit Blasaggregat n
amount of blowing agent Treibmittelmenge f
automatic bottle blowing machine Flaschenblasautomat m
bottle blowing compound Flaschencompound m,n
bottle blowing line Flaschenproduktionslinie f
bottle blowing machine Flaschen(blas)maschine f
film blowing Folienblasen n, Folienblasverfahren n, Schlauchfolienblasen n, Schlauchfolienextrusion f, Schlauchfolienfertigung f, Schlauchfolienherstellung f, Blasextrusion f
film blowing die Schlauchformeinheit f,

Schlauchwerkzeug n, Schlauch(extrusions)düse f, Schlauch(spritz)kopf m, Blasfolienschlauchkopf m, Folienblaskopf m, Schlauchfoliendüse f, Schlauchfolien(extrusions)werkzeug n, Schlauchfolienkopf m, Blaswerkzeug n, Blasdüse f
film blowing line Folienblasanlage f, Schlauchfolienanlage f, Schlauchfolienblasanlage f, Schlauchfolienextrusionsanlage f
film blowing machine Folienblaseinheit f
free from blowing agent treibmittelfrei
mechanical blowing (process) [method of making foam] Direktbegasungsverfahren n, Begasungsverfahren n
mechanical blowing unit Direktbegasungsanlage f, Begasungsanlage f
twin-die film blowing head Doppelblaskopf m
twin-die film blowing line Doppelkopf-Blasfolienanlage f
two-layer film blowing line Zweischicht-Folienblasanlage f
type of blowing agent Treibmittelart f
blown blasgeformt, getrieben, geblasen
  blown container geblasener/blasgeformter Hohlkörper m, Blaskörper m, Blas(form)teil n
  blown film Blasfolie f, Schlauchfolie f
  blown film coextrusion die Koextrusionsblaskopf m, Koextrusionsblaswerkzeug n
  blown film coextrusion line coextrudierende Schlauchfolienanlage f, Coextrusions-Blasfolienanlage f
  blown film cooling system Schlauchfolienkühlung f
  blown film die Schlauchfoliendüse f, Schlauchfolienwerkzeug n, Blaswerkzeug n, Blasfolienschlauchkopf m, Folienblaskopf m, Schlauchfolienextrusionswerkzeug n, Schlauch(extrusions)düse f, Schlauchwerkzeug n, Schlauchformeinheit f, Schlauchfolienkopf m, Schlauch(spritz)kopf m, Foliendüse f, Folienkopf m, Folien(spritz)werkzeug n
  blown film die design Blaskopfkonzeption f
  blown film extruder Schlauchfolienextruder m
  blown film (extrusion) line Blasanlage f
  blown film extrusion (process) Blasprozeß m, Folienblasen n, Blasen n, Blasverarbeitung f, Blasvorgang m, Blasverfahren n, Blasextrusion f, Schlauchfolienfertigung f, Schlauchfolienherstellung f, Schlauchfolienverfahren n, Folienblasverfahren n, Folienblasen n, Schlauchfolienextrusion f, Schlauchfolienblasen n
  blown film line Schlauchfolien(blas)anlage f
  blown film orientation (process) Schlauchstreckverfahren n
  blown film plant Folienblasbetrieb m
  blown film stretching (process) Schlauchstreckverfahren n
  blown foam [using gas as opposed to blowing agent] Begasungsschaum m
  blown part removal device Blaskörperentnahme f
  blown polyethylene film Polyethylenschlauchfolie f
  center-fed blown film die [GB: centre] zentral angespritzter Blaskopf m, Dornhalterblaskopf m
  chemically blown chemisch getrieben
  chemically blown foam chemischer Schaum m, Treibmittelschaumstoff m, Begasungsschaum m
  four-die blown film extrusion line Vierfach-Schlauchfolien-Extrusionsanlage f
  high capacity blown film line Hochleistungs-Schlauchfolienanlage f, Hochleistungsblasfolienanlage f
  mechanically blown [expanded by introducing gas as opposed to chemical blowing using a blowing agent] direktbegast, mechanisch getrieben
  mechanically blown foam mechanischer Schaum m, Schlagschaum m
  paste for making mechanically blown foam Schlagschaumpaste f
  side-fed blown film die seitlich eingespeister Folienblaskopf m, stegloser Folienblaskopf m, Umlenkblaskopf m, Pinolenblaskopf m
  side-gussetted blown film Seitenfaltenschlauchfolie f
  spider-type blown film die Stegdorn(halter)blaskopf m, Steg(dorn)halterkopf m, Stegdornhalterwerkzeug n
  spiral mandrel blown film die Schmelzewendelverteilerwerkzeug n, Schmelzewendelverteilerkopf m, Spiraldorn(blas)kopf m, Wendelverteilerwerkzeug n, Wendelverteilerkopf m
  three-layer blown film die Dreischicht-Folienblaskopf m
  two-layer blown film die Zweischicht-Folienblaskopf m
blue blau
  blue-collar workers gewerbliche Mitarbeiter mpl
  cobalt blue Kobaltblau n
  phthalocyanine blue Phthalocyaninblau n
  toolmaker's blue Tuschierfarbe f
blushing [of paint film, due to excessive moisture] Weißanlauf m
board Brett n, Bohle f
  collapsing boards Flachlegebretter npl, Flachlegeeinrichtung f, Flachlegebleche npl, Leitbleche npl, Leitplatten fpl, Flachlegung f
  printed circuit board Leiterplatte f, Elektrolaminat n, Elektroschichtpreßstoff m
  skirting board Fußbodenleiste f, Sockelleiste f
boat Schiff n, Boot n
  combustion boat Verbrennungsschiffchen n
boatbuilding Bootsbau m
bobbin Spule f, Trommel f, Bobine f
  cheese bobbin Kreuzspule f
body Körper m
  body-compatible körperverträglich
  car body Karosserie f

    die body Düsenkörper m, Werkzeugkörper m
    screen changer body Siebwechslerkörper m
    vehicle body Karosserie f
boiler Sieder m, (Dampf-)Kessel m, Boiler m
    steam boiler Dampfkessel m
boiling siedend
    boiling point Siedepunkt m, Siedetemperatur f
    boiling range Siedeintervall n
    boiling test Kochversuch m
    boiling water Kochwasser n
    low boiling niedrigsiedend
    resistant to boiling water kochfest
bolt Bolzen m
    bolt-type mandrel support Bolzendornhalterung f
    angled bolt Schrägbolzen m
    ejector bolt Auswerferbolzen m
    guide bolt Führungsbolzen m
    locking bolt Riegelbolzen m, Sperrbolzen m
    retaining bolt Haltebolzen m
bond/to (ver)kleben
    bond strength Bindefestigkeit f, Klebfestigkeit f, Klebwert m, Trennfestigkeit f
    atomic bond Atombindung f
    double bond Doppelbindung f
    double bond content Doppelbindungsgehalt m, Doppelbindungsanteil m
    easy to bond klebefreundlich
    primary valency bond Hauptvalenzbindung f
    single bond Einfachbindung f
bondable verklebbar
bonded geklebt
    bonded alumin(i)um joint Aluminiumverklebung f
    bonded area Verklebungsfläche f
    bonded glass joint Glas(ver)klebung f
    bonded joint Klebeverbindung f, Verklebung f
    bonded plastics joint Kunststoffverklebung f
    bonded steel joint Stahl(ver)klebung f
bonding Verklebung f
    bonding agent Klebemittel n
    bonding of glass Glas(ver)klebung f
    bonding of plastics Kunststoffverklebung f
    bonding of steel Stahl(ver)klebung f
    hot bonding Warmkleben n
    hot melt bonding Heißschmelzverklebung f
bonnet Fronthaube f, Frontklappe f, Motorhaube f
bookbinding film Bucheinbandfolie f
booklet Broschüre f
booster pump Druckerhöhungspumpe f
boot Kofferraum m
    boot liner Kofferraumauskleidung f
booth Bude f, Zelle f, Kabine f
    paint spraying booth Spritzkabine f
borate Borat n
    calcium borate Calciumborat n
    zinc borate Zinkborat n
borax Borax n
borderline case Grenzfall m

bore Bohrung f (Seitlich am Zylinder angebrachte Bohrungen dienen zur Aufnahme von Thermofühlern / holes drilled into the side of the barrel serve to accomodate thermocouples); Bohrungsdurchmesser m
    ejector bore Auswerferbohrung f
    pipe bore Rohrinnenfläche f
bored gebohrt
    accurately bored hole Paßbohrung f
boric acid Borsäure f
boron compound Borverbindung f
boron fiber [GB: fibre] Borfaser f
boron trifluoride Bortrifluorid n
borrowed capital Fremdkapital n
borrowed funds Fremdmittel npl
bottle Flasche f
    bottle blow mo(u)lding Blasformteil n, blasgeformter Hohlkörper m, geblasener Hohlkörper m
    bottle blowing compound Flaschencompound m
    bottle blowing line Flaschenproduktionslinie f
    bottle blowing machine Flaschenblasmaschine f
    bottle blowing plant Flaschenblasanlage f
    bottle cap Flaschenverschluß m
    bottle closure Flaschenverschluß m
    bottle crate Flaschenkasten m
    bottle gripping unit Behältergreifstation f
    bottle manufacturer Hohlkörperhersteller m
    bottle testing machine Flaschenprüfstand m
    automatic bottle blowing machine Flaschenblasautomat m
    beverage bottle Getränkeflasche f
    disposable bottle Einwegflasche f
    windscreen wash bottle Scheibenwaschbehälter m
bottom Boden m, Grund m
    bottom force Unterwerkzeug n
    bottom half Werkzeugunterteil n
    bottom part Unterteil n
    bottom roll Unterwalze f
    bottom weld Bodenquetschnaht f, Bodenschweißnaht f
    bottom-weld bag Bodennahtbeutel m
bound gebunden
    synthetic resin bound kunstharzgebunden
boundary Grenze f, Grenzlinie f
    boundary condition Grenzbedingung f, Randbedingung f
    boundary layer Grenzschicht f
    boundary zone Grenzzone f
    phase boundary Phasengrenze f, Phasengrenzfläche f
box Behälter m, Kasten m
    box-shaped kastenförmig
    fish box Fischkasten m
    grass box Grasfangkorb m
    switch box Schaltkasten m
    tool box Werkzeugkasten m
bracing Versteifen n
bracket Träger m
    bumper bracket Biegeträger m, Stoßfängerträger m
braid Geflecht n

copper braid Kupfergeflecht n
brake Bremse f
  brake band Bremsband n
  brake fluid Bremsflüssigkeit f
  brake lining resin Bremsbelagharz m
braking Bremsen n
  braking device Bremsvorrichtung f
  braking mechanism Bremssystem n
  braking system Bremssystem n
  braking valve Bremsventil n
branch Zweig m, Ast m, Zweigstelle f
  branch (office) Niederlassung f
  branch of industry Industriesparte f, Industriezweig m, Branche f
  branch point Verzweigungsstelle f
branched verzweigt
  partly branched teilverzweigt
branching Verzweigung f
  branching reaction Verzweigungsreaktion f
  degree of branching Verzweigungsgrad m
  long-chain branching Langkettenverzweigung f
  short-chain branching Kurzkettenverzweigung f
break Bruch m
  break down/to [e.g. solids] zerteilen
  break in production Verarbeitungsunterbrechung f, Produktionsstörung f, Produktionsunterbrechung f
  break resistant bruchsicher
  bending/flexural stress at break Bruchbiegespannung f
  change in elongation at break Bruchdehnungsänderung f
  deflection at break Durchbiegung f beim Bruch
  easy to break down (e.g. PVC particles) aufschließbar
  elongation at break Bruchdehnung f, Reißdehnung f
  flexural stress at break Bruchbiegespannung f
  limiting elongation at break Grenzbruchdehnung f
  residual elongation at break Restbruchdehnung f, Restreißdehnung f
  retained elongation at break Restbruchdehnung f, Restreißdehnung f
  tensile stress at break Bruchspannung f
breakdown Versagen n, Störung f [e.g. of production figures] Aufschlüsselung f
  breakdown field strength Durchschlagfeldstärke f
  breakdown of the power supply Stromausfall m
  breakdown process Versagensprozeß m
  breakdown test Durchschlagversuch m
  breakdown voltage Durchschlagspannung f
  machine breakdown Maschinenausfall m
  market breakdown Marktaufteilung f, Marktgliederung f
breaker Brecher m
  breaker plate Siebträgerscheibe f, Brecherplatte f, Extruderlochplatte f, Lochscheibe f, Lochring m, Siebstützplatte f, Sieblochplatte f, Stütz(loch)platte f, Siebträger m
  breaker plate-type die Siebkorbkopf m
  breaker plate-type mandrel support Loch(scheiben)dornhalter m, Lochtragring m, Siebkorbhalterung f, Sieb(korb)dornhalter m, Lochdornhalterwerkzeug n
  automatic circuit breaker Sicherungsautomat m
  die with breaker plate-type mandrel support Lochdornhalterkopf m
breaking Bruch m, Riß m
  breaking-down effect [of solids] Zerteileffekt m
  breaking length Reißlänge f
  breaking limit Bruchgrenze f
  breaking strength [e.g. of a blow mo(u)lded container] Bruchfestigkeit f
  breaking stress Bruchlast, Reißkraft f
  number of breaking stress cycles Bruchlastspielzahl f
  short-term breaking stress Kurzzeitbruchlast f
  tensile breaking stress Zugbruchlast f
breathing Atmen n
  breathing apparatus Atemschutzgerät n
  mo(u)ld breathing Lüften n, Werkzeugatmung f
brickwork Ziegelmauerwerk n
bridge/to überbrücken
bridge Brücke f
  bridge bearing pad Brückenlager n
  dimethylene ether bridge Dimethylenetherbrücke f
  ether bridge Etherbrücke f
  hydrogen bridge H-Brücke f, Wasserstoffbrücke f
  hydrogen bridge linkage Wasserstoffbrückenbindung f
  particle bridge Partikelbrücke f
  polymethylene bridge Polymethylenbrücke f
  transducer bridge Aufnehmer-Meßbrücke f
  urea bridge Harnstoffbrücke f
  Wheatstone bridge Wheatstone-Meßbrücke f
brief description Kurzbeschreibung f
bright annealed blankgeglüht
brightness Helligkeit f
Brinell hardness Brinellhärte f
brittle spröde
  become brittle/to verspröden, brüchig werden
  brittle failure Trennbruch m, Sprödbruch m
  brittle fracture Trennbruch m, Sprödbruch m
  brittle-tough transition Spröd/Zäh-Übergang m
brittleness Sprödigkeit f
  low-temperature brittleness point Kältesprödigkeitspunkt m
brochure Broschüre f
brominated bromiert, bromhaltig
bromine compound Bromverbindung f
bromocresol green Bromkresolgrün n
Brookfield viscometer Brookfield-Viskosimeter n
brown braun
  brown discolo(u)ration Braunfärbung f
  brown iron oxide Eisenoxidbraun n
  electrical-grade brown elektrobraun
Brownian movement Brownsche Bewegung f

brush Bürste f, Pinsel m
  brush cleaner Pinselreiniger m
  air brush Luftbürste f
  steel wire brush Stahldrahtbürste f
  wire brush Drahtbürste f
brushing Streichen n, Bürsten n
bubble Blase f
  bubble circumference film Schlauchumfang m
  bubble collapsing angle Flachlegewinkel m
  bubble contour Schlauchkontur f
  bubble cooling system Innenluftkühlsystem n
  bubble expansion Schlauchaufweitung f
  bubble expansion zone [The part between die and frost line, where the extruded tube is inflated into] Schlauchbildungszone f, Aufblaszone f, Aufweitungszone f
  bubble film Luftpolsterfolie f
  bubble-free blasenfrei
  bubble guide Schlauchführung f, Leitblech n, Leitplatte f
  bubble neck Folienhals m
  bubble pack Bubblepackung f
  air bubble Luftblase f
  film bubble Blasschlauch m, Schlauchblase f, Filmballon m, Folienballon m, Folienblase f, Folienschlauch m
  film bubble contours Blasenkontur f
  film bubble cooling system Folienkühlung f, Schlauchkühlvorrichtung f, Schlauchkühlung f
  film bubble diameter Schlauchdurchmesser m
  film bubble radius Folienschlauchhalbmesser m, Folienschlauchradius m
  film bubble stability Schlauchblasenstabilität f, Blasenstabilität f
  film bubble tube Schlauch m
  internal bubble cooling system Folieninnnenkühlung f, Blaseninnenkühlung f, Blaseninnenkühlsystem n
  lateral bubble guide Seitenführung f
bucket conveyor Becherwerk n
buckling Beule f, Knickung f
  buckling pressure Beuldruck m
  buckling resistance Beulsteifigkeit f, Knickfestigkeit f
buffered gepuffert
buffing wheel Schwabbelscheibe f
build/to bauen, errichten
  ability to build up pressure Druckaufbauvermögen n
  capacity to build up pressure Druckaufbauvermögen n
  clamping force build-up Schließkraftaufbau m
  edge build-up [resulting from uneven reeling of film] Randwülste mpl
  heat build-up Wärmeaufbau m
  high build system Dickschichtsystem n
  pressure build-up Druckaufbau m
  pressure build-up phase Druckaufbauphase f

temperature build-up Temperatureinschwingen n
building Gebäude n
  building conservation Bautenschutz m
  building industry Bauindustrie f, Bauwesen n, Bauwirtschaft f
  conservation of buildings Bautenschutz m
  road building Straßenbau m
bulk Umfang m, Größe f, Füllung f, Ladung f
  bulk factor Füllfaktor m, Verdichtungsgrad m
  bulk mo(u)lding compound (BMC) Feuchtpreßmasse f
  bulk plastics Konsumkunststoffe mpl, Massenkunststoffe mpl
  bulk polymer Massepolymerisat n, Schmelzpolymerisat n, Blockpolymerisat n
  bulk polymerization Blockpolymerisation f, Massepolymerisation f, Substanzpolymerisation f
  bulk polymerization process Massepolymerisationsverfahren n
  bulk PVC Masse-PVC n
  bulk storage Bevorratung f
  bulk storage unit Massenspeicher m
  compacted bulk density Stampfdichte f
  compacted bulk volume Stampfvolumen n
  polyester bulk mo(u)lding compound (polyester BMC) Feuchtpolyester m
bulkhead lamp Ovalleuchte f
bumper Stoßfänger m, Stoßstange f
  bumper apron Stoßfängerschürze f
  bumper bracket Biegeträger m, Stoßfängerträger m
  bumper cover Stoßstangenabdeckung f
  front bumper Frontstoßfänger m
  rear bumper Heckstoßfänger m
burette Bürette f
buried [e.g. pipes] erdverlegt
burning 1. brennend; 2. Brennen n, Brand m
  burning rate Brenngeschwindigkeit f
  burning test Brandversuch m
  large scale burning test Brandgroßversuch m
bursting Bersten n, berstend
  bursting disc Berstscheibe f
  bursting pressure Berstdruck m
  bursting strength [particularly of roofing sheet] Schlitzdruckfestigkeit f, Bruchfestigkeit f
  bursting stress Berstspannung f
  bursting test Berstversuch m
  accelerated bursting test Kurzzeit-Berstversuch m
bus Sammelschiene f, Stromschiene f, Bus-Leitung f
  address bus Adreßbus m
  control bus Steuerbus m
  data bus Datenbus m
  instruction bus Befehlsbus m
busbar Sammelschiene f, Stromschiene f
bush(ing) Buchse f
  ante-chamber sprue bush Vorkammer(anguß)buchse f
  barrel bushing Zylinderbuchse f
  bearing bush Lagerbuchse f, Lagerschale f

    cent(e)ring bush Zentrierbuchse f, Zentrierbund m, Zentrierhülse f
    connecting bush Verbindungshülse f
    ejector bush Auswerferhülse f
    feed bush Einlaufhülse f, Einzugsbuchse f, Trichterstückbuchse f
    flow restriction bush Staubüchse f
    grooved feed bush Einzugsnutbuchse f
    guide bush Führungsbuchse f
    sprue bush Angußbuchse f, Anschnittbuchse f
    sprue puller bush Angußziehbuchse f
    stripper bush Abstreifhülse f
    threaded bush Gewindebuchse f
butadiene Butadien n
    butadiene-acrylonitrile copolymer Butadien-Acrylnitril-Copolymer n
    butadiene-acrylonitrile rubber Butadien-Acrylnitril-Kautschuk m
    butadiene-styrene rubber Butadien-Styrol-Kautschuk m
    butadiene rubber Butadienkautschuk m
butane Butan n
butanediol adipate Butandioladipat n
butanediol dimethacrylate Butandioldimethacrylat n
butanol Butanol n
butt Stoß m, strumpfes Ende n
    butt joint Stoßverklebung f, Stumpfstoß m
    butt welded stumpfgeschweißt
    butt welded joint Stumpfschweißverbindung f, Stumpfnaht f
    butt welding Stumpfschweißen n
    hot plate butt welding Heizelementstumpfschweißen n
    machine butt welding Maschinenstumpfschweißen n
    manual butt welding Handstumpfschweißen n
button Knopf m, Taste f
    emergency cut-out button [switch/device] Notausschalter m
    push button Druckknopf m, Drucktaste f
    start button Startknopf m, Starttaste f
    "stop" button Stoptaste f
butyl Butyl-
    butyl acetate Butylacetat n, Essigsäurebutylester m
    butyl acrylate Butylacrylat n, Acrylsäurebutylester m,
    butyl alcohol Butylalkohol m
    butyl benzyl adipate Butylbenzyladipat n
    butyl benzyl phthalate (BBP) Butylbenzylphthalat n
    butyl diglycol acetate Butyldiglykolacetat n
    butyl glycol Butylglykol m
    butyl glycolate Glykolsäurebutylester m
    butyl hydroperoxide Butylhydroperoxid n
    butyl peroctoate Butylperoctoat n
    butyl rubber Butylkautschuk m
    butyl titanate Butyltitanat n
    butyl-tin carboxylate Butylzinncarboxylat n
    butyl-tin compound Butylzinn-Verbindung f
    butyl-tin mercaptide Butylzinnmercaptid n
    butyl-tin stabilizer Butylzinnstabilisator m
    benzyl butyl phthalate (BBP) Benzylbutylphthalat n
    chlorinated butyl rubber Chlorbutylkautschuk m
    di-tertiary butyl peroxide (DTBB) Di-tert.-butylperoxid n
    ethylene glycol butyl ether Ethylenglykolmonobutylether m
    isobutylene-isoprene rubber Isobutylen-Isopren-Kautschuk m
butylglycidyl ether Butylglycidylether m
butylphenylglycidyl ether Butylphenylglycidylether m
butyral Butyral-
    polyvinyl butyral Polyvinylbutyral n
butyrate Butyrat-
    cellulose acetate butyrate Celluloseacetobutyrat n
    polyvinyl butyrate Polyvinylbutyrat n
butyric acid Buttersäure f
butyric anhydride Buttersäureanhydrid n
by-pass Umlenkung f
by-product Nebenprodukt n
cabinet Schrank m
    circulating air drying cabinet Umlufttrockenschrank m
    conditioning cabinet Klimaschrank m, Klimazelle f
    control cabinet Elektrokontrollschrank m, Elektroschaltschrank m, Kontrollschrank m, Regelschrank m, Schaltschrank m, Steuerschrank m
    drying cabinet Trockenschrank m
    vacuum cabinet Vakuumschrank m
    vacuum drying cabinet Vakuumtrockenschrank m
cable Kabel n
    cable compound Kabelgranulat n, Kabelmischung f, Kabelisolationsmischung f, Kabelmasse f, Kabeltype f, Mantelmasse f, Mantelmischung f
    cable compound formulation Kabelrezeptur f
    cable conduit Elektroschutzrohr n, Kabelkanal m, Kabelschutzrohr n
    cable factory Kabelwerk n
    cable industry Kabelindustrie f
    cable insulating compound Kabelgranulat n, Kabelmischung f, Kabelisoliermasse f, Kabelisolationsmischung f, Kabelmasse f, Kabeltype f, Mantelmasse f, Mantelmischung f
    cable insulation Kabelisolation f
    cable jointing compound Kabelvergußmasse f
    cable reclaiming plant Kabelaufbereitungsanlage f
    cable sector Kabelsektor m
    cable sheathing Kabelmantel m, Kabelummantelung f
    cable sheathing die Ummantelungswerkzeug n, Kabelkopf m, Kabelspritzkopf m, Kabelummantelungsdüse f, Kabelummantelungswerkzeug n
    cable sheathing extruder Kabelummantelungsextruder m, Ummantelungsextruder m
    cable sheathing line Ummantelungsanlage f, Kabelummantelungsanlage f

coaxial cable Koaxialkabel n
connecting cable Anschlußkabel n, Verbindungskabel n
extension cable Verlängerungskabel n
fiber-optic cable [GB: fibre] Glasfaserkabel n
high temperature cable Hochtemperaturkabel n
input cable Eingabekabel n
mains cable Netzkabel n
optical cable Lichtwellenleiterkabel n
power cable Starkstromkabel n
power supply cable Zuführungskabel n
telecommunication cable Fernmeldekabel n, Fernsprechkabel n
transducer cable Aufnehmerkabel n
CAD [computer-aided design] rechnerunterstütztes Konstruieren n, CAD-Rechentechnik f
cadmium Cadmium (Cd) n
  cadmium emission Cadmiumabgabe f
  cadmium laurate Cadmiumlaurat n
  cadmium orange Cadmiumorange n
  cadmium pigment Cadmiumpigment n
  cadmium pigmented cadmiumpigmentiert
  cadmium red Cadmiumrot n
  cadmium yellow Cadmiumgelb n
cage Käfig m
  cage pallet Gitterboxpalette f
  ball bearing cage Kugellagerkäfig m
  calibrating cage Kalibrierkorb m, Korb m
  sizing cage Kalibrierkorb m, Korb m
  squirrel cage motor Kurzschlußankermotor m, Kurzschlußläufermotor m
cake Kuchen m
  filter cake Filterkuchen m
caking Versintern n, Zusammenbacken n
calcination Calzinierung f, Kalzinierung f
calcined calziniert, geglüht, kalziniert
calcium Calcium n
  calcium borate Calciumborat n
  calcium carbonate Calciumcarbonat n
  calcium montanate Calciummontanat n
  calcium oxide Calciumoxid n
  calcium soap Calciumseife f
  calcium stearate Calciumstearat n
  calcium-zinc stabilizer Calcium-Zink-Stabilisierer m
  calcium-zinc stearate Calcium-Zink-Stearat n
calculated berechnet, rechnerisch
  calculated shot volume rechnerisches Spritzvolumen n
  calculated value Rechenwert m
calculation Berechnung f, Kalkulation f
  cost-benefit calculation Kosten-Nutzenrechnung f
  design calculation Auslegungsrechnung f
  energy calculation Energieabrechnung f
  profitability calculation Wirtschaftlichkeitsberechnung f
calculator Rechner m
  pocket calculator Taschenrechner m
calender Kalander m

calender downstream equipment Kalandernachfolge f
calender flow Kalanderströmung f
calender heating Kalanderbeheizung f
calender nip Kalander(walzen)spalt m
calender roll Kalanderwalze f
calender speed Kalandergeschwindigkeit f
calender take-off roll Kalanderabzugswalze f
contact laminating calender Doublierkalander m
embossing calender Prägekalander m
five-roll calender Fünfwalzenkalander m
five-roll L-type calender Fünfwalzen-L-Kalander m
four-roll calender Vierwalzenkalander m
four-roll inclined S-type calender Vierwalzen-S-Kalander m, S-Form f
four-roll inverted F-type calender Vierwalzen-F-Kalander m, F-Form f
four-roll L-type calender Vierwalzen-L-Kalander m
four-roll offset calender Vierwalzen-I-Kalander m mit schräggestellter Oberwalze
four-roll vertical calender Vierwalzen-I-Kalander m
four-roll Z-type calender Vierwalzen-Z-Kalander m, Z-Form-Kalander m
four-type inclined Z-type calender Vierwalzen-Z-Kalander m
laboratory calender Laborkalander m
laminating calender Laminierkalander m, Schmelzwalzenkalander m
polishing calender Glättkalander m
production calender Betriebskalander m
sheeting calender Folienziehkalander m
strips for feeding to the calender Fütterstreifen m
three-roll calender Dreiwalzenkalander m
three-roll offset calender Dreiwalzenkalander m [Schrägform]
three-roll superimposed calender Dreiwalzen-Kalander m
three-roll vertical calender Dreiwalzen-Kalander m
two-roll calender Zweiwalzenkalander m
two-roll superimposed calender Zweiwalzen-I-Kalander m
two-roll vertical calender Zweiwalzen-I-Kalander m
calendered film Kalanderfolie f
calendered sheeting Kalanderfolie f
calendering Folienziehen n, Kalanderverfahren n, Kalandrieren n (PVC-Folien werden zu 90% nach dem Kalanderverfahren hergestellt / 90% of PVC film and sheeting is made by calendering)
calendering line Kalanderstraße f, Kalanderlinie f, Kalanderanlage f
calendering (process) Kalandrierprozeß m
calendering speed Kalandriergeschwindigkeit f
ease of calendering: Kalandrierbarkeit f (Das Kriterium für die Kalandrierbarkeit einer Kautschukmischung / the criterion for the ease with which a rubber compound can be calendered)

high temperature calendering (process) HT-Verfahren n, Hochtemperaturverfahren n
low temperature calendering (process) LT-Verfahren n
sheet calendering Plattziehen n
calibrate/to kalibrieren
calibrating [also: sizing] Kalibrieren n
   calibrating air Kalibrierluft f
   calibrating air holes Stützluftbohrungen fpl
   calibrating basket Rollenkorb m, Führungskorb m, Kalibrierkorb m
   calibrating current Kalibrierstrom m
   calibrating device/unit Kalibriereinrichtung f, Kalibrator m, Kalibriervorrichtung f
   calibrating die Kalibrierdüse f, Kalibrierwerkzeug n
   calibrating mandrel Kalibrierdorn m [nicht zu verwechseln mit Kalibrierblasdorn]
   calibrating nip Kalibrierspalt m
   calibrating plate Kalibrierplatte f, Kalibratorscheibe f, Kalibrierscheibe f
   calibrating plate assembly Scheibenpaket n
   calibrating pressure Kalibrierdruck m
   calibrating ring Kalibrierring m
   calibrating section Kalibrierstrecke f
   calibrating sleeve Kalibrierhülse f, Kalibrierbüchse f
   calibrating system Kalibriersystem n
   calibrating table Kalibriertisch m
   calibrating tube Kalibrierrohr n
   calibrating unit Kalibriereinheit f
   compressed air calibrating sleeve Druckluftkalibrierhülse f
   compressed air calibrating unit Druckluftkalibriereinheit f
   high speed calibrating system Hochgeschwindigkeitskalibriersystem n
   internal air pressure calibrating unit Überdruckkalibrator m
   neck calibrating device Halskalibrierung f
   pipe calibrating (unit) Rohrkalibrierung f
   profile calibrating (unit) Profilkalibrierung f
   vacuum calibrating tank Vakuumkalibrierbecken n
   vacuum calibrating (unit) Vakuumkalibrierung f, Vakuumtank-Kalibrierung f
   waterbath vacuum calibrating unit Vakuum-Kühltank-Kalibrierung f
calibration Eichung f, Kalibrierung f
   calibration unit design Kalibrierkonzept n
   compressed air calibration Druckkalibrierung f, Druckluftkalibrierung f
   external calibration Außenkalibrieren n
   friction calibration Rutschkalibrierung f
   internal air pressure calibration (unit) Stützluftkalibrierung f, Überdruckkalibrierung f
   large-bore pipe calibration Großrohrkalibrierung f
   neck calibration Halskalibrierung f
   pipe calibration Rohrkalibrierung f
   profile calibration Profilkalibrierung
   thickness calibration (unit) Dickenkalibrierung f
   vacuum calibration Vakuumkalibrieren n, Vakuumtank-Kalibrierung f
   vacuum calibration section Vakuumkalibrierstrecke f
   vacuum calibration system Vakuumkalibriersystem n
calibrator Kalibrier-, -Kalibrator m
   calibrator bore Kalibrierbohrung f
   compressed air calibrator (unit) Druckkalibrierung f
   draw plate calibrator Blendenkalibrator m
   low-friction calibrator Gleitkalibrator m
   profile calibrator Profilkalibrierung f
   steel calibrator Stahlkalibrierung f
call up/to abrufen
calorimetry Kalorimetrie f
   differential calorimetry Differentialkalorimetrie f
   diffential scanning calorimetry DSC-Methode f
cam Nocke f, Steuernocke f
   cam disc Nockenscheibe f
   cam system Nockensystem n
   cam-type rotor Nockenrotor m
   actuating cam Betätigungsnocken m
CAM [computer aided manufacture] CAM-Rechentechnik f, rechnerunterstützte Fertigung f
candle Kerze f
   candle filter Filterkerze f
   candle filter assembly Filterkerzenpaket n
canister blow mo(u)lding machine Kanisterblasmaschine f
cantilever arm Kragarm m, freistehender Arm m
cantilevered [e.g. the clamping unit of an injection mo(u)lding machine, fixed at one end and free at the other] freistehend
cap Verschluß m
   bottle cap Flaschenverschluß m
   cap nut Überwurfmutter f
   fuel tank cap Tankverschluß m, Tankdeckel m
   hub cap Radkappe f, Rad(zier)blende f
   petrol tank cap Tankverschluß m, Tankdeckel m
   screw cap Schraubverschluß m
capable fähig, imstande, tüchtig
   capable of addition additionsfähig
   capable of being centered zentrierbar [Die Düse ist leicht zentrierbar / the die can be easily centered]
   capable of being high frequency welded hochfrequenzverschweißbar [Die Folie ist hochfrequenzverschweißbar / the film can be high frequency welded]
   capable of being polished polierfähig [Kunststoffe sind polierfähig / plastics can be polished]
   capable of being printed out protokollierbar
   capable of being processed verarbeitungsfähig
   capable of being swivelled upwards hochklappbar
   capable of being taken apart zerlegbar
   capable of being tilted upwards hochklappbar
capacitor [use version only in an electrical context] Kondensator m
   capacitor film Kondensatorfolie f

capacitor plates Kondensatorplatten fpl
capacity Fassungsvermögen n, Kapazität f, Füllinhalt m
  capacity increase, increase in capacity Kapazitätsausweitung f, Kapazitätszuwachs m, Kapazitätsausbau m
  capacity reduction Kapazitätsabbau m
  capacity to build up pressure Druckaufbauvermögen n
  capacity utilization Kapazitätsauslastung f
  absorptive capacity Aufnahmevermögen n
  annual world capacity Weltjahreskapazität f
  automatic high capacity blow mo(u)lding machine Hochleistungsblasformautomat m
  barrel heating capacity Zylinderheizleistung f
  connected heating capacity installierte Heizleistung f
  conveying capacity rate Förderleistung f
  cooling capacity Kühlkapazität f, Kühlleistung f (Hier spielt besonders die Kühlleistung eine Rolle / here, efficient cooling is particularly important)
  dirt capacity (of a filter) Schmutzspeichervermögen n [Filter]
  effective capacity Nutzinhalt m
  energy-absorbing capacity Arbeitsaufnahmevermögen n
  extrusion capacity Extrusionskapazität f
  feed capacity Einzugsvermögen n
  flexural load-carrying capacity Biegebelastbarkeit f
  handling capacity Hantierungskapazität f
  heat capacity Wärmekapazität f
  heating capacity Heizleistung f
  hopper capacity Fülltrichterinhalt m, Trichterinhalt m
  impact energy absorptive capacity Stoßenergieaufnahmevermögen n
  increase in capacity Kapazitätsausweitung f, Kapazitätsaufstockung f, Kapazitätszuwachs m
  injection capacity Einspritzleistung f, Spritzleistung f
  installed heating capacity installierte Heizleistung f
  installed injection capacity installierte Einspritzleistung f
  load-bearing capacity Belastbarkeit f, Belastungsgrenze f, Tragfähigkeit f, Lastaufnahmefähigkeit f, Lastaufnahmevermögen n
  load-carrying capacity maximum memory capacity maximaler Speicherausbau m
  memory capacity Speicherkapazität f
  nominal capacity Nenninhalt m
  plasticizing capacity Plastifizierkapazität f, Plastifizierleistung f, Plastifiziermenge f, Verflüssigungsleistung f, Weichmachungsvermögen n, Plastifiziervolumen n
  production capacity Produktionskapazität f, Produktionsleistung f
  pump capacity Pumpenleistung f
  shot capacity mögliches Schußvolumen n, Spritzkapazität f, Schußleistung f
  storage capacity Speicherkapazität f
  total heating capacity Gesamtheizleistung f

  with identical capacities kapazitätsidentisch
capillary Kapillare f
  capillary rheometer Kapillarrheometer n
  capillary thermostat Kapillarrohrthermostat m
  capillary viscometer Kapillarviskosimeter n
capital Kapital n
  capital assets Anlagevermögen n
  capital expenditure Investitionsausgaben fpl
  capital goods Investitionsgüter npl
  capital holding Kapitalbeteiligung f
  capital-intensive kapitalintensiv
  capital investment Investitionsausgaben fpl
  capital market Kapitalmarkt m
  capital outlay Investitionspreis m, Kapitaleinsatz m, Investitionsaufwand m
  capital stock Grundkapital n
  borrowed capital Fremdkapital n
  need for capital investment Investitionsbedarf m
  ordinary capital Stammkapital n
  own capital (of a company) Eigenkapital n
  share capital Aktienkapital n, Grundkapital n
caprolactam Caprolactam n
  caprolactam ring Caprolactamring m
car Wagen m, Auto(mobil) n
  car construction Automobilbau m
  car industry Auto(mobil)industrie f
carbamic acid Carbamidsäure f
carbazole Carbazol n
  polyvinyl carbazole Polyvinylcarbazol n
  vinyl carbazole Vinylcarbazol n
carbide Karbid n
  carbide tipped hartmetallbestückt
  carbide tipped tool Hartmetallwerkzeug n
  silicon carbide Siliciumkarbid n
carbody Karosserie f, Karosse f
  carbody paint Karosserielack m
carbon Kohlenstoff m
  carbon atom Kohlenstoffatom n
  carbon black Farbruß m, Rußpigment n, Kohlenstoffpigment n, Ruß m
  carbon black concentrate Rußpaste f
  carbon black-filled rußgefüllt
  carbon-carbon linkage Kohlenstoff-Kohlenstoff-Bindung f
  carbon chain Kohlenstoffkette f
  carbon dioxide Kohlendioxid n
  carbon fiber [GB: fibre] C-Faser f, Kohlefaser f, Kohlenstoff-Faser f
  carbon fiber reinforced kohlen(stof)faserverstärkt
  carbon fiber reinforcement Kohlenstofffaserverstärkung f, Kohlenfaserfüllung f
  carbon-like kohleartig
  carbon monoxide Kohlenmonoxid n
  carbon residues Kohlenstoffrückstände mpl, Kohlenstoffreste mpl
  carbon tetrachloride Tetrachlorkohlenstoff m
  active carbon black Aktivruß m

chopped carbon fiber Kohlenstoff-Kurzfaser f
carbonate Carbonat n
  alkaline earth carbonate Erdalkalicarbonat n
  calcium carbonate Calciumcarbonat n
  lead carbonate Bleicarbonat n
  lithium carbonate Lithiumcarbonat n
carbonated kohlensäurehaltig
carbonized track [caused by leakage current] Kriechspur f (Veränderungen durch Kriechspuren und Erosion / changes due to tracking and erosion)
carbonization Verkoken n
carbonyl Carbonyl-
  carbonyl compound Carbonylverbindung f
  carbonyl group Carbonylgruppe f
carboxyl group Carboxylgruppe f
  terminal carboxyl group Carboxylendgruppe f
carboxylate Carboxylat n
  alkyl tin carboxylate Alkylzinncarboxylat n
  butyl tin carboxylate Butylzinncarboxylat n
  octyl tin carboxylate Octylzinncarboxylat n
  tin carboxylate Zinncarboxylat n
carboxylated carboxyliert
carboxylic acid Carbonsäure f, Carboxylsäure f
  carboxylic acid anhydride Carbonsäureanhydrid n
  alkyl carboxylic acid Alkylcarbonsäure f
  aryl carboxylic acid Arylcarbonsäure f
carboxymethylcellulose Carboxymethylcellulose f
carburetor [GB: carurettor] Vergaser m, Verdampfer m
  carburet(t)or float Vergaserschwimmer m
  carburet(t)or housing Verdampfergehäuse n
card Karte f
  controlled by punched cards lochkartengesteuert
  filter card Filterkarte f
  input card Eingangskarte f
  magnetic card Magnetkarte f
  magnetic card reader Magnetkartenleser m
  mark-sense card Markierungsbeleg m
  mo(u)ld setting record card Formeinrichterkarte f
  output card Ausgangskarte f
  plug-in card Steckkarte f
  punched card Lochkarte f
  punched card control system Lochkartensteuerung f
  punched card reader Lochkartenleser m
cardan Kardan-
  cardan joint Kardangelenk n
  cardan shaft Gelenkwelle f, Kardanwelle f
cardboard Pappe f, Kartonpapier n
  cardboard container Pappdose f
  cardboard drum Papptrommel f
careful sanft, vorsichtig
  careful closing of the mo(u)ld sanfter Werkzeugschluß m
  careful treatment of the mo(u)ld Werkzeugschonung f (Werkzeugschonung durch sanften Werkzeugschluß / thanks to the careful clamping mechanism, the mo(u)ld is not damaged)
carousel [also: corrousel] Karussell n

carousel-type design Drehtischbauweise f
carousel-type injection mo(u)lding machine Revolverspritzgießmaschine f, Spritzgießrundtischanlage f, Drehtisch-Spritzgußmaschine f
carousel-type machine Drehtischmaschine f, Revolvermaschine f
carousel-type rotomo(u)lder kreisförmige Rotationsschmelzanlage f
carousel unit Revolvereinheit f
automatic carousel-type injection mo(u)lding machine Revolverspritzgießautomat m
carpet Teppich m
  carpet backing Teppichrückseitenbeschichtung f, Teppichrückenbeschichtung f
  carpet tiles Teppichfliesen fpl
carriage Schlitten m, Wagen m, Laufwerk n
  extruder carriage Extruderfahrwagen m
  mo(u)ld carriage Werkzeugschlitten m
  moving carriage Arbeitsschlitten m
  nozzle carriage Düsenhalteplatte f
carrier Träger m
  carrier bag Tragetasche f
  carrier gas Trägergas n
  data carrier Datenträger m
  mo(u)ld carrier Werkzeugträger m, Formenträger m, Formträgerplatte f
carrousel s. carousel
cartridge Kartusche f, Patrone f
  cartridge filter Filterkorb m, Filterkartusche f, Kerzenfilter m, Filterkerze f
  cartridge heater patronenförmiger Heizkörper m, Heizpatrone f, Patronenheizkörper m
  adhesive cartridge Klebstoffpatrone f
  heavy duty cartridge heater Hochleistungsheizpatrone f
cascade Kaskade f
  cascade arrangement Huckepack-Anordnung f, Kaskadenanordnung f
  cascade control Kaskadenregelung f
  cascade extruder Extruderkaskade f, Kaskadenextruder m, Tandemextruder m
  cascade temperature control Kaskadentemperaturregelung f
  cascade-type melt recirculation extruder Kaskadenumlaufextruder m
case 1. Fall m; 2. Gehäuse n, Kasten m; 3. Zarge f
  case-hardened einsatzgehärtet
  case-hardened steel Einsatzstahl m
  borderline case Grenzfall m
  door case Türzarge f
  in case of fire im Brandfall m
  in case of malfunction im Störfall m
casein Kasein n
casement Fensterflügel m, Flügelrahmen m
  casement profile Flügelprofil n
  casement window Flügelfenster n
  exterior casement (of window) Außenflügel m

cash subscription Bareinlage f
casing Gehäuse n
cassette Kassette f
  cassette memory Kassettenspeicher m
  cassette recorder Kassettenrecorder m,
  Bandkassettengerät n, Kassettenbandgerät n,
  Magnetband-Kassettengerät n, Kassettengerät n
  cassette screen changer Kassettensiebwechsler m
  magnetic tape cassette Magnetbandkassette f
  screen cassette Siebkassette f
cast/to (i.e. make an object from a mo(u)ld by pouring in resin) abformen, gießen
cast gegossen
  cast alumin(i)um Alu-Guß m, Aluminiumguß m
  cast alumin(i)um housing Alu-Gußgehäuse n
  cast alumin(i)um mo(u)ld Aluminiumguß-Werkzeug n
  cast epoxy insulator Epoxidgießharz-Isolator m
  cast film Chillrollfolie f, Gießfolie f
  cast film extrusion Kühlwalzenverfahren n, Extrusionsgießverfahren n, Chillroll-Verfahren n, Foliengießen n
  cast film extrusion line Chillroll-Filmgießanlage f, Chillroll-Folienanlage f, Chillroll-Anlage f, Chillroll-Flachfolien-Extrusionsanlage f, Foliengießanlage f, Gießfolienanlage f,
  cast polyamide Gußpolyamid n
  cast resin Gießharzformstoff m
  cast resin housing Gießharz-Gehäuse n
  cast sheet Gießplatte f
  cast steel Stahlguß m
  centrifugally cast pipe Schleuderrohr n
casting Gießharzteil n, Gießling m, Gießkörper m, Gießharzformstoff m, Formkörper m
  casting compound Gießmischung f, Gießmasse f
  casting machine Gießmaschine f
  casting paste [PVC] Gießpaste f
  casting plant Gießanlage f
  casting process Gießverfahren n
  casting resin Gießharz m
  casting resin technology Gießharztechnik f
  casting roll [used in chill casting of film] Gießwalze f
  casting roll unit Gießwalzeneinheit f
  centrifugal casting alloy Schleuderlegierung f
  centrifugal casting machine Schleudermaschine f
  centrifugal casting plant Schleuderanlage f
  centrifugal casting (process) Schleuder(guß)verfahren n, Schleudergießen n
  chill roll casting (process) Chillroll-Verfahren n, Kühlwalzenverfahren n, Extrusionsgießverfahren n
  chill roll casting line Chillroll-Filmgießanlage f, Chillroll-Anlage f, Chillroll-Folienanlage f, Chillroll-Flachfolien-Extrusionsanlage f
  cured casting resin Gießharzformstoff m
  epoxy casting resin Epoxidgießharz n, Epoxygießharz n
  face casting Frontguß m
  film casting Foliengießen n
  film casting line [for making, say, cellulose acetate film from a solution] Foliengießanlage f, Gießfolienanlage f
  silicone casting resin Silicongießharz n
  solid casting Vollguß m
  synthetic resin casting Gießharzkörper m, Kunststoffgießharzkörper m
  two-part casting compound Zweikomponenten-Gießharzmischung f
  zinc casting alloy Zink-Gießlegierung f
castor oil Rizinusöl n
catalyze/to katalysieren
catalyzed katalysiert
  catalyzed epoxy resin EP-Reaktionsharzmasse f
  catalyzed lacquer Reaktionslack m
  catalyzed polyester resin UP-Reaktionsharzmasse f
  catalyzed resin Reaktionsmasse f, Reaktionslacksystem n, Reaktionsharz n, Reaktionsharzmasse f
  catalyzed surface coating system Reaktionslacksystem n
  platinum catalyzed platinkatalysiert
catalyst Katalysator m, Härter m
  catalyst paste Härterpaste f
  catalyst poison Katalysatorgift n
  catalyst poisoning Katalysatorvergiftung f
  catalyst residues Katalysatorrückstände mpl, Katalysatorreste mpl
  catalyst surface Katalysatoroberfläche f
  acid anhydride catalyst Säureanhydridhärter m
  acid catalyst Säurekatalysator m
  amine catalyst Aminkatalysator m
  hot curing catalyst Heißhärter m
  oxidation catalyst Oxidationskatalysator m
  platinum catalyst Platinkatalysator m
  titanium catalyst Titankatalysator m
  Ziegler catalyst Ziegler-Katalysator m
catalytic katalytisch
catastrophic failure katastrophaler Bruch m, katastrophales Versagen n
categorization Typisierung f
categorized typisiert
caterpillar take-off (unit) Raupenabzug m, Abzugsraupe f
cathode ray oscillograph Kathodenstrahloszillograph m
cation Kation n
cationic kationisch, kationaktiv
cause/to verursachen
  causing wear verschleißverursachend
cause of malfunctioning Störungsursache f
caused bedingt
  caused by extrusion [e.g. stresses] extrusionsbedingt
  caused by flow strömungsbedingt
causes Ursachen fpl
  causes of ag(e)ing Alterungsursachen fpl

caustic 1. Alkali n, Beize f, 2. Ätz-
   caustic potash solution Kalilauge f
   caustic soda solution Natriumhydroxidlösung f, Natronlauge f
cavitate/to kavitieren
cavity Höhlung f, Loch n, Kavität f, Hohlraum m
   cavity insert Nesteinsatz m, Gesenkeinsatz m
   cavity plate Formnestblock m, Formnestplatte f, düsenseitige Formplatte f, Gesenkplatte f, Konturplatte f
   cavity pressure Form(nest)innendruck m, Form(nest)druck m, Kavitäteninnendruck m, Werkzeug(innen)druck m, Innendruck m
   cavity pressure-dependent werkzeuginnendruckabhängig, (form)innendruckabhängig
   cavity pressure-independent (form)innendruckunabhängig, werkzeuginnendruckunabhängig
   cavity pressure profile Werkzeug(innen)druckverlauf m
   cavity pressure transducer Werkzeuginnendruckaufnehmer m, Werkzeuginnendruck-Meßwertaufnehmer m
   cavity surface Form(nest)oberfläche f, Werkzeughohlraumoberfläche f
   cavity surface temperature Form(nest)oberflächentemperatur f
   cavity volume Formnest-Füllvolumen n, Formfüllvolumen n, Füllvolumen n, Spritzteil-Füllvolumen n
   cavity wall Formnestwand(ung) f, Werkzeug(hohl)raumwand(ung) f
   cavity wall temperature Nestwandtemperatur f, Werkzeugwandtemperatur f
   blow mo(u)ld cavity Blasgesenk n
   depending on the cavity pressure werkzeuginnendruckabhängig
   experimental cavity Versuchsformnest n
   in the (mo(u)ld) cavity im Werkzeuginneren n
   independent of cavity pressure innendruckunabhängig
   individual cavity cent(e)ring device Einzelformnest-Zentrierung f
   mo(u)ld cavity Formeinarbeitung f, Werkzeugeinarbeitung f, Formkavität f, Form(nest)kontur f, Formnesthohlraum m, Gesenk n, Gesenkhöhlung f, Werkzeuggesenk n, Nest n, Werkzeughöhlung f, Werkzeughohlraum m, Werkzeuginnenraum m, Werkzeugkavität f, Werkzeugkontur f, Werkzeugnest n, Werkzeugraum m,
   mo(u)ld cavity depth Formhohlraumtiefe f, Werkzeughohlraumtiefe f
   mo(u)ld cavity venting Formnestentlüftung f [Möglichkeiten zur Formnestentlüftung / methods of venting the mo(u)ld cavity]
   mo(u)ld cavity volume Formhohlraumvolumen n, Werkzeughohlraumvolumen n
   number of cavities Formnestzahl f, Kavitätenzahl f

   test cavity Versuchsformnest n
ceiling (Zimmer-)Decke f
   sound absorbent ceiling tile Schallschluckdeckenplatte f
cell Zelle f, Pore f
   cell content Poreninhalt m
   cell-regulating zellregulierend
   cell regulator Zellregulierungsmittel n
   cell size Porengröße f, Zellgröße f
   cell size distribution Porengrößenverteilung f, Zellgrößenverteilung f
   cell structure Porenbild n, Porenstruktur f, Zellengefüge n, Zell(en)struktur f
   cell wall Zellwand f
   photoelectric cell Photozelle f
cellophane [Although the word started life as a trade name, it has now become part of everyday English] Zellglas n, Cellophan f
cellular zellenförmig, zellig
   cellular concrete Gasbeton m
   cellular polyethylene Zellpolyethylen n
celluloid Celluloid n
cellulose Cellulose f, Zellstoff m
   cellulose acetate Celluloseacetat n
   cellulose acetate butyrate Celluloseacetobutyrat n
   cellulose derivative Cellulosederivat n
   cellulose ester Celluloseester m
   cellulose film Zellglas n
   cellulose hydrate Cellulosehydrat n
   cellulose nitrate Cellulosenitrat n
   cellulose propionate Cellulosepropionat n
   cellulose triacetate Cellulosetriacetat n
   acetyl cellulose Acetylcellulose f
   benzyl cellulose Benzylcellulose f
   ethyl cellulose Ethylcellulose f
   regenerated cellulose Regeneratcellulose f
cement Zement m
   cement surfacer Zementspachtel m
   asbestos cement Asbestzement m
central zentral, Zentral-
   central computer Zentralrechner m, Zentralcomputer m, Leitrechner m
   central cooling water manifold Zentralkühlwasserverteilung f
   central data processor BDE-Zentrale f
   central ejector Mittenauswerfer m, Zentralauswerfer m
   central ejector pin Zentralausdrückstift m
   central film gate mittiger Bandanschnitt m
   central gating Zentralanguß m, axiale Anspritzung f
   central grease lubricating system Fettzentralschmierung f
   central hydraulic system Zentralhydraulikanlage f
   central lubricating system Zentralschmierung f, Zentralschmieranlage f
   central lubricating unit Zentralschmieranlage f, Zentralschmierung f

central part Kernstück n, Herzstück n (Das Herzstück des Laminators ist die Kühlwalze / the most important part of the laminating line is the cooling roll)
central processing unit zentrales Rechenwerk n, zentrale Prozeßeinheit f, Zentraleinheit f
central processor Zentralprozessor m
central program unit Programmzentraleinheit f
central shaft Zentralwelle f
central unit Zentraleinheit f
central winder Zentrumswickler m
centralized [GB: centralised] zentral
center [GB: centre] Zentrum n, Mittelpunkt m
  center distance Mittenabstand m
  center-drive winder Direktwickler m, Achswickler m
  center feed Zentral(ein)speisung f, Zentralanguß m
  center height (height from floor level to centre line of extruder barrel) Extrudierhöhe f, Extrusionhöhe f
  center line Mittellinie f
  center of gravity Schwerpunkt m
  center roll Mittelwalze f
  center winder Zentralwickler m
  computer center Rechenzentrum n
  fluid center flüssige Seele f
  roll face center Walzenballenmitte f
center-fed zentralangespeist, zentral (an)gespeist, axial angeströmt, zentralumströmt, zentral eingespeist
  center-fed blown film die zentral angespritzter Blaskopf m, Dornhalterblaskopf m
  center-fed die Pinolenkopf m, Dornhalter(spritz)kopf m, Dornhalterwerkzeug n, zentralgespeister Dornhalterkopf m, Torpedokopf m
  center-fed parison die Dornhalterschlauchkopf m
cent(e)red zentrierbar
centrifugal Schleuder-, zentrifugal
  centrifugal casting Schleudergießen n
  centrifugal casting alloy Schleuderlegierung f
  centrifugal casting machine Schleudermaschine f
  centrifugal casting plant Schleuderanlage f
  centrifugal casting process Schleuder(guß)verfahren n
  centrifugal drier Zentrifugaltrockner m
  centrifugal force Fliehkraft f, Zentrifugalkraft f
  centrifugal pump Zentrifugalpumpe f
  centrifugal separator Zentrifugalabscheider m
centrifugally cast pipe Schleuderrohr n
centrifuge/to zentrifugieren
cent(e)ring Zentrier-, Zentrieren n
  cent(e)ring bush Zentrierbuchse f, Zentrierbund m, Zentrierhülse f
  cent(e)ring cone Zentrierkonus m
  cent(e)ring device Zentrierung f, Zentriervorrichtung f, Zentriergerät n
  cent(e)ring element Zentrierelement n
  cent(e)ring hole Zentrierbohrung f
  cent(e)ring screw Zentrierschraube f
  die cent(e)ring device Düsenzentrierung f
  individual cavity cent(e)ring device Einzelformnest-Zentrierung f

  mo(u)ld cent(e)ring device Werkzeugzentrierung f
  mo(u)ld cent(e)ring hole Werkzeugzentrierbohrung f
ceramic keramisch
  ceramic(-insulated) band heaters Keramikbänder npl, Keramikheizung f, Keramikheizkörper m
certainty Sicherheit f
  statistic certainty statistische Sicherheit f
certificate Zeugnis n
  test certificate Prüfzeugnis n
chain Kette f
  chain branching Kettenverzweigung f
  chain conveyor Kettenförderer m
  chain degradation Kettenabbau m
  chain-degrading kettenabbauend
  chain drive Kettentrieb m
  chain end Kettenende n
  chain entanglement Kettenverschlaufung f
  chain extender Kettenverlängerer m
  chain extension Kettenverlängerung f
  chain fission Kettenspaltung f, Kettenbruch m
  chain initiation Kettenstart m
  chain initiator Kettenstarter m
  chain length Kettenlänge f
  chain-like kettenartig, kettenförmig
  chain mobility Kettenbeweglichkeit f
  chain molecule Kettenmolekül n
  chain propagation Kettenwachstum n
  chain reaction Kettenreaktion f
  chain scission Kettenspaltung f, Kettenbruch m
  chain segment Kettenbaustein m, Kettenglied n, Kettensegment n, Kettenstück n, Kettenteil n
  chain stiffness Kettensteifigkeit f
  chain structure Kettenaufbau m, Kettengebilde n, Kettenstruktur f
  chain termination Kettenabbruch m
  chain transfer Kettenübertragung f
  chain transfer reaction Kettenübertragungsreaktion f
  alcohol chain Alkoholkette f
  carbon chain Kohlenstoffkette f
  ethylene chain Ethylenkette f
  free-radical chain Radikalkette f
  free-radical chain growth mechanism Radikalkettenmechanismus m
  hydrocarbon chain Kohlenwasserstoffkette f
  main chain Hauptkette f
  main polymer chain Polymerhauptkette f
  molecule chain Molekülkette f
  molecule chain length Molekülkettenlänge f
  polydialkyl siloxane chain Polydialkylsiloxankette f
  polyethylene chain Polyethylenkette f
  polymer chain Polymerkette f
  propylene chain Propylenkette f
  side chain Seitenkette f
  sprocket chain Zahnkette f
  sprocket chain drive Zahnkettentrieb m
chalk Kreide f
  natural chalk Naturkreide f

chalking Auskreidung f, Kreidung f
  chalking effect Kreidungseffekt m
  resistance to chalking Kreidungsbeständigkeit f, Kreidungsresistenz f
chamber Kammer f, Raum m
  accumulator chamber (space in front of the screw) Speicherraum m, Sammelraum m
  constant temperature chamber Temperierkammer f
  cutting chamber Schneidkammer f, Schneidraum m, Mahlraum m
  cutting chamber hood Granulierhaube f
  degassing chamber Entgasungskammer f
  devolatilizing chamber Entgasungskammer f
  filtration chamber Filterkammer f
  flash chamber Butzenkammer f, Quetschtasche f
  low-temperature test chamber Kältekammer f, Kälteschrank m
  mixing chamber Knetkammer f, Knetraum m
  outer chamber Außenkammer f
  plasticating chamber Plastifizierkammer f
  pump displacement chamber Pumpenverdrängerraum m
  steam chamber Dampfkammer f
  transfer chamber Füllraum m
chamfer/to abkanten, (an)fasen
chamfer Fase f
change Änderung f
  change in colo(u)r Farb(ton)änderung f, Farbumschlag m, Farbveränderung f, Farbübergang m
  change in diameter Querschnittsveränderung f
  change in dimensions Dimensionsänderung f
  change in elongation at break Bruchdehnungsänderung f
  change in flow resistance Fließwiderstandsänderung f
  change in length Längenänderung f
  change in molecular weight Molekulargewichtsänderung f
  change in position Wegänderung f
  change in properties Eigenschaftsänderung f
  change in Shore hardness Shore-Änderung f
  change in speed Drehzahländerung f
  change in temperature Temperaturwechsel m
  change in tensile strength Zugfestigkeitsänderung f
  change in viscosity Viskositätsänderung f
  change in volume Volumen(ver)änderung f
  change in weight Gewichts(ver)änderung f
  change of concentration Konzentrationsänderung f
  change of direction Richtungsänderung f
  change of supplier Lieferantenwechsel m
  change over/to umschalten
  change-over Umschaltung f
  change-over from injection to holding pressure Druckumschaltung f
  change-over point Umschalt(zeit)punkt m, Umschaltniveau n, Umschaltschwelle f
  change-over pressure Umschaltdruck m
  change-over threshold Umschaltschwelle f
  change-over time Umschaltzeit f
  change-over time interval Pausenzeit f
  change-over to hold(ing) pressure Nachdruckumschaltung f
  changes in pressure Druckverlauf m
  changes in temperature Temperaturverlauf m [Die zeitabhängig gemessenen Temperaturverläufe / changes of temperature measured over a period of time]
  dimensional change Formänderung f, Maßänderung f
  flow resistance changes Fließwiderstandsänderungen fpl
  pressure change-over Druckumschaltung f
  pressure change-over time Druckumschaltzeit f
  speed change-over point Geschwindigkeitsumschaltpunkt m
  structural change Gefügeänderung f, Strukturveränderung f
  surface change Oberflächenveränderung f
  viscosity changes Viskositätsänderungen fpl
changeable veränderlich
changer Wechsler m
  cassette screen changer Kassettensiebwechsler m
  high speed screen changer Schnellsiebwechsel-Einrichtung f
  quick action screen changer Sieb-Schnellwechseleinrichtung f
  screen changer Siebwechseleinrichtung f, Siebwechselkassette f, Siebwechsler m, Siebscheibenwechsler m, Siebwechselgerät n
  screen changer body Siebwechslerkörper m
changing [part of a machine, e.g. a mo(u)ld] Umrüsten n
  changing the screw Schneckenaustausch m, Schneckenwechsel m [... die den Schneckenwechsel erleichtern / ...which make it easier to change the screw]
  automatic changing mechanism Wechselautomatik f
  mo(u)ld changing time Werkzeugwechselzeit f [Die Werkzeugwechselzeit ist 65 Minuten / the time required to change the mo(u)ld is 65 minutes]
  reel changing mechanism/system/unit Rollenwechselsystem n
  resistance to changing climatic conditions Klimawechselbeständigkeit f
  speed changing device Geschwindigkeitsumschaltung f
channel Kanal m
  channel depth ratio Kanaltiefenverhältnis n
  cooling channel Kühlbohrung f, Kühlkanal m
  cooling channel layout Kühlkanalanordnung f
  cooling channel walls Kühlkanalwandungen fpl
  distance between the cooling channel Kühlkanalabstand m
  feed channel Einspeisekanal m, Zuströmkanal m
  feed channel length Anspritztiefe f
  feed channel runner manifold Zuführ(ungs)kanal m,

Anströmkanal m, Angußtunnel m, Anschnittkanal m
flow channel Strömungskanal m, Rundlochkanal m
flow channel design Fließkanalgestaltung f
flow channel runner Flußkanal m, Fließkanal m, Kanal m, Fließquerschnitt m
heating-cooling channel Temperierbohrung f, Temperierkanal m
heating-cooling channel layout Temperierkanalauslegung f
heating-cooling channel system Temperierkanalsystem n
input channel Eingangskanal m
input-output channel Ein-Ausgabekanal m
narrowing flow channels Querschnittsverengungen fpl
outlet channel Abströmkanal m
output channel Ausgangskanal m
rectangular flow channel [general term] Rechteckkanal m
screw channel Schnecken(gang)kanal m
screw channel depth Schneckenkanaltiefe f
screw channel profile Schneckengangprofil n, Schneckenkanalprofil n, Gangprofil n
screw channel volume Gangvolumen n, Schneckengangvolumen n, Schneckenkanalvolumen n
screw channel width Gangbreite f
slender cooling channels [method of cooling long core inserts in injection mo(u)lding] Fingerkühlung f
spiral channel Wendelkanal m
venting channel Entlüftungskanal m
char/to verkohlen
tendency to char Verkohlungsneigung f
character Charakter m
dipole character Dipolcharakter m
character row Schriftzeile f, Textzeile f
fine-particle character Feinteiligkeit f
long-chain character Langkettigkeit f
modular character Modularität f
open-cell character Offenporigkeit f
characteristic 1. Eigenschaft f, Charakteristik f; 2. charakteristisch, Kenn-
characteristic features kennzeichnende Eigenschaften fpl
characteristic value Kennwert m, Kenngröße f
die characteristic (curve) Düsenkennlinie f, Werkzeugkennlinie f
extruder characteristic (curve) Extruderkennlinie f
screw characteristic Schneckenkennzahl f, Schneckenkennlinie f
characteristics Kenndaten pl, Eigenschaftsmerkmale npl
abrasion characteristics Abriebverhalten n
ag(e)ing characteristics Alterungsverhalten n
damping characteristics Dämpfungseigenschaften fpl
dissolving characteristics Lösungsverhalten n
drying characteristics Trockenverhalten n
film surface characteristics Filmoberflächeneigenschaften fpl
filtration characteristics Filtrationsverhalten n

flow characteristics Fließeigenschaften fpl, Fließverhalten n
foaming characteristics Verschäumbarkeit f
fusion characteristics Gelierverhalten n
melting characteristics Aufschmelzverhalten n
particle characteristics Kornbeschaffenheit f
performance characteristics Gebrauchseigenschaften fpl
PID characteristics PID-Verhalten n
principal characteristics Hauptmerkmale npl
processing characteristics Verarbeitungsmerkmale npl, Verarbeitungsverhalten n
punching characteristics Stanzbarkeit f, Stanzfähigkeit f
screw characteristics Schneckeneigenschaften fpl
setting characteristics Abbindeverhalten n
solubility characteristics Löseverhalten n, Löslichkeitsverhalten n, Löslichkeitseigenschaften fpl
surface slip characteristics Gleiteigenschaften fpl
swelling characteristics Schwellverhalten n
true-running characteristics Rundlaufgenauigkeit f
vulcanizing characteristics Vulkanisierverhalten n
wear characteristics Verschleißverhalten n
weathering characteristics Bewitterungseigenschaften fpl
wetting characteristics Benetzungsvermögen n
characterization Charakterisierung f, Kennzeichnung f, Typisierung f
characterized typisiert
charge/to 1. aufladen [elektrisch]; 2. eindosieren, einspeisen, beladen
charge Ladung f
charge amplifier Ladungsverstärker m
charge distribution Ladungsverteilung f
charge monitor Ladungsmonitor m
charge signal Ladungssignal n
input charge Eingangsladung f
charged geladen, beladen
compound being charged to a machine Aufgabegut n
charger Ladegerät n
battery charger Batterieladegerät n
charging 1. Lade-; 2. Auflagung f [elektrisch]
electrostatic charging elektrostatische Aufladung f
Charpy impact tester Charpygerät n
charred [polymer, during processing] thermisch geschädigt
charring Verkohlung f
signs of charring Verbrennungserscheinungen fpl
chart Tabelle f, Tafel f, Diagramm n
chart recorder Linienschreiber m
four-channel chart recorder Vierkanal-Linienschreiber m
trouble-shooting chart Fehlersuchliste f
chassis Fahrgestell n, Fahrwerk n
check Kontrolle f, Untersuchung f, Check m
check list Checkliste f

check nut Kontermutter f
check plates Begrenzungsbacken fpl, Dichtbacken fpl
check valve Sperrventil n, Rückschlagventil n
check weigher Ausfallwaage f
  plausibility check Plausibilitätskontrolle f
  routine check Routinekontrolle f
checking fixture Kontrollehre f
cheese Kreuzspule f
chelating agent Komplexbildner m, Chelatbildner m, Chelator m
chelator Chelatbildner m, Chelator m, Komplexbildner m
chemical chemisch
  chemical attack Chemikalienangriff m
  chemical embrittlement Chemikalienversprödung f
  chemical equation (chemische) Reaktionsgleichung f
  chemical equipment construction Chemieapparatebau m
  chemical plant construction Chemieanlagenbau m
  chemical reaction chemische Reaktion f, chemische Umsetzung f, chemischer Umsatz m
  chemical resistance chemische Beständigkeit f, Chemikalienresistenz f
  chemical resistant chemikalienresistent, chemikalienbeständig, chemikalienfest
  exposed to chemical attack chemikalienbeansprucht
  exposure to chemical attack Chemikalienbeanspruchung f
  long-term chemical resistance Chemikalien-Zeitstandverhalten n, Chemikalien-Tauglichkeit f, Chemikalienbeständigkeit f
  resistance to chemical attack Widerstandsfähigkeit f gegen Chemikalieneinwirkung
  subjected to chemical attack chemikalienbeansprucht
chemically chemisch
  chemically blown chemisch getrieben
  chemically blown foam chemischer Schaum m, Treibmittelschaumstoff m
chemicals Chemikalien fpl
  exposure to chemicals Chemikalienbelastung f
  effect of chemicals Chemikalieneinfluß m
chemiluminescence Chemilumineszenz f
chemisorption Chemisorption f
chemist Chemiker m
chest Kasten m, Kiste f
  steam chest Dampfkasten m
child-proof kindersicher
chill-cast roll Kokillen-Hartgußwalze f
chill-roll Kühlwalze f
  chill roll casting/extrusion Chillroll-Verfahren n, Kühlwalzenverfahren n
  chill roll casting/extrusion line Chillroll-Filmgießanlage f, Chillroll-Anlage f, Chillroll-Flachfolien-Extrusionsanlage f, Chillroll-Folienanlage f
  chill roll casting/extrusion process, cast film extrusion Extrusionsgießverfahren n
  chill roll unit Chillroll-Walzengruppe f
chimney effect Kaminwirkung f

chip Stückchen n, Splitter m, Chip m
  silicon chip Silicium-Einkristall-Plättchen n, Siliciumchip m
chipboard Spanplatte f
chloride Chlorid n
  acid chloride Säurechlorid n
  allyl chloride Allylchlorid n
  ethylene chloride Ethylenchlorid n
  hydrogen chloride Chlorwasserstoff m
  mono-alkyl tin chloride Monoalkylzinnchlorid n
  organo-tin chloride Organozinnchlorid n
  plasticized polyvinyl chloride Polyvinylchlorid n weich
  polyvinyl chloride Polyvinylchlorid n, PVC n
  polyvinylidene chloride Polyvinylidenchlorid n
  residual vinyl chloride (monomer) content VC-Restmonomergehalt m
  rigid/unplasticized PVC Polyvinylchlorid n hart
  unplasticized PVC Polyvinylchlorid n hart
  vinyl chloride Vinylchlorid n
  vinyl chloride copolymer VC-Copolymerisat n
  vinyl chloride homopolymer Vinylchlorid-Homopolymer n
  vinyl chloride polymer Vinylchloridpolymerisat n
  vinyl chloride-vinyl acetate copolymer Polyvinylchloridacetat n, Vinylchlorid-Vinylacetat-Copolymerisat n
  vinylidene chloride Vinylidenchlorid n
  zinc chloride Zinkchlorid n
chlorinated chlorhaltig, chloriert
  chlorinated butyl rubber Chlorbutylkautschuk m
  chlorinated hydrocarbon Chlorkohlenwasserstoff m
  chlorinated paraffin Chlorparaffin n
  chlorinated polyethylene Chlorpolyethylen n
  chlorinated rubber Chlorkautschuk m
  chlorinated rubber coating/finish Chlorkautschukanstrich m
  chlorinated rubber paint Chlorkautschukfarbe f, Chlorkautschuklack m
chlorination Chlorierung f
  degree of chlorination Chlorierungsgrad m
chlorine Chlor n
  chlorine atom Chloratom n
  chlorine-containing chlorhaltig
  chlorine content Chlorgehalt m, Chlorierungsgrad m
  chlorine derivative Chlorderivat n
chlorobenzene Chlorbenzol n
chloroprene rubber Chloroprenkautschuk m
chlorosilane Chlorsilan n
chlorosulphonated chlorsulfoniert
choice Auswahl f
  choice of material Werkstoffauswahl f
  choice of supplier Lieferantenwahl f
chopped [e.g. fibers] kurzgeschnitten, schnitzelförmig
  chopped carbon fiber Kohlenstoff-Kurzfaser f
  chopped glass fiber Glasseiden-Kurzfaser f
  chopped (glass) strands Glaskurzfasern fpl

chopped fibers [if referring to fibers other than those made of glass] Kurzfasern fpl
chopped strand (glass fiber) mat Glasseiden-Schnittmatte f
chopped strand mat Glasfaservliesstoff m, Faserschnittmatte f
chopped strand prepreg Kurzfaser-Prepreg m
chopped strand reinforced kurzglasfaserverstärkt
chopped strands geschnittenes Textilglas n, geschnittene Glasfasern fpl, Kurz(glas)fasern fpl, Schnittglasfasern fpl
chopping rovings Schneidrovings mpl
chromate Chromat n
   lead chromate pigment Bleichromatpigment n
   zinc chromate Zinkchromat n
   zinc chromate primer Zinkchromatgrundierung f
chromatogram Chromatogramm n
   gas chromatogram Gaschromatogramm n
   gel chromatogram Gel-Chromatogramm n
chromatographic chromatographisch
   gas chromatographic gaschromatographisch
chromatography Chromatographie f
   gas chromatography Gaschromatographie f
   gel chromatography Gel-Chromatographie f
   gel permeation chromatography Gel-Permeationschromatographie f
   high pressure liquid chromatography Hochdruckflüssigkeitschromatographie f
   liquid chromatography Flüssigkeitschromatographie f
   paper chromatography Papierchromatographie f
   thin-layer chromatography Dünnschichtchromatographie f
chrome Chrom m
   chrome green Chromoxidgrün n
   chrome pigment Chromatpigment n
   chrome plated verchromt
   chrome yellow Chromgelb n
chroming Verchromen n
   hard chroming Hartverchromen n
chromophore Chromophor m
chromophoric chromophor
chute Rutsche f, Schurre f
   delivery chute Ausfallrutsche f, Ausfallschacht m
   feed chute Aufgabeschurre f
   large-volume chute Großvolumenschurre f
   ordinary chute Normalschurre f
   soundproof chute Schalldämmschurre f
   standard feed chute Serienaufgabeschurre f
   tangential chute Tangentialschurre f
   vibrating chute Vibrationsrinne f
circle Kreis m
   bed knife cutting circle Statormesserkreis m
   rotor knife cutting circle Rotormesserkreis m
   vicious circle Teufelskreis m
circuit Kreislauf m
   circuit block diagram Blockschaltbild n
   circuit diagram Schaltbild n, Schaltplan m, Schaltschema n
   automatic circuit breaker Sicherungsautomat m
   barrel heating circuit Zylinderheizkreis m
   closed-loop control circuit geschlossener Regelkreis m
   control circuit Steuerkreis m, Regelkreis m
   cooling circuit Kühl(kanal)kreis m, Kühlkreislauf m
   cooling water circuit Kühlwasserkreislauf m
   drive switch circuit Antriebsschaltkreis m
   electric circuit Stromkreis m
   illuminated circuit diagram Leuchtschaltbild n
   input circuit Eingangskreis m
   logic circuit Logikschaltkreis m
   oil circuit Ölkreislauf m
   open-loop control circuit offener Steuerkreis m, offener Regelkreis m
   position control circuit Lageregelkreis m
   pressure control circuit Druckregelkreis m, Drucksteuerkreis m
   printed circuit gedruckte Schaltung f
   printed circuit board Leiterplatte f, Elektrolaminat n, Elektroschichtpreßstoff m
   pump circuit Pumpenkreis m
   separate control circuit Einzelregelkreis m
   short circuit Kurzschluß m
   single-parameter control circuit Einzelparameter-Regelkreis m
   switch circuit Schaltkreis m
   temperature circuit Temperierkreis(lauf) m
   temperature control circuit Temperaturregelkreis m, Temperierkreis(lauf) m
circular kreisförmig, rund, kreisrund
   circular area Kreisfläche f
   circular cutting device Kreisschneidevorrichtung f
   circular loom Rundwebmaschine f, Rundwebstuhl m
   circular runner Ringkanal m
   circular saw Kreissäge f
   circular surface Kreisfläche f
   circular-woven fabric Rundgewebe n
circulate/to zirkulieren
circulating zirkulierend, Umlauf-
   circulating air drying cabinet Umlufttrockenschrank m, Umluftwärmeschrank m
   circulating air (drying) oven Umluftofen m
   circulating hot water heating system Heißwasser-Umlaufheizung f
   circulating medium Umlaufmedium n
   circulating oil heating system Ölumlaufheizung f
   circulating oil lubricating system Ölumlaufschmierung f, Umlaufölung f, Umlaufschmierung f
   circulating oil temperature control (system) Ölumlauftemperierung f
   circulating pump Umwälzpumpe f
   circulating water Umlaufwasser n
   circulating water temperature control unit Wasserumlauftempriergerät n
   air circulating oven Luftumwälzofen m

air circulating system Luftumwälzungssystem n
positive displacement circulating pump Umlaufverdrängerpumpe f
water circulating unit Wasserumwälzung f, Wasserumwälzeinheit f
circulation Zirkulation f, Umwälzung f
   circulation cooling (unit/system) Umwälzkühlung f
   air circulation Luftzirkulation f
   air circulation rate Luftzirkulationsrate f
   air circulation unit Luftumwälzvorrichtung f
   forced air circulation Luftdurchwirbelung f, Zwangsluftzirkulation f
   forced circulation Zwangsströmung f, Zwangsführung f
   forced circulation heater Zwangsumlauferhitzer m
   forced circulation system Zwangsumlaufsystem n
   melt circulation Schmelzeumlagerung f
   water circulation (unit) Wasserumwälzung f
circumference Umfang m
   die circumference Düsenumfang m
   parison circumference Schlauchumfang m
   pipe circumference Rohrumfang m
   screw circumference Schneckenumfang m
circumferential direction Umfangsrichtung f
cis-trans isomerism Cis-Trans-Isomerie f
citrate Zitronensäureester m
citric acid Zitronensäure f
civil engineering Tiefbau m
cladding Verkleidung f
   cladding panel Fassadenplatte f, Verkleidungsplatte f
   copper cladding Kupferbelag m, Kupferkaschierung f
   wall cladding Fassadenverkleidung f, Wandverkleidung f
clamp Einspannklemme f, Klemme f, Klemmpratze f, Spannklemme f, Spannzange f
   clamp ram Schließkolben m
   clamp(ing) force Schließzahl f, Werkzeugschließkraft f
   clamp(ing) unit Schließeinheit f, Schließhälfte f, Werkzeugschließeinheit f
   distance between clamps [e.g. in tensile tests] Klemmabstand m
   double toggle clamp unit Doppelkniehebel-Schließeinheit f
   fixing clamp Befestigungsklemme f
   four-column clamp unit Vierholmschließeinheit f
   mo(u)ld clamp delaying mechanism Formschließverzögerung f
   quick acting clamp Schnellverschluß m
   screw clamp Schraubzwinge f
   single-speed clamp(ing) unit Eingeschwindigkeitsschließeinheit f
   three-plate clamp(ing) unit Dreiplatten-Schließeinheit f
   toggle clamp machine Kniehebelmaschine f
   toggle clamp mechanism Kniehebelschließsystem n, Kniehebelverschluß m
   toggle clamp system Kniehebelschließsystem n, Kniehebelverschluß m
   two-tie bar clamp unit Zweiholmenschließeinheit f
clamped eingespannt
clamping Schließ -
   clamping cylinder Werkzeugschließzylinder m, Schließzylinder m, Formschließzylinder m
   clamping cylinder ram Schließkolben m
   clamping device [for test specimen] Einspannvorrichtung f, Niederhalter m, Spannvorrichtung f
   clamping distance [of test specimen] Einspannlänge f
   clamping element Schließglied n, Aufspannelement n
   clamping force Formzufahrkraft f, Niederhaltekraft f, Schließkraft f, Formschließkraft f, Zufahrkraft f
   clamping force build-up Schließkraftaufbau m
   clamping force control device Schließkraftregelung f
   clamping force distribution Schließkraftverteilung f
   clamping force profile Schließkraftverlauf m
   clamping force transducer Schließkraftmeßplatte f
   clamping frame Spannrahmen f
   clamping mechanism Schließeinrichtung f, Schließmechanismus m, Werkzeugschließsystem n, Formschlußm, Formschließsystem n
   clamping pressure Einspanndruck m, Niederhaltedruck m, Schließdruck m
   clamping ring Spannring m
   clamping stroke Formschließhub m
   clamping unit Formschließeinheit f, Schließseite f
   clamping unit ram Schließkolben m
   double toggle clamping system/mechanism Doppelkniehebel-Schließsystem n
   gentle mo(u)ld clamping mechanism sanfter Werkzeugschluß m
   high speed clamping cylinder Schnellschließzylinder m
   high speed mo(u)ld clamping mechanism/device Werkzeugschnellspannvorrichtung f
   housing containing the mo(u)ld clamping mechanism Formschließgehäuse n
   mo(u)ld clamping control Formschließregelung f
   mo(u)ld clamping frame Schließgestell n
   reduced mo(u)ld clamping pressure [to prevent damage to the mo(u)ld] Werkzeugsicherungsdruck m
   three-plate clamping mechanism Dreiplatten-Schließsystem n
   toggle mo(u)ld clamping unit Kniehebel-Formschließaggregat n
clarity Klarheit f
   crystal clarity Glasklarheit f
classification Klassifizierung f
   flammability classification Brennbarkeitsklasse f
cleaner Reinigungsmittel n
   brush cleaner Pinselreiniger m
cleaning Reinigung f
   cleaning bath Reinigungsbad n
   amount of cleaning Reinigungsaufwand m (Der Reinigungsaufwand nach Produktionsende ist sehr

gering / very little cleaning is necessary when production has been completed)
    mo(u)ld cleaning device Werkzeugreinigungsvorrichtung f
    time required for cleaning Reinigungszeit f
cleansing cream Reinigungscreme f
clear klar
    clear lacquer Klarlack m, Lasur f
    crystal clear glasklar
    crystal clear film Glasklarfolie f
    clear varnish Klarlack m, Lasur f
clearance Spiel n, Spielraum m
    axial clearance Axialspiel n
    clearance angle Freiwinkel m, Hinterschliffwinkel m
    flight clearance Grundspiel n, Schneckengrundspalte f, Kopfspiel n
    flight land clearance Flankenspiel n, Flankenabstand m, Schneckenspalt m, Walzenspalt m
    inter-screw clearance [clearance between the flight lands of twin screws] Schneckenspalt m, Flankenabstand m, Flankenspiel n, Schneidspalt m
    play clearance Betriebsspiel n
    radial clearance radiales Spiel n
    (radial) screw clearance Scherspalt m, Schneckenscherspalt m, Schneckenspiel n
clearly defined definiert
climatic conditions Klimabedingungen fpl
    ag(e)ing under standard climatic conditions Klimaraumlagerung f
    resistance to changing climatic conditions Klimawechselbeständigkeit f
clock signal generator Taktgenerator m
clogging Verstopfung f
close-mesh engmaschig
close-tolerance [e.g. mo(u)ldings] engtoleriert
closed geschlossen, Schließ -
    closed-cell geschlossenporig, geschlossenzellig
    closed-circuit cooling system/unit geschlossener Kühlkreislauf m, Rückkühlung f, Rückkühlaggregat n, Rückkühlwerk n
    closed-loop control Regeln n, Regelung f
    closed-loop control circuit geschlossener Regelkreis m
    closed-loop control instrument Regelgerät n
    closed-loop control system Regel(ungs)system n
    closed-loop process control Prozeßregelung f
    closed position Schließstellung f
    open and closed loop controls Steuer-, Schalt- und Regelelemente npl
closely eng
    closely cross-linked engvernetzt, engmaschig vernetzt
    closely intermeshing (twin screws) engkämmend, volleingreifend, dichtkämmend, vollständig kämmend
closing (of mo(u)ld) Zufahren n
    closing the mo(u)ld bei Werkzeugschluß m
    closing and opening movements Schließ- und Öffnungsbewegungen fpl
    closing movement Formschließbewegung f, Schließbewegung f, Schließvorgang m, Werkzeugschließbewegung f
    closing speed Schließgeschwindigkeit f, Zufahrgeschwindigkeit f
    closing time Schließzeit f
    opening and closing cycle Öffnungs- und Schließspiel n
    opening and closing valve Öffnungs- und Schließventil n
closure Verschluß m
    ring closure Ringschluß m
cloth Gewebe n
    abrasive cloth Schleifleinen n, Schmirgelleinen n, Schmirgeltuch n, Schleifgewebe n
    asbestos cloth Asbestgewebe n
    emery cloth Schleifleinen n, Schmirgelleinen n, Schmirgeltuch n, Schleifgewebe n
    epoxy glass cloth laminate Epoxidharz-Glashartgewebe n
    filter cloth Filtertuch n
    glass cloth Glasgewebe n, Glasseidengewebe n, Textilglasgewebe n, Glasfilamentgewebe n
    glass cloth covering Glasgewebeabdeckung f
    glass cloth laminate Glasgewebelaminat n, Glashartgewebe n
    glass cloth laminate pipe Glashartgeweberohr n
    glass cloth laminate sheet Glashartgewebeplatte f
    glass cloth reinforcement Glasgewebeverstärkung f
    glass cloth tape Glasgewebeband n
    roving cloth Glasrovinggewebe n
    test filter cloth Prüfsiebgewebe n
    varnished glass cloth Lackglasgewebe n
    wire cloth Drahtgewebe n
clothing Bekleidung f, Kleidung f
    clothing industry Bekleidungssektor m, Bekleidungsindustrie f
    protective clothing Schutz(be)kleidung f
cloud Wolke f
    cloud point Trübungspunkt m
    electron cloud Elektronenwolke f
cloudiness Trübung f
cloudy trübe
clutch Kupplung f
    clutch fluid Kupplungsflüssigkeit f
    clutch lining Kupplungsbelag m
co-condensate Cokondensat n
co-extruded flat film Breitschlitzverbundfolie f
co-monomer Comonomer m
co-precipitate Copräzipitat n
co-precipitation Copräzipitieren n
co-rotating gleichläufig, gleichlaufend, gleichsinnig
    co-rotating twin screw Gleichdrall-Doppelschnecke f
    co-rotating twin screw compounder/plasticator Gleichdralldoppelschneckenkneter m
    co-rotating twin screw extruder Gleichdrall-Schneckenextruder m

co-rotating twin screw system Gleichdrallsystem n
co-stabilizer Costabilisator m
coagulation Koagulation f
coal tar Steinkohlenteer m
coalescence Zusammenlaufen n
coarse grob
   coarse cotton fabric Baumwollgrobgewebe n
   coarse fabric Grobgewebe n
   coarse granules Grobgranulat n
   coarse material Grobgut n
   coarse mesh grobmaschig
   coarse-particle grobkörnig, grobteilig
   coarse particles Grobbestandteile mpl
coarsely crystalline grobkristallin
coarsely ground grob gemahlen
coat Schicht f
   gel coat Feinschicht f, Gelcoatierung f, Gelcoatschicht f, Harzfeinschicht f, Reinharzschicht f, Deckschicht f, Schutzschicht f, Oberflächenschicht f
   gel coat resin Deckschichtharz m
   gel coat surface Deckschichtoberfläche f
   tie coat Haftgrundierung f
   top coat [e.g. of a PVC paste] Deckschicht f, Schlußstrich m
coated oberflächenbeschichtet
   coated products Streichartikel mpl
   primer coated primerlackiert, grundiert
   (PVC) coated fabric Gewebekunstleder n, Kunstleder n
   synthetic rubber coated [e.g. rolls] kunstkautschukbeschichtet
coater Streichmaschine f, Auftragmaschine f, Beschichtungsmaschine f
   fluidized bed coater Wirbelsintergerät n
   hot-melt coater Schmelzwalzenmaschine f
   knife-(over-)blanket coater Gummituchrakel m
   knife-roll coater Walzenrakel m
   reverse roll coater Umkehrwalzenbeschichter m
coathanger Kleiderbügel m
   coathanger die Kleiderbügeldüse f
   coathanger manifold Kleiderbügelkanal m, Kleiderbügel-Verteilerkanal m
coating Beschichtung f, Beschichten n, Überzug m, Belag m
   coating compound Belegemischung f, Beschichtungsmasse f, Beschichtungsmittel n
   coating defect Beschichtungsfehler m
   coating extruder Beschichtungsextruder m
   coating line Beschichtungsanlage f
   coating machine Beschichtungseinrichtung f, Beschichtungsmaschine f
   coating plant Beschichtungsanlage f
   coating quality Beschichtungsqualität f
   coating system Beschichtungssystem n
   coating thickness Beschichtungsdicke f, Schichtstärke f, Schichtdicke f
   coating trials Beschichtungsversuche mpl
   coating unit Beschichtungseinheit f, Beschichtungseinheit f, Beschichtungseinrichtung f, Beschichtungsmaschine f
   coating weight Beschichtungsgewicht n
   coating with primer Primerlackierung f
   anchor coating [for PVC paste] Kupplungsschicht f, Grundstrich m, Haftstrich m
   anchor coating paste Grundierungspaste f
   anti-corrosive coating Korrosionsschutzanstrich m, Rostschutzanstrich m
   anti-drumming coating Antidröhnbeschichtung f
   anti-fouling coating Antifouling-Anstrich m
   barrier coating Barriereschicht f
   catalyzed surface coating system Reaktionslacksystem n
   chlorinated rubber coating Chlorkautschukanstrich m
   coextrusion coating line Coextrusionsbeschichtungsanlage f
   coextrusion coating (process) Coextrusionsbeschichtungsverfahren n
   coil coating paint Coil-Coating-Decklack m, Walzlack m
   coil coating primer Coil-Coating-Grundierung f
   dip coating Tauchbeschichtung f
   dip coating bath Tauchbad n
   dip coating compound Tauchmasse f
   electrostatic fluidized bed coating elektrostatisches Wirbelsintern n
   epoxy coating system Epoxidharzanstrichsystem n
   extrusion coating Extrusionsbeschichten n
   extrusion coating line Extrusionsbeschichtungsanlage f
   fabric coating Gewebebeschichtung f
   flow coating Fluten n
   fluidized bed coating Wirbelsintern n
   foam coating Schaumstrich m
   hard face coating Panzerung f, Oberflächenpanzerung f, Panzerschicht f
   heat sealable coating Heißsiegelbeschichtung f, Siegelschicht f
   hot-melt coating machine Schmelzwalzenmaschine f
   laboratory coating machine/line Laborbeschichtungsanlage f
   lacquer coating [e.g. for PVC leather cloth] Schlußlack m
   non-skid coating Rutschfestbeschichtung f
   non-stick coating Abhäsivbeschichtung f, Antihaftbelag m, Antihaftüberzug m
   plastics coating plant/line Kunststoffbeschichtungsanlage f
   plastisol coating Plastisolschicht f
   polyester surface coating resin Lackpolyester m
   powder coating Pulverbeschichtung f, Pulverlack m
   primer coating Grundlackierung f, Haftbrücke f, Primerschicht f, Haftgrundierung f
   protective coating Schutzanstrich m, Schutzbeschichtung f, Schutzüberzug m,

Schutzschicht f
reverse roll coating Umkehrbeschichtung f
spread coating Streichbeschichten n
spread coating compound Streichmasse f
spread coating machine Streichmaschine f
spread coating paste Streichpaste f
spread coating plant Streichanlage f
spread coating process Streichverfahren n
surface coating resin Lackbindemittel n, Lackharz m, Lackrohstoff m
surface coating system Lacksystem n
top coating Deck(an)strich m, Decklackierung f
wear resistant coating Oberflächenpanzerung f, Panzerschicht f, Panzerung f
coaxial cable Koaxialkabel n
cobalt Kobalt n
  cobalt accelerated Co-beschleunigt, kobaltbeschleunigt
  cobalt accelerator Co-Beschleuniger m, Kobaltbeschleuniger m
  cobalt blue Kobaltblau n
  cobalt cure Kobalthärtung f
  cobalt drier Kobaltsikkativ n
  cobalt naphthenate Kobaltnaphthenat n
  cobalt octoate Kobaltoctoat n
coconut oil fatty acid Kokosölfettsäure f
code Code m
  code pen Lesestift m
  code letter Kennbuchstabe m
  bar code Balkencode m, Barcode m, Strichcode m
  binary code Binär-Code m
  product code Typenbezeichnung f
coded codiert, kodiert
  binary coded binär kodiert
coding Kodierung f
  colo(u)r coding device [used for impressing identifying marks into extruded pipe] Signiergerät n
  colo(u)r coding machine Farbkennzeichnungsmaschine f
coefficient Koeffizient m
  coefficient of cubical/volume expansion kubischer Ausdehnungskoeffizient m, (thermischer) Volumenausdehnungskoeffizient m
  coefficient of expansion Ausdehnungskoeffizient m
  coefficient of friction Reibwert m, Reibungszahl f, Reibungsbeiwert m, Reibungskoeffizient m
  coefficient of linear expansion linearer Ausdehnungskoeffizient m, linearer Wärmeausdehnungskoeffizient m, thermischer Längenausdehnungskoeffizient m
  coefficient of sliding friction Gleitreibungszahl f, Gleitreibungskoeffizient m
  coefficient of static friction Haftreibungszahl f, Haftreibungskoeffizient m
  coefficient of thermal expansion [usually, but not necessarily, linear expansion] Wärmeausdehnungskoeffizient m, Wärmedehn(ungs)zahl f, thermischer Ausdehnungskoeffizient m
  absorption coefficient Absorptionskoeffizient m
  correlation coefficient Korrelationskoeffizient m
  diffusion coefficient Diffusionskoeffizient m, Permeationskoeffizient m, Permeationswert m
  drag coefficient cw-Wert m, Luftwiderstandsbeiwert m
  expansion coefficient Ausdehnungskoeffizient m
  extinction coefficient Extinktionskoeffizient m
  heat transfer coefficient Wärmedurchgangskoeffizient m, Wärmeübergangszahl f, Wärmeübergangskoeffizient m
  permeability coefficient Durchlässigkeitswert m, Permeabilitätswert m, Permeabilitätskoeffizient m
coextrudable coextrudierbar
coextruded coextrudiert
coextrusion Coextrudieren n, Vielschichtextrusion f, Coextrusionsverfahren n [auch: Ko-]
  blown film coextrusion die Coextrusionsblaskopf m, Coextrusionsblaswerkzeug n
  blown film coextrusion line coextrudierende Schlauchfolienanlage f, Coextrusions-Blasfolienanlage f, Coextrusionsblasanlage f
  coextrusion blow mo(u)lding Coextrusionsblasen n
  coextrusion blow mo(u)lding line Coextrusions-Blas(form)anlage f
  coextrusion blow mo(u)lding machine Coextrusionsblasformmaschine f
  coextrusion blow mo(u)lding plant Coextrusions-Blas(form)anlage f
  coextrusion blow mo(u)lding process Coextrusions-Blasformverfahren n
  coextrusion blow mo(u)lding trials Coextrusionsblasversuche mpl
  coextrusion coating Coextrusionsbeschichtungsverfahren n
  coextrusion coating line Coextrusionsbeschichtungsanlage f
  coextrusion die Coextrusionswerkzeug n
  coextrusion die head Coextrusionskopf m
  coextrusion line Coextrusionsanlage f
  coextrusion process Coextrusions-Plastformverfahren n
  flat film coextrusion line (coextrudierende) Flachfolienanlage f, Coextrusions-Breitschlitzfolienanlage f
  parison coextrusion die Coextrusionsschlauchkopf m
  three-layer coextrusion Dreischicht-Coextrusion f
  two-layer coextrusion Zweischicht-Coextrusion f
  two-layer coextrusion die Zweischichtdüse f
coffeemaker Kaffeeautomat m
cog Zahn m [Zahnrad], Knagge f, Nase f
  kneading cog Knetzahn m
coherent kohärent
cohesion Kohäsion f
cohesive kohäsiv
  cohesive force Zusammenhaltskraft f
  cohesive fracture Kohäsivbruch m, Kohäsionsbruch m, Kohäsionsversagen n

cohesive strength Kohäsionsfestigkeit f, Kohäsivfestigkeit f
coil Rolle f, Bund m, Windung f, Wendel f; Spule f [Elektro]; Rohrschlange f
  coil coating paint/enamel/lacquer Walzlack m, Coil-Coating-Decklack m
  coil coating primer Coil-Coating-Grundierung f
  cooling coil Kühlschlange f, Kühlwendel f
  spiral coil vaporizer Schlangenrohrverdampfer m
cold 1. kalt; 2. Kälte f
  cold bend test Kältebiegeversuch m
  cold bending properties Kaltbiegeeigenschaften fpl
  cold bonding Kaltkleben n
  cold crack behavio(u)r Kältebruchverhalten n
  cold crack resistance [e.g. of PVC coated fabrics] Kältebruchfestigkeit f
  cold crack temperature [e.g. of PVC coated fabrics] Kältebruchtemperatur f
  cold-cure foam Kaltschaumstoff m
  cold-cure mo(u)lded foam Kaltformschaumstoff m
  cold-curing kalthärtend, Kalthärtung f
  cold-curing catalyst Kalthärter m
  cold-curing mo(u)lding compound Kaltpreßmasse f
  cold-curing system Kalthärtungssystem n
  cold dipping (process) Kalttauchverfahren n
  cold feed extruder Kaltfütterextruder m
  cold flow kalter Fluß m, Kaltfluß m
  cold forming Kalt(um)formen n
  cold forming techniques/methods Kaltformmethoden fpl
  cold mixing Kaltmischung f
  cold press mo(u)lding Kaltpressen n
  cold press mo(u)lding tool Kaltpreßwerkzeug n
  cold runner Kalt(kanal)verteiler m
  cold runner feed system Kaltkanalangußsystem n
  cold runner mo(u)ld Kaltkanalform f, Kaltkanalwerkzeug n
  cold setting kaltabbindend
  cold slug kalter Pfropfen m
  cold slug retainer Pfropfenhalterung f
  cold water Kaltwasser n
  cold-worked steel Kaltarbeitsstahl m
collapse/to zusammenfallen, flachlegen
  bubble collapsing angle Flachlegewinkel m
  collapsing boards Flachlegebretter npl, Flachlegeeinrichtung f, Flachlegebleche npl, Flachlegung f
collapsible core Faltkern m
collecting Sammel-
  data collecting system Datensammelsystem n
  domestic refuse collecting Hausmüllsammlung f
  production data collecting system BDE-System n
  production data collecting unit Produktionsdaten-Erfassungsgerät n
collection Sammlung f
  data collection Datensammlung f, Datenabnahme f
  machine data collection Maschinendatenerfassung f
  production data collection Betriebsdatenerfassung f
collodion cotton Collodiumwolle f
colloid Kolloid n
  colloid mill Kolloidmühle f
  protective colloid Schutzkolloid n
colloidal kolloidal
colophony Kolophonium n
  colophony ester Kolophoniumester m
colorimetric farbmetrisch
color/to [GB: **colour/to**] einfärben
color [GB: **colour**] Farbe f
  colo(u)r change Farbveränderung f, Farbumschlag m, Farbtonänderung f
  colo(u)r coding device [used for impressing identifying marks into extruded pipe] Signiergerät n
  colo(u)r coding machine Farbkennzeichnungsmaschine f
  colo(u)r fast farbecht, farbtonbeständig
  colo(u)r fastness Farbechtheit f, Farbtonbeständigkeit f
  colo(u)r Hazen units Hazenfarbzahl f
  colo(u)r intensity Farbintensität f, Farbstärke f
  colo(u)r printing unit Farbdruckvorrichtung f
  colo(u)r scale Farbskala f
  colo(u)r screen Farbbildschirm m
  colo(u)r stability Farbbeständigkeit f
  colo(u)r variations Farbschwankungen fpl
  colo(u)r visual display unit Farbsichtgerät n
  absence of colo(u)r Farblosigkeit f
  change in colo(u)r Farbtonänderung f, Farbübergang m, Farbumschlag m, Farbveränderung f
  contrasting colo(u)r Kontrastfarbe f
  deepening in colo(u)r Farbvertiefung f
  difference in colo(u)r Farbunterschied m
  Gardner colo(u)r standard number Gardner Farbzahl f
  Hazen colo(u)r scale Hazen-Farbskala f
  imparting colo(u)r Farbgebung f
  inherent colo(u)r Eigenfarbe f
  iodine colo(u)r scale Jodfarbskala f
  preferred colo(u)r Vorzugsfarbe f
  standard colo(u)r range Standardfarben-Palette f
  uniform colo(u)r Farbhomogenität f
  variations in colo(u)r Farbschwankungen fpl
  with mo(u)lded-in colo(u)r durchgefärbt
colo(u)rant [which can be a pigment or a dye] Farbmittel n
colo(u)red (ein)gefärbt
  colo(u)red pigment [when a distinction is made between Farbpigmente and Weißpigmente] Farbpigment n, Buntpigment n
  colo(u)red streaks Farbschlieren fpl
  anthracite colo(u)red anthrazitfarbig
  dark colo(u)red dunkelfarbig
colo(u)rfast farbstabil
colo(u)ring Farbgebung f, Einfärbung f
  colo(u)ring agent [which can be a pigment or a dye] Farbmittel n

in-plant colo(u)ring [i.e. the addition of pigments by the mo(u)lder rather than by the mo(u)lding compound manufacturer] Selbsteinfärben n
colo(u)rless farblos
colo(u)rlessness Farblosigkeit f
column Säule f, Holm m
   liquid column Flüssigkeitssäule f
   mercury column Quecksilbersäule f
   separating column [in chromatography] Trennsäule f
combinable kombinierbar
combination Kombination f
   combination of properties Eigenschaftskombination f
   barrel-screw combination Zylinder-Schneckensystem n
   binder combination Bindemittelkombination f
combined 1. kombiniert; 2. -Kombination f
   combined extrusion-thermoforming line Extrusionsformanlage f
   combined take-off and wind-up unit Abzugswicklerkombination f, Abzugswerk-Wickler-Kombination f
combustibility Brennbarkeit f
combustible brennbar
combustion Verbrennen n, Verbrennung f
   combustion boat Verbrennungsschiffchen n
   combustion process Verbrennungsprozeß m
   combustion product Verbrennungsprodukt n
comfort Komfort m
   driving comfort Fahrkomfort m
   operator comfort Bedienungskomfort m
command Kommando n, Befehl m
commercial handelsüblich
   commercial product Handelsprodukt n
   commercial vehicle Nutzfahrzeug n
   in commercial quantitites in Handelsmengen fpl
commitment liability Verbindlichkeit f
commodity plastic Standardkunststoff m
common market countries EWG-Länder npl
common name [of a chemical compound] Trivialname m
commutator motor Kommutatormotor m
compact/to komprimieren, verdichten
compact unit/instrument Kompaktgerät n
compacted verdichtet, komprimiert
   compacted apparent density Stopfdichte f
   compacted bulk density Stampfdichte f
   compacted bulk volume Stampfvolumen n
compaction Verdichtung f
   data compaction Datenverdichtung f
compactly designed/constructed film extrusion line Folienkompaktanlage f
company Gesellschaft f
   associated company Beteiligungsgesellschaft f
   company funds/reserves Gesellschaftsmittel n
   company literature Firmenliteratur f
   company policy Firmenstrategie f
   company's contribution [e.g. to welfare, insurance etc.] Firmenleistung f
   developed by our company von unserem Haus entwickelt
   parent company Muttergesellschaft f
comparability Vergleichbarkeit f
comparative vergleichbar, Vergleichs-
   comparative figure Vergleichszahl f, Vergleichswert m
   comparative measurement Vergleichsmessung f
   comparative table Vergleichstabelle f
   comparative test Vergleichstest m, Vergleichsversuch m
comparison of properties Eigenschaftsvergleich m
compartment Kammer f, Raum m
   engine compartment Motorraum m
   feed compartment Dosierkammer f
   glove compartment Handschuhfach n, Handschuhkasten m
   grinding compartment Mahlkammer f
   weighing compartment Wägeraum m
compatibility Kompatibilität f, Verträglichkeit f
   compatibility characteristics Verträglichkeitseigenschaften fpl
   compatibility limit Verträglichkeitsgrenze f
   compatibility problems Verträglichkeitsschwierigkeiten fpl
   compatibility test Verträglichkeitsprüfung f, Verträglichkeitsuntersuchung f
   degree of compatibility Verträglichkeitsgrad m
   pigment compatibility Pigmentverträglichkeit f
compatible verträglich
   compatible with fillers füllstoffverträglich
compensating arm sensor Tänzerarmfühler m
compensating roller Tänzerwalze f
compensation Kompensation f, Ausgleich m
   compensation for thermal expansion Wärmeausdehnungsausgleich m
   shrinkage compensation Schrumpfungskompensation f
competition Konkurrenz f, Wettbewerb m
competitive konkurrenzfähig, wettbewerbsfähig
competitiveness Konkurrenzfähigkeit f, Wettbewerbsfähigkeit f
compiler Compiler m
   compiler language Compilersprache f
complete komplett, voll, vollständig
   complete plant Komplettanlage f
   complete vulcanization Ausvulkanisieren n, Durchvulkanisation f
completely komplett, voll, vollständig
   completely etherified vollveretert
   completely soluble klarlöslich
   completely transparent hochtransparent
complex 1. Komplex m; 2. aufwendig, komplex
   complex in design konstruktiv aufwendig
   complex ion Komplexion n
   heavy metal complex Schwermetallkomplex m

of complex shape formkompliziert
component Komponente f, Teil n, Bestandteil m
  alkyl component Alkylkomponente f
  ancillary component Zusatzbauteil n
  basic components Grundbauteile npl
  electrical insulating component Elektroisolierteil n
  formulation components Rezepturbestandteile npl, Rezepturkomponenten fpl
  machine components Maschinenelemente npl
  mo(u)ld component Werkzeugbauteil n, Formenbauteil n
  screw assembly components Schneckensatzelemente npl
composite zusammengesetzt, Verbund-
  composite control system Regelverbund m
  composite film Verbundfolie f
  composite flat film Kombinationsflachfolie f
  composite material Verbund(werk)stoff m
  composite panel Verbundplatte f
  composite sheet Verbundplatte f
  composite system Verbundsystem n
  continuous strand composite Endlosfaserverbund m
  fiber composite Faserverbundwerkstoff m
  flexible composite Weichverbund m
  rigid composite Hartverbund m
  two-layer composite Zweischichtverbund m
composition Zusammensetzung f, Komposition f
composting Kompostierung f
compound Verbindung f
  aliphatic compound Aliphat n
  alkyl tin compound Alkylzinn-Verbindung f
  alkylol compound Alkylolverbindung f
  amino mo(u)lding compound Aminoplast-Preßmasse f
  antimony compound Antimonverbindung f
  aromatic compound Aromat m
  azo compound Azoverbindung f
  blow mo(u)lding compound Blas(form)masse f, Hohlkörper-Thermoplast n, Hohlkörpercompound m,n
  boron compound Borverbindung f
  bottle blowing compound Flaschencompound m,n
  bromine compound Bromverbindung f
  butyl-tin compound Butylzinn-Verbindung f
  cable insulating compound Kabelgranulat n, Kabelisolationsmischung f, Kabel(isolier)masse f, Mantelmasse f
  cable jointing compound Kabelvergußmasse f
  carbonyl compound Carbonylverbindung f
  casting compound Gießmasse f, Gießmischung f, Gießharzmasse f, Gießharzsystem n
  coating compound Belegemischung f, Beschichtungsmasse f, Beschichtungsmittel n
  cold-curing mo(u)lding compound Kaltpreßmasse f
  cyclic compound Ringverbindung f
  di-organotin compound Diorganozinn-Verbindung f
  dialkyl compound Dialkylverbindung f
  dialkyltin compound Dialkylzinnverbindung f

  diallyl phthalate mo(u)lding compound Diallylphthalatpreßmasse f
  dibutyltin mercapto compound Dibutylzinnmerkaptoverbindung f
  diglycidyl compound Diglycidylverbindung f
  dihydroxy compound Dihydroxyverbindung f
  dip coating compound Tauchmasse f
  dough mo(u)lding compound [DMC] Feuchtpreßmasse f
  electrical insulating compound Elektroisoliermasse f
  embedding compound Einbettmasse f, Einbettmaterial n, Vergußmasse f
  encapsulating compound Einbettmasse f, Einbettmaterial n, Vergußmasse f
  epoxy compound Epoxidverbindung f
  epoxy mo(u)lding compound Epoxidformmasse f, Epoxidharz-Preßmasse f, Epoxidpreßmasse f
  extrusion compound Extrusionsmasse f, Extrusionsmischung f, Strangpreßmasse f
  free from aromatic compounds aromatenfrei
  frictioning compound Friktionsmischung f
  glycidyl compound Glycidylverbindung f
  halogen compound Halogenverbindung f
  high impact compound Schlagzähmischung f
  hydrocarbon compound Kohlenwasserstoffverbindung f
  injection compound Spritzgießmasse f, Spritzgießmaterial n
  injection mo(u)lding compound Spritzteilmasse f
  jointless flooring compound Bodenausgleichmasse f, Bodenspachtelmasse f, Fußbodenausgleichmasse f, Estrichmasse f
  lead compound Bleiverbindung f
  low pressure mo(u)lding compound Niederdruckformmasse f
  mono-alkyl tin compound Monoalkylzinnverbindung f
  mono-organotin compound Monoorganozinn-Verbindung f
  mo(u)lding compound Formmasse f, Rohstoff m
  mo(u)lding compound producer Formmassehersteller m
  octyl tin compound Octylzinnverbindung f
  one-component/-pack compound Einkomponentenmasse f
  organo-tin compound Organozinnverbindung f
  peroxy compound Peroxyverbindung f
  phenolic mo(u)lding compound Phenolharz-Preßmasse f, Phenoplast-Formmasse f, Phenoplast-Preßmasse f
  phosphorus compound Phosphorverbindung f
  phosphorus-halogen compound Phosphor-Halogen-Verbindung f
  pipe extrusion compound Rohrgranulat n, Rohrware f, Rohrwerkstoff m
  plasticized compound Weich-Granulat n, Weich-Masse f, Weichcompound m,n Weichmischung f

plastics mo(u)lding compound Kunststoffmasse f
polyester dough mo(u)lding compound [polyester DMC] Feuchtpolyester m
polyester mo(u)lding compound Polyesterharz-Formmasse f, Polyesterharz-Preßmasse f, Polyesterpreßmasse f
polyester sheet mo(u)lding compound Polyesterharzmatte f
polyethylene mo(u)lding compound Polyethylen-Formmasse f
polyhydroxy(l) compound Polyhydroxyverbindung f
polyolefin mo(u)lding compound Polyolefin-Formmasse f
polystyrene mo(u)lding compound Polystyrol-Formmasse f
purging compound Reinigungscompound m,n Reinigungsgranulat n
PVC cable compound Kabel-PVC n
range of thermoset mo(u)lding compounds Duroplastformmassen-Sortiment n
sheet mo(u)lding compound [SMC] Harzmatte f
sheeted-out compound Fell n (Das gelieferte Material wird auf einem Walzwerkzeug zum Fell ausgezogen / the gelled compound is sheeted out on the mill)
silicone resin mo(u)lding compound Siliconharzpreßmasse f
silicone rubber compound Siliconkautschukmischung f
special-purpose mo(u)lding compound Spezialformmasse f
spread coating compound Streichmasse f
standard injection mo(u)lding compound Standardspritzgußmarke f
sulphur compound Schwefelverbindung f
tetra-alkyl tin compound Tetraalkyzinnnverbindung f
tetra-aryl tin compound Tetraarylzinnverbindung f
thermoset mo(u)lding compound Duroplast-Formmasse f, Duroplastmasse f
trialkyl tin compound Trialkylzinnverbindung f
triorgano-tin compound Triorganozinnverbindung f
two-pack/component compound Zweikomponentenmasse f
two-part casting compound Zweikomponenten-Gießharzmischung f
tyre sidewall compound Reifenseitenwandmischung f
unplasticized compound Hart-Granulat n, Hartcompound m,n
unplasticized/rigid PVC (mo(u)lding) compound Hart-PVC-Formmasse f, Hart-PVC-Compound m,n
urea mo(u)lding compound Harnstoff-Preßmasse f
virgin compound Neugranulat n
window profile compound Fensterformulierung f
wire covering compound Ädermischung f
compounder Kneter m, Compoundiermaschine f
  co-rotating twin-screw compounder Gleichdralldoppelschneckenkneter m
  plastics compounder Kunststoffaufbereitungsmaschine f
  production compounder Produktionskneter m
  screw compounder Knetscheiben-Schneckenpresse f, Schneckenkneter m
  single-screw compounder Einschneckenkneter m, Einschnecken-Plastifizieraggregat n, einwelliger Schneckenkneter m
  twin screw compounder Doppelschneckenmaschine f, Zweischneckenmaschine f, Zweiwellenkneter m, zweiwellige Schneckenmaschine f, Doppelschneckencompounder m, Doppelschneckenkneter m
compounding Compoundieren n, Aufbereitungsverfahren n
  compounding extruder Aufschmelzextruder m, Aufbereitungsextruder m, Compoundierextruder m, Schmelzehomogenisierextruder m
  compounding line Compoundierstraße f, Aufbereitungsanlage f, Aufbereitungsstraße f
  compounding plant Compoundieranlage f
  compounding screw unit Knetschneckeneinheit f
  compounding section Aufbereitungsteil m
  compounding system Aufbereitungssystem n
  compounding unit Aufbereitungsaggregat n, Aufbereitungsmaschine f, Compoundiermaschine f, Plastifizierteil n
  high-capacity compounding line/plant Groß-compoundieranlage f
  single screw compounding extruder Plastifizier-Einschneckenextruder m
  twin compounding unit Aufbereitungs-Doppelanlage f
  twin screw compounding extruder Zweiwellen-Knetscheiben-Schneckenpresse f
compress/to verdichten, zusammendrücken, komprimieren
compressed Druck-, Press-, verdichtet, komprimiert
  compressed air Druckluft f, Preßluft f
  compressed air calibrating/sizing sleeve Druckluftkalibrierhülse f
  compressed air calibrating/sizing (unit) Druck-(luft)kalibrierung f
  compressed air cylinder Druckluftzylinder m
  compressed air forming Druckluftformung f, Druckluftverformen n
  compressed air forming machine Druckluftformmaschine f
  compressed air jet Preßluftstrahl m, Preßluftstrom m
  compressed air supply Druckluftnetz n, Preßluftnetz n
  compressed gas cylinder Druckgasbehälter m
  compressed wood Preßholz n
compressibility Kompressibilität f
compression Kompression f, Stauchung f, Verdichtung f
  compression blow mo(u)lding Kompressionsblasen n
  compression blow mo(u)lding process Kompressionsblasformverfahren n
  compression mo(u)ld Preßwerkzeug n
  compression mo(u)lded (form)gepreß t

compression mo(u)lding Formpressen n, Kompressionsformen n, Preßformen n, Preßverfahren n
compression ratio Schneckenkompression f, Volumenkompressionsverhältnis n, Kompressionsverhältnis n, Verdichtungsverhältnis n
compression resistance Druckbeanspruchbarkeit f
compression screw Kompressionsschnecke f, Verdichtungsschnecke f
compression section Kompressionsbereich m, Komprimierzone f, Kompressionszone f, Verdichtungszone f
compression set Druckverformungsrest m, Verformungsrest m
compression spring Druckfeder f
compression test Druckprüfung f, Druckversuch m, Stauchdruckprüfung f
automatic compression mo(u)lding machine Preßautomat m
data compression Datenkompression f
modulus of elasticity in compression Druck-E-Modul m
oil compression Ölkompression f
rate of compression Stauchungsgeschwindigkeit f
resistant to compression druckfest
three-plate compression mo(u)ld Dreiplattenpreßwerkzeug n
compressive Druck-
  compressive behavio(u)r Druckverhalten n
  compressive creep test Zeitstanddruckversuch m
  compressive force Druckkraft f, Stauchdruck m, Stauchkraft f
  compressive modulus Druckmodul m, Kompressionsmodul m
  compressive strength Stauchfestigkeit f, Stauchhärte f
  compressive stress Druckbelastung f, Druckbeanspruchung f, Drucklast f, Druckspannung f, Stauchbelastung f
  compressive strength Druckfestigkeit f
  compressive strength test machine Druckfestigkeitsprüfmaschine f
  development of compressive strength Druckfestigkeitsentwicklung f, Druckfestigkeitsverlauf m
  internal compressive stresses Druckeigenspannungen fpl
  under compressive stress druckbeansprucht, druckbelastet
compressor Kompressor m
  air compressor Luftkompressor m
compromise solution Kompromißlösung f
computation Berechnen n, Berechnung f
computer Computer m, Rechner m, Recheneinheit f
  computer aided computerunterstützt, rechnergestützt, rechnerunterstützt
  computer aided design [CAD] CAD-Rechentechnik f, rechnerunterstütztes Konstruieren n
  computer aided manufacture [CAM] CAM-Rechentechnik f, rechnerunterstützte Fertigung f

computer center Rechenzentrum n
computer controlled rechneransteuerbar, rechnergesteuert, computergesteuert
computer print-out Computer-Ausdruck m
computer program EDV-Rechenprogramm n, Rechnerprogramm n, EDV-Programm n, Rechenprogramm n, Berechnungsprogramm n
computer with integral screen Bildschirmcomputer m
central computer Leitrechner m, Zentralcomputer m, Zentralrechner m
control computer Kontrollrechner m, Steuerungsrechner m
desk-top computer Tischcomputer m, Tischrechner m
digital computer Digitalrechner m
master computer Leitrechner m, Zentralcomputer m, Zentralrechner m
process computer Prozeßrechner m
real-time computer Realzeitrechner m
supervisory computer Supervisorrechner m
conceived konzipiert
concentrate Konzentrat n
  antistatic concentrate Antistatikumkonzentrat n
  blowing agent concentrate Treibmittelkonzentrat n
  carbon black concentrate Rußpaste f
  pigment concentrate Farbkonzentrat n
concentrated konzentriert
concentration Konzentration f
  concentration gradient Konzentrationsgefälle n
  concentration ratio Konzentrationsverhältnis n
  concentration-related konzentrationsbezogen
  active substance concentration Wirkstoffkonzentration f
  change of concentration Konzentrationsänderung f
  curing agent concentration Vernetzerkonzentration f
  filler concentration Füllstoffkonzentration f
  final concentration Endkonzentration f
  initial concentration Ausgangskonzentration f
  limiting concentration Grenzkonzentration f
  maximum allowable concentration [MAC] MAK-Wert m, maximale Arbeitsplatzkonzentration f
  monomer concentration Monomerkonzentration f
  original concentration Ausgangskonzentration f
  peroxide concentration Peroxidkonzentration f
  pigment volume concentration [p.v.c.] Pigmentvolumenkonzentration f
  plasticizer concentration Weichmacherkonzentration f
  polymer concentration Polymerkonzentration f
  radical concentration Radikalkonzentration f
  range of concentration Konzentrationsbereich m
  saturation concentration Sättigungskonzentration f
  stress concentration Spannungskonzentration f
  sudden increase in concentration Konzentrationssprung m
  volume concentration Volumenkonzentration f
  wetting agent concentration Netzmittelkonzentration f
concentric konzentrisch

concentricity Konzentrizität f
concept Konzept n, Idee f, Vorstellung f
  basic concept Grundkonzept n
  model concept Modellvorstellung f
conclusion Schlußfolgerung f
  wrong conclusions Fehlschlüsse mpl
concrete Beton m
  aerated concrete Gasbeton m
  cellular concrete Gasbeton m
  epoxy concrete Epoxidharzbeton m
  fresh concrete Frischbeton m
  lightweight concrete Leichtbeton m
  old concrete Altbeton m
  polyester concrete Polyesterbeton m
  polymer concrete Kunstharzbeton m, Polymerbeton m
  reinforced concrete construction Stahlbetonbau m
condensate Kondensat n
  cresol-formaldehyde condensate Kresol-Formaldehyd-Kondensat n
  phenol-formaldehyde condensate Phenol-Formaldehyd-Kondensat n
  urea-formaldehyde condensate Harnstoff-Formaldehyd-Kondensat n
condensation Kondensation f
  condensation-curing 1. kondensationshärtend, kondensationsvernetzend; 2. Kondensationsvernetzung f
  condensation polymer Kondensationspolymer n
  condensation polymerization Kondensationspolymerisation f
  condensation product Kondensationsprodukt n
  condensation reaction Kondensationsreaktion f
  condensation resin Kondensationsharz n
  continued condensation Weiterkondensation f
  degree of condensation Kondensationsgrad m
  solid phase condensation Festphasenkondensation f
condense/to kondensieren
condensed kondensiert, Kondensations-
  condensed moisture Kondenswasser n, Schwitzwasser n
  condensed moisture atmosphere Schwitzwasserklima n
  condensed moisture test Schwitzwasserprüfung f
condenser Kondensator m [Wasser]
  reflux condenser Rückflußkühler m
condition/to klimatisieren, konditionieren, regeln, abbinden, härten, tempern
condition Kondition f, Bedingung f, Voraussetzung f, Zustand m
  condition of the melt Schmelzezustand m
  conditions of exposure Expositionsbedingungen fpl
  conditions of use Einsatzbedingungen fpl, Beanspruchungsbedingungen fpl, Beanspruchungsverhältnisse npl
  ag(e)ing conditions Alterungsbedingungen fpl, Lagerungsbedingungen fpl
  ag(e)ing under standard climatic conditions Klimaraumlagerung f
  ag(e)ing under warm/humid conditions Feucht-Warm-Lagerung f
  air flow conditions Luftströmungsverhältnisse npl
  ambient conditions Umgebungsbedingungen fpl
  approximate processing conditions Verarbeitungsrichtwerte mpl
  baking conditions Einbrennbedingungen fpl
  basic condition Grundvoraussetzung f
  blow-up conditions Aufblasbedingungen fpl
  boundary conditions Randbedingungen fpl, Grenzbedingungen fpl
  climatic conditions Klimabedingungen fpl
  constant production conditions Produktionskonstanz f
  cooling conditions Abkühlbedingungen fpl, Abkühlungsverhältnisse npl, Kühlungsverhältnisse npl
  crystallizing conditions Kristallisationsbedingungen fpl
  curing conditions Aushärtungsbedingungen fpl, Härtungsbedingungen fpl, Vernetzungsbedingungen fpl
  degassing conditions Entgasungsbedingungen fpl
  devolatilization conditions Entgasungsbedingungen fpl
  driving conditions Fahrbedingungen fpl
  drying conditions Trockenbedingungen fpl
  environmental conditions Umgebungsbedingungen fpl
  experimental conditions Experimentierbedingungen fpl
  exposure conditions Expositionsbedingungen fpl
  extrusion condition processing conditions Extrudierbedingungen fpl, Extrusionsbedingungen fpl
  feed conditions Einzugsverhältnisse npl
  flow conditions Fließbedingungen fpl, Strömungsbedingungen fpl, Strömungsverhältnisse npl
  injection conditions Einspritzbedingungen fpl
  laboratory conditions Laborbedingungen fpl
  market conditions Marktverhältnisse npl, Konjunktur f
  mo(u)lding conditions Formgebungsbedingungen fpl, Spritzbedingungen fpl
  operating conditions Arbeitsbedingungen fpl, Betriebsbedingungen fpl, Betriebszustände mpl
  outdoor conditions Freiluftklima n
  polymerizing conditions Polymerisationsbedingungen fpl
  post-curing conditions Nachhärtungsbedingungen fpl
  practical conditions Praxisbedingungen fpl
  pre-set processing conditions Sollverarbeitungsbedingungen fpl
  precipitation conditions Fällungsparameter mpl
  process conditions Prozeßbedingungen fpl
  processing conditions verfahrenstechnische Bedingungen fpl, Fahrbedingungen fpl, Verarbeitungsbedingungen fpl

production conditions Herstellungsparameter mpl, Fertigungsparameter mpl, Produktionsbedingungen fpl, Produktionsverhältnisse npl
reaction conditions Reaktionsbedingungen fpl, Reaktionsparameter mpl
resistance to changing climatic conditions Klimawechselbeständigkeit f
resistant to tropical conditions tropenbeständig, tropenfest
service conditions Beanspruchungsbedingungen fpl, Beanspruchungsverhältnisse npl
standard conditions Normbedingungen fpl, Serienbedingungen fpl, Standardbedingungen fpl
starting conditions Ausgangsbedingungen fpl
storage conditions Lagerungsbedingungen fpl
stoving conditions Einbrennbedingungen fpl
technical conditions maschinentechnische Gegebenheiten fpl
test conditions Prüfbedingungen fpl, Versuchsbedingungen fpl
tropical conditions Tropenbedingungen fpl, Tropenklima n
vulcanizing conditions Vulkanisationsbedingungen fpl
weathering conditions Bewitterungsbedingungen fpl, Witterungsbedingungen fpl
welding conditions Schweißbedingungen fpl
conditioned konditioniert, klimatisiert, normaltrocken, getempert
conditioning Konditionierung(slagerung) f, Temperaturbehandlung f, Temperung f
  conditioning cabinet Klimaschrank m, Klimazelle f
  conditioning oven Temperofen m
  conditioning period Temperzeit f
  conditioning section Temperstrecke f
  conditioning temperature Temper-Temperatur f
  air conditioning equipment Klimaanlage f
  air conditioning unit Klimagerät n
  DIN standard conditioning atmosphere [This is usually followed by figures such as 23/50, which means 23°C and 5% relative humidity] Normklima n, DIN-Klima n, Normalklima n
  standard conditioning atmosphere Normklima n
conduction of heat Wärmeleitung f
conductive leitend, leitfähig
  thermally conductive nozzle Wärmeleitdüse f
  thermally conductive torpedo Wärmeleittorpedo m
conductivity Leitfähigkeit f
  surface conductivity Oberflächenleitfähigkeit f
  thermal conductivity Wärmeleitfähigkeit f
  thermal conductivity coefficient Wärmeleitzahl f
  thermal conductivity equation Wärmeleitungsgleichung f
conductometric konduktometrisch
conductor (of electricity) (elektrischer) Leiter m
  conductor of heat Wärmeleiter m
  conductor temperature Leitertemperatur f
conduit Röhre f, Rohr n, Kanal m, Installationsrohr n, Kabelrohr n
  cable conduit Elektroschutzrohr n
  electrical conduit Elektroisolierrohr n
cone Konus m
  cone and plate viscometer Kegel-Platte-Viskosimeter n
  cent(e)ring cone Zentrierkonus m
configuration Konfiguration f
  roll configuration Walzenanordnung f
  screw configuration Schneckenbauform f, Schneckenkonfiguration f, Schneckengestalt(ung) f, Schneckengeometrie f
conical konisch, kegel(förmig)
  conventional conical [screw] einfachkonisch, normalkonisch
conjugated [e.g. double bond] konjugiert
connect/to ankuppeln, koppeln, verbinden
connected angekoppelt, angeschlossen, verbunden, Anschluß -
  connected heating capacity installierte Heizleistung f
  connected load (elektrischer) Anschlußwert m, Anschlußleistung f, installierte elektrische Gesamtleistung f
  connected voltage Anschlußspannung f
  ready to be connected anschlußfertig
  total connected load gesamtelektrischer Anschlußwert m, installierte Gesamtleistung f, Gesamtanschlußwert m
connecting Anschluß-, Verbindungs-
  connecting bush Verbindungshülse f
  connecting cable Anschlußkabel n, Verbindungskabel n
  connecting flange Anschlußflansch m
  connecting nipple Anschlußnippel m
  connecting pipe Verbindungsrohr n
  connecting plug Anschlußstecker m
  connecting rod Pleuel m, Pleuelstange f
  connecting tube Verbindungsrohr n
connection Anschluß m, Verbindung f
  cooling water connection Kühlwasseranschluß m
  extruder connection Extruderanschluß m
  plug-in connection Steckanschluß m, Steckverbindung f
  threaded connection Gewindeanschluß m
  vacuum connection Vakuumanschluß m
conservation Schutz m, Konservierung f
  conservation of buildings Bautenschutz m
considerations Überlegungen fpl
  cost-benefit considerations Kosten-Nutzen-Überlegungen fpl
  ecological considerations Umweltaspekte mpl
  economic considerations Rationalisierungsgründe mpl, Wirtschaftlichkeitsbetrachtungen fpl, Wirtschaftlichkeitsüberlegungen fpl
  technical considerations verfahrenstechnische Überlegungen fpl
consistency Konsistenz f

final consistency Endkonsistenz f
spraying consistency Spritzviskosität f
consisting of separate units [machine] offene Bauweise f
console Konsole f
  console typewriter Protokollschreibmaschine f
  display console Anzeigefeld n, Anzeigetafel f
  push-button console Druckknopf-Station f
consolidate/to komprimieren
consolidated konsolidiert
consolidation Verdichtung f
constant 1. Konstante f, Kennwert m, Kenngröße f; 2. konstant, gleichbleibend
  constant current Dauerstrom m, Konstantstrom m
  constant delivery pump Konstantpumpe f
  constant production conditions Produktionskonstanz f
  constant quality Qualitätskonstanz f
  constant speed Drehzahlkonstanz f
  constant taper screw [screw with constantly increasing root diameter] Kernprogressivschnecke f, kernprogressive Schnecke f
  constant temperature Temperaturkonstanz f
  constant temperature chamber Temperierkammer f
  constant temperature medium Temperierflüssigkeit f, Temperiermedium n, Temperiermittel n
  constant temperature rolls Temperierwalzen fpl
  constant temperature zone Temperierzone f
  constant test atmosphere Konstantklima n
  constant weight Gewichtskonstanz f (Die Gewichtskonstanz ist besser / a more constant weight is achieved)
  constant width Breitenkonstanz f (Enge Breitenkonstanz der Folie / very constant film width)
  at constant pressure isobar
  at constant volume isochor
  attenuation constant Abschwächungsbeiwert m
  die constant Düsenkennzahl f
  dielectric constant (relative) Dielektrizitätskonstante f, Dielektrizitätskonstante f, Dielektrizitätszahl f, dielektrische Konstante f
  diffusion constant Diffusionskonstante f
  dissociation constant Dissoziationskonstante f
  gas constant Gaskonstante f
  having constant viscosity viskositätsstabil
  material constant Stoffkonstante f, Stoffkennwert m, Werkstoffkennwert m, Werkstoffkenngröße f, Stoffkenndaten npl
  more constant quality bessere Qualitätskonstanz f
  of constant quality qualitätskonstant
  product constant Produktkennzahl f
  source of constant voltage Konstantspannungsquelle f
  velocity constant Geschwindigkeitskonstante f
  with constant torque drehmomentkonstant
constitution Konstitution f, Zusammensetzung f
  Law on the Constitution of Business Betriebsverfassungsgesetz n
constriction Einschnürung f
constructed konstruiert, bebaut
  constructed from units/modules in Segmentbauweise f
  simply constructed einfache Bauart f
construction Aufbau m, Konstruktionsaufbau m, Ausführung f, Ausbau m
  aircraft construction Flugzeugbau m
  car construction Automobilbau m
  chemical equipment construction Chemieapparatebau m
  chemical plant construction Chemieanlagenbau m
  construction guidelines Konstruktionshinweise mpl
  construction possibilities Konstruktionsmöglichkeiten fpl
  construction principles Konstruktionsprinzipien npl
  four-column construction Vierholm-Bauweise f
  framework construction Rahmenbauweise f
  hot runner unit construction Heißkanalblock-Ausführung f
  lightweight construction Leichtbau(weise) m(f)
  low construction Niedrigbau(weise) m(f)
  method of construction Bauweise f (Die Maschine zeichnet sich durch eine sehr einfache Bauweise aus / the machine is very simply constructed)
  modular construction Blockaufbau m, Modulbauweise f, Baukastenbauweise f, Elementebauweise f
  narrow construction Schmalbauweise f (Einstationenmaschinen in Schmalbauweise / narrowly constructed single-station machines)
  reinforced concrete construction Stahlbetonbau m
  sandwich construction Depotverfahren n, Sandwich-Verbundweise f
  shell-type construction Schalenbauweise f
  socket(-type) construction Muffenkonstruktion f
  transformer construction Trafobau m
  type of construction Konstruktionsform f
  unit construction (system) Baukastenbauweise f, Elementebauweise f, Modulbauweise f
  vehicle construction Fahrzeugbau m, Kraftfahrzeugbau m
  welded steel construction Stahlschweißkonstruktion f
constructional engineering Ingenieurbau m
constructional features konstruktiver Aufbau m
consume/to [The word has many other nontechnical meanings which should be looked up in a general dictionary] aufnehmen
consumer Konsument m, Verbraucher m
  consumer goods Bedarfsgegenstände mpl, Konsumartikel mpl, Konsumgüter npl, Verbrauchsgüter npl, Gebrauchsgüter npl
  consumer goods for food contact applications Lebensmittelbedarfsgegenstände mpl
  consumer goods regulations Bedarfsgegenständegesetz n
  consumer-oriented verbrauchernah
  consumer unit Verbraucher m [Maschine]

cooling water consumer Kühlwasserverbraucher m
energy consumer Energieverbraucher m
heat consumer Wärmeverbraucher m
injection mo(u)lded consumer goods Konsumspritzteile npl
injection mo(u)lding of consumer goods Konsumentenspritzguß m
consumption Verbrauch m
  continuous power consumption Dauerleistungsbedarf m
  drop in consumption Verbrauchseinbuße f
  electricity consumption Stromverbrauch m (der tatsächliche Stromverbrauch / the actual amount of electricitiy used)
  energy consumption Energieverbrauch m, Leistungsbedarf m
  fuel consumption Kraftstoffverbrauch m
  increase in consumption Verbrauchssteigerung f, Verbrauchszuwachs m
  per capita consumption Pro-Kopf-Verbrauch m
  petrol consumption Benzinverbrauch m
  power consumption Leistungsbedarf m, Energieverbrauch m
  total consumption Gesamtverbrauch m, Gesamtbedarf m
  total power consumption Gesamtleistungsbedarf m
  world consumption Weltverbrauch m
contact Kontakt m
  contact adhesive Haftklebstoff m, Kaltsiegelkleber m, Kontaktkleber m, Kontaktklebstoff m
  contact angle Berührungswinkel m
  contact cooling drum/roll Kontaktkühlwalze f
  contact cooling (unit) Kontaktkühlung f
  contact heating Kontaktheizung f
  contact laminating calender Doublierkalander m
  contact laminator/laminating unit Doubliereinrichtung f
  contact mo(u)lding (process) Kontaktverfahren n
  contact point Kontaktstelle f
  contact pressure Anpreßkraft f, Anpreßdruck m, Berührungsdruck m, Kontaktdruck m
  contact surface Anlagefläche f, Berührungsfläche f, Kontaktfläche f
  contact thermometer Berührungsthermometer n, Kontaktthermometer n
  contact zone Berührungszone f
  angle of contact Berührungswinkel m
  area of contact Kontaktfläche f
  consumer goods for food contact applications Lebensmittelbedarfsgegenstände mpl
  continuous contact Dauerkontakt m
  hot melt contact adhesive Schmelzhaftklebstoff m
  limit contact Grenz(wert)kontakt m
  long-term contact Langzeitkontakt m
  nozzle contact Düsenanlage f
  nozzle contact pressure Düsenanpressung f, Düsenanlagekraft f, Düsenanpreßkraft f, Düsenanpreßdruck m, Düsenanlagedruck m
  permanent/continuous contact Dauerkontakt m
  skin contact Hautkontakt m
  suitable for food contact applications lebensmittelecht
contacting electrode Haftelektrode f
contactor Schütz n
container Behälter m, Behältnis n, Gebinde n, Packmittel n, Packungshohlkörper m
  blown container geblasener Hohlkörper m, blasgeformter Hohlkörper m, Blaskörper m, Blas(form)teil n
  disposable container Wegwerfgefäß n, Einwegbehälter m
  injection mo(u)lding of packaging containers Verpackungsspritzgießen n
  original container Originalgebinde n
  packaging container Verpackungsbehälter m, Verpackungsgebinde n, Verpackungshohlkörper m, Verpackungsmittel n, Verpackungsteil n
  production of large containers Großhohlkörperfertigung f
  single-layer expanded polystyrene container Einschicht-PS-Schaumhohlkörper m
  small cardboard container Pappdose f
  stacking container Stapelbehälter m
  transit container Transportbehälter m
  weighing container Wägebehälter m
containing -haltig, beinhaltend
  containing electrolyte elektrolythaltig
  containing heavy metals schwermetallhaltig
  containing hydroxyl groups hydroxylgruppenhaltig
  containing lead bleihaltig
  containing phosphorus phosphorhaltig
  containing stabilizer stabilisatorhaltig
  containing sulphur schwefelhaltig
  containing water wasserhaltig
  housing containing the mo(u)ld clamping mechanism Formschließgehäuse n
  not containing lubricant gleitmittelfrei
contaminant [any material introduced from an outside source into a mo(u)lding compound for example] Schmutzstoff m, Fremdmaterial n
contaminant separator Fremdkörperabscheider m
  surface contaminants Oberflächenverunreinigungen fpl
contaminated verunreinigt, verschmutzt
contamination Verunreinigung f, Verschmutzung f
  contamination monitor Verschmutzungsüberwachung f
  degree of contamination Verschmutzungsgrad m, Verunreinigungsgrad m
  risk of contamination Verunreinigungsrisiko n, Verunreinigungsgefahr f
content Anteil m, Gehalt m
  acid content Säuregehalt m
  active oxygen content Aktivsauerstoffgehalt m

active substance content Wirkstoffgehalt m
air content [e.g. in foam] Luftbeladung f
allophanate linkage content Allophanatbindungsanteil m
aromatic content Aromatengehalt m
ash content Aschegehalt m
binder content Bindemittelgehalt m
biuret linkage content Biuretbindungsanteil m
blowing agent content Treibmittelanteil m,
Treibmittelgehalt m
cell content Poreninhalt m
chlorine content Chlorgehalt m, Chlorierungsgrad m
data content Dateninhalt m
double bond content Doppelbindungsanteil m,
Doppelbindungsgehalt m
dry solids content Trockengehalt m
emulsifier content Emulgatorgehalt m
energy content Arbeitsinhalt m
ester content Estergehalt m
fiber content Faseranteil m
filler content Füllstoffmenge f, Füllstoffgehalt m,
Füllungsgrad m, Füllstoffanteil m
final moisture content Endfeuchte f
gas content [e.g. in foam production] Gasbeladung f
gel content Gel-Gehalt m, Gelanteil m
glass content Glasgehalt m, Glasanteil m
glass fiber content Glasfasergehalt m,
Glasfaserkonzentration f, Glasfaseranteil m (Ohne
Glasfaseranteil ist eine nur sehr geringe
Verschleißwirkung vorhanden / only very little wear is
apparent in the absence of glass fibers)
halogen content Halogengehalt m
heat content Wärmeinhalt m
lead content Bleigehalt m
low fish eye content Stippenarmut f
lubricant content Gleitmittelanteil m
memory content Speicherinhalt m
moisture content Feuchtwert m, Feuchtigkeitsgehalt m,
Feuchtegehalt m, Feuchtigkeitsniveau n
moisture content of the atmosphere Luftfeuchtegehalt m
original acrylonitrile content Ausgangsacrylnitrilgehalt m
original moisture content Ausgangsfeuchte f
original monomer content
Ausgangsmonomergehalt m
original styrene content Ausgangsstyrolgehalt m
paint solids content Lackfeststoffgehalt m
pigment content Farbanteil m, Pigmentierungshöhe f
plasticizer content Weichmachergehalt m,
Weichmacheranteil m
residual acid content Restsäuregehalt m
residual acrylonitrile content Acrylnitrilrestgehalt m
residual epoxy group content Rest-Epoxidgruppengehalt m
residual ethylene content Restethylengehalt m
residual moisture content Restfeuchte f,
Restfeuchtigkeit(sgehalt) f(m)
residual monomer content Restmonomergehalt m
residual solvent content Lösungsmittelrestgehalt m
residual styrene content Reststyrolanteil m,
Reststyrolgehalt m, Styrolrestgehalt m
resin content Harzanteil m, Harzgehalt m
silicone content Siliconanteil m, Silicongehalt m
solids content Festkörperanteil m, Festkörpergehalt m,
Feststoffanteil m, Fest(stoff)gehalt m
solvent content Lösemittelgehalt m,
Lösungsmittelanteil m
total lead content Gesamtbleigehalt m
total pigment content Gesamtpigmentierung f
urea linkage content Harnstoffbindungsanteil m
urethane linkage content Urethanbindungsanteil m
UV content UV-Anteil m
volatile content Flüchtegehalt m
volume content Volumenanteil m, Volumengehalt m
water content Wassergehalt m, Wasserinhalt m
weight content Gewichtsanteil m
with a high butadiene content hochbutadienhaltig
with a high plasticizer content weichmacherreich
with a high resin content harzreich
with a low emulsifier content emulgatorarm
with a low filler content füllstoffarm
with a low fish eye content stippenarm
with a low monomer content monomerarm
with a low pigment content niedrigpigmentiert,
schwachpigmentiert
with a low plasticizer content weichmacherarm
with a low resin content harzarm
contents [of a package, bottle etc.] Füllgut n
(Verpackungen für fetthaltige Füllgüter / containers for
fat-containing products)
ante-chamber contents Vorkammerkegel m (damit der
Vorkammerkegel nicht einfriert / so that the contents of
the hot well do not solidify)
display contents Bildschirminhalt m
hopper contents Fülltrichterinhalt m, Trichterinhalt m
hot well contents Vorkammerkegel m
mo(u)ld contents Forminhalt m
original contents Originalfüllung f
screw flight contents Schneckengangfüllung f
continued condensation Weiterkondensation f
continuous kontinuierlich, durchgezogen [Kurve]
continuous control unit Stetigregler m
continuous cooling Dauerkühlung f
continuous extrusion stetige Extrusion f
continuous filament Endlosfaser f
continuous film Folien- und Bahnenware f
continuous flow cooling (system) Durchflußkühlung f
continuous glass fiber Glasseidenfaser f,
Langglasfasern fpl
continuous (glass) strands Endlosglasfasern fpl
continuous load Dauerlast f
continuous mixer Durchlaufmischer m
continuous operation permanenter Betrieb m,

Dauereinsatz m
continuous painting line Durchlauflackieranlage f
continuous plant Kontianlage f
continuous power consumption Dauerleistungsbedarf m
continuous process Fließprozeß m
continuous production Dauerproduktion f
continuous service temperature Dauerbetriebstemperatur f
continuous sheeting Folien- und Bahnenware f
continuous strand composite Endlosfaserverbund m
continuous strand (glass fiber) mat Glasseiden-Endlosmatte f
continuous strand mat Endlos(faser)matte f
continuous strand-reinforced langfaserverstärkt
continuous temperature Dauertemperatur f
continuous working temperature Dauergebrauchstemperatur f
in continuous use/operation in Dauerbetrieb m
maximum continuous operating temperature Dauerbetriebsgrenztemperatur f
reinforced with continuous glass strands langfaserverstärkt
contour Kontur f
  bubble contour Schlauchkontur f
  flight contours Stegkontur f
  film bubble contours Blasenkontur f
  screw flight contours Stegkontur f
  surface contours Oberflächenkonturen fpl
contoured fassoniert
contra gegen
  contra-extruder (Gegendrall-)Kontraextruder m
  contra-screw Kontraschnecke f
contradictory widersprüchlich
contrasting colo(u)r Kontrastfarbe f
contribution Beitrag m (e.g. another contribution towards greater efficiency and increased profits / Ein weiterer Beitrag zur Erhöhung der Wirtschaftlichkeit)
  company's contribution [e.g. to welfare, insurance etc.] Firmenleistung f
  non-cash contribution Sacheinlage f
  social (insurance) contributions Sozialaufwand m
control Kontrolle f, Regulierung f, Regelung f, Steuerung f, Regel-, Steuer-
  control accuracy Regelgüte f
  control algorithm Regelalgorithmus m
  control amplifier Regelverstärker m
  control bus Steuerbus m
  control cabinet Elektroschaltschrank m, Elektrokontrollschrank m, Kontrollschrank m, Regelschrank m, Schaltschrank m, Steuerschrank m
  control circuit Steuerkreis m, Regelkreis m
  control computer Kontrollrechner m, Steuerungsrechner m
  control console Bedienungspult n, Bedienungskonsole f, Prozeßwarte f, Maschinenpult n, Schaltwarte f, Steuerpult n, Steuerwarte f, Regiepult n
  control desk Bedienungspult n, Bedienungskonsole f, Prozeßwarte f, Maschinenpult n, Schaltwarte f, Steuerpult n, Steuerwarte f, Regiepult n
  control elements Bedienungselemente npl, Funktionselemente npl
  control equipment Reguliereinrichtung f, Steuereinrichtung f
  control function Steuer(ungs)funktion f, Regelfunktion f, Steueraufgaben fpl
  control input Steuer(ungs)eingang m, Regeleingang m
  control instruction Steuerkommando n, Regelsteuerkommando n
  control instrument Steuergerät n, Regelgerät n, Kontrollgerät n
  control key Bedienungstaste f
  control keyboard Bedienungstastatur f
  control mechanism Schaltgetriebe n
  control options Regelmöglichkeiten fpl, Steuervarianten fpl
  control output Steuer(ungs)ausgang m, Regelausgang m
  control panel Fronttafel f, Schalttableau n, Überwachungstafel f, Bedienungstafel f, Bedienungstableau n, Bedienungsfeld n, Bedienungsfront f, Frontplatte f
  control probe Reguliersonde f
  control quality Regelgüte f
  control sensor Reguliersonde f
  control signal Kontrollsignal n, Steuersignal n, Regelsignal n, Steuerbefehl m
  control solenoid Regelmagnet m
  control status Schaltzustand m
  control system Schalt(ungs)system n, Steuersystem n, Regelsystem n
  control task Regel(ungs)aufgabe f, Steuer(ungs)aufgabe f
  control test Überwachungsprüfung f
  control thermocouple Regelthermoelement n
  control unit Regeleinrichtung f, Regelinstrument n, Steueraggregat n, Steueranlage f, Steuerblock m, Steuereinheit f
  control valve Regulierventil n, Schaltventil n, Steuerventil n, Regelventil n
  control voltage Steuerspannung f
  adaptive control AC-Regelung f
  anti-drift control mechanism Antidrift-Regelung f
  automatic speed control system/mechanism/device Drehzahl-Steuerautomatik f
  barrel temperature control (system) Zylindertemperierung f
  batch control Chargenprüfung f, Partiekontrolle f, Partieprüfung f
  cascade control (unit) Kaskadenregelung f
  cascade temperature control (unit) Kaskadentemperaturregelung f
  circulating oil temperature control (system) Ölumlauftemperierung f

circulating water temperature control unit Wasserumlauftemperiergerät n
clamping force control (device/system/mechanism) Schließkraftregelung f
closed-loop control circuit geschlossener Regelkreis m
closed-loop process control Prozeßregelung f
composite control system Regelverbund m
continuous control unit Stetigregler m
cooling water control valve Kühlwasserregulierventil n
core puller control (mechanism) Kernzugsteuerung f
cushion control Polsterregelung f
cycle control Taktsteuerung f
cylinder temperature control (system) Zylindertemperierung f
deadline control Terminüberwachung f
digital control (unit) Digitalsteuerung f
digital volume control unit Digitalmengenblock m
direct digital control DDC-Regelung f
direct digital temperature control DDC-Temperaturregelung f
direct digital temperature control circuit DDC-Temperaturregelkreis m
directional control valve Wegeventil n
ejection control Ausfallüberwachung f
ejector control mechanism Auswerfersteuerung f
electric control system Elektrosteuerung f, Elektroregelung f
electronic control (system) Elektroniksteuerung f, Elektronikregelung f
electronic measuring and control system MSR-Elektronik f
feedback control system Regelungssystem n
film edge control Bahnkantensteuerung f
film width control (mechanism) Folienbreitenregelung f
flow control agent Verlaufmittel n
flow control (device) Durchsatzregulierung f
flow control (mechanism/system) Flußregulierung f (...welches ebenfalls zur Flußregulierung dient / which likewise helps to control flow)
flow control valve Durchflußsteuerventil n, Strom(regel)ventil n
in-house control Eigenkontrolle f
incoming goods control Wareneingangskontrolle f
incoming raw material control Rohstoffeingangskontrolle f
individual control unit Einzelregler m
injection control unit Einspritzsteuereinheit f
level control Niveauüberwachung f, Niveaukontrolle f, Niveauregelung f
limiting value control Grenzwertüberwachung f
machine control Maschinenüberwachung f
machine control program Maschinensteuerprogramm n
machine cycle control Maschinentaktsteuerung f

machine speed control system Fahrgeschwindigkeitsüberwachung f
main control room Hauptsteuerwarte f
measuring and control instruments MSR-Geräte npl
mo(u)ld clamping control Formschließregelung f
mo(u)ld temperature control (system) Formentemperierung f, Werkzeugtemperierung f, Werkzeugtemperaturregelung f
mo(u)lding cycle control Zykluskontrolle f
oil level control (unit) Ölstandkontrolle f
oil temperature control system Öltemperierung f
open-loop control circuit offener Steuerkreis m
outside control Fremdüberwachung f
parison diameter control Schlauchdurchmesserregelung f
parison length control (mechanism) Vorformlingslängenregelung f
parison wall thickness control Schlauchdickenregelung f
phase interface control Phasenanschnittsteuerung f
pneumatic control valve Luftsteuerventil n
pollution control equipment Umweltschutzanlage f
pollution control measures Umweltschutzmaßnahmen fpl
pollution control regulations Umweltschutzbedingungen fpl, Umweltschutzgesetzgebung f
position control circuit Lageregelkreis m
pressure control Druckführung f
pressure control device Druckregelgerät n
pressure control valve Druckregelventil n, Druck(regulier)ventil n
process control Prozeßregelung f, Prozeßführung f, Prozeßsteuerung f, Programmsteuerung f, Arbeitsablaufsteuerung f, Prozeßkontrolle f, Ablaufsteuerung f
process control equipment/system Prozeßsteueranlage f, Prozeßführungssystem n, Prozeßführungseinrichtung f, Prozeßleitsystem n
process control unit Prozeßführungsinstrument n
process data control unit Prozeßdatenüberwachung f
production control (system) Fabrikationskontrolle f, Fabrikationsprüfung f, Fabrikationsüberwachung f, Fertigungskontrolle f, Produktionskontrolle f, Produktionsüberwachung f, Fertigungssteuerung f
program control unit Programmregler m, Programmsteuerung f
proportional control valve Proportionalstellventil n
punched card control (system) Lochkartensteuerung f
push-button control (unit) Drucktastensteuerung f, Druckknopfsteuerung f
quality control Güteüberwachung f, Qualitätskontrolle f, Qualitätsüberwachung f
real time control Echtzeitsteuerung f
relay control Relais-Steuerung f, Schützensteuerung f
remote control adjusting/setting mechanism Ferneinstellung f
remote control (unit) Fernsteuerung f, Fernkontrolle f

remote speed control (mechanism) Geschwindigkeitsfernsteuerung f
screw root temperature control (system) Schneckenkerntemperierung f
screw temperature control (system) Schneckentemperierung f
separate control circuit Einzelregelkreis m
sequence control (unit) Folgeregelung f, Schrittfolgesteuerung f, Ablaufsteuerung f
single-parameter control circuit Einzelparameter-Regelkreis m
slide-in temperature control module Temperatur-Regeleinschub m
speed control (system/mechanism/device) Drehzahlregulierung f, Drehzahlsteuerung f, Drehzahlregelung f
stored-program control PC-Steuerung f
stroke control Wegesteuerung f
temperature control circuit Temperaturregelkreis m
temperature control instrument Temperaturregler m, Temperaturregelgerät n
temperature control medium Temperiermedium n, Temperierflüssigkeit f, Temperiermittel n
temperature control (system) Temperaturkontrolle f, Temperaturregelung f, Temperatursteuerung f, Temperaturregulierung f, Temperaturüberwachung f, Temperiersystem n
temperature control unit Temperiergerät n, Temperatursteuereinheit f, Temperaturüberwachung f
tension control (mechanism) [for film as it comes off the calender] Spannungskontrolle f
three-speed control mechanism Dreigang-Schaltgetriebe n
three-way flow control (unit) Dreiwegestromregelung f
three-way flow control valve Dreiwege-Mengenregelventil n
throttle control Drosselsteuerung f
thyristor control unit Thyristorregler m
two-way flow control (unit) Zweiwegestromregelung f
very accurate controls große Regelgenauigkeit f
visual control Sichtkontrolle f
volume control (unit/device) Volumensteuerung f, Volumenstromregler m
wall thickness control (system) Wanddickenkontrolle f, Wanddickensteuerung f, Wanddickenregulierung f
wall thickness control mechanism Wanddickenregulierungssystem n
water-fed temperature control unit Wassertemperiergerät n
web tension control (unit/mechanism) Bahnspannungskontrolle f, Bahnspannungsregelung f, Zugspannungsregelung f
width control (mechanism) Breitensteuerung f, Breitenregelung f, Breitenregulierung f
works control Betriebskontrolle f
controllable kontrollierbar, steuerbar, regelbar

thermostatically controllable thermostatisierbar
controlled gesteuert, geregelt
controlled by punched cards lochkartengesteuert
controlled variable Regelgröße f
computer controlled rechneransteuerbar, computergesteuert, rechnergesteuert
digitally controlled numerisch gesteuert
laser controlled lasergesteuert
process controlled prozeßgeführt
push-button controlled drucktastengesteuert
separately controlled fremdgesteuert
thermostatically controlled thermostatisch geregelt, thermostatisiert
thyristor controlled thyristorgespeist, thyristorgesteuert
controller Regler m, Steuerwerk n
back pressure controller Staudruckregler m
flow controller Verlaufsmittel n
individual controller Einzelregler m
level controller Niveauwächter m, Niveauregler m
parison diameter controller Schlauchdurchmesserregler m
parison length controller Schlauchlängenregler m
parison wall thickness controller Schlauchdickenregler m
pressure controller Druckregelgerät n
program controller Programmregler m
temperature controller Temperaturregelgerät n, Temperaturregler m, Thermowächter m
three-point controller Dreipunktregler m
time controller Zeitregler m
two-point controller Zweipunktregler m
controls Bedienungsgeräte npl, Bedienungsorgane npl, Einstellelemente npl, Regelelemente npl, Stelleinrichtungen fpl, Steuerelemente npl, Schalt- und Steuergeräte npl
electric controls Elektrosteuerung f
electronic controls Steuerelektronik f, Steuerungselektronik f
hydraulic controls Regelhydraulik f
solid-state controls kontaktlose Steuerung f
supplementary controls Zusatzsteuerungen fpl
convection Konvektion f
heat losses due to convection Konvektionsverluste mpl
conventional herkömmlich, konventionell, gängig
conventional conical [screw] einfachkonisch, normalkonisch
conventional injection mo(u)lding Kompaktspritzen n, Normalspritzguß m, Kompaktspritzguß m
conventional injection mo(u)lding machine Kompaktspritzgießmaschine f
conventional screw Normalschnecke f
converging zulaufend
conversion/turnover Umformung f, Umwandlung f, Umsetzung f, Konversion f

conversion equipment Konfektionierungsmaschine f
conversion factor Umrechnungsfaktor m
conversion rate Umsatzrate f
conversion reaction Umsetzung f
conversion unit Umbausatz m
  degree of conversion Umsetzungsgrad m, Umsetzungskennzahl f
  energy conversion Energieumsatz m
  energy conversion process Energieumwandlungsprozeß m
  film conversion Folienweiterverarbeitung f
  machine conversion unit Maschinenumbausatz m
  percentage conversion [e.g. of monomer into polymer] prozentualer Umsatz m
  polymerization conversion Polymerisationsumsatz m
  rate of conversion Umsatzrate f
convert/to umwandeln
converter Wandler m, Weiterverarbeiter m
  analog-digital converter A/D-Wandler m, Analog-Digital-Wandler m
  digital-analog converter D/A-Wandler m, Digital-Analog-Wandler m
  sonic converter Schallwandler m
convex grinding Bombage f, Walzenbombage f
convex ground bombiert
conveyance [e.g. melt in an extruder barrel] Fördern n, Zufuhr f
conveyed gefördert
  amount conveyed by the pump Pumpenfördermenge f
conveying fördernd, Förder-
  conveying capacity Förderleistung f
  conveying efficiency Förderwirkungsgrad m, Förderwirksamkeit f
  conveying pressure Förderdruck m
  conveying process Fördervorgang m
  conveying rate capacity Förderleistung f
  conveying section [of screw] Transportteil m
  conveying unit Transportanlage f, Förderaggregat n
  forced conveying Zwangsförderung f
  forced conveying effect Zwangsförderungseffekt m
  plasticating extruder with a forced conveying section Plastifizierextruder m mit förderwirksamer Einzugszone f
  solids conveying Feststoff-Förderung f
  with forced/positive conveying action förderwirksam
conveyor Förderer m
  conveyor belt Förderband n, Fördergurt m, Transportband n, Fließband n
  bucket conveyor Paternosteranlage f, Becherwerk n
  feed conveyor Zuführband n
  paternoster-type conveyor Paternosteranlage f
  pneumatic conveyor Druckluftfördergerät n
  screw conveyor Schneckenförderer m, Schneckenfördergerät n
  suction conveyor Saugfördergerät n
  vacuum conveyor Vakuumfördergerät n

cool down/to abkalten, erkalten
coolant Kühlmittel n, Kühlmedium n, Kältemittel n
cooled gekühlt, Kühl-
  cooled die Kühldüsenwerkzeug n, Kühldüse f
  air cooled luftgekühlt
  water cooled wassergekühlt
  water cooled feed section/zone Naßbüchse f
cooling Abkühlung f, Kühlung f
cooling air Kühlluft f
cooling air blower Kühlluftgebläse n
cooling air delivery rate Kühlluftaustrittsgeschwindigkeit f
cooling air impingement angle Anblaswinkel m, Kühlluftanblaswinkel m
cooling air ring Luftkühlring m, Kühl(luft)ring m
cooling air stream/flow Kühlluftströmung f
cooling bath Kühlbad n
cooling by means of long, slender cooling channels [method of cooling long core inserts in injection mo(u)lding] Fingerkühlung f
cooling capacity/efficiency Kühlleistung f, Kälteleistung f, Kühlkapazität f
(Hier spielt besonders die Kühlleistung eine Rolle / here efficient colling is particularly important)
cooling channel Kühlkanal m, Kühlbohrung f
cooling channel layout Kühlkanalanordnung f
cooling channel walls Kühlkanalwandungen fpl
cooling circuit Kältekreis m, Kühlkanalkreis m, Kühlkreis(lauf) m
cooling coil Kühlschlange f, Kühlwendel m
cooling conditions Abkühlbedingungen fpl, Abkühlungsverhältnisse npl, Kühlungsverhältnisse npl
cooling drum Kühltrommel f
cooling due to evaporation Verdampfungskühlung f
cooling effect Kühleffekt m
cooling fan Kühlventilator m, Gebläsekühlung f
cooling fins Kühlrippen fpl
cooling in a waterbath Wasserbadkühlung f
cooling intensity Kühlintensität f
cooling jacket Kühlmantel m
cooling mandrel Kühldorn m
cooling medium Kühlmedium n
cooling mixer Kühlmischer m
cooling of film bubble Schlauchkühlung f
cooling phase Abkühlphase f
cooling pin Kühlstift m
cooling plant Kühlanlage f
cooling process Abkühlvorgang m
cooling rate Abkühlgeschwindigkeit f
cooling ring lips Kühlringlippen fpl
cooling ring orifice (Kühlring-)Lippenspalt m
cooling roll Kühlwalze f
cooling section Abkühlstrecke f, Kühlkanal m, Kühlstrecke f, Kühlzone f, Kühlabschnitt m, Kühlpartie f
cooling station Kühlstation f
cooling stresses Abkühlspannungen fpl

cooling system Kühlsystem n, Kühlung f
cooling take-off unit Kühlabzugsanlage f
cooling time Erstarrungszeit f, Kühlzeit f, Erstarrungsdauer f, Abkühlzeit f, Abkühldauer f
cooling tower Kühlturm m
cooling trough Kühltrog m, Kühlwanne f
cooling tunnel Kühltunnel m
cooling unit Kühlgerät n
cooling water Kühlwasser n
cooling water circuit Kühlwasserkreislauf m
cooling water connection Kühlwasseranschluß m
cooling water consumer Kühlwasserverbraucher m
cooling water control valve Kühlwasserregulierventil n
cooling water flow rate Kühlwasserdurchflußmenge f
cooling water inlet Kühlwasserzulauf m, Kühlwasserzufuhr f
cooling water inlet temperature Kühlwasserzulauftemperatur f
cooling water line Kühlwasserleitung f
cooling water manifold Kühlwasserverteiler m
cooling water outlet Kühlwasserablauf m
cooling water outlet temperature Kühlwasserablauftemperatur f
cooling water recovery/recycling unit Kühlwasserrückgewinnungsanlage f
cooling water supply Kühlwasserzulauf m, Kühlwasserzufuhr f
cooling water throughput Kühlwasserdurchflußmenge f
cooling zone Temperierzone f
air cooling Luftkühlung f
air cooling system Kühlluftsystem n
air cooling unit Luftkühlaggregat n, Luftkühlgerät n
amount of cooling air Kühlluftmenge f
amount of external cooling air Außenkühlluftstrom m
amount of internal cooling air Innenkühlluftstrom m
barrel cooling sections Zylinderkühlzonen fpl
barrel cooling unit Zylinderkühlaggregat n
blown film cooling (system) Schlauchfolienkühlung f
central cooling water manifold Zentralkühlwasserverteilung f
circulation cooling (unit/system) Umwälzkühlung f
closed-circuit cooling system geschlossener Kühlkreislauf m
closed-circuit cooling unit Rückkühlaggregat n, Rückkühlwerk n, Rückkühlung f
contact cooling drum/roll Kontaktkühlwalze f
contact cooling (unit) Kontaktkühlung f
continuous cooling Dauerkühlung f
continuous flow cooling (system) Durchflußkühlung f
counterflow oil cooler Gegenstromölkühler m
counterflow oil cooling system Gegenstromölkühler m
distance between the cooling channels Kühlkanalabstand m

due to cooling abkühlbedingt
emergency cooling Notkühlung f
emergency cooling system Notkühlsystem n
even cooling Kühlgleichmäßigkeit f
external air cooling (system) Außenluftkühlung f
external cooling air (stream) Außenkühlluftstrom m
fan cooling Gebläsekühlung f
film bubble cooling system Folienkühlung f, Schlauchkühlvorrichtung f
intensive cooling Zwangskühlung f
intermediate cooling Zwischenabkühlung f
internal air cooling (system) Innenluftkühlung f, Innenkühlluftsystem n
internal bubble cooling system Blaseninnenkühlung f, Folieninnenkühlung f, Folieninnenkühlvorrichtung f, Innenluftkühlsystem n
internal cooling stresses Abkühleigenspannungen fpl
internal cooling (system) Innenkühlung f
internal screw cooling (system) Schneckeninnenkühlung f
liquid nitrogen cooling (system) Flüssig-Stickstoffkühlung f
main cooling system Primärkühlung f
mo(u)ld cooling (channel) system Werkzeugkühlkanalanlage f, Werkzeugkühlung f, Werkzeugkühlsystem n
mo(u)ld cooling medium Werkzeugkühlmedium n
oil cooling (system) Ölkühlung f
oil cooling unit Ölkühler m
pellet cooling unit Granulatkühlvorrichtung f
rate of cooling Abkühlgeschwindigkeit f
recirculating cooling Umlaufkühlung f
recirculating oil cooling (system) Umlaufölkühlung f
screw cooling (system) Schneckenkühlung f
secondary cooling system/unit Sekundärkühlung f
single-circuit cooling system Einkreis-Kühlsystem n
spray cooling section Sprühkühlstrecke f
spray cooling (unit) Sprühkühlung f
sudden cooling Schockkühlung f
water cooling ring Wasserkontaktkühlung f
water cooling (unit/system) Wasserkühlung f
waterbath cooling (unit) Wasserbadkühlung f
copolyester Copolyester m
copolymer Copolymer(isat) n
copolymer-based film Copolymerfolie f
acetal copolymer Acetal-Copolymerisat n
acrylonitrile copolymer Acrylnitril-Copolymerisat n
acrylonitrile-methyl-methacrylate copolymer Acrylnitril-Methylmethacrylat-Copolymer n
block copolymer Segmentpolymer(isat) n, Blockpolymer(isat) n
butadiene-acrylonitrile copolymer Butadien-Acrylnitril-Copolymer n
ethylene-tetrafluoroethylene copolymer Ethylen-Tetrafluorethylen-Copolymer n
ethylene-vinyl acetate copolymer [EVA] Ethylen-Vinylacetat-Copolymerisat n

ethylene-vinyl alcohol copolymer Ethylen-Vinylalkohol-Copolymerisat n
fluorinated ethylene-propylene copolymer [FEP] Fluorethylenpropylen n
graft copolymer Pfropfcopolymer(isat) n, Pfropfencopolymer n
styrene-acrylonitrile copolymer [SAN] Styrol-Acrylnitril-Copolymerisat n
styrene-butadiene copolymer Styrol-Butadienpfropfpolymerisat n
styrene copolymer Styrolcopolymerisat n
suspension copolymer Suspensionscopolymerisat n
suspension graft copolymer Suspensionspfropfcopolymerisat n
vinyl chloride copolymer VC-Copolymerisat n
vinyl chloride-vinyl acetate copolymer Polyvinylchloridacetat n, Vinylchlorid-Vinylacetat-Copolymerisat n
vinylidene fluoride-tetrafluoroethylene copolymer Vinylidenfluorid-Tetrafluorethylen-Copolymerisat n
copolymeric copolymer
copolymerization Copolymerisation f
  block copolymerization Block-Copolymerisation f
  graft copolymerization Pfropf(co)polymerisation f
copolymerize/to copolymerisieren
  graft copolymerized pfropfcopolymerisiert
copper Kupfer n
  copper braid Kupfergeflecht n
  copper-clad kupferkaschiert
  copper cladding Kupferbelag m, Kupferkaschierung f
  copper foil Kupferfolie f
  copper wire mesh Kupfergeflecht n
copy/to kopieren
  copy-milling machine Kopierfräsmaschine f
  copy-milling model Kopier(fräs)modell n
  hard copy printer Protokolldrucker m
  production of hard copy Protokollerstellung f
core Kernschicht f, Kern m, Wickelhülse f
  core box Kernbüchse f, Kernkasten m
  core displacement Kernversatz m (um einen Kernversatz zu vermeiden.... / to prevent the core getting out of alignment)
  core insert Kerneinsatz m
  core inserts eingelegte Kerne mpl
  core layer [of a laminate] Innenlage f
  core material Kernwerkstoff m, Kernmaterial n
  core misalignment Kernversatz m (um einen Kernversatz zu vermeiden.... / to prevent the core getting out of alignment)
  core pattern Kernmodell n
  core pin Bohrungskern m, Bohrungsstift m, Kernstift m
  core plate schließseitige Formplatte f, Kernplatte f, Kernträger m, Kernträgerplatte f, Stempelplatte f, Patrize f
  core puller Kernzug(vorrichtung) m(f)
  core puller control (mechanism) Kernzugsteuerung f
  core pulling Kernziehen n
  core pulling mechanism Kernzug(vorrichtung) m(f)
  core rotating device/mechanism Kerndrehvorrichtung f
  core unscrewing device/mechanism Kernausschraubvorrichtung f
  core wall temperature Kernwandtemperatur f
  collapsible core Faltkern m
  foam core Schaumstoffkern m
  foundry core Gießereikern m
  honeycomb core Wabenkern m
  integral foam core Integralschaumstoffkern m
  mandrel core Werkzeugdorn m
  mo(u)ld core Spritzgießkern m, Formkern m, Werkzeugkern m, Stempel m
  rigid foam core Hartschaumkern m
  separate cores Einzelkerne mpl
  solid core [of material] Feststoffkern m
  threaded core Gewindekern m, Schraubkern m
  unscrewing core Ausschraubkern m
corner Ecke f
  corner joint Eckverbindung f
  corner protector Kantenschutz m
  corner strength [of a PVC window frame] Eckfestigkeit f
corona Corona f, Korona f
  corona discharge Coronaentladung f
  corona pretreating unit Corona-Vorbehandlungsanlage f
  corona pretreatment Coronavorbehandlung f, Korona-Vorbehandlung f
  corona resistance Coronabeständigkeit f, Koronafestigkeit f
  corona treatment Coronabehandlung f
correct/to korrigieren
correction Korrektur f
  correction factor Korrekturwert m, Korrekturfaktor m
  fault correction Mängelkorrektur f
  flow correction Fließkorrektur f, Flußkorrektur f
  mo(u)ld correction factor Formenkorrekturmaß n
  pressure correction Druckkorrektur f
  zero correction Nullpunktkorrektur f
correctly designed richtig konstruiert
correctly timed zeitoptimal
correlation Korrelation f
  correlation coefficient Korrelationskoeffizient m
corrosion Korrosion f
  corrosion damage [Note that the word damage is always used in its singular form in this context. The plural is something entirely different] Korrosionsschäden mpl
  corrosion inhibiting korrosionsverhindernd, korrosionshemmend
  corrosion inhibitor Korrosionsinhibitor m
  corrosion near the edges Kantenkorrosion f
  corrosion resistance Korrosionsbelastbarkeit f, Korrosionsbeständigkeit f

corrosion resistant korrosionsfest, korrosionsbeständig
edge corrosion Kantenkorrosion f
electrolytic corrosion elektrolytische Korrosion f
freedom from corrosion Korrosionsfreiheit f
protected against corrosion korrosionsgeschützt
protecting against corrosion korrosionsschützend
protection against corrosion Korrosionsschutz m
risk of corrosion Korrosionsgefahr f
signs of corrosion Korrosionserscheinungen fpl
surface corrosion Oberflächenkorrosion f
susceptibility to corrosion Korrosionsanfälligkeit f
susceptible to corrosion korrosionsanfällig
corrosive ätzend, korrodierend, korrosiv
  corrosive effect Korrosivwirkung f
corrugate/to riffeln, riefen, wellen
  corrugated pipe extrusion line Wellrohranlage f
  longitudinal corrugating device Längswellvorrichtung f
  longitudinal corrugating machine Längswellmaschine f
  transverse corrugating machine Querwellmaschine f
corundum Korund n
  corundum slurry Korundschlämme f
cosmetics Kosmetika npl (Kosmetikum n)
cost [also: costs] Preis m, Kosten pl, Kostenaufwand m
  cost advantage Kostenvorteil m, Preisvorteil m
  cost analysis Kostenanalyse f
  cost-benefit calculation Kosten-Nutzenrechnung f
  cost-benefit considerations Kosten-Nutzen-Überlegungen fpl
  cost-benefit ratio Preis-Wirkungs-Relation f, Kosten-Nutzen-Verhältnis n
  cost factor Kostenfaktor m
  cost increase Kostensteigerung f
  cost-intensive kostenintensiv
  cost of spare parts Ersatzteilkosten pl
  cost-performance factor Preis-Leistungs-Index m
  cost-performance ratio Preis-Leistungs-Verhältnis n, Aufwand-Nutzen-Relation f, Preis-Durchsatz-Verhältnis n
  cost reduction Kostensenkung f
  cost saving Kosteneinsparung f, Kostenersparnis f
  cost-service life factor Kosten-Standzeit-Relation f, Preis-Lebensdauer-Relation f
  cost structure Kostenstruktur f
  administrative costs Verwaltungskosten pl
  capital cost Kapitaleinsatz m
  development costs Entwicklungsaufwand m
  electricity costs Stromkosten pl
  estimated cost Schätzkosten pl
  extra cost Aufpreis m, Zusatzkosten pl
  finishing costs Nacharbeitungskosten pl
  initial tooling costs Erstwerkzeugkosten pl
  itemized cost Einzelkosten pl
  labo(u)r costs Lohnkosten pl, Personalkosten pl
  machine costs Maschinenkosten pl
  machine rebuilding costs Maschinenumbaukosten pl
  maintenance costs Unterhaltskosten pl, Wartungskosten pl
  manufacturing costs Fertigungkosten pl
  material costs Werkstoffkosten pl, Stoffkosten pl
  operating costs Unterhaltskosten pl, Betriebsaufwand m, Betriebskosten pl
  production costs Produktionskosten pl, Fertigungskosten pl
  raw material costs Rohmaterialkosten pl, Stoffkosten pl
  reasonable cost Preiswürdigkeit f
  repair costs Reparaturaufwand m, Reparaturkosten pl
  research costs Forschungsaufwand m
  running costs Betriebskosten pl
  tooling costs Werkzeug(herstellungs)kosten pl
  total cost Gesamtkosten pl
  total electricity costs Gesamtstromkosten pl
  transport costs Transportkosten pl
costly kostenaufwendig, kostspielig
cotton Baumwolle f
  cotton fabric Baumwollgewebe n
  cotton flock Baumwollflocken fpl
  cotton linters Baumwollinters pl
  coarse cotton fabric Baumwollgrobgewebe n
  collodion cotton Collodiumwolle f
  extremely fine cotton fabric Baumwollfeinstgewebe n
  fine cotton fabric Baumwollfeingewebe n
cottonseed oil fatty acid Baumwollsamenölfettsäure f
coumarone Cumaron n
  coumarone-indene resin Cumaron-Indenharz n
counter 1. gegenläufig; Gegen-, 2. Zählwerk n
  counter-rotating entgegengesetzt rotierend, gegenläufig
  counter-rotating twin screw Gegendrallschnecke f, Gegendrall-Doppelschnecke f
  counter-rotating twin screw extruder Gegendrallmaschine f, Gegendrall-Schneckenextruder m
  counter-rotating twin screw system Gegendrallsystem n
  cycle counter Zykluszähler m
  decade counter Dekadenzähler m
  hours counter Stundenzähler m
  instruction counter Befehlszähler m
  machine hours counter Betriebsstundenzähler m
  parts counter Stückzähler m
  speed counter Drehzahlmesser m
counterflow Gegenstrom m
  counterflow oil cooler Gegenstromölkühler m
counterpressure Gegendruck m
counting 1. Zählung f; 2. Zähl-
  counting function Zählfunktion f
  counting mechanism Zählwerk n
  parts counting unit/device Stückzähler m
country Land n
  country of the European Free Trade Association [EFTA country] EFTA-Staat m

common market country [EEC country] EWG-Land n
neighbo(u)ring country Nachbarland n
oil-producing country Ölland n, ölproduzierendes Land n
underdeveloped country Entwicklungsland n
couple/to koppeln, ankuppeln, anlenken
coupled angekoppelt
  coupled operation Verbundbetrieb m
coupling 1. Kopplung f, Kupplung f, Ankopplung f, Haftvermittlung f; 2. haftvermittelnd
  coupling agent Haftverbesserer m, Haftvermittler m
  coupling mechanism Kopplungsmechanismus m
  coupling reaction Kupplungsreaktion f
  coupling system Kopplungssystem n
  bayonet coupling Bajonettverschluß m
  flange coupling Flanschanschluß m
  high speed coupling Schnellkupplung f, Schnellverbindung f
  plug-in coupling Steckkupplung f
  quick-action coupling mechanism/system Schnellspannvorrichtung f, Schnellspannsystem n
  screw coupling Schraubenanschluß m, Verschraubung f
course of the process Prozeßablauf m
covalent kovalent
cover Abdeckung f
  bumper cover Stoßstangenabdeckung f
  safety cover Schutzverdeck n
  seat cover Sitzbezug m
  transparent cover Klarsichtdeckel m
covered bedeckt, abgedeckt
  rubber covered gummiert
  rubber covered (back-up) roll Gummiwalze f
  rubber covered pressure roll Anpreßgummiwalze f
covering Abdeckung f, Belag m, Bezug m
  glass cloth covering Glasgewebeabdeckung f
  roll surface covering Walzenmantel m, Walzenbezug m
  wall covering Wandbelag m, Wandbekleidung f
  wire covering compound Adermischung f
  wire covering die Drahtummantelungs-Düsenkopf m, Drahtmantelungskopf m
  wire covering line Drahtummantelungsanlage f, Drahtisolierlinie f
  wire covering torpedo Drahtummantelungspinole f
crack Riß m
  crack direction Rißrichtung f
  crack formation Rißbildung f
  crack front Rißfront f
  crack growth Rißvergrößerung f, Rißverlängerung f, Rißfortschritt m, Rißwachstum n, Rißausbreitung f, Rißfortpflanzung f, Bruchausbreitung f, Bruchfortschritt m
  crack growth direction Rißwachstumsrichtung f
  crack growth rate Rißausbreitungsgeschwindigkeit f, Rißerweiterungsgeschwindigkeit f, Rißwachstumsrate f, Rißwachstumsgeschwindigkeit f, Riß(fortpflanzungs)geschwindigkeit f
  crack-initating rißauslösend
  crack initiation Rißauslösung f, Brucheinleitung f, Rißinitiierung f
  crack length Rißlänge f
  crack propagation Rißvergrößerung f, Rißverlängerung f, Rißfortschritt m, Rißwachstum n, Rißausbreitung f, Rißfortpflanzung f, Bruchausbreitung f, Bruchfortschritt m
  crack propagation force Rißausbreitungskraft f
  crack propagation rate Rißausbreitungsgeschwindigkeit f, Rißerweiterungsgeschwindigkeit f, Rißwachstumsrate f, Rißwachstumsgeschwindigkeit f, Riß(fortpflanzungs)geschwindigkeit f
  crack propagation resistance Rißausbreitungswiderstand m
  crack propagation test Rißfortpflanzungsversuch m
  crack resistance Rißbildungsresistenz f, Rißwiderstand m
  crack speed Riß(fortpflanzungs)geschwindigkeit f [s.a. crack growth rate]
  crack surface Rißfläche f
  crack tip Rißspitze f
  fatigue crack Ermüdungsriß m
  fatigue crack propagation Ermüdungsrißausbreitung f [s.a. crack propagation]
  hairline crack Haarriß m
  liability to crack Bruchanfälligkeit f
  shrinkage crack Schwundriß m
  stress crack Spannungs(korrosions)riß m
  tendency to crack Rißbildungsneigung f
cracking Reißen n, Rißbildung f, Spalten n, Kracken n
  environmental stress cracking umgebungsbeeinflußte Spannungsrißbildung f
  fatigue cracking Ermüdungsrißbildung f
  resistant to stress cracking spannungsrißbeständig
  stress cracking behavio(u)r Spannungskorrosionsverhalten n
  stress cracking resistance Spannungsriß(korrosions)beständigkeit f
  susceptibility to cracking Rißanfälligkeit f
  susceptibility to stress cracking Spannungsrißanfälligkeit f
  susceptible to stress cracking spannungsrißempfindlich
crash Krach m, Bruch m, Zusammenstoß m
  crash helmet Schutzhelm m, Sturzhelm m
  crash test Auffahrversuch m
crate Kiste f, Steige f
  bottle crate Flaschenkasten m
cratering Kraterbildung f, Auskohlung f
  freedom from cratering Kraterfreiheit f
craze Craze f
  craze formation Craze-Bildung f, Trübungszonenbildung f
  craze-initiating Craze-auslösend

craze initiation Craze-Initiierung f
craze propagation Craze-Wachstum n
craze zone Craze-Zone f, Crazefeld n, Fließzone f, Trübungszone f
craze zone structure Fließzonenstruktur f
shear craze Scher-Craze f
tensile craze Zugspannungs-Craze f
crazed material Fließzonenmaterial n
crazing Craze-Bildung f, Trübungszonenbildung f, Fließzonenbildung f
cream Creme f
   cream time [in PU foaming] Startzeit f
   barrier cream Hautschutzcreme f, Schutzcreme f
   cleansing cream Reinigungscreme f
crease-free faltenfrei, faltenlos
creel Gatter n
   pay-off creel Abspulgatter n
creep Kriechen n
   creep behavio(u)r Zeitstandverhalten n, Kriechverhalten n, Zeitdehnverhalten n
   creep curve Dehngrenzlinie f, Kriechkurve f, Zeitdehnlinie f, Zeitspannungslinie f, Zeitstandkurve f
   creep deformation Kriechverformung f
   creep diagram Zeitstanddiagramm n, Zeitstandschaubild n
   creep elongation Kriechdehnung f
   creep factor Kriechfaktor m
   creep modulus Kriechmodul m
   creep modulus curve Kriechmodulkurve f
   creep rate Kriechgeschwindigkeit f, Kriechrate f
   creep resistance Kriechwiderstand m
   creep rupture curve Zeitbruchkurve f, Zeitbruchlinie f
   creep rupture strength Zeitstandfestigkeit f, Kriechfestigkeit f
   creep strain Kriechdehnung f
   creep strength Zeitstandfestigkeit f, Kriechfestigkeit f
   creep strength curve Zeitstandfestigkeitskurve f, Zeitstandfestigkeitslinie f
   creep stress Kriechbeanspruchung f, Kriechspannung f, Zeit(dehn)spannung f, Zeitstandbeanspruchung f
   creep test Langzeitprüfung f, Langzeittest m, Langzeituntersuchung f, Langzeitversuch m, Kriechversuch m, Zeitstandprüfung f, Zeitstandversuch m
   creep test machine Zeitstandanlage f
   compressive creep test Zeitstanddruckversuch m
   flexural creep modulus Biege-Kriechmodul m
   flexural creep strength Zeitstandbiegefestigkeit f
   flexural creep test Zeitstandbiegeversuch m, Biegekriechversuch m
   tendency to creep Kriechneigung f
   tensile creep modulus Zug-Kriechmodul m
   tensile creep strength Zeitstandzugfestigkeit f
   tensile creep stress Zeitstand-Zugbeanspruchung f
   tensile creep test Zeitstandzugversuch m
cresol Kresol n
cresol-formaldehyde condensate Kresol-Formaldehyd-Kondensat n
cresol novolak Kresolnovolak n
cresol resin Kresolharz n
cresol resol Kresolresol n
cresyl glycidyl ether Kresylglycidylether m
criteria Kriterien npl (Singular: Kriterium)
   design criteria Auslegungskriterien npl, Auslegekriterien npl
   quality criteria Qualitätskriterien npl
   selection criteria Auswahlkriterien npl
criterion (plural: criteria) Kriterium n
   selection criterion Auswahlkriterium n
critical kritisch
cross Kreuz n, Kreuzung f, Kreuz(rohr)stück n
   cross-axis roll adjustment Walzenschrägstellung f, Walzenschrägverstellung f, Schrägverstellung f, Walzenschrägeinstellung f
   cross-breaking strength Biegefestigkeit f
   cross-hatch adhesion test Gitterschnitttest m
   cross-section Querschnitt m
   cross-sectional area Querschnittfläche f
   cross twill weave Kreuzköperbindung f
   cross-wound bobbin Kreuzspule f
   closely cross-linked engmaschig vernetzt, engvernetzt
   complete cross-section Gesamtquerschnitt m
   entire cross-section Gesamtquerschnitt m
   initial cross-section Anfangsquerschnitt m
   mo(u)lded-part cross-section Formteilquerschnitt m
   narrowing cross-section Querschnittsverengungen fpl
   original cross-section Anfangsquerschnitt m
   part cross-section Formteilquerschnitt m
   runner cross-section Verteilerkanalquerschnitt m
crossbar distributor Kreuzschienenverteiler m
crosshead (die) Quer(spritz)kopf m, Umlenkkopf m, Umlenkwerkzeug n, Winkel(spritz)kopf m, Schrägkopf m
crossing Kreuzung f, Schrägverstellung f
   axis/roll crossing Schrägverstellung f, Walzenschrägeinstellung f, Walzenschräg(ver)stellung f
crosslink Vernetzung f
   crosslink density Vernetzungsdichte f
   crosslink point Vernetzungsstelle f
   crosslink structure Vernetzungsstruktur f
crosslinkability Vernetzbarkeit f (Vernetzbarkeit mit reaktiven Lackrohstoffen ist gegeben / can be crosslinked with reactive surface coating resins)
crosslinkable vernetzbar
crosslinkage Vernetzung f, Härtung f
   addition crosslinkage Additionsvernetzung f
   peroxide crosslinkage peroxidische Vernetzung f
   radiation crosslinkage Strahlenvernetzung f
crosslinked vernetzt
   closely crosslinked engmaschig vernetzt
   densely crosslinked stark vernetzt
   loosely crosslinked schwachvernetzt, schwach vernetzt, weitmaschig vernetzt

partly crosslinked teilvernetzt, teilweise vernetzt
radiation crosslinked strahlenvernetzt
crosslinking Vernetzung f, Vernetzen n
  crosslinking efficiency Vernetzungseffizienz n, Vernetzungswirksamkeit f
  crosslinking mechanism Vernetzungsmechanismus m
  crosslinking reaction Verknüpfungsreaktion f
  crosslinking process Vernetzungsprozeß m
  crosslinking reaction Vernetzungsreaktion f
  crosslinking time Vernetzungszeit f
  addition crosslinking additionsvernetzend
  degree of crosslinking Vernetzungsgrad m
crosspiece Steg m
crude rubber Rohkautschuk m
crumb form [e.g. a rubber] Krümelform f
crumbs Krümel mpl
crumple zone [in a car] Verformungsbereich m, Knautschzone f
cryostat Kryostat m
crystal Kristall m
  crystal clarity Glasklarheit f
  crystal clear glasklar
  crystal clear film Glasklarfolie f
  crystal growth Kristallwachstum n
  crystal lattice Kristallgitter n
  liquid crystal display [LCD] Flüssigkristall-Anzeige f, LCD-Display n
  rate of crystal growth Kristallwachstumsgeschwindigkeit f
crystalline kristallin
  crystalline state Kristallisationszustand m
  coarsely crystalline grobkristallin
  finely crystalline feinkristallin
  partially crystalline teilkristallin
crystallinity Kristallinität f
  degree of crystallinity Kristallinitätsgrad m, Kristallisationsgrad m
  partial crystallinity Teilkristallinität f
crystallite Kristallit m
  crystallite orientation Kristallitorientierung f
crystallizability Kristallisationsfähigkeit f
crystallization Kristallisation f, Kristallisierung f
  crystallization temperature Kristallisationstemperatur f
  rate of crystallization Kristallisationsgeschwindigkeit f
  water of crystallization Hydrat(ions)wasser n, Kristallwasser n, chemisch gebundenes Wasser n
crystallize/to out (aus)kristallisieren
  tendency to crystallize Kristallisationsneigung f
crystallizing conditions Kristallisationsbedingungen fpl
cube-shaped würfelförmig, kubisch
cubic(al) kubisch, würfelförmig
  coefficient of cubical expansion thermischer Volumenausdehnungskoeffizient m, kubischer Ausdehnungskoeffizient m
culminating point Kulminationspunkt m

cumol hydroperoxide Cumolhydroperoxid n
cumulation Kumulation f
cup Becher m, Tasse f, Verpackungsbecher m
  disposable cup Einwegtasse f
  flow cup Auslaufbecher m
  yog(h)urt cup Yoghurtbecher m
cure/to [crosslinking of polymer chains to impart elastomeric properties] aushärten
cure Aushärtung f, Durchhärtung f, Vulkanisation f
  cure temperature Härtungstemperatur f
  cure time Härtungszeit f, Aushärtungszeit f
  accelerated cure Schnellhärtung f
  amine cure Aminhärtung f
  anhydride cure Anhydridhärtung f, Anhydridvernetzung f
  cobalt cure Kobalthärtung f
  degree of cure Aushärtungsgrad m, Härtungsgrad m
  full cure Durchhärtung f
cured [resin, rubber, etc.] ausgehärtet, gehärtet, vernetzt, vulkanisiert
  cured casting resin Gießharzformstoff m
  cured epoxy resin Epoxidharz-Formstoff m
  cured polyester resin Polyesterharzformstoff m, UP-Harzformstoff m
  cured resin Preßstoff m, Harzformstoff m
  cured UP resin UP-Harzformstoff m
curing Vernetzung f, Härtung f
  curing agent Vernetzungsmittel n, Vernetzungshilfe f, Vernetzungskomponente f, Vernetzer m
  curing agent concentration Vernetzerkonzentration f
  curing agent decomposition products Vernetzerspaltprodukte npl
  curing agent paste Vernetzerpaste f
  curing behavio(u)r Härtungsverhalten n, Aushärteverhalten n, Vernetzungsverhalten n
  curing characteristics Härtungsverhalten n, Aushärteverhalten n, Vernetzungsverhalten n
  curing conditions Aushärtungsbedingungen fpl, Härtungsbedingungen fpl, Vernetzungsbedingungen fpl
  curing mechanism Härtungsmechanismus m
  curing oven Aushärteofen m
  curing pattern Härtungsverlauf m
  curing rate Aushärtungsgeschwindigkeit f, Vernetzungsrate f
  curing reaction Aushärtungsreaktion f, Härtungsreaktion f, Härtungsablauf m
  curing schedule Härtungsprogramm n
  curing shrinkage Härtungsschrumpf m, Härtungsschwund m
  curing speed Härtungsgeschwindigkeit f
  curing stage Härtungsstufe f
  curing system Härtungssystem n, Vernetzungssystem n
  curing temperature Aushärtungstemperatur f, Härtetemperatur f, Vernetzungstemperatur f
  curing time Härtezeit f
  acid curing säurehärtend
  amount of curing agent Vernetzermenge f,

Vernetzungsmitteldosierung f
electron beam curing EB-Verfahren n, Elektronenstrahlhärtung f
fast curing rasch aushärtbar, rasch härtend, schnellhärtend
heat curing 1. heißhärtend, wärmehärtend; 2. Heißhärtung f, Warmhärtung f
hot curing catalyst Heißhärter m
low-temperature curing Tieftemperaturhärtung f
radiation curing Strahlungshärtung f
room temperature curing Normalklimahärtung f, Normaltemperatur-Aushärtung f, Raumtemperaturhärtung f
type of curing agent Vernetzerart f
current 1. Strom m [elektrischer]; 2. laufend, gegenwärtig, (allgemein)gültig
current assets Umlaufvermögen n
current input Stromaufnahme f
current intensity Stromstärke f
current-limiting resistance Strombegrenzungswiderstand m
current transformer Stromwandler m
alternating current [a.c.] Wechselstrom m
constant current Dauerstrom m, Konstantstrom m
direct current [d.c.] Gleichstrom m
leakage current Ableitstrom m, Kriechstrom m
output current Ausgangsstrom m
three-phase current Drehstrom m
cursor Bildschirmmarke f, Cursor m
curtain rail Gardinengleiter m
curvature Krümmung f
angle of curvature Krümmungswinkel m
radius of curvature Krümmungsradius m
curve Kurve f, Kurvenzug m, Kennlinie f
acoustic emission curve Schallemissionskurve f
actual value curve Ist(wert)kurve f
creep curve Dehngrenzlinie f, Kriechkurve f, Zeitdehnlinie f, Zeitspannungslinie f, Zeitstandkurve f
creep modulus curve Kriechmodulkurve f
creep rupture curve Zeitbruchkurve f, Zeitbruchlinie f
creep strength curve Zeitstandfestigkeitskurve f, Zeitstandfestigkeitslinie f
family of curves Kurvenschar f
flow curve Fließkurve f
force-deflection curve Kraft-Durchbiegungskurve f
group of curves Kurvenschar f
hysteresis curve Federkennlinie f, Hysteresekurve f
molecular weight distribution curve Molekulargewichtsverteilungskurve f
operating curve Arbeitskennlinie f
output rate curve Ausstoßkennlinie f
pressure curve Druckkennlinie f, Druckverlaufkurve f
reference curve Bezugskurve f, Sollkurve f
reference pressure curve Solldruckkurve f
relaxation curve Relaxationskurve f
setpoint curve Soll(wert)kurve f
sine curve Sinuskurve f
stress-strain curve Kraft-Längenänderungs-Diagramm n
tensile stress-elongation curve Zug-Dehnungs-Diagramm n, Zugspannungs-Dehnungs-Kurve f
velocity curve Geschwindigkeitskennlinie f
WLF curve WLF-Kurve f
Wöhler curve Wöhlerlinie f
curved gekrümmt
cushion Polster n, Kissen n
cushion control Polsterregelung f
air cushion Luftkissen n
melt cushion Massepolster n, Schmelzpolster n
cushioning material [e.g. foam for packaging] Polstermaterial n
customer's requirements Kundenbedürfnisse npl
customer's sample Kundenvorlage f
cut/to schneiden, trennen
cut into lengths/to [extruded pipe, profiles or sheets] ablängen
cut to size/to formatschneiden
cut up/to zerkleinern
cut with a plasma arc plasmageschnitten
cut Schnitt m
cut blank Plattenzuschnitt m
cut-off frequency Grenzfrequenz f
cut surface Schnittfläche f
cut-to-size piece of sheet Plattenzuschnitt m
cut-to-size pieces Zuschnitte mpl
cut-to-size prepregs Harzmattenzuschnitte mpl
automatic cut-out Sicherungsautomat m
emergency cut-out Not-Auseinrichtung f, Notausschalter m
hot cut pelletizing system Direktabschlagsystem n
longitudinal cut Längsschnitt m
machine cut-out Maschinenabschaltung f
pressure cut-out Druckabschneidung f
cutter Schneidewerkzeug n, Schneider m
automatic pipe cutter Rohrtrennautomat m
high speed cutter Hochgeschwindigkeitsschneider m
impact cutter Schlagmesser n, Schlagschere f
incandescent wire cutter Glühbandabschneider m
longitudinal cutter Längsschneideeinrichtung f, Längsschneidevorrichtung f, Längstrenneinrichtung f
milling(-type) cutter Fräsrotor m, Walzenfräser m
parison cutter Schlauchabschneider m
rotary transverse cutter Rotationsquerschneider m
transverse cutter Querschneidemaschine f, Querschneider m, Quertrenneinrichtung f
cutting Schneiden n
cutting accuracy Schnittpräzision f
cutting angle Schnittwinkel f
cutting chamber Schneidraum m, Schneidkammer f, Mahlraum m, Mahlkammer f
cutting chamber hood Granulierhaube f
cutting die Stanzwerkzeug n
cutting edge [of knife] Schneidkante f
cutting force Schnittkraft f

cutting oil Bohröl n, Schneidöl n
cutting quality Schnittgüte f
cutting speed Schnittgeschwindigkeit f
cutting tool Schnittwerkzeug n
automatic device for cutting plastics profiles into lengths Kunststoffprofil-Ablängeautomat m
bed knife cutting circle Statormesserkreis m
circular cutting device Kreisschneidevorrichtung f
gear cutting Verzahnung f
incandescent wire cutting Glühdrahtschneiden n
laser beam cutting Laserstrahlschneiden n
laser beam cutting process Laserschneidverfahren n
pipe cutting device Rohrtrennvorrichtung f
rotor knife cutting circle Rotormesserkreis m
water jet cutting Wasserstrahlschneiden n
cyanacrylate Cyanacrylat n
cyanacrylate adhesive Cyanacrylatklebstoff m
ethyl cyanacrylate Ethylcyanacrylat n
cyanurate Cyanurat n
triallyl cyanurate Triallylcyanurat n
cycle Zyklus m, Takt m
cycle control Taktsteuerung f, Zykluskontrolle f
cycle counter Zykluszähler m
cycle-independent zyklusunabhängig
cycle memory Zyklusgedächtnis n
cycle monitor Taktüberwachung f
cycle sequence Zyklusablauf m
cycle time Taktzeit f, Zyklusdauer f, Zykluszeit f
cycle time reduction Zykluszeitverkürzung f, Zykluszeitreduzierung f (zum Zweck der Zykluszeitreduzierung / to reduce cycle time)
dry cycle time Taktzeit f im Trockenlauf, Taktzeit f, Trockenlauf-Zykluszeit f, Trockentaktzeit f, Trockenlaufzeit f
end of cycle Zyklusablauf m, Zyklusende n (nach Zyklusablauf / after completion of the cycle)
factor influencing cycle time Zykluszeitfaktor m (Einer der Zykluszeitfaktoren ist die Werkzeugwanddicke / one of the factors which influences cycle time is mo(u)ld wall thickness)
feed(ing) cycle Dosiertakt m, Förderzyklus m
independent of the mo(u)lding cycle zyklusunabhängig
interruption of the mo(u)lding cycle Zyklusunterbrechung f (Abschalten der Werkzeugkühlung bei Zyklusunterbrechung / switching off of mo(u)ld cooling if the mo(u)lding cycle is interrupted)
load cycle Lastspiel n, Lastwechsel m, Lastzyklus m, Belastungszyklus m
load cycle frequency Lastspielfrequenz f
machine cycle Maschinentakt m
machine cycle control Maschinentaktsteuerung f
mo(u)lding cycle Arbeitstakt m, Arbeitszyklus m, Fertigungszyklus m, Formzyklus m, Preßzyklus m, Spritztakt m, Spritzzyklus m, zyklischer Prozeßablauf m
mo(u)lding cycle control Zykluskontrolle f

mo(u)lding cycle monitoring (system) Zyklusüberwachung f
mo(u)lding cycle, production cycle Fertigungszyklus m
mo(u)ld opening and closing cycle Formenöffnungs- und Schließspiel n
number of breaking stress cycles Bruchlastspielzahl f
number of dry cycles [usually per minute] Trockenlaufzahl f, Anzahl der Leerlaufzyklen mpl
number of (mo(u)lding) cycles Taktzahl(en) f(pl)
number of stress/load cycles Lastspielzahl f, Lastwechselzahl f
operating in cycles periodisch arbeitend
processing cycle Verarbeitungszyklus m
production cycle Fertigungsfluß m, Produktionsablauf m, Produktionszyklus m, Fertigungszyklus m
stress cycle Lastspiel n, Lastwechsel m, Lastzyklus m, Belastungszyklus m
stress cycle frequency Lastspielfrequenz f
total cycle time Zyklusgesamtzeit f, Gesamtzykluszeit f
cyclic ringförmig, zyklisch
cyclic compound Ringverbindung f
cyclic/ring structure Ringstruktur f
cycling Taktablauf m, Ablauf m des Arbeitszyklus, Dauerschwingbeanspruchung f
automatic fast cycling injection mo(u)lding machine Schnellspritzgießautomat m
fast cycling rasche Taktfolge f
fast cycling machine Schnelläufer m
cyclization Cyclisierung f, Zyklisierung f, Vernetzung f, Ringbildung f
cycloaliphatic cycloaliphatisch
cycloaromatic cycloaromatisch
cyclohexane Cyclohexan n
cyclohexanol Cyclohexanol n
cyclohexanone Cyclohexanon n
cyclohexanone peroxide Cyclohexanonperoxid n
cyclohexanone resin Cyclohexanonharz n
cyclone separator Zyklonabscheider m
cycloolefin Cycloolefin n
cylinder [also: barrel] Zylinder m
cylinder/barrel dimensions Zylinderabmessungen fpl
cylinder/barrel head Zylinderkopf m
cylinder/barrel heater Zylinderheizelement n
cylinder/barrel heaters Zylinderbeheizung f, Zylinderheizung f
cylinder/barrel heating capacity Zylinderheizleistung f
cylinder/barrel heating circuit Zylinderheizkreis m
cylinder/barrel length Zylinderlänge f
cylinder/barrel liner Zylinderinnenfläche f, Zylinderauskleidung f, Zylinderinnenwand f
cylinder/barrel temperature Zylindertemperatur f
cylinder/barrel wall Zylinderwand f
cylinder/barrel wall temperature Zylinder-

wandtemperatur f
cylinder/barrel wear Zylinderverschleiß m
accumulator cylinder Tauchkammerzylinder m, Akkumulierzylinder m, Speicherzylinder m
bimetallic cylinder/barrel Bimetallzylinder m
clamping cylinder Formschließzylinder m, Schließzylinder m, Werkzeugschließzylinder m, Schließkolben m
compressed air cylinder Druckluftzylinder m, Druckluftbehälter m
compressed gas cylinder Druckgasbehälter m
ejector cylinder Ausstoßzylinder m
feed cylinder Füllzylinder m, Dosierzylinder m
high speed clamping cylinder Schnellschließzylinder m
high speed cylinder Eilgangszylinder m
high speed (injection) cylinder Schnellfahrzylinder m
hydraulic injection cylinder Hydraulikspritzzylinder m
injection cylinder Einspritzzylinder m, Spritzzylinder m, Maschinenzylinder m
locking cylinder Arretierzylinder m, Zuhaltezylinder m
metering cylinder Dosierzylinder m
mo(u)ld opening cylinder Formaufdrückzylinder m
nozzle advance cylinder Düsenanpreßzylinder m
outer cylinder/barrel surface Zylinderaußenfläche f
plasticizing cylinder/barrel Plastifizierzylinder m
plunger injection cylinder Kolbenspritzzylinder m
plunger plasticizing cylinder Kolbenplastifizierzylinder m
pneumatic cylinder Pneumatikzylinder m
preplasticizing cylinder/barrel Schneckenzylinder m
reciprocating-screw cylinder Schneckenschubzylinder m
screw injection cylinder Schneckenspritzzylinder m
screw plasticizing cylinder Schneckenplastifizierungszylinder m
sprue ejector cylinder Anguß-Ausstoßzylinder m
vented cylinder/barrel Entgasungszylinder m
cylindrical zylinderförmig, zylindrisch
  cylindrical and/or diced pellets Kaltgranulat n
  cylindrical pellets Stranggranulat n, Zylindergranulat n
dairy products Molkereiprodukte npl
damage [Note that the word damage is always used in its singular form in this context. The plural is something entirely different] Schädigung f
  damage due to wear Verschleißschäden mpl
  damage to the mo(u)ld Werkzeugbeschädigungen fpl (...um Werkzeugbeschädigungen zu vermeiden / to prevent the mo(u)lds being damaged)
  amount of damage Schädigungsgrad m
  corrosion damage Korrosionsschäden mpl
  polymer damage Polymerenschädigung f
  risk of damage Beschädigungsgefahr f
  structural damage Gefügeschädigung f

surface damage Oberflächenschädigung f
  without (suffering) damage schädigungslos
damaged area Schadstelle f, örtliche Schadstellen fpl
damp 1. feucht, naß; 2. Feuchte f, Feuchtigkeit f
damping Dämpfung f, Anfeuchtung f, Benetzung f
  damping behavio(u)r Dämpfungsverhalten n
  damping characteristics Dämpfungseigenschaften fpl
  damping element Dämpfungselement n
  damping fluid Dämpfungsflüssigkeit f, Dämpfungsmedium n
  damping maximum Dämpfungsmaximum n
  damping mechanism Dämpfungseinrichtung f
  damping medium Dämpfungsflüssigkeit f, Dämpfungsmedium n
  damping modulus Dämpfungsmodul m
  ejector damping (mechanism) Auswerferdämpfung f
danger to health Gesundheitsgefährdung f
dark colo(u)red dunkelfarbig, dunkelpigmentiert
darkening Dunkelwerden n
dart Ankerstift m
  falling dart test Falldorntest m
dashboard Instrumenttafel f, Armaturenbrett n, Armaturentafel f
dashed [curve] gestrichelt [Linie]
data Daten npl
  data acquisition device/unit/equipment Datenerfassungsgerät n, Datenerfassungsstation f, Datenerfassungsanlage f
  data analyzer Datenanalysator m
  data bank Datenbank f, Datenspeicherung f, Datenspeicher m
  data bus Datenbus m, Informationssammelschiene f
  data carrier Datenträger m
  data collecting system Datensammelsystem n, Datenabnahme f, Datensammlung f
  data compaction Datenverdichtung f
  data compression Datenkompression f
  data content Dateninhalt m
  data display unit Datenanzeige f, Datensichtgerät n
  data documentation Datendokumentation f
  data exchange Datenaustausch m
  data file Datei f, Datenfile n
  data filing Datenarchivierung f
  data flow Datenfluß m
  data gathering Datenerfassung f
  data gathering device/unit/equipment Datenerfassungsgerät n [s.a. data acquisition device]
  data generator Datengeber m
  data input Dateneingabe f
  data interface Datenschnittstelle f
  data link Datenverbund m
  data loss Datenverlust m
  data management Datenverwaltung f
  data output Datenausgabe f
  data preparation Datenaufbereitung f
  data presentation Datendarstellung f

data processing Datenverarbeitung f
data processing facilities Datenverarbeitungsmöglichkeiten fpl
data processing unit Datenverarbeitungsanlage f
data protection Datensicherung f
data reduction Datenreduktion f
data register Datenregister n
data retention Datenhaltung f
data retrieval Datenabfrage f
data station Datenstation f
data storage Daten(ab)speicherung f
data store Datenvorrat m
data transfer Datentransfer m, Datenübertragung f
data transmission Datentransfer m,
  Datenübertragung f
additional data Zusatzdaten npl
allocation data Zuordnungsdaten npl
analog(ue) data Analogwerte mpl
basic data Grunddaten npl
central data processor BDE-Zentrale f
design data Auslegungsdaten npl, Entwurfsdaten npl
formulation data file Rezeptdatei f
heat transfer data/values Wärmeübertragungskennwerte mpl
input data Eingabedaten npl
job data Auftragsdaten npl
job data input Auftragsdateneingabe f
machine control data file Maschinensteuerungsdatei f
machine data Maschinendaten npl
machine data collection/acquisition Maschinendatenerfassung f
output data Ausgabedaten npl, Leistungsangaben fpl
  [z.B. Maschine]
performance data [of a machine] Leistungsdaten npl
permeability data Permeabilitätsdaten npl
process data/variables Prozeßdaten npl,
  Produktionswerte mpl
process data control (unit) Prozeßdatenüberwachung f
process data gathering/acquisition (unit) Prozeßdatenerfassung f
product data sheet Typenmerkblatt n
production data Betriebsdaten npl,
  Betriebsgrößen fpl, Produktionsdaten npl
production data acquisition (unit) Produktionsdatenerfassung f
production data collecting system BDE-System n
production data collecting/acquisition (unit)
  Produktionsdaten-Erfassungsgerät n,
  Betriebsdatenerfassung f
production data record Betriebsdatenprotokoll n,
  Produktionsdatenprotokoll n
production data terminal BDE-Terminal n,
  Betriebsdatenerfassungsstation f
raw material data file Rohstoffdatei f
record/list of mo(u)ld data Werkzeugdatenkatalog m
safety data Sicherheitskennzahlen fpl

test data Prüfdaten npl
weighing data Wägedaten npl
date Datum n
  date display Datumsanzeige f
  date identification Datumsidentifikation f
  production/manufacturing date Produktionsdatum n
day Tag m
  day shift Tagesschicht f
  amount of raw material required for a day's
    production Rohstofftagesmenge f
  silo containing day's supply of mo(u)lding
    compound Tagessilo m, Tagesbehälter m
daylight Tageslicht n
  maximum daylight between platens größte lichte
    Weite f zwischen den Platten,
    Werkzeugplattenabstand m
DBP [dibutyl phthalate] Dibutylphthalat n,
  Phthalsäuredibutylester m
DBS [dibutyl sebacate] Dibutylsebacat n
d.c. [direct current, DC] Gleichstrom m
  d.c. amplifier Gleichstromverstärker m
  d.c. drive Gleichstromantrieb m
  d.c. motor Gleichstrommotor m
  d.c. permanent magnet motor Gleichstrom-
    Drehankermagnet-Motor m
  d.c. shunt motor Gleichstrom-Nebenschlußmotor m
  d.c. signal Gleichstromsignal n
  d.c. source Gleichstromquelle f
  d.c. voltage Gleichspannung f
  shunt-wound geared DC motor Nebenschluß-
    Gleichstrom-Getriebemotor m
deactivate/to desaktivieren
dead spots tote Ecken fpl, fließtote Räume mpl, tote
  Stellen fpl, Toträume mpl, tote Zonen fpl
deadline control Terminüberwachung f
deaerate/to entlüften
deaerated [compound or melt from which volatiles
  such as moisture, solvent or unreacted monomer
  have been removed] entgast
deaeration Entlüftung f
  paste deaeration Pastenentlüftung f
deagglomeration Desagglomeration f
deal Handel m
  package deal Paketlieferung f
debts Forderungen fpl, Schulden fpl
decade Dekade f
  decade counter Dekadenzähler m
  decade switch Dekadenschalter m
decahydronaphthalene Dekahydronaphthalin n
decarboxylation Decarboxylierung f
decentralized dezentral
decimal dezimal
  floating decimal point Fließkomma n
decision Entscheidung f
  decision aid Entscheidungshilfe f
  decision-making features Entscheidungsmerkmale npl

decision table Entscheidungstabelle f
decline Neigung f, Abfall m, Einbruch m
   decline in performance Leistungsabfall m
decoder Decoder m, Dekodierer m
   instruction decoder Befehlsdecoder m
decomposition Zerfall m, Zersetzung f
   decomposition behavio(u)r Zersetzungsverhalten n
   decomposition product Spaltprodukt n, Zersetzungsprodukt n, Zerfallsprodukt n
   decomposition range Zersetzungsbereich m
   decomposition rate Zersetzungsgeschwindigkeit f, Zerfallgeschwindigkeit f
   decomposition reaction Zersetzungsreaktion f, Zerfallsreaktion f
   decomposition temperature Zersetzungspunkt m, Zersetzungstemperatur f
   curing agent decomposition products Vernetzerspaltprodukte npl
   peroxide decomposition Peroxidzerfall m, Peroxidzersetzung f
   signs of decomposition Zersetzungserscheinungen fpl
decompression Kompressionsentlastung f, Druckkompressionsentlastung f, Dekompressions-Entspannung f
   decompression section Dekompressionszone f
   melt decompression (system) Schmelzedekompression f
decorative dekorativ, Zier-
   decorative film Dekorfilm m
   decorative laminate Dekorlaminat n, Dekorschichtstoff m
   decorative paper Dekorpapier n
decrease Abfall m, Abnahme f, Verminderung f
   decrease in hardness Härteabfall m
   decrease in pressure Druckabfall m
   decrease in strength Festigkeitsminderung f, Festigkeitsabfall m
   pressure decrease Druckabfall m
   temperature decrease Temperaturabsenkung f, Temperaturabfall m
   volume decrease Volumenabnahme f, Volumenverminderung f
decreasing abnehmend, zurückgehend
   decreasing pitch screw steigungsdegressive Schnecke f
   pressure decreasing phase Druckabfallphase f (während der Druckabfallphase / while the pressure is decreasing)
   with decreasing pitch [screw] steigungsdegressiv
decrement Dekrement n, stufenweise Abnahme f
dedicated zugeordnet (zugeordneter Rechner / dedicated computer)
deep embossing Tiefprägen n
deep-flighted [screw] tiefgängig, tiefgeschnitten
deep-freezer Gefriertruhe f
deepening in colo(u)r Farbvertiefung f
defect [in a material] Fehler m, Defekt m, Mangel m

coating defect Beschichtungsfehler m
film defects Filmstörungen fpl
flaw defect Fehlstelle f
list of defects Fehlerkatalog m
mo(u)ld defects Werkzeugmängel mpl
paint film defect Lackierungsdefekt m
structural defect Gefügestörung f, Strukturfehler m
surface defects Oberflächenstörungen fpl
defective fehlerhaft, mangelhaft
defined definiert
definite bestimmt, eindeutig, definitiv, endgültig
deflashing Entgraten n, Entbutzen n
   deflashing device Entgratungsvorrichtung f, Entbutzeinrichtung f, Butzenabschlag-Einrichtung f, Entbutzvorrichtung f, Abreißbacken fpl
   deflashing station Entbutzstation f
   deflashing unit Entgratungsstation f
   automatic deflashing unit Entgratautomat m
   base deflashing device Bodenentgrateinrichtung f
deflection Auslenkung f, Durchbiegung f, Ablenkung f
   deflection angle Umlenkwinkel m
   deflection at break Durchbiegung f beim Bruch
   deflecting roller Umlenkwalze f
   deflection temperature Formbeständigkeitstemperatur f, Wärmeformbeständigkeit f
   journal deflection Zapfendurchbiegung f
   mo(u)ld deflection Werkzeugdurchbiegung f
   platen deflection Plattendurchbiegung f
   roll deflection Walzendurchbiegung f, Walzenverbiegung f
   roll journal deflection Zapfendurchbiegung f
deflector Ablenkvorrichtung f, Leitblech n, Prallwand f, Deflektor m
   melt deflector Schmelzeumlenkventil n
defoamer Entschäumer m
deform/to deformieren
deformability Deformierbarkeit f
deformable deformierbar
deformation Deformation f, Verformung f
   deformation behavio(u)r Verformungsverhalten n
   deformation energy Verformungsenergie f
   deformation history Deformationsgeschichte f
   deformation limit Deformationsgrenze f, Verformungsgrenze f
   deformation process Verformungsvorgang m
   deformation resistance Deformationswiderstand m, Verformungsstabilität f
   deformation through internal pressure Innendruckverformung f
   deformation under load Verformung f unter Last
   creep deformation Kriechverformung f
   heat of deformation Verformungswärme f
   nozzle deformation Düsendeformation f
   permanent deformation Dauerverformung f
   rate of deformation Deformationsgeschwindigkeit f, Verformungsgeschwindigkeit f
   resistance to deformation through impact Schlag-

unverformbarkeit f
  shear deformation Scherdeformation f
  state of deformation Verformungszustand m
  tensile deformation Zugdeformation f
  total deformation Gesamtdeformation f, Gesamtverformung f
deforming amplitude Verformungsamplitude f
degassed entgast
degassing [also: devolatilizing or venting] Entgasung f
  degassing chamber Entgasungskammer f
  degassing conditions Entgasungsbedingungen fpl
  degassing effect Entgasungseffekt m, Entgasungsergebnis n
  degassing efficiency Entgasungsleistung f
  degassing intensity Entgasungsintensität f
  degassing silo Entgasungssilo m
  degassing stages Entgasungsstufen fpl
  degassing unit Entgasungsanlage f
  melt degassing (system/unit) Schmelzeentgasung f
  vented hopper degassing unit Vakuumtrichter-Entgasungsanlage f
degating device Angußabtrennvorrichtung f
degradability Abbaubarkeit f
degradable abbaubar
degradation Abbau m, Zersetzung f, Degradation f, Verkracken n
  degradation mechanism Zersetzungsmechanismus m
  degradation process Abbauprozeß m
  degradation product Abbauprodukt n
  degradation reaction Abbaureaktion f
  polymer degradation Kunststoffabbau m
  thermal degradation thermische Schädigung f
degrade/to abbauen, zersetzen, vermindern
degraded verkräckt
degrease/to entfetten
degree Grad m, Ausmaß n
  degree of accuracy Genauigkeitsgrad m
  degree of agglomeration Agglomerationsgrad m
  degree of automation Automatisierungsgrad m, Automatisationsgrad m, Automatisierungsaufwand m (Ein höherer Automatisierungsaufwand kann vorgesehen werden / a greater degree of automation can be provided)
  degree of branching Verzweigungsgrad m
  degree of chlorination Chlorierungsgrad m
  degree of compatibility Verträglichkeitsgrad m
  degree of condensation Kondensationsgrad m
  degree of contamination Verschmutzungsgrad m, Verunreinigunggrad m
  degree of conversion Umsetzungsgrad m, Umsetzungskennzahl f
  degree of crosslinking Vernetzungsgrad m
  degree of crystallinity Kristallisationsgrad m, Kristallinitätsgrad m
  degree of cure Aushärtungsgrad m, Härtungsgrad m
  degree of dispersion Verteilungsgrad m
  degree of freedom [e.g. of a control system] Freiheitsgrad m
  degree of fusion Gelier(ungs)grad m
  degree of gelation Gelier(ungs)grad m
  degree of grafting Pfropfungsgrad m
  degree of hardness Härtegrad m
  degree of influence Einflußhöhe f
  degree of integration Integrationsgrad m
  degree of orientation Orientierungsgrad m
  degree of oxidation Oxidationsgrad m
  degree of packing Verdichtungsgrad m
  degree of penetration Eindringrad m
  degree of plasticization Weichmachungsgrad m, Plastifiziergrad m
  degree of polymerization Polymerisationsgrad m
  degree of saturation Sättigungsgrad m
  degree of separation Abscheidegrad m
  degree of solvation [PVC in plasticizer] Geliergrad m
  degree of unsaturation Ungesättigtheitsgrad m
dehumidification Entfeuchtung f
dehydrated castor acid Ricinensäure f
dehydrating dehydratisierend
dehydration Dehydra(ta)tion f, Dehydratisierung f
dehydrochlorination Dehydrochlorierung f
dehydrogenating dehydrierend
dehydrogenation Dehydrierung f
deionized entionisiert, deionisiert
delaminating force Spaltkraft f, Spaltlast f
delamination Aufblätterung f, Delaminierung f
  delamination test Spaltversuch m
delay Verzögerung f
  delay time Verzögerungszeit f
  without delay verzögert
delayed verzögert
  delayed feed Dosierverzögerung f
delaying 1. Verzögerungs-; 2. -verzögerung f
  delaying mechanism Verzögerungseinrichtung f
  delaying the reaction Reaktionsverzögerung f
  mo(u)ld clamp delaying mechanism Formschließverzögerung f
delicate empfindlich
delivery Lieferung f, Zustellung f
  delivery chute Ausfallrutsche f
  delivery line [of hydraulic system] Druckleitung f
  delivery rate [e.g. of an extrudate or melt from a die] Austrittsgeschwindigkeit f, Fördergeschwindigkeit f
  delivery rolls Auslaufwalzen fpl
  delivery screw Auspreßschnecke f
  at the delivery/discharge end auslaufseitig
  constant delivery pump Konstantpumpe f
  cooling air delivery rate Kühlluftaustrittsgeschwindigkeit f
  extrudate delivery pressure Ausstoßdruck m
  extrudate delivery temperature Düsenaustrittstemperatur f
  melt delivery/discharge Schmelzeaustrag m (um

einen Schmelzeaustrag innerhalb von 20-30 Minuten zu gewährleisten / to ensure that the melt is discharged within 20-30 minutes)
melt delivery screw Schmelzeaustragsschnecke f
parison delivery speed/rate Schlauchaustrittsgeschwindigkeit f, Schlauchausstoßgeschwindigkeit f
part delivery safety mechanism Ausfallsicherung f
temperature at the delivery end [e.g. extrudate temperature at the die] Außstoßtemperatur f
uniform delivery [of extrudate] Ausstoßgleichmäßigkeit f
variable delivery pump Regelpumpe f
demand Nachfrage f (Angebot und Nachfrage / supply and demand)
  demands Forderungen fpl
  extreme demands Extremforderungen fpl
  increased demand Nachfragebelebung f
  rise in demand Bedarfsanstieg m
  technical demands/requirements verfahrenstechnische Forderungen fpl
demineralized entmineralisiert
demo(u)ld/to entformen
demo(u)ldability Entformbarkeit f (Die Entformbarkeit der Formteile ist ausgezeichnet / the parts are extremely easy to demo(u)ld)
demo(u)lding Entformung f
  demo(u)lding aid Entformungshilfe f
  demo(u)lding pressure Entformungsdruck m
  demo(u)lding problems Entformungsschwierigkeiten fpl
  demo(u)lding robot Entnahmeautomat m, Entnahmegerät n, Entnahmevorrichtung f, Entnahmeroboter m, Formteilentnahmegerät n
  demo(u)lding stroke Entformungsweg m
  demo(u)lding system Entformungssystem n
  demo(u)lding temperature Entformungstemperatur f
  demo(u)lding without damaging the mo(u)lded part spritzteilschonendes Entformen n
  direction of demo(u)lding Entformungsrichtung f
  method of demo(u)lding Entformungsprinzip n
  moment of demo(u)lding Entformungszeitpunkt m
  stiffness on demo(u)lding Entformungssteifigkeit f, Formsteifigkeit f
dense dicht
densely crosslinked stark vernetzt
density Dichte f, Raumdichte f, Raumgewicht n, Rohdichte f
  apparent density Fülldichte f, Schüttdichte f, Schüttgewicht n
  compacted apparent density Stopfdichte f
  compacted bulk density Stampfdichte f
  crosslink density Vernetzungsdichte f
  foam density Schaumdichte f
  high density polyethylene Polyethylen n hart, Niederdruck-Polyethylen n, Polyethylen n hoher Dichte
  low density polyethylene [LDPE] Polyethylen n weich, Polyethylen n niederiger Dichte, PE weich

  packing density Packungsdichte f, Rütteldichte f, Rüttelgewicht n
  smoke density Rauch(gas)dichte f, Rauchgaswert m
denting Einbeulen n
DEP [diethyl phthalate] Diethylphthalat n
department Abteilung f
  goods control (department) Wareneingangskontrolle f
  mo(u)ld design (department) Werkzeugauslegung f
  production monitoring (department) Fertigungsüberwachung f
  production planning (department) Fertigungsplanung f, Fertigungssteuerung f
  scheduling department Terminbüro n
  toolmaking department Werkzeugbau m
dependability Betriebssicherheit f, Funktionssicherheit f
dependable betriebssicher
dependence Abhängigkeit f
  dependence on imports Importabhängigkeit f
  temperature dependence Temperaturabhängigkeit f (Die Temperaturabhängigkeit der Viskosität / the dependence of viscosity on temperature)
  time dependence Zeitabhängigkeit f
dependent variable Zielgröße f
depending (on something) abhängig (von etwas)
  depending on the cavity pressure werkzeuginnendruckabhängig
  depending on the formulation rezepturabhängig
  depending on the molecular weight molekulargewichtsabhängig
  depending on the number of strokes hubzahlabhängig
  depending on the output leistungsabhängig
  depending on the plunger/ram stroke kolbenwegabhängig
  depending on the pressure druckabhängig
  depending on the product produktabhängig
  depending on the screw stroke schneckwegabhängig
  depending on the temperature temperaturabhängig
  depending on (the) wall thickness wanddickenabhängig
depolymerization Depolymerisation f
deposit Belag m, Ablagerung f, Vorkommen n
  deposits formed in the die Düsenablagerungen fpl
  size deposits Schlichteablagerungen fpl
  surface deposit Oberflächenbelegung f
depreciation Abschreibung f
  sum set aside for depreciation Abschreibungsvolumen n
depressant drückendes Schwimmittel n, Drücker m
  viscosity depressant Viskositätserniedriger m
depression Vertiefung f
depth Tiefe f
  cavity depth Formhohlraumtiefe f, Werkzeughohlraumtiefe f
  channel depth Schneckenkanaltiefe f
  depth filter Tiefenfilter m
  depth filtration Innenfiltration f

depth of abrasion Abtragtiefe f
depth of draw Umformungstiefe f, Ziehtiefe f
depth of indentation [when measuring ball indentation hardness] Eindringtiefe f, Eindrucktiefe f
depth of wear Verschleißtiefe f
feed section flight depth Einzugsgangtiefe f
flight depth Gangtiefe f, Schneckengangtiefe f (mit großer Gangtiefe / deep-flighted, with deep flights)
flight depth ratio Gangtiefenverhältnis n
mo(u)ld cavity depth Formhohlraumtiefe f, Werkzeughohlraumtiefe f
screw channel depth Schneckenkanaltiefe f
screw depth Schneckentiefe f
with gradually increasing flight depth [screw] gangprogressiv
derivative 1. Derivat n; 2. Ableitung f [Funktion]
cellulose derivative Cellulosederivat n
chlorine derivative Chlorderivat n
fatty acid derivative Fettsäurederivat n
urea derivative Harnstoffderivat n
DES [diethyl sebacate] Diethylsebacat n
desalinated entsalzt
description Beschreibung f
description of products/materials/grades Typenbeschreibung f
brief description Kurzbeschreibung f
outline description Kurzbeschreibung f
desiccator Exsiccator m, Exsikkator m
design Gestaltung f, konstruktive Gestaltung f, konstruktive Auslegung f, Formgebung f
design aids Auslegungshilfen fpl
design calculation Auslegungsrechnung f
design criteria Auslegekriterien npl, Konstruktionsprinzipien npl
design data Auslegungsdaten npl, Entwurfsdaten npl
design details Konstruktionseinzelheiten fpl
design drawing Konstruktionszeichnung f
design error Auslegefehler m
design fault Auslegefehler m
design features Konstruktionsmerkmale npl
design freedom Gestaltungsfreiheit f
design guidelines Gestaltungsrichtlinien fpl, **design improvements** konstruktive Verbesserungen fpl
design modification Konzeptionsänderung f
design possibilities Konstruktionsmöglichkeiten fpl, Gestaltungsmöglichkeiten fpl
design principles Auslegungskriterien npl, Konstruktionsprinzipien npl
design stage Konstruktionsphase f, Konzeptphase f (bereits in der Konzeptphase / already at the design stage)
basic design Grundkonstruktion f, Basisausführung f
blown film die design Blaskopfkonzeption f
calibration unit design Kalibrierkonzept n
carousel-type design Drehtischbauweise f
complex in design konstruktiv aufwendig
computer-aided design [CAD] rechnerunterstütztes Konstruieren n, CAD-Rechentechnik f
die design Düsenauslegung f, Werkzeuggestaltung f, Werkzeugkonzept n, Düsenkonstruktion f, Düsenbauart f
die head design Kopfkonstruktion f [Düse]
drive (unit) design Antriebsauslegung f, Getriebekonstruktion f
extruder design Extruderkonstruktion f, Extruderkonzeption f
extruder head design Kopfkonstruktion f [Düse]
flow channel design Fließkanalgestaltung f
four-column design Viersäulenkonstruktion f
gate design Angußgestaltung f, Angußkonstruktion f
hot runner (unit) design Heißkanalblock-Ausführung f, Heißkanal-Konstruktionen fpl
machine design Maschinenkonzeption f
mandrel support design Dornhalterkonstruktion f
mo(u)ld design Formenkonstruktion f, Werkzeugkonzept n, Werkzeugauslegung f, Düsenauslegung f, Werkzeuggestaltung f, (bei der Werkzeuggestaltung / when designing the mo(u)ld/die)
mo(u)lded-part design Formteilgestaltung f, Formteilkonstruktion f
multi-daylight design Etagenbauweise f (Spritzgießwerkzeug in Etagenbauweise / multi-daylight injection mo(u)ld)
new design Neukonstruktion f (eine weitere Neukonstruktion / another recent design)
nozzle design Düsenkonstruktion f
optimizing the design of single-screw extruders Optimierung f der verfahrenstechnischen Auslegung von Einschneckenextrudern
original design Ausgangsbauart f
overall design Gesamtkonzept n
part design Formteilkonstruktion f, Formteilgestaltung f
pipe die design Rohrwerkzeugkonstruktion f
plant design Anlagenkonzept n (Anlagenkonzept zur Herstellung coextrudierter Blasfolie / plant for making coextruded blown film)
product design Produktgestaltung f
recent design Neukonstruktion f (eine weitere Neukonstruktion / another recent design)
rotary table design Drehtischbauweise f
runner design Fließkanalgestaltung f
screw design Schneckenentwurf m, Schneckenkonstruktion f, Schneckenkonzept n, Schneckenkonzeption f, Schneckenausbildung f, Schneckenauslegung f, Schneckenausführung f (verschiedene Schneckenausführungen / different types of screw)
special-purpose design Einzweckausführung f
standard design Serienausführung f, Normalausführung f, Standardausführung f (Aufgabeschurren in Serienausführung oder Sonderkonstruktion / standard or specially designed feed chutes)

technical design verfahrenstechnische Auslegung f
three-station design Dreiplatzanordnung f
twin screw design Doppelschneckenausführung f
two-tie bar design Zweiholmenausführung f
wind-up design Wickelkonzeption f
designed konzipiert
  badly designed falsch ausgelegt
  compactly designed film extrusion line Folienkompaktanlage f
  correctly designed richtig konstruiert, richtig ausgelegt
  wrongly designed falsch ausgelegt
designer Konstrukteur m, Designer m, Gestalter m
  mo(u)ld designer Formgestalter m, Werkzeugkonstrukteur m
  mo(u)lded-part designer Formteilkonstrukteur m
  part designer Formteilkonstrukteur m
de-sized entschlichtet
desk Tisch m
  desk-top computer Tischcomputer m, Tischrechner m
  control desk/panel/console Maschinenpult n, Prozeßwarte f, Regiepult n, Schaltwarte f, Schaltpult n, Steuerwarte f, Steuerpult n, Bedienungskonsole f, Bedienungspult n
desorption Desorption f
destaticizing unit Entstatisierungseinrichtung f
destruction Zerstörung f
destructive zerstörend
  destructive test Zerrüttungsprüfung f, Zerrüttungsuntersuchung f
detach/to ablösen
detachable abnehmbar
detail Detail n, Einzelheit f
  accurate reproduction of detail Abbildegenauigkeit f
  design details Konstruktionseinzelheiten fpl
  faithful (to the original) in every detail detailgetreu
  standard platen details Lochbild n, Werkzeugaufspannzeichnung f, Aufspannplan m, Bohrbild n
  structural details Konstruktionseinzelheiten fpl
  tolerance details Toleranzangaben fpl
detection Nachweis m, Erkennung f
  fire detection Brandentdeckung f
detector Detektor m, Nachweisgerät n, Suchgerät n
  pore detector Lochsuchgerät n, Porensuchgerät n
detergent Detergens n, Waschmittel n
  detergent resistance Detergentienfestigkeit f, Beständigkeit f gegen Reinigungsmittel
  detergent solution Detergentienlösung f, Waschmittellauge f
  liquid detergent Flüssigwaschmittel n
deteriorated verschlechtert
deterioration Verschlechterung f
  deterioration of properties Werteabfall m
determination Bestimmung f, Ermittlung f, Erfassung f
  absolute determination Absolutmessung f
  method of determination Bestimmungsmethode f
  relative determination Relativmessung f
  routine determination Routinemessung f
  separate determinations Einzelmessungen fpl
  weight loss determination Gewichtsabnahmebestimmung f
determine/to bestimmen, ermitteln, erfassen
develop/to entwickeln, erschließen
  newly/recently developed neuentwickelt (our machines are constantly being developed and improved / unsere Maschinen werden ständig weiterentwickelt)
developing Entwicklung f, Entwickeln n
  developing a formulation Rezepturgestaltung f, Rezepturentwicklung f
  developing/underdeveloped countries Entwicklungsländer npl
development Entwicklung f
  development costs Entwicklungsaufwand m
  development of compressive strength Druckfestigkeitsentwicklung f
  development phase Entwicklungsphase f
  development product Entwicklungsprodukt n
  development trends Entwicklungstendenzen fpl
  development work Entwicklungsarbeit(en) f(pl)
  basic development Grundentwicklung f
  program development Programmentwicklung f
  recent development Neuentwicklung f
  technical development(s) verfahrenstechnische Entwicklung f
deviation Abweichung f
  dimensional deviations Maßabweichungen fpl
  setpoint deviations Sollwertabweichungen fpl
  standard deviation Standardabweichung f
device Gerät n, Vorrichtung f
  device for measuring pipe wall thickness Rohrwanddicken-Meßanlage f
  device for preventing screw fracture Schneckenbruchsicherung f
  device to prevent screw retraction Schneckenrückdrehsicherung f
  device which cuts up the sheeted-out compound into strips and returns these to the mill Fellwendevorrichtung f
  device which divides the melt stream Schmelzestromteiler m
  air cooling device Luftdusche f
  automatic device for cutting plastics profiles into lengths Kunststoffprofil-Ablängeautomat m
  automatic mo(u)ld safety device Automatikwerkzeugschutz m
  base deflashing device Bodenentgrateinrichtung f
  blowing device Blasvorrichtung f
  braking device Bremsvorrichtung f
  cent(e)ring device Zentriergerät n, Zentrierung f, Zentriervorrichtung f
  circular cutting device Kreisschneidevorrichtung f
  clamping device Einspannvorrichtung f, Spannvorrichtung f, Niederhalter m
  clamping force control device Schließkraftregelung f
  colo(u)r coding device Signiergerät n

core rotating device Kerndrehvorrichtung f
core unscrewing device Kernausschraubvorrichtung f
data gathering device Datenerfassungsgerät n
deflashing device Entgratungsvorrichtung f, Butzenabschlag-Einrichtung f, Entbutzeinrichtung f, Entbutzvorrichtung f, Abreißbacken fpl
degassing device Entlüftungsvorrichtung f, Entgasungsvorrichtung f
degating device Angußabtrennvorrichtung f
devolatilizing device Entlüftungsvorrichtung f, Entgasungsvorrichtung f
die cent(e)ring device Düsenzentrierung f
edge trimming/cutting device Seitenschneidvorrichtung f
ejector monitoring device Auswerfer-Überwachungssteuerung f
ejector stroke limiting device Auswerferhubbegrenzung f
excess-pressure safety device Überdrucksicherung f
feed device Einzugsvorrichtung f
film edge control device Bahnkantensteuerung f
film stretching device Folienreckvorrichtung f
flow restriction device Drossel f, Drosselorgan n, Drosselvorrichtung f
handling device Handlinggerät n
individual cavity cent(e)ring device Einzelformnest-Zentrierung f
injection device Einspritzvorrichtung f
input device Eingabegerät n
input-output device Ein-Ausgabeeinrichtung f
lateral feed device Seitenspeisevorrichtung f
level control device Niveauwächter m
lifting device Hebeeinrichtung f
load removal device Entlastungsvorrichtung f
longitudinal corrugating device Längswellvorrichtung f
low pressure mo(u)ld safety device Niederdruckwerkzeugschutz m
marking device [used for impressing identifying marks into extruded pipe] Signiergerät n
monitoring device Überwachungsorgan n
monitoring devices, controls Überwachungsarmaturen fpl, Überwachungselemente npl
mo(u)ld cent(e)ring device Werkzeugzentrierung f
mo(u)ld cleaning device Werkzeugreinigungsvorrichtung f
mo(u)ld opening stroke limiting device Werkzeugöffnungshub-Begrenzung f
neck calibrating device Halskalibrierung f
opening stroke limiting device Öffnungswegbegrenzung f
overload prevention device Überlastschutz m, Überlast(ungs)sicherung f
pipe cutting device Rohrtrennvorrichtung f
pressure adjusting device Druckeinstellorgan n
pressure recording device Druckschreiber m
recording device Registriereinrichtung f

safety device Schutzeinrichtung f, Schutzvorrichtung f, Sicherheitseinrichtung f, Betriebssicherung f
side gussetting device Seitenfalteinrichtung f
slitting device [for making film tape from film] Schneidvorrichtung f, Schneidaggregat n [zum Schneiden von Streifen], Schlitzvorrichtung f
smear device [used to obliterate spider lines in the melt flow] Stegverwischungseinrichtung f, Verwischgewinde n
speed control device Drehzahlregler m
stacking device Stapelvorrichtung f
stress application device Belastungseinrichtung f
stress removal device Entlastungsvorrichtung f
stuffing device Stopfvorrichtung f
temperature measuring device Temperaturmeßvorrichtung f, Temperaturmeßgerät n
test piece insertion device Probeneinlegevorrichtung f
threaded mandrel unscrewing device Gewindedorn-Ausdrehvorrichtung f
timing device Zeitgeber m, Zeitmesser m, Zeituhr f
venting device Entlüftungsvorrichtung f, Entgasungsvorrichtung f
wall thickness programming device Wanddicken-Programmiergerät n
warning device Warneinrichtung f
water spraying device Beregnungsvorrichtung f
web monitoring device Bahnwächter m
web tension measuring device Bahnzugkraftmeßvorrichtung f
web tensioning device Folienbahnspannung f, Waren(bahn)spannung f
devolatilization Entspannungsvorgang m, Entspannung f
   hopper devolatilization Trichterentgasung f
   melt devolatilization Schmelzedekompression f
devolatilize/to flüchtige Bestandteile entfernen, entgasen
devolatilizer Entgasungsmaschine f, Entgasungseinheit f, Entgasungsaggregat n
   screw devolatilizer Schneckenverdampfer m
devolatilizing extruder Entgasungsextruder m
devolatilizing section Entgasungsschuß m, Entgasungsbereich m, Zylinderentgasungszone f, Entgasungszone f, Entspannungszone f, Dekompressionszone f
devolatilizing unit Entgasungsaggregat n, Entgasungseinheit f
dew point Taupunkt(temperatur) m(f)
di-2-ethylhexyl adipate [DOA] Di-2-ethylhexyladipat n
di-2-ethylhexyl azelate [DOZ] Di-2-di-2-ethylhexylazelat n
di-2-ethylhexyl fumarate Di-2-ethylhexylfumarat n
di-2-ethylhexyl maleinate Di-2-ethylhexylmaleinat n
di-2-ethylhexyl phthalate [DOP] Di-2-ethylhexylphthalat n
di-2-ethylhexyl sebacate [DOS] Di-2-ethylhexylsebacat n
di-2-ethylhexyl terephthalate [DOTP] Di-2-ethylhexylterephthalat n

di-acid Disäure f
diacetone alcohol Diacetonalkohol m
diagnostic Diagnose f
  diagnostic function Diagnosefunktion f
  diagnostic text Diagnosetext m
  diagnostics program Diagnoseprogramm n
  diagnostics system Diagnosesystem n
  error diagnostics Fehlerdiagnose f
  error diagnostics program/routine Fehlerdiagnoseprogramm n
  malfunction diagnostics Störungsdiagnose f
diagram Diagramm n, schematische Darstellung f
  block diagram Blockdarstellung f, Blockdiagramm n
  circuit diagram Schaltbild n, Schaltplan m, Schaltschema n
  creep diagram Zeitstanddiagramm n, Zeitstandschaubild n
  exploded-view diagram Explosionsdarstellung f
  extruder diagram [output vs. pressure curve] Extruderdiagramm n
  flow diagram Ablaufdiagramm n, Fließbild n, Fließschema n, Flußdiagramm n, Strömungsprofil n
  force-deformation diagram Kraft-Verformungsdiagramm n
  functional diagram Funktionsplan m
  illuminated circuit diagram Leuchtschaltbild n
  logic diagram Funktionsplan m
  mo(u)ld fixing diagram Werkzeugaufspannzeichnung f, Aufspannplan m, Lochbild n, Bohrbild n
  stress-strain diagram Spannungs-Dehnungs-Diagramm n, Kraft-Dehnungs-Diagramm n
dialkyl compound Dialkylverbindung f
dialkyl peroxide Dialkylperoxid n
dialkyl peroxide group Dialkylperoxidgruppe f
dialkyltin-bis-mercaptide Dialkylzinnbismerkaptid n
dialkyltin compound Dialkylzinnverbindung f
dialkyltin dichloride Dialkylzinndichlorid n
dialkyltin dimercaptide Dialkylzinndimerkaptid n
dialkyltin laurate Dialkylzinnlaurat n
dialkyltin maleate Dialkylzinnmaleat n
dialkyltin mercaptide Dialkylzinnmerkaptid n
dialkyltin mercaptochloride Dialkylzinnmerkaptochlorid n
dialkyltin stabilizer Dialkylzinnstabilisator m
dialkyltin thioglycolate Dialkylzinnthioglycolsäureester m
diallyl phthalate Diallylphthalat n, Phthalsäurediallylester m
  diallyl phthalate mo(u)lding compound Diallylphthalatpreßmasse f
diameter Durchmesser m
  changes in diameter Querschnittsveränderungen fpl
  die diameter Düsendurchmesser m
  external screw diameter Schneckenaußendurchmesser m
  inside pipe diameter Rohrinnendurchmesser m
  internal screw diameter Schneckeninnendurchmesser m
  nominal diameter Nenndurchmesser m
  outside diameter Außendurchmesser m
  outside pipe diameter Rohraußendurchmesser m
  parison diameter Schlauchdurchmesser m
  parison diameter controller Schlauchdurchmesserregler m
  reel diameter Wickeldurchmesser m
  required diameter Solldurchmesser m
  roll diameter Walzendurchmesser m
  screw diameter Schneckendurchmesser m
  with constantly increasing root diameter [screw] kernprogressiv
diamine Diamin n
  phenylene diamine Phenylendiamin n
diamino-thiodiazole Diaminothiodiazol n
diamond-shaped rautenförmig
dianhydride Dianhydrid n
diaphragm Scheidewand f, Diaphragma n, Membran f, Blende f
  diaphragm gate Pilzanguß m, Scheibenanguß m, Schirmanschnitt m, Schirmanguß m, Telleranguß m
  fixed diaphragm Festblende f
  steel diaphragm Stahlmembrane f
diatomaceous earth Kieselgur n, Diatomenerde f
diazodicarbonamide Diazodicarbonamid n
dibasic dibasisch, zweibasisch
dibenzyl adipate Dibenzyladipat n
dibutyl adipate Dibutyladipat n
dibutyl phthalate [DBP] Dibutylphthalat n, Phthalsäuredibutylester m
dibutyl sebacate [DBS] Dibutylsebacat n
dibutyltin dilaurate Dibutylzinndilaurat n
dibutyltin maleinate Di-n-butylzinnmaleinat n, Dibutylzinnmaleinat n
dibutyltin mercaptide Dibutylzinnmerkaptid n
dibutyltin mercapto compound Dibutylzinnmerkaptoverbindung f
dibutyltin stabilizer Dibutylzinnstabilisator m
dibutyltin thioacetate Dibutylzinnthioacetat n
dibutyltin thioglycolate Dibutylzinnthioglykolat n
dicarboxylate Dicarbonsäureester m
dicarboxylic acid Bicarbonsäure f, Dicarbonsäure f
  dicarboxylic acid anhydride Dicarbonsäureanhydrid n
  dicarboxylic acid ester Dicarbonsäureester m
  ethylene dicarboxylic acid Ethylendicarbonsäure f
diced würfelförmig
  diced granules Würfelgranulat n
dicer Bandschneider m, Würfelschneider m, Bandgranulator m, Kaltgranulator m, Kaltgranuliermaschine f
  stair step dicer Bandgranulator m mit Stufenschnitt
dicetylperoxy dicarbonate Dicetylperoxydicarbonat n
dichloropolyether Dichloropolyether m
dicresylphenyl phosphate Dikresylphenylphosphat n
dicumyl peroxide Dicumylperoxid n
dicyandiamide Dicyandiamid n

dicyclohexyl phthalate Dicyclohexylphthalat n
dicyclohexylperoxy dicarbonate
Dicyclohexyl-Peroxydicarbonat n
dicyclopentadiene Dicyclopentadien n
DIDA [di-isodecyl adipate] Diisodecyladipat n
DIDP [di-isodecyl phthalate] Diisodecylphthalat n
die Düse f, formgebendes Werkzeug n,
Verformungswerkzeug n, Werkzeugkopf m,
Düsenkopf m
die adjusting mechanism Düsenjustiereinrichtung f,
Düseneinstellung f
die assembly Düsensatz m
die back pressure Werkzeugrückdruck m,
Werkzeuggegendruck m
die body Düsenkörper m, Werkzeugkörper m
die cent(e)ring device Düsenzentrierung f
die characteristic (curve) Werkzeugkennlinie f,
Düsenkennlinie f
die circumference Düsenumfang m
die constant Düsenkennzahl f
die design Düsenauslegung f, Werkzeugkonzept n,
Düsenkonstruktion f, Düsenbauart f
die diameter Düsendurchmesser m
die dimensions Düsenabmessung fpl
die-face [This prefix is used in connection with pelletizers where strands are extruded and cut into pellets at the die face, whilst still hot] Heißabschlag-
die-face pelletizer Heißgranuliervorrichtung f,
Heißabschlageinrichtung f, Heißabschlaggranulator m,
Heißabschlagvorrichtung f,
Heißabschlaggranuliereinrichtung f
die face pelletizing system, hot cut pelletizing system Direktabschlagsystem n
die gap Austrittsspalt m, Austrittsöffnung f,
Düsen(lippen)spalt m, Düsenaustrittsspalt m,
Düsen(austritts)öffnung f, Mundstück(ring)spalt m,
Düsenmund m, Düsenbohrung f
die gap adjusting device Düsenspaltverstellung f
die gap width/thickness Düsenspaltweite f,
Düsenspaltbreite f, Austrittsspaltweite f, Spaltbreite f,
Spaltweite f
die geometry Werkzeuggeometrie f
die head Spritzkopf m, Werkzeugkopf m,
Extruder(düsen)kopf m, Extrusionskopf m
die head design Kopfkonstruktion f
die head pressure [melt pressure as material enters the die] Düseneintrittdruck m
die head temperature Kopftemperatur f
die insert Düseneinsatz m
die land Bügellänge f, Bügelstrecke f, Bügelzone f,
Parallelteil n, Parallelzone f, Schlupflänge f,
Werkzeugparallelführung f (Die Parallelführung der beweglichen Aufspannplatten ist dann gewährleistet... / parallel travel of the moving platens is assured if...)
die lip adjusting mechanism Lippenverstellung f,
Lippenverstellmöglichkeit f

die lips Austrittslippen fpl, Düsenlippen fpl,
Lippenpaar n, Lippenpartie f, Werkzeuglippen fpl,
Blaslippen fpl
die material Düsenwerkstoff m, Düsenbaustoff m
die mount Düsenhalterung f
die opening Düsenausgang m
die opening force Düsenöffnungsdruck m,
Auftreibkraft f, Werkzeugauftriebskraft f,
Aufreißkraft f
die orifice Lippenspalt m
die plate Düsenteller m, Werkzeughalteplatte f,
Düsenplatte f
die pressure Düsendruck m
die resistance Düsenwiderstand m,
Spritzkopfwiderstand m, Werkzeugwiderstand m
die ring Düsenring m, Zentrierring m
die section Düsenteil n
die swell Schwellverhalten n, Düsenquellung f,
Quellfaktor m, Strangaufweitung f,
Strangaufweitungsverhältnis n
die width Düsenbreite f, Werkzeugbreite f
die with breaker plate-type mandrel support
Lochdornhalterkopf m, Lochdornhalterwerkzeug n
amount of melt passing through the die
Düsenfluß m
angled extrusion die head Schrägkopf m
annular die Ringspaltwerkzeug n,
Ringschlitzdüsenwerkzeug n, Ringschlitzdüse f,
Ring(spalt)düse f
blown film coextrusion die Koextrusionsblaskopf m,
Koextrusionsblaswerkzeug n
blown film die Schlauchkopf m,
Blasfolienschlauchkopf m, Folienblaskopf m,
Schlauchfolien(extrusions)werkzeug n,
Schlauchfolienkopf m, Schlauch(folien)düse f,
Schlauchspritzkopf m,
Schlauchwerkzeug n, Schlauchkopf m,
Schlauchformeinheit f, Schlauchextrusionsdüse f,
Blasdüse f, Blaswerkzeug n, Foliendüse f,
Folienkopf m, Folien(spritz)werkzeug n
blown film die design Blaskopfkonzeption f
breaker plate-type die Siebkorbkopf m
cable sheathing die Ummantelungswerkzeug n,
Kabel(sprit)zkopf m, Kabelummantelungswerkzeug n,
Kabelummantelungsdüse f
center-fed blown film die zentral angespritzter
Blaskopf m, Dornhalterblaskopf m, Dornhalterkopf m,
zentralgespeister Dornhalterkopf m,
Dornhalterwerkzeug n, Torpedokopf m
center-fed parison die Dornhalterschlauchkopf m
coathanger die Kleiderbügeldüse f
coextrusion die Koextrusionswerkzeug n
coextrusion die head Koextrusionskopf m
cooled die Kühldüse f, Kühldüsenwerkzeug n
cutting die Stanzwerkzeug n
deposits formed in the die Düsenablagerungen fpl
distance between die and nip rolls Abzugshöhe f

double-manifold die Zweikanalwerkzeug n
due to the die werkzeugbedingt
experimental die Prüfwerkzeug n, Versuchswerkzeug n
extrusion die Ausformdüse f, Düsenwerkzeug n, Extruderwerkzeug n, Extrudierwerkzeug n, Extrusionsdüse f, Extrusionswerkzeug n
extrusion die assembly Extruder-Düseneinheit f
film blowing die Blaswerkzeug n, Schlauchformeinheit f, Blasfolienschlauchkopf m, Folienblaskopf m, Schlauchfoliendüse f, Schlauchfolien(extrusions)werkzeug n, Schlauchfolienkopf m
fishtail die Fischschwanzdüse f
flat film (extrusion) die Flach(folien)düse f, Flachfolienwerkzeug n, Breitschlitzdüse f, Breitschlitzwerkzeug n, Schlitzdüse f
flat-profile die Flachprofildüse f
four-parison die Vierfachschlauchkopf m
in-line die Geradeaus(extrusions)werkzeug n, Geradeaus(spritz)kopf m, Längenspritzkopf m
large-bore pipe die Großrohrwerkzeug n
manifold-type die Verteilerkanaldüse f, Verteilerwerkzeug n
monofilament die Monofilwerkzeug n
multi-orifice die Vielfachwerkzeug n
multi-stage die Stufenwerkzeug n
multi-strand die Vielfachstrangdüse f
near the die in Düsennähe f
parison coextrusion die Coextrusionsschlauchkopf m
parison die Vorformlingswerkzeug n
pelletizing die Granulierdüse f, Granulier(loch)platte f, Granulierwerkzeug n, Lochdüse f
pipe die Rohr(düsen)kopf m, Rohr(extrusions)werkzeug n, Rohrspritzkopf m
pipe die design Rohrwerkzeugkonstruktion f
pipe die head Rohr(düsen)kopf m, Rohr(extrusions)werkzeug n, Rohrspritzkopf m
pipe sheathing die [This is a die used to extrude a polyethylene sleeve on to a steel pipe] Rohrbeschichtungswerkzeug n
possibility of adjusting the die lips Lippenverstellmöglichkeit f
precision die Präzisionswerkzeug n
production die Produktionswerkzeug n
profile die Profildüse f, Profil(spritz)kopf m, Profilwerkzeug n
prototype die Prototypwerkzeug n
rectangular profile die Rechteckprofildüse f
removal of the die Werkzeugausbau m
replacing the die Düsenwechsel m (bei einem erforderlichen Düsenwechsel / if the die needs changing)
ring-shaped die Kreisprofildüse f, Extrusionsrunddüse f, Rundstrangdüsenkopf m, Rund(strang)düse f
round profile die Kreisprofildüse f, Extrusionsrunddüse f, Rundstrangdüsenkopf m, Rund(strang)düse f
round section die Kreisprofildüse f, Extrusionsrunddüse f, Rundstrangdüsenkopf m, Rund(strang)düse f
sheet die Platten(-Breitschlitz)düse f, Plattenwerkzeug n, Plattenkopf m
side-fed blown film die seitlich eingespeister Folienblaskopf m, stegloser Folienblaskopf m, Pinolenblaskopf m, Umlenkblaskopf m
side-fed die Krümmerkopf m
side-fed parison die seitlich angeströmter Pinolenkopf m, Pinolenschlauchkopf m, Pinolenspritzkopf m, Pinolenwerkzeug n
single-layer die Einschichtwerkzeug n, Einschichtdüse f
single-parison die Einfach-Schlauchkopf m
slit die Schlitzdüse f, Breitschlitzdüse f, Breitschlitzwerkzeug n
slit die extrusion (process) Breitschlitz(düsen)verfahren n
slit die film extrusion Breitschlitzfolienverfahren n, Breitschlitzfolienextrusion f
slit die film extrusion line Breitschlitzfolienanlage f, Folienextrusionsanlage f
spider-type blown film die Stegdorn(halter)-blaskopf m
spider-type die Steg(dorn)halterkopf m, Stegdornhalterwerkzeug n
spiral mandrel (blown film) die Spiralverteilerblaskopf m, Wendelverteilerkopf m, Wendelverteilerwerkzeug n, Schmelzewendelverteilerkopf m, Schmelzewendelverteilerwerkzeug n
straight-through die Geradeauswerkzeug n, Geradeausspritzkopf m, Längsspritzkopf m, Geradeausextrusionswerkzeug n, Längenspritzwerkzeug n
strand die Strangdüse f, Strangwerkzeug n, Vollstrangdüse f
strand die head Strangdüsenkopf m
three-layer blown film die Dreischicht-Folienblaskopf m
three-layer (coextrusion) die Dreischichtdüse f
triple-manifold die Dreikanalwerkzeug n
twin die (extruder) head Zweifachextrusionskopf m, Zweifachkopf m, Doppel(spritz)kopf m, Doppelwerkzeug n, Zweifachwerkzeug n
twin-manifold die Zweiverteilerwerkzeug n
twin-orifice die Doppelstrangwerkzeug n
twin parison die Doppelschlauchkopf m, Zweifachblaskopf m, Zweifachschlauchkopf m
two-layer blown film die Zweischichtfolienblaskopf m
two-layer coextrusion die Zweischichtdüse f
two-layer slit die Zweischicht-Breitschlitzwerkzeug n
type of die Düsenart f, Düsenform f
vertical die head Senkrechtspritzkopf m

water-cooled die face pelletization Wasserringgranulierung f
when designing the die bei der Werkzeuggestaltung f
wire covering die Drahtummantelungs(-Düsen)kopf m
diecasting mo(u)ld Druckgußform f
(pressure) diecasting mo(u)ld Druckgußwerkzeug n
dielectric 1. Dielektrikum n; 2. dielektrisch
   dielectric constant (relative) Dielektrizitätskonstante f, dielektrischer Konstant m, Dielektrizitätszahl f
   dielectric strength Durchschlagfestigkeit f
   long-term dielectric strength Dauerdurchschlagfestigkeit f, Langzeit-Durchschlagfestigkeit f
   short-term dielectric strength Kurzzeit-Durchschlagfestigkeit f
dienol fatty acid Dienolfettsäure f
diesel fuel Dieselkraftstoff m
diester Diester m
diethyl aniline Diethylanilin n
diethyl phthalate [DEP] Diethylphthalat n
diethyl sebacate [DES] Diethylsebacat n
diethylene glycol Diethylenglykol n
diethylene glycol adipate Diethylenglykoladipat n
diethylene triamine Diethylentriamin n
diethyldioctyl phthalate [DOP] Phthalsäurediethylhexylester m
diethylhexyl phthalate [DOP] Phthalsäurediethylhexylester m
difference Differenz f
   difference in colo(u)r Farbunterschied m
   difference in pressure Druckdifferenz f, Druckunterschied m
   differences in wall thickness Wanddickenunterschiede mpl
   pressure difference Druckdifferenz f, Differenzdruck m
   pressure differences Druckunterschiede mpl
   processing differences Verarbeitungsunterschiede mpl
   shrinkage difference Schwindungsunterschied m
   structural differences Strukturunterschiede mpl
   temperature difference Temperaturdifferenz f
   viscosity differences Viskositätsunterschiede mpl
differential Differenz-, Differential-, Ausgleichs-
   differential calorimetry Differentialkalorimetrie f
   differential pressure Differenzdruck m
   differential pressure transducer Differenzdruckumformer m
   differential scales Differentialwaage f
   differential scanning calorimetry DSC-Methode f
   differential thermal analyzer Differentialthermoanalysengerät n
   differential thermoanalysis [DTA] Differentialthermoanalyse f
   differential-thermoanalytical differentialthermoanalytisch
   temperature differential Temperaturdifferenz f
differentiation Differenzierung f
difficult schwierig, aufwendig
difficult to disperse dispergierhart
difficult to dissolve schwer löslich
diffuse/to diffundieren
   diffuse into/to [a material] eindiffundieren
   diffuse out/to [of something] ausdiffundieren
diffuser Streuscheibe f
   headlamp diffuser Scheinwerferstreuscheibe f
diffusion Diffusion f, Permeation f
   diffusion behavio(u)r Permeationsverhalten n
   diffusion coefficient Diffusionskoeffizient m, Permeationskoeffizient m, Permeationswert m
   diffusion constant Diffusionskonstante f
   diffusion into [a material] Eindiffusion f
   diffusion process Diffusionsvorgang m
   diffusion pump Diffusionspumpe f
   diffusion rate Diffusionsgeschwindigkeit f, Permeationsrate f
   diffusion resistance Diffusionswiderstand m
   oil diffusion pump Öldiffusionspumpe f
   rate of diffusion Diffusionsgeschwindigkeit f, Permeationsrate f
   vapo(u)r diffusion resistance Dampfdiffusionswiderstand m
   water vapo(u)r diffusion Wasserdampfdiffusion f
diffusivity Diffusionsfähigkeit f, Diffusionskoeffizient m
digital digital
   digital-analog(ue) converter D/A-Wandler m, Digital-Analog-Wandler m
   digital computer Digitalrechner m
   digital control unit Digitalsteuerung f, Digitalregler m
   digital display (unit) digitale Anzeige f, Digitalanzeige f
   digital hydraulic system Digitalhydraulik f
   digital input Digitaleingang m
   digital output Digitalausgang m
   digital printer Digitaldrucker m
   digital setting Digitaleinstellung f
   digital signal Digitalsignal n
   digital switch Digitalschalter m
   digital timer Digital-Zeitmeßgerät n, Digitalzeituhr f
   digital valve Digitalventil n
   digital volume control unit Digitalmengenblock m
   direct digital control DDC-Regelung f
   direct digital temperature control circuit DDC-Temperaturregelkreis m
   grows of digital switches Digitalschalterreihen fpl
digitally adjustable digital einstellbar
digitally controlled numerisch gesteuert
digitization Digitalisierung f
digitizer Digitalisierer m
diglycidyl compound Diglycidylverbindung f
   diglycidyl ether Diglycidylether m
   diglycidyl hexahydrophthalate Hexahydrophthalsäurediglycidylester m
   diglycidyl phthalate Phthalsäurediglycidylester m
   neopentyl diglycidyl ether Neopentyldiglycidylether m

diglycol acetate Diglykolacetat n
  butyl diglycol acetate Butyldiglykolacetat n
dihexyl adipate Dihexyladipat n
dihydric [alcohol] zweiwertig [Alkohol]
dihydroquinoline Dihydroquinolin n
dihydroxy compound Dihydroxyverbindung f
dihydroxypolydimethyl siloxane Dihydroxypolydimethylsiloxan n
di-isodecyl adipate [DIDA] Diisodecyladipat n
di-isodecyl phthalate [DIDP] Diisodecylphthalat n
di-isononyl phthalate [DINP] Diisononylphthalat n
di-isooctyl phthalate [DIOP] Diisooctylphthalat n
di-isotridecyl phthalate Diisotridecylphthalat n
dilatancy Dilatanz f
dilatant dilatant
dila(ta)tion Dilatation f
dilatometer Dilatometer n
dilatometric dilatometrisch
dilauroyl peroxide Dilauroylperoxid n
diluent Verdünner m, Verdünnungsmittel n
dilutability Verdünnbarkeit f
  dilutability with ethanol Ethanolverdünnbarkeit f
dilutable verdünnbar
dilute/to verdünnen
dilute(d) verdünnt
dilution Verdünnung f
dimension Dimension f, Maß n, Ausmaß n, Abmessung f
  barrel dimensions Zylinderabmessungen fpl
  change in dimensions Dimensionsänderung f
  cylinder dimensions Zylinderabmessungen fpl
  die dimensions Düsenabmessungen fpl
  external dimensions Außenmaße npl
  inside dimensions Stichmaße npl
  maximum dimension Größtmaß n
  minimum dimension Kleinstmaß n
  mo(u)ld dimensions Formenmaße npl, Werkzeugmaße npl
  mo(u)ld fixing dimensions Formaufspannmaße npl, Aufspannmaße npl
  mo(u)ld impression dimensions formgebende Werkzeugmaße npl
  mo(u)ld mounting dimensions Werkzeugaufspannmaße npl, Werkzeugeinbaumaße npl, Werkzeuganschlußmaße npl
  mo(u)ld-dependent dimensions [These are dimensions formed entirely in one mo(u)ld part, i.e. dimensions not affected by the mating parts of the mo(u)ld] werkzeuggebundene Maße npl
  mo(u)ld-independent dimensions [These are dimensions formed across the mo(u)ld parting line, in the direction of mo(u)lding] nicht werkzeuggebundene Maße npl
  mo(u)lded-part dimensions Formteilmaße npl
  nominal dimensions Nennmaße npl
  nozzle dimensions Düsenabmessungen fpl
  original dimensions Ausgangsmaße npl
  outside dimensions Außenmaße npl
  overall dimensions Gesamtabmessungen fpl
  part dimensions Formteilmaße npl
  platen dimensions Werkzeugaufspannmaße npl, Werkzeuganschlußmaße npl, Werkzeugeinbaumaße npl
  required dimension Sollmaß n
  test piece dimensions Probenabmessungen fpl, Probendimensionen fpl
dimensional Form-, Dimensions-, Maß-
  dimensional accuracy Formtreue f
  dimensional changes Formänderungen fpl, Maßänderungen fpl
  dimensional deviation Abmaß n, Maßabweichungen fpl (Schwindung ist eine Maßabweichung in 3 Richtungen / shrinkage is a change in dimensions in three directions)
  dimensional stability Formbeständigkeit f, Formänderungsfestigkeit f, Dimensionsbeständigkeit f, Dimensionsstabilität f, Formstabilität f, Maßbeständigkeit f
  dimensional strength Gestaltfestigkeit f
  dimensional tolerance Formattoleranz f
  dimensional variations Abmessungsschwankungen fpl
dimensionally accurate maßgenau
dimensionally stable dimensionsbeständig, dimensionsstabil, formbeständig, formstabil
  dimensionally stable at elevated temperatures warmformbeständig
dimensioned bemessen, dimensioniert
  generously dimensioned großzügig dimensioniert, großzügig bemessen
dimensionless dimensionslos
dimer Dimer m, Dimerisat n
dimeric dimer
dimerization Dimerisierung f, Dimerisation f
  dimerization reaction Dimerisierungsvorgang m
dimerized dimerisiert
dimethacrylate Dimethacrylat n
  butanediol dimethacrylate Butandioldimethacrylat n
  ethylene glycol dimethacrylate Ethylenglykoldimethacrylat n
dimethoxyethyl phthalate Dimethoxyethylphthalat n
dimethyl Dimethyl-
  dimethyl aniline Dimethylanilin n
  dimethyl benzylamine Dimethylbenzylamin n
  dimethyl formamide Dimethylformamid n
  dimethyl glycol phthalate Dimethylglykolphthalat n
  dimethyl glyoxime Dimethylglyoxim n
  dimethyl phenol Dimethylphenol n
  dimethyl phthalate [DMP] Dimethylphthalat n
  dimethyl polysiloxane Dimethylpolysiloxan n
  dimethyl succinate Bernsteinsäuredimethylester m
  dimethyl sulphate Dimethylsulfat n
  dimethyl sulphoxide Dimethylsulfoxid n
  dimethyl terephthalate Dimethylterephthalat n
dimethylamine Dimethylamin n
  benzyl dimethylamine Benzyldimethylamin n
dimethylcyclohexyl phthalate Dimethylcylco-

hexylphthalat n
dimethylene ether bridge Dimethylenetherbrücke f
dimethylol urea Dimethylolharnstoff m
di-n-butyltin maleinate Di-n-butylzinnmaleinat n
di-n-butyltin thioglycolate Di-n-butylzinnthioglykolat n
DIN standard conditioning atmosphere DIN-Klima n
dinonyl adipate [DNA] Dinonyladipat n
DINP [di-isononyl phthalate] Diisononylphthalat n
dioctyl Dioctyl-
  dioctyl adipate [di-2-ethylhexyl adipate, DOA] Di-2-ethylhexyladipat n, Dioctyladipat n
  dioctyl azelate [di-2-ethylhexyl azelate, DOZ] Di-2-ethylhexylazelat n, Dioctylazelat n
  dioctyl fumarate [di-2-ethylhexyl fumarate] Di-2-ethylhexylfumarat n
  dioctyl maleinate [di-2-ethylhexyl maleinate] Di-2-ethylhexylmaleinat n
  dioctyl phthalate [di-2-ethylhexyl phthalate, DOP] Di-2-ethylhexylphthalat n, Dioctylphthalat n
  dioctyl sebacate [di-2-ethylhexyl sebacate, DOS] Di-2-ethylhexylsebacat n, Dioctylsebacat n
  dioctyl terephthalate [di-2-ethylhexyl terephthalate, DOTP] Di-2-ethylhexylterephthalat n
dioctyltin mercaptide Dioctylzinnmerkaptid n
dioctyltin thioglycolate Dioctylzinnthioglykolat n
diode Diode f
  light-emitting diode [LED] Leuchtdiode f
diol Diol n
DIOP [di-isooctyl phthalate] Diisooctylphthalat n
di-organotin compound Diorganozinn-Verbindung f
di-substituted disubstituiert
di-tertiary butyl peroxide [DTBB] Di-tert.-butylperoxid n
diorganosilane Diorganosilan n
dioxide Dioxid n
  carbon dioxide Kohlendioxid n
  titanium dioxide Titandioxid n
dip/to tauchen
  dip blow mo(u)lder Tauchblasmaschine f
  dip blow mo(u)lding Tauchblas(form)en n
  dip blow mo(u)lding (process) Tauchblasverfahren n
  dip coating Tauchbeschichtung f
  dip coating bath Tauchbad n
  dip coating compound Tauchmasse f
  dip impregnation Tauchimprägnierung f
diphenyl Diphenyl-
  diphenyl 2-ethylhexyl phosphate [diphenyloctyl phosphate] Diphenyl-(2-ethylhexyl-)phosphat n
  diphenyl ester Diphenylester m
  diphenyl thiourea Diphenylthioharnstoff m
  diphenyl urea Diphenylharnstoff m
diphenylamine Diphenylamin n
diphenylcresyl phosphate Diphenylkresylphosphat n
diphenylmethane-di-isocyanate [MDI] Diphenylmethandiisocyanat n
diphenyloctyl phosphate Diphenyloctylphosphat n, Diphenyl-(2-ethylhexyl-)phosphat n

diphenylsulphone group Diphenylsulfongruppe f
dipole Dipol m
  dipole character Dipolcharakter m
  dipole-dipole attraction Dipol-Dipol-Anziehung f
  dipole forces Dipolkräfte fpl
  dipole linkage Dipolbindung f
  dipole molecule Dipolmolekül n
  dipole moment Dipolmoment n
dipping Tauchen n, Tauchverfahren n
  dipping bath Tauchbad n
  dipping equipment Tauchapparatur f
  dipping lacquer Tauchlack m
  dipping mandrel Tauchdorn m
  dipping paint Tauchlack m
  dipping paste Tauchpaste f
  dipping tank Tauchwanne f
  dipping varnish Tauchlack m
  hot dipping (process) Warmtauchverfahren n
  paste dipping (process) Pastentauchverfahren n
diproylene glycol Diproylenglykol n
direct direkt
  direct addition Direktdosierung f
  direct current [d.c., D.C.] Gleichstrom m
  direct digital control DDC-Regelung f
  direct digital temperature control circuit DDC-Temperaturregelkreis m
  direct feed angußlose Direktanspritzung f, Direktanspritzung f
  direct gating angußlose Direktanspritzung f, Direktanspritzung f
  direct screw injection system Schneckendirekteinspritzung f
  direct titration Direkttitration f
  ante-chamber direct feed injection Vorkammerdurchspritzverfahren n
direction Richtung f
  direction-dependent richtungsabhängig
  direction of demo(u)lding Entformungsrichtung f
  direction of flame spread Flammenausbreitungsrichtung f
  direction of flow Fließrichtung f, Strömungsrichtung f
  direction of greatest stress Hauptspannungsrichtung f
  direction of impact Schlagrichtung f
  direction of loading Belastungsrichtung f
  direction of mo(u)lding Spritzrichtung f
  direction of movement Bewegungsrichtung f
  direction of orientation Orientierungsrichtung f
  direction of rotation Drehsinn m, Drehrichtung f
  direction of screw rotation Schneckendrehrichtung f, Schneckendrehsinn m
  direction of tensile stress Zugbeanspruchungsrichtung f
  axial direction Achsrichtung f
  change of direction Richtungsänderung f
  circumferential direction Umfangsrichtung f
  crack direction Rißrichtung f
  crack growth direction Rißwachstumsrichtung f

extrusion direction Auspreßrichtung f, Extrusionsrichtung f
flow direction Fließrichtung f, Strömungsrichtung f
in machine direction in Folienlaufrichtung f
longitudinal direction Längsrichtung f, Arbeitsrichtung f, Warenlaufrichtung f
machine direction Längsrichtung f, Arbeitsrichtung f, Warenlaufrichtung f
main direction of movement Hauptbewegungsrichtung f
mo(u)ld opening direction Formöffnungsrichtung f
orientation in machine direction Längsorientierung f
rotating in opposite directions [twin screws] gegenläufig, gegeneinanderlaufend
rotating in the same direction [twin screws] gleichläufig, gleichlaufend, gleichsinnig
stress direction Beanspruchungsrichtung f
take-off direction Abzugsrichtung f
tangential direction Tangentenrichtung f
transport direction Förderrichtung f
transverse direction Querrichtung f
weft direction Schußrichtung f [Textil]
directional gerichtet, Richtungs-
  directional control valve Wegeventil n
  directional servo-valve Wege-Servoventil n
directors Direktoren mpl
  board of directors Aufsichtsrat m, Verwaltungsrat m [Vorstand + Aufsichtsrat]
  managing directors Vorstand m
  member of the board of directors Aufsichtsratsmitglied n
dirt Schmutz m, Verunreinigung f
  dirt capacity [of a filter] Schmutzspeichervermögen n
  dirt repellent schmutzabweisend
  amount of dirt [e.g. in an oil filter or a mo(u)lding compound] Verschmutzungsgrad m
  surface dirt Oberflächenverunreinigung f
disadvantage Nachteil m
  processing disadvantages verfahrenstechnische Nachteile mpl
disc s. disk
discharge Entleerung f, Abfluß m, Auslaufen n, (elektrische) Entladung f, Entsorgung f
  discharge line Entsorgungsleitung f, Abflußleitung f
  discharge monitor Entleerüberwachung f
  discharge opening Entleerungsstutzen m, Entleerungsöffnung f
  discharge rate Entladungsgeschwindigkeit f
  discharge rolls Auslaufwalzen fpl
  discharge screw Auspreßschnecke f, Ausstoßschnecke f, Austragsschnecke f
  discharge section Austragsteil n, Auslaufgehäuse n
  discharge temperature [e.g. extrudate temperature at the die] Ausstoßtemperatur f
  discharge valve Entladeschieber m, Auslaßventil n
  corona discharge Coronaentladung f
  electric discharge machined funkenerodiert
  electric discharge machining [EDM] Funkenerosion f
  glow discharge Glimmentladung f
  partial discharge Teilentladung f
discolo(u)ration Verfärbung f
  brown discolo(u)ration Braunfärbung f
discontinous diskontinuierlich, chargenweise, stoßweise
  discontinuous operation (plant) Batchanlage f, intermittierender Betrieb m
  discontinuous process Stückprozeß m
disentangled entknäuelt, entwirrt
dishwasher Spülmaschine f
disk Scheibe f, Platte f
  disk drive Plattenlaufwerk n
  disk filter Scheibenfilter m
  disk spring Tellerfeder f
  bursting disk Berstscheibe f
  cam disk Nockenscheibe f
  exchangeable disk Wechselplatte f
  filter disk Filterscheibe f, Siebronde f, Siebscheibe f
  fixed disk Festplatte f
  fixed magnetic disk Festplattenspeicher m
  floppy disk drive Floppy-Disk(etten)-Laufwerk n
  kneading disks Knetelemente npl, Knetscheiben fpl
  magnetic disk Magnetplattenspeicher m, Magnetplatte f, Magnetträger m
  magnetic disk drive Magnetplattenlaufwerk n
  perforated disk Ringlochplatte f
  screw flights and kneader disks Schnecken- und Knetelemente npl
  spacer disk Distanzscheibe f
  toothed disk mill Zahnscheibenmühle f
  twin floppy disk drive Doppel-Floppy-Disk(etten)-Laufwerk n
  video disk Videoplatte f
  Winchester disk Winchesterplatte f
diskette Diskette f
  diskette management Diskettenverwaltung f
Diskpack plasticator/processor [trade name of a machine made by Farrel machinery group] Scheibenextruder m, Scheibenplastifizieraggregat n
dismantle/to ausbauen, demontieren
dismantled ausgebaut, demontiert
dismantling Demontage f
disordered [molecular structure] ungeordnet
disorientation Desorientierung f
disperse/to dispergieren, verteilen
  difficult to disperse dispergierhart
  finely dispersed feinverteilt
disperse dispers, dispergiert, Dispersions-
  disperse phase dispergierte Phase f, disperse Phase f
dispersibility Dispergierfähigkeit f, Dispergierbarkeit f, Verteilbarkeit f
dispersible dispergierbar
  water dispersible wasserdispergierbar
dispersing Dispergier-, Dispersions-
  dispersing agent Dispergier(hilfs)mittel n, Dispersionsmittel n, Dispersionshilfsstoff m

dispersing effect Dispersierwirkung f, Verteileffekt m
dispersing efficiency Dispergierleistung f, Verteilungsgüte f
dispersing problems Dispergierprobleme npl
dispersion Dispergierung f, Dispersion f, Verteilung f
  dispersion adhesive Dispersionskleber m
  dispersion medium Dispersionsträger m
  dispersion problems Verteilungsprobleme npl
  degree of dispersion Verteilungsgrad m
  efficiency of dispersion Verteilungsgüte f
  pigment dispersion Farbverteilung f, Pigmentverteilung f
  polymer dispersion Kunststoffdispersion f, Polymerdispersion f
displacement Verschiebung f, Verlagerung f, Fördermenge f, Verdrängung f
  axial displacement axiale Verschiebung f
  fixed displacement pump Konstantpumpe f
  liquid displacement Flüssigkeitsverdrängung f
  nozzle displacement/misalignment Düsenversatz m (um einen Düsenversatz zu verhindern / ...to prevent the nozzle becoming misaligned)
  positive displacement circulating pump Umlaufverdrängerpumpe f
  positive displacement pump Verdrängerpumpe f
  positive displacement vacuum pump Verdränger-Vakuumpumpe f
  pump displacement chamber Pumpenverdrängerraum m
  rotary positive displacement pump Rotationsverdrängerpumpe f
  variable displacement pump Verstellpumpe f
display Anzeige f
  display console Anzeigetafel f, Anzeigefeld n
  display contents Bildschirminhalt m
  display line Textzeile f, Schriftzeile f, Anzeigezeile f
  display module Anzeigebaugruppe f
  display page Bildschirmseite f
  display terminal Bildschirmterminal n
  actual value display Istwertanzeige f
  alphanumeric display Alphanumerik-Bildschirm m
  bar display Balkenanzeige f
  colo(u)r visual display unit [colo(u)r VDU] Farbsichtgerät n
  data display Datenanzeige f
  data display unit Datensichtgerät n
  digital display digitale Anzeige f, Digitalanzeige f, Ziffernanzeige f
  function display Funktionsanzeige f
  graphic(s) display Grafik-Bildschirm m, Bildschirmgrafik f
  LED display LED-Leuchtanzeige f, Leuchtziffernanzeige f
  liquid crystal display [LCD] Flüssigkristall-Anzeige f, LCD n
  monochrome visual display unit S/W-Bildschirm m, Schwarz-Weiß-Monitor m
  numeric display Ziffernanzeige f, Digitalanzeige f
  operating sequence display Funktionsablaufanzeige f
  plain language display Klartextanzeige f
  setpoint display Sollwertanzeige f
  signal status display Signalzustandsanzeige f
  status display Zustandsanzeige f
  temperature display Temperaturanzeige f
  video display unit Sichtgerät n
  visual display Bildschirmanzeige f
  visual display unit [VDU] Datensichtgerät n
displayable anzeigbar
disposable wegwerfbar, Einweg-
  disposable article [plural: disposables] Einwegartikel m, Wegwerfartikel m
  disposable bottle Einwegflasche f
  disposable container Wegwerfgefäß n, Einwegbehälter m
  disposable cup Einwegtasse f
  disposable drum Einwegfaß n
  disposable gloves Wegwerfhandschuhe mpl
  disposable towel Wegwerfhandtuch n
disposal Entsorgung f
  disposal of fixed assets Anlagenabgänge mpl
  disposal of sprues Angußabführung f
disproportionation Disproportionierung f
disproportioning reaction Disproportionierungsreaktion f
dissimilar [materials, e.g. PE and PA] artfremd
dissipation Dissipation f, Vernichtung f, Verbrauch m, Ableitung f [Wärme]
  dissipation factor (dielektrischer) Verlustfaktor m
  energy dissipation Energiedissipation f
  heat dissipation Wärmeableitung f
  heat dissipation losses Wärmeableitungsverluste mpl
  heat of dissipation Dissipationswärme f
dissociation Dissoziation f
  dissociation constant Dissoziationskonstante f
dissolve/to lösen
  difficult to dissolve schwer löslich
  partly dissolve/to anlösen
dissolved (auf)gelöst
dissolving characteristics Lösungsverhalten n
dissolving process Lösungsvorgang m
distance Distanz f, Abstand m
  distance between clamps [e.g. in tensile tests] Klemmabstand m
  distance between die and nip rolls Abzugshöhe f
  distance between supports [of test piece] Stützweite f, Widerlagerabstand m
  distance between the cooling channels Kühlkanalabstand m
  distance between tie bars lichter Holmabstand m, lichter Abstand m zwischen den Säulen
  distance block Distanzblock m
  distance piece Abstandhalter m, Abstandstück n, Distanzbolzen m, Distanzhalter m, Distanzstück n
  center distance Mittenabstand m

clamping distance [of test specimen] Einspannlänge f
distillation apparatus Destillationsgerät n
distilled destilliert
   double distilled bidestilliert
distinguishing features Unterscheidungsmerkmale npl, Unterscheidungskriterien npl
distortion Deformation f, Verformung f, Verzerrung f, Distorsion f
   distortion temperature Formbeständigkeitstemperatur f
   Martens heat distortion temperature Formbeständigkeit f (in der Wärme) nach Martens
distribute/to verteilen
   randomly distributed glass fibers wirre Glasfaserverteilung f
distribution Verteilung f
   distribution (of dividends) Ausschüttung f
   distribution of residual welding stresses Schweißrestspannungsverteilung f
   distribution system [e.g. for cooling air runner system/manifold system/feed system] Verteilersystem n
   cell size distribution Porengrößenverteilung f, Zellgrößenverteilung f
   charge distribution Ladungsverteilung f
   clamping force distribution Schließkraftverteilung f
   glass fiber distribution Glasfaserverteilung f
   heat distribution Wärmeverteilung f
   melt distribution Schmelzeverteilung f (Isolierkanal zur Schmelzeverteilung / insulated runner for distributing the melt)
   melt temperature distribution Massentemperaturverteilung f
   molecular weight distribution Molekulargewichtsverteilung f, Molmassenverteilung f
   molecular weight distribution curve Molekulargewichtsverteilungskurve f
   mo(u)ld temperature distribution Werkzeugtemperaturverteilung f
   orientation distribution Orientierungsverteilung f
   particle size distribution Kornzusammensetzung f, Partikelgrößenverteilung f, Korn(größen)verteilung f, Teilchen(größen)verteilung f
   phase distribution Phasenverteilung f
   pressure distribution Druckverteilung f
   stress distribution Spannungsverteilung f
   temperature distribution Temperaturverteilung f
   velocity distribution Geschwindigkeitsverteilung f
   wall thickness distribution Wanddickenverteilung f
distributor Verteiler m
   distributor drive Verteilgetriebe n
   crossbar distributor Kreuzschienenverteiler m
   laminar flow distributor Laminarflußverteiler m
   ring-type spiral distributor Ring(-Wendel)verteiler m
   spiral mandrel (melt) distributor Schmelzewendelverteiler m
   star-type distributor Sternverteiler m
   star-type spiral distributor Stern-Wendelverteiler m
disturbing störend, Stör(ungs)-
   disturbing influence Störgröße f, Störfaktor m, Störeinflüsse mpl
   unaffected by disturbing influence störgrößenunabhängig
divalent zweiwertig, zweibindig
diverging auseinanderlaufend
divided geteilt, getrennt
   finely divided feinverteilt
dividing (ver)teilend
   torque dividing drive Drehmomentverteilergetriebe n
   melt dividing schmelzeteilend (schmelzeteilende Hindernisse / obstacles which divide the melt stream)
division Teilung f, Division f
   scale division Skalenteilung f
DMC [dough mo(u)lding compound] Feuchtpreßmasse f
   polyester DMC Feuchtpolyester m
DMP [dimethylphthalate] Dimethylphthalat n
DNA [dinonyl adipate] Dinonyladipat n
DOA [dioctyl adipate, di-2-ethylhexyl adipate] Di-2-ethylhexyladipat n, Dioctyladipat n
doctor knife Rakelmesser n
document/to protokollieren
documentation Dokumentieren n, Dokumentation f, Protokollierung f
   data documentation Datendokumentation f
dodecalactam Dodecalactam n
dodecanoic/lauric acid Dodecansäure f
dodecanoic/lauric anhydride Dodecansäureanhydrid n
dodecyl benzene Dodecylbenzol n
dodecyl monoethanolamine Dodecylmonoethanolamin n
dodecyl succinic anhydride Dodecylbernsteinsäureanhydrid n
dolly [This is a rolled-up piece of hot compound cut from milled strip for passing to the calender feed] Puppe f, Wickel m
dolomite Dolomit n
domain Domäne f
dome light Lichtkuppel f
domestic häuslich, inländisch, Inlands-, Haus-
   domestic holdings Inlandsbeteiligungen fpl
   domestic refuse Hausmüll m
   domestic refuse collecion Hausmüllsammlung f
   domestic turnover Inlandsumsatz m
   domestic waste pipe Hausabflußrohr n
donor Donator m, Spender m
   hydrogen donor Wasserstoffdonator m
   radical donor Radikalspender m
door Tür f
   door case Türzarge f
   door seal Türabdichtung f
   door sill Schwelle f
   door trim Türinnenverkleidung f
   guard door Schutztür f
   guard door interlock system/mechanism Schutz-

türsicherung f, Schutzgittersicherung f
  operator's guard door Bedienungsschutztür f
  sliding guard door Schiebegitter n, Schiebeschutzgitter n, Sicherheitsschiebetür f
DOP [dioctyl phthalate, di-2-ethylhexyl phthalate] Di-2-ethylhexylphthalat n, Dioctylphthalat n, Phthalsäurediethylhexylester m
DOS [dioctyl sebacate, di-2-ethylhexyl sebacate] Di-2-ethylhexylsebacat n, Dioctylsebacat n
dosage Einsatzmenge f
  radiation dosage Strahlungsdosis f, Bestrahlungsdosis f
dose Dosis f
  energy dose Energiedosis f
dot-dash [curve] strichpunktiert
DOTP [dioctyl terephthalate, di-2-ethylhexyl terephthalate] Di-2-ethylhexylterephthalat n
dots Bildpunkte mpl
dotted [curve] punktiert
double doppelt, Doppel-
  double bond Doppelbindung f
  double bond content Doppelbindungsanteil m, Doppelbindungsgehalt m
  double-conical [screw] doppelkonisch
  double-conical screw Doppelkonusschnecke f
  double-daylight mo(u)ld Dreiplattenwerkzeug n, Zweietagenwerkzeug n
  double-daylight mo(u)lding Zweietagenspritzen n
  double distilled bidestilliert
  double flighted zweigängig, doppelgängig
  double glazing Doppelverglasung f
  double-manifold die Zweikanalwerkzeug n
  double refraction Doppelbrechung f
  double runner Doppelverteilerkanal m
  double-shell zweischalig
  double-shell roof Kaltdach n
  double sided doppelseitig
  double toggle Doppelkniehebel m
  double toggle clamp unit Doppelkniehebel-Schließeinheit f
  double toggle clamping system/mechanism Doppelkniehebel-Schließsystem n
  double toggle machine Doppelkniehebelmaschine f
  double toggle system Doppelkniehebelsystem n
  double-V notch Doppel-V-Kerbe f
  double-walled doppelwandig
  double-walled jacket Doppelmantel m
  five-point double toggle Fünfpunkt-Doppelkniehebel m
  free from double bonds doppelbindungsfrei
dough mo(u)lding compound [DMC] Feuchtpreßmasse f
downpipe Fallrohr n, Regenfallrohr n
downstream stromabwärts, nachfolgend, nachgeschaltet
  downstream degassing Stromabwärtsentgasung f
  downstream devolatilizing Stromabwärtsentgasung f
  downstream machine Folgemaschine f, Nachfolgemaschine f
  downstream process/operation Nachfolgeprozeß m
  downstream unit Folgeeinrichtung f, Nachfolgevorrichtung f, Folgeaggregat n, Folgegerät n, Nachfolgeeinheit f, Nachfolgeaggregat n
  downstream venting Stromabwärtsentgasung f
  extruder downstream equipment Extrudernachfolge(maschine) f
  profile extrusion downstream equipment Profilnachfolge f
downstroke Abwärtshub m
  downstroke press Oberkolbenpresse f
  automatic downstroke press Oberkolbenpreßautomat m
  automatic downstroke transfer mo(u)lding press Oberkolbenspritzpreßautomat m
downtime Ausfallzeit f, Fehlzeit f, Ruhezeit f, Leerlaufzeit f, Abstellzeit f, Stillstandzeit f, Nebenzeit f, Totzeit f, Stehzeit f
  downtime minimization Stillstandminimierung f (damit können keine Maßnahmen zur Stillstandminimierung getroffen werden / no measures can therefore be taken to minimize downtime)
  machine downtime Maschinenausfallzeit f, Maschinenstillstandzeit f
  production downtime Produktionsausfall m
downward abwärts
  downward extrusion Abwärtsextrusion f
  downward movement Abwärtsbewegung f
  downward pressure on prices Preisdruck m
DOZ [dioctyl azelate] Dioctylazelat n
draft Entformungsschräge f, Entformungskonizität f, Konizität f, Ausformschräge f
drag Ziehen n, Schleppen n, Strömungswiderstand m, Hemmung f
  drag coefficient/factor cw-Wert m, Luftwiderstandsbeiwert m
  drag flow Schleppfluß m, Schleppströmung f
  drag link Schleppstange f
  aerodynamic drag Luftwiderstand m
drain/to entleeren, ablassen, ablaufen
drainpipe Dränrohr n, Drainagerohr n
draw Zug m, Ziehvorgang m, Entformungskonizität f, Entformungsschräge f, Konizität f
  draw-down Unterziehen n, Unterzug m
  draw-down ratio Unterzugsverhältnis n
  draw plate assembly Blendenpaket n
  draw plate calibrator Blendenkalibrator m
  depth of draw Umformungstiefe f, Ziehtiefe f
drawdown Auslängen n, Durchhang m, Durchhängen n, Aushängen n
drawing [e.g. of film] 1. Strecken n, Verstrecken n; 2. Zeichnung f
  customer's drawing Kundenzeichnung f
  design drawing Konstruktionszeichnung f
  mo(u)ld drawing Werkzeugzeichnung f
  mo(u)lded-part drawing Formteilzeichnung f,

Produktionsteilzeichnung f
schematic drawing Schemaskizze f, Prinzipskizze f
sectional drawing Schnittbild n, Schnittdarstellung f
specification drawing Zeichnungsvorschrift f
dried getrocknet
  spray dried sprühgetrocknet
drier Sikkativ n, Trockner m
  drier [in a paint] Trockenstoff m
  centrifugal drier Zentrifugaltrockner m
  cobalt drier Kobaltsikkativ n
  hair drier Haartrockner m
  high performance pellet drier Hochleistungsgranulattrockner m
  high speed drier Schnelltrockner m
  hopper drier Trichtertrockner m
  hot air drier Warmlufttrockner m
  machine-side drier Beistelltrockner m
  tumble drier Taumeltrockner m
drift Drift f, Abweichung f, Drall m
  drift-free driftfrei
drill Bohrer m
  drill jig Bohrlehre f
  twist drill Spiralbohrer m
drilled roll Bohrungswalze f
drilling Bohren n
drink Getränk n
drinking water pipe Trinkwasserrohr n
drinking water supply Trinkwasserversorgung f
drip tray Auffangwanne f
drive/to fahren
drive Antrieb m, Laufwerk n, Treiber m, Getriebe n
  drive motor Antriebsmotor m
  drive power Antriebsleistung f
  drive shaft Antriebswelle f, Getriebewelle f
  drive switch circuit Antriebsschaltkreis m
  drive torque Antriebsdrehmoment n
  drive unit Antriebseinheit f
  drive (unit) design Antriebsauslegung f, Getriebekonstruktion f
  d.c. drive Gleichstromantrieb m
  disk drive Plattenlaufwerk n
  distributor drive Verteilergetriebe n
  electric drive motor Elektro-Antriebsmotor m
  extruder drive Extruderantrieb m, Extrudergetriebe n
  floppy disk drive Floppy-Disk(etten)-Laufwerk n
  gear drive Zahnradgetriebe n
  hollow drive shaft Getriebehohlwelle f
  hydraulic drive (unit) Antriebshydraulik f
  magnetic disk drive Magnetplattenlaufwerk n
  main drive shaft Hauptantriebswelle f
  main drive (unit) Hauptantrieb m
  manual drive Handgetriebe n
  modular drive Bausteingetriebe n
  motor drive motorischer Antrieb m
  piston drive mechanism Kolbentriebwerk n
  pump drive power Pumpenantriebsleistung f
  rack-and-pinion drive Zahnstangenantrieb m
  screw drive Schneckenantrieb m, Schneckengetriebe n
  screw drive mechanism Schneckenantriebssystem n
  screw drive motor Schneckenantriebsmotor m
  screw drive power Schneckenantriebsleistung f
  screw drive shaft Schneckenantriebswelle f
  separate drive Einzelantrieb m (Moderne Kalander haben für jede Walze Einzelantrieb / in modern calenders, each roll is driven separately)
  solenoid (operated) valve driver Magnetventiltreiber m
  sprocket chain drive Zahnkettenantrieb m
  torque dividing drive Drehmomentverteilergetriebe n
  total drive power Gesamtantriebsleistung f
  twin drive Doppellaufwerk n
  twin floppy disk drive Doppel-Floppy-Disk(etten)-Laufwerk n
  type of screw drive Schneckenantriebsart f
  V-belt speed reducing drive Keilriemenuntersetzung f
driving fahrend, Fahr-, Antriebs-
  driving comfort Fahrkomfort m
  driving conditions Fahrbedingungen fpl
  driving rain Schlagregen m
  driving wheel Antriebsrad n
  resistance to driving rain Schlagregensicherheit f
drooling Nachtropfen n, Tropfenbildung f
drop/to fallen
drop Tropfen m, Fallen n, Sinken n, Einbruch m
  drop hammer tool Schlaghammerwerkzeug n
  drop height Fallhöhe f
  drop impact strength Fallbruchfestigkeit f
  drop in consumption Verbrauchseinbuße f
  drop in power Leistungsabfall m
  drop in pressure Druckeinbruch m, Druckabfall m
  drop in profits Gewinnrückgang m
  drop in sales Absatzrückgang m
  drop in temperature Temperaturabsenkung f, Temperaturabfall m
  drop in turnover Umsatzeinbruch m, Umsatzeinbuße f, Umsatzrückgang m
  drop in viscosity Viskositätsminderung f, Viskositätsabsenkung f, Viskositätsabfall m, Viskositätserniedrigung f
droplet Tröpfchen n
  monomer droplet Monomertröpfchen n
dropping weight Fallhammer m
drum Faß n, Hobbock m, Trommel f
  drum mixer Trommelmischer m
  drum wind-up unit Trommelwickler m
  drum winder Trommelwickler m
  cardboard drum Papptrommel f
  contact cooling drum Kontaktkühlwalze f, Kontakttrommel f
  cooling drum Kühltrommel f
  disposable drum Einwegfaß n
  lidded drum Deckelfaß n
  machine for blow mo(u)lding drums Faßblasmaschine f

powered drum Antriebstrommel f
preheating drum Vorheiztrommel f
wind-up drum Wickeltrommel f
dry/to trocknen
dry trocken
  dry adhesive Trockenklebstoff m
  dry air Trockenluft f
  dry blend Dry-Blend-Mischung f, pulverförmige Formmasse f, Pulvercompound m, Pulvermischung f, Heißmischung f
  dry blend extruder Pulverextruder m
  dry blend extrusion Pulverextrusion f
  dry blend processing Pulververarbeitung f
  dry blend screw Pulverschnecke f
  dry cycle Leerhub m
  dry cycle time Trockenlauf-Zykluszeit f, Trockentaktzeit f, Trockenlaufzeit f, Taktzeit im Trockenlauf m
  dry film Trockenfilm m
  dry film thickness Trockenfilmdicke f, Trockenschichtstärke f
  dry grinding Trockenschliff m
  dry handle trockener Griff m
  dry residue Trockenrückstand m
  dry resin content Trockenharzgehalt m
  dry solids content Trockengehalt m
  dry state Trockenzustand m
  dry strength Trockenfestigkeit f
  dry weight Trockengewicht n
  dust dry time Staubtrockenzeit f
  number of dry cycles [usually per minute] Anzahl der Leerlaufzyklen mpl, Trockenlaufzahl f
  resistant to dry sliding friction trockengleitverschleißarm
drying trocknend, Trocken-
  drying agent Trockenmittel n
  drying air Trockenluft f, Trocknungsluft f
  drying cabinet Trockenschrank m, Wärmeschrank m
  drying characteristics Trockenverhalten n, Trocknungsverhalten n
  drying conditions Trockenbedingungen fpl, Trocknungsbedingungen fpl
  drying hopper Trocknungstrichter m
  drying oven Trockenofen m, Wärmeschrank m
  drying oven/cabinet, laboratory oven Wärmeschrank m
  drying process Trocknungsablauf m
  drying properties Trocknungsverhalten n, Trockenverhalten n
  drying section Trockenstrecke f
  drying temperature Trocknungstemperatur f
  drying time Trockenzeit f, Trocknungszeit f
  drying tunnel Trockenkanal m, Tunnelofen m
  drying unit Trockner m, Trocknungsanlage f, Trocknungsgerät n, Trockengerät n
  air drying 1. lufttrocknend; 2. Lufttrocknung f
  fast drying schnelltrocknend

  circulating air drying cabinet/oven Umlufttrockenschrank m, Umluftwärmeschrank m
  high speed drying oven Schnelltockner m
  physical drying physikalische Trocknung f
  shrinkage on drying Trocknungsschrumpf m
  spray drying Sprühtrocknen n
  vacuum drying cabinet Vakuumtrockenschrank m
dryness Trockengehalt m, Trockenheit f
  dust dryness Staubtrockenheit f
DTA [differential thermoanalysis] Differentialthermoanalyse f
DTBB [di-tertiary butyl peroxide] Di-tert.-butylperoxid n
dual dual, doppelt, zweifach
  dual-circuit hydraulic system Zweikreis-Hydrauliksystem n
  dual-circuit system Zweikreissystem n
  dual-circuit unit Zweikreisgerät n
duct Kanal m
  air duct Luft(führungs)kanal m
ductile duktil
  ductile failure/fracture Verformungsbruch m, Verstreckungsbruch m, Zähbruch m
ductility Duktilität f
due to bedingt
  due to cooling abkühlbedingt
  due to flow strömungsbedingt
  due to friction friktionsbedingt, reibungsbedingt
  due to the die werkzeugbedingt
  due to the machine maschinenbedingt
  due to the mo(u)ld werkzeugbedingt
  due to wear verschleißbedingt
  cooling due to evaporation Verdampfungskühlung f
  damage due to wear Verschleißschäden mpl
  heat losses due to convection Konvektionsverluste mpl
dumbbell Hantel f
  dumbbell-shaped hantelförmig
  dumbbell-shaped tensile test piece Schulterzugstab m
  dumbbell-shaped test piece Schulterstab m, Schulterprobe f, Hantelstab m
  standard dumbbell-shaped test piece/specimen Norm-Schulterprobe f
duplicate model Duplikatmodell n, Modellduplikat n
durability Dauerhaftigkeit f, Haltbarkeit f
durable dauerhaft
duration Dauer f
  duration of blowing Blaszeit f
  stress duration Laststandzeit f, Beanspruchungsdauer f, Belastungsdauer f, Belastungszeit f
  test duration Prüfdauer f, Versuchdauer f
dust Staub m
  dust dry time Staubtrockenzeit f
  dust dryness Staubtrockenheit f
  dust-free staubfrei
  dust formation Staubbildung f

dust particles Staubteilchen npl
dust removal Entstauben n
dust removal system Entstaubungssystem n
dust separator Staubabscheider m
fine dust Feinstaub m
precautions against inhalation of dust Staubschutz-Vorsichtsmaßnahmen fpl
zinc dust Zinkstaub m
dwell/to (an)schwellen
  dwell pressure Formgebungsdruck m
  dwell pressure phase Formgebungsphase f
dynamic dynamisch
  dynamic stress zügige Belastung f, zügige Beanspruchung f
  dynamic viscosity dynamische Viskosität f
Dynstat apparatus Dynstatgerät n
early früh
  in the early stages im Frühstadium n
  in the early years in den Anfangsjahren npl
earning power Ertragskraft f
earnings Einnahmen fpl, Verdienst m
earth Erde f
  alkaline earth carbonate Erdalkalicarbonat n
  alkaline earth metal Erdalkalimetall n
  alkaline earth oxide Erdalkalioxid n
  alkaline earth sulphate Erdalkalisulfat n
  diatomaceous earth Kieselgur n, Diatomenerde f
ease Leichtigkeit f, Bequemlichkeit f
  ease of access Zugänglichkeit f
  ease of calendering Kalandrierbarkeit f (Das Kriterium für die Kalandrierbarkeit einer Kautschukmischung / the criterion for the ease with which a rubber compound can be calendered)
  ease of demo(u)lding Entformbarkeit f (Die Entformbarkeit der Formteile ist ausgezeichnet / the parts are extremely easy to demo(u)ld)
  ease of flow Fließfähigkeit f
  ease of maintenance Servicefreundlichkeit f
  ease of operation Bedienungsfreundlichkeit f, Bedienungskomfort m
  ease of servicing Servicefreundlichkeit f
easily leicht, einfach
  easily accessible leicht zugänglich, griffgünstig
  easily pigmented leicht pigmentierbar
  easily scratched kratzempfindlich
easy leicht
  easy access bequemer Zugang m
  easy-care pflegeleicht
  easy-flow leichtfließend, weichfließend
  easy-process [e.g. PVC] leichtlaufend
  easy to apply leicht applizierbar
  easy to automate automatisierungsfreundlich
  easy to bond klebefreundlich
  easy to break down [e.g. PVC particles] aufschließbar
  easy to fit montagefreundlich, montageleicht
  easy to handle handhabungsfreundlich

easy to install montagefreundlich, montageleicht
easy to maintain wartungsfreundlich, servicefreundlich
easy to operate leichtbedienbar (Diese Maschine zeichnet sich durch leichte Bedienbarkeit aus / this machine is easy to operate)
easy to process verarbeitungsfreundlich
easy to repair reparaturfreundlich
easy to service wartungsfreundlich, servicefreundlich
easy to stick klebefreundlich
easy to use benutzerfreundlich
eccentric unrund laufend, exzentrisch
  eccentric rolls unrund laufende Walzen fpl
  eccentric stroke Exzenterhub m
ecological ökologisch
  ecological considerations Umweltaspekte mpl
  ecological problems Umwelt(schutz)probleme npl
ecologically safe ökologisch unbedenklich, umweltfreundlich
economic (volks)wirtschaftlich
  economic boom Hochkonjunktur f
  economic climate Wirtschaftklima n
  economic considerations Wirtschaftlichkeitsüberlegungen fpl, Wirtschaftlichkeitsbetrachtungen fpl, Rationalisierungsgründe mpl
  economic policy Konjunkturpolitik f
  economic reasons Wirtschaftlichkeitsgründe mpl
  economic recovery Konjunkturerholung f
  economic situation Konjunktur(lage) f
economical preisgünstig, kostengünstig, rationell
economy measures Rationalisierungsmaßnahmen fpl
edge Kante f
  edge build-up [resulting from uneven reeling of film] Randwülste mpl
  edge corrosion Kantenkorrosion f
  edge coverage [of a paint] Kantenbedeckung f
  edge gating seitliches Anspritzen n
  edge scanner Kantenabtaster m
  edge trim [also: trimmings] Folienrandstreifen m, Randstreifen m
  edge trim reclaim unit Folienrandstreifenaufbereitung f
  edge trim recycling Randstreifenrückspeisung f (die unmittelbare Randstreifenrückspeisung in den Extruder / the immediate return of edge trims to the extruder)
  edge trim recycling unit Randstreifenrückspeiseeinrichtung f, Randstreifenrückführanlage f, Randstreifenrückführsystem n, Randstreifenrezirkulierung f
  edge trim removal unit Randstreifenabsauganlage f
  edge trim shredder Randstreifenzerhacker m
  edge trim waste Randstreifenabfall m
  edge trim wind-up (unit) Randstreifenaufwicklung f, Abfallaufwicklung f
  edge trimmer [of film or sheeting] Kantenschneider m, Randbeschneidung f, Randbeschnitt m, Kantenbeschnitt m, Randstreifen-Schneidvor-

richtung f, Seitenschneider m, Seitenbeschneidung f
edge trimming unit Randstreifen-Schneidvorrichtung f, Seitenschneidvorrichtung f
corrosion near the edges Kantenkorrosion f
cutting edge [of knife] Schneidkante f
film edge control device Bahnkantensteuerung f
flight land edge Kammkante f
leading edge of flight treibende Flanke f, vordere Flanke f, aktive Flanke f, aktive Gewindeflanke f, aktive Schneckenflanke f, treibende Schneckenflanke f, Schubflanke f, aktive Stegflanke
pinch-off edges Quetschkanten fpl
sealing edges Dichtkanten fpl
sharp edged scharfkantig
trailing edge of flight passive Gewindeflanke f, hintere Flanke f, passive Flanke f, hintere Schneckenflanke f, passive Schneckenflanke f, hintere Stegflanke f, passive Stegflanke f
edible eßbar, Speise-
edible fat Speisefett n
edible oil Speiseöl n
editing Editierung f
EDM [electric discharge machining] Funkenerosionsbearbeitung f
EEC countries EWG-Länder npl
effect Wirkung f, Effekt m
effect of ag(e)ing Alterungseinfluß m, Alterungsauswirkung f
effect of chemicals Chemikalieneinfluß m
effect of moisture Feuchteeinfluß m
effect of plasticizer Weichmachereinfluß m
effect of processing Verarbeitungseinfluß m
abrasive effect Abrasionswirkung f, Abrasivwirkung f, Abriebwirkung f
accumulator effect Akku-Wirkung f
adhesive effect Haftwirkung f, Klebwirkung f
anti-blocking effect Antiblockeffekt m, Antiblockwirkung f
anti-corrosive effect Korrosionsschutz m, Rostschutzwirkung f
barrier effect Sperrwirkung f
breaking-down effect [of solids] Zerteileffekt m
conveying effect Förderwirkung f
cooling effect Kühleffekt m
corrosive effect Korrosivwirkung f
degassing effect Entgasungseffekt m
devolatizing effect Entgasungseffekt m
dispersing effect Dispersierwirkung f, Verteileffekt m
external lubricating effect Außengleitwirkung f
filtration effect Abscheidewirkung f
flame retardant effect Flammschutzeffekt m, Flammschutzwirkung f
flexibilizing effect Flexibilisierungseffekt m
flow restriction effect Drosselwirkung f
forced conveying effect Zwangsförderungseffekt m
gelling effect Gelierwirkung f
heat effect Wärmewirkung f

insulating effect Dämmeffekt m, Dämmwirkung f
leather grain effect Ledernarbung f
light stabilizing effect Lichtschutzeffekt m, Lichtstabilisatorwirkung f, Lichtstabilisierwirkung f
long-term effect Langzeitwirkung f
lubricating effect Schmierkraft f, Schmier(mittel)wirkung f, Gleitmittelverhalten n, Gleitwirkung f, Schmiereffekt m
memory effect Erinnerungseffekt m
mo(u)ld release effect Formtrennwirkung f
non-stick effect Antiklebewirkung f, Trenn(mittel)wirkung f
nucleating effect Nukleierungswirkung f
orange peel effect Apfelsinenschaleneffekt m
orientation effect Ausrichteffekt m
piezo-electric effect Piezoeffekt m
plasticizer effect Weichmachereffekt m
release effect Trenn(mittel)wirkung f
reinforcing effect Verstärkungseffekt m, Verstärkungswirkung f
screening effect Abschirmeffekt m
separation effect Abscheidewirkung f
shear effect Scherwirkung f
side effect Nebeneffekt m
stabilizing effect Stabilisierwirkung f
stick-slip effect Haft-Gleit-Effekt m
stiffening effect Versteifungswirkung f
stress-whitening effect Weißbrucheffekt m
swelling effect Quellwirkung f
thickening effect Eindickungsaktivität f
thixotropic effect Thixotropiewirkung f
venting effect Entgasungswirkung f
effective effektiv, wirksam, tatsächlich
effective capacity Nutzinhalt m
effective filter area freie Durchgangsfläche f [Filter]
effective length Arbeitslänge f, Funktionslänge f
effective motor power Motorwirkleistung f
effective power input aufgenommene Wirkleistung f, Nutzleistung f
effective screen area freie Siebfläche f
effective screw length Schneckenarbeitslänge f
effective volume Nutzvolumen n
effective width Arbeitsbreite f
effectiveness Wirksamkeit f
efficiency Wirkungsgrad m, Leistungsfähigkeit f (...hat einen höheren Wirkungsgrad als... / ..is more efficient than..)
conveying efficiency Förderwirkungsgrad m
crosslinking efficiency Vernetzungseffizienz f, Vernetzungswirksamkeit f
degassing efficiency Entgasungsergebnis n, Entgasungsleistung f
devolatizing efficiency Entgasungsergebnis n, Entgasungsleistung f
dispersing efficiency Dispergierleistung f
extruder efficiency Extruderleistung f
filtration efficiency Abscheideleistung f

low-temperature plasticizer efficiency Kälteelastifizierungsvermögen n
machine efficiency Maschineneffizienz f
plasticizer efficiency Weichmacherwirksamkeit f
stabilizer efficiency Stabilisatorwirksamkeit f
thermal efficiency thermische Leistung f
venting efficiency Entgasungsqualität f, Entgasungsleistung f
efficient leistungsfähig, leistungsstark, funktionstüchtig, rationell
efflorescence Salzausblühungen fpl
effluent Ausfluß m, Abwasser n [industriell]
  effluent pipe Abwasserrohr n
EFTA countries [European Free Trade Association] EFTA-Staaten mpl
eight-cavity injection mo(u)ld Achtfach-Spritzwerkzeug n
eight-cavity mo(u)ld Achtfachwerkzeug n
eight-runner arrangement Achtfachverteilerkanal m
ejection Ausstoßen n, Auswerfen n
  ejection control Ausfallüberwachung f
  space for part ejection Ausfallöffnung f
ejector Auswerfer m
  ejector bolt Auswerferbolzen m
  ejector bore Auswerferbohrung f
  ejector bush Auswerferhülse f
  ejector control mechanism Auswerfersteuerung f
  ejector cylinder Ausstoßzylinder m
  ejector damping mechanism Auswerferdämpfung f
  ejector end position Auswerferendstellung f
  ejector force Ausstoßdruck m, Auswerferkraft f
  ejector forward movement Auswerfervorlauf m
  ejector forward speed Auswerfervorlaufgeschwindigkeit f
  ejector halt Auswerferseite f, auswerferseitige Werkzeughälfte f
  ejector mark Auswerfermarkierung f
  ejector mechanism Auswerfereinrichtung f, Auswerfersystem n
  ejector monitoring device Auswerfer-Überwachung f
  ejector movement Auswerferbewegung f
  ejector pin Auswerferstempel m, Auswerferstift m, Ausstoßstift m
  ejector plate (assembly) Auswerfer(halte)platte f, Plattenauswerfer m, Auswerferteller m
  ejector plate return pin Rückdrückstift m
  ejector plate safety mechanism Auswerferplattensicherung f
  ejector release (mechanism) Auswerferfreistellung f
  ejector retaining plate Auswerfergrundplatte f
  ejector retraction force Auswerferrückzugkraft f
  ejector return movement Auswerferrücklauf m
  ejector return speed Auswerferrücklaufgeschwindigkeit f
  ejector ring Auswerferring m
  ejector rod Auswerferstange f, Auswerferkolben m, Auswerferstößel m, Stangenauswerfer m
  ejector speed Auswerfergeschwindigkeit f
  ejector spindle Austriebsspindel f
  ejector stroke Auswerferweg m, Auswerferhub m, Ausstoßweg m
  ejector stroke limitation Auswerferhubbegrenzung f
  central ejector Mittenauswerfer m, Zentralauswerfer m
  central ejector pin Zentralausdrückstift m
  on the ejector side of the mould auswerfseitig
  pneumatic ejector Druckluftauswerfer m, Luftauswerfer m
  side ejector Seitenauswerfer m
  sprue ejector Angußauswerfer m, Angußauswerfvorrichtung f
  sprue ejector cylinder Anguß-Ausstoßzylinder m
  sprue ejector pin Angußausdrückstift m
  two-stage ejector Zweistufenauswerfer m
  valve ejector Ventilauswerfer m
elastic elastisch, dehnbar
  elastic limit Dehnungsgrenze f, Elastizitätsgrenze f
  elastic modulus Elastizitätsmodul m, E-Modul m
  elastic recovery elastische Rückfederung f, elastische Rückdeformation f
  to make elastic elastifizieren
elasticity Dehnfähigkeit f, Dehnbarkeit f, Elastizität f
  elasticity number Elastizitätszahl f
  energy elasticity Energieelastizität f
  entropy elasticity Entropieelastizität f
  melt elasticity Schmelzeelastizität f
  modulus of elasticity Elastizitätsmodul m, E-Modul m
  modulus of elasticity in bending Biegeelastizitätsmodul m, Biege-E-Modul m
  modulus of elasticity in compression Druck-E-Modul m
  modulus of elasticity in tension Zug-E-Modul m
  rubber-like elasticity Kautschukelastizität f
elastomer Elastomer m
  elastomer blend Elastomerblend m
  elastomer-modified elastomermodifiziert
  elastomer phase Elastomerphase f
  silicone elastomer Siliconelastomer m
elastomeric elastomer
elbow Bogenstück n, Krümmer m, Winkel m [Rohr]
electric elektrisch
  electric control system Elektrosteuerung f, Elektroregelung f
  electric discharge machined funkenerodiert
  electric discharge machining [EDM] Funkenerosionsbehandlung f
  electric drive motor Elektro-Antriebsmotor m
  electric field strength elektrische Feldstärke f
  electric filament lamp Glühlampe f
  electric heating Elektro(be)heizung f
  electric iron Bügeleisen n
  electric motor E-Motor m, Elektromotor m
electrical elektrisch [s.a. electric]
  electrical appliance Elektrogerät n

electrical appliance industry Elektrogeräteindustrie f
electrical applications Elektroanwendungen fpl
electrical conduit Elektroisolierrohr n
electrical engineering Elektrotechnik f
electrical equipment Elektroausrüstung f
electrical-grade brown elektrobraun
electrical industry Elektroindustrie f
electrical insulating component Elektroisolierteil n, Elektroisoliermasse f, Elektroisolierstoff m
electrical insulating properties Elektroisolierverhalten n
electrical insulating resin Elektroisolierharz n
electrical insulating varnish Elektroisolierlack m
electrical loading elektrische Beanspruchung f
electrical properties elektrische Werte mpl
electrically erasable programmable read only memory EEPROM n
electrically interlocked [e.g. safety guard] elektrisch abgesichert
electricity Elektrizität f
   electricity consumption Stromverbrauch m [der tatsächliche Stromverbrauch / the actual amount of electricity used]
   electricity costs Stromkosten pl
   total electricity costs Gesamtstromkosten pl
electrochemical elektrochemisch
electrode Elektrode f
   electrode arrangement Elektrodenanordnung f
   contacting electrode Haftelektrode f
   plate electrode Plattenelektrode f
   spherical electrode Kugelelektrode f
electroforming Galvanoformung f
electrokinetic elektrokinetisch
electrolyte Elektrolyt m
   containing electrolyte elektrolythaltig
electrolytic elektrolytisch
   electrolytic corrosion elektrolytische Korrosion f
electromagnetic elektromagnetisch
electron Elektron n
   electron beam Elektronenstrahl m
   electron beam curing EB-Verfahren n, Elektronenstrahlhärtung f
   electron beam oscillograph Elektronenstrahloszillograph m
   electron cloud Elektronenwolke f
   electron irradiation Elektronenbestrahlung f
   electron micrograph elektronenmikroskopische Abbildung f, elektronenmikroskopische Aufnahme f
   electron microscope Elektronenmikroskop n
   electron pair Elektronenpaar n
   electron shell Elektronenschale f
   scanning electron micrograph rasterelektronenmikroskopische Aufnahme f, REM-Aufnahme f
   scanning electron microscope Rasterelektronenmikroskop n
electronegative elektronegativ
electronegativity Elektronegativität f

electronic elektronisch
   electronic control (system) Elektroniksteuerung f
   electronic controls Steuer(ungs)elektronik f
   electronic industry Elektronikindustrie f
   electronic measuring and control system MSR-Elektronik f
   electronic module Elektronikbaustein m
   electronic system Elektronik f
   main electronic system Primärelektronik f
   modular electronic system Bausteinelektronik f
electronics Elektronik f
electrophoretic elektrophoretisch
electroplatable galvanisierbar
electroplated [metals and plastics] galvanisiert
electroplating plant/unit Galvanikanlage f
electroplating solution Galvanisierungslösung f
electrostatic elektrostatisch
   electrostatic charging elektrostatische Aufladung f
   electrostatic fluidized bed coating elektrostatisches Wirbelsintern n
   electrostatic spraying Elektrostatikspritzen n
element Element n
   additional logic element Zusatzlogik f
   air distributing elements Luftverteilungselemente npl
   air supply elements Luftzuführungselemente npl
   cent(e)ring element Zentrierelement n
   clamping element Schließglied n, Aufspannelement n
   control elements Bedienungselemente npl, Funktionselemente npl, Stellglieder mpl
   damping element Dämpfungselement n
   filter element Filterelement n, Siebkörper m
   filter element support Filterelementhalteplatte f
   finite element method FEM-Verfahren n
   flow sensing element Durchflußfühler m
   granulating elements Zerkleinerungselemente npl
   guide element Führungselement n
   hydraulic control element [plural: hydraulic controls] Hydraulikstellglied n
   lip elements Lippenelemente npl
   logic element Logikelement n
   operating elements Funktionselemente npl
   plasticizing/plasticating element Plastifizierorgan n
   shear element Scherelement n
   soundproofing elements Schallschutzelemente npl
   static mixing element Statikmischelement n
   structural element Strukturelement n
   transducer element Aufnehmerelement n
elementary fiber Elementarfaser f
elevated temperature erhöhte Temperatur f
   dimensionally stable at elevated temperatures warmformbeständig
elevation Aufriß m
eliminate/to ausschalten, ausregeln, eliminieren
elimination Eliminierung f, Beseitigung f
   elimination of faults [e.g. in a mo(u)lding process] Pannenbehebung f
   elimination of water [e.g. during a chemical

reaction] Water elimination f
elongation Dehnung f, Streckdehnung f, Zugdehnung f
  elongation at break Bruchdehnung f, Reißdehnung f
  elongation at yield Dehnung f bei Streckgrenze, Zugdehnung f bei Streckgrenze
  elongation at yield stress Dehnung f bei Streckspannung
  change in elongation at break Bruchdehnungsänderung f
  irreversible elongation bleibende Dehnung f
  limiting elongation at break Grenzbruchdehnung f
  nominal elongation Nenndehnung f
  permanent elongation bleibende Dehnung f
  rate of elongation Dehn(ungs)geschwindigkeit f
  residual elongation bleibende Dehnung f
  residual elongation at break Restreißdehnung f, Restbruchdehnung f
  retained elongation Restdehnung f
  retained elongation at break Restreißdehnung f, Restbruchdehnung f
  short-term elongation Kurzzeitlängung f
  total elongation Gesamtdehnung f
eluant [used in chromatography] Fließmittel n
elution Elution f
embedded eingebettet, eingekapselt
embedding Verguß m, Einbettung f
  embedding compound Einbettmaterial n, Einbettmasse f
emboss/to prägen, bossieren
embossing Prägen n
  embossing calender Prägekalander m
  embossing machine Prägemaschine f
  embossing nip Prägespalt m
  embossing press Prägepresse f
  embossing pressure Prägedruck m
  embossing roller Prägewalze f
  embossing unit Prägevorrichtung f, Prägeeinrichtung f, Prägewerk n
  deep embossing Tiefprägen n
  high frequency embossing Hochfrequenzprägen n
  twin embossing unit Doppelprägewerk n
embrittled versprödet
embrittlement Verspröden n, Versprödung f
  chemical embrittlement Chemikalienversprödung f
  local(ized) embrittlement Lokalversprödung f
  surface embrittlement Oberflächenversprödung f
emergency Not f, Gefahr f
  emergency cooling Notkühlung f
  emergency cooling system Notkühlsystem n
  emergency cut-out Notausschalter m
  emergency power supply unit Notstromaggregat n
  emergency program Notprogramm n
  emergency stop button Notausschalter m
emery Schmirgel m
  emery cloth Schleifleinen n, Schmirgelleinen n, Schmirgeltuch n
  emery paper Schleifpapier n, Schmirgelpapier n

emission Emission f, Ausstoß m
  emission rate Abgaberate f, Emissionsrate f
  acoustic emission Schallemission f
  acoustic emission analysis Schallemissionsanalyse f
  acoustic emission curve Schallemissionskurve f
  cadmium emission Cadmiumabgabe f
  lead emission Bleiabgabe f
  sound emission value [in a scientific context] Lärmemissionswert m
emphasis Schwerpunkt m (der Schwerpunkt bei diesen Entwicklungsarbeiten war... / the emphasis in this development work was on... OR: this development centered on...)
empirical empirisch
  empirical formula Summenformel f
employees Belegschaft f, Angestellte mpl, Personal n
  number of employees Beschäftigtenzahl f
  total number of employees Gesamtbelegschaft f, Gesamtbeschäftigtenzahl f
employment [of machines, labo(u)r etc.] Einsatz m
  employment situation Beschäftigungslage f
empty/to entleeren
emulsifiable emulgierbar
emulsifier Emulgator m, Emulsionshilfsmittel n
  emulsifier blend Emulgatorgemisch n
  emulsifier content Emulgatorgehalt m
  emulsifier residues Emulgatorreste mpl
  emulsifier system Emulgatorsystem n, Emulgiersystem n
  amount of emulsifier Emulgatormenge f
  type of emulsifier Emulgatorart f
  with a low emulsifier content emulgatorarm
emulsify/to emulgieren
emulsifying agent Emulgator m, Emulsionshilfsmittel n
emulsion Emulsion f
  emulsion homopolymer Emulsionshomopolymerisat n
  emulsion paint Dispersionslack m, Dispersionsfarbe f, Dispersionsanstrichmittel n
  emulsion polymer Emulsionspolymerisat n
  emulsion polymerization Emulsionspolymerisation f, Emulsionsverfahren n
  emulsion polymerization process Emulsionspolymerisationsverfahren n
  emulsion PVC Emulsions-PVC n
enamel Email n, Emaille f
  stoving enamel Einbrennlack m
  wire enamel Drahtlack m
enamelled emailliert
  stove enamelled einbrennlackiert
encapsulated eingekapselt
  encapsulated by injection [mo(u)lding inserts] umspritzen
encapsulating compound Vergußmasse f
encapsulating resin Vergußharz m
encapsulation Verguß m, Verkapselung f
enclosed [part of a machine] gekapselt
  totally enclosed [machine] in geschlossener Bauart f,

in geschlossener Bauweise f, in Kompaktbauweise f
encouraging [results, developments etc.] erfreulich
end Ende n
   end group Endgruppe f
   end of cycle Zyklusende n
   end of injection Einspritzende n (unmittelbar vor dem Einspritzende / immediately before injection has been completed)
   end of plasticization Plastifizierende n (bei Plastifizierende / when plasticization has been completed)
   end-of-travel damping mechanism Endlagendämpfung f
   end product Fertigteil n, Fertigprodukt n, Fertigerzeugnis n, Fertigartikel m, Enderzeugnis n, Endprodukt n
   end product properties Endprodukteigenschaften fpl, Fertigprodukteigenschaften fpl
   end properties Endeigenschaften fpl
   end use Endanwendung f, Einsatzzweck m
   at the delivery/discharge end einlaufseitig
   ejector end position Auswerferendstellung f
   feed end Einzug m
   front end Frontende n
   melt temperature at the feed end [e.g. of an extruder] Einlaufmassetemperatur f, Eingangsmassetemperatur f
   shaft end Wellenstummel m
   temperature at the end [e.g. extrudate temperature at the die] Ausstoßtemperatur f
endothermic endotherm
endurance Dauer(haftigkeit) f, Ertragen n, Aushalten n, Zeitfestigkeit f, Dauerfestigkeit f
   endurance limit Zeitwechselfestigkeit f
   folding endurance [e.g. of PVC leathercloth] Knickfestigkeit f
   folding endurance test Dauer-Knickversuch m
   thermal endurance Wärmebelastbarkeit f
energy Energie f
   energy absorbent energieabsorbierend
   energy absorber Energieabsorber m
   energy-absorbing capacity Arbeitsaufnahmevermögen n
   energy balance Energiebilanz f, Energiehaushalt m, Leistungsbilanz f
   energy calculation Energieabrechnung f
   energy consumer Energieverbraucher m, Energieabnehmer m
   energy-consuming energieverzehrend
   energy consumption Energieverbrauch m
   energy content Arbeitsinhalt m
   energy conversion Energieumsatz m
   energy conversion process Energieumwandlungsprozeß m
   energy-dissipating energiedissipierend
   energy dissipation Energiedissipation f
   energy dose Energiedosis f
   energy-elastic energieelastisch, stahlelastisch
   energy elasticity Energieelastizität f, Stahlelastizität f
   energy input Energieeinsatz m, Arbeitsaufnahme f, aufgenommene Energie f, Energieaufnahme f
   energy input rate Energieanlieferungsrate f
   energy-intensive energieaufwendig
   energy loss Energieverlust m
   energy recovery Energierückgewinnung f
   energy release Energiefreisetzung f
   energy release rate Energiefreisetzungsrate f
   energy requirements Energie(mehr)aufwand m (der Energiemehraufwand beträgt etwa 15% / about 15% more energy is required)
   energy-rich energiereich
   energy saving 1. energiesparend; 2. Energieeinsparung f (Dies entspricht einer Energieeinsparung von etwa 45% / this is equivalent to a saving in energy of about 45%)
   energy shortage Energieknappheit f
   energy source Energiequelle f
   energy supply Energieversorgung f
   energy transfer Energieübertragung f
   energy used aufgenommene Energie f, Energieaufnahme f
   absorption of energy Arbeitsaufnahme f
   activation energy Aktivierungsenergie f
   amount of energy Energiestrom m
   amount of energy required Energieaufwand m
   deformation energy Verformungsenergie f
   dissociation energy Dissoziationsenergie f
   fracture energy Brucharbeit f, Bruchenergie f, Schädigungsarbeit f, Schädigungsenergie f
   from the energy point of view energetisch (betrachtet)
   impact energy Schlagarbeit f, Schlagenergie f, Schlagkraft f, Stoßarbeit f, Stoßenergie f
   impact energy absorptive capacity Stoßenergieaufnahmevermögen n
   increased energy requirements erhöhter Energiebedarf m
   kinetic energy kinetische Energie f
   light energy Lichtenergie f
   outside energy Fremdenergie f
   penetration energy Durchstoßarbeit f
   requiring a lot of energy energieaufwendig
   separation energy Trennungsenergie f
   shear energy Scherenergie f
   source of energy Energiequelle f
   surface energy Oberflächenenergie f
   tensile energy Zugarbeit f
   thermal energy Wärmeenergie f
   total energy input Gesamtenergieaufnahme f
   total energy used Gesamtenergieaufnahme f
   ultrasonic energy Ultraschallenergie f
   vibrational energy Schwingungsenergie f
engaged eingerastet
engine compartment Motorraum m

engine sound shield Geräuschkapsel f, Schallverkleidung f
engineer Ingenieur m, Techniker m
  service engineer Wartungstechniker m
engineering Technik f, Ingenieurtechnik f
  engineering plastic technischer Kunststoff m
  engineering thermoplastics technische Thermoplaste mpl
  aerospace engineering Raumfahrttechnik f
  civil engineering Tiefbau m
  electrical engineering Elektrotechnik f
  high frequency engineering Hochfrequenztechnik f
  hydraulic engineering Wasserbau m
  low-frequency engineering Niederfrequenztechnik f
  mechanical engineering Maschinenbau m
  phenolic engineering resin technisches Phenolharz n
  precision engineering Feinmechanik f, Feinwerktechnik f
enhanced verstärkt, erhöht, gesteigert
enlarge/to erweitern, ausdehnen, ausbauen
entangled [molecule] verknäuelt, geknäuelt
entanglement [of molecule chains] Verschlaufung f, Knäuelung f, Verknäuelung f
enthalpy Enthalpie f
  reaction enthalpy Reaktionsenthalpie f
entire cross-section Gesamtquerschnitt m
entrapped air Lufteinschlüsse mpl
entropy Entropie f
  entropy-elastic entropieelastisch
  entropy-elasticity Entropieelastizität f
envelop/to umhüllen
envelope Umhüllung f, Hülle f, Hüllkurve f
environment Umwelt f
  pollution of the environment Umweltverschmutzung f, Umweltbelastung f
  protection of the environment Umweltschutz m
  reaction environment Reaktionsmilieu n
environmental Umgebungs-, Umwelt-
  environmental conditions Umgebungsbedingungen fpl
  environmental considerations Umweltfaktoren mpl
  environmental factors Umweltfaktoren mpl
  environmental influences Umgebungseinflüsse mpl, Umwelteinflüsse mpl, Umwelteinwirkungen fpl
  environmental problems Umweltprobleme npl
  environmental protection Umweltschutz m
  environmental stress cracking umgebungsbeeinflußte Spannungsrißbildung f
EPDM [ethylene-propylene-diene rubber] Ethylen-Propylen-Dien-Kautschuk m
epichlorhydrin Epichlorhydrin n
  epichlorhydrin rubber Epichlorhydrinkautschuk m
epitaxial epitaktisch
epoxidation Epoxidation f
epoxide Epoxid n
  residual epoxide group content Rest-Epoxidgruppengehalt m

epoxidized epoxidiert
epoxy 1. Epoxid n; 2. Epoxy-
  epoxy adhesive Epoxidharzklebstoff m
  epoxy-based knifing filler EP-Spachtel m
  epoxy-based mortar EP-Mörtel m
  epoxy-based paint EP-Anstrichmittel n
  epoxy-based stopper EP-Spachtel m
  epoxy blend Epoxyverschnitt m
  epoxy casting resin Epoxidgießharz n, Epoxygießharz n
  epoxy coating system Epoxidharzanstrichsystem n
  epoxy concrete Epoxidharzbeton m
  epoxy compound Epoxidverbindung f
  epoxy equivalent Epoxidäquivalent(gewicht) n
  epoxy-functional epoxidfunktionell
  epoxy glass cloth laminate Epoxidharz-Glashartgewebe n
  epoxy group Epoxygruppe f, Epoxidgruppe f
  epoxy laminate Epoxidharz-Schichtstoff m
  epoxy mo(u)lding compound Epoxidformmasse f, Epoxidharz-Preßmasse f, Epoxidpreßmasse f
  epoxy paint Epoxidharzanstrich(system) m(n)
  epoxy paper laminate Epoxidharzhartpapier n
  epoxy plasticizer Epoxidweichmacher m
  epoxy resin EP-Harz n, Epoxidharz n, EP-Reaktionsharz n
  epoxy resin solution Epoxidharzlösung f
  epoxy value Epoxidwert m, Epoxidgehalt m
  catalyzed epoxy resin EP-Reaktionsharzmasse f
  cured epoxy resin Epoxidharz-Formstoff m
  typical of epoxy resins epoxidtypisch
epoxysilane Epoxysilan n
EPROM [abbr. of erasable and programmable read-only memory] Eprom-Speicher m, EPROM n
equalize/to ausgleichen
equation Gleichung f
  equation of state Zustandsgleichung f
  chemical equation (chemische) Reaktionsgleichung f
  mathematical equation Rechenansatz m, mathematische Gleichung f
  numerical equation Zahlenwertgleichung f
  thermal conductivity equation Wärmeleitungsgleichung f
  viscosity equation Viskositätsansatz m
  WLF equation [the initials stand for the names of the three people who first proposed the equation Williams, Landel and Ferry] WLF-Gleichung f
equilibrium Gleichgewicht n
  equilibrium reaction Gleichgewichtsreaktion f
  equilibrium system Gleichgewichtssystem n
  phase equilibrium Phasengleichgewicht n
  state of equilibrium Gleichgewichtszustand m
  stress equilibrium Spannungsgleichgewicht n
  temperature equilibrium Temperaturgleichgewicht n
  thermal equilibrium thermisches Gleichgewicht n
equimol(ecular) äquimolar
equipment Einrichtung f, maschinelle Ausrüstung f,

Maschinenausrüstung f, Maschineneinrichtung f
ancillary equipment Nebeneinrichtungen fpl, Zubehör n, Zubehöreinrichtungen fpl, Zubehörteile npl, Zusatzausrüstung f, Zusatzeinrichtung f
automatic feeding equipment Beschickungsautomat m
automatic processing equipment Verarbeitungsautomat m
basic equipment Basisausrüstung f, Grundausstattung f
calender downstream equipment Kalandernachfolge f
chemical equipment construction Chemieapparatebau m
control equipment [plural: controls] Reguliereinrichtung f, Steuereinrichtung f
conversion equipment Konfektionierungsmaschine f
dipping equipment Tauchapparatur f
electrical equipment Elektroausrüstung f
experimental equipment Versuchsausrüstung f
extraction equipment Absauganlage f
extruder downstream equipment Extrudernachfolge(maschine) f
feed equipment Dosieranlage f, Dosiereinrichtung f, Zuführeinrichtung f, Beschick(ungs)einrichtung f
machine and equipment manufacture Maschinen- und Apparatebau m
metering equipment Dosiereinrichtung f, Dosieranlage f
monitoring equipment Überwachungseinrichtung f
office equipment Bürogeräte npl, Büromaschinen fpl
optional equipment Wahlausrüstung f
original equipment Erstausrüstung f
peripheral equipment Peripheriegerät n
pollution control equipment Umweltschutzanlage f
post-extrusion equipment Extrudernachfolgemaschine f
process control equipment Prozeßsteueranlage f
processing equipment Verarbeitungsgeräte npl
production equipment Fertigungseinrichtungen fpl, Fertigungsmittel n
profile extrusion downstream equipment Profilnachfolge f
range of equipment Ausrüstungsumfang m
sports equipment Sportgeräte npl
spray lay-up equipment Faserspritzanlage f
standard equipment serienmäßige Ausrüstung f, Normalausrüstung f, Standardeinrichtung f
stereo equipment Stereoausrüstung f
weighing equipment Verwiegeanlage f, Wägeeinrichtungen fpl
wind-up equipment Aufwickelmaschinen fpl
erase head Löschkopf m
ergonomic ergonomisch
Erichsen indentation Erichsentiefung f
Erlenmeyer flask Erlenmeyerkolben m
eroded erodiert

erosion Erosion f
  erosion resistance Erosionsbeständigkeit f, Erosionsfestigkeit f
  amount of erosion Erosionsgrad m
  spark erosion Funkenerosion f
  surface erosion Oberflächenerosion f
erosive erosiv
error Fehler m
  error diagnostics Fehlerdiagnose f
  error diagnostics program/routine Fehlerdiagnoseprogramm n
  error-free fehlerfrei
  design error Auslegefehler m
  experimental error Versuchsfehler m
  input error(s) Eingabefehler m
  limit of error Fehlergrenze f
  margin of error Fehlergrenze f
  setting error Einstellfehler m
  source of error Fehlerquelle f
  test error Prüffehler m
  true-running error Rundlauffehler m
  weighing error Wägefehler m
ester Ester m
  ester alcohol radical Esteralkoholrest m
  ester-carbonyl group Estercarbonylgruppe f
  ester content Estergehalt m
  ester group Estergruppe f
  ester plasticizer Esterweichmacher m
  ester-soluble esterlöslich
  acrylic acid ester Acrylsäureester m
  adipic acid ester Adipinsäureester m
  alkyl sulphonic acid ester Alkylsulfonsäureester m
  azelaic acid ester Azelainsäureester m
  cellulose ester Celluloseester m
  colophony ester Kolophoniumester m
  dicarboxylic acid ester Dicarbonsäureester m
  diphenyl ester Dyphenylester m
  fatty acid ester Fettsäureester m
  glycidyl ester Glycidylester m
  maleic acid ester Maleinsäureester m
  monocarboxylic acid ester Monocarbonsäureester m
  montanic acid ester Montansäureester m
  oleic acid ester Ölsäureester m
  partial ester Partialester m
  pentaerythritol ester Pentaerythritester m
  phosphoric acid ester Phosphorsäureester m
  phosphorous acid ester Phosphorigsäureester m
  phthalic acid ester Phthalsäureester m
  polyether ester Polyetherester m
  polyglycol fatty acid ester Polyglykolfettsäureester m
  rosin ester Kolophoniumester m
  salicylic acid ester Salicylsäureester m
  sebacic acid ester Sebacinsäureester m
  succinic acid ester Bernsteinsäureester m
  sulphonic acid ester Sulfonsäureester m
  trimellitic acid ester Trimellitsäureester m
  vinyl ester Vinylester m

vinyl ester resin Vinylesterharz n
esterification Veresterung f
esterified verestert
estimate/to abschätzen,
   estimated cost Schätzkosten pl
estimate Abschätzung f
estimating Abschätzung f
   method of estimating Abschätzmethode f
etch/to ätzen
etched angeätzt
ethane Ethan n
   tetraphenyl ethane Tetraphenylethan n
ethanol Ethanol n
   ethanol solution ethanolische Lösung f
   ethanol-soluble ethanollöslich
ethanolic ethanolisch
ether Ether m
   ether alcohol Etheralkohol m
   ether amine Etheramin n
   ether bridge Etherbrücke f
   ether extract Etherextrakt m
   ether group Ethergruppe f
   ether linkage Etherbindung f
   allyl ether Allylether m
   bisphenol-A-diglycidyl ether Bisphenol-A-Diglycidylether m
   butylglycidyl ether Butylglycidylether m
   butylphenylglycidyl ether Butylphenylglycidylether m
   cresyl glycidyl ether Kresylglycidylether m
   diglycidyl ether Diglycidylether m
   diethyl ether Diethylether m
   dimethylene ether bridge Dimethylenetherbrücke f
   ethylene glycol butyl ether Ethylenglykolmonobutylether m
   ethylene glycol ether Ethylenglykolether m
   ethylene glycol ethyl ether Ethylenglykolmonoethylether m
   ethylhexyl glycidyl ether Ethylhexylglycidylether m
   glycidyl ether Glycidylether m
   glycol ether Glykolether m
   glycol ether acetate Glykoletheracetat n
   hexanediol diglycidyl ether Hexandioldiglycidylether m
   neopentyl diglycidyl ether Neopentyldiglycidylether m
   petroleum ether Petrolether m
   polyaryl ether Polyarylether m
   polyvinyl ether Polyvinylether m
   propylene glycol ether Propylenglykolether m
   propylene glycol methyl ether Propylenglykolmonomethylether m
   silyl ether Silylether m
   vinyl ether Vinylether m
etherification Veretherung f
etherified verethert
   completely etherified vollverethert
   partly etherified partiellverethert

ethoxy group Ethoxygruppe f
ethoxylated ethoxyliert, oxethyliert
ethyl Ethyl-, Essig-
   ethyl acetate Essig(säureethyl)ester m, Ethylacetat n
   ethyl acetate test Essigestertest m
   ethyl acrylate Acrylsäureethylester m, Ethylacrylat n
   ethyl alcohol Ethylalkohol m
   ethyl cellulose Ethylcellulose f
   ethyl cyanacrylate Ethylcyanacrylat n
   ethyl glycol Ethylglykol n
   ethyl glycol acetate Ethylglykolacetat n
   ethyl group Ethylgruppe f
   ethyl silicate Ethylsilikat n
ethylene Ethylen(gas) n
   ethylene chain Ethylenkette f
   ethylene chloride Ethylenchlorid n
   ethylene dicarboxylic acid Ethylendicarbonsäure f
   ethylene glycol Ethylenglykol n
   ethylene glycol butyl ether Ethylenglykolmonobutylether m
   ethylene glycol dimethacrylate Ethylenglykoldimethacrylat n
   ethylene glycol ether Ethylenglykolether m
   ethylene glycol ethyl ether Ethylenglykolmonoethylether m
   ethylene glycol polyadipate Ethylenglykolpolyadipat n
   ethylene oxide Ethylenoxid n
   ethylene oxide group Ethylenoxidgruppe f
   ethylene-propylene-diene rubber [EPDM] Ethylen-Propylen-Dien-Kautschuk n
   ethylene-propylene rubber Ethylen-Propylen-Kautschuk n, Ethylen-Propylen-Elastomer n
   ethylene-tetrafluoroethylene copolymer Ethylen-Tetrafluorethylen-Copolymer n
   ethylene-vinyl acetate [EVA] EVAC n
   ethylene-vinyl acetate copolymer [EVA] Ethylen-Vinylacetat-Copolymerisat n
   ethylene-vinyl acetate rubber Ethylen-Vinylacetat-Kautschuk n
   ethylene-vinyl alcohol copolymer Ethylen-Vinylalkohol-Copolymerisat n
   fluorinated ethylene-propylene copolymer [FEP] Fluorethylenpropylen n
   residual ethylene content Restethylengehalt m
ethylhexyl glycidyl ether Ethylhexylglycidylether m
EVA [ethylene-vinyl acetate copolymer] Ethylen-Vinylacetat-Copolymerisat n
evacuate/to evakuieren, entlüften
   partly evacuated teilevakuiert
evaluation [e.g. of test results] Auswertung f
evaporation Abdampfen n, Abdunsten n, Ausdampfen n, Verdampfung f, Verdunstung f
   evaporation index Verdunstungszahl f
   evaporation loss Verdampfungsverlust m
   evaporation rate Verdampfungsrate f
   cooling due to evaporation Verdampfungskühlung f

heat of evaporation Verdampfungswärme f
loss through evaporation, evaporation loss Verdampfungsverlust m, Verdunstungsverlust m
   plasticizer evaporation Weichmacherverdampfung f
   solvent evaporation Lösemittelverdunstung f, Lösungsmittelabdunstung f, Lösungsmittelabgabe f
evaporator Verdampfer m
even cooling Kühlgleichmäßigkeit f
even temperature Temperaturgleichmäßigkeit f, Temperaturhomogenität f (damit eine ausreichende Temperaturhomogenität gewährleistet ist / so that the temperature is sufficiently uniform throughout)
everyday use Alltagsbetrieb m
evolution (zeitliche) Entwicklung f, Evolution f
   gas evolution Gasabspaltung f
evolution of heat Wärmeentwicklung f
exact fit Paßgenauigkeit f
exacting anspruchsvoll
example Beispiel n
   examples of hot runner systems Heißkanalbeispiele npl
   application examples Anwendungsbeispiele npl
   mathematical example Rechenbeispiel n
   numerical example Zahlenbeispiel n
excess 1. Überschuß m; 2. überschüssig
   excess hardener Härterüberschuß m
   excess oxygen Sauerstoffüberschuß m
   excess pressure Überdruck m
   excess-pressure safety device Überdrucksicherung f
   excess water pressure Wasserüberdruck m
excessive shear Überscherung f
excessive temperature Übertemperatur f
exchange Austausch m, Wechsel m
   exchange of information Informationsaustausch m
   data exchange Datenaustausch m
   heat exchange Wärmeaustausch m
exchangeable auswechselbar
   exchangeable disk Wechselplatte f
exchanger Tauscher m, Austauscher m
   heat exchanger Wärme(aus)tauscher m, Wärmeüberträger m
   heat exchanger pipe Wärmetauschrohr n
   plate heat exchanger Plattenwärmeaustauscher m
exchanging austauschend
   heat exchanging fluid wärmeaustauschende Flüssigkeit f
exclude/to ausschließen, ausschalten
exclusion Ausschluß m
   exclusion of air Luftausschluß m
   under the exclusion of oxygen unter Sauerstoffausschluß m
exothermic exotherm
expand aufblähen, erweitern, expandieren
expandable aufschäumbar, ausbaufähig, erweiterungsfähig, expandierbar, schäumbar, schäumfähig, verschäumbar
   expandable paste [PVC] Schaumpaste f

expanded getrieben, expandiert, aufgebläht, aufgeschäumt
   expanded film extruder Schaumfolienextruder m
   expanded film extrusion line Schaumfolienanlage f
   expanded polystyrene Polystyrolschaum(stoff) m, Schaumpolystyrol n, Polystyrol-Hartschaum m
   expanded polystyrene pack Polystyrolschaumstoff-Verpackung f
   expanded polystyrene production line Polystyrol-Schaumanlage f
   single-layer expanded polystyrene container Einschicht-PS-Schaumhohlkörper m
expansion Ausdehnung f, Expansion f, Dilatation f
   expansion coefficient Ausdehnungskoeffizient m
   expansion joint Bauwerksfuge f, Bewegungsfuge f, Dehnfuge f
   expansion phase Expansionsphase f
   expansion tank Ausdehnungsgefäß n, Ausgleichsbehälter m
   bubble expansion Schlauchaufweitung f
   coefficient of cubical/volume expansion kubischer Ausdehnungskoeffizient m
   coefficient of expansion Ausdehnungskoeffizient m
   coefficient of linear expansion linearer Ausdehnungskoeffizient m, linearer Wärmeausdehnungskoeffizient m, Längenausdehnungskoeffizient m
   coefficient of thermal expansion thermischer Ausdehnungskoeffizient m, Wärmedehn(ungs)zahl f
   compensation for thermal expansion Wärmeausdehnungsausgleich m
   linear expansion Längenausdehnung f
   longitudinal expansion Längs(aus)dehnung f
   market expansion Marktausweitung f
   thermal expansion Wärme(aus)dehnung f
   thermal expansion piece Wärmedehnbolzen m
   transverse expansion Querdehnung f
   volume/volumetric expansion Volumendilatation f, Volumen(aus)dehnung f
expenditure Aufwendungen fpl
   personnel expenditure Personalaufwand m
expense Kosten pl, Aufwand m
expensive kostenaufwendig, kostspielig, teuer
experiment Experiment n, Versuch m
experimental Versuchs-
   experimental conditions Experimentierbedingungen fpl
   experimental die Versuchswerkzeug n, Prüfwerkzeug n
   experimental equipment Versuchsausrüstung f
   experimental error Versuchsfehler m
   experimental extruder Versuchsextruder m
   experimental mo(u)ld Prüfwerkzeug n, Versuchswerkzeug n
   experimental plant Versuchsanlage f
   experimental procedure Versuchsdurchführung f
   experimental product Versuchsprodukt n
   experimental quantities Versuchsmengen fpl

experimental screw Versuchsschnecke f
experimental set-up Versuchsaufbau m
experimental stage Versuchsstadium n
exploded-view diagram Explosionsdarstellung f
explosion-proof ex-geschützt, explosionsgeschützt
explosive explosiv, explosionsartig
   explosive limit Explosionsgrenze f
   potentially explosive explosionsgefährlich
   representing an explosive hazard explosionsgefährlich
export business Exportgeschäft n
export share Exportquote f
exposed exponiert, freigelegt
   exposed melt surface [i.e. during degassing] Entgasungsoberfläche f
   exposed to chemical attack chemikalienbeansprucht
   side exposed to view Sichtseite f
exposure Belastung f, Beanspruchung f
   exposure conditions Expositionsbedingungen fpl
   exposure to acids Säurebelastung f
   exposure to alternating temperatures Temperaturwechselbeanspruchung f
   exposure to chemical attack Chemikalienbelastung f, Chemikalienbeanspruchung f
   exposure to chemicals Chemikalienbelastung f, Chemikalienbeanspruchung f
   exposure to high temperatures Wärmebeanspruchung f, Temperaturbeanspruchung f
   exposure to high voltage Hochspannungsbelastung f
   exposure to hot water Heißwasserbelastung f
   exposure to light Belichtung f
   exposure to steam Wasserdampflagerung f
   exposure to water Wasserbelastung f
   conditions of exposure Expositionsbedingungen fpl
   light exposure test Belichtungsprüfung f
   low temperature exposure Kältebelastung f
   period of exposure [e.g. of a test specimen] Expositionszeit f, Expositionsdauer f, Einwirkungszeit f
   time of exposure [e.g. of a test specimen] Expositionszeit f, Expositionsdauer f, Einwirkungszeit f
   time of exposure to heat thermische Belastungszeit f
extend/to verlängern, ausbauen
extended nozzle verlängerte Düse f, Tauchdüse f
extender Extender m, Streck(ungs)mittel n
   extender plasticizer Verschnittkomponente f, Verschnittmittel n
   extender polymer Fremdharz n
   extender PVC Extender-PVC n
   extender resin Verdünnerharz n
   paste extender resin Pastenverschnittharz n
   plasticizer extender Extenderweichmacher m
extensibility Dehnbarkeit f, Dehnfähigkeit f, Streckbarkeit f
   melt extensibility Dehnviskosität f, Schmelzedehnung f

extensible dehnbar, streckbar
extension Ausdehnung f, Dehnung f, Verlängerung f
   extension cable Verlängerungskabel n
exterior außen
   exterior casement [of window] Außenflügel m
   exterior paint Fassadenfarbe f
   exterior use Außeneinsatz m, Außenverwendung f
external äußerlich, außen
   external air cooling (system) Außenluftkühlsystem n, Außenluftkühlung f
   external calibration Außenkalibrieren n
   external cooling air stream Außenkühlluftstrom m
   external dimensions Außenmaße npl
   external lubricant äußeres Gleitmittel n
   external lubricating effect Außengleitwirkung f
   external lubrication äußere Gleitwirkung f
   external screw diameter Schneckenaußendurchmesser m
   external sizing Außenkalibrieren n
   external store Externspeicher m
   external stress Fremdspannung f
   external thread Außengewinde n
externally heated außenbeheizt
extinction coefficient Extinktionskoeffizient m
extra 1. zusätzlich, extra; 2. Sonderausrüstung f (Eine elektronische Steuerung kann als Sonderausrüstung geliefert werden / electronic controls can be supplied as an extra)
   extra cost Aufpreis m, Zusatzkosten pl
   extra function Zusatzfunktion f
   extra heater band Zusatzheizband n
   extra large überdimensioniert, übergroß
   extra module Zusatzbaustein m
   extra output Zusatzausgang m
   extra part Zusatzbauteil n
   extra unit Zusatzeinheit f
extract Auszug m, Extrakt m
   aqueous extract wäßriger Auszug m
   benzene extract Benzolextrakt m
   ether extract Etherextrakt m
extractability Extrahierbarkeit f
extractable extrahierbar
extractant Extraktionsflüssigkeit f, Extraktionsmittel n
extracted extrahiert
extraction Extraktion f
   extraction behavio(u)r Extraktionsverhalten n
   extraction equipment Absauganlage f
   extraction hood Absaughaube f
   extraction line Absaugleitung f
   extraction pump Austragspumpe f
   extraction rate Extraktionsgeschwindigkeit f
   extraction resistance Auslaugebeständigkeit f, Extraktionsbeständigkeit f
   extraction resistant extraktionsbeständig
   platicizer extraction Weichmacherextraktion f
extractor Absauger m
   extractor fan Absaugvorrichtung f, Absauggebläse n

plasticizer vapo(u)r extractor
Weichmacherdampfabsaugung f
extrapolated extrapoliert
extrapolation Extrapolation f
extreme demands Extremforderungen fpl
extremely extrem, äußerst
  extremely fine adjustment Feinstregulierung f
  extremely fine cotton fabric Baumwollfeinstgewebe n
  extremely fine fabric Feinstgewebe n
  extremely fine-mesh oil filter Feinstölfilter n,m
  extremely fine particle Feinstpartikel m
  extremely hard wearing hochverschleißwiderstandsfähig, hochverschleißfest
  extremely short runs Kleinstserien fpl
  extremely small amount Kleinstmenge f
  extremely small mo(u)ld Kleinstwerkzeug n
  extremely small quantity Kleinstmenge f
  extremely stable hochstabil
  extremely wear resistant hochverschleißfest,
hochverschleißwiderstandsfähig
extrudability Extrudierbarkeit f, Spritzbarkeit f
extrudable extrudierbar
extrudate Extrudat n
  extrudate delivery pressure Ausstoßdruck m
  extrudate delivery temperature Düsenaustrittstemperatur f
  extrudate melt strand Massestrang m,
Schmelzestrang m
  extrudate strand Strang m
extrude/to spritzen, extrudieren
extruded extrudiert, gespritzt
  extruded film Extruderfolie f
  extruded product Extrusionsartikel m
  extruded sheet Breitschlitzplatte f,
Breitschlitzdüsenplatte f
extruder Extruder m, Extrudereinheit f,
Extrusionseinheit f, Extrusionsmaschine f,
Strangextrusionsmaschine f, Strangpresse f
  extruder accessories Extruderzubehör n
  extruder carriage Extruderfahrwagen m
  extruder characteristic (curve) Extruderkennlinie f
  extruder connection Extruderanschluß m
  extruder design Extruderkonzeption f,
Extruderkonstruktion f, Extruderkonzept n
  extruder diagram [output vs. pressure curve]
Extruderdiagramm n
  extruder downstream equipment Extrudernachfolge f
  extruder drive Extruderantrieb m, Extrudergetriebe n
  extruder efficiency Extruderleistung f
  extruder feed Extruderspeisung f
  extruder head Extrusionskopf m,
Extruder(düsen)kopf m, Düsenkopf m,
Werkzeugkopf m
  extruder head design Kopfkonstruktion f
  extruder length Extruder(bau)länge f,
Maschinenlänge f
  extruder manufacturer Extruderbaufirma f
  extruder output equation
Extruderkennliniengleichung f
  extruder output (rate) Extruderausstoß m,
Extruderleistung f, Extrusionsleistung f
  extruder performance Extruderverhalten n
  extruder range Extruderbaureihe f
  extruder settings Extrudereinstellwerte mpl
  extruder start-up waste Extrusionsabfälle mpl,
Extruderanfahrfladen m
  extruder suspension (unit) Extruderaufhängung f
  extruder system Extrudersystem n
  ancillary extruder Beistellextruder m,
Nebenextruder m, Seitenextruder m,
Sekundärextruder m, Zusatzextruder m
  auxiliary twin screw extruder
Doppelschnecken-Seitenextruder m
  blown film extruder Schlauchfolienextruder m
  cable sheathing extruder Kabelummantelungs-
extruder m, Ummantelungsextruder m
  cascade extruder Extruderkaskade f,
Tandemextruder m
  co-rotating twin screw extruder Gleichdrall-
Schneckenextruder m
  coating extruder Beschichtungsextruder m
  cold feed extruder Kaltfütterextruder m
  compounding extruder Aufbereitungsextruder m,
Aufschmelzextruder m, Compoundierextruder m,
Schmelzehomogenisierextruder m,
Plastifizierextruder m
  counter-rotating twin screw extruder Gegendrall-
Schneckenextruder m
  devolatilizing extruder Entgasungsextruder m
  dry blend extruder Pulverextruder m
  expanded film extruder Schaumfolienextruder m
  experimental extruder Versuchsextruder m
  feed extruder Speiseextruder m
  film tape extruder Folienbandextruder m
  flat film extruder Flachfolien-Extrusionseinheit f,
Flachfolienextruder m, Flachfolienmaschine f
  general-purpose extruder Allzweckextruder m,
Universalextruder m
  general-purpose extruder head Universalkopf m
  grooved-barrel extruder Nutbuchsenextruder m
  high speed extruder Hochleistungsextruder m,
Schnelläufer-Extruder m
  high speed single screw extruder Hochleistungseinschneckenextruder m
  high speed twin screw extruder Hochleistungs-
doppelschneckenextruder m
  horizontal extruder Waagerechtextruder m
  hot feed extruder Warmfütterextruder m
  hot melt extruder Schmelzeaustragsextruder m,
Schmelzeextruder m
  laboratory extruder Labor(meß)extruder m
  laboratory single-screw extruder Labor-
Einwellenextruder m
  laboratory twin screw extruder Doppelschnecken-

Laborextruder m
main extruder Hauptextruder m, Primärextruder m
melt fed extruder Heißschmelzextruder m
monofilament extruder Spinnextruder m
multi-screw extruder Vielschneckenmaschine f
planetary gear extruder Walzenextruder m, Planetwalzenextruder m
plastics extruder Kunststoffschneckenpresse f
principal extruder Hauptextruder m, Primärextruder m
production extruder Fertigungsextruder m, Produktionsextruder m
ram extruder Kolbenextruder m, Ramextruder m, Kolbenmaschine f, Kolbenstrangpresse f
reciprocating extruder reversierender Extruder m
reciprocating-screw extruder Schubschneckenextruder m
reclaim extruder Regranulierextruder m
rotary extruder Drehextruder m
short single-screw extruder Einschnecken-Kurzextruder m
short-screw extruder Kurzschneckenextruder m
single-die extruder head Einfach(extrusions)kopf m, Einfachwerkzeug n
single screw compounding/plasticating extruder Plastifizier-Einschneckenextruder m
single-screw extruder Einschneckenaggregat n, Einschnecken-Plastifizierextruder m, Einschneckenanlage f, Einschneckenextruder m, Einschneckenextrusionsanlage f, Einschneckenpresse f, Einwellenextruder m
sinter plate extruder Schmelztellerextruder m
special-purpose extruder Einzweckextruder m, Sonderextruder m
subsidiary extruder Beispritzextruder m, Beistellextruder m
triple screw extruder Dreischneckenextruder m
twin screw compounding extruder Zweiwellen-Knetscheiben-Schneckenpresse f
twin screw extruder Doppelschneckenextruder m, Zweischneckenextruder m, Zweiwellenextruder m, Doppelschneckenpresse f
two-stage extruder Zweistufenextruder m
two-stage twin screw extruder Zweistufen-Doppelschneckenextruder m
type of extruder Extrudertyp m
vented barrel extruder Zylinderentgasungsextruder m
vented extruder Vakuumextruder m
vented single-screw extruder Einschnecken-Entgasungsextruder m
vented twin screw extruder Doppelschnecken-Entgasungsextruder m
vertical extruder Senkrechtextruder m, Vertikalextruder m
vertical extruder head Senkrechtspritzkopf m
vertical twin screw extruder Vertikaldoppelschneckenextruder m

extrusion Extrudieren n, Extrusion f, Extruderverarbeitung f, Strangpressen n, Verspritzen n, Extrusionsverfahren n, Extruderprozeß m, Extrusionsvorgang m, Strangpreßverfahren n
extrusion blow mo(u)lded extrusionsgeblasen
extrusion blow mo(u)lder Extrusionsblas(form)maschine f
extrusion blow mo(u)lding Extrusionsblasen n, Extrusionsblasformen n, Blasextrusion f
extrusion blow mo(u)lding line Blasextrusionsanlage f, Extrusionsblas(form)anlage f
extrusion blow mo(u)lding machine Extrusionsblas(form)maschine f
extrusion blow mo(u)lding of expandable thermoplastics Thermoplast-Extrusions-Schaumblasen n
extrusion blow mo(u)lding plant Blasextrusionsanlage f, Extrusionsblas(form)anlage f
extrusion capacity Extrusionskapazität f
extrusion coating Extrusionsbeschichten n
extrusion coating line Extrusionsbeschichtungsanlage f, Extrusionsbeschichtungslinie f
extrusion coating plant Extrusionsbeschichtungsanlage f, Extrusionsbeschichtungslinie f
extrusion compound Extrusionsmasse f, Extrusionsmischung f, Strangpreßmasse f
extrusion conditions Extrusionsbedingungen fpl, Extrudierbedingungen fpl, Verarbeitungsbedingungen fpl
extrusion direction Auspreßrichtung f, Extrusionsrichtung f
extrusion grade [of compound] Extrusionsqualität f, Extrusionstyp m, Extrusionsmarke f
extrusion height Extrusionshöhe f, Extruderhöhe f, Extrudierhöhe f
extrusion line Extruderanlage f, Extruderstraße f, Extrusionslinie f, Extrusionsstraße f
extrusion of expandable thermoplastics Thermoplast-Schaumextrusion f
extrusion of granular compound [as opposed to the extrusion of dry blends] Granulatextrusion f
extrusion of plasticized PVC Weich-Extrusion f, PVC-Extrusion f
extrusion of unplasticized PVC Hart-PVC-Extrusion f, Hartextrusion f
extrusion operation Extrusionsverlauf m
extrusion performance Extrusionsverhalten n, Extrudierbarkeit f
extrusion plastimeter Extrusionsviskosimeter n
extrusion pressure Extrusionsdruck m, Spritzdruck m
extrusion process Extrusionsverlauf m
extrusion quality Extrusionsqualität f
extrusion rate Extrudiergeschwindigkeit f, Extrusionsgeschwindigkeit f, Spritzgeschwindigkeit f
extrusion rheometer Extrusionsviskosimeter n
extrusion scrap Extrusionsabfälle mpl
extrusion shop Spritzerei f, Extruderhalle f, Extrusionshalle f, Produktionsstätte f, Fabrikationshalle f

extrusion speed Extrudiergeschwindigkeit f, Extrusionsgeschwindigkeit f, Spritzgeschwindigkeit f
extrusion stretch blow mo(u)lding Extrusions-Streckblasformen n, Extrusions-Streckblasen n
extrusion system Extrusionssystem n
extrusion tasks Extrusionsaufgaben fpl
extrusion trial Extrusionsversuch m
extrusion welding Extrusionsschweißen n
extrusion welding unit Extrusionsschweißgerät n
angled extrusion die head Schräg(spritz)kopf m
blown film extrusion Folienblasverfahren n, Folienblasen n, Schlauchfolienherstellung f, Schlauchfolienfertigung f, Blasverfahren n, Blasvorgang m, Blasverarbeitung f, Blasextrusion f
blown film extrusion process Blasprozeß m
cast film extrusion Chillroll-Verfahren n, Extrusionsgießverfahren n, Kühlwalzenverfahren n
cast film extrusion line Chillroll-Filmgießanlage f, Chillroll-Folienanlage f, Chillroll-Anlage f, Chillroll-Flachfolien-Extrusionsanlage f, Foliengießanlage f, Gießfolienanlage f
caused by extrusion [e.g. stresses] extrusionsbedingt
combined extrusion-thermoforming line Extrusionsformanlage f
compactly designed/constructed extrusion line Folienkompaktanlage f
continuous extrusion stetige Extrusion f
corrugated pipe extrusion line Wellrohranlage f
downward extrusion Abwärtsextrusion f
dry blend extrusion Pulverextrusion f
expanded film extrusion line Schaumfolienanlage f
fiber extrusion line Faser(extrusions)anlage f
film extrusion Breitschlitzfolienverfahren n, Breitschlitzfolienextrusion f
flat (film) extrusion Flachfolienextrusion f, Breitschlitzfolienextrusion f
flat film extrusion line Breitschlitz(flach)-folienanlage f, Flachfolienanlage f, Folienextrusionsanlage f
flat film extrusion process Flachfolienverfahren n
four-die blown film extrusion line Vierfach-Schlauchfolien-Extrusionsanlage f
high capacity/performance extrusion blow mo(u)lding plant Hochleistungsblasextrusionsanlage f
high speed extrusion Hochgeschwindigkeitsextrusion f
high speed extrusion line Hochleistungsextrusionsanlage f
high speed pipe extrusion line Hochleistungs-Rohrextrusionslinie f
high speed single-screw extrusion line Einschnecken-Hochleistungsextrusionsanlage f
intermittent extrusion taktweise Extrusion f
large-bore pipe extrusion line Großrohrstraße f
monofilament extrusion line Monofilanlage f
paste extrusion Pastenextrusion f
pipe extrusion compound Rohrgranulat n, Rohrwerkstoff m, Rohrware f
pipe extrusion head Rohr(spritz)kopf m
pipe extrusion line Rohrextrusionsanlage f, Rohr(fertigungs)straße f, Rohrherstellungsanlage f
profile extrusion downstream equipment Profilnachfolge f
profile extrusion line Profilanlage f, Profil(extrusions)straße f
rate of extrusion Extrusionsgeschwindigkeit f
sheet extrusion Plattenextrusion f, Breitschlitzplattenextrusion f
sheet extrusion head Plattenkopf m
sheet extrusion line Platten-Extrusionsanlage f, Plattenanlage f, Plattenextrusionslinie f, Plattenstraße f
single-die extrusion line Einkopfanlage f
single-layer extrusion Einschichtextrusion f
single-screw extrusion Einschneckenextrusion f
single-screw extrusion line Einschneckenanlage f
single-stage extrusion stretch blow mo(u)lding Einstufen-Extrusionsstreckblasen n
single-station extrusion blow mo(u)lding machine Einstationenextrusionsblasformanlage f
slit die extrusion Breitschlitzextrusion f, Breitschlitzverfahren n
slit die extrusion line Breitschlitzextrusionsanlage f
slit die extrusion process Breitschlitzdüsenverfahren n
slit die film extrusion line Breitschlitzfolienanlage f, Folienextrusionsanlage f
spinneret extrusion line Extrusionsspinnanlage f
spinneret extrusion (process) Extrusionsspinnprozeß m
spinneret extrusion unit Extrusionsspinneinheit f
thermoforming sheet extrusion line Tiefziehfolien-Extrusionsanlage f
twin screw extrusion Doppelschneckenextrusion f
two-layer sheet extrusion Zweischicht-Tafelherstellung f
two-layer thermoforming film extrusion line Zweischicht-Tiefziehfolienanlage f
two-stage extrusion blow mo(u)lding Zweistufen-Extrusionsblasformen n
wide sheet extrusion line Platten-Großanlage f
exudation [of plasticizer] Ausschwitzen n
eye Auge n
   fish eye Stippe f, Fischauge n [Materialfehler]
   free from fish eyes stippenfrei
   low fish eye content Stippenarmut f
   with a low fish eye content stippenarm
fabric Gewebe n, Textilgewebe n
   fabric-based laminate Hartgewebe n
   fabric-based laminate sheet Hartgewebetafel f
   fabric coating Gewebebeschichtung f
   fabric reinforced textilverstärkt
   fabric softener Textilweichmacher m
   backing fabric Trägergewebe n
   circular-woven fabric Rundgewebe n
   coarse cotton fabric Baumwollgrobgewebe n

coarse fabric Grobgewebe n
extremely fine cotton fabric Baumwollfeinstgewebe n
extremely fine fabric Feinstgewebe n
filter fabric Siebgewebe n, Filtergewebe n
fine cotton fabric Baumwollfeingewebe n
fine fabric Feingewebe n
flat-woven fabric Flachgewebe n
glass cloth fabric Glasfilamentgewebe n
glass (fiber) fabric Textilglasgewebe n
industrial fabrics technische Gewebe npl
knitted fabric Gewirke n, Gestrick n, Maschenware f
(PVC) coated fabric Kunstleder n
PVC leathercloth fabric Gewebekunstleder n
raschel-knit fabric Raschelgewirke n
reinforcing fabric Armierungsgewebe n, Verstärkungsgewebe n
synthetic fabric Chemiefasergewebe n
varnished fabric Lackgewebe n
woven fabric Textilgewebe n
fabricator Verarbeiter m, Weiterverarbeiter m
facade Fassade f
face Gesicht n, Stirnfläche f
  face casting Frontguß m
  face mask Atem(schutz)maske f, Filtermaske f, Gesichtsmaske f, Schutzmaske f, Staubmaske f
  die face pelletizing system Direktabschlagsystem n
  front face of flight Schubflanke f, aktive Flanke f, treibende Flanke f, Gewindeflanke f
  grinding face Mahlbahn f
  hard face coating Oberflächenpanzerung f, Panzerschicht f, Panzerung f
  rear face of flight hintere Flanke f, passive Schneckenflanke f, Gewindeflanke f
  roll face Ballen m, Walzenballen m,
  roll face center Walzenballenmitte f
  roll face width Ballenbreite f, Ballenlänge f, Walzenballenbreite f, Walzenballenlänge f
  sealing face Dichtfläche f
  thrust face of flight vordere Flanke f, aktive Flanke f, treibende Stegflanke f, Schneckenflanke f
  vertical flash face Tauchkante f, Werkzeugtauchkante f
  water-cooled die face pelletization Wasserringgranulierung f
faceted facettiert
  faceted mixing section [of screw] Rautenmischteil n
  faceted smear section [of screw] Flächenscherteil n
facility Einrichtungen fpl, Anlagen fpl
  data processing facilities Datenverarbeitungsmöglichkeiten fpl
  output facility Ausgabemöglichkeit f
  production facility Fabrikationsstätte f, Fertigungsstätte f
  research facilities Forschungseinrichtungen fpl
facing Frontguß m
factor Faktor m
  factor influencing cycle time Zykluszeitfaktor m (Einer der Zykluszeitfaktoren ist die Werkzeugwanddicke / one of the factors which influences cycle time is the mo(u)ld wall thickness)
  factors influencing machine performance Maschineneinflußgrößen fpl
  factors influencing the process Verfahrenseinflußgrößen fpl
  bulk factor Füllfaktor m
  conversion factor Umrechnungsfaktor m
  correction factor Korrekturfaktor m
  cost-benefit factor Kosten-Nutzen-Verhältnis n
  cost factor Kostenfaktor m
  cost-performance factor Preis-Leistungs-Index m
  cost-service life factor Kosten-Standzeit-Relation f
  creep factor Kriechfaktor m
  dissipation factor dielektrischer Verlustfaktor m
  environmental factors Umweltfaktoren mpl
  influencing factor Einflußfaktor m, Einflußparameter m, Einflußgröße f
  long-term welding factor Langzeitschweißfaktor m
  loss factor (mechanischer) Verlustfaktor m
  mo(u)ld correction factor Formenkorrekturmaß n
  quality factor Gütefaktor m
  safety factor Sicherheitsbeiwert m, Sicherheitsfaktor m, Sicherheitskoeffizient m, Sicherheitszahl f
  stress intensity factor Spannungsintensitätsfaktor m
  time factor Zeitfaktor m
  welding factor Schweißfaktor m
factory Fabrik f, Fertigungsstätte f, Fabrikationsstätte f, Fertigungsbetrieb m, Produktionsstätte f
  factory gate Werkstor n
failure Bruch m, Versagen n, Störung f
  failure-initiating versagensauslösend
  failure mechanism Bruchmechanismus m, Versagensmechanismus m
  brittle failure Trennbruch m
  catastrophic failure katastrophaler Bruch m, katastrophales Versagen n
  ductile failure Verformungsbruch m
  fatigue failure Dauerbruch m
  long-term failure test under internal hydrostatic pressure [test used for plastic pipes] Zeitstand-Innendruckversuch m
  material failure Werkstoffversagen n
  number of vibrations to failure Bruch-Schwingspielzahl f
  power failure Spannungsausfall m, Stromausfall m
  shear stress at failure Abscherspannung f
faithful (to the original) in every detail detailgetreu, originalgetreu
falling fallend, Fall-
  falling ball test Kugelfallversuch m
  falling dart test Falldorntest m
  falling film vaporizer Fallfilmverdampfer m
  falling sphere viscometer Kugelfallviskosimeter n
  falling weight Fallbolzen m, Fallgewicht n
  falling weight test Bolzenfallversuch m,

Fallbolzenversuch m
falling weight tester Fallbolzenprüfgerät n
false start Fehlstart m
falsification [e.g. of test results] Verfälschung f
fan Lüfter m, Ventilator m
  fan blade Lüfterflügel m
  fan cooling Gebläsekühlung f
  fan housing Gebläsegehäuse n, Lüfterzarge f
  fan shroud Gebläsegehäuse n, Lüfterzarge f
  cooling fan Kühlventilator m
  extractor fan Absaugvorrichtung f
  (waste air) exhaust fan Abluftventilator m
fast 1. im Eilgang, schnell; 2. fest, beständig
  fast curing rasch aushärtbar, raschhärtend, schnellhärtend, schnellabbindend
  fast cycling hohe Bewegungsgeschwindigkeit f, rasche Taktfolge f
  fast cycling injection mo(u)lding machine Schnelläufer-Spritzgießmaschine f, Hochleistungsspritzgießmaschine f
  fast-cycling machine Schnelläufer m
  fast-drying schnelltrocknend
  fast-gelling raschplastifizierend, schnellgelierend
  fast-reacting hardener Schnellhärter m
  fast-setting schnellabbindend
  fast-setting adhesive Schnellkleber m
  fast-solidifying schnellerstarrend
  fast-solvating schnellgelierend
  automatic fast-cycling injection mo(u)lding machine Schnellspritzgießautomat m
  colo(u)r fast farbtonbeständig
fastening lugs Befestigungslaschen fpl
fastness 1. Festigkeit f, Beständigkeit f; 2. Schnelligkeit f
  colo(u)r fastness Farbechtheit f, Farbtonbeständigkeit f
  rub fastness Reibechtheit f
  weathering fastness of pigments Wetterechtheit f
fat Fett n
  edible fat Speisefett n
fatigue Ermüdung f
  fatigue crack Ermüdungsriß m
  fatigue crack propagation Ermüdungsrißausbreitung f
  fatigue cracking Ermüdungsrißbildung f
  fatigue endurance Zeitfestigkeit f
  fatigue failure Dauerbruch m
  fatigue limit Wechselfestigkeit f, Dauer(stand)festigkeit f
  fatigue strength Schwingfestigkeit(sverhalten) f(n)
  fatigue stress Ermüdungsbelastung f, Schwellbeanspruchung f
  fatigue test Dauerfestigkeitsversuch m, Dauerschwingversuch m, Zeitwechselfestigkeitsversuch m
  flexural fatigue strength Biegeschwingfestigkeit f, (Zeit-)Biegewechselfestigkeit f
  flexural fatigue stress Biegewechselbelastung f, Biegewechselbeanspruchung f
  flexural fatigue test Biegeschwingversuch m, Wechselbiegeversuch m
  tensile fatigue strength Zugschwingungsversuch m
fatty acid Fettsäure f
  fatty acid amide Fettsäureamid n
  fatty acid derivative Fettsäurederivat n
  fatty acid ester Fettsäureester m
  fatty acid mono-ester Fettsäuremonoester m
  fatty acid soap Fettsäureseife f
  coconut oil fatty acid Kokosölfettsäure f
  cottonseed oil fatty acid Baumwollsamenölfettsäure f
  dienol fatty acid Dienolfettsäure f
  linseed oil fatty acid Leinölfettsäure f
  polyglycol fatty acid ester Polyglykolfettsäureester m
  soya bean oil fatty acid Sojaölfettsäure f
  tall oil fatty acid Tallölfettsäure f
fault Fehler m
  fault analysis Fehleranalyse f
  fault correction Mängelkorrektur f
  fault localization Fehlerlokalisierung f
  fault location Fehlerort m, Pannensuche f
  fault location indicator Fehlerortsignalisierung f, Störungsortsignalisierung f
  fault location program Fehlersuchprogramm n
  elimination of faults Pannenbehebung f
  if there is a fault im Störungsfall m
  mo(u)lding fault Spritzfehler m
faulty fehlerhaft, defekt
  faulty adhesion Haftungsschäden fpl
  faulty operation Fehlbedienung f
feasibility Machbarkeit f
  feasibility study Machbarkeitsstudie f
feasible durchführbar, ausführbar, machbar
  technically feasible verfahrenstechnisch machbar
feature Kennzeichen n
  as a standard feature serienmäßig
  basic features Grundmerkmale npl
  constructional features konstruktiver Aufbau m
  decision-making features Entscheidungsmerkmale npl
  design features konstruktive Auslegung f
  distinguishing features Unterscheidungsmerkmale npl, Unterscheidungskriterien npl
  performance features Leistungsmerkmale npl
  structural features Konstruktionsmerkmale npl
feed Dosierung f, Zuführung f, Speisung f
feed/to eindosieren, einspeisen, beschicken
  feed bush Einlaufhülse f, Einzugsbuchse f
  feed capacity Einzugsvermögen n
  feed channel Einspeisekanal m, Zuströmkanal m, Zuführungskanal m
  feed chute Aufgabeschurre f
  feed compartment Dosierkammer f
  feed conditions Einzugsverhältnisse npl
  feed conveyor Zuführband n
  feed cycle Förderzyklus m
  feed cylinder Dosierzylinder m, Füllzylinder m
  feed device Einzugsvorrichtung f
  feed end Einzug m

feed equipment Beschick(ungs)einrichtung f, Dosieranlage f, Dosiereinrichtung f, Zuführeinrichtung f
feed extruder Speiseextruder m
feed line Anströmleitung f, Versorgungsleitung f
feed mechanism Einzugswerk n
feed mill Speisewalzwerk n
feed opening Einspeisebohrung f, Zuführöffnung f, Maschineneinlauf m, Zuführungsbohrung f
feed performance Einziehverhalten n, Einzugsverhalten n
feed pipe Einlaufrohr n, Zuleitungsrohr n
feed plate Anguß(verteiler)platte f, Verteilerplatte f
feed pocket Einzugstasche f
feed point Einspeisepunkt m, Einspeisestelle f, Einspritzpunkt m
feed port Zugabeöffnung f, Einfüllschacht m, Zulauföffnung f
feed problems Einspeisungsschwierigkeiten fpl, Einzugsschwierigkeiten fpl
feed pump Dosierpumpe f, Speisepumpe f
feed ram Dosierkolben m
feed rate Einzugsgeschwindigkeit f
feed roll Dosierwalze f, Einlaufwalze f, Einzugswalze f, Speisewalze f, Zuführ(ungs)walze f
feed screw Aufgabeschnecke f, Beschickungsschnecke f, Einspeiseschnecke f, Einzugsschnecke f, Füllschnecke f, Speiseschnecke f, Dosierschnecke f, Zuführschnecke f
feed screw unit Speiseschneckeneinheit f
feed section Förderzone f, Einfüllbereich m, Einfüllteil m, Trichterzone f, Einlaufteil m, Einzugszonenabschnitt m, Einzugszonenbereich m, Einfüllabschnitt m, Förderlänge f, Einzugszone f, Einlaufstück n, Eingangsteil n, Extrudereinzugsbereich m, Eingangszone f, Extruder-Einfüllbereich m
feed section flight depth Einzugsgangtiefe f
feed side Beschickseite f
feed station Dosierstation f
feed system Angußsystem n, Einspeisevorrichtung f, Beschickungssystem n, Dosiersystem n, Einspeisung f
feed tank Vorlaufbehälter m
feed temperature [e.g. of melt] Einlauftemperatur f
feed throat Beschickungsöffnung f, Aufgabeöffnung f, Aufgabeschacht m, Einfüllöffnung f, Einzugsöffnung f, Extrudereinfüllöffnung f, Einlauföffnung f
feed twin screws Zuführschneckenpaar n
feed unit Dosieraggregat n, Dosiereinheit f, Dosierelement n, Dosiergerät n, Dosierwerk n, Speiseaggregat n, Speisegerät n, Speisungsanlage f, Vorschubeinheit f, Vorschubgerät n, Zuführvorrichtung f
feed unit equipment Beschickung f, Beschickungsanlage f, Speiseeinrichtung f
feed zone Einlaufteil m, Einzugszonenabschnitt m, Einzugszonenbereich m, Einfüllabschnitt m, Förderlänge f, Einzugszone f, Einlaufstück n, Eingangsteil n, Extrudereinzugsbereich m, Eingangszone f, Extruder-Einfüllbereich m, Förderzone f, Einfüllbereich m, Einfüllteil m, Trichterzone f, Zylindereinzugsteil n, Zylindereinzugszone f, Einzugs(zonen)teil n, Zylindereinzug m, Füllzone f
ante-chamber direct feed injection Vorkammerdurchspritzverfahren n
ante-chamber feed system Vorkammer-Angießtechnik f, Vorkammeranguß m
at the feed end einlaufseitig
automatic feed unit Vorschubautomat m, Zuführungsautomat m
center feed Zentraleinspeisung f
cold runner feed system Kaltkanalangußsystem n
delayed feed Dosierverzögerung f
direct feed direkte Anspritzung f
extruder feed Extruderspeisung f (Da die Extruderspeisung über einen Tichter erfolgt.... / since the extruder is fed from a hopper....)
forced feed Zwangsbeschickung f, Zwangsfütterung f
forced feed mechanism Zwangszuführeinrichtung f
forced feed (system) Zwangsbeschickung f, Zwangsfütterung f
forced feed unit Zwangsdosiereinrichtung f
granule feed unit Granulatdosiergerät n
gravity feed Gravitationsdosierung f
grooved feed bush Einzugsnutbuchse f
grooved feed section/zone Nuteneinzugszone f
heart-shaped groove type of feed system Herzkurveneinspeisung f
horizontal feed system Horizontalbeschickung f
hot feed extruder Warmfütterextruder m
hot runner feed system Heißkanalanguß m, Heißkanal-Angußsystem n, Heißkanal-Verteileranguß m
insulated runner feed system Isolierkanal-Angußplatte f
lateral feed device Seitenspeisevorrichtung f
material feed Masseinzug m
melt feed (system) Schmelzeeinspeisung f, Schmelzezufluß m
melt temperature at the feed end [e.g. of an extruder] Eingangsmassetemperatur f, Einlaufmassetemperatur f
on the feed side angußseitig
raw material feed unit Rohstoffzuführgerät n
rotary feed unit Rotationsdosiereinrichtung f
separate feed (unit) Fremddosierung f
side feed seitliches Anspritzen n
standard feed chute Serienaufgabeschurre f
twin feed screw Doppel-Dosierschnecke f, Doppel-Einlaufschnecke f
twin screw feed section Zweischneckeneinzugszone f
vibratory feed hopper Vibrationseinfülltrichter m
vibratory feed (unit) Vibratordosierung f

water cooled feed section/zone Naßbüchse f
feedback Rückmeldung f
  feedback control system Regelungssystem n
  feedback information Informationsausgabe f
  feedback system Rückführsystem n
  electronic feedback elektronische Rückführung f
  position feedback Lagerückmeldung f
feeder Zuführvorrichtung f [s.a. feed]
  automatic feeder Dosierautomat m, Zuführungsautomat m, Beschickungsautomat m
  ram feeder Kolbenstopfaggregat n
  screw feeder Schneckendosiereinheit f, Schneckendosierer m, Schneckendosiergerät n, Schneckendosiervorrichtung f, Schneckenspeiseeinrichtung f
  vertical feeder Vertikalspeiseapparat m
  volumetric feeder Volumendosieraggregat n, Volumendosierung f
  weigh feeder Dosierwaage f, Gewichtsdosiereinrichtung f, Gewichtsdosierung f, Waagedosierung f
feeding Einziehen n, Beschickung f
  feeding aid Einzugshilfe f
  automatic feeding equipment Beschickungsautomat m
  starve feeding Unterdosierung f
  strips for feeding to the calender Fütterstreifen m
  vibratory feeding Vibratordosierung f
  volumetric feeding Volumendosierung f
  weigh feeding Waagedosierung f, Gewichtsdosierung f
feedstock Aufgabegut n, Aufgabematerial n
female forming Negativverfahren n, Negativformen n
female tool Negativform f, Negativwerkzeug n
fender Kotflügel m
FEP [fluorinated ethylene-propylene copolymer] Fluorethylenpropylen n
ferrous metal FE-Metall n
few wenig
fiber [GB: fibre] Faser f
  fiber composite Faserverbundwerkstoff m
  fiber content Faseranteil m
  fiber extrusion line Faser(extrusions)anlage f
  fiber-optic cable Glasfaserkabel n
  fiber orientation Faseranordnung f, Faserausrichtung f, Faserorientierung f, Faserverlauf m
  fiber reinforced faserverstärkt
  fiber strand Faserbündel n
  fiber strands Faserstränge mpl
  aramid fiber Aramidfaser f
  asbestos fiber Asbestfaser f
  boron fiber Borfaser f
  carbon fiber C-Faser f, Kohlenstoff-Faser f, Kohlefaser f
  carbon fiber reinforced kohlenstoffaserverstärkt
  carbon fiber reinforcement Kohlenfaserfüllung f, Kohlenstoffaserverstärkung f
  chopped carbon fiber Kohlenstoff-Kurzfaser f
  chopped fibers [if referring to fibers other than those made of glass] Kurzfasern
  chopped glass fiber Glasseiden-Kurzfaser f
  continous glass fiber Glasseidenfaser f, Langglasfasern fpl
  elementary fiber Elementarfaser f
  film fibers Spleißfasern fpl
  glass fiber Glasfaser f, Textilglasfaser f
  glass fiber content Glasfasergehalt m, Glasfaserkonzentration f
  glass fiber distribution Glasfaserverteilung f
  glass fiber filled glasfaserhaltig, glasfasergefüllt
  glass fiber laminate Glaslaminat n
  glass fiber mat Glasfasermatte f
  glass fiber material Textilglas n
  glass fiber materials/products Textilglaserzeugnisse npl, Textilglasprodukte npl
  glass fiber orientation Glasfaserorientierung f
  glass fiber reinforced glasfaserverstärkt, textilglasverstärkt
  glass fiber reinforced plastic [GRP] Glasfaserkunststoff m
  glass fiber reinforcement Glasseidenverstärkung f, Textilglasverstärkung f
  graphite fiber Graphitfaser f
  high modulus fiber HM-Faser f, Hochmodulfaser f
  high strength fiber Hochleistungsfaser f
  high tensile fiber HT-Faser f
  man-made fiber Chemiefaser f, Synthesefaser f
  milled glass fibers Mahlglasfasern fpl
  optic fiber cable Glasfaserkabel n
  optical fiber Lichtleitfaser f
  outer fiber Außenfaser f, Randfaser f
  outer fiber strain Randfaserdehnung f
  outer fiber zone Randfaserbereich m
  reinforcing fibers Verstärkungsfasern fpl
  staple fiber Stapelfaser f
  staple glass fiber Glasstapelfaser f
  synthetic fiber Chemiefaser f, Synthesefaser f
  synthetic fiber reinforced chemiefaserarmiert
  textile fiber Textilfaser f
  vulcanized fiber Vulkanfiber n
  woven staple fiber Glasstapelfasergewebe n
fibril Fibrille f, Elementarfaser f
fibrillate/to spleißen, fibrillieren
fibrillated film Spleißfasern fpl
fibrillation [a method of making film fibers from film tape, consisting essentially of splitting the tape lengthways] Fibrillieren n, Aufspleißen n
fibrillator Fibrillator m
  pin roller fibrillator Nadelwalzenfibrillator m
fibrous faserförmig, fas(e)rig, fibrogen
field Feld n, Bereich m, Fachgebiet n
  field strength Feldstärke f
  field trial Feldversuch m
  fields of application Anwendungsgebiete npl
  breakdown field strength Durchschlagfeldstärke f
  electric field strength elektrische Feldstärke f
  high frequency field Hochfrequenzfeld n

main field Hauptanwendungsgebiet n, Haupteinsatzgebiet n
figure Ziffer f, Zahl f, Größe f, Wert m, Zahlenwert m
  approximate figure [to be taken as a guide] Richtgröße f
  comparative figure Vergleichswert m, Vergleichszahl f
  original figure Ausgangswert m
  production figures Produktionszahlen fpl
  sales figures Absatzzahlen fpl
  WLF figure WLF-Wert m
filament Faden m, Faser f
  filament winding Faserwickelverfahren n, Wickelverfahren n, Wickeltechnik f
  filament winding resin Wickelharz n
  filament winding robot Wickelroboter m
  filament wound pipe Wickelrohr n
  filament wound section Wickelschuß m
  continous filament Endlosfaser f
  electric filament lamp Glühlampe f
  glass filament Textilglasspinnfaden m
  GRP filament winding machine GFK-Wickelmaschine f
  precision filament winding Präzisionswickeltechnik f
  rovings for filament winding Wickelrovings mpl
file Akte f, Ordner m, Datei f, File n
  data file Datei f, Daten(bank)datei f, Datenfile n
  formulation data file Rezeptdatei f
  job file Auftragsdatei f
  machine control data file Maschinensteuerungsdatei f
  master file Stammdatei f
  raw material data file Rohstoffdatei f
filing Archivierung f
  data filing Datenarchvierung f
filled füllstoffhaltig, gefüllt
  filled with carbon black rußgefüllt
  glass fiber filled glasfasergefüllt
  highly filled füllstoffreich
filler Füllstoff m, Füllsubstanz f, Harzträger m, Zusatzstoff m
  filler concentration Füllstoffkonzentration f
  filler content Füllstoffmenge f, Füllungsgrad m, Füllstoffgehalt m, Füllstoffanteil m
  filler loading Füllgrad m
  filler particle Füllstoffteilchen n, Füllstoffkorn n
  filler particle orientation Füllstofforientierung f
  addition of filler Füllstoffzugabe f
  compatible with fillers füllstoffverträglich
  jointing filler Fugenspachtel m(f), Fugenmasse f, Fugenmörtel m, Fugenfüller m
  knifing epoxy-based filler EP-Spachtel m(f)
  knifing filler Spachtelmasse f, Spachtel m(f), Ausgleichsmasse f
  light weight filler Leichtfüllstoff m
  powdered filler Pulverfüllstoff m
  reinforcing filler Verstärkerstoff m, Verstärkungsadditiv n
  silicate filler Silikatfüllstoff m
  type of filler Füllstoffsorte f
  with a low filler content füllstoffarm
fillet Hohlkehle f, Leiste f, Steg m
filling Füllung f
  filling pressure Fülldruck m
  filling speed Füllgeschwindigkeit f
  filling station Füllstation f
  filling study Füllstudie f
  automatic filling machine Abfüllautomat m, Füllautomat m
  blow mo(u)lding and filling machine Blas- und Füllmaschine f
  mo(u)ld filling monitoring system Formfüllüberwachung f
  mo(u)ld filling operation Formfüllprozeß m, Werkzeugfüllvorgang m, Formenfüllung f
  mo(u)ld filling phase Formfüllphase f, Füllphase f
  mo(u)ld filling pressure Formfülldruck m
  mo(u)ld filling speed Formfüllgeschwindigkeit f, Werkzeugfüllgeschwindigkeit f
  mo(u)ld filling time Einspritzzeit f, Füllzeit f, Spritzzeit f, Formfüllzeit f, Werkzeugfüllzeit f
film Film m, Folie f
  film adhesive Klebefilm m
  film appearance Filmaussehen n
  film blowing Blasen n, Blasextrusion f, Blasverarbeitung f, Blasverfahren n, Blasvorgang m, Folienblasen n, Folienblasverfahren n, Schlauchfolienextrusion f, Schlauchfolienherstellung f, Schlauchfolienblasen n
  film blowing die Blasfolienschlauchkopf m, Folienblaskopf m, Schlauchfolien(extrusions)werkzeug n, Schlauchfolienkopf m, Schlauchfoliendüse f, Blaswerkzeug n, Schlauchformeinheit f, Blasdüse f
  film blowing line Schlauchfolienblasanlage f, Schlauchfolien(extrusions)anlage f, Folienblasanlage f
  film blowing machine Folienblaseinheit f
  film blowing unit Blasaggregat n
  film bubble Blasschlauch m, Filmballon m, Folienblase f, Folienschlauch m, Folienballon m, Schlauchblase f
  film bubble circumference Schlauchumfang m
  film bubble contours Blaskontur f
  film bubble cooling system Schlauchkühlvorrichtung f, Folienkühlung f
  film bubble diameter Schlauchdurchmesser m
  film bubble radius Folienschlauchhalbmesser m, Folienschlauchradius m
  film bubble stability Schlauchblasenstabilität f
  film casting line Foliengießanlage f, Gießfolienanlage f
  film conversion (plant) Folienweiterverarbeitung f
  film defects Filmstörungen fpl
  film edge control (device) Bahnkantensteuerung f
  film extruder Folienmaschine f

film extrusion Folienextrusion f
film fibers Spleißfasern fpl
film formation Filmbildung f
film former Filmbildner m
film-forming filmbildend
film-forming aid Verfilmungshilfsmittel n
film gate Bandanguß m, Bandanschnitt m, Delta-Anguß m, Filmanguß m, Filmanschnitt m
film-gated bandangeschnitten
film ga(u)ge equalizing unit [this equalizes ga(u)ge variations across the width of the roll to avoid local diameter build-ups when operating with a stationary die head or winder] Folienverlegegerät n, Folienverlegeeinheit f
film ga(u)ge measuring instrument Foliendickenmeßgerät n
film ga(u)ge variations Filmdickenunterschiede mpl, Foliendickenabweichungen fpl
film guide rolls Folienführungswalzen fpl
film hardness Filmhärte f
film inspection unit Folienbeobachtungsstand m
film laminating plant Folienkaschieranlage f
film pretreating instrument Folienvorbehandlungsgerät n
film production machine Folienproduktionsmaschine f
film properties Filmeigenschaften fpl, Folienbeschaffenheit f
film quality Folienbeschaffenheit f, Filmqualität f
film scrap Folienabfall m
film scrap re-processing plant Folienaufbereitungsanlage f
film stretching device Folienreckvorrichtung f
film stretching plant Folienreckanlage f
film surface Filmoberfläche f, Beschichtungsoberfläche f
film surface characteristics Filmoberflächeneigenschaften fpl
film take-off (unit) Folienabziehwerk n, Folienabzug m
film tape Bändchen n, Folienband n, Folienbändchen n
film tape extruder Folienbandextruder m
film tape production line Folienbändchenanlage f
film tape stretching plant Bändchenreckanlage f, Folienbandreckanlage f, Folienbandstreckwerk n
film tape stretching unit Bändchenreckanlage f, Folienbandreckanlage f, Folienbandstreckwerk n
film thickness Filmdicke f
film transport Folientransport m
film web Folienbahn f
film web guide Folien(bahn)führung f
film width control Folienbreitenregelung f
film width control mechanism Folienbreitenregelung f
film winder Folienwickelmaschine f, Folienwickler m, Folienaufwicklung f

film winding system/mechanism/arrangement Folienwickelsystem n
film wind-up (unit) Folienaufwickler m, Folienaufwicklung f
film yarn Bändchen n, Folienbändchen n, Folienband n
adhesive film Kleb(stoff)schicht f
adhesive film thickness Klebschichtstärke f
agricultural film Landwirtschaftsfolie f, Agrarfolie f
automatic film slitter Folienschneidautomat m, Folienschneidaggregat n
barrier film layer Sperrschicht f
blown film coextrusion die Koextrusionsblaskopf m, Koextrusionsblaswerkzeug n
blown film coextrusion line Coextrusions-Blasfolienanlage f, coextrudierende Schlauchfolienanlage f
blown film cooling (system) Schlauchfolienkühlung f
blown film die Blaswerkzeug n, Folienblaskopf m, Schlauchwerkzeug n, Blasfolienschlauchkopf m, Blasdüse f, Schlauch(extrusions)düse f, Folienblaskopf m, Foliendüse f, Folienkopf m, Folien(spritz)werkzeug n, Schlauchfolien(extrusions)werkzeug n, Schlauch(folien)kopf m, Schlauchfoliendüse f, Schlauchformeinheit f, Schlauchspritzkopf m
blown film die design Blaskopfkonzeption f
blown film extruder Schlauchfolienextruder m
blown film extrusion Folienblasen n, Schlauchfolienfertigung f, Folienblasverfahren n, Schlauchfolienextrusion f, Schlauchfolienherstellung f, Blasvorgang m, Schlauchfolienblasen n, Blasextrusion f, Blasverarbeitung f, Blasverfahren n
blown film (extrusion) line Blasanlage f
blown film plant Folienblasanlage f
blown film stretching (process) Schlauchstreckverfahren n
blown polyethylene film Polyethylenschlauchfolie f
bookbinding film Bucheinbandfolie f
bubble film Luftpolsterfolie f
capacitor film Kondensatorfolie f
cast film Chillrollfolie f
cast film extrusion Foliengießen n, Extrusionsgießverfahren n, Kühlwalzenverfahren n
cast film extrusion line Chillroll-Anlage f, Chillroll-Filmgießanlage f, Chillroll-Flachfolien-Extrusionsanlage f, Chillroll-Folienanlage f, Gießfolienanlage f, Foliengießanlage f
center-fed blown film die zentral angespritzter Blaskopf m, Dornhalterblaskopf m
central film gate mittiger Bandanschnitt m
co-extruded flat film Breitschlitzverbundfolie f
compactly designed/constructed film extrusion line Folienkompaktanlage f
composite film Verbundfolie f
composite flat film Kombinationsflachfolie f
continuous film Folien- und Bahnenware f
cooling of film bubble Schlauchkühlung f
copolymer-based film Copolymerfolie f
crystal clear film Glasklarfolie f

decorative film Dekorfilm m
dry film Trockenfilm m
dry film thickness Trockenfilmdicke f, Trockenschichtstärke f
expanded film extruder Schaumfolienextruder m
expanded film extrusion line Schaumfolienanlage f
extruded film Extruderfolie f
fabrication of rigid film Hartfolienverarbeitung f (Emulsions-PVC für die Hartfolienverarbeitung / emulsion PVC for making rigid film)
falling film vaporizer Fallfilmverdampfer m
fibrillated film Spleißfasern fpl
flat film Flachfolie f
flat film coextrusion line coextrudierende Flachfolienanlage f, Coextrusions-Breitschlitzfolienanlage f
flat film die Folienwerkzeug n, Folienkopf m, Flach(folien)düse f, Flachfolienwerkzeug n, Flachdüse f, Schlitzdüse f, Breitschlitzwerkzeug n, Breitschlitzdüse f, Foliendüse f, Folienspritzwerkzeug n
flat film extruder Flachfolienmaschine f, Flachfolienextruder m, Flachfolien-Extrusionseinheit f
flat film extrusion Breitschlitzfolienextrusion f, Breitschlitzfolienverfahren n, Flachfolienextrusion f
flat film extrusion line Breitschlitz(flach)-folienanlage f, Flachfolienanlage f
flat film extrusion process Flachfolienverfahren n
flat film orientation (process) Flachfolienstreckverfahren n
flexible film Weichfolie f
flexible PVC film PVC-Weichfolie f, Weich-PVC-Folie f
food packaging film Lebensmittelverpackungsfolie f
four-die blown film extrusion line Vierfach-Schlauchfolien-Extrusionsanlage f
guiding of the film bubble Folienführung f
heavy-duty sack film Schwergutsackfolie f
heavy-ga(u)ge film winder Schwerfolienwickler m
high capacity/performance blown film plant Hochleistungsblasfolienanlage f
high molecular weight polyethylene film HM-Folie f
HM-HDPE blown film plant HM-Anlage f
lateral film gate seitlicher Bandanschnitt m
liquid film Flüssigfolie f
lubricant film Gleitmittelfilm m, Schmierfilm m
multi-layer flat film Kombinationsflachfolie f
oxide film Oxidschicht f
packaging film Verpackungsfolie f
paint film Anstrich(film) m, Lackfilm m, Lackschicht f, Lackierung f
paint film composition Anstrichaufbau m
paint film defect Lackierungsdefekt m
paint film quality Lackierungsqualität f
paint film structure Anstrichaufbau m
paint film surface Lackoberfläche f
paper-like film papierähnliche Folie f, Papierfolie f
paper-like polyethylene film HM-Folie f

plastics film Kunststoff-Folie f
release film Mitläuferfolie f, Trennfolie f, Trennbeschichtung f, Trennfilm m
rigid film Hartfolie f
rigid PVC film Hart-PVC-Folie f, PVC-Hartfolie f
shrink wrapping film Schrumpffolie f
side-fed blown film die seitlich eingespeister Folienblaskopf m, stegloser Folienblaskopf m, Pinolenblaskopf m, Umlenkblaskopf m
side-gussetted blown/tubular film Seitenfaltenschlauchfolie f
single-layer film Einschichtfolie f
slit die film extrusion Breitschlitzfolienextrusion f, Breitschlitzfolienverfahren n
slit die film extrusion line Breitschlitzfolienanlage f
spider-type blown film die Stegdorn(halter)-blaskopf m, Steghalterkopf m, Stegdornhalterwerkzeug n
spiral mandrel blown film die Schmelzewendelverteilerkopf m, Schmelzewendelverteilerwerkzeug n, Wendelverteilerwerkzeug n
stretch wrapping film Dehnfolie f, Streckfolie f
thin film Dünnfolie f
thin film vaporizer Dünnschichtverdampfer m
three-layer blown film die Dreischicht-Folienblaskopf m
total film thickness Gesamtschichtdicke f
transparent film Klarsichtfolie f
twin-die film blowing head Doppelblaskopf m
two-layer blown film die Zweischichtfolienblaskopf m
two-layer film blowing line Zweischicht-Folienblasanlage f
two-layer thermoforming film extrusion line Zweischicht-Tiefziehfolienanlage f
wet film Naßfilm m
wet film thickness Naßfilmdicke f, Naßschichtdicke f
wet film viscosity Naßfilmviskosität f
width of film after trimming Folienfertigbreite f
wrapping film Einschlagfolie f
filter Filter n,m
filter cake Kuchen m, Filterkuchen m
filter card Filterkarte f
filter cloth Filtertuch n, Siebgewebe n
filter disk Filterscheibe f, Siebronde f, Siebscheibe f
filter element support Filterelementhalteplatte f
filter fabric Filtergewebe n
filter insert Filtereinsatz m
filter medium Filtermedium n, Filtermittel n
filter plate Filterplatte f
filter press Filterpresse f
filter ribbon Siebband n
filter screen Gewebesieb n, Filtergewebe n, Siebgewebe n
filter unit Filter n,m
air filter Luftfilter n,m
automatic filter unit Filterautomatik f
candle filter Filterkerze f

candle filter assembly Filterkerzenpaket n
cartridge filter Filterkartusche f, Filterkorb m, Filterkerze f
depth filter Tiefenfilter n,m
disk filter Scheibenfilter n,m
effective filter area freie Durchgangsfläche f
extremely fine-mesh oil filter Feinstölfilter n,m
fluted filter Faltenfilter n,m
large-area melt filter Großflächen-Schmelzefilter n,m
long-life filter Langzeitfilter n,m
low-pass filter Tiefpaßfilter n,m
melt filter Schmelzefilter n,m
pressure filter Druckfilter n,m
quick-change filter unit Schnellwechselfilter n,m
stainless steel filter screen Edelstahlmaschengewebe n
suction filter Ansaugfilter n,m, Saugfilter n,m
surface filter Flächenfilter n,m
test filter cloth Prüfsiebgewebe n
filtrability Filtrierbarkeit f
filtration Filterung f, Filtrieren n, Filtration f
 filtration aid Filterhilfsmittel n
 filtration chamber Filterkammer f
 filtration characteristics Filtrationsverhalten n
 filtration efficiency Abscheideleistung f
 filtration performance Abscheideleistung f
 filtration residue Filterrückstand m
 filtration temperature Filtertemperatur f
 filtration unit Filterapparatur f, Filtergerät n
 fine filtration unit Feinfilter-Siebeinrichtung f
 melt filtration Schmelzefiltrierung f
 melt filtration unit Rohmaterialfilter m, Schmelzefilter n,m, Schmelzenfilterung f, Schmelzenfiltereinrichtung f
 oil filtration unit Ölfilterung f
 polymer melt filtration Kunststoffiltration f
 surface filtration Siebfiltration f
fin Rippe f, Kühlrippe f
 cooling fins Kühlrippen fpl
final letzt, endlich, schließlich, End...
 final concentration Endkonzentration f
 final consistency Endkonsistenz f
 final foaming Fertigschäumen n
 final melt temperature Schmelzeendtemperatur f
 final moisture content Endfeuchte f
 final position Endposition f
 final strength Endfestigkeit f
 final temperature Endtemperatur f
financial finanziell
 financial reasons Kostengründe mpl
 financial strength Finanzkraft f
 financial year Geschäftsjahr n
fine fein
 fine adjustment Feineinstellung f, Feinjustierung f, Feinregulierung f, Feinverstellung f
 fine-cell [foam] feinporig, feinzellig
 fine cotton fabric Baumwollfeingewebe n
 fine dust Feinstaub m
 fine fabric Feingewebe n
 fine filtration unit Feinfilter-Siebeinrichtung f
 fine grinding Fein(ver)mahlen n
 fine grinding machine Feinmahlaggregat n
 fine grooves Feinnuten fpl
 fine-mesh feinmaschig
 fine-mesh screen Feinsieb n
 fine-particle feindispers, feinteilig, hochdispers
 fine-particle character Feinteiligkeit f
 extremely fine adjustment Feinstregulierung f
 extremely fine cotton fabric Baumwollfeinstgewebe n
 extremely fine fabric Feinstgewebe n
 extremely fine-mesh oil filter Feinstölfilter n,m
 extremely fine particle Feinstpartikel m
finely fein
 finely crystalline feinkristallin
 finely dispersed feinverteilt
 finely divided feinverteilt
 finely grooved feingenutet
 finely powdered feinpulverig
 very finely ground feinstgemahlen
finish Oberflächenbeschaffenheit f, Oberflächenbehandlung f, Appretur f, Lackierung f
 alkyd finish Alkydharzlackierung f
 automotive finish Autolack m
 baked finish Einbrennlackierung f
 hammer finish paint Hammerschlaglack m
 polished to a mirror finish spiegelhochglanzpoliert
 stoved finish Einbrennlackierung f
 with a woodgrain finish holzgemasert
finished Fertig...
 finished article Fertigteil n, Fertigprodukt n, Fertigartikel m, Fertigerzeugnis n
 finished products Fertigwaren fpl, Fertigartikel mpl
finishing [e.g. mo(u)lding] Nach(be)arbeitung f
 finishing agent [textile] Appreturmittel n
 finishing costs Nacharbeitungskosten pl
 surface finishing [i.e. the application of special colo(u)r or texture effects to coated products to improve their appearance] Veredeln n
finite element method [FEM] FEM-Verfahren n
fire Feuer n
 fire behavio(u)r Brandverhalten n, Brennverhalten n
 fire detection Brandentdeckung f
 fire fighting Brandbekämpfung f
 fire precautions Brandschutzmaßnahmen fpl
 fire protection Brandschutz m
 fire resistant feuerbeständig
 fire resistance Brandsicherheit f, Feuerwiderstandsfähigkeit f
 fire retardant feuerhemmend
 fire risk Brandgefahr f, Feuerrisiko n
 in case of fire im Brandfall m
firm 1. fest, standfest; 2. Firma f, Haus n [Firma]
firmly fixed festverankert
first zuerst, erst
 first order reaction Reaktion f erster Ordnung

first-time user Erstanwender m
fish box Fischkasten m
fish eye Stippe f, Fischauge n
  low fish eye content Stippenarmut f
  with a low fish eye content stippenarm
fishing rod Angelrute f
fishtail die Fischschwanzdüse f
fishtail manifold Fischschwanzkanal m
fit/to anpassen, montieren
  easy to fit montagefreundlich, montageleicht
fit Passung f
  accurate fit Paßgenauigkeit f, Passergenauigkeit f
  exact fit Paßgenauigkeit f, Passergenauigkeit f
fitting Formstück n, Armatur f, Fitting n
  fitting instructions Einbaurichtlinien fpl
  furniture fittings Möbelausrüstungsteile npl, Möbelbeschläge mpl
  injection mo(u)lded fittings Spritzguß-Fittings npl
  pipe fitting Rohrfitting n, Rohrleitungsteil n
  sanitary fittings Sanitärarmaturen fpl
  tight fitting spielfrei
five fünf
  five-point double toggle Fünfpunkt-Doppelkniehebel m
  five-point gating Fünffachanguß m
  five-point toggle Fünfpunkt-Kniehebel m
  five-roll calender Fünfwalzenkalander m
  five-roll L-type calender Fünfwalzen-L-Kalander m
  five-runner arrangement Fünffachverteilerkanal m
fixed stationär, fix, ortsfest
  fixed amount Festmenge f
  fixed diaphragm Festblende f
  fixed disk Festplatte f
  fixed displacement pump Konstantpumpe f
  fixed lip feste Düsenlippe f, feste Lippe f
  fixed knife Statorscheibe f
  fixed magnetic disk Festplattenspeicher m, Magnetplattenspeicher m
  fixed mo(u)ld half Einspritzseite f, Düsenseite f
  fixed platen düsenseitige Formaufspannplatte f, Düsenplatte f, feststehende Aufspannplatte f, spritzseitige Aufspannplatte f, spritzseitige Werkzeughälfte f, düsenseitige Werkzeughälfte f, feststehende Werkzeughälfte f, Gesenkseite f, Spritzseite f, Werkzeugdüsenplatte f
  fixed point Fixpunkt m
  fixed roll Festwalze f
  fixed stop Festanschlag m
  fixed value Festwert m
  disposal of fixed assets Anlagenabgänge mpl
  firmly fixed festverankert
  on the fixed mo(u)ld half spritzseitig, düsenseitig
fixing Befestigung f
  fixing clamp Befestigungsklemme f
  fixing hole Befestigungsbohrung f
  fixing pin Fixierstift m
  fixing screw Befestigungsschraube f

mo(u)ld fixing details Bohrbild n, Lochbild n, Werkzeugaufspannzeichnung f
mo(u)ld fixing dimensions Formaufspannmaße npl, Aufspannmaße npl
fixture Lehre f, Haltevorrichtung f
  checking fixture Kontrollehre f
  welding fixture Schweißaufnahme f
flakes Schuppen fpl
flaky blättrig
flame Flamme f
  flame polisher Flammpoliergerät n
  flame polishing Flammpolieren n
  flame resistance Flammwidrigkeit f, Schwerbrennbarkeit f, Schwerentflammbarkeit f
  flame resistant flammfest, flammsicher, flammwidrig, schwerentflammbar
  flame retardancy Flammhemmung f
  flame retardant 1. flammhemmend, flammgehemmt, schwerbrennbar; 2. Antiflammittel n, Brandschutzmittel n, Flammhemmer m, Flammschutzkomponente f, Flammschutzmittel n, Flammschutzsystem n, Flammschutz-Additiv n, Flammschutzmittel-Zusatz m
  flame retardant effect Flammschutzeffekt m, Flammschutzwirkung f
  flame retardant properties Flammschutz-eigenschaften fpl, Schwerentflammbarkeit f
  flame retardants flammhemmende Additive npl
  flame spraying Flammspritzen n
  flame spread Brandausbreitung f, Brandweiterleitung f, Feuerweiterleitung f, Flammenausbreitung f
  flame treating unit Beflammstation f
  flame treatment [e.g. of a polyethylene surface to make it suitable for printing] Beflammen n, Abflammen n, Flammstrahlen n
  flame treatment station Beflammstation f
  flame/fire proofing Flammschutz m, Flammfestausrüsten n
  flame/fire resistance Flammfestigkeit f, Nichtbrennbarkeit f
  direction of flame spread Flammen-ausbreitungsrichtung f
  oxyhydrogen flame Knallgasflamme f
  rate of flame spread Flammen-ausbreitungsgeschwindigkeit f
flameproof(ed) feuerbeständig, flammsicher, flammschutzausgerüstet, brandgeschützt, brandschutzausgerüstet
flammability Entflammbarkeit f, Entzündlichkeit f, Brennbarkeit f
  flammability classification Brennbarkeitsklasse f
  flammability test Brandversuch m, Brennbarkeitstest m
flammable entflammbar, feuergefährlich, brennbar
  highly flammable leichtentflammbar, leichtentzündlich
  normally flammable normalentflammbar

flange/to bördeln, flanschen
flange Flansch m, Wulst m, Steg m
   flange connection Flanschanschluß m
   flange coupling Flanschanschluß m
   flange joint Flanschverbindung f
   flange mounting Flanschbefestigung f
   connecting flange Anschlußflansch m
   hinged flange Klappflansch m
   mounting flange Befestigungsflansch m, Flanschbefestigung f
flanged eingebördelt
   flanged connection Flanschanschluß m
flank Flanke f, Stegflanke f
   screw flank Schneckenflanke f
   screw flight flank Schneckenflügelflanke f, Stegflanke f
flap valve Klappenventil n
flash/to blinken
flash 1. Aufblitzen n, 2. Breite f, Grat m, Preßgrat m, Schwimmhaut f, Spritzgrat m, Spritzhaut f, Abquetschgrat m, Abfallbutzen m, Butzen m, Butzenabfall m, Butzenmaterial n, Blas(teil)butzen m
   flash chamber Butzenkammer f, Quetschtasche f
   flash-free gratfrei
   flash ignition temperature Fremdentzündungstemperatur f
   flash mo(u)ld Abquetschwerkzeug n, Überlaufwerkzeug n
   flash-over Überschlag m
   flash-over resistance Überschlagfestigkeit f
   flash point Flammpunkt m
   flash removal Entbutzen n, Butzenbeseitigung f
   flash removal device Abfallentfernung f
   flash trimmer Butzenabtrennvorrichtung f, Butzentrenner m
   flash trimming mechanism Butzenabtrennung f
   flash vaporization Entspannungsverdampfung f, Flashverdampfung f
   base flash Bodenabfall m, Bodenbutzen m
   minimizing the amount of flash produced Butzenminimierung f
   neck and base flash Abquetschlinge mpl
   neck flash Halsabfälle mpl, Halsbutzen m, Halsüberstände mpl
   shoulder flash Schulterbutzen m
   tail flash Bodenabfall m, Bodenbutzen m
   vertical flash face Tauchkante f, Werkzeugtauchkante f
flashing light signal Blinksignal n
flashing light warning signal Blink-Störungslampe f, Störungsblinkanzeige f
flask Kolben m
   Erlenmeyer flask Erlenmeyerkolben m
   flat-bottom flask Stehkolben m
   long-neck flat-bottom flask Langhals-Stehkolben m
   long-neck round-bottom flask Langhals-Rundkolben m
   round-bottom flask Rundkolben m

flat flach, flächenartig
   flat film Flachfolie f
   flat film coextrusion line coextrudierende Flachfolienanlage f, Breitschlitzfolienanlage f
   flat film extruder Flachfolienextruder m, Flachfolien-Extrusionseinheit f, Flachfolienmaschine f
   flat film extrusion Breitschlitzfolienverfahren n, Breitschlitzfolienextrusion f, Flachfolienextrusion f
   flat film (extrusion) die Flach(folien)düse f, Foliendüse f, Folien(spritz)werkzeug n, Flachfolienwerkzeug n, Folienkopf m, Breitschlitzwerkzeug n, Breitschlitzdüse f
   flat film extrusion line Flachfolienanlage f, Breitschlitz(flach)folienanlage f, Folienextrusionsanlage f
   flat film extrusion process Flachfolienverfahren n
   flat film orientation (process) Flachfolienstreckverfahren n
   flat material [e.g. paper, film, fabric etc.] Flächengebilde n
   flat-profile die Flachprofildüse f
   flat roof Flachdach n
   flat-woven fabric Flachgewebe n
   co-extruded flat film Breitschlitzverbundfolie f
   composite flat film Kombinationsflachfolie f
   long-neck flat-bottom flask Langhals-Stehkolben m
   multi-layer flat film Kombinationsflachfolie f
flattened screw tip Schneckenballen m
flaw Fehlstelle f, Fehler m
flex/to falzen, durchbiegen
   mandrel flex test Dornbiegeversuch m
flexibility Flexibilität f, Weichheitsgrad m, Biegsamkeit f, Elastizität f
   initial flexibility Anfangsflexibilität f
   low temperature flexibility Tieftemperaturflexibilität f
   mandrel flexibility Dornbiegefestigkeit f
   permanent flexibility Dauerelastizität f
flexibilization Flexibilisierung f
flexibilizer [used with epoxy resin] Flexibilisator m
flexibilizing effect Flexibilisierungseffekt m
flexible flexibel, weich, elastisch, biegsam
   flexible composite Weichverbund m
   flexible film Weichfolie f
   flexible foam Weichschaum(stoff) m
   flexible PVC film/sheeting PVC-Weichfolie f, Weich-PVC-Folie f
   flexible PVC product/article Weich-PVC-Artikel m
   moderately flexible mittelflexibel
   permanently flexible dauerelastisch, dauerplastisch
flexographic printing Flexodruck m
flexographic printing ink Flexodruckfarbe f
flexural biegbar
   flexural behavio(u)r Biegeverhalten n
   flexural creep modulus Biege-Kriechmodul m
   flexural creep strength Zeitstandbiegefestigkeit f
   flexural creep test Biegekriechversuch m, Zeitstandbiegeversuch m

flexural fatigue strength Biegeschwingfestigkeit f, (Zeit-)Biegewechselfestigkeit f, Biegewechselbelastung f, Biegewechselbeanspruchung f, Biegeschwingversuch m, Wechselbiegeversuch m
flexural impact behavio(u)r Schlagbiegeverhalten n
flexural impact strength Schlagbiegezähigkeit f, Schlagbiegefestigkeit f
flexural impact stress Schlagbiegebeanspruchung f
flexural impact test Schlagbiegeversuch m
flexural impact test piece Schlagbiegestab m
flexural load-carrying capacity Biegebelastbarkeit f
flexural modulus Biegemodul m
flexural rigidity Biegesteifigkeit f
flexural specimen Biegestab m
flexural strength Biegefestigkeit f
flexural stress Biegebelastung f, Biegebeanspruchung f
flexural test Biegeversuch m
flexural test piece Biegestab m
long-term flexural strength Dauerbiegefestigkeit f
residual flexural strength Restbiegefestigkeit f
retained flexural strength Restbiegefestigkeit f
flight 1. Flug m; 2. Flügel m, Schneckenflügel m, Schneckensteg m, Treppenflügel m
flight clearance Kämmspalt m, Kopfspiel n, Grundspiel n, Schneckengrundspalte f
flight contours Stegkontur f
flight depth Gangtiefe f, Schneckengangtiefe f
flight depth ratio Gangtiefenverhältnis n
flight land Kamm m, Schneckenkamm m, Schneckenstegfläche f, Stegkopf m, Stegoberfläche f
flight land clearance Walzenspalt m, Schneckenspalt m, Flankenabstand m, Flankenspiel n
flight land edge Kammkante f
barrier flight Sperrsteg m
feed section flight depth Einzugsgangtiefe f
interrupted flights Gangdurchbrüche mpl
leading edge of flight aktive Flanke f, treibende Flanke f, vordere Flanke f, aktive Gewindeflanke f, aktive Schneckenflanke f, Schubflanke f
mixing flight [of a screw] Knetgang m, Knetsteg m
number of flights Gangzahl f, Schneckengangzahl f
rear face of flight passive Flanke f, hintere Flanke f, passive Gewindeflanke f
screw flight Schneckenelement n, Schneckenwendel m
screw flight contents Schneckengangfüllung f
screw flight flank Schneckenflügelflanke f
screw flights and kneader disks Schnecken- und Knetelemente npl
thrust face of flight aktive Flanke f, treibende Flanke f, treibende Stegflanke f, vordere Schneckenflanke f, aktive Gewindeflanke f
trailing edge of flight passive Gewindeflanke f, passive Schneckenflanke f, hintere Flanke f, passive Flanke f
with hardened flights [screw] steggepanzert
flighted screw tip Schneckenförderspitze f, Förderspitze f

float 1. Schwimmer m; 2. Glättscheibe f
carburettor float Vergaserschwimmer m
petrol/fuel tank float Tankschwimmer m
floating 1. Ausschwimmen n; 2. fliegend angeordnet, Fließ-
floating bung Schleppstopfen m
floating decimal point Fließkomma n
floating plug Schleppstopfen m
floating roller Pendelwalze f
floating saw Pendelsäge f
floating screed schwimmender Estrich m
flocculation Flockulation f
flock Flocke f
flock spraying Beflocken n
cotton flock Baumwollflocken fpl
flocking Beflocken n
floor Boden m
floor area Bodenfläche f
floor screed Fußbodenausgleichmasse f, Estrich m
floor space Bodenfläche f, Aufstellfläche f
floor space requirement(s) Stellflächenbedarf m, Grundflächenbedarf m
at floor level bodeneben
floorcovering Fußbodenbelag m, Bodenbelag m
flooring Fußbodenbelag m
jointless flooring compound Bodenausgleichmasse f, Bodenspachtelmasse f, Fußbodenausgleichmasse f
floppy disk drive Floppy-Disk(etten)-Laufwerk n
twin floppy disk drive Doppel-Floppy-Disk(etten)-Laufwerk n
flour Mehl n
silica flour Quarzmehl n
synthetic silica flour Quarzgutmehl n
flow Strömung f, Fluß m, Fließen n, Durchflußstrom m, Fließfähigkeit f, Fließprozeß m, Verlauf m
flow anomalies Fließanomalien fpl
flow channel Fließkanal m, Flußkanal m, Fließquerschnitt m, Strömungskanal m
flow channel design Fließkanalgestaltung f
flow characteristics Fließverhalten n, Fließeigenschaften fpl, Verlaufseigenschaften fpl
flow coating Fluten n, Flow-Coating n
flow conditions Fließbedingungen fpl, Strömungsverhältnisse npl, Strömungsbedingungen fpl
flow control Durchsatzregulierung f, Flußregulierung f
flow control agent Verlaufmittel n
flow control device Durchsatzregulierung f
flow control valve Durchflußsteuerventil n, Strom(regel)ventil n
flow controller Verlaufmittel n, Durchsatzregler m
flow correction Fließkorrektur f, Flußkorrektur f
flow cup Auslaufbecher m
flow curve Fließkurve f
flow-dependent strömungsabhängig
flow diagram Ablaufdiagramm n, Fließbild n, Fließschema n, Flußdiagramm n, Strömungsprofil n
flow direction Fließrichtung f

flow distance Fließweg(länge) m(f)
flow front Fließfront f
flow index Fließexponent m
flow indicator Durchlaufanzeige f
flow law index Fließgesetzexponent m
flow length Fließweg(länge) m(f)
flow length-wall thickness ratio Fließweg-Wanddickenverhältnis n
flow line Bindenaht f, Strömunglinie f, Fließlinie f
flow marks Fließmarkierungen fpl
flow monitor Strömungswächter m
flow mo(u)lding (process) Fließgießverfahren n, Fließguß m
flow path Strömungsweg m, Fließweg m
flow pattern Fließfiguren fpl
flow problems Verlaufschwierigkeiten fpl, Verlaufstörungen fpl
flow processes Strömungsvorgänge mpl
flow profile Strömungsprofil n
flow promoting verlaufsfördernd
flow properties Verlaufeigenschaften fpl, Fließverhalten n, Fließeigenschaften fpl
flow rate Durchflußgeschwindigkeit f, Durchflußleistung f, Durchflußrate f, Durchflußstrom m, Fördermenge f, Strömungsgeschwindigkeit f, Fließgeschwindigkeit f, Durchlaufmenge f, Durchfluß m, Durchflußmenge f
flow rate transducer Strömungsgeschwindigkeitsaufnehmer m
flow rating [of pump] Schluckvolumen n
flow resistance Durchflußwiderstand m, Fließwiderstand m, Strömungswiderstand m
flow resistance changes Fließwiderstandsänderungen fpl
flow restriction bush Staubüchse f
flow restriction device Drossel f, Drosselvorrichtung f, Drosselorgan n, Drosselkörper m
flow restriction effect Drosselwirkung f
flow restrictor Drossel f, Drosselvorrichtung f, Drosselorgan n, Drosselkörper m
flow restrictor gap Drosselspalt m
flow restrictor grid Drosselgitter n
flow sensing element Durchflußfühler m
flow sensor Durchflußfühler m
flow time [in viscosity determinations] Durchflußzeit f, Auslaufzeit f
flow variations Durchflußschwankungen fpl
flow velocity [liquids in pipes] Fließgeschwindigkeit f
air flow conditions Luftströmungsverhältnisse npl
axial flow Axialströmung f
caused by flow strömungsbedingt
changes in flow resistance Fließwiderstandsänderungen fpl
cold flow kalter Fluß m
continuous flow cooling system Durchflußkühlung f
cooling water flow rate Kühlwasserdurchflußmenge f
data flow Datenfluß m

direction of flow Fließrichtung f, Strömungsrichtung f
drag flow Schleppfluß m, Schleppströmung f
due to flow strömungsbedingt
ease of flow Fließfähigkeit f
forced flow principle Zwangslaufprinzip n
frontal flow Quellfließen n, Quellfluß m, Quellströmung f
heat flow Wärmefluß m
hydraulic oil flow Hydrauliköstrom m
laminar flow laminare Strömung f, laminarer Fluß m, Schichtenströmung f, laminares Fließen n
laminar flow distributor Laminarflußverteiler m
leakage flow Verlustströmung f, Leckströmung f
longitudinal flow Längsströmung f
material flow Massefluß m, Produktstrom m
measurement of flow rate Durchflußmessung f
melt flow Massefluß m, Schmelzefluß m
melt flow index [MFI] Fließindex m, Schmelzindex m
melt flow path Strömungsweg m, Fließweg m
melt flow rate Schmelzestrom m, Massedurchsatz m
melt flow-way Schmelzeleitung f, Massekanal m, Schmelze(führungs)bohrung f, Schmelze(führungs)kanal m, Strömungsweg m, Strömungskanal m, Fließweg m
melt flow-way system Schmelzeleitsystem n
narrowing flow channels Querschnittsverengungen fpl
Newtonian flow Newtonsches Fließen n
oil flow regulator Ölstromregler m
particle flow Teilchenfluß m
plug flow Pfropfenströmung f
power law flow [polymer melt] Potenzgesetzverhalten n
pressure flow Druckfluß m, Druckströmung f, Stauströmung f
pressure flow-drag flow ratio Drosselquotient m, Drosselkennzahl f
rectangular flow channel Rechteckkanal m
restricted flow zone Dammzone f, Drosselstelle f, Drosselfeld n, Stauzone f
reverse flow Rückfluß m, Rückströmung f
reverse melt flow Schmelzerückfluß m
round-section flow channel Rundlochkanal m
signal flow chart Signalflußplan m
solid flow Blockströmung f
spiral flow Wendelströmung f
spiral flow length Spirallänge f
spiral flow test Spiraltest m
stretching flow Dehnströmung f
tangential flow Tangentialströmung f
three-way flow control Dreiwegestromregelung f
three-way flow control unit Dreiwegestromregler m
three-way flow control valve Dreiwege-Mengenregelventil n
transverse flow Querströmung f, Schichtenströmung f, Transversalströmung f
turbulent flow turbulente Strömung f
two-way flow control unit Zweiwegestromregler m

viscous flow viskoses Fließen n
volume(tric) flow rate Volumendurchsatz m, Volumenstrom m, volumetrischer Durchsatz m
water flow-way Wasserfließweg m, Wasserführungsrohr n
with poor flow [e.g. a mo(u)lding compound] hartfließend, schwerfließend
flowability Fließfähigkeit f, Fließvermögen n
flowmeter Durchflußmeßgerät n, Durchflußmengenmesser m, Strömungsmesser m
   ultrasonic flowmeter Ultraschalldurchflußmeßgerät n
fluctuation Schwankung f
   holding pressure fluctuations Nachdruckpulsationen fpl
   output fluctuations Ausstoßschwankungen fpl
   pressure fluctuations Druckpulsationen fpl
   temperature fluctuations Temperaturschwankungen fpl
   viscosity fluctuations Viskositäts-Inhomogenitäten fpl
   voltage fluctuations Spannungsschwankungen fpl
fluid 1. flüssig; 2. Flüssigkeit f
   fluid center flüssige Seele f, plastische Seele f
   fluid heating system Flüssigkeitsheizung f
   fluid mixer Fluidmischer m
   brake fluid Bremsflüssigkeit f
   clutch fluid Kupplungflüssigkeit f
   fluorosilicone fluid Fluorsilikonöl n
   heat exchanging fluid wärmeaustauschende Flüssigkeit f
   hydraulic fluid Druckflüssigkeit f, Druckmedium n
   phenylmethyl silicone fluid Phenylmethylsiliconöl n
   power law fluid Potenzgesetz-Flüssigkeit f, Potenzgesetzstoff m
   silicone fluid Siliconöl n
fluidize/to fluidisieren, wirbeln
fluidity Fluidität f
   fluidized bed Wirbelbett n
   fluidized bed coater Wirbelsintergerät n
   fluidized bed coating Wirbelsintern n
   fluidized pouring Wirbelschütten n
   electrostatic fluidized bed coating elektrostatisches Wirbelsintern n
fluorescence Fluoreszenz f
fluorescent lamp Leuchtstofflampe f
fluorescent pigment Fluoreszenz-Pigment n
fluoride Fluorid n
   polyvinyl fluoride Polyvinylfluorid n
   polyvinylidene fluoride [PVDF] Polyvinylidenfluorid n
   vinylidene fluoride Vinylidenfluorid n
   vinylidene fluoride-tetrafluoroethylene copolymer Vinylidenfluorid-Tetrafluoroethylen-Copolymerisat n
fluorinated fluoriert
   fluorinated ethylene-propylene copolymer [FEP] Fluorethylenpropylen n
   fluorinated hydrocarbon Fluorkohlenwasserstoff m
fluorocarbon plastic/polymer Fluorkunststoff m, Fluorkunstharz n, Fluorpolymer n
fluorocarbon rubber Fluorcarbonkautschuk m
fluoroplastic Fluorkunstharz n, Fluorkunststoff m
fluoropolymer Fluorpolymer n, Fluorkunstharz n, Fluorkunststoff m, Fluorpolymer n
fluororubber Fluorelastomer m, Fluorkautschuk m
fluorosilicone Fluorsilikon n
   fluorosilicone fluid Fluorsilikonöl n
   fluorosilicone rubber Fluorsilikonkautschuk m
flush bündig
   flush with the surface oberflächenbündig
flush/to spülen
   back flushing Rückspülung f
   melt back flushing Schmelzenrückspülung f
fluted filter Faltenfilter n,m
flying fliegend, Flieg-
   flying knife Ablängvorrichtung f
   protection against flying stones Steinschlagschutzwirkung f
   resistance to flying stones Steinschlagfestigkeit f
foam/to aufschäumen
foam Schaum(stoff) m
   foam blow mo(u)lding (process) Schaumblasverfahren n
   foam coating Schaumstrich m
   foam core Schaumstoffkern m
   foam density Schaumdichte f
   foam mo(u)lding Form(teil)schäumen n, Formverschäumung f
   foam mo(u)lding machine Schäummaschine f
   foam mo(u)lding plant Schäumanlage f
   chemically blown foam chemischer Schaum m, Treibmittelschaumstoff m
   cold-cure mo(u)lded foam Kaltformschaumstoff m
   flexible foam Weichschaum(stoff) m
   hot-cured mo(u)lded foam Heißformschaumstoff m
   in-situ foam Ortsschaum m
   integral foam Strukturschaum(stoff) m
   mechanically blown foam mechanischer Schaum m, Schlagschaum m
   mo(u)lded foam Formschaumstoff m
   multi-component structural foam mo(u)lding TSG-Mehrkomponentenverfahren n
   multi-component structural foam mo(u)lding machine TSG-Mehrkomponenten-Rundläufermaschine f
   one-component structural foam mo(u)lding (process) Einkomponenten-Schaumspritzgießverfahren n
   paste for making mechanically blown foam Schlagschaumpaste f
   phenolic foam Phenolharzschaumstoff m, Phenolschaum m
   plastics foam Schaumkunststoff m
   polyethylene foam Polyethylenschaumstoff m, Zellpolyethylen n
   polyurethane integral/structural foam Polyurethan-

Integralschaum m, Polyurethan-Struktur-
schaum(stoff) m
PVC foam Schaum-PVC n
reaction foam mo(u)lding (process) Reaktions-
Schaumgießverfahren n
rigid foam Hartschaum(stoff) m
rigid foam core Hartschaumkern m
rigid foam insulating material Hartschaumisolierstoff m
rigid polyurethane foam Hart-Urethanschaum m
slabstock foam Blockschaumstoff m, Blockware f,
Schaumstoffblock m, Schaumstoffblockmaterial n
structural foam Strukturschaum(stoff) m
structural foam mo(u)ld TSG-Werkzeug n
[TSG = Thermoplast-Schaum-Guß]
structural foam mo(u)lding TSG-Verarbeitung f,
TSG-Verfahren n, Schaumspritzgießen n
structural foam mo(u)lding machine Strukturschaummaschine f, Thermoplast-Schaumgießmaschine f, TSG-Spritzgießmaschine f,
TSG-Maschine f
urea-formaldehyde foam Harnstoffharzschaum m
fusion Fusion f
fusion [of PVC paste] Fertiggelieren n
socket fusion welding Muffenschweißung f
gage s. gauge
galvanized verzinkt, galvanisiert
hot galvanized feuerverzinkt
gamma irradiation Gammabestrahlung f
gamma rays Gammastrahlen mpl
gap Spalt m
gap-filling fugenfüllend
air gap Luftspalte f
die gap Austrittsspalt m, Austrittsöffnung f,
Düsenbohrung f, Düsenl(ippen)spalt m,
Düsenaustrittsspalt m, Düsen(austritts)öffnung f,
Mundstück(ring)spalt m, Düsenmund m
die gap adjusting device Düsenspaltverstellung f
die gap adjustment Düsenspaltverstellung f
die gap thickness Düsenspaltbreite f,
Düsenspaltweite f, Spaltbreite f, Austrittsspaltweite f
die gap width Düsenspaltbreite f, Düsenspaltweite f,
Spaltbreite f, Austrittsspaltweite f
flow restrictor gap Drosselspalt m
mixing gap [of a mixing screw] Knetspalt m
garden chair Gartenstuhl m
garden hose Gartenschlauch m
Gardner colo(u)r standard number Gardner Farbzahl f
gas Gas n
gas chromatogram Gaschromatogramm n
gas chromatographic gaschromatographisch
gas chromatography Gaschromatographie f
gas constant Gaskonstante f
gas content [e.g. in foam production] Gasbeladung f
gas evolution Gasabspaltung f
gas heated gasbeheizt
gas impermeability Gasdichtheit f,

Gasundurchlässigkeit f
gas nitriding Gasnitrierung f
gas permeability Gasdurchlässigkeit f
gas pipe Gasrohr n
gas pressure Gasdruck m
gas-producing gasabspaltend
gas supply Gasversorgung f
gas-tight gasdicht
gas yield Gasausbeute f
automatic hot gas welding unit Warmgas-Schweißautomat m
blowing gas Treibgas n
carrier gas Trägergas n
compressed gas cylinder Druckgasbehälter m
inert gas Edelgas n
internal gas pressure Gasinnendruck m
waste gas purification plant Abgasreinigungsanlage f
welding gas supply Schweißgasversorgung f
gaseous gasförmig
gaseous phase Gasphase f
gasket Dichtring m
gasoline Benzin n
gate/to (into) anbinden, anspritzen, anschneiden
gate 1. Tor n; 2. Anschnittbereich m, Einspritzpunkt m,
Füllquerschnitt m, Anbindung f, Angußbohrung f,
Angußloch n, Angußöffnung f, Anschnittbohrung f,
Anschnittöffnung f
gate area Anschnittpartie f
gate cross-section Anschnittquerschnitt m
gate design Angußgestaltung f, Angußkonstruktion f
gate land Angußsteg m
gate location Angußlage f
gate mark Angußmarkierung f, Anschnittmarkierung f
gate opening time Versiegelungszeit f
gate sealing Angußversiegeln n
gate sealing point Siegelzeitpunkt m
gate size Angußgröße f, Anschnittgröße f
antechamber-type pin gate Punktanguß m mit
Vorkammer, Vorkammerpunktanguß m,
Stangenpunktanguß m
away from the gate [if referring to the mo(u)ld]
angußfern, in Angußferne f
central film gate mittiger Bandanschnitt m
diaphragm gate Pilzanguß m, Scheibenanguß m,
Schirmanguß m, Schirmanschnitt m, Telleranguß m
edge gate seitlicher Anschnitt m
factory gate Werkstor n
film gate Bandanguß m, Bandanschnitt m,
Delta-Anguß m, Filmanguß m, Filmanschnitt m
four-point pin gate Vierfach-Punktanschnitt m
lateral film gate seitlicher Bandanschnitt m
multi-point pin gate Reihenpunktanschnitt m
multiple pin gate Reihenpunktanschnitt m
near the gate (area) im Anschnittbereich m, in
Angußnähe f, angußnah
overlap gate Überlappungsanschnitt m
pin gate Punktanguß m, Punktanschnitt(kanal) m,

punktförmiger Anschnittkanal m
pin gate nozzle Punktangußdüse f, Punktanschnittdüse f
pin gate sprue Punktangußkegel m
positioning of the gate Angußpositionierung f
ring gate Ringanschnitt m, ringförmiger Bandanschnitt m
side gate seitlicher Anschnitt m
single pin gate Einzelpunktanschnitt m
sprue gate Stangenanguß m, Stangenanschnitt m
sprue puller gate Abreißanschnitt m
sprue puller pin gate Abreißpunktanschnitt m
submarine gate Tunnelanguß m, Tunnelanschnitt m
tunnel gate Tunnelanguß m, Tunnelanschnitt m, Abscheranschnitt m
tunnel gate with pin point feed Tunnelanguß m mit Punktanschnitt
type of gate Angußart f
gated angespritzt, angegossen
  film gated bandangeschnitten
  valve gated mo(u)ld Nadelventilwerkzeug n
gathering [of data] Erfassung f [von Daten]
  data gathering Datenerfassung f
  process data gathering unit Prozeßdatenerfassung(seinheit) f
gating (technique/method) Angußtechnik f, Einspritzung f
  gating at the (mo(u)ld) parting line Trennlageneinspritzung f
  gating surface Anschnittebene f
  gating system Angußsystem n
  ante-chamber gating Vorkammer-Angießtechnik f
  central gating Zentralanguß m, axiale Anspritzung f
  direct gating (angußlose) Direktanspritzung f
  five-point gating Fünffachanguß m
  hot runner multiple gating Heißkanal-Mehrfachanguß m
  multi-point gating mehrfaches Anbinden n
  side gating seitliches Anspritzen n
  two-point gating Zweifachanspritzung f
  valve gating system Nadelventilangußsystem n
gauge [also: gage] 1. Meßgerät n; 2. Stärke f, Dicke f
  ga(u)ge tolerance [of film] Dickentoleranz f
  ga(u)ge variations [of film] Dickenschwankungen fpl, Dickenabweichungen fpl
  film ga(u)ge equalizing unit Folienverlegegerät n, Folienverlegeeinheit f, Folienverlegung f, Verlegeeinheit f
  film ga(u)ge measuring instrument Foliendickenmeßgerät n
  film ga(u)ge variations Foliendickenunterschiede mpl, Foliendickenabweichungen fpl
  fuel ga(u)ge Tankgeber m
  hydraulic system pressure ga(u)ge Hydrauliköl-Druckanzeige f
  isotope thickness ga(u)ge Isotopendickenmeßgerät n
  petrol ga(u)ge Tankgeber m

pressure ga(u)ge Druckanzeigegerät n, Druckmeßeinrichtung f, Druckmeßgeber m, Druckmeßgerät n, Manometer n
strain ga(u)ge Dehn(ungs)meßstreifen m
thickness ga(u)ge Dickenmeßeinrichtung f, Dickenmeßgerät n
wall thickness ga(u)ge Wanddickenmeßgerät n, Wanddickenmessung f
gauze Gaze f
  wire gauze Drahtgewebe n
  wire gauze screen/filter Drahtgewebefilter n,m
gear/to ineinandergreifen [Zahnräder], mit Zahnrädern fpl verbinden
  geared motor Getriebemotor m
  shunt-wound geared DC motor Nebenschluß-Gleichstrom-Getriebemotor m
  three-phase geared motor Drehstromgetriebemotor m
gear 1. Getriebe n, Zahnrad n, Gang m; 2. Gerät n
  gear cutting Verzahnung f
  gear drive Zahnradgetriebe n
  gear housing Getriebegehäuse n
  gear motor Zahnradmotor m
  gear pump Zahnradpumpe f
  gear reduction (unit) Getriebeuntersetzung f, Zahnraduntersetzung f
  gear tooth system Verzahnung f
  gear train Getriebetransmission f
  gear transmission Getriebetransmission f
  gear wheel Zahnrad n
  gear wheel blank Zahnradrohling m
  lifting gear Hebezeug n
  nip adjusting gear Walzenanstellung f
  planetary gear Planetengetriebe n
  planetary gear extruder Walzenextruder m
  reducing gear Reduziergetriebe n, Zahnraduntersetzungsgetriebe n
  reduction gear Reduziergetriebe n, Zahnraduntersetzungsgetriebe n
  speed reduction gear Untersetzungsgetriebe n
  spur gear Stirn(zahn)rad n
  spur gear speed reduction mechanism Stirnrad-Untersetzungsgetriebe n
  toggle gear Kniehebelgetriebe n
  worm gear Schneckengetriebe n, Schneckenvorgelege n
gearing Übersetzung f [Zahnrad], Verzahnung f
  helical gearing Schrägverzahnung f
gel Gel n
  gel chromatogram Gel-Chromatogramm n
  gel chromatography Gel-Chromatographie f
  gel coat Feinschicht f, Gelcoatschicht f, Gelcoatierung f, Harzfeinschicht f, Feinharzschicht f, Deckschicht f [GFK], Oberflächenschicht f [GFK], Schutzschicht f [GFK]
  gel coat resin Deckschichtharz n [GFK]
  gel coat surface Deckschichtoberfläche f [GFK]
  gel content Gel-Gehalt n, Gelanteil m

gel-like gelartig, gelförmig
gel particle Gelpartikel m, Gelteilchen n
gel permeation chromatography Gel-Permeationschromatographie f
gel point Geliertemperatur f, Gelierpunkt m
gel time Gelierzeit f, Gelierdauer f
gel time tester Gelzeitprüfgerät n
automatic gel time tester Gelierzeitautomat m
readiness to gel Gelierfreudigkeit f
silica gel Silikagel n
tendency to gel Gelierneigung f
with a low gel point niedriggelierend
gelation Gelieren n, Gelbildung f, Vergelung f, Ausgelierung f, Solvatation f
gelation resistance Gelierungsbeständigkeit f
gelation speed Geliergeschwindigkeit f
gelation temperature [PVC paste] Vorgeliertemperatur f [PVC]
degree of gelation Gelier(ungs)grad m
partial gelation [PVC paste] Angelieren n, Vorgelierung f
pre-gelation temperature [UP, EP resin] Vorgeliertemperatur f [UP-, EP-Harz]
pressure gelation (process) Druckgelierverfahren n
gelcoat Oberflächenharzschicht f
gelcoat resin Oberflächenharz n
gelled ausgeliert
partly gelled angeliert, vorgeliert
gelling solvatisierend, gelierend
gelling aid Gelierhilfe f
gelling behavio(u)r Gelierverhalten n
gelling effect Gelierwirkung f
gelling oven Gelierofen m
gelling power Geliervermögen n
gelling tunnel Gelierkanal m
fast gelling raschplastifizierend
signs of gelling Gelierungserscheinungen fpl
general allgemein, gewöhnlich, Haupt-
general meeting [of shareholders etc.] Hauptversammlung f
general performance Gesamtverhalten n
general-properties Allgemeineigenschaften fpl
general-purpose [machine] universell einsetzbar, Mehrzweck-, Allzweck-, Universal-
general-purpose catalyst Allzweckvernetzungsmittel n
general-purpose extruder Universalextruder m, Allzweckextruder m
general-purpose extruder head Universalkopf m
general-purpose granulator Universal-Zerkleinerer m
general-purpose machine Universalmaschine f
general-purpose mixer Universalmischer m
general-purpose model [of a machine] Universaltyp m
general-purpose rubber Allzweckkautschuk m
general-purpose screw Allzweckschnecke f, Standardschnecke f, Universalschnecke f

general reaction Reaktionsschema n
general survey Übersicht f
general table Übersichtstabelle f
generated-rotor pump Wälzkolbenpumpe f
generation of steam Dampferzeugung f
generator Generator m
clock signal generator Taktgenerator m
data generator Datengeber m
function generator Funktionsgenerator m
signal generator Signalgeber m
ultrasonic generator Ultraschall-Generator m
generously dimensioned großzügig dimensioniert
gentle sanft, mild
gentle mo(u)ld clamping mechanism sanfter Werkzeugschluß m
gentle treatment Schonung f
geometry Geometrie f
die geometry Werkzeuggeometrie f
gerotor pump Wälzkolbenpumpe f
giant molecule Riesenmolekül n
given vorgegeben
given setpoints vorgegebene Sollwerte mpl
glass Glas n
glass beads Glaskugeln fpl
glass cloth Glas(filament)gewebe n, Glasseidengewebe n
glass cloth covering Glasgewebeabdeckung f
glass cloth laminate Glashartgewebe n, Glasgewebelaminat n
glass cloth laminate pipe Glashartgeweberohr n
glass cloth laminate sheet Glashartgewebeplatte f
glass cloth reinforcement Glasgewebeverstärkung f
glass cloth tape Glasgewebeband n
glass content Glasanteil m, Glasgehalt m
glass fabric Glasfilamentgewebe n, Textilglasgewebe n
glass fiber Glasfaser f, Textilglasfaser f, Glasseide f
glass fiber content Glasfasergehalt m, Glasfaserkonzentration f, Glasfaseranteil m
glass fiber distribution Glasfaserverteilung f
glass fiber fabric Glasfilamentgewebe n, Textilglasgewebe n
glass fiber filled/reinforced glasfasergefüllt, glasfaserhaltig
glass fiber laminate Glaslaminat n
glass fiber mat Glasfasermatte f, Textilglasvlies n, Textilglasmatte f
glass fiber material Textilglas n
glass fiber materials/products Textilglasprodukte npl, Textilglaserzeugnisse npl
glass fiber orientation Glasfaserorientierung f
glass fiber reinforced glasfaserverstärkt, textilglasverstärkt
glass fiber reinforced plastic Glasfaserkunststoff m, GFK
glass fiber reinforcement Glasfaserverstärkung f, Glasseidenverstärkung f, Textilglasverstärkung f

glass filament Textilglasspinnfaden m
glass mat Glasmatte f, Glaseidenmatte f, Glasvlies n
glass mat reinforced glasmattenverstärkt
glass mat reinforced polyester laminate GF-UP-Mattenlaminat n
glass paper Glaspapier n
glass reinforcement Glasverstärkung f
glass spheres Glaskugeln fpl
glass stopper Glasstopfen m
glass transition [i.e. the transition of a polymer from the glassy to the viscoelastic state] Glasübergang m, Glasumwandlung f
glass transition range Einfrierbereich m
glass transition temperature Tg-Wert m, Glas(umwandlungs)punkt m, Glas(umwandlungs)temperatur f, Einfrier(ungs)temperatur f, Glasübergangstemperatur f
glass transition temperature range Glasumwandlungstemperaturbereich m
glass transition zone Glasumwandlungsbereich m, Glasübergangsbereich m, Glasübergangsgebiet n
addition of glass fibers Glasfaserzusatz m
bonded glass joint Glas(ver)klebung f
bonding of glass Glas(ver)klebung f
chopped glass fiber Glasseiden-Kurzfaser f
continous glass fiber Langglasfaser f
epoxy glass cloth laminate Epoxidharz-Glashartgewebe n
high alkali glass A-Glas n
incorporation of glass fibers Glasfaserzusatz m
low-alkali glass E-Glas n
milled glass fibers Mahlglasfasern fpl
randomly distributed glass fibers wirre Glasfaserverteilung f
reinforced with continous glass strands langglasverstärkt
safety glass Sicherheitsglas n, Verbundsicherheitsglas n
sight glass Einblickfenster n, Schauglas n, Sichtfenster n, Sichtglas n, Sichtscheibe f
staple glass fiber Glasstapelfaser f
varnished glass cloth Lackglasgewebe n
woven glass tape Glasgewebeband n
glassy glasartig, spröd-hart
glassy state Glaszustand m
isolated glassy zones Glasinseln fpl
glaze Glasur f
glazed glasiert
glazing Verglasung f, Glasierung f, Glasieren n
glazing bead Glasleiste f
glazing material Verglasungsmaterial n
acrylic glazing sheet/material Acrylglas n
double glazing Doppelverglasung f
gloss Glanz m
gloss retention Glanzbeständigkeit f, Glanzerhalt m, Glanzhaltung f, Glanzhaltevermögen n
reduction in gloss Glanz(grad)verlust m, Glanzminderung f
surface gloss Oberflächenglanz m
gloves Handschuhe mpl
glove compartment Handschuhfach n, Handschuhkasten m
disposable gloves Wegwerfhandschuhe mpl
protective gloves Schutzhandschuhe mpl
rubber gloves Gummihandschuhe mpl
glow/to glimmen
glow discharge Glimmentladung f
glue Leim m
glue applicator Leimauftragungsgerät n
glue applicator roll Beleimungswalze f
assembly glue Montageleim m
glueline Klebespalte f, Klebefuge f
glutaric acid Glutarsäure f
glutaric anhydride Glutarsäureanhydrid n
glycerin mono-oleate Glycerinmonooleat n
glyceryl phthalate Glycerylphthalat n
glycidyl... Glycidyl-
glycidyl compound Glycidylverbindung f
glycidyl ester Glycidylester m
glycidyl ether Glycidylether m
glycidyl group Glycidylgruppe f
glycidyl radical Glycidylrest m
cresyl glycidyl ether Kresylglycidylether m
ethylhexyl glycidyl ether Ethylhexylglycidylether m
glycol Glykol n
glycol diacetate Glykoldiacetat n
glycol ether Glykolether m
glycol ether acetate Glykoletheracetat n
glycol group Glykolgruppe f
butyl glycol Butylglykol n
diethylene glycol Diethylenglykol n
dieythylene glycol adipate Diethylenglykoladipat n
dimethyl glycol phthalate Dimethylglykolphthalat n
dipropylene glycol Dipropylenglykol n
ethylene glycol Ethylenglykol n
ethylene glycol butyl ether Ethylenglycolmonobutylether m
ethylene glycol dimethacrylate Ethylenglykoldimethacrylat n
ethylene glycol ether Ethylenglycolether m
ethylene glycol ethyl ether Ethylenglykolmonoethylether m
ethylene glycol polyadipate Ethylenglycolpolyadipat n
ethyl glycol Ethylglykol n
ethyl glycol acetate Ethylglykolacetat n
neopentyl glycol Neopentylglykol n
polyether glycol Polyetherglykol n
polyethylene glycol Polyethylenglykol n
polypropylene glycol Polypropylenglykol n
polytetramethylene glycol Polytetramethylenglykol n
propylene glycol Propylenglykol n
propylene glycol ether Propylenglycolether m
propylene glycol maleate Propylenglycolmaleat n

propylene glycol methyl ether Propylenglykolmonomethylether m
glyoxime Glyoxim n
   dimethyl glyoxime Dimethylglyoxim n
godet roll stretch unit Galettenstreckwerk n
godets Galetten fpl
goggles Schutzbrille f
goods Waren fpl, Artikel mpl, Gegenstände mpl
   bakery goods Backwaren fpl
   consumer goods Bedarfsgegenstände mpl, Gebrauchsgüter npl
   consumer goods for food contact applications Lebensmittelbedarfsgegenstände mpl
   consumer goods regulations Bedarfsgegenständeverordnung f
   household goods Haushaltsartikel mpl, Haushaltswaren fpl, Konsumartikel mpl, Konsumgüter npl, Verbrauchsgüter npl
   incoming goods control (department/section) Wareneingangskontrolle f
   incoming goods quality/properties Wareneingangsparameter mpl
   incoming goods section Wareneingangsbereich m
grade/to [powders] klassieren, klassifizieren
grade Grad m, Rang m, Qualität f
   grading unit Klassiereinheit f
   blow mo(u)lding grade Blasmarke f
   electrical grade presspahn Elektropreßspan m
   extrusion grade [of compound] Extrusionstyp m, Extrusionsqualität f, Extrusionsmarke f
   general-purpose grade Universaltyp m
   injection mo(u)lding grade [of compound] Spritzgießmarke f
   monofilament grade [of compound] Monofilament-Typ m
   paste-making grade Pastentyp m [PVC]
   range of grades Typenpalette f, Typenprogramm n, Typensortiment n, Typenübersicht f
   special-purpose grade Sonderqualität f
   standard grade Standardqualität f, Standardtype f, Grundtyp m
gradient Gradient m, Gefälle n
   concentration gradient Konzentrationsgefälle n
   pressure gradient Druckgefälle n, Druckgradient m
   temperature gradient Temperaturgradient m, Temperaturgefälle n
   velocity gradient Geschwindigkeitsgefälle n
graft/to (auf)pfropfen, aufpolymerisieren, anpolymerisieren
graft Transplantat n, Pfropf-
   graft copolymer Pfropfencopolymer n
   graft copolymerization Pfropfcopolymerisation f
   graft polymer Pfropfpolymer(isat) n
   graft polymerized pfopfpolymerisiert
   suspension graft copolymer Suspensionspfropfcopolymerisat n
grafted (auf)gepfropft

grafting Pfropfung f
   degree of grafting Pfropfungsgrad m
grain Korn n, Körnung f, Struktur f, Maserung f
   leather grain effect Ledernarbung f
grained körnig, gekörnt, genarbt
granular granulatförmig, granulös, körnig, kornartig
   granular form Granulatform f
granulate/to zerkleinern, granulieren
   granulating elements Zerkleinerungselemente npl
   granulating line Granulieranlage f
granulate Granulat n
granulation Granulierung f
   hot melt granulation Heißzerkleinern n, Heißgranulierung f
   preliminary granulation Vorzerkleinern n
granulator Granuliereinrichtung f, Granuliervorrichtung f, Schneidgranulator m, Granulator m, Mühle f, Granuliermaschine f, Haufwerkschneider m, Schneidmühle f, Abfallgranulator m, Zerkleinerer m, Zerkleinerungsmaschine f, Zerkleinerungsmühle f, Zerkleinerungsaggregat n, Zerkleinerungsanlage f, Abfallzerkleinerer m, Abfall(zerkleinerungs)mühle f
   granulator range Mühlenbaureihe f
   general-purpose granulator Universal-Zerkleinerer m
   heavy-duty granulator Großschneidmühle f, Großraumschneidgranulator m
   hot melt granulator Heißgranuliervorrichtung f, Heißgranuliermaschine f
   machine-side granulator Beistellgranulator m
   pipe scrap granulator Rohrschneidmühle f
   precision granulator Präzisionsschneidmühle f
   preliminary granulator Vorzerkleinerer m, Vorzerkleinerungsmühle f
   press-side granulator Beistellgranulator m
   vertical granulator Vertikalschneidmühle f
   web scrap granulator Stanzgittermühle f
granule feed unit Granulatdosiergerät n
granule metering unit Granulatdosiergerät n
granules Granulatkörner npl, Granulat n
   alumin(i)um granules Aluminiumgrieß m
   cubed granules Würfelgranulat n
   coarse granules Grobgranulat n
   diced granules Würfelgranulat n
   polystyrene granules Polystyrol-Granulat n
graph graphische Darstellung f
   graph paper Diagrammpapier n
   bar graph Balkendiagramm n, Blockdarstellung f, Blockdiagramm n
graphics Grafik f
   graphics capability Grafikfähigkeit f
   graphics display Grafik-Bildschirm m, Bildschirmgrafik f
graphite Graphit m
   graphite fiber Graphitfaser f
   powdered graphite Graphitpulver n
graphitization Graphitierung f

gravimetric gravimetrisch
gravity Schwerkraft f, Gravitation f
  gravity feed Gravitationsdosierung f
  centre of gravity Schwerpunkt m
  force of gravity Schwerkraft f
  specific gravity spezifisches Gewicht n
gravure printing ink Tiefdruckfarbe f
gray [GB: grey] grau
  gray cast iron Grauguß m, GG
  gray scale Graumeßstab m
grease Fett n
  grease-dissolving fettlösend
  grease gun Fett(schmier)presse f
  grease residues Fettreste mpl
  grease resistance Fettbeständigkeit f
  grease resistant fettbeständig
  grease solvent Fettlösungsmittel n
  anti-friction bearing grease Wälzlagerfett n
  central grease lubricating system Fettzentralschmierung f
  free from grease fettfrei
  low temperature lubricating grease Tieftemperaturschmierfett n
  lubricating grease Schmierfett n
greaseproof fettdicht
green grün
  bromocresol green Bromkresolgrün n
  chrome green Chromoxidgrün n
  phthalocyanine green Phthalocyaningrün n
greenhouse Gewächshaus n
grey s. gray
grid Gitter n
  flow restrictor grid Drosselgitter n
grind/to schleifen, mahlen, abreiben
grinder Mahlaggregat n
  abrasive belt grinder Bandschleifmaschine f
grinding Schleifen n, Vermahlung f
  grinding compartment Mahlkammer f, Mahlraum m
  grinding face Mahlbahn f
  grinding rolls Mahlwalzwerk n
  grinding unit Mahlaggregat n
  grinding wheel resin Schleifscheibenharz n
  convex grinding Bombage f, Walzenbombage f
  dry grinding Trockenschliff m
  fine grinding Fein(ver)mahlen n
  fine grinding machine Feinmahlaggregat n
  impact grinding machine Prallmahlanlage f
  pigment grinding Pigmentvermahlen n
  precision grinding Präzisionsschleifen n
  roll grinding Walzenschliff m
  wet grinding Naßschliff m
gripper Entnahmemaske f, Greifer m
  gripper arrangement Greifervorrichtung f, Greiferanordnung f
  gripper unit Greifereinheit f
  parison gripper Schlauchgreifer m
grippers Klemmbacken fpl

gripping unit/mechanism Greifvorrichtung f
  bottle gripping unit Flaschengreifstation f
  parison gripping mechanism Schlauchgreifvorrichtung f
grit Grobsand m, Schrot n,m, Strahlkies m, Körnung f
  grit blasting Schrotstrahlen n
  grit blasting unit Strahlanlage f
groove Nut f, Rille f, Riefe f
  annular groove Ringkanal m
  fine grooves Feinnuten fpl
  longitudinal grooves Axialnuten fpl, Längsnuten fpl
  mandrel with a heart-shaped groove Herzkurvendorn m
  mandrel with a ring-shaped groove Ringrillendorn m
  plasticizing grooves Plastifizierkanäle mpl
  ring-shaped groove Ringnut
  spiral grooves Wendelnuten fpl
  vent groove Entlüftungsnut f, Entlüftungsschlitz m, Entlüftungsspalt m
grooved genutet, geriffelt
  grooved barrel Nutenzylinder m
  grooved-barrel extruder Nutenbuchsenextruder m
  grooved bushing Nutenbuchse f
  grooved feed bush Einzugsnutbuchse f
  grooved feed section/zone Nuteneinzugszone f
  grooved shear section Nutenscherteil n
  grooved smear head Nutenscherteil n
  grooved torpedo Nutentorpedo m
  grooved torpedo screw Nutentorpedoschnecke f
  finely grooved feingenutet
  longitudinally grooved längsgenutet
  spirally grooved spiralförmig genutet, spiralgenutet
gross Brutto-
  gross national product Bruttosozialprodukt n
ground 1. gemahlen, geschliffen; 2. Grund m, Boden m, Erde f
  ground water Grundwasser n
  ground-glass joint Schliffverbindung f
  ground-glass stopper Schliffstopfen m
  coarsely ground grob gemahlen
  convex ground bombiert
  laid above ground [e.g. pipes] freiverlegt
  material being ground Mahlgut n
  precison-ground feingeschliffen
  very finely ground feinstgemahlen
group Gruppe f
  group of atoms Atomgruppe f
  group of curves Kurvenschar f
  group of materials Werkstoffgruppe f
  acetate group Acetatgruppe f
  acetyl group Acetylgruppe f
  acid amide group Säureamidgruppe f
  acrylonitrile group Acrylnitril-Gruppe f
  aldehyde group Aldehydgruppe f
  alkenyl group Alkenylgruppe f
  alkoxymethylene group Alkoxymethylengruppe f
  alkyl group Alkylgruppe f

allylidene group Allylidengruppe f
amine group Amingruppe f
amino group Aminogruppe f
anhydride group Anhydridgruppe f
aryl group Arylgruppe f
carbonyl group Carbonylgruppe f
carboxyl group Carboxylgruppe f
containing hydroxyl groups hydroxylgruppenhaltig
dialkyl peroxide group Dialkylperoxidgruppe f
diphenylsulphone group Diphenylsulfongruppe f
end group Endgruppe f
epoxy/epoxide group Epoxidgruppe f, Epoxygruppe f
ester-carbonyl group Estercarbonylgruppe f
ester group Estergruppe f
ether group Ethergruppe f
ethoxy group Ethoxygruppe f
ethyl group Ethylgruppe f
ethylene oxide group Ethylenoxidgruppe f
glycidyl group Glycidylgruppe f
glycol group Glykolgruppe f
ignition group Zündgruppe f
nitrile group Nitrilgruppe f
organic group Organogruppe f
oxymethylene group Oximethylengruppe f
peroxide group Peroxidgruppe f
phenyl group Phenylgruppe f
product group Produktgruppe f
residual epoxide/epoxy group content Rest-Epoxidgruppengehalt n
side group Seitengruppe f
silanol group Silanolgruppe f
siloxane group Siloxangruppe f
sulphone group Sulfongruppe f
terminal carboxyl group Carboxyl-Endgruppe f
terminal hydroxyl group Hydroxyl-Endgruppe f
urethane group Urethangruppe f
vinyl group Vinylgruppe f
grout Fugenfüller m, Fugenmasse f, Fugenmörtel m, Fugenspachtel m
growth Wachstum n
crack growth Rißfortschritt m
crack growth direction Rißwachstumsrichtung f
crack growth rate Rißwachstumsrate f, Rißwachstumsgeschwindigkeit f, Rißausbreitungsgeschwindigkeit f, Rißerweiterungsgeschwindigkeit f, Rißvergrößerung f, Rißverlängerung f
crystal growth Kristallwachstum n
free-radical chain growth mechanism Radikalkettenmechanismus m
rate of crystal growth Kristallwachstumsgeschwindigkeit f
rate of growth Wachstumsgeschwindigkeit f, Steigerungsrate f
rate of spherulite growth Sphärolithwachstumsgeschwindigkeit f
GRP [glass fiber reinforced plastic] Glasfaserkunststoff m, GFK
GRP filament winding machine GFK-Wickelmaschine f
GRP sheet GF-UP-Platte f
guaranteed garantiert
guaranteed output Garantieausstoß m, Leistungsgarantie f
guaranteed shelf life Lagerfähigkeitsgarantie f
guard Schranke f, Betriebssicherung f
guard door Schutztür f, Sicherheitstür f
guard door interlock system/mechanism Schutzgittersicherung f, Schutztürsicherung f
light beam guard Lichtschranke f, Lichtvorhang m
operator's guard door Bedienungsschutztür f
safety guard Schutzgitter n
screen guard Sicherheitsgitter n
sliding guard door Sicherheitsschiebetür f, Schiebe(schutz)gitter n
sliding safety guard Sicherheitsschiebetür f, Schiebe(schutz)gitter n
guide Führung f, Richtwert, Anhaltspunkt m
guide bar Führungsschiene f, Leitstange f
guide bolt Führungsbolzen m
guide bush Führungsbuchse f
guide element Führungselement n
guide hole Führungsloch n
guide pillar Führungssäule f, Führungsstift m, Führungsholm m
guide pillar system Holmführung f
guide pin Führungssäule f, Führungsstift m, Führungsholm m
guide price Richtpreis m
guide rails Führungsschienen fpl
guide ring Führungsring m
guide rod Führungsstange f
guide roll Führungsrolle f, Führungswalze f, Leitwalze f
guide rolls Rollenführung f
bubble guide Schlauchführung f
bubble guides Leitstangen fpl, Leitplatten fpl, Leitbleche npl
film guide rolls Folienführungswalzen fpl
film web guide Folien(bahn)führung f
lateral (film) bubble guide Seitenführung f
metal guide plate Leitblech n
pattern guides Modellführungen fpl
web guide Bahnführung f, Bahnsteuereinrichtung f, Warenbahnführung f, Warenbahnsteuerung f
guideline Richtlinie f
application guidelines Anwendungsrichtlinien fpl
construction guidelines Konstruktionshinweise mpl
design guidelines Gestaltungsrichtlinien fpl
formulation guideline Rezepturhinweis m
formulation guidelines Formulierungshinweise mpl
processing guidelines Verarbeitungshinweise mpl, Verarbeitungsrichtlinien fpl
quality guidelines Güterichtlinie fpl,

Qualitätsrichtlinien fpl
safety guidelines Sicherheitsrichtlinien fpl
works safety guidelines Arbeitsschutz-Richtlinien fpl
guideway Führungsbahn f
gun Pistole f, Presse f, Spritze f, Spritzpistole f
  grease gun Fett(schmier)presse f
gutter Rinne f, Rinnstein m
  roof gutter Dachrinne f
gyration Kreisbewegung f
  radius of gyration Trägheitshalbmesser m, Gyrationsradius m
hair Haar n
  hair crack Harriß m
  hair drier Haartrockner m
hairline crack Haarriß m
half Hälfte f
  half life period Halbwert(s)zeit f
  bottom mo(u)ld half Werkzeugunterteil n
  ejector (mo(u)ld) half auswerferseitige Werkzeughälfte f
  fixed mo(u)ld half Gesenkseite f, Spritzseite f, Einspritzseite f
  mo(u)ld half Formhälfte f, Werkzeughälfte f
  moving mo(u)ld half Auswerferseite f, auswerferseitige Werkzeughälfte f, Kernseite f, schließseitige Werkzeughälfte f, Schließseite f
  on the moving mo(u)ld half schließseitig
  on the statinary mo(u)ld half düsenseitig
  stationary mo(u)ld half Düsenseite f, düsenseitige Werkzeughälfte f, spritzseitige Werkzeughälfte f
  top mo(u)ld half Werkzeugoberteil n
halogen Halogen n
  halogen compound Halogenverbindung f
  halogen-containing halogenhaltig
  halogen content Halogengehalt m
  halogen-free halogenfrei
  halogen hydracid Halogenwasserstoffsäure f
  halogen hydrin Halogenhydrin n
  halogen radical Halogenradikal n
halogenated halogenhaltig, halogeniert
hammer Hammer m
  hammer finish paint Hammerschlaglack m
  hammer mill Hammermühle f
  drop hammer tool Schlaghammerwerkzeug n
  water hammer Druckstoß m
hand Hand f
  hand lay-up Handauflegen n, Handlaminieren n
  hand lay-up laminate Handlaminat n
  hand lay-up process Hand(laminier)verfahren n, Handauflegeverfahren n
  hand mixer Handrührgerät n
  hand operated handbetrieben, handbetätigt, handbedient, manuell bedient
  hand spraying Handspritzen n
  hand welding instrument Handschweißgerät n
handle/to handhaben, benutzen
  easy to handle handhabungsfreundlich
  safe to handle handhabungssicher
handle Handgriff m, Griff m, Klinke f
  dry handle trockener Griff m
  tool handle Werkzeuggriff m
handling Handhabung f
  handling capacity Hantierungskapazität f
  handling characteristics Handlingeigenschaften fpl
  handling device Handhabeeinrichtung f, Handlinggerät n
  handling instructions Handhabungsrichtlinien fpl
  handling properties Handlingeigenschaften fpl
  part handling Formteilhandhabung f
handrail profile Handlaufprofil n
hard hart
  hard and brittle spröd-hart
  hard and tough zähhart
  hard chroming Hartverchromen n
  hard copy Protokoll n, Hardcopy n [Rechnerausdruck auf Papier]
  hard copy printer Protokolldrucker m
  hard-elastic starr-elastisch
  hard-face/to panzern
  hard-face coating Oberflächenpanzerung f, Panzerung f, Panzerschicht f
  hard-faced gepanzert
  hard-facing Panzerung f
  hard resin Hartharz n
  hard rubber [rubber cured with 1% to 40% of added sulphur] Hartgummi m,n
  hard wearing verschleißfest, verschleißwiderstandsfähig, verschleißarm
  hard wearing properties Verschleißarmut f
  hard-wired festverdrahtet, verbindungsprogrammierbar
  extremely hard wearing hochverschleißwiderstandsfähig, hochverschleißfest
  production of hard copy Protokollerstellung f
  semi-hard halbhart
hardboard Hartfaserplatte f
harden/to härten, erstarren, abbinden
hardened gehärtet, gepanzert
  hardened on the inside [barrel] innengepanzert
  hardened steel Hartstahl m
  hardened steel (outer) layer Hartstahlmantel m
  case hardened einsatzgehärtet
  surface hardened oberflächenvergütet, oberflächengehärtet
  with hardened flights [screw] steggepanzert
hardener Härter m, Härterkomponente f, Härtungsmittel n
  amine hardener Aminhärter m
  anhydride hardener Anhydridhärter m
  excess hardener Härterüberschuß m
  fast-reacting hardener Schnellhärter m
  high speed hardener Schnellhärter m
  less hardener Härterunterschuß m
  polyamine hardener Polyaminhärter m
  polyaminoamide hardener Polyaminoamidhärter m

polyol hardener Polyolhärter m
hardening shop Härterei f
surface hardening Oberflächenvergütung f
hard-face coating Panzerung f
hardness Härte f
  hardness test Härteprüfung f
  hardness tester Härteprüfgerät n, Härteprüfer m
  ball indentation hardness Kugel(ein)druckhärte f
  Brinell hardness Brinellhärte f
  change in Shore hardness Shore-Änderung f
  decrease in hardness Härteabfall m
  degree of hardness Härtegrad m
  film hardness Filmhärte f
  intermediate Shore hardness Zwischenshorehärte f
  original hardness Ausgangshärte f
  pendulum hardness Pendelhärte f
  residual pendulum hardness Restpendelhärte f
  retained pendulum hardness Restpendelhärte f
  Rockwell hardness Rockwellhärte f
  scratch hardness Ritzhärte f
  shore-A hardness Shore-Härte A f
  shore-D hardness Shore-Härte D f
  shore hardness Shore-Härte f
  surface hardness Oberflächenhärte f
  Sward hardness Schenkelhärte f
  Sward hardness test Schenkelhärteprüfung f
  Vickers hardness Vickershärte f
  with a high shore hardness hochshorig
hardware 1. Hardware f [Computer]; 2. Eisenwaren fpl, Stahlwaren fpl, Beschläge mpl
  hardware module Hardwarebaustein m
  hardware shop/store Haushaltsgeschäft n
harmful substance Schadstoff m
haul-off ratio Abzugsverhältnis n
hazard Gefahr f
  health hazard Gesundheitsgefährdung f
  representing an explosive hazard explosionsgefährlich
haze [in film testing] Trübung f
Hazen colo(u)r scale Hazen-Farbskala f [Öl]
Hazen colo(u)r units Hazenfarbzahl f
HDPE [high density polyethylene] Niederdruck-Polyethylen n, PE n hart, Polyethylen n hart, Polyethylen n hoher Dichte
head Kopf m
  head rest Kopfstütze f
  accumulator head Akku-Kopf m, Speicherkopf m, Staukopf m
  accumulator head blow mo(u)lding (process) Staukopfverfahren n
  accumulator head plasticizing unit Staukopfplastifiziergerät n
  barrel head Zylinderkopf m
  coextrusion die head Koextrusionskopf m
  cylinder head Zylinderkopf m
  die head Extruder(düsen)kopf m, Extrusionskopf m, Spritzkopf m, Werkzeugkopf m, Düsenkopf m
  die head design Kopfkonstruktion f
  die head pressure Düseneintrittdruck m
  die head temperature Kopftemperatur f
  erase head Löschkopf m, Löscheinheit f
  extruder head Extruder(düsen)kopf m, Extrusionskopf m
  four-die (extruder) head Vierfachkopf m
  general-purpose extruder head Universalkopf m
  grooved shear head Nutenscherteil n
  pipe extrusion head Rohr(spritz)kopf m
  scanning head Abtastkopf m
  screw head Schraubenkopf m
  sheet extrusion head Plattenkopf m
  single-die extruder head Einfach(extrusions)kopf m, Einfachwerkzeug n
  smear head Schertorpedo m, Torpedoscherteil m, Schmierkopf m, Scherkopf m
  strand die head Strangdüsenkopf m
  triple-die (extruder) head Dreifachkopf m
  tubular ram accumulator head Ringkolbenspeicherkopf m
  twin die (extruder) head Zweifach(extrusions)kopf m, Doppel(spritz)kopf m, Zweifachwerkzeug n, Doppelwerkzeug n
  twin-die film blowing head Doppelblaskopf m
  twin mixing head Doppelmischkopf m
  vertical die head Senkrechtspritzkopf m
  vertical extruder head Senkrechtspritzkopf m
  write head Schreiber m, Aufnahmekopf m, Schreibkopf m
headlamp Scheinwerfer m
  headlamp diffuser Scheinwerferstreuscheibe f
  headlamp housing Scheinwerfergehäuse n
  headlamp reflector Scheinwerferreflektor m
health Gesundheit f
  health hazard Gesundheitsgefährdung f
  health risk Gesundheitsrisiko n
  danger to health Gesundheitsgefährdung f
  Federal German Public Health Office Bundesgesundheitsamt n
  injurious to health gesundheitsschädlich
heart Herz n
  heart-shaped groove Herzkurvenkanal m
  heart-shaped groove type of feed system Herzkurveneinspeisung f
  mandrel with a heart-shaped groove Herzkurvendorn m
heat Wärme f
  heat absorption Wärmeabsorption f
  heat accumulation Wärmestau m
  heat activation [of dry adhesive film] Hitzeaktivierung f
  heat ag(e)ing Hitzealterung f, Hitzelagerung f, thermische Alterung f, Wärmealterung f, Wärmebehandlung f, Wärmelagerung f
  heat ag(e)ing behavio(u)r Wärmealterungsverhalten n
  heat ag(e)ing period Wärmelagerungszeit f

heat ag(e)ing properties Wärmealterungswerte mpl
heat ag(e)ing resistance Wärmealterungsbeständigkeit f
heat ag(e)ing temperature Warmlagerungstemperatur f
heat ag(e)ing test Wärmealterungsversuch m, Warmlagerungsversuch m
heat balance Wärmehaushalt m, Wärme(strom)bilanz f
heat barrier Wärmesperre f
heat bonding Heißverklebung f
heat build-up Wärmeaufbau m
heat capacity Wärmekapazität f
heat-conducting wärmeleitend
heat consumer Wärmeverbraucher m
heat content Wärmeinhalt m
heat curing 1. Heißhärtung f, Warmhärtung f; 2. wärmehärtend, warmhärtend, heißhärtend
heat dissipation Wärmeableitung f
heat dissipation loss Wärmeableitungsverlust m
heat distortion temperature Formbeständigkeitstemperatur f
heat distribution Wärmeverteilung f
heat effect Wärmetönung f
heat evolution Wärmeentwicklung f
heat exchange Wärmeaustausch m
heat exchanger Wärme(aus)tauscher m
heat exchanger pipe Wärmetauschrohr n
heat exchanging fluid wärmeaustauschende Flüssigkeit f
heat exposure Wärmeeinwirkung f
heat flow Wärmefluß m
heat impulse welding Wärmeimpulsschweißen n, Wärmeimpulssiegelung f
heat input zugeführte Heizleistung f, Heizleistungsaufnahme f
heat insulating wärmedämmend, wärmeisolierend
heat insulating jacket Wärmeschutzmantel m
heat insulating layer Wärmedämmschicht f
heat insulating material Wärmedämmstoff m
heat insulating properties Wärmeisoliereigenschaften fpl
heat laminating plant Thermokaschieranlage f
heat lamination Heißkaschierung f
heat loss Wärmeabfluß m, Wärmeverlust m
heat loss due to radiation Strahlungsverlust m
heat losses due to convection Konvektionsverluste mpl
heat of deformation Verformungswärme f
heat of dissipation Dissipationswärme f
heat of evaporation Verdampfungswärme f
heat of polymerization Polymerisationswärme f
heat of reaction Reaktionswärme f
heat of transformation Umwandlungswärme f
heat pipe Heizrohr n, Wärmerohr n
heat produced Wärmetönung f
heat pump Wärmepumpe f
heat radiation Wärmestrahlung f

heat recovery Wärmerückgewinnung f
heat reflecting surface Wärmeabstrahlfläche f
heat requirements Wärmebedarf m
heat resistance Hitzebelastbarkeit f, Hitzebeständigkeit f, Temperatur(stand)festigkeit f, thermische Beständigkeit f, Wärmebeständigkeit f, Wärme(stand)festigkeit f, Warmfestigkeit f
heat resistant hitzebeständig, hitzefest, wärme(form)beständig, wärmestabil, wärmestandfest, temperaturbeständig, hitzebelastbar
heat sealability Heißsiegelbarkeit f
heat sealable heißsiegelbar, heißsiegelfähig
heat sealable coating Heißsiegelschicht f, Heißsiegelbeschichtung f
heat sealing instrument Heißsiegelgerät n
heat sealing lacquer Heißsiegellack m
heat sensitive hitzeempfindlich, temperaturempfindlich, wärmeempfindlich
heat setting Thermofixierung f, Wärmestabilisierung f
heat shield Hitzeschild m
heat sink paste Wärmeleitpaste f
heat source Wärmequelle f
heat stabilization Hitzestabilisierung f, Thermostabilisierung f, Wärmestabilisierung f
heat stabilized hitzestabilisiert, wärmestabilisiert
heat stabilizer Hitzestabilisator m, Thermostabilisator m, Wärmestabilisator m
heat stabilizing thermostabilisierend
heat stabilizing properties Wärmestabilisierungsvermögen n
heat sterilizable hitzesterilisierbar
heat sterilized hitzesterilisiert
heat transfer Wärmeübertragung f, Wärmeübergang m
heat transfer coefficient Wärmeübergangskoeffizient m, Wärmeübergangszahl f, Wärmedurchgangskoeffizient m
heat transfer data Wärmeübertragungskennwerte mpl
heat transfer medium Wärmeübertragungsmittel n, Wärmeüberträgermittel n, Wärmeübertragungsmedium n, Wärmeträger m
heat transfer oil Wärmeträgeröl n
heat transfer resistance Wärmedurchgangswiderstand m
heat transfer values Wärmeübertragungskennwerte mpl
heat treatment Temperaturbehandlung f, Warmbehandlung f, Wärmebehandlung f
affected by heat hitzeempfindlich, temperaturempfindlich, wärmeempfindlich
amount of heat Wärmemenge f
automatic heat impulse welding instrument Wärmeimpulsschweißautomat m, Wärmeimpulsschweißmaschine f
automatic heat sealer Heißsiegelautomat m
conduction of heat Wärmeleitung f
conductor of heat Wärmeleiter m

convection heat loss Lüftungswärmeverlust m
evolution of heat Wärmeentwicklung f
frictional heat Friktionswärme f, Reibungswärme f
long-term heat ag(e)ing Dauerwärmelagerung f
long-term heat resistance Dauerwärmebelastbarkeit f, Dauerwärmebeständigkeit f, Dauerwärmestabilität f, Dauertemperaturbeständigkeit f
Martens heat distortion temperature Formbeständigkeit f (in der Wärme) nach Martens
molar heat Molwärme f
plate heat exchanger Plattenwärmeaustauscher m
radiation heat loss Strahlungs(wärme)verlust m
removal of heat Wärmeentzug m
specific heat spezifische Wärme f
the heat put into the system die Wärmezufuhr f an das System
transmission heat loss Transmissionswärmeverlust m
unaffected by heat wärmeunempfindlich, hitzeunempfindlich
waste heat recycling Abwärmenutzung f
heatable beheizbar
heated beheizt, geheizt, Heiz-
   heated tool Heizelement n, Heizkeil m
   heated tool pipe welding line Heizelement-Rohrschweißanlage f
   heated tool welding Heizkeilschweißen n, Heizelementschweißen n
   externally heated außenbeheizt
   gas heated gasbeheizt
   oil heated ölbeheizt, öltemperiert
   partly heated teilbeheizt
   steam heated dampfbeheizt
heater Heizkörper m, Heizelement n
   heater band Heizband n, Temperierband n
   heater controls Heizungssteuerung f
   heater voltage Heizspannung f
   band heater Temperierband n
   barrel heater Zylinderheizelement n
   barrel heaters Zylinderheizung f, Zylinderbeheizung f
   cartridge heater patronenförmiger Heizkörper m, Patronenheizkörper m
   ceramic(-insulated) band heaters Keramik(heiz)-bänder npl, Keramikheizung f, Keramikheizkörper mpl
   cylinder heater Zylinderheizelement n
   cylinder heaters Zylinderheizung f
   extra heater band Zusatzheizband n
   forced circulation heater Zwanglauferhitzer m
   heavy duty cartridge heater Hochleistungsheizpatrone f
   heavy duty heater Hochleistungsheizkörper m
   immersed heater Tauchheizung f
   mica-insulated heater bands Glimmerheizbänder npl
   mo(u)ld heater Formheizgerät n
   nozzle heater Düsenheizung f, Düsenheizkörper m
   nozzle heater band Düsenheizband n
   platen heaters Plattenheizung f
   radiant heaters Strahlerheizung f, Strahlungsheizung f
   resistance band heater Widerstandsheizband n
   resistance heater Widerstandsheizelement n, Widerstandsheizkörper m, Widerstandsheizung f
   ring heater Ringheizelement n
   separate heaters Fremdbeheizung f
   storage heater Wärmespeicherofen m
   tubular heater Rohrheizkörper m
heating Heizen n [s.a. heater]
   heating area Heizfläche f
   heating bath Heizbad n
   heating capacity Heizleistung f
   heating channel Heizbohrung f, Heizkanal m
   heating circuit Heizkreis(lauf) m, Heiz(strom)kreis m
   heating coil Heizschlange f, Heizwendel f
   heating-cooling channel Temperierbohrung f, Temperierkanal m
   heating-cooling channel layout Temperierkanalauslegung f
   heating-cooling channel system Temperierkanalsystem n
   heating-cooling circuit Temperierkreis(lauf) m
   heating-cooling collar Heiz-Kühlmanschette f
   heating-cooling medium Temperiermittel n, Temperierflüssigkeit f, Temperiermedium n
   heating-cooling mixer Heiz-Kühlmischer m
   heating-cooling section Heiz-Kühlzone f
   heating-cooling system Heiz-Kühlsystem n
   heating-cooling unit Heiz-Kühleinrichtung f, Heiz- und Kühlaggregat n
   heating cylinder Heizzylinder m
   heating energy Heizenergie f
   heating jacket Heizmantel m
   heating medium Heizmedium n, Temperiermittel n, Temperiermedium n, Temperierflüssigkeit f
   heating mixer Heizmischer m
   heating platen [of press] Heizplatte f
   heating rod Stabheizkörper m
   heating rolls Temperierwalzen fpl
   heating section Heizstrecke f, Temperierstrecke f
   heating sleeve Heizmanschette f
   heating station Heizstation f
   heating surface Heizfläche f
   heating system Temperiersystem n
   heating tunnel Heizkanal m, Temperierkanal m
   heating tunnel layout Temperierkanalauslegung f
   heating unit Heizgerät n, Temperiergerät n
   heating unit consumption Heizleistung f
   heating-up period Aufheizzeit f
   heating-up rate Aufheizgeschwindigkeit f, Aufheizrate f
   heating-up station Aufheizstation f
   heating-up times Erwärmungszeiten fpl
   heating zone Temperierzone f, Heizzone f
   barrel heating capacity Zylinderheizleistung f
   barrel heating circuit Zylinderheizkreis m
   barrel heating zones Zylinderheizzonen fpl
   calender heating system Kalanderbeheizung f

circulating hot water heating (system) Heißwasser-Umlaufheizung f
circulating oil heating system Ölumlaufheizung f
connected heating capacity installierte Heizleistung f
contact heating (unit) Kontaktheizung f
cylinder heating capacity Zylinderheizleistung f
electric heating Elektrobeheizung f
fluid heating (system) Flüssigkeitsheizung f
hot water heating unit Heißwasser-Heizanlage f
internal oil heating system Ölinnentemperierung f
method of heating Aufheizmethode f
mo(u)ld heating Werkzeugbeheizung f
mo(u)ld heating medium Werkzeugheizmedium n
mo(u)ld heating system Werkzeugbeheizung f
oil heating unit Öltemperiergerät n
platen heating (system) Plattenheizung f
rate of heating Aufheizrate f
total heating capacity Gesamtheizleistung f
type of heating Heizungsart f
underfloor (central) heating Fußbodenheizung f
heavy schwer, zäh(flüssig), hart, fest, hochsiedend
　heavy-duty hochbelastbar, Hochleistungs-
　heavy-duty blade Hochleistungsschneide f
　heavy-duty cartridge heater Hochleistungsheizpatrone f
　heavy-duty granulator Großschneidmühle f, Großraumschneidgranulator m, Hochleistungs-Schneidmühle f
　heavy-duty heater Hochleistungsheizkörper m
　heavy-duty high speed mixer Hochleistungsschnellmischer m
　heavy-duty internal mixer Hochleistungsinnenmischer m
　heavy-duty kneader Hochleistungskneter m
　heavy-duty mixer Hochleistungsmischer m
　heavy-duty mo(u)ld Hochleistungswerkzeug n, Hochleistungsform f
　heavy-duty plasticator Hochleistungsplastifizieraggregat n
　heavy-duty sack Schwergutsack m
　heavy-duty sack film Schwergutsackfolie f
　heavy-duty screw Hochleistungsschnecke f
　heavy-duty window profile Fensterhochleistungsprofil n
　heavy-gauge film winder Schwerfolienwickler m
　heavy metal Schwermetall n
　heavy metal complex Schwermetallkomplex m
　heavy metal salt Schwermetallsalz n
　containing heavy metals schwermetallhaltig
heel Schuhabsatz m
height Höhe f
　adjustable in height höhenverstellbar
　drop height Fallhöhe f
　frost line height Einfriergrenzenabstand m, Erstarrungslinienhöhe f
　mo(u)ld height Formeinbauhöhe f, Formeinbauraum m, Formaufspannhöhe f

mo(u)ld height adjusting mechanism Werkzeughöhenverstelleinrichtung f, Werkzeughöhenverstellung f
roughness height Rauhtiefe f
screw center height Extrudierhöhe f
helical 1. schraubenförmig, schneckenförmig; 2. schrägverzahnt [Zahnrad]
　helical flow Helixströmung f
　helical gearing Schrägverzahnung f
helix angle Drallwinkel m, Gangsteigungswinkel m, Steigungswinkel m
helmet Helm m
　crash helmet Schutzhelm m, Sturzhelm m
heptane Heptan n
HET acid [hexachloroendomethylene tetrahydrophthalic acid] Hetsäure f
heterocyclic heterozyklisch
heterogeneous heterogen
hexafluoropropylene Hexafluoropropylen n
hexagonal screw Sechskantschraube f
hexagonal-section pipe Sechskantrohr n
hexagonal-section solid rod Sechskant-Vollstab m
hexahydrophthalic acid Hexahydrophthalsäure f
hexahydrophthalic anhydride [HHPSA] Hexahydrophthalsäureanhydrid n
hexamethoxymethyl melamine Hexamethoxymethylmelamin n
hexamethylene diamine Hexamethylendiamin n
hexamethylene tetramine Hexamethylentetramin n
hexanediol diglycidyl ether Hexandioldiglycidylether m
hexaphenyl ethane Hexaphenylethan n
HF welding line HF-Schweißanlage f
hiding power [of pigment] Deckkraft f, Deckfähigkeit f, Deckvermögen n, Deckungsgrad m
hierarchal hierarchisch
hierarchic(al) hierarchisch
high hoch
　high-alkali glass A-Glas n
　high-boiling hochsiedend
　high-boiling solvent Hochsieder m
　high-build system Dickschichtsystem n
　high-capacity blown film line Hochleistungs-Schlauchfolien-anlage f, Hochleistungsblasformanlage f, Hochleistungsblasfolienanlage f
　high-capacity blow mo(u)lding line Großblasformanlage f
　high-capacity blow mo(u)lding machine Großblasformmaschine f
　high-capacity compounding line Großcompoundieranlage f
　high-capacity extrusion blow mo(u)lding plant Hochleistungsblasextrusionsanlage f
　high-capacity plant Hochleistungsanlage f
　high-contrast kontrastreich
　high-density polyethylene [HDPE] Polyethylen n hoher Dichte, Polyethylen n hart,

Niederdruck-Polyethylen n
high-energy hochenergetisch
high-frequency hochfrequent
high-frequency embossing Hochfrequenzprägen n
high-frequency engineering Hochfrequenztechnik f
high-frequency field Hochfrequenzfeld n
high-frequency welding Hochfrequenzschweißen n, Hochfrequenzschweißverfahren n
high-frequency welding machine Hochfrequenzschweißmaschine f
high-gloss hochglänzend
high-impact hochschlagfest, hochschlagzäh, hochstoßfest, erhöht schlagzäh
high-impact compound Schlagzähmischung f
high-melting hochschmelzend
high-modulus fiber HM-Faser f, Hochmodulfaser f
high-molecular weight hochmolekular
high-molecular weight polyethylene film HM-Folie f
high-output leistungsstark, leistungsfähig
high-performance leistungsfähig, leistungsstark
high-performance pellet drier Hochleistungsgranulattrockner m
high-polish chromium plated hochglanzverchromt
high polymer Hochpolymer n
high-polymeric hochpolymer
high pressure Hochdruck m
high-pressure blower Hochdruckgebläse n
high-pressure injection Hochdruckinjektion f
high-pressure laminate Hochdrucklaminat n
high-pressure liquid chromatography [HPLC] Hochdruckflüssigkeitschromatographie f
high-pressure metering unit Hochdruckdosieranlage f
high-pressure mixing unit Hochdruckmischanlage f
high-pressure plasticator Hochdruckplastifiziergerät n
high-pressure plastic(iz)ation Hochdruckplastifizierung f
high-pressure polymerization Hochdruckpolymerisation f
high-pressure process Hochdruckverfahren n
high-pressure pump Hochdruckpumpe f
high-pressure pump unit Hochdruckpumpenaggregat n
high-pressure reactor Hochdruckreaktor m
high-pressure spraying plant Hochdruckspritzanlage f
high-pressure valve Hochdruckventil n
high-quality (qualitativ) hochwertig
high-reactivity hochaktiv
high-resolution hochauflösend
high-solids (fest)körperreich
high-solids [paint] bindemittelreich, lösemittelarm
high-speed hochtourig, Schnell-, Hochgeschwindigkeits-
high-speed cable sheathing plant Hochleistungs-Kabelummantelungsanlage f
high-speed calibrating system Hochgeschwindigkeitskalibriersystem n
high-speed clamping cylinder Schnellschließzylinder m
high-speed coupling Schnellverbindung f, Schnellkupplung f
high-speed cutter [especially for film] Hochgeschwindigkeitsschneider m
high-speed cylinder Eilgangzylinder m
high-speed drier Schnelltrockner m
high-speed drying oven Schnelltrockner m
high-speed extruder Hochleistungsextruder m, Schnelläufer-Extruder m, Großextruder m
high-speed extrusion Hochgeschwindigkeitsextrusion f, Hochleistungsextrusion f
high-speed extrusion line Hochleistungsextrusionsanlage f
high-speed hardener Schnellhärter m
high-speed hot air welding Warmgasschnellschweißen n
high-speed injection Schnellspritzen n
high-speed (injection) cylinder Schnellfahrzylinder m
high-speed injection mo(u)lding Leistungsspritzguß m, Hochleistungsspritzgießen n
high-speed injection unit Hochleistungsspritzeinheit f
high-speed machine Schnelläufer m
high-speed mixer Schnellmischer m
high-speed mo(u)ld changing system Werkzeugschnellwechselsystem n
high-speed mo(u)ld clamping mechanism Werkzeugschnellspannvorrichtung f
high-speed nozzle retraction mechanism Düsenschnellabhebung f
high-speed pipe extrusion line Hochleistungs-Rohrextrusionslinie f
high-speed plant Hochgeschwindigkeitsanlage f
high-speed printer Schnelldrucker m
high-speed production line Hochgeschwindigkeitsanlage f
high-speed screen changer Sieb-Schnellwechseleinrichtung f, Schnellsiebwechsel-Einrichtung f
high-speed single screw extruder Hochleistungseinschneckenextruder m, Einschnecken-Hochleistungsmaschine f
high-speed single-screw extrusion line Einschnecken-Hochleistungsextrusionsanlage f
high-speed slitter [especially for film] Hochgeschwindigkeitsschneider m
high-speed steel Schnellarbeitsstahl m
high-speed stirrer Dissolver m, Schnellrührer m
high-speed test Hochgeschwindigkeitsversuch m
high-speed twin screw extruder Hochleistungsdoppelschneckenextruder m
high-speed welding Schnellschweißen n
high-strength hochfest
high-strength fiber Hochleistungsfaser f
high-temperature applications Hochtemperatur-

anwendungen fpl
high-temperature behavio(u)r Hochtemperaturverhalten n
high-temperature cable Hochtemperaturkabel n
high-temperature calendering (process) HT-Verfahren n, Hochtemperaturverfahren n
high-temperature packing Hochtemperaturpackung f
high-temperature process HT-Verfahren n, Hochtemperaturverfahren n
high-temperature processing Hochtemperaturverarbeitung f
high-temperature properties Hochtemperatureigenschaften fpl
high-temperature resistant hochwärme(form)beständig, hochtemperaturbeständig, hochhitzebeständig
high-temperature stabilized hochwärmestabilisiert
high-temperature thermostat Hochtemperaturthermostat m
high-temperature vulcanization Hitzevulkanisation f
high-temperature vulcanizing heißvulkanisierend
high temperatures Hochtemperaturen fpl
high-tensile fiber HT-Faser f
high-tension insulator Hochspannungsisolator m [s.a. high-voltage]
high vacuum Hochvakuum n
high-vacuum metallization Hochvakuumbedampfen n
high-viscosity 1. Dickflüssigkeit f, Zähflüssigkeit f; 2. dickflüssig
high voltage Hochspannung f
high-voltage range Hochspannungsbereich m
high-voltage switch Hochspannungsschalter m
high-voltage test Hochspannungsprüfung f
high-voltage test bed Hochspannungprüffeld n
exposure to high temperatures Temperaturbeanspruchung f, Wärmebeanspruchung f, Temperaturbelastung f
exposure to high voltage Hochspannungsbelastung f
heavy-duty-high speed mixer Hochleistungsschnellmischer m
subject to high temperatures hitzebeansprucht
ultra-high vacuum Ultrahochvakuum n
with a high plasticizer content weichmacherreich
with a high resin content harzreich
with a high Shore hardness hochshorig
highly hoch
highly abrasion resistant hochabriebfest
highly concentrated hochkonzentriert
highly condensed hochkondensiert
highly corrosive hochaggressiv
highly effective hochwirksam
highly elastic hochdehnfähig
highly filled füllstoffreich, hochfüllstoffhaltig, hochgefüllt
highly flammable leichtentzündlich, leichtentflammbar
highly flexible hochelastisch
highly integrated hochintegriert

highly lubricated hochgeschmiert
highly oriented hochorientiert
highly pigmented hochpigmentiert
highly polished feinstpoliert, hochglanzpoliert
highly polished roll Hochglanzwalze f
highly reactive hochreaktiv, hochaktiv
highly stressed hochbeansprucht
highly toxic hochgiftig
highly unsaturated hoch-ungesättigt
with a highly polished surface oberflächenpoliert
hinge Scharnier n
hinge-mounted klappbar, schwenkbar
integral hinge Filmscharnier n
hinged (auf)klappbar, schwenkbar
hinged flange Klappflansch m
history Geschichte f
deformation history Deformationsgeschichte f
previous history [i.e. the thermal or mechanical history of a test piece or mo(u)lded part] Vorgeschichte f, Fließgeschichte f
thermal history Temperaturgeschichte f
HM-HDPE blown film plant HM-Anlage f
hold(ing) Halten n, Halt m
hold(ing) pressure Nachdruckniveau n, Nachdruck m, Haltedruck m, Druckstufe II f, Formgebungsdruck m
holding pressure fluctuations Nachdruckpulsationen fpl
hold(ing) pressure phase Nachdruckphase f, Formgebungsphase f
hold(ing) pressure profile Nachdruckverlauf m
hold(ing) pressure program Nachdruckprogramm n
hold(ing) pressure stages Nachdruckstufen fpl
hold(ing) pressure time Nachdruckdauer f, Nachdruckzeit f, Druckhaltezeit f
change-over from injection to holding pressure Druckumschaltung f
change-over to hold(ing) pressure Nachdruckumschaltung f
mandrel holding plate Dornhalteplatte f
hole Loch n, Bohrung f
accurately bored hole Paßbohrung f
blind hole Sackloch n, Sacklochbohrung f
calibrating air holes Stützluftbohrungen fpl
cent(e)ring hole Zentrierbohrung f
fixing hole Befestigungsbohrung f
guide hole Führungsloch n
locating holes Positionierungslöcher npl
mo(u)ld attachment holes Aufspannbohrungen fpl
mo(u)ld cent(e)ring hole Werkzeugzentrierbohrung f
pouring hole Eingußloch n
punched holes Stanzlöcher npl
suction hole Saugloch n
threaded hole Gewindebohrung f, Gewindeloch n
vacuum suction holes Vakuumsauglöcher npl
hollow 1. Höhlung f, 2. hohl
hollow-bored hohlbebohrt
hollow drive shaft Getriebehohlwelle f,

Hohlwellengetriebe n
hollow rod Hohlstab m
hollow section Hohlprofil n
hollow shaft Hohlwelle f
holographic holographisch
holography Holographie f
home 1. inländisch; 2. Heim n
  home business Inlandsgeschäft n
  home entertainment industry Unterhaltungselektronikindustrie f
  home market Binnenmarkt m
homogeneity Homogenität f
  melt homogeneity Schmelzehomogenität f
homogeneous homogen
homogenization Homogenisierung f
homogenize/to homogenisieren
homogenizing homogenisierend, Homogenisier-
  homogenizing aid Homogenisierhilfe f
  homogenizing effect Homogenisierwirkung f
  homogenizing efficiency Homogenisierleistung f
  homogenizing section Aufschmelzbereich m
  homogenizing zone Ausgleichszone f
homolytic homolytisch
homopolymer Homopolymer n, Homopolymersiat n, Reinpolymerisat n
  homopolymer-based film Homopolymerfolie f
  emulsion homopolymer Emulsionshomopolymerisat n
  styrene homopolymer Styrol-Homopolymerisat n
  suspension homopolymer Suspensionshomopolymer n
  vinyl chloride homopolymer Vinylchlorid-Homopolymer n
homopolymeric homopolymer
homopolymerized homopolymerisiert
honed gehont
honeycomb Wabe(nstruktur) f
  honeycomb core Wabenkern m
  honeycomb structure Wabenstruktur f
honing Kurz(hub)honen n, Schwingschleifen n, Ziehschleifen n
  honing machine Schwingschleifer m
hood Absaughaube f, Haube f
  cutting chamber hood Granulierhaube f
  extraction hood Absaughaube f
  safety hood Schutzabdeckung f, Schutzhaube f
  soundproof hood Schallschutzhaube f
Hooke's Law Hookesches Gesetz n
hoop stress [of pipes] Rohrwandbeanspruchung f, Wandbeanspruchung f, Vergleichsspannung f
hoops Kolbenringe mpl [Fehler bei der Folienherstellung]
  free from hoops kolben(ring)frei
hopper [also: feed/material hopper] Trichter m, Aufgabetrichter m, Aufsatztrichter m, Dosiertrichter m, Fullguttrichter m, Einfüllgehäuse n, Einfülltrichter m, Einlaufgehäuse n, Einlauftrichter m, Extruder(einfüll)trichter m, Extrusionstrichter m, Formmassetrichter m, Fülltrichter m, Granulattrichter m, Maschinentrichter m, Massebehälter m, Speisetrichter m, Vorratstrichter m, Zugabetrichter m
  hopper capacity Fülltrichterinhalt m, Trichterinhalt m
  hopper contents Fülltrichterinhalt m, Trichterinhalt m
  hopper devolatilization Trichterentgasung f
  hopper drier Trichtertrockner m
  drying hopper Trocknungstrichter m
  single-compartment hopper Einkammertrichter m
  twin vacuum hopper Doppelvakuumtrichter m, Vakuum-Doppeltrichter m
  twin vacuum hopper assembly Vakuum-Doppeltrichteranlage f
  two-compartment hopper Zweikammertrichter m
  vacuum hopper Entgasungstrichter m, Vakuumspeisetrichter m
  vented hopper Entgasungstrichter m, Vakuumtrichter m
  vented hopper degassing unit Vakuumtrichter-Entgasungsanlage f
  vented hopper system Trichterentgasung f
  vibratory feed hopper Vibrationseinfülltrichter m
horizontal waag(e)recht, horizontal
  horizontal adjustment Horizontalverstellung f
  horizontal displacement Horizontalverschiebung f
  horizontal extruder Horizontalextruder m, Waagerechtextruder m
  horizontal feed system Horizontalbeschickung f
  horizontal feeding Horizontalbeschickung f
  horizontal granulator Horizontalschneidmühle f
  horizontal in construction flache Bauweise f, horizontale Bauweise f
horizontally constructed in Horizontalbauweise, in Horizontalausführung f
horn 1. Horn n, Hupe f; 2. Sonotrode f [Schweißen]
hose Schlauch m
  garden hose Gartenschlauch m
  pressure hose Druckschlauch m
  suction hose Saugschlauch m
hot heiß
  hot air Heißluft f, Warmluft f
  hot air ag(e)ing Heißluftalterung f
  hot air blower Warmluftgebläse f
  hot air drier Warmlufttrockner m
  hot air oven Heißluftofen m, Heißluftschrank m
  hot air resistance Heißluftbeständigkeit f
  hot air tunnel Heißluft(düsen)kanal m
  hot air welding Heißgasschweißen n, Warmgasschweißen n
  hot bending Warmbiegen n
  hot bonding Warmkleben n
  hot-cured foam Heißschaumstoff m
  hot-cured mo(u)lded foam Heißformschaumstoff m
  hot curing warmhärtbar (warmhärtbare Preßmasse / thermoset mo(u)lding compound)
  hot curing catalyst Heißhärter m
  hot-cut [this prefix is used in connection with

pelletizers where strands are extruded and cut into pellets at the die face, whilst still hot] Heißabschlag m
hot-cut pelletizer Heißgranuliervorrichtung f
hot cut pelletizing system Direktabschlagsystem n
hot-cut water-cooled pelletization Heißabschlag-Wassergranulierung f
hot-cut water-cooled pelletizer Heißabschlag-Wassergranulierung f
hot-dip galvanized feuerverzinkt
hot dipping (process) Warmtauchverfahren n
hot embossing Heißprägen n
hot feed extruder Warmfütterextruder m
hot melt adhesive Schmelzkleber m, Schmelzklebstoff m
hot melt adhesive applicator Schmelzklebstoff-Auftragsgerät n
hot melt bonding Heißschmelzverklebung f
hot-melt coater Schmelzwalzenmaschine f
hot-melt coating machine Schmelzwalzenmaschine f
hot melt contact adhesive Schmelzhaftklebstoff m
hot melt extruder Schmelze(austrags)extruder m, Heißschmelzextruder m
hot melt granulation Heißzerkleinern n, Heißgranulierung f
hot melt granulator Heißgranuliervorrichtung f, Heißgranuliermaschine f
hot-melt roller application [of an adhesive] Walzenschmelzverfahren n
hot melt sealant Schmelzmasse f
hot mixing Heißmischung f
hot plate Heizplatte f
hot plate butt welding Heizelementstumpfschweißen n
hot plate welding (process) Spiegel(schweiß)-verfahren n
hot press mo(u)lded laminate Warmpreßlaminat n
hot press mo(u)lding Warm(form)pressen n, Heißpreßverfahren n, Warmpreßverfahren n, Heißpressen n
hot runner Heißkanalverteiler m, Heißkanal m, Heißläufer m
hot runner design Heißkanal-Konstruktion f
hot runner feed system Heißkanalanguß m, Heißkanal-Angußsystem n, Heißkanal-Verteileranguß m
hot runner injection mo(u)ld Heißkanal-Spritzgießwerkzeug n
hot-runner injection mo(u)lding Durchspritzverfahren n
hot runner injection mo(u)lding system Spritzgieß-Heißkanalsystem n
hot runner manifold Heißkanal(verteiler)platte f
hot runner manifold block Heißkanalverteilerblock m, Heizblock m, Heißkanalblock m, Querverteiler m
hot runner mo(u)ld Heißkanalform f, Heißkanalwerkzeug n
hot runner multi-daylight mo(u)ld Heißkanal-Etagenwerkzeug n
hot runner multiple gating Heißkanal-Mehrfachanguß m
hot runner needle shut-off mechanism Heißkanal-Nadelverschlußsystem n
hot runner needle valve Heißkanal-Nadelventil n
hot runner nozzle Heißkanaldüse f
hot runner plate Verteilerplatte f
hot runner stack mo(u)ld Heißkanal-Etagenwerkzeug n
hot runner system Heißkanal-Verteilersystem n, Heißkanalsystem n, Heißkanal-Rohrsystem n
hot runner tooling Heißkanal-Formenbau m
hot runner two-cavity mo(u)ld Heißkanal-Doppelwerkzeug n
hot runner unit Heißkanalelement n, Heißkanalverteilerbalken m, Querverteiler m, Heizblock m, Heißkanal(verteiler)block m
hot runner unit construction Heißkanalblock-Ausführung f
hot runner unit design Heißkanalblock-Ausführung f
hot-setting warmabbindend
hot water Heißwasser n
hot water heating unit Heißwasser-Heizanlage f
hot water resistant heißwasserbeständig
hot well Vorkammerbohrung f, Vorkammerraum m, Vorkammer f
hot well contents Vorkammerkegel m
hot wire welding Trennahtschweißen n
hot worked steel Warmarbeitsstahl m
automatic hot gas welding unit Warmgas-Schweißautomat m
circulating hot water heating system Heißwasser-Umlaufheizung f
examples of hot runner systems Heißkanal-beispiele npl
exposure to hot water Heißwasserbelastung f
high-speed hot air welding Warmgas-schnellschweißen n
immersion in hot water Heißwasserbelastung f
partial hot runner Teilheißläufer m
standard hot runner unit Standard-Heißkanalblock m
types of hot runner Heißkanalbauarten fpl
hotplate Heizplatte f
hour Stunde f
hours of operation Betriebsstunden fpl
hours of sunshine Sonnenstunden fpl
machine hour rate Maschinenstundensatz m
machine hours counter Betriebsstundenzähler m
output per hour Stundenleistung f
production hours Produktionsstunden fpl
throughput per hour Stundendurchsatz m
hourly stündlich
hourly output Stundenleistung f
hourly throughput Stundendurchsatz m

household Haushalt m
  household appliance paint Haushaltsgerätelack m
  household appliances Haushaltsgeräte npl
  household goods Haushaltswaren fpl, Haushaltsartikel mpl
housing Gehäuse n
  housing containing the mo(u)ld clamping mechanism Formschließgehäuse n
  bearing housing Lagergehäuse n
  carburet(t)or housing Verdampfergehäuse n
  cast (resin) housing Gießharz-Gehäuse n
  cast alumin(i)um housing Alu-Gußgehäuse n
  fan housing Lüfterzarge f
  gear housing Getriebegehäuse n
  headlamp housing Scheinwerfergehäuse n
  kneader housing Knetergehäuse n
  light metal housing Leichtmetallgehäuse n
  rear light housing Rückleuchtengehäuse n
  spring housing Federgehäuse n, Federhaus n
  transducer housing Aufnehmergehäuse n
hub Nabe f
  hub cap Radkappe f, Rad(zier)blende f
humanization Humanisierung f
humid feucht
  ag(e)ing under warm, humid conditions Feucht-Warm-Lagerung f
humidity Feuchtigkeitsgehalt m, Feuchtigkeitsniveau n, Feuchtigkeit f, Feuchte f, Feuchtgehalt m
  atmospheric humidity Luftfeuchtigkeit f
  relative humidity relative Luftfeuchte f
hybrid composite (material) Hybrid-Verbundwerkstoff m
hybrid material Hybridwerkstoff m
hydrate/to hydratisieren
hydrate Hydrat n
  cellulose hydrate Cellulosehydrat n
hydration Hydra(ta)tion f, Hydratisation f
hydraulic Hydraulik-, hydraulisch
  hydraulic accumulator Druckmittelakkumulator m, Druckmittelspeicher m, Druckölspeicher m, Hydrospeicher m, Hydraulikspeicher m
  hydraulic circuit Hydraulikkreis m
  hydraulic control element [plural: hydraulic controls] Hydraulikstellglied n
  hydraulic controls Regelhydraulik f
  hydraulic cylinder Hydraulikzylinder m
  hydraulic drive (unit) Antriebshydraulik f, Hydraulikantrieb m
  hydraulic engineering Wasserbau m
  hydraulic fluid Druckflüssigkeit f, Druckmedium n, Hydraulikflüssigkeit f
  hydraulic injection cylinder Hydraulikspritzzylinder m
  hydraulic injection unit Einspritzhydraulik f
  hydraulic line Hydraulikleitung f
  hydraulic monitoring package Hydraulik-Überwachungspaket n
  hydraulic motor Hydraulikmotor m, Hydromotor m, Ölmotor m
  hydraulic oil Drucköl n, Hydrauliköl n
  hydraulic oil flow Hydraulikölstrom m
  hydraulic oil requirements Hydraulikölbedarf m
  hydraulic oil reservoir Hydrauliköltank m
  hydraulic press ölhydraulische Presse f
  hydraulic pressure Hydraulikdruck m, hydraulischer Systemdruck m
  hydraulic pressure-dependent hydraulikdruckabhängig
  hydraulic pressure-inpendent hydraulikdruckunabhängig
  hydraulic pull-back system Vorspanneinrichtung f, Walzenvorspanneinrichtung f, Walzenvorspann m, Vorspannung f
  hydraulic pump Hydraulikpumpe f, Hydropumpe f
  hydraulic ram Hydraulikkolben m
  hydraulic system Hydrauliksystem n
  hydraulic system pressure ga(u)ge Hydrauliköl-Druckanzeige f
  hydraulic unit Hydraulikaggregat n, Hydraulikblock m
  hydraulic valve Hydraulikventil n
  amount of hydraulic oil Hydraulikölstrom m
  central hydraulic system Zentralhydraulikanlage f
  digital hydraulic system Digitalhydraulik f
  dual-circuit hydraulic system Zweikreis-Hydrauliksystem n
  fully hydraulic(ally) voll(öl)hydraulisch
  proportional hydraulic system Proportionalhydraulik f
hydraulics Hydrauliksystem n, Hydraulik(anlage) f
  digital hydraulics Digitalhydraulik f
  proportional hydraulics Proportionalhydraulik f
hydrazine derivative Hydrazin-Derivat n
hydrazo compound Hydrazoverbindung f
hydrin Hydrin n
  halogen hydrin Halogenhydrin n
hydrocarbon Kohlenwasserstoff m
  hydrocarbon compound Kohlenwasserstoffverbindung f
  hydrocarbon radical Kohlenwasserstoffrest m
  hydrocarbon resin Kohlenwasserstoffharz n
  hydrocarbon wax Kohlenwasserstoffwachs n
  aliphatic hydrocarbon aliphatischer Kohlenwasserstoff m, Paraffinkohlenwasserstoff m
  aromatic hydrocarbon aromatischer Kohlenwasserstoff m
  chlorinated hydrocarbon Chlorkohlenwasserstoff m
  fluorinated hydrocarbon Fluorkohlenwasserstoff m
  terpene hydrocarbon Terpenkohlenwasserstoff m
hydrochloric acid Salzsäure f, Chlorwasserstoffsäure f
hydrogen Wasserstoff m
  hydrogen acceptor Wasserstoff-Akzeptor m
  hydrogen atom Wasserstoffatom n
  hydrogen bridge H-Brücke f, Wasserstoffbrücke f
  hydrogen bridge linkage

Wasserstoffbrückenbindung f
hydrogen chloride Chlorwasserstoff(gas) m(n)
hydrogen donor Wasserstoffdonator m
hydrogen-free wasserstoffrei
hydrogen halide Halogenwasserstoff m
hydrogen peroxide Wasserstoffsuperoxid n, Wasserstoffperoxid n
removal of hydrogen Wasserstoff-Abstraktion f
hydrogenated hydriert
hydrogenation Hydrierung f
hydrolysis Hydrolyse f
hydrolysis resistance Hydrolysebeständigkeit f, Hydrolyseresistenz f
not resistant to hydrolysis hydrolyseinstabil
subject to hydrolysis hydrolyseanfällig
hydrolytic hydrolytisch
hydrolyzable hydrolysierbar
hydrolyze/to hydrolysieren
hydrometer Aräometer n
hydroperoxide Hydroperoxid n
hydroperoxide group Hydroperoxidgruppe f
alkyl hydroperoxide Alkylhydroperoxid n
butyl hydroperoxide Butylhydroperoxid n
cumol hydroperoxide Cumolhydroperoxid n
hydroperoxy radical Hydroperoxyradikal n
hydrophilic hydrophil, wasserfreundlich
hydrophilic properties Hydrophilie f
hydrophobic hydrophob, wasserfeindlich
hydrophobic properties Hydrophobie f
hydroquinone Hydrochinon n
hydrostatic hydrostatisch
hydrostatic pressure hydrostatischer Druck m
long-term failure test under internal hydrostatic pressure [test used for plastics pipes] Zeitstand-Innendruckversuch m
long-term resistance to internal hydrostatic pressure [of plastics pipes] Innendruck-Zeitstandwert m
hydroxide Hydroxid n
alkali hydroxide Alkalihydroxid n
alumin(i)um hydroxide Aluminiumhydroxid n, Tonerdehydrat n
lithium hydroxide Lithiumhydroxid n
sodium hydroxide solution Natriumhydroxidlösung f
hydroxy Hydroxy-
hydroxy equivalent Hydroxyäquivalent n
hydroxy ester Hydroxyester m
hydroxy stearic acid Hydroxystearinsäure f
hydroxyfunctional hydroxyfunktionell
hydroxyl Hydroxyl-
hydroxyl compound Hydroxylverbindung f
hydroxyl content Hydroxylgehalt m
hydroxyl group Hydroxylgruppe f
hydroxyl number Hydroxylwert m, Hydroxylzahl f
hydroxyl value OH-Zahl f, OHZ
hydroxylamine hydrochloride Hydroxylaminhydrochlorid n

hydroxymethyl group Hydroxymethylgruppe f
hygroscopic hygroskopisch, wasseranziehend
hygroscopicity Hygroskopizität f, Hygroskopie f
hysteresis Hysterese f
hysteresis curve Federkennlinie f, Hysteresekurve f
hysteresis loop Hystereseschleife f
identification Identifizierung f
date identification Datumsidentifikation f
identifying mark Identifizierungskennzeichen n
ignition Inbrandsetzen n, Zündung f
ignition group Zündgruppe f
ignition point Entzündungspunkt m, Zündpunkt m, Zündtemperatur f
ignition test Zündversuch m
flash ignition temperature Fremdentzündungstemperatur f
loss on ignition Glühverlust m
source of ignition Entzündungsquelle f, Zündquelle f
illuminated circuit diagram Leuchtschaltbild n
imide Imid n
polyamide imide Polyamidimid n
polyester imide Polyesterimid n
polyether imide Polyetherimid n
immediately verzögerungsfrei, sofort
immersed versenkt, eingetaucht
immersion Eintauchen n, Untertauchen n, Einlagerung f
immersion in hot water Heißwasserbelastung f
immersion in water Wasserbelastung f, Wasser(ein)lagerung f
immersion period Lagerdauer f, Lagerungszeit f, Lagerungsdauer f
immersion temperature Lagerungstemperatur f
immersion test Einlagerungsversuch m, Immersionsversuch m, Tauchversuch m, Lagerungsversuch m
prolonged immersion [of test piece in a liquid] Langzeitlagerung f, Dauerlagerung f
prolonged immersion in water Langzeitwasserlagerung f
short-term immersion Kurzzeitlagerung f
solder bath immersion Lötbadlagerung f
time of immersion Lagerungszeit f, Einlagerungszeit f, Lagerdauer f, Lagerungsdauer f
impact Schlag m, Stoß m
impact behavio(u)r Schlagverhalten n
impact cutter Schlagmesser n, Schlagschere f
impact energy Schlagarbeit f, Schlagenergie f, Schlagkraft f, Stoßarbeit f, Stoßenergie f
impact energy absorptive capacity Stoßenergieaufnahmevermögen n
impact force 1. Auftreffkraft f, Schlagintensität f, Stoßkraft f; 2. Zerkleinerungswucht f [Granulator]
impact grinding machine Prallmahlanlage f
impact mill Prallmühle f
impact modification Schlagzähmodifizierung f
impact modified schlagzähmodifiziert
impact modifier Schlagzähkomponente f, Schlagzähmacher m, Schlagzähmodifikator m,

Schlagzähmodifiziermittel n, Schlagzähmodifizierharz n, Schlagfestmacher m, Schlagzähigkeitsverbesserer m, Schlagzähmodifier m
impact resistance Schlagbeständigkeit f
impact resistant stoßfest, stoßsicher
impact speed Stoßgeschwindigkeit f, Schlaggeschwindigkeit f, Aufprallgeschwindigkeit f
impact strength Schlagfestigkeit f, Schlagzähigkeit f, Stoßfestigkeit f, Stoßzähigkeit f
impact strength reduction Schlagzähigkeitsminderung f
impact stress Schlagbeanspruchung f, Stoßbeanspruchung f, Stoßbelastung f
impact surface Stoßfläche f
impact test Schlagprüfung f, Schlagversuch m, Stoßversuch m
impact tester Schlagwerk n
angle of impact Schlagwinkel m
behavio(u)r on impact Schlagverhalten n
Charpy impact tester Charpygerät n
direction of impact Schlagrichtung f
drop impact strength Fallbruchfestigkeit f
Dynstat impact tester Dynstatgerät n
flexural impact behavio(u)r Schlagbiegeverhalten n
flexural impact strength Schlagbiegefestigkeit f, Schlagbiegezähigkeit f
flexural impact stress Schlagbiegebeanspruchung f
flexural impact test Schlagbiegeversuch m
flexural impact test piece Schlagbiegestab m
high-impact erhöht schlagzäh
high-impact compound Schlagzähmischung f
low temperature impact strength Tieftemperatur-Schlagzähigkeit f
making impact resistant Schlagfestmachen n
medium-impact mittelzäh
normal-impact normalschlagfest, normal(schlag)zäh, normalstoßfest
notched tensile impact strength Kerbschlagzugzähigkeit f
pendulum impact speed Pendelhammergeschwindigkeit f, Schlagpendelgeschwindigkeit f
pendulum impact test Pendelschlagversuch m
pendulum impact tester Pendelhammergerät n, Pendelschlagwerk n
point of impact Auftreffpunkt m, Auftreffstelle f
resistance to impact Schlagwiderstand m
resistivity to deformation through impact Schlagunverformbarkeit f
resistant to deformation through impact schlagunverformbar
sensitivity to impact Stoßempfindlichkeit f
speed of impact Aufprallgeschwindigkeit f, Schlaggeschwindigkeit f, Stoßgeschwindigkeit f
tensile impact strength Schlagzugzähigkeit f
tensile impact test Schlagzugversuch m
under impact stress schlagbeansprucht
with enhanced impact resistance erhöht schlagzäh

impart/to verleihen, vermitteln
  imparting colo(u)r Farbgebung f
  imparting thixotropy Thixotropierung f
  imparting water repellency Hydrophobierung f
impedance Impedanz f
  input impedance Eingangsimpedanz f
  output impedance Ausgangsimpedanz f
impeller Kreisel m, Laufrad n, Rührwerkzeug n, Schleuderrad n
  impeller shaft Rührwelle f
impermeability Undurchlässigkeit f, Impermeabilität f
  impermeability to light Lichtundurchlässigkeit f
  impermeability to water Wasserdichtheit f
  gas impermeability Gasdichtheit f, Gasundurchlässigkeit f
  vapo(u)r impermeability Dampfdichtheit f, Dampfundurchlässigkeit f
  water impermeability Wasserdichtheit f
impermeable undurchlässig
  impermeable to light lichtundurchlässig
  impermeable to moisture feuchtigkeitsundurchlässig
  impermeable to X-rays röntgenstrahlundurchlässig
imperviousness Dichtigkeit f
impingement Stoß m, Anprall m, Aufprall m
  cooling air impingement angle Kühlluftanblaswinkel m
implementation Ausführung f, Verwirklichung f, Implementierung f
  program implementation Programmimplementierung f
import duty Einfuhrzoll m
important wichtig, wesentlich, relevant
  important applications Anwendungsschwerpunkt m
  important from the processing point of view verfahrenstechnisch wesentlich
  technically important parameters verfahrenstechnisch relevante Parameter mpl
impossible unmöglich
  technically impossible verfahrenstechnisch nicht möglich
impregnant Tränkmittel n, Imprägniermittel n
impregnate/to imprägnieren, tränken
impregnating Imprägnier-, Tränk-
  impregnating agent Imprägniermittel n, Tränkmittel n
  impregnating bath Tränkbad n
  impregnating conditions Imprägnierparameter mpl
  impregnating machine Imprägniermaschine f
  impregnating paste Tränkpaste f
  impregnating period Tränkungszeit f
  impregnating plant Tränkanlage f
  impregnating resin Imprägnierharz n, Tränkharz n
  impregnating time Tränkungszeit f
impregnated durchtränkt, getränkt, imprägniert
  resin impregnated harzgetränkt, harzimprägniert
impregnation Imprägnierung f, Tränkung f
  dip impregnation Tauchimprägnierung f
  vacuum impregnation (process)

Vakuumtränkverfahren n
impression Forminnenraum m, Formnest n, Formhöhlung f
  mo(u)ld impression dimensions formgebende Werkzeugmaße npl
improved verfeinert, verbessert, weiterentwickelt, fortentwickelt
  improved adhesion Haftungssteigerung f
improvement Verbesserung f, Fortentwicklung f, Weiterentwicklung f
  improvement in quality Qualitätsverbesserung f
  improvement in sales Absatzverbesserung f
  design inprovements konstruktive Verbesserungen fpl
impulse Impuls m
  impulse sequence Impulsfolge f
  impulse welding Impulsschweißen n
  automatic heat impulse welding instrument Wärmeimpulsschweißautomat m
  heat impulse welding Wärmeimpulssiegelung f, Wärmeimpulsschweißen n
  heat impulse welding machine Wärmeimpulsschweißmaschine f
impurity Verunreinigung f, Fremdmaterial n
in-house betriebseigen, firmenintern, betriebsintern
  in-house control Eigenkontrolle f
in-line die Geradeaus(extrusions)werkzeug n, Geradeaus(spritz)kopf m, Längsspritzkopf m
in-mo(u)ld coating IMC-Beschichtung f, IMC-Verfahren n
in-mo(u)ld labelling IML-Technik f
in-plant betriebsintern, innerbetrieblich
  in-plant colo(u)ring Selbsteinfärben n
  in-plant waste Produktionsabfälle mpl
in-situ foam Ortsschaum m
inaccuracy Ungenauigkeit f
inaccurate metering ungenaue (unexakte) Dosierung f
incandescence resistance Glutbeständigkeit f
incandescent (weiß)glühend, Glüh-
  incandescent mandrel test Glühdornprüfung f
  incandescent wire cutter Glühbandabschneider m, Glühdrahtschneider m
  incandescent wire test Glühdrahtprüfung f
incision Einschnitt m
incline/to sich neigen
  four-roll inclined Z-type calender Vierwalzen-S-Kalander m
including einschließlich
incoming hereinkommend, ankommend, einfallend, Eingangs-
  incoming air Zuluft f
  incoming goods control (department) Wareneingangskontrolle f, Eingangskontrolle f
  incoming goods quality Wareneingangsparameter mpl
  incoming goods section Wareneingangsbereich m
  incoming orders Auftragseingang m
  incoming raw materials control Rohstoffeingangskontrolle f

incompatibility Unverträglichkeit f
incompatible unverträglich, inkompatibel
incompressibility Inkompressibilität f, Nichtkomprimierbarkeit f
incompressible inkompressibel, nichtkomprimierbar
incorporate/to beimengen, einmischen, zudosieren, einarbeiten
incorporating of small amounts anteilige Mitverwendung f
incorporation Beigabe f, Einarbeitung f, Zudosierung f
  incorporation of glass fibers Glasfaserzusatz m
  incorporation of small amounts anteilige Mitverwendung f
incorrect fehlerhaft, unrichtig, Fehl-
increase Anstieg m, Erhöhung f, Steigerung f
  increase in capacity Kapazitätsaufstockung f, Kapazitätsausweitung f, Kapazitätszuwachs m
  increase in consumption Verbrauchszuwachs m, Verbrauchssteigerung f
  increase in length Verlängerung f
  increase in orientation Orientierungszuwachs m
  increase in pressure Druckanstieg m
  increase in production Produktionszuwachs m, Produktionserhöhung f
  increase in volume Volumenerhöhung f, Volumenvergrößerung f, Volumenzunahme f
  increase in weight Gewichtszunahme f
  cost increase Kostensteigerung f
  net increase Netto-Zugang m
  price increase Preissteigerung f, Preiserhöhung f, Preisanhebung f
  production increase Produktionszuwachs m
  productivity increase Produktivitätssteigerung f
  rate of increase Steigerungsrate f
  rate of speed increase Drehzahlsteigerungsrate f
  speed increase Drehzahlerhöhung f
  sudden increase in concentration Konzentrationssprung m
  sudden pressure increase Druckstoß m
  temperature increase Temperaturzunahme f, Temperaturanstieg m, Temperaturerhöhung f
  torque increase Drehmomentanstieg m
  turnover increase Umsatzausweitung f, Umsatzplus n, Umsatzzuwachs m
  viscosity increase Viskositätsaufbau m, Viskositätserhöhung f, Viskositätsanstieg m
  volume increase Volumenvergrößerung f, Volumenzunahme f, Volumenerhöhung f
  weight increase Gewichtszunahme f
increased erhöht
  increased demand Nachfragebelebung f
  increased energy requirements Energiemehraufwand m
  increased output Leistungserhöhung f, Leistungssteigerung f
increasing steigend, anwachsend, sich vermehrend
  increasing pitch screw Progressivspindel f

increasing sales Absatzausweitung f
with constantly increasing root diameter [screw] kernprogressiv
with increasing pitch [screw] steigungsprogressiv
increment Inkrement n
indene Inden n
indentation Einbeulen n, Einkerbung f, Eindruck m, Vertiefung f
  indentation resistance Eindruckwiderstand m
  indentation test Kerbversuch m; Eindrückversuch m
  ball indentation hardness Kugel(ein)druckhärte f
  ball indentation test Kugeldruckprüfung f, Kugeleindruckverfahren n
  depth of indentation [when measuring ball indentation hardness] Eindringtiefe f, Eindrucktiefe f
  Erichsen indentation Erichsentiefe f
indentor Eindringkörper m
  Vicat indentor Vicatnadel f
independence of pressure Druckunabhängigkeit f
independent eigenständig, unabhängig
  independent of cavity pressure innendruckunabhängig
  independent of speed drehzahlunabhängig
  independent of the mo(u)lding cycle zyklusunabhängig
  independent of the temperature temperaturunabhängig
  independent of time zeitunabhängig
  independent variable Einflußgröße f
index Index m, Kennziffer f, Beiwert m, Faktor m
  evaporation index Verdunstungszahl f
  flow index Fließexponent m
  flow law index Fließgesetzexponent m
  loss index dielektrische Verlustzahl f, Verlustziffer f
  melt flow index [MFI] Schmelzindex m, Fließindex m
  oxygen index Sauerstoffindex m
  power law index Potenzgesetzexponent m
  refractive index Brechungsindex m, Lichtbrechungsindex m, Brechungszahl f
  yellowness index Vergilbung f, Vergilbungsgrad m, Vergilbungszahl f, YI-Wert m
indicator Anzeigegerät n, Anzeigeinstrument n, Indikator m
  fault location indicator Fehlerortsignalisierung f, Störungsortsignalisierung f
  flow indicator Durchlaufanzeige f
  level indicator Füllstand(an)geber m, Füllstandanzeige(r) f(m), Füllstandmelder m, Niveaumelder m
  light signal indicator panel Leuchtanzeigetableau n
  malfunction indicator Störungsanzeige f, Störungsmeldegerät n
  melt pressure indicator Massedruckanzeige(r) f(m)
  melt temperature indicator Massetemperaturanzeige(r) f (m)
  oil level indicator Ölstandanzeige(r) f (m)
  speed indicator Drehzahlanzeige(r) f (m)
  temperature indicator Temperaturanzeigegerät n
indifferent to unempfindlich
individual individuell, einzeln
  individual cavity cent(e)ring device Einzelformnest-Zentrierung f
  individual controller Einzelregler m
  individual control unit Einzelregler m
  individual units Einzelgeräte npl
induction Induktion f
  induction period Induktionszeit f, Induktionsperiode f
  induction phase Induktionsphase f
inductive induktiv
industrial industriell, Industrie-
  industrial atmosphere Industrieklima n
  industrial bag Industriesack m
  industrial cleaner Industriereinigungsmittel n
  industrial fabrics technische Gewebe npl
  industrial floor screed Industriebodenbelag m, Industrieestrich m
  industrial laminates technische Schichtstoffplatten fpl
  industrial paint Industrielack m
  industrial pipe Industrierohr n
  industrial robot Industrieroboter m
  industrial-scale großtechnisch
  industrial sheets Industrieplatten fpl
  industrial shelving Industrieregale npl
industrialized countries Industrienationen fpl, Industrieländer npl, Industriestaaten mpl
industry Industrie f, Branche f
  adhesives industry Klebstoffindustrie f
  aerospace industry Raumfahrtindustrie f
  aircraft industry Luftfahrtindustrie f
  automotive industry Automobilindustrie f
  branch of industry Industriesparte f, Industriezweig m
  building industry Bauwirtschaft f, Bauindustrie f
  car industry Automobilindustrie f
  clothing industry Bekleidungssektor m
  electrical appliance industry Elektrogeräteindustrie f
  electrical industry Elektroindustrie f
  electronics industry Elektronikindustrie f
  food industry Nahrungsmittelindustrie f
  foundry industry Gießereiindustrie f
  furniture industry Möbelindustrie f
  home entertainment industry Unterhaltungselektronikindustrie f
  leisure industry Freizeitsektor m
  motor industry Kraftfahrzeugindustrie f
  packaging industry Verpackungswesen n, Emballageindustrie f
  paint industry Farbenindustrie f, Lackindustrie f
  petroleum industry Erdölindustrie f
  pharmaceutical industry Pharmaindustrie f
  plastics industry Kunststoffbranche f
  printing industry Druckindustrie f
  shipbuilding industry Schiffsbauindustrie f
  shoe industry Schuhindustrie f
  transport industry Transportwesen n

ineffective unwirksam, ineffektiv
inefficient leistungsschwach
inert inaktiv, inert, indifferent, reaktionsträge
   inert gas Edelgas n, Inertgas n
   physiologically inert physiologisch indifferent
inertia Massenträgheit f
   force of inertia Massenträgheitskraft f
   moment of inertia [body/area] Trägheitsmoment n [Masse/Fläche]
inertness Trägheit f, Massenwiderstand m
inexpensive preiswert, preisgünstig, preiswürdig, kostengünstig, billig
infinite unendlich, stufenlos
infinitely variable stufenlos einstellbar, stufenlos regelbar (variabel)
inflammability Entflammbarkeit f
inflammable entflammbar, entzündlich
inflate/to aufblasen
inflating mandrel Blaspinole f, Blasdorn m, Spritzdorn m
inflating pressure Aufblasdruck m
inflation Aufblähung f, Aufblasen n
   inflation air Blasluft f
   inflation needle Blasnadel f, Injektionsblasnadel f, Blasstift m
influence Einfluß m
   atmospheric influences atmosphärische Einflüsse mpl, Atmosphärilien fpl
   degree of influence Einflußhöhe f
   disturbing influence Störfaktor m, Störgröße f
   disturbing influences Störeinflüsse mpl
   environmental influences Umgebungseinflüsse mpl, Umwelteinwirkungen fpl
   unaffected by disturbing influences störgrößenunabhängig
   weathering influences Witterungseinflüsse mpl
influencing Einfluß -
   influencing factor Einflußfaktor m, Einflußparameter m, Einflußgröße f
   factors influencing machine performance Maschineneinflußgrößen fpl
   factors influencing the process Verfahrenseinflußgrößen fpl
information Information f
   information content Informationsgehalt m
   information interface Informationsschnittstelle f
   information output Informationsausgabe f
   information sheet Informationsschrift f
   additional information Zusatzinformation f
   advance information Vor(ab)information f
   exchange of information Informationsaustausch m
   source of information Informationsquelle f
informative value Aussagekraft f, Aussagewert m
infra-red ultrarot, infrarot, IR
   infra-red differential spectrometry Infrarot-Differenzspektrometrie f
   infra-red heater Infrarotstrahler m

   infra-red heating tunnel Infrarottunnel m
   infra-red oven Infrarotofen m
   infra-red range Infrarotbereich m
   infra-red sensor Infrarotmeßfühler m
   infra-red spectrographic infrarotspektrographisch
   infra-red spectroscopy Infrarotspektroskopie f, IR-Spektroskopie f
   infra-red spectrum Infrarotspektrum n
infrasonic range Infraschallbereich m
infusible unschmelzbar
inherent anhaftend, eigen, Eigen-
   inherent colo(u)r Eigenfarbe f
   inherent smell Eigengeruch m
   inherent tack Eigenklebrigkeit f
inhibited phlegmatisiert, gehemmt
inhibiting hemmend, inhibierend, Sperr-, verzögernd
   corrosion inhibiting korrosionsverhindernd
inhibition Hemmung f, Inhibierung f
inhibitor Inhibitor m, Phlegmatisierungsmittel n, Verzögerer m
   corrosion inhibitor Korrosionsinhibitor m
inhomogeneous inhomogen
initial anfänglich, Anfangs-
   initial adhesion Anfangshaftung f
   initial cross-section Anfangsquerschnitt m
   initial flexibility Anfangsflexibilität f
   initial injection pressure Anfangsspritzdruck m
   initial orientation Anfangsorientierung f
   initial position Anfangsposition f
   initial reaction Startreaktion f
   initial speed Startdrehzahl f, Eingangsdrehzahl f
   initial strength Anfangsfestigkeit f
   initial temperature Anfangstemperatur f
   initial tooling costs Erstwerkzeugkosten pl
   initial value Anschaffungswert m
   initial viscosity Anfangsviskosität f
   initial volume Anfangsvolumen n
initiate/to initiieren, auslösen
initiation Einleitung f, Auslösung f, Initiierung f
   crack initiation Brucheinleitung f, Rißinitiierung f
   craze initiation Craze-Initiierung f
   tear initiation force Anreißkraft f
initiator Initiator m, Initiierungsmittel n
   initiator radical Startradikal n
   initiator suspension Initiator-Suspension f
   azo initiator Azoinitiator m
   low temperature initiator Tieftemperaturinitiator m
   polymerization initiator Polymerisationsstarter m, Polymerisations-Initiator m
inject/to einspritzen, injizieren
injection Einspritzvorgang m, Einspritzung f
   injection blow mo(u)ld Spritzblaswerkzeug n
   injection blow mo(u)lded spritzgeblasen
   injection blow mo(u)lding Spritzblasformen n, Spritzgieß-Blasformen n, Spritzblasen n, Spritzblasverfahren n
   injection blow mo(u)lding machine Spritzblas-

maschine f
injection capacity Einspritzleistung f, Spritzleistung f
injection-compression mo(u)lding Prägespritzen n, Spritzprägen n
injection conditions Einspritzbedingungen fpl
injection control unit Einspritzsteuereinheit f
injection cylinder Einspritzzylinder m, Spritzzylinder m, Maschinenzylinder m
injection device Einspritzvorrichtung f
injection force Einspritzkraft f, Spritzkraft f
injection module Einspritzbaustein m
injection mo(u)ld/to spritzen
injection mo(u)ld Spritz(gieß)werkzeug n, Spritz(gieß)form f
injection mo(u)ldable spritzgießfähig, spritzgießbar
injection mo(u)lded gespritzt, spritzgegossen
injection mo(u)lded consumer goods Konsumspritzteile npl
injection mo(u)lded fittings Spritzguß-Fittings mpl
injection mo(u)lded part Spritzgieß(form)teil n
injection mo(u)lded rubber article Gummi-Spritzgußteil n
injection mo(u)lder Spritzgießaggregat n, Spritzgießer m [Person]
injection mo(u)lding Spritzgießverfahren n, Spritzguß m, Spritzgießfertigung f, Spritzgießen n, Formling m, Spritz(gieß)(form)teil n
injection mo(u)lding formulation Spritzgießrezeptur f
injection mo(u)lding grade [of mo(u)lding compound] Spritzgießmarke f
injection mo(u)lding line Spritzgießanlage f
injection mo(u)lding machine Spritzgießeinheit f, Spritzgießaggregat n
injection mo(u)lding of consumer goods Konsumentenspritzguß m
injection mo(u)lding of packaging containers Verpackungsspritzgießen n
injection mo(u)lding of plasticized PVC Weichspritzguß m, Weich-PVC-Spritzguß m
injection mo(u)lding of unplasticized PVC Hartspritzguß m, Hart-PVC-Spritzguß m
injection mo(u)lding process Spritzgießvorgang m
injection mo(u)lding resin Spritzgießharz n
injection mo(u)lding shop Spritzgießbetrieb m, Spritzgießfabrik f
injection mo(u)lding tool Spritzgießform f, Spritz(gieß)werkzeug n
injection nozzle Einspritzdüse f, Injektionsdüse f, Spritzdüse f
injection options Einspritzmöglichkeiten fpl
injection phase Einspritzphase f, Formfüllphase f, Füllphase f
injection plunger Injektionskolben m, Einspritzkolben m
injection point Einspritzpunkt m
injection pressure Einspritzdruck m, Fülldruck m, Druckstufe I f, Spritzdruck m

injection pressure regulator Spritzdruckregler m
injection pressure stage Spritzdruckstufe f
injection pressure time Spritzdruckzeit f
injection program Einspritzprogramm n
injection rate Einspritzgeschwindigkeit f, Einspritzmenge f
injection screw Spritzgießmaschinenschnecke f
injection speed profile Einspritzgeschwindigkeitsverlauf m
injection stretch blow mo(u)lding Spritzstreckblasen n
injection stretch blow mo(u)lding plant Spritzstreckblasanlage f
injection stretch blow mo(u)lding machine Spritzgieß-Streckblasmaschine f, Spritzstreckblasmaschine f
injection stroke Spritzhub m, Einspritzbewegung f, Einspritzweg m, Einspritzhub m
injection temperature Einspritztemperatur f
injection time Formfüllzeit f, Werkzeugfüllzeit f, Spritzzeit f, Einspritzzeit f
injection unit Einspritzeinheit f, Spritzeinheit f, Einspritzaggregat n, Spritzaggregat n, Einspritzseite f, Spritzseite f
injection volume Spritzvolumen n, Einspritzvolumen n
automatic carousel-type injection mo(u)lding machine Revolverspritzgießautomat m, Drehtisch-Spritzgußautomat m
automatic fast cycling injection mo(u)lding machine Schnellspritzgießautomat m
automatic injection blow mo(u)lder Spritzblasautomat m
automatic injection mo(u)lding machine Spritz(gieß)automat m
automatic reciprocating-screw injection mo(u)lding machine Schneckenspritzgießautomat m
automatic rotary injection mo(u)lding machine Rotations-Spritzgußautomat m
automatic two-colo(u)r injection mo(u)lding machine Zweifarben-Spritzgußautomat m
carousel-type injection mo(u)lding machine Drehtisch-Spritzgußmaschine f, Revolverspritzgießmaschine f
change-over from injection to holding pressure Druckumschaltung f
conventional injection mo(u)lding Normalspritzguß m, Kompaktspritzen n
conventional injection mo(u)lding machine Kompaktspritzgießmaschine f
direct feed injection Vorkammerdurchspritzverfahren n
direct screw injection system Schneckendirekteinspritzung f
eight-cavity injection mo(u)ld Achtfach-Spritzwerkzeug n
encapsulated by injection mo(u)lding [insert] umspritzt
fast cycling injection mo(u)lding machine

Hochleistungsspritzgießmaschine f,
Schnelläufer-Spritzgießmaschine f
high-pressure injection Hochdruckinjektion f
high-speed injection Schnelleinspritzen n
high-speed injection mo(u)lding Hochleistungsspritzgießen n, Leistungsspritzguß m
high-speed injection unit Hochleistungsspritzeinheit f
hot runner injection mo(u)ld Heißkanal-Spritzgießwerkzeug n
hot runner injection mo(u)lding Durchspritzverfahren n
hot runner injection mo(u)lding system Spritzgieß-Heißkanalsystem n
hydraulic injection unit Einspritzhydraulik f
initial injection pressure Anfangsspritzdruck m
installed injection capacity installierte Einspritzleistung f
low pressure injection mo(u)lding machine Niederdruckspritzgießmaschine f
modular injection mo(u)lding machine Baukasten-Spritzgießmaschine f
multi-daylight injection mo(u)ld Etagenspritzgießwerkzeug n, Stockwerk-Spritzgießwerkzeug n
multi-daylight injection mo(u)lding Etagenspritzen n
on the injection side spritzseitig
plastics injection mo(u)lding shop Kunststoffspritzerei f
plunger injection cylinder Kolbenspritzzylinder m
plunger injection mechanism Kolbenspritzsystem n
plunger injection mo(u)lding machine Kolben(spritzgieß)maschine f
plunger injection system Kolbeneinspritzsystem n
plunger injection unit Kolbenspritzeinheit f
precision injection mo(u)lded part Präzisionsspritzgußteil n
precision injection mo(u)lding Präzisionsspritzgießverarbeitung f, Präzisionsspritzguß m, Präzisionsspritzgußteil n, Qualitätsspritzguß m
range of injection mo(u)lders Spritzgießmaschinenbaureihe f, Spritzgießmaschinenprogramm n
rate of injection Einspritzgeschwindigkeit f, Spritzgeschwindigkeit f, Einspritzstrom m, Einspritz-Volumenstrom m, Einspritzrate f
reaction injection mo(u)lding [RIM] Reaktionsguß m, Reaktionsgießen n, Reaktionsgießverarbeitung f, Reaktionsspritzgießen n
reaction injection mo(u)lding machine Reaktionsgießmaschine f
ready for injection einspritzfertig
reciprocating-screw injection Schneckenkolbeninjektion f, Schneckenkolbeneinspritzung f
reciprocating-screw injection mo(u)lding machine Schneckenkolbenspritzgießmaschine f, Schubschneckenspritzgießmaschine f
reciprocating-screw injection unit Schneckenkolbeneinspritzaggregat n

rotary table injection mo(u)lding machine Spritzgießrundtischanlage f, Drehtisch-Spritzgußmaschine f
rubber injection mo(u)lding machine Gummi-Spritzgießmaschine f
screw injection cylinder Schneckenspritzzylinder m
screw injection mo(u)lding machine Schneckenspritzgießmaschine f
screw injection unit Schneckenspritzeinheit f, Schneckeneinspritzaggregat n
screw-plunger injection Schneckenkolbeninjektion f
screw-plunger injection system Schneckenkolbeneinspritzsystem n
single-cavity injection mo(u)ld Einfachspritzgießwerkzeug n
single-screw injection mo(u)lding machine Einschnecken-Spritzgießmaschine f
sprueless injection mo(u)lding angußloses Spritzen n, Vorkammerdurchspritzen n, Durchspritzverfahren n
standard injection mo(u)lding compound Standardspritzgußmarke f
thermoset injection mo(u)lding machine Duroplast-Spritzgießmaschine f
thermoset injection mo(u)lding (process) Duroplast-Spritzgießverfahren n
three-cavity injection mo(u)ld dreifaches Spritzgießwerkzeug n
tubular ram injection Ringkolbeninjektion f
tubular ram injection mo(u)lding Ringkolbeninjektionsverfahren n
twelve-cavity injection mo(u)ld Zwölffach-Spritzwerkzeug n
two-cavity injection blow mo(u)ld Zweifachspritzblaswerkzeug n
two-cavity injection mo(u)ld Zweifachspritzgießwerkzeug n
two-component injection mo(u)lding Zweikomponenten-Spritzgießen n
two-component low-pressure injection mo(u)lding machine Zweikomponenten-Niederdruck-Spritzgießmaschine f
two-stage injection blow mo(u)lding Zweistufenspritzblasen n
two-stage injection stretch blow mo(u)lding Zweistufen-Spritzstreckblasverfahren n
two-stage injection unit Zweistufen-Einspritzaggregat n
vacuum injection mo(u)lding Vakuumeinspritzverfahren n
vented injection mo(u)lding machine Entgasungsmaschine f
injector Injektor m
　high-pressure injector Hochdruckinjektionsgerät n, Hochdruckinjektor m
　ram injector unit Kolbeninjektor m
injurious to health gesundheitsschädlich
ink Tinte f, Farbe f

flexographic printing ink Flexodruckfarbe f
gravure printing ink Tiefdruckfarbe f
magnetic ink document Magnetbeleg m
printing ink Druckfarbe f
printing ink binder Druckfarbenbindemittel n
inlet Einlaß m, Eingang m
 inlet port Einlaufstutzen m
 inlet temperature Eingangstemperatur f, Eintrittstemperatur f, Einlauftemperatur f
 air inlet Lufteintritt m, Luftzufuhr f
 cooling water inlet Kühlwasserzufluß m, Kühlwasserzufuhr f
 cooling water inlet temperature Kühlwasserzulauftemperatur f
 oil inlet Ölzufluß m
 water inlet Wasseranschluß m, Wassereintritt m, Wasserzulauf m, Wasserzufuhr f, Wasserzufluß m
 water inlet port Wasserzugabestutzen m
inner zone Innenbereich m
inorganic anorganisch
input Eingang m, Eingabe f, Input m
 input cable Eingangskabel n
 input card Eingangskarte f
 input channel Eingangskanal m
 input charge Eingangsladung f
 input circuit Eingangskreis m
 input data Eingabedaten npl
 input device Eingabegerät n
 input error(s) Eingabefehler m(pl)
 input impedance Eingangsimpedanz f
 input keyboard Eingabetastatur f
 input mask Eingabemaske f
 input medium Eingabemedium n
 input module Eingabebaugruppe f, Eingabebaustein m, Eingangsmodul m, Eingabestein m
 input options Eingabemöglichkeiten fpl
 input-output channel Ein-Ausgabekanal m
 input-output device Ein-Ausgabeeinrichtung f
 input-output module Ein-Ausgabestein m
 input-output unit Ein-Ausgabeeinheit f
 input resistance Eingangswiderstand m
 input sensitivity Eingangsempfindlichkeit f
 input side Eingabeseite f, Eingangseite f
 input signal Eingangssignal n
 input station Eingabestation f
 input transistor Eingangstransistor m
 input unit Eingabeeinheit f
 input value eingebbarer Wert m, Eingabewert m
 input variable Eingangsgröße f, Eingabewert m
 input voltage Eingangsspannung f
 alarm input Alarmeingang m
 analog input Analogeingang m, Analogeingabe f
 control input Steuer(ungs)eingang m
 current input Stromaufnahme f
 data input Dateneingabe f
 digital input Digitaleingang m
 effective power input aufgenommene Wirkleistung f
 energy input Arbeitsaufnahme f, Energieaufnahme f, Energieeinleitung f
 energy input rate Energieanlieferungsrate f
 instruction input Befehlseingabe f
 job data input Auftragsdateneingabe f
 power input aufgenommene Leistung f
 setpoint input Sollwerteingabe f, Sollwertvorgabe f
 signal input Signaleingang m
 total energy input Gesamtenergieaufnahme f
insert/to einsetzen, einführen, einlegen
insert Einbetteil m, Einlage f, Einlegeteil n, Einsatz m
 insert-placing robot Einlegeroboter m, Einlegeautomat m
 insert plate Einsatzplatte f
 insert ring Einsatzring m
 cavity insert Nesteinsatz m, Gesenkeinsatz m, Kontureinsatz m, Werkzeugeinsatz m
 die insert Düseneinsatz m
 filter insert Filtereinsatz m
 mo(u)ld insert Formeinsatz m
 pin inserts eingelegte Stifte mpl
 sliding insert Schiebeeinsatz m
 sprue insert Angußeinsatz m
 steel insert Stahleinsatz m
 thread insert Gewindeeinsatz m
insertion Einsetzung f, Einlegung f
 test piece insertion device Probeneinlegevorrichtung f
inside innen, Innen-
 inside diameter [of pipes] Innendurchmesser m
 inside dimensions Stichmaße npl
 inside pipe diameter Rohrinnendurchmesser m
 inside the mo(u)ld im Werkzeuginneren n
 air inside the (film) bubble Blaseninnenluft f
 amount of air inside [e.g. a film bubble] Innenluftstrom m
 hardened on the inside [barrel] innengepanzert
insolubility Unlöslichkeit f
insoluble unlöslich
 insoluble in acids säureunlöslich
 water insoluble wasserunlöslich
inspection Beobachtung f, Prüfung f, Inspektion f
 film inspection unit Folienbeobachtungsstand m
 sample for inspection [e.g. of a mo(u)lding] Anschauungsmuster n
 visual inspection Sichtprüfung f
 visual inspection unit Sichtprüfstrecke f, Sichtprüfstelle f
instability [of a chemical compound] Labilität f
install/to einbauen, montieren
installation Einbau m, Installation f, Montage f, Anlage f
 installation costs Installationskosten pl
 installation dimensions [of a machine] Einbaumaße npl
 for underground installation für die Erdverlegung f
 method of installation Einbaumethode f
 ready for installation einbaufertig
 silo installation Siloanlage f

installed installiert
  installed heating capacity installierte Heizleistung f
  installed injection capacity installierte Einspritzleistung f
  installed load installierte Leistung f
  total installed motor power gesamte installierte Motorenleistung f
instantaneous value Momentanwert m
instruction Befehl m, Instruktion f, Kommando n
  instruction bus Befehlsbus m
  instruction counter Befehlszähler m
  instruction decoder Befehlsdecoder m
  instruction input Befehlseingabe f
  instruction register Befehlsregister n
  assembly instruction Assemblerbefehl m
  control instruction Steuerkommando n
  fitting instructions Einbaurichtlinien fpl
  handling instructions Handhabungsrichtlinien fpl
  operating instruction Bedienungsbefehl m
  operating instructions Betriebsanleitung f, Betriebsanweisung f
  program instruction Programmbefehl m
  safety instructions Sicherheitsauflagen fpl, Sicherheitshinweise mpl, Sicherheitsratschläge mpl
  servicing instructions Wartungsanweisungen fpl
  setting-up instructions Einrichtblätter npl
instrument Gerät n, Instrument n
  instrument front panel Gerätefrontplatte f
  instrument panel Instrumententafel f
  accelerated weathering instrument Kurzbewitterungsgerät n
  automatic heat impulse welding instrument Wärmeimpulsschweißautomat m
  control instrument Kontrollgerät n, Regelgerät n, Steuergerät n
  film ga(u)ge measuring instrument Foliendickenmeßgerät n
  film pretreating instrument Folienvorbehandlungsgerät n
  friction welding instrument Reibschweißgerät n
  hand welding instrument Handschweißgerät n
  heat sealing instrument Heißsiegelgerät n, Siegelwerkzeug n
  inside the instrument geräteintern
  measuring and control instruments MSR-Geräte npl
  monitoring instrument Überwachungsgerät n
  outside the instrument geräteextern
  pipe welding instrument Rohrschweißgerät n
  programming instrument Programmiergerät n
  recording instrument Aufzeichnungsgerät n, Registriergerät n
  sealing instrument Siegelgerät n
  temperature control instrument Temperaturregler m, Temperaturregelgerät n
  test instrument Prüfgerät n
  ultrasonic spot welding instrument Ultraschall-Punktschweißgerät n
  weathering instrument Bewitterungsgerät n
  welding instrument Schweißgerät n
instrumentation Geräteausrüstung f, Geräteausstattung f, Gerätetechnik f
insufficient torque Drehmoment-Defizit n
insulated isoliert, gedämmt
  insulated runner Isolier(verteiler)kanal m, Isolierverteiler m
  insulated runner feed system Isolierkanal-Angußplatte f, Isolierkanalanguß m
  insulated runner mo(u)ld Isolierkanalform f, Isolierkanalwerkzeug n
  insulated runner system Isolierverteiler-Angießsystem n, Isolierkanal-Angußplatte f, Isolierkanalanguß m
  thermally insulated thermoisoliert, wärmegedämmt
insulating isolierend, Isolier-, dämmend, Dämm-
  insulating bush Isolierbuchse f
  insulating effect Dämmeffekt m, Dämmwirkung f, Isolierwirkung f
  insulating jacket Isoliermantel m
  insulating layer Dämmschicht f, Isolationsschicht f, Isolierschicht f
  insulating material Dämmaterial n, Dämmstoff m, Isolier(werk)stoff m
  insulating plate Isolierplatte f, Temperierschutzplatte f
  insulating properties Isoliereigenschaften fpl, Isolationswerte mpl, Isolationseigenschaften fpl, Isolierwerte mpl, Isolationsvermögen n
  insulating sheet Dämmplatte f
  insulating sleeve Isolierhülse f
  insulating tube Isolierrohr n
  insulating varnish Isolierlack m
  electrical insulating component Elektroisolierteil n
  electrical insulating compound Elektroisoliermasse f
  electrical insulating material Elektroisolierstoff m
  electrical insulating properties Elektoisolierverhalten n
  electrical insulating resin Elektroisolierharz n
  electrical insulating varnish Elektroisolierlack m
  footfall sound insulating material Trittschallisoliermaterial n
  heat insulating wärmedämmend, wärmeisolierend
  heat insulating layer Wärmedämmschicht f
  heat insulating properties Wärmeisoliereigenschaften fpl
  rigid foam insulating material Hartschaumisolierstoff m
  solid-borne sound insulating material Körperschallisoliermaterial n
  sound insulating schalldämmend
  sound insulating material Schalldämmstoff m
  thermal insulating material Wärmedämmstoff m
  thermal insulating properties Wärmedämmeigenschaften fpl
insulation Isolierung f
  insulation resistance Isolierwiderstand m,

Widerstand m zwischen Stöpseln
roof insulation Dachdämmung f, Dachisolierung f
sound insulation Schallschutz m, Schalldämmung f
thermal insulation Wärmedämmung f,
Wärmeisolation f, Wärmeisolierung f
wire insulation Drahtisolierung f
insulator Isolator m
  cast epoxy insulator Epoxidgießharz-Isolator m
  high-tension insulator Hochspannungsisolator m
  outdoor insulator Freiluftisolator m
  post insulator Stützisolator m
  suspended insulator Hängeisolator m
  synthetic resin insulator Kunstharzisolator m
intaglio printing Tiefdruck m
integrable integrationsfähig
integral integral
  integral foam Integral(hart)(schaumstoff) m
  integral foam core Integralschaumstoffkern m
  integral hinge Filmscharnier n
  computer with integral screen Bildschirmcomputer m
  polyurethane integral foam Polyurethan-Strukturschaumstoff m, Polyurethan-Integralschaum m
integrated integriert
integration Integration f
  degree of integration Integrationsgrad m
intelligent intelligent
  intelligent terminal intelligentes Terminal n
intensity Intensität f, Stärke f
  colo(u)r intensity Farbintensität f, Farbstärke f
  cooling intensity Kühlintensität f
  current intensity Stromstärke f
  degassing intensity Entgasungsintensität f
  light intensity Lichtintensität f
  radiation intensity Bestrahlungsintensität f, Strahlungsintensität f
  sound pressure intensity Schalldruckstärke f
  stress intensity Spannungsintensität f
  stress intensity amplitude Spannungsintensitätsamplitude f
  stress intensity factor Spannungsintensitätsfaktor m
intensive intensiv
  intensive cooling Zwangskühlung f
  intensive kneader Intensivkneter m
  intensive mixer Intensivmischer m
inter-layer adhesion Zwischenlagenhaftung f, Zwischenschichthaftung f
inter-screw clearance Flankenspiel n, Schneckenspalt m, Flankenabstand m, Walzenspalt m
interaction Wechselwirkung f
interactive mode Dialogverkehr m, Dialogführung f, Dialogbetrieb m
interception Abfangen n, Einfangen n
  radical interception Radikaleinfang m
interceptor Abfänger m
  radical interceptor Radikalfänger m
interchangeability Austauschbarkeit f

interchangeable austauschbar, auswechselbar
interconnected [e.g. processes] ineinandergreifend (z.B. Prozesse)
interest Verzinsung f, Zinsertrag m
interesting from the processing point of view verfahrenstechnisch interessant
interface Grenzfläche f, Schnittstelle f, Interface n
  interface routine Anschlußprogramm n
  data interface Datenschnittstelle f
  particle interface Partikelgrenzfläche f
  phase interface control Phasenanschnittsteuerung f
interfacial (surface) tension Grenzflächenspannung f
interference Störung f, Interferenz f
  interference lines Interferenzlinien fpl
  interference pattern Interferenz-Grundmuster n
interferogram Interferogramm n
interferometry Interferometrie f
interior innen, Innen-
  interior casement Innenflügel m
  interior fitments Innenausstattung f
  interior paint Innenfarbe f
  interior use Inneneinsatz m
interlaminar interlaminar
  interlaminar adhesion Lagenbindung f
interlock Verriegelung f
  electrically interlocked [e.g. safety guard] elektrisch abgesichert
  guard door interlock system Schutzgittersicherung f, Schutztürsicherung f
  safety interlock system Sicherheitsverriegelung f
intermediate 1. Zwischenstufe f; 2. Zwischen-
  intermediate cooling Zwischen(ab)kühlung f
  intermediate reaction Zwischenreaktion f
  intermediate Shore hardness Zwischenshorehärte f
  intermediate stage Zwischenstufe f, Zwischenstadium n
intermeshing [screws] eingreifend, kämmend, ineinandergreifend
  intermeshing screws with a self-sealing/-wiping profile Dichtprofilschnecken fpl
  intermeshing zone [between twin screws] Zwickel(bereich) m, Eingriffsbereich m
  closely intermeshing [screws] dichtkämmend, einkämmend, voll eingreifend
  screw intermeshing Schneckeneingriff m
intermittent intermittierend, pulsierend, stoßweise, im Taktverfahren n
  intermittent extrusion taktweise Extrusion f
  intermittent operation intermittierender Betrieb m
intermolecular intermolekular, zwischenmolekular
internal innerlich, Innen-
  internal air cooling (system) Innenluftkühlung f, Innenkühlluftsystem n
  internal air pressure Innenluftdruck m
  internal air pressure calibrating unit Überdruckkalibrator m, Stützluftkalibrator m
  internal air pressure calibration Überdruck-

kalibrierung f, Stützluftkalibrierung f
internal air stream Innenluftstrom m
internal bubble cooling (system) Folieninnen-kühlvorrichtung f, Folieninnenkühlung f, Blaseninnenkühlung f, Blaseninnenkühlsystem n
internal calibration Innenkalibrierung f
internal compressive stresses Druckeigen-spannungen fpl
internal cooling Innenkühlung f
internal cooling air Innenkühlluft f
internal cooling air stream Innenkühlluftstrom m
internal cooling stresses Abkühleigenspannungen fpl, innere Abkühlspannungen fpl
internal cooling system Innenkühlung f, Innenkühlsystem n
internal cooling with liquid nitrogen Stickstoff-innenkühlung f
internal dimensions Innenmaße npl
internal friction innere Reibung f
internal gas pressure Gasinnendruck m
internal gear pump Innenzahnradpumpe f
internal lubricant inneres Gleitmittel n
internal lubrication innere Gleitwirkung f, Innenschmierung f
internal memory interner Speicher m, Kernspeicher m
internal mixer Innenkneter m, Innenmischer m, Stempelkneter m
internal oil heating system Ölinnentemperierung f
internal plasticization innere Weichmachung f
internal pressure Innendruck m
internal screw cooling (system) Schneckeninnen-kühlung f
internal screw diameter Schneckeninnen-durchmesser m
internal standard (specification) Hausnorm f, firmeninterne Norm f
internal stress Eigenspannung f
internal stresses Eigenspannungsfeld n
internal tensile stress Zugeigenspannung f
internal thread Innengewinde n
amount of internal cooling air Innenkühlluftstrom m
deformation through internal pressure Innen-druckverformung f
heavy-duty internal mixer Hochleistungsinnen-mischer m
long-term failure test under internal hydrostatic pressure [test used for plastics pipes] Zeitstand-Innendruckversuch m
long-term internal pressure test for pipes Rohrinnendruckversuch m
long-term resistance to internal hydrostatic pressure [of plastics pipes] Innendruck-Zeitstandwert m
resistance to internal pressure Innendruckfestigkeit f
state of internal stress Eigenspannungszustand m
internally innen
   internally hardened innengepanzert

internally heated innenbeheizt
internally plasticized innerlich weichgemacht
interplay Zusammenspiel n
interpretation [e.g. of test results] Auswertung f, Interpretation f
interrogate/to abfragen
interrogation [of data] Abfrage f
interrupted flights Gangdurchbrüche mpl
interruption of the mo(u)lding cycle Zyklus-unterbrechung f, Zyklusstörung f
intersection Kreuzung f, Schnittpunkt m
   point of intersection Kreuzungspunkt m, Schnittpunkt m
interstice Zwischenraum m
interval Zeitabschnitt m, Zeitspanne f, Intervall n, Zeitintervall n, Zeitraum m
   interval timing mechanism Pausenzeituhr f, Intervallzeitgeber m
intramolecular innermolekular, intramolekular
intrinsic viscosity Grenzviskosität f, Staudinger-Index m
introduction Einleitung f
   introduction of shear forces Scherkrafteinleitung f
intrusion (process) Intrusionsverfahren n
inventory Vorrat m, Bestand m; Inventur f, Bestandsaufnahme f
inverted invertiert
   four-roll inverted L-type calender Vierwalzen-L-Kalander m
investment Kapitalanlage f, Investition f
   investment program Investitionsprogramm n
   investment volume Investitionsvolumen n
   foreign investments Auslandsinvestitionen fpl
   need for capital investment Investitionsbedarf m
   rate of investment Investitionsrate f
   total investments Gesamtinvestitionen fpl
iodine colo(u)r value Jodfarbzahl f
iodometric jodometrisch
ion Ion n
   ion complex Ionenkomplex m
   ion interceptor Ionenfänger m
   complex ion Komplexion n
   sulphate ion Sulfation n
ionic ionisch
   ionic bond Ionen(ver)bindung f
   ionic character Ionogenität f
   ionic lattice structure Ionengitter n
ionitrided ionitriert
ionitriding Ionitrieren n
ionization Ionisierung f, Ionisation f
   ionization resistance Ionisationsbeständigkeit f
ionizing ionisierend
ionogenic ionogen
ionomer (resin) Ionomerharz n
iron Eisen n
   iron-constantan thermocouple Fe-Ko-Thermo-element n, Eisen-Konstantan-Thermoelement n
   iron-on [interlinings] einbügelbar

iron sheet Eisenblech n
brown iron oxide Eisenoxidbraun n
electric iron Bügeleisen n
gray cast iron Grauguß m
micaceous iron ore Eisenglimmer m
red iron oxide Eisenoxidrot n
irradiated bestrahlt
irradiation Bestrahlung f
   electron irradiation Elektronenbestrahlung f
   gamma irradiation Gammabestrahlung f
   long-term irradiation Langzeitbestrahlung f
   solar irradiation Sonnenbestrahlung f
irregular unregelmäßig
irrespective of the number of items stückzahlunabhängig
irreversible irreversibel
irrigation pipe Bewässerungsrohr n, Irrigationsrohr n
irritant reizend, ätzend, Reiz-
iso-compound Isoverbindung f
isobar Isobare f [Luftdruck]; Isobar m [Kernphysik]
isobaric isobar
isobutyl stearate Isobutylstearat n
isobutylene-isoprene rubber Isobutylen-Isopren-Kautschuk m
isochore Isochore f, Linie f konstanten Volumens
isochoric isochor
isochromatic photograph Isochromatenaufnahme f
isochronous isochron
isocyanate Isocyanat n
   isocyanate group Isocyanatgruppe f
   isocyanate resin Isocyanatharz n
isocyanurate Isocyanurat n
   triglycidyl isocyanurate Triglycidylisocyanurat n
isododecane Isododecan n
isolated glassy zones Glasinseln fpl
isomer Isomer n
isomeric isomer
isomerism Isomerie f
   cis-trans isomerism Cis-Trans-Isomerie f
isophorone Isophoron n
   isophorone diisocyanate Isophorondiisocyanat n
isophthalic acid Isophthalsäure f
isoprene Isopren n
   isoprene rubber Isoprenkautschuk m
isopropanol Isopropanol n
isopropyl acetate Isopropylacetat n
isostatic mo(u)lding isostatisches Pressen n
isotactic isotaktisch
isothermal isotherm
isotope thickness ga(u)ge Isotopendickenmeßgerät n
isotropic isotrop
isotropy Isotropie f
itemized cost Einzelkosten pl
Izod notched impact strength Izod-Kerbschlagzähigkeit f
jacket Mantel m
   cooling jacket Kühlmantel m
   double-walled jacket Doppelmantel m
   heat insulating jacket Wärmeschutzmantel m
jacketed thermocouple Mantelthermoelement n
jam/to blockieren, verklemmen, festfressen
jerrycan Kanister m
jet Strahl m
   jet black tiefschwarz
   jet fuel Düsentreibstoff m
   jet of material freier Strahl m
   compressed air jet Preßluftstrahl m, Preßluftstrom m
   water jet cutting Wasserstrahlschneiden n
jetting Würstchenspritzguß m, Freistrahlbildung f
jig Lehre f, werkzeugsteuernde Vorrichtung f, Spannvorrichtung f
   bending jig Biegeschablone f
   drill jig Bohrlehre f
   test jig Prüflehre f
   welding jig Schweißlehre f
job 1. Auftrag m; 2. Job [EDV]
   job (data) file Auftragsdatei f
   job data input Auftragsdateneingabe f
   job number Auftragsnummer f
   job status Auftragszustand m
   (injection) mo(u)lding jobs Spritzgießaufgaben fpl
join/to zusammenfügen
joining Füge-, Verbindungs-
   joining pressure Fügedruck m
   method of joining Verbindungsverfahren n
   parts to be joined Fügeteile npl
   surface to be joined Fügeflächen fpl
joint Fuge f, Stoß m
   bonded alumin(i)um joint Aluminium(ver)klebung f
   bonded glass joint Glas(ver)klebung f
   bonded joint Klebeverbindung f
   bonded plastics joint Kunststoff(ver)klebung f
   bonded steel joint Stahl(ver)klebung f
   butt joint Stumpfstoß m, Stumpfverbindung f
   butt-welded joint Stumpfnaht f, Stumpfschweißverbindung f
   corner joint Eckverbindung f
   expansion joint Bauwerksfuge f, Bewegungsfuge f, Dehnfuge f
   flange(d) joint Flanschverbindung f
   ground-glass joint Schliffverbindung f
   lap joint Überlappstoß m, Überlappungsverklebung f
   lap welded joint Überlappnaht f
   permanent joint unlösbare Verbindung f
   pipe joint Rohrverbindung f
   snap-in joint Schnappverbindung f
   socket joint Muffenverbindung f
   temporary joint lösbare Verbindung f
   universal joint Kreuzgelenk(kupplung) f
   welded joint Schweißfuge f, Schweißverbindung f
   welded socket joint Muffenschweißverbindung f
jointing Verfugen n
   jointing filler Fugenmasse f, Fugenspachtel m, Fugenmörtel m, Fugenfüller m

jointless fugenlos, nahtlos
   jointless flooring compound Bodenausgleichmasse f, Fußbodenausgleichmasse f, Bodenspachtelmasse f
journal Lagerzapfen m, Wellenzapfen m
   journal deflection Zapfendurchbiegung f
   roll journal Walzenzapfen m
justifiable vertretbar
K-value K-Wert m
kaolin(e) Kaolin n
keto acid Ketosäure f
keto-carbonyl group Ketocarbonylgruppe f
keto group Ketogruppe f
ketone Keton n
   ketone hydroperoxide Ketonhydroperoxid n
   ketone peroxide Ketonperoxid n
   ketone resin Ketonharz n
key Schlüssel m, Taste f
   key in/to einlesen
   key-operated switch Schlüsselschalter m
   key position Schlüsselstellung f
   key product Schlüsselprodukt n
   key role Schlüsselrolle f
   by pressing a key auf Tastenaufdruck m
   control key Bedientaste f
   function key Funktionstaste f
   numeric key Zahlentaste f
   program key Programmtaste f
   program selection key Programmwahltaste f
   setpoint selector key Sollwertauswahltaste f
   symbolic key Symboltaste f
keyboard Tastatur f, Tastaturfeld n, Tastenfeld n, Tastenplatte f
   character keyboard Klarschrifttastatur f, Klartext-Tastatur f
   control keyboard Bedienungstastatur f
   input keyboard Eingabetastatur f
   touch-sensitive keyboard Folientastatur f
keypad kleine Tastatur f
   numeric keypad Zifferntastatur f
kick-off temperature [of a catalyst] Anspringtemperatur f
kicker Kicker m; Zersetzungsbeschleuniger m [Schaum]
kieselgu(h)r Diatomenerde f, Kieselgur f
kinematic viscosity kinematische Viskosität f, kinematische Zähigkeit f
kinetic kinetisch
   kinetic energy kinetische Energie f
kinetics Kinetik f
kitchenware Küchengeschirr n
kneadable knetbar
kneader Kneter m
   kneader blade Knetarm m
   kneader housing [of a mixer or kneader] Knetergehäuse n
   heavy duty kneader Hochleistungskneter m
   screw flights and kneader discs Schnecken- und Knetelemente npl

kneading Kneten n
   kneading block Knetblock m
   kneading block assembly Knetblockanordnung f
   kneading cog Knetzahn m
   kneading disks Knetelemente fpl, Knetscheiben fpl
   kneading section Knetgehäuse n
knife Messer n
   knife application Rakel(messer)auftrag m, Rakeln n, Spachteln n
   knife blade Rakelblatt n
   knife-blanket coater Gummituchrakel n
   knife-roll coater Walzenrakel n
   air knife Luftmesser n, Luftrakel n
   all-steel knife Ganzstahlmesser n
   bed knife Gegenmesser n, Festmesser n
   bed knife block Statormesserbalken m, Statormesserblock m
   bed knife cutting circle Statormesserkreis m
   doctor knife Rakelmesser n
   fixed knife Statorscheibe f
   flying knife Ablängvorrichtung f
   granulator knife Granulier(messer)flügel m
   pelletizer knife Granulier(messer)flügel m
   putty knife Spachtel f, Ziehklinge f
   rotor knife Kreismesser n, Rotormesser n, Rotorscheibe f
   rotor knife block Rotormesserbalken m
   rotor knife cutting circle Rotormesserkreis m, Kreismesserwelle f
   spreading knife Streichmesser n
   square-section knife Vierkantmesser n
   stationary knife Festmesser n, Gegenmesser n, Statormesser n
knifing filler Spachtel f, Spachtelmasse f, Ausgleichsmasse f
   epoxy-based knifing filler EP-Spachtel f
knitted fabric Gewirke n
knives Zerkleinerungselemente npl
knob Drehknopf m
knock-out pin Auswerferstift m, Auswerferstempel m
known value Erfahrungswert m
knurled mixing section [of screw] Igelkopf m, Nockenmischteil n
L-type L-Form f
   five-roll L-type calender Fünfwalzen-L-Kalander m
   four-roll inverted L-type calender Vierwalzen-L-Kalander m
laboratory Labor n
   laboratory calender Laborkalander m
   laboratory coating machine Laborbeschichtungsanlage f
   laboratory conditions Laborbedingungen fpl
   laboratory extruder Labor(meß)extruder m
   laboratory mixing rolls Labormischwalzwerk n
   laboratory oven Wärmeschrank m, Laborschrank m
   laboratory scale Labormaßstab m
   laboratory single-screw extruder Labor-Einwellen-

extruder m
laboratory test Laborversuch m
laboratory thermoforming machine Laborziehgeräte npl
laboratory twin screw extruder Doppelschnecken-Laborextruder m
laboratory two-roll mill Zweiwalzen-Labormischer m
material testing laboratory Werkstoffprüflabor n
labo(u)r costs Lohnkosten pl, Personalkosten pl
labo(u)r-intensive arbeitsaufwendig, personalaufwendig, personalintensiv
lack of oxygen Sauerstoffmangel m
lacquer Firnis m, Lack m
  lacquer coating [e.g. for PVC leather cloth] Schlußlack m
  lacquer film Lackfilm m
  catalyzed lacquer Reaktionslack m
  clear lacquer Klarlack m, Lasur f
  heat sealing lacquer Heißsiegellack m
  nitrocellulose lacquer Nitrocelluloselack m
  wood lacquer Harzfirnis m
lactam Lactam n
ladder polymer Leiterpolymer n
laid above ground [e.g. pipes] freiverlegt
laitance Schlempeschicht f, Zementschlämme f, Zementschlamm m
lambswool polishing wheel Lammfellscheibe f
lamella [plural: lamellae] Lamelle f
  venting lamella Entlüftungslamelle f
lamellar lamellar, lamellenartig
lamellar structure Lamellenstruktur f
laminar laminar, schichtförmig
  laminar flow laminares Fließen n, laminarer Fluß m, Schichtenströmung f, laminare Strömung f
  laminar flow distributor Laminarflußverteiler m
laminate/to doublieren, kaschieren, laminieren
laminate Laminat n, Schicht(preß)stoff m
  laminate pipe Schichtpreßstoffrohr n
  laminate sheet Schichtpreßstoffplatte f, Schichtpreßstofftafel f
  laminate structure Laminataufbau m
  laminate surface Laminatoberfläche f
  laminate web Laminatband m
  decorative laminate Dekorlaminat n, Dekorschichtstoff m
  epoxy glass cloth laminate Epoxidharz-Glashartgewebe n
  epoxy laminate Epoxharz-Schichtstoff m
  epoxy paper laminate Epoxidharzhartpapier n
  fabric-based laminate Hartgewebe n
  fabric-based laminate sheet Hartgewebetafel f
  glass cloth laminate Glasgewebelaminat n, Glashartgewebe n
  glass cloth laminate pipe Glashartgeweberohr n
  glass cloth laminate sheet Glashartgewebeplatte f
  glass fiber laminate Glaslaminat n
  glass mat reinforced polyester laminate GF-UP-Mattenlaminat n
  hand lay-up laminate Handlaminat n
  high-pressure laminate Hochdrucklaminat n
  hot press mo(u)lded laminate Warmpreßlaminat n
  industrial laminates technische Schichtstoffplatten fpl
  paper-based laminate Hartpapier(laminat) n, Papierschichtstoff m
  phenolic laminate Phenolharz-Schichtstoff m
  phenolic paper laminate Phenolharz-Hartpapier n
  press-mo(u)lded laminate Preßlaminat n
  roving laminate Rovinglaminat n
  spray lay-up laminate Spritzlaminat n
  spray-up laminate Faserspritzlaminat n
  synthetic resin laminate Kunstharzlaminat n
  urea laminate Harnstoffharz-Schichtstoff m
laminating Kaschier-, Laminier-
  laminating adhesive Kaschierklebstoff m, Laminierkleber m
  laminating calender Laminierkalander m, Schmelzwalzenkalander m
  laminating coating Kaschierstrich m
  laminating machine Laminiermaschine f
  laminating mo(u)ld Laminierform f
  laminating paste Kaschierpaste f, Laminierpaste f
  laminating plant Kaschieranlage f, Laminieranlage f
  laminating pressure Laminierdruck m
  laminating process Laminierverfahren n
  laminating resin Laminierharz n
  laminating roll Kaschierwalze f
  laminating unit Kaschierwerk n, Kaschiervorrichtung f, Laminator m
  contact laminating calender Doublierkalander m
  film laminating plant Folienkaschieranlage f
  heat laminating plant Thermokaschieranlage f
laminator Kaschiermaschine f, Laminiermaschine f
  contact laminator unit Doubliereinrichtung f
lamp Lampe f
  lamp black Flammruß m, Lampenruß m
  lamp socket Lampensockel m, Lampenfassung f
  electric filament lamp Glühlampe f
  fluorescent lamp Leuchtstoffflampe f
  fog lamp Nebelleuchte f
  xenon lamp Xenonbogenstrahler m
lampshade Lampenschirm m
language Sprache f
  assembly language Assemblersprache f
  compiler language Compilersprache f
  machine language Maschinensprache f
  plain language [i.e. plain English or plain German] Klartext m
  plain language display Klartextanzeige f
  plain language print-out Klartextausdruck m
  plain language signal Klartextmeldung f
  program language Programm(ier)sprache f
lap Überlappung f
  lap joint Überlappstoß m, Überlappungsverklebung f
  lap welded joint Überlappnaht f, überlappt

geschweißte Naht f
lap welding Überlappschweißen n
large groß(dimensioniert), weit, geräumig
   large-area großflächig
   large-area melt filter Großflächen-Schmelzefilter n,m
   large-bore pipe Großrohr n
   large-bore pipe calibration Großrohrkalibrierung f
   large-bore pipe die Großrohrwerkzeug n
   large-bore pipe extrusion line Großrohrstraße f
   large-bore pipe production line Großrohrstraße f
   large-capacity silo Großsilo n
   large extruder Großextruder m
   large mo(u)ld Großwerkzeug n
   large mo(u)lding Großformteil n
   large numbers große Stückzahlen fpl
   large-scale großtechnisch
   large-scale burning test Brandgroßversuch m
   large-scale production Großproduktion f, Großserienfertigung f
   large-scale production unit Großproduktionsmaschine f
   large-scale trial Großversuch m
   large-size großformatig
   large-volume großvolumig
   large-volume chute Großvolumenschurre f
   extra large überdimensioniert, übergroß
   machine for making large blow mo(u)ldings Großhohlkörper-Blasmaschine f
   production of large-container production unit Großhohlkörperfertigung f
   production of large containers Großhohlkörperfertigung f
largely soluble weitgehend löslich
laser Laser m
   laser beam Laserlicht n, Laserstrahl m
   laser beam cutting Laserstrahlschneiden n
   laser beam cutting process Laserschneidverfahren n
   laser controlled lasergesteuert
   laser pyrolysis Laserpyrolyse f
   laser scanning analyzer Laser-Abtastgerät n
lasting dauerhaft
lateral lateral, seitlich
   lateral feed device Seitenspeisevorrichtung f
   lateral (film) bubble guide Seitenführung f
   lateral film gate seitlicher Bandanschnitt m
latex [water emulsion of rubber polymer] Latex m (Gummi-Wasser-Emulsion)
   latex particle Latexteilchen n
   dewatered latex [e.g. for making solid parts] wasserfreier Latex m
   nitrile rubber latex Nitrillatex m
   synthetic resin latex Kunststofflatex m
latitude Breite f, Weite f, Spielraum m, Umfang m
   processing latitude Verarbeitungsbreite f, Verarbeitungsspielraum m
lattice Gitter n
   lattice structure Gitterstruktur f

crystal lattice Kristallgitter n
molecular lattice structure Molekülgitter n
laurate Laurat n
   cadmium laurate Cadmiumlaurat n
   dialkyltin laurate Dialkylzinnlaurat n
   vinyl laurate Vinyllaurat n
lauric acid Laurinsäure f
laurinlactam Laurinlactam n
lauroyl peroxide Lauroylperoxid n
law Gesetz n
   Law on the Constitution of Business Betriebsverfassungsgesetz n
   flow law index Fließgesetzexponent m
   model law Modellgesetz n
   Ohm's law Ohmsches Gesetz n
   power law Potenzansatz m, Potenz(fließ)gesetz n
   power law behavio(u)r Potenzgesetzverhalten n
   power law flow Potenzgesetzverhalten n
   power law fluid Potengesetz-Flüssigkeit f, Potenzgesetzstoff m
   power law index Potenzgesetzexponent m
layer Schicht f
   barrier layer Sperrschicht f
   boundary layer Grenzschicht f
   hardened steel (outer) layer Hartstahlmantel m
   heat insulating layer Wärmedämmschicht f
   insulating layer Dämmschicht f
   melt layer Schmelzeschicht f
   outer layer Randschicht f
   surface layer Oberflächenschicht f
   top layer [e.g. of a sandwich structure] Deckschicht f
   wear resistant layer coating Verschleißschicht f
layflat width flachgelegte Breite f, Flachlegebreite f, Flachliegebreite f, flachgelegte Folienbreite f, Schlauchliegebreite f
layout Auslegung f, Anordnung f
   cooling channel layout Kühlkanalanordnung f
   heating-cooling channel layout Temperierkanalauslegung f
   heating tunnel layout Temperierkanalauslegung f
   technical layout verfahrenstechnische Auslegung f
lay-up Auflegen n
   hand lay-up Handauflegen n, Handlaminieren n
   hand lay-up laminate Handlaminat n
   hand lay-up process Hand(auflege)verfahren n, Handlaminerverfahren n
   spray lay-up Faser(harz)spritzen n
   spray lay-up equipment Faserspritzanlage f
   spray lay-up laminate Spritzlaminat n
   spray lay-up plant Faser(harz)spritzanlage f
LCD [liquid crystal display] Flüssigkristall-Anzeige f, LCD-Display n
locking mechanism Verriegelungssystem n
LDPE [low density polyethylene] Hochdruck-Polyethylen n
leach out/to auslaugen

lead Blei n
  lead carbonate Bleicarbonat n
  lead chromate pigment Bleichromatpigment n
  lead compound Bleiverbindung f
  lead-containing bleihaltig
  lead content Bleigehalt n
  lead emission Bleiabgabe f
  lead maleate Bleimaleat n
  lead phosphite Bleiphosphit n
  lead phthalate Bleiphthalat n
  lead pigment Bleipigment n
  lead salicylate Bleisalicylat n
  lead shot Bleischrot n
  lead soap Bleiseife f
  lead stabilized bleistabilisiert
  lead stabilizer Bleistabilisator m
  lead stearate Bleistearat n
  lead sulphate Bleisulfat n
  containing lead bleihaltig
  total lead content Gesamtbleigehalt m
  white lead Bleiweiß n
leading edge of flight treibende Flanke f, (vordere) Stegflanke f, aktive Flanke f, Schneckenflanke f, Gewindeflanke f, aktive Stegflanke f
leaf spring Blattfeder f
leaflet Druckschrift f, Prospekt n
leak Undichtheit f, Leckage f, Leck n
  leak-proof leckagefrei, dicht
  leak tester Leckprüfgerät n
leakage current Ableitstrom m, Kriechstrom m
leakage flow Verlustströmung f, Leckströmung f
leakproof leckagefrei, dicht
leather grain effect Ledernarbung f
leathercloth Lederkleidung f
  foam leathercloth Schaumkunstleder n
  PVC leathercloth Gewebekunstleder n, Kunstleder n
LED [light-emitting diode] Leuchtdiode f
  LED adaptor Leuchtdioden-Adapter m
  LED display LED-Leuchtanzeige f, Leuchtziffernanzeige f
  LED matrix Leuchtdiodenmatrix f
left-hand thread Linksgewinde n
legislation Gesetzgebung f
legislative measures Gesetzgebungsmaßnahme f
leisure industry Freizeitsektor m
length Länge f
  barrel length Zylinderlänge f
  breaking length Reißlänge f
  change in length Längenänderung f
  crack length Rißlänge f
  cut into lengths/to [extruded pipes, profiles or sheets] ablängen
  cylinder length Zylinderlänge f
  effective length Arbeitslänge f, Funktionslänge f
  effective screw length Schneckenarbeitslänge f
  extruder length Extruderbaulänge f
  flow length Fließweglänge f
  flow length-wall thickness ratio Fließweg-Wanddickenverhältnis n
  increase in length Verlängerung f
  initial measured length Ausgangs(meß)länge f
  machine length Maschinenlänge f
  molecule chain length Molekülkettenlänge f
  molecule length Moleküllänge f
  of equal length gleichlang
  oil length Öllänge f
  original length Ausgangslänge f
  parison length Vorformlingslänge f
  parison length control (mechanism) Vorformlingslängenregelung f, Schlauchlängenregelung f
  parison length controller Schlauchlängenregelung f, Vorformlingslängenregelung f
  parison length programming (device) Schlauchlängen-Programmierung f
  screw length Schneckenbaulänge f
  spiral flow length Spirallänge f
  standard length Fixlänge f, Standardlänge f
  total length Gesamtlänge f
  word length Daten(wort)breite f, Wortlänge f
Leno weave Dreherbindung f [Textil]
  mock Leno wave Scheindreherbindung f
less kleiner, geringer
  less hardener Härterunterschuß m
  using less energy energiegünstiger
letter Buchstabe m
  self-adhesive letter Klebefolienbuchstabe m, Selbstklebebuchstabe m
letterpress printing Buchdruck m
level Niveau n, Höhe f, Pegel m, Füllstand m
  level controller Niveaukontrolle f, Niveauwächter m, Niveauregelung f, Niveauüberwachung f
  level indicator Füllstandangeber m, Füllstandmelder m, Füllstandanzeige(r) f(m), Niveaumelder m
  level monitoring device Niveauwächter m, Niveauregelung f, Niveaukontrolle f, Niveauüberwachung f, Füllstandüberwachung f
  level monitoring unit Niveauwächter m, Niveauregelung f, Niveaukontrolle f, Niveauüberwachung f, Füllstandüberwachung f
  at floor level bodeneben
  noise level Schallpegel m, Geräuschniveau n
  oil level Ölstand m, Ölniveau n, Ölspiegel m
  oil level control (unit) Ölstandkontrolle f
  oil level indicator Ölstandanzeiger m
  oil level monitoring Ölstandüberwachung f
  overall noise level Gesamtschallpegel m
  sound pressure level Schalldruckstärke f
  stress level Beanspruchungshöhe f, Belastungshöhe f
  temperature level Temperaturniveau n
  water level Wasserstand m
lever Hebel m
  actuating lever Betätigungshebel m

operating lever Betätigungshebel m
toggle lever Einfachkniehebel m
Lewis acid Lewissäure f
liability Verbindlichkeit f
   liability to crack Bruchanfälligkeit f
liable to give trouble störanfällig, störempfindlich
lidded drum Deckelfaß n
life Leben n
   life of the die Düsenstandzeit f
   cost-service life factor Kosten-Standzeit-Relation f, Preis-Lebensdauer-Relation f
   cost-service life ratio Kosten-Standzeit-Relation f, Preis-Lebensdauer-Relation f
   guaranteed shelf life Lagerfähigkeitsgarantie f
   half life period Halbwertzeit f
   mo(u)ld life Werkzeugstandzeit f
   pot life Topfzeit f, Verarbeitungsperiode f, Verarbeitungsspielraum m, Verarbeitungszeit f, Gebrauchsdauer f
   required pot life Verarbeitungsbedarf m
   shelf life Lägerfähigkeit f
   shelf life problems Lagerstabilitätsschwierigkeiten fpl
   storage life Lagerbeständigkeit f, Lager(ungs)eigenschaften fpl
lift/to heben
lifting hebend, Hebe-
   lifting device Hebeeinrichtung f
   lifting gear Hebezeug n
   lifting stroke Hebehub m
ligands Liganden mpl
light 1. Licht n, Fenster n; 2. leicht
   light absorption Lichtabsorption f
   light ag(e)ing Lichtalterung f
   light beam Lichtstrahl m
   light beam guard Lichtschranke f, Lichtvorhang m
   light beam oscillograph Lichtstrahloszillograph m
   light bulb Glühlampe f
   light colo(u)red hellfarbig
   light-emitting diode [LED] Leuchtdiode f
   light energy Lichtenergie f
   light exposure test Belichtungsprüfung f
   lightfast lichtstabil, lichtecht
   lightfastness Lichtechtheit f
   light gray hellgrau
   light-induced lichtinduziert
   light intensity Lichtintensität f
   light metal housing Leichtmetallgehäuse n
   light microscope Auflichtmikroskop n, Lichtmikroskop n
   light resistance Lichtbeständigkeit f
   light resistant lichtbeständig
   light scattering Lichtstreuung f
   light sensitive lichtempfindlich
   light sensitivity Lichtempfindlichkeit f
   light signal Lichtsignal n, Leuchtanzeige f
   light signal indicator panel Leuchtanzeigetableau n
   light source Lichtquelle f
   light spot line recorder Lichtpunktlinienschreiber m
   light stability Lichtstabilität f
   light stabilization Lichtschutz m, Lichtstabilisierung f
   light stabilized lichtstabilisiert
   light stabilizer Lichtschutzzusatz m, Lichtschutzmittel n
   light stabilizing lichtstabilisierend
   light stabilizing effect Lichtstabilisatorwirkung f, Lichtschutzeffekt m, Lichtstabilisierwirkung f
   light stabilizing system Lichtschutzsystem n
   light transmission Lichtdurchlässigkeit f, Lichtdurchgang m, Lichttransmission f
   dome light Lichtkuppel f
   exposure to light Belichtung f
   flashing light signal Blinksignal n
   flashing light warning signal Blink-Störungslampe f, Störungsblinkanzeige f
   impermeability to light Lichtundurchlässigkeit f
   impermeable to light lichtundurchlässig
   indicator light Anzeigelampe f
   percentage light transmission Lichtdurchlässigkeitszahl f
   pilot light Kontrollampe f, Überwachungskontrollampe f
   rear light Rücklicht n
   rear light housing Rücklichtgehäuse n
   warning light Alarmlampe f, Warnlampe f, Warnleuchte f
lightweight leicht(gewichtig), Leicht-
   lightweight concrete Leichtbeton m
   lightweight construction Leichtbau(weise) m(f)
   lightweight filler Leichtfüllstoff m
   lightweight material Leicht(bau)werkstoff m
limestone Kalkstein m
limit Grenze f, Grenzwert m, Schwellenwert m
   limit contact Grenz(wert)kontakt m
   limit of error Fehlergrenze f
   limit switch Endschalter m, Grenzwertschalter m
   breaking limit Bruchgrenze f
   compatibility limit Verträglichkeitsgrenze f
   deformation limit Deformationsgrenze f, Verformungsgrenze f
   elastic limit Dehnungsgrenze f, Elastizitätsgrenze f
   endurance limit Zeitwechselfestigkeit f, Dauer(schwell)festigkeit f, Dauerstandfestigkeit f, Langzeitfestigkeit f, Schwellfestigkeit f, Wechselfestigkeit f
   explosive limit Explosionsgrenze f
   fatigue limit Zeitwechselfestigkeit f, Dauer(schwell)festigkeit f, Dauerstandfestigkeit f, Langzeitfestigkeit f, Schwellfestigkeit f, Wechselfestigkeit f
   long-term temperature limit Dauertemperaturgrenze f
   lower limit unterer Grenzwert m, Untergrenze f
   lower temperature limit Temperatur-Untergrenze f
   performance limit Leistungsgrenze f

pressure limit Drucklimit n
proportional limit Proportionalitätsgrenze f
stroke limit (switch) Hubbegrenzung f
switch limit Schaltschwelle f
temperature limit Temperaturgrenze f, Temperaturgrenzwert m
tolerance limit Toleranzgrenze f
tolerance limits Toleranzbänder npl
top limit oberer Grenzwert m, Obergrenze f
upper limit oberer Grenzwert m, Obergrenze f
upper temperature limit Temperatur-Obergrenze f
viscosity limit Viskositätsgrenze f
limitation Begrenzung f
ejector stroke limitation Auswerferhubbegrenzung f
pressure limitation Druckbegrenzung f
technical limitations verfahrenstechnische Nachteile mpl
limited begrenzt
limited resistance [material in the presence of chemicals] bedingt beständig
limiting begrenzend, Grenz-
limiting concentration Grenzkonzentration f
limiting device Begrenzung f
limiting elongation at break Grenzbruchdehnung f
limiting locking force Grenzzuhaltekraft f
limiting molecular weight Grenzmolekulargewicht n
limiting pressure Grenzdruck m
limiting screw speed Grenzdrehzahl f [Schnecke]
limiting shear stress Grenzschubspannung f
limiting speed Grenzdrehzahl f
limiting strain Grenzdehnung f
limiting stress Grenzspannung f
limiting temperature Grenztemperatur f, Temperaturgrenzwert m
limiting value Grenzwert m
limiting value control Grenzwertüberwachung f
limiting value signal Grenzwertmeldung f, Grenzwertsignal n
limiting viscosity number Staudinger-Index m, Grenzviskosität f, Viskositätsgrenze f
ejector stroke limiting device Auswerferhubbegrenzung f, Öffnungswegbegrenzung f
mo(u)ld opening stroke limiting device Werkzeugöffnungshub-Begrenzung f
opening stroke limiting device Öffnungswegbegrenzung f
line 1. Linie f, Reihe f, Zeile f, Markierung f; 2. Anlage f
air supply line Luftzuführungsschlauch m
assembly line Fließband n
assembly line production Fließbandfertigung f
blow mo(u)lding line Hohlkörperblasanlage f, Blasanlage f, Blaslinie f
blown film coextrusion line coextrudierende Schlauchfolienanlage f, Coextrusions-Blasfolienanlage f, Schlauchfolienextrusionsanlage f
blown film line Folienblasanlage f
bottle blowing line Flaschenproduktionslinie f

cable sheathing line Kabelummantelungsanlage f, Ummantelungsanlage f
cast film line Gießfolienanlage f
center line Mittellinie f
chill roll casting line Chillroll-Anlage f, Chillroll-Filmgießanlage f, Chillroll-Flachfolien-Extrusionsanlage f
coating line Beschichtungsanlage f
coextrusion coating line Coextrusionsbeschichtungsanlage f
coextrusion line Coextrusionsanlage f, Koextrusionsanlage f
combined extrusion-thermoforming line Extrusionsformanlage f
compactly designed film extrusion line Folienkompaktanlage f
compounding line Aufbereitungsstraße f, Aufbereitungsanlage f, Compoundierstraße f
continuous painting line Durchauflackieranlage f
cooling water line Kühlwasserleitung f
corrugated pipe extrusion line Wellrohranlage f
delivery line [of hydraulic system] Druckleitung f
discharge line Entsorgungsleitung f
display line Schriftzeile f, Textzeile f, Anzeigezeile f
exhaust line Absaugleitung f
expanded film extrusion line Schaumfolienanlage f
extrusion blow mo(u)lding line Blasextrusionsanlage f
extrusion coating line Extrusionsbeschichtungslinie f, Extrusionsbeschichtungsanlage f
extrusion line Extruderanlage f, Extruderstraße f, Extrusionslinie f, Extrusionsstraße f
feed line Versorgungsleitung f, Anströmleitung f
fiber extrusion line Faser(extrusions)anlage f
film blowing line Folienblasanlage f, Schlauchfolienanlage f, Foliengießanlage f
film tape production line Folienbändchenanlage f
flat film coextrusion line coextrudierende Flachfolienanlage f, Coextrusions-Breitschlitzfolienanlage f, Breitschlitzeflachfolienanlage f
flow line Bindenaht f, Fließnaht f, Strömungslinie f
frost line Einfrierbereich m, Einfriergrenze f, Erstarrungslinie f, Frostgrenze f, Frostlinie f, Frostzone f
frost line height Einfrierungsgrenzabstand m, Erstarrungslinienhöhe f
fuel line Kraftstoffleitung f
gating at the (mo(u)ld) parting line Trennlageneinspritzung f
heated tool pipe welding line Heizelement-Rohrschweißanlage f
HF welding line HF-Schweißanlage f
high-capacity blow mo(u)lding line Großblasformanlage f
high-capacity blown film line Hochleistungs-Schlauchfolienanlage f, Hochleistungsblasformanlage f
high-capacity compounding line Großcompoundier-

anlage f
high-speed extrusion line Hochleistungsextrusionsanlage f
high-speed pipe extrusion line Hochleistungs-Rohrextrusionslinie f
high-speed production line Hochgeschwindigkeitsanlage f
high-speed single-screw extrusion line Einschnecken-Hochleistungsextrusionsanlage f
hydraulic line Systemleitung f [Hydraulik]
injection mo(u)lding line Spritzgießanlage f
large-bore pipe production line Großrohrstraße f
light spot line recorder Lichtpunktlinienschreiber m
monofilament extrusion line Monofilanlage f
(mo(u)ld) parting line Formtrennlinie f
packaging line Verpackungsstraße f
parting line Formtrennaht f, Teilungslinie f, Trennaht f
pelletizing line Granulieranlage f
pipe extrusion line Rohrextrusionsanlage f
pipe production line Rohr(fertigungs)straße f, Rohrherstellungsanlage f
production line Fertigungsanlage f, Fertigungslinie f, Fertigungsstraße f, Produktionsanlage f, Produktionsstraße f
profile extrusion line Profilanlage f, Profil(extrusions)straße f
reprocessing line Aufbereitungsanlage f, Aufbereitungsstraße f
scrap repelletizing line Regranulieranlage f
sheet extrusion line Platten(-Extrusions)anlage f, Plattenextrusionslinie f, Plattenstraße f
sheet thermoforming line Plattenformmaschine f
single-die extrusion line Einkopfanlage f
slit die (film) extrusion line Folienextrusionsanlage f, Breitschlitzextrusionsanlage f, Breitschlitzfolienanlage f
spinneret extrusion line Extrusionsspinnanlage f
strand pelletizing line Stranggranulieranlage f
strip pelletizing line Bandgranulierstraße f
suction line Ansaugleitung f, Saugleitung f
thermoforming sheet extrusion line Tiefziehfolien-Extrusionsanlage f
twin-die film blowing line Doppelkopf-Blasfolienanlage f
twin screw pelletizing line Doppelschnecken-Granulieranlage f
two-layer film blowing line Zweischicht-Folienblasanlage f
two-layer thermoforming film extrusion line Zweischicht-Tiefziehfolienanlage f
weld line Schweißnaht f, Zusammenflußlinie f
wide sheet extrusion line Platten-Großanlage f
wire covering line Drahtummantelungsanlage f, Drahtisolierlinie f
linear linear
 linear expansion Längenausdehnung f
 linear motor Linearmotor m

linear phthalate Linearphthalat n
coefficient of linear expansion linearer Ausdehnungskoeffizient m, thermischer Längenausdehnungskoeffizient m, thermischer Wärmeausdehnungskoeffizient m
linearity Linearität f
lined ausgekleidet
liner Futter n, Auskleidung f, Liner m
 barrel liner Gehäuseinnenwandung f, Zylinderauskleidung f, Zylinderinnenwand f, Zylinderinnenfläche f
 boot liner Kofferraumauskleidung f
 car roof liner Autodachhimmel m
 lines of numbers Zahlenzeilen fpl
 nitrided liner [of an extruder barrel] nitriergehärtete Bohrung f
 polyethylene liner PE-Innensack m
 prefabricated roof liner Fertighimmel m
 refrigerator liner Kühlschrank-Innengehäuse n, Kühlschrankinnenverkleidung f
 roof liner Dachhimmel m
 spider lines Stegdornmarkierungen fpl, Dornhaltermarkierungen fpl, Dornstegmarkierungen fpl, Fließmarkierungen fpl, Fließschatten mpl, Längsmarkierungen fpl, Linienmarkierungen fpl, Steg(halter)markierungen fpl, Strömungsschatten mpl
 wear resistant liner [of an extruder barrel] Innenpanzerung f
lining Auskleidung f, Belag m [s.a. liner]
 brake lining resin Bremsbelagharz n
 clutch lining Kupplungsbelag m
 friction lining Reibbelag m
 tank lining Tankauskleidung f
link Glied n, Verbindung(sglied) f(n)
 data link Datenverbund m
 drag link Schleppstange f
linkage Bindung f, Kopplung f, Verknüpfung f
 allophanate linkage Allophanatbindung f
 allophanate linkage content Allophanatbindungsanteil m
 biuret linkage Biuretbindung f
 biuret linkage content Biuretbindungsanteil m
 carbon-carbon linkage Kohlenstoff-Kohlenstoff-Bindung f
 dipole linkage Dipolbindung f
 ether linkage Etherbindung f
 hydrogen bridge linkage Wasserstoffbrückenbindung f
 possible linkages Verknüpfungsmöglichkeiten fpl
 secondary linkage Sekundärbindung f
 silicon-oxygen linkage Silicium-Sauerstoffbindung f
 siloxane linkage Siloxanbindung f
 urea linkage Harnstoffbindung f
 urea linkage content Harnstoffbindungsanteil m
 urethane linkage Urethanbindung f
 urethane linkage content Urethanbindungsanteil m

linked verknüpft, ineinandergreifend
linseed oil Leinöl n
   linseed oil fatty acid Leinölfettsäure f
linters Linters pl
   cotton linters Baumwollinters pl
lip Lippe f
   lip angel Keilwinkel m
   lip elements Lippenelemente npl
   adjustable lip biegsame Lippe f, einstellbare Düsenlippe f
   cooling ring lips Kühlringlippen fpl
   die lip adjusting mechanism Lippenverstellmöglichkeit f, Lippenverstellung f
   die lips Blaslippen fpl, Austrittslippen fpl, Düsenlippen fpl, Lippenpaar n, Lippenpartie f, Werkzeuglippen fpl
   fixed lip feste Düsenlippe f, feste Lippe f
   flexible lip einstellbare Lippe f
   possibility of adjusting the die lips Lippenverstellmöglichkeit f
liquid 1. Flüssigkeit f; 2. flüssig
   liquid additive Flüssigadditiv n
   liquid assets flüssige Mittel npl
   liquid bath Flüssigkeitsbad n
   liquid chromatograpy Flüssigkeitschromatographie f
   liquid column Flüssigkeitssäule f
   liquid crystal display [LCD] LCD-Display n, Flüssigkristall-Anzeige f
   liquid detergent Flüssigwaschmittel n
   liquid displacement Flüssigkeitsverdrängung f
   liquid film Flüssigfolie f, Flüssigkeitsfilm m
   liquid nitrogen cooling system Flüssig-Stickstoffkühlung f
   liquid phase Flüssigkeitsphase f
   liquid phenolic resin Phenolflüssigharz n
   liquid resin Flüssigharz n
   liquid resin blend Flüssigharzkombination f
   liquid resin press mo(u)lding Naßpreßverfahren n
   liquid-ring pump Wasserringpumpe f
   liquid-ring vacuum pump Wasserringvakuumpumpe f
   liquid rubber Flüssigkautschuk m
   liquid silicone rubber Flüssigsilikonkautschuk m
   absorption of liquid Flüssigkeitsaufnahme f
   high-pressure liquid chromatography [HPLC] Hochdruckflüssigkeitschromatographie f
   internal cooling with liquid nitrogen Stickstoffinnenkühlung f
   Newtonian liquid Newtonsche Flüssigkeit f
   non-Newtonian liquid Nicht-Newtonsche Flüssigkeit f
   test liquid Prüfflüssigkeit f
   two-pack liquid silicone rubber Zweikomponentenflüssigsilikonkautschuk m
liquify/to verflüssigen
   liquifying agent Verflüssigungsmittel n
list Liste f, Katalog m, Verzeichnis n
   list of defects Fehlerkatalog m
   list of requirements Anforderungskatalog m, Lastenheft n
   list of spare parts Ersatzteilliste f
   list of suppliers Lieferantenverzeichnis n
   check list Checkliste f
literature Literatur f
   company literature Firmenliteratur f
   manufacturers' literature Firmenmerkblätter npl
   technical literature Fachliteratur f
lithium Lithium n
   lithium carbonate Lithiumcarbonat n
   lithium hydroxide Lithiumhydroxid n
   lithium oxide Lithiumoxid n
load Last f; (elektrische) Leistung f; Belastung f, Beanspruchung f
   load-bearing capacity Lastaufnahmefähigkeit f, Belastbarkeit f, Lastaufnahmevermögen n, Tragverhalten n, Tragfähigkeit f
   load-carrying capacity Lastaufnahmefähigkeit f, Belastbarkeit f, Lastaufnahmevermögen n, Tragverhalten n, Tragfähigkeit f
   load-compensated lastkompensiert
   load pressure Lastdruck m
   load removal device Entlastungsvorrichtung f
   applications of load(s) Krafteinleitung f
   applied load Prüfkraft f
   axial loads Axialkräfte fpl
   connected load Anschlußleistung f, (elektrischer) Anschlußwert m
   continuous load Dauerlast f
   deformation under load Verformung f unter Last
   drive load Antriebsleistung f
   flexural load-carrying capacity Biegebelastbarkeit f
   installed load installierte Leistung f
   thermal load Wärmebeanspruchung f
   total connected load gesamtelektrischer Anschlußwert m
   total installed load Gesamtanschlußwert m
loaded belastet
   spring-loaded federbelastet
loader Ladegerät n, Zuführmaschine f
loading Laden n, Beladen n; Beanspruchung f, Belastung f [s.a. load]
   direction of loading Belastungsrichtung f
   electrical loading elektrische Beanspruchung f
   filler loading Füllgrad m
   load(ing)-bearing Belastungsgrenze f
   loading stages Belastungsschritte mpl
   loading station Beladestation f
   loading test Belastungsprüfung f, Belastungsversuch m
   loading time Beanspruchungsdauer f, Belastungsdauer f, Belastungszeit f
   mechanical loading mechanische Beanspruchung f
   thermal loading thermische Beanspruchung f
local lokal, örtlich
   local overheating Punktüberheizung f, örtliche Überhitzung f
   local(ized) embrittlement Lokalversprödung f

localization Lokalisierung f
   fault localization Fehlerlokalisierung f
   locating holes Positionierungslöcher npl
location Standort m
   fault location Fehlerort m, Pannensuche f
   fault location indicator Fehlerortsignalisierung f, Störungsortsignalisierung f
   fault location program Fehlersuchprogramm n
   gate sprue location Angußlage f
lock/to arretieren, feststellen, verriegeln
lock Feststellvorrichtung f, Verschluß m, Verriegelung f
   lock nut Kontermutter f
   safety lock Sicherheitsverschluß m
   toggle lock mechanism Kniehebelverriegelung f
lockable arretierbar, verschließbar
locked verriegelt, verschlossen
locking Riegel-, Sperr-
   locking bolt Riegelbolzen m, Sperrbolzen m
   locking cylinder Arretierzylinder m, Zuhaltezylinder m
   locking force Werkzeugschließkraft f, Zuhaltedruck m, Werkzeugzuhaltedruck m, Zuhaltekraft f, Verriegelkraft f, Werkzeugzuhaltekraft f, Formzuhaltekraft f, Maschinenzuhaltekraft f, Maschinenzuhaltung f, Schließzahl f, Zufahrkraft f
   locking mechanism Arretiervorrichtung f, Verriegelungs(einrichtung) f, Zuhalteeinrichtung f, Maschinenzuhaltung f, Zuhaltemechanismus m
   locking screw Feststellschraube f
   automatic locking mechanism Zwangsverriegelung f
   clamping locking force Formschließkraft f
   limiting locking force Grenzzuhaltekraft f
   mo(u)ld locking mechanism Formzuhaltung f
   reserve locking force Zuhaltekraftreserve f
log/to protokollieren, registrieren, aufzeichnen
log Protokoll n
   shift log Schichtprotokoll n
logarithm Logarithmus m
logarithmic logarithmisch
logging Registrierung f, Protokollierung f
logic 1. Logik f; 2. logisch
   logic circuit Logikschaltkreis m
   logic diagram Funktionsplan m, Logikdiagramm n
   logic element Logikelement n
   logic function Logikfunktion f
   logic module Logikbaustein m
   logic output Logikausgang m
   logic signal Logiksignal n
   additional logic element Zusatzlogik f
long lang
   long-chain langkettig
   long-chain branching Langkettenverzweigung f
   long-chain character Langkettigkeit f
   long-compression zone screw Langkompressionsschnecke f
   long-lasting dauerhaft, langandauernd
   long-life filter Langzeitfilter n,m

long-neck flat-bottom flask Langhals-Stehkolben m
long-neck round-bottom flask Langhals-Rundkolben m
long-oil langölig
long-oil alkyd (resin) fettes Alkydharz n, Langölalkydharz n
long-reach nozzle verlängerte Düse f
long runs große Serien fpl
long-stroke langhubig
long-term langfristig, langzeitig
long-term ag(e)ing Langzeitlagerung f, Dauerlagerung f
long-term behavio(u)r Zeitstandverhalten n
long-term chemical resistance Chemikalien-Zeitstandverhalten n
long-term contact Langzeitkontakt m
long-term dielectric strength Dauerdurchschlagfestigkeit f, Langzeit-Durchschlagfestigkeit f
long-term effect Langzeitwirkung f
long-term failure test under internal hydrostatic pressure [used for plastics pipes] Innendruck-Zeitstanduntersuchung f, Innendruck-Zeitstandversuch m, Zeitstand-Innendruckversuch m
long-term flexural strength Dauerbiegefestigkeit f
long-term heat ag(e)ing Dauerwärmelagerung f
long-term heat resistance Dauerwärmestabilität f, Dauerwärmebeständigkeit f, Dauertemperaturbelastungsbereich m, Dauertemperaturbeständigkeit f, Dauerwärmebelastbarkeit f
long-term internal pressure test for pipes Rohrinnendruckversuch m
long-term irradiation Langzeitbestrahlung f
long-term milling test Dauerwalztest m
long-term performance Langzeitverhalten n, Dauergebrauchseigenschaften fpl
long-term properties Langzeiteigenschaften fpl
long-term service temperature Dauergebrauchstemperatur f
long-term stabilizer Langzeitstabilisator m
long-term stress Dauerbelastung f, Langzeitbeanspruchung f, Langzeitbelastung f
long-term temperature limit Dauertemperaturgrenze f
long-term temperature resistance Dauertemperaturbelastungsbereich m
long-term tensile stress Zeitstand-Zugbeanspruchung f
long-term test Langzeitprüfung f, Langzeittest m, Langzeituntersuchung f, Langzeitversuch m
long-term test voltage Dauerprüfspannung f
long-term thermal stability Dauertemperaturbeständigkeit f, Dauerwärmebeständigkeit f, Dauerwärmestabilität f
long-term torsional bending stress Dauertorsionsbiegebeanspruchung f

long-term tracking resistance Kriechstromzeitbeständigkeit f
long-term weathering resistance Dauerwitterungsstabilität f
long-term welding factor Langzeitschweißfaktor m
equally long gleichlang
having a long shelf life lagerstabil
longitudinal longitudinal, Längs-, in Richtung der Längsachse
  longitudinally adjustable längsverschiebbar
  longitudinal corrugating device Längswellvorrichtung f
  longitudinal corrugating machine Längswellmaschine f
  longitudinal cut Längsschnitt m
  longitudinal cutter Längsschneidevorrichtung f, Längstrenneinrichtung f, Längsschneideeinrichtung f
  longitudinal direction Längsrichtung f
  longitudinal expansion Längsdehnung f
  longitudinal flow Längsströmung f
  longitudinally grooved längsgenutet
  longitudinal grooves Axialnuten fpl, Längsnuten fpl
  longitudinal machine axis Maschinenlängsachse f
  longitudinal marks Linienmarkierungen fpl, Längsmarkierungen fpl
  longitudinal mixing Axialvermischung f, Längsmischung f
  longitudinal mo(u)ld axis Werkzeuglängsachse f
  longitudinal movement Längsbewegung f
  longitudinally oriented längsgerichtet, längsverstreckt
  longitudinal section Längsschnitt m
  longitudinal shear Längsscherung f
  longitudinal shrinkage Längsschwindung f, Längenschwund m, Längsschrumpf(ung) m(f), Längenschrumpf m, Längskontraktion f
  longitudinal stress Längsspannung f
  longitudinal stretching Längs(ver)streckung f
  longitudinal stretching unit Längsstreckmaschine f, Längsstreckwerk n
  barrel vented through longitudinal slits Längsschlitz-Entgasungsgehäuse n
loom 1. Webstuhl m; 2. Isolierschlauch m
  circular loom Rundwebmaschine f, Rundwebstuhl m
loose locker, lose
  loose-laid [e.g. roofing sheet] freiverlegt
  loosely crosslinked schwachvernetzt, weitmaschig vernetzt
loss Verlust m
  loss angle Verlustwinkel m
  loss factor (mechanischer) Verlustfaktor m, (dielektrischer) Verlustfaktor m
  loss index (dielektrische) Verlustzahl f, Verlustziffer f
  loss modulus Verlustmodul m
  loss of adhesion Haftungseinbuße f
  loss of pressure Druckverlust m, Druckabfall m
  loss of quality Qualitätseinbuße f
  loss of strength Festigkeitseinbuße f

  loss of weight Gewichtsabnahme f, Gewichtsverlust m
  loss on ignition Glühverlust m
  loss through evaporation Verdampfungsverlust m, Verdunstungsverlust m
  convection heat loss Lüftungswärmeverlust m
  data loss Datenverlust m
  energy loss Energieverlust m
  evaporation loss Verdampfungsverlust m, Verdunstungsverlust m
  frictional loss Reibungsverlust m
  heat dissipation losses Wärmeableitungsverluste mpl
  heat loss Wärmeabfluß m, Wärmeverlust m
  heat loss due to radiation Strahlungsverlust m
  heat losses due to convection Konvektionsverluste mpl
  moisture loss Feuchtigkeitsverlust m
  plasticizer loss Weichmacherverlust m
  power loss Verlustleistung f
  pressure loss Druckverlust m
  profit and loss account Gewinn- und Verlustrechnung f
  radiation heat loss Strahlungs(wärme)verlust m
  rate of plasticizer loss Weichmacher-Verlustrate f
  solvent loss Lösungsmittelverlust m
  transmission heat loss Transmissionswärmeverlust m
  weight loss Gewichtsverlust m, Gewichtsabnahme f
  weight loss determination Gewichtsabnahmebestimmung
low niedrig, tief; gering, schwach
  low-alkali glass E-Glas n, alkaliarmes Glas n
  low-boiling niedrigsiedend
  low-boiling solvent Niedrigsieder m
  low-carbon [e.g. steel] kohlenstoffarm
  low-compression screw Niedrigkompressionsschnecke f
  low-cost kostengünstig, preiswert, preisgünstig, preiswürdig, billig
  low construction Niedrigbauweise f
  low-density niedrigdicht
  low density polyethylene [LDPE] Hochdruck-Polyethylen n, Polyethylen n niedriger Dichte
  low-emission emissionsarm
  low-emulsifier emulgatorarm
  low-energy energiearm, niedrigenergetisch
  low fish eye content Stippenarmut f
  low-frequency engineering Niedrigfrequenztechnik f
  low-friction friktionsarm
  low-friction calibrator Gleitkalibrator m
  low-inerta trägheitsarm
  low-maintenance wartungsarm
  low melting niedrigschmelzend
  low-molecular weight niedrigmolekular, niedermolekular
  low-noise geräuscharm, lärmarm
  low-odo(u)r gerucharm
  low-pass filter Tiefpaßfilter n,m
  low-pressure injection mo(u)lding machine Niedrig-

druckspritzgießmaschine f
low-pressure mo(u)ld safety device Niederdruckwerkzeugschutz m
low-pressure mo(u)lding compound Niederdruckformmasse f
low-pressure polymerization Niederdruckpolymerisation f
low-pressure process Niederdruckverfahren n
low-pressure reactor Niederdruckreaktor m
low-pressure scanning (device) Niederdruckabtastung f
low-pressure separator Niederdruckabscheider m
low-profile [especially used for UP resins] schwindungsarm, schrumpfarm
low-reactivity niedrigaktiv, niedrigreaktiv, reaktionsträg
low-resin harzarm
low-shear scherungsarm
low-shrinkage [general term] schrumpfarm, schwindungsarm [s.a. low-profile]
low-solvent lösungsmittelarm
low-stress spannungsarm
low-temperature ag(e)ing Kältelagerung f
low-temperature applications Tieftemperaturanwendungen fpl
low-temperature brittleness point Kältesprödigkeitspunkt m
low-temperature brittleness test Kältesprödigkeitsversuch m
low-temperature calendering (process) LT-Verfahren n
low-temperature curing Tieftemperaturhärtung f
low-temperature exposure Kältebelastung f
low-temperature flexibility Kälteelastizität f, Kälteflexibilität f, Tieftemperaturflexibilität f
low-temperature folding [endurance test] Kältefalzversuch m
low-temperature impact resistant kaltschlagzäh
low-temperature impact strength Kälteschlagwert m, Kälteschlagzähigkeit f, Kälteschlagbeständigkeit f, Tieftemperatur-Schlagzähigkeit f
low temperature initiator Tieftemperaturinitiator m
low-temperature insulating material Kältedämmstoff m
low-temperature lubricating grease Tieftemperaturschmierfett n
low-temperature performance Kälteverhalten n, Tieftemperaturverhalten n
low-temperature peroxide Tieftemperaturperoxid n
low-temperature plasticizer efficiency Kälteelastifizierungsvermögen n
low-temperature process LT-Verfahren n
low-temperature properties Tieftemperatureigenschaften fpl, Kälteeigenschaften fpl
low-temperature resistance Tieftemperaturbeständigkeit f, Tieftemperaturfestigkeit f, Kältebeständigkeit f, Kälte(stand)festigkeit f

low-temperature resistant kältezäh, kältefest
low temperature shock resistance Kälteschockfestigkeit f
low-temperature test chamber Kälteschrank m, Kältekammer f
low temperature thermostat Kryostat m, Kältethermostat m
low temperatures Tieftemperaturen fpl
low vacuum Grobvakuum n
low-viscosity dünnflüssig, niederviskos, niedrigviskos
low viscosity primer Einlaßgrund(ierung) m [speziell für starkabsorbierende Flächen wie z.B. Beton]
low-volatility 1. niedrigflüchtig, schwerflüchtig; 2. Schwerflüchtigkeit f
low-voltage Niederspannung f
low-voltage range Niederspannungsbereich m
low-warpage verzugsarm
low-warpage properties Verzugsarmut f
resistant to high and low temperatures temperaturbeständig
two-component low-pressure injection mo(u)lding machine Zweikomponenten-Niederdruck-Spritzgießmaschine f
with a low emulsifier content emulgatorarm
with a low filler content füllstoffarm
with a low fish eye content stippenarm
with a low fusion point [PVC] niedriggelierend
with a low gel point niedriggelierend
with a low monomer content monomerarm
with a low pigment content niedrigpigmentiert, schwachpigmentiert
with a low plasticizer content weichmacherarm
with a low resin content harzarm
lower limit unterer Grenzwert m, Untergrenze f
lower temperature limit Temperatur-Untergrenze f
lowering Reduzierung f
lubricant Gleitmittel n, Gleitsubstanz f, Schmiermittel n, Schmierstoff m
  lubricant blend Gleitmittelgemisch n, Gleitmittelkombination f, Gleit(mittel)system n, Kombinationsgleitmittel n
  lubricant content Gleitmittelanteil m
  lubricant film Gleitmittelfilm m
  lubricant performance Gleitmittelverhalten n
  amount of lubricant Gleitmittelanteil m
  external lubricant äußeres Gleitmittel n
  not containing lubricant gleitmittelfrei
lubricated geschmiert
lubricating schmierend, Schmier-, Gleit-
  lubricating effect Gleitmittelverhalten n, Gleitwirkung f, Schmierkraft f, Schmier(mittel)wirkung f, Schmiereffekt m
  lubricating film Schmierfilm m
  lubricating grease Schmierfett n
  lubricating oil Schmieröl n
  lubricating points Schmierstellen fpl
  lubricating properties Schmiereigenschaften fpl

lubricating properties [of lubricants] Gleiteigenschaften fpl
  lubricating system Schmiersystem n
  central grease lubricating system Fettzentralschmierung f
  central lubricating system Zentralschmierung f
  central lubricating unit Zentralschmieranlage f
  circulating oil lubricating system Ölumlaufschmierung f, Umlaufölung f, Umlaufschmierung f
  external lubricating effect Außengleitwirkung f
  pressurized oil lubricating system Drucköllung f
lubrication Schmierung f
  lubrication oil reservoir Schmieröltank m
  circulating oil lubrication Ölumlaufschmierung f, Umlaufölung f. Umlaufschmierung f
  external lubrication äußere Gleitwirkung f
  internal lubrication innere Gleitwirkung f
  pressurized oil lubrication Drucköllung f
lug Nase f, Ansatz m, Vorsprung m, Öse f
  fastening lugs Befestigungslaschen fpl
lump Klumpen m, Brocken m
  free from lumps klümpchenfrei, klumpenfrei
  tendency to form lumps Verklumpungsneigung f
MAC [maximum allowable concentration] MAK-Wert m, maximale Arbeitsplatzkonzentration f
machinability Bearbeitbarkeit f, Zerspanbarkeit f
machinable bearbeitbar, (zer)spanbar
machine/to bearbeiten
machine Maschine f, Fertigungsmittel n
  machine and equipment manufacture Maschinen- und Apparatebau m
  machine base Maschinenbett n, Gestell n, Maschinengestell n, Maschinenrahmen m, Maschinenständer m
  machine breakdown Maschinenausfall m
  machine built up from modules Baukastenmaschine f
  machine butt welding Maschinenstumpfschweißung f
  machine components Maschinenelemente npl
  machine control Maschinenüberwachung f, Maschinensteuerung f
  machine control data file Maschinensteuerungsdatei f
  machine control program Maschinensteuerprogramm n
  machine conversion unit Maschinenumbausatz m
  machine cost(s) Maschinenkosten pl
  machine cut-out Maschinenabschaltung f
  machine cycle Maschinentakt m
  machine cycle control Maschinentaktsteuerung f
  machine data Maschinendaten npl
  machine data collection Maschinendatenerfassung f
  machine design Maschinenkonzeption f
  machine dimensions [e.g. in a diagram of a machine with all dimensions indicated] Raumbedarfsplan m
  machine direction Arbeitsrichtung f, Warenlaufrichtung f
  machine downtime Maschinenausfallzeit f, Maschinenstillstandzeit f
  machine efficiency Maschineneffizienz f
  machine exports Maschinenausfuhr f
  machine for blow mo(u)lding drums Faßblasmaschine f
  machine for blow mo(u)lding fuel oil storage tanks Heizöltankmaschine f, Heizöltank-Blasmaschine f
  machine for making large blow mo(u)ldings Großhohlkörper-Blasmaschine f
  machine frame Gestell n, Maschinenbett n, Maschinengestell n, Maschinenrahmen m, Maschinenständer m
  machine functions Maschinenfunktionen fpl
  machine hour rate Maschinenstundensatz m
  machine hours counter Betriebsstundenzähler m
  machine language Maschinensprache f
  machine length Maschinenlänge f
  machine life Maschinenlebensdauer f
  machine malfunction Betriebsstörung f, Maschinenstörung f
  machine manufacturer Maschinenhersteller m
  machine movements Bewegungsabläufe mpl, Maschinenbewegungen fpl
  machine operating program Maschinenprogramm n
  machine operation Maschinenbetrieb m
  machine operator Maschinenbediener m, Maschinenführer m
  machine operators Maschinen(bedien)personal n, Bedienungsmannschaften fpl, Bedienungspersonal n
  machine-oriented maschinennah
  machine output Maschinenleistung f
  machine parameter Maschinenparameter m
  machine performance Maschinenverhalten n
  machine productivity Maschinennutzungsgrad m
  machine purgings Anfahrbrocken m, Anfahrfladen m
  machine-readable maschinenlesbar
  machine rebuilding costs Maschinenumbaukosten pl
  machine setter Einrichter m, Maschineneinsteller m
  machine setting Maschineneinstellgröße f, Maschineneinstellparameter m
  machine setting program Maschineneinstellprogramm n
  machine setting store Einstelldatenabspeicherung f
  machine settings Maschineneinstellungen fpl, Maschineneinstellwerte mpl, Maschineneinstelldaten npl
  machine-side drier Beistelltrockner m
  machine-side granulator Beistellgranulator m
  machine speed Maschinengeschwindigkeit f, Fahrgeschwindigkeit f
  machine status Maschinenzustand m
  machine stoppage Maschinenstillstand m
  machine supplier Maschinenlieferant m
  machine support Maschinenlagerung f
  machine tool Werkzeugmaschine f
  machine utilization Maschinennutzung f
  machine variable Maschinenstellwert m,

Maschinenstellgröße f
machine wear Maschinenverschleiß m
machine's lifetime Maschinenlebensdauer f
accumulator-type machine Speichermaschine f
automatic bag-making machine Beutelautomat m
automatic bag-welding machine Beutelschweißautomat m
automatic blow mo(u)lding-filling machine Hohlkörper-Blas- und Füllautomat m
automatic blow mo(u)lding machine Blas(form)automat m, Hohlkörperblasautomat m
automatic bottle blowing machine Flaschenblasautomat m
automatic carousel-type injection mo(u)lding machine Revolverspritzgießautomat m
automatic compression mo(u)lding machine Preßautomat m
automatic fast cycling injection mo(u)lding machine Schnellspritzgießautomat m
automatic filling machine Abfüllautomat m, Füllautomat m
automatic heat sealing machine Heißsiegelautomat m
automatic high capacity blow mo(u)lding machine Hochleistungsblasformautomat m
automatic injection blow mo(u)lding machine Spritzblasautomat m
automatic injection mo(u)lding machine Spritz(gieß)automat m
automatic machine Automat m
automatic mo(u)lding machine Formteilautomat m
automatic oriented PP stretch blow mo(u)lding machine OPP-Streckblasautomat m
automatic rotary(-table) injection mo(u)lding machine Revolverspritzgießautomat m, Rotations-Spritzgußautomat m
automatic rotary-table machine Rundläuferautomat m
automatic thermoforming machine Thermoform-Automat m, Tiefziehautomat m, Warmformautomat m, Formautomat m
automatic transfer mo(u)lding machine Spritzautomat m
automatic two-colo(u)r injection mo(u)lding machine Zweifarben-Spritzgußautomat m
automatic vacuum forming machine Vakuumformautomat m
back of the machine Maschinenrückseite f
bag-making machine Beutelmaschine f
basic machine Basismaschine f, Grundmaschine f
blister packaging machine Blisterformmaschine f
blow mo(u)lding and filling machine Blas- und Füllmaschine f
blow mo(u)lding machine Blas(form)maschine f, Hohlkörperblas(form)maschine f
bottle blowing machine Flaschen(blas)maschine f
bottle testing machine Flaschenprüfstand m
canister blow mo(u)lding machine Kanisterblasmaschine f
carousel-type injection mo(u)lding machine Spritzgießrundtischanlage f, Revolverspritzgießmaschine f, Drehtisch-Spritzgußmaschine f
carousel-type machine Drehtischmaschine f, Rundtischanlage f, Rundtischmaschine f, Revolvermaschine f
casting machine Gießmaschine f
centrifugal casting machine Kunststoffschleudermaschine f, Schleudermaschine f
coating machine Beschichtungseinrichtung f, Beschichtungsmaschine f
coextrusion blow mo(u)lding machine Coextrusionsblasformmaschine f
colo(u)r coding machine Farbkennzeichnungsmaschine f
compressed air forming machine Druckluftformmaschine f
compressive strength test machine Druckfestigkeitsprüfmaschine f
conventional injection mo(u)lding machine Kompaktspritzgießmaschine f
copy milling machine Kopierfräsmaschine f
creep test machine Zeitstandanlage f
damage to a machine Maschinenbeschädigungen fpl
dip blow mo(u)lding machine Tauchblasmaschine f
double toggle machine Doppelkniehebelmaschine f
downstream machine Folgemaschine f, Nachfolgemaschine f
due to the machine maschinenbedingt
embossing machine Prägemaschine f
extrusion blow mo(u)lding machine Extrusionsblas(form)maschine f
factors influencing machine performance Maschineneinflußgrößen fpl
fast cycling injection mo(u)lding machine Hochleistungsspritzgießmaschine f, Schnelläufer-Spritzgießmaschine f
fast cycling machine Schnelläufer m
film blowing machine Folienblaseinheit f
film production machine Folienproduktionsmaschine f
fine grinding machine Feinmahlaggregat n
foam mo(u)lding machine Schäummaschine f
form-fill-seal machine Form-Füll-Siegelmaschine f
friction welding machine Reibschweißmaschine f
fully automatic machine Vollautomat m
general-purpose machine Universalmaschine f
GRP filament winding machine GFK-Wickelmaschine f
heat impulse welding machine Wärmeimpulsschweißmaschine f
high-capacity blow mo(u)lding machine Großblasformmaschine f
high-frequency welding machine Hochfrequenzschweißmaschine f
high-speed machine Schnelläufer m

honing machine Schwingschleifer m, Ziehschleifmaschine
hot-melt coating machine Schmelzwalzenmaschine f
impact grinding machine Prallmahlanlage f
in machine direction in Laufrichtung f
injection blow mo(u)lding machine Spritzblasmaschine f
injection mo(u)lding machine Spritzgießeinheit f, Spritzgießmaschine f, Spritzgießaggregat n
injection stretch blow mo(u)lding machine Spritzgieß-Streckblasmaschine f
laboratory coating machine Laborbeschichtungsanlage f
laboratory thermoforming machine Laborziehgerät n
laminating machine Kaschiermaschine f, Laminiermaschine f
longitudinal corrugating machine Längswellmaschine f
longitudinal machine axis Maschinenlängsachse f
low-pressure injection mo(u)lding machine Niederdruckspritzgießmaschine f
melt accumulator machine Kopfspeichermaschine f
modular injection mo(u)lding machine Baukasten-Spritzgießmaschine f
modular machine Baukastenmaschine f
multi-component structural foam mo(u)lding machine TSG-Mehrkomponenten-Rundläufermaschine f [TSG = Thermoplast-Schaum-Guß]
orientation in machine direction Längsorientierung f
oriented polypropylene blow mo(u)lding machine OPP-Maschine f
pelleting machine Tablettenmaschine f
pilot plant machine Technikummaschine f
pilot plant-scale machine Pilotmaschine f
plastics processing machine Kunststoffmaschine f
plunger injection mo(u)lding machine Kolbeninjektions-Spritzgießmaschine f, Kolben(spritz)gießmaschine f
production-scale machine Produktionsmaschine f
range of automatic blow mo(u)lding machines Blasformautomaten-Baureihe f
range of machines Maschinenreihe f, Maschinenprogramm n, Anlagenspektrum n, Anlagenprogramm n
raschel machine Raschelmaschine f
reaction injection mo(u)lding machine [RIM machine] Reaktionsgießmaschine f
reciprocating screw injection mo(u)lding machine Schubschneckenspritzgießmaschine f, Schneckenkolbenspritzgießmaschine f
reciprocating-screw machine Schubschneckenmaschine f, Schneckenschubmaschine f
rotary table injection mo(u)lding machine Spritzgießrundtischanlage f, Revolverspritzgießmaschine f
rotary table machine Drehtischmaschine f, Rundläuferanlage f, Rundläufer(maschine) f, Revolvermaschine f

rotational mo(u)lding machine Rotationsgießmaschine f
rubber injection machine Gummi-Spritzgießmaschine f
screw injection mo(u)lding machine Schneckenspritzgießmaschine f
screw-plunger machine Schneckenkolbenmaschine f
sealing machine Verschließanlage f
semi-automatic machine Halbautomat m
single-screw injection mo(u)lding machine Einschnecken-Spritzgießmaschine f
single-screw machine Einschneckenmaschine f, Einwellenmaschine f
single-station automatic blow mo(u)lding machine Einstationenblasformautomat m
single-station blow mo(u)lding machine Einstationenblas(form)maschine f
single-station extrusion blow mo(u)lding machine Einstationenextrusionsblasformanlage f
single-station machine Einstationenanlage f, Einstationenmaschine f
sliding table machine Schiebetischmaschine f
skin packaging machine Skinformmaschine f
slow-speed machine Langsamläufer m
special purpose machine Einzweckausführung f, Sondermaschine f
spread coating machine Streichmaschine f
stopping of the machine Maschinenstopp m, Maschinenstillstand m
storage of machine settings Einstelldatenabspeicherung f
stretch blow mo(u)lding machine Streckblas(form)maschine f
structural foam mo(u)lding machine Strukturschaummaschine f, Thermoplast-Schaumgießmaschine f, TSG-Maschine f, TSG-Spritzgießmaschine f
tensile testing machine Zerreiß(prüf)maschine f, Zugprüfmaschine f
thermoforming machine Tiefziehmaschine f, Warmformmaschine f, Warmformanlage f
thermoset injection mo(u)lding machine Duroplast-Spritzgießmaschine f
toggle clamp machine Kniehebelmaschine f
top part of the machine Maschinenoberteil n
transverse corrugating machine Querwellmaschine f
transverse stretching machine Breitreckmaschine f
two-component low-pressure injection mo(u)lding machine Zweikomponenten-Niederdruck-Spritzgießmaschine f
two-station blow mo(u)lding machine Zweistationenblasmaschine f
two-station machine Zweistationenmaschine f
ultrasonic welding machine Ultraschallschweißmaschine f, Ultraschallschweißanlage f
vacuum forming machine Vakuumformmaschine f, Vakuumtiefziehmaschine f

vented injection mo(u)lding machine Entgasungsmaschine f
washing machine Waschmaschine f
16-station rotary table machine Sechzehn-Stationen-Drehtischmaschine f
machined herausgearbeitet, bearbeitet
electric discharge machined funkenerodiert
machining spanabhebende Bearbeitung f, spanende Bearbeitung f, mechanische Bearbeitung f, spangebende Bearbeitung f, Zerspanung f
  machining properties Bearbeitbarkeit f
  machining tolerances Bearbeitungstoleranzen fpl
  electric discharge machining [EDM] Funkenerosionsbearbeitung f
macrocrack Makroriß m
macromolecular makromolekular
macromolecule Makromolekül n
macroradical Makroradikal n
macroscopic makroskopisch
macrostructure Makrostruktur f
made into a slurry aufgeschlämmt
made worse verschlechtert
magnesium Magnesium n
  magnesium alloy Magnesiumlegierung f
  magnesium oxide Magnesiumoxid n
  magnesium silicate Magnesiumsilikat n
  magnesium stearate Magnesiumstearat n
magnet Magnet m
  d.c. permanent magnet motor Gleichspannungs-Drehankermagnet-Motor m
magnetic magnetisch, Magnet-
  magnetic armature Magnetanker m
  magnetic card Magnetkarte f
  magnetic card reader Magnetkartenleser m
  magnetic disk Magnetplatte f, Magnetplattenspeicher m, Magnetträger m, Plattenspeicher m
  magnetic disk drive Magnetplattenlaufwerk n
  magnetic ink document Magnetbeleg m
  magnetic tape Magnetträger m, Magnetband n
  magnetic tape cassette Magnetbandkassette f
  magnetic tape unit Magnetbandgerät n
  fixed magnetic disk Festplattenspeicher m
main Haupt-, hauptsächlich
  main applications Anwendungsschwerpunkte mpl
  main chain Hauptkette f
  main control room Hauptsteuerwarte f
  main cooling system Primärkühlung f
  main direction of movement Hauptbewegungsrichtung f
  main drive shaft Hauptantriebswelle f
  main drive (unit) Hauptantrieb m
  main electronics system Primärelektronik f
  main extruder Primärextruder m, Hauptextruder m
  main field of use Hauptanwendungsgebiet n, Haupteinsatzgebiet n
  main motor Hauptmotor m
  main polymer chain Polymerhauptkette f
  main runner Hauptverteilerkanal m, Hauptverteilersteg m; Anströmkanal m, Anschnittkanal m, Angußtunnel m [bei Mehrfachwerkzeugen]
  main screw Hauptschnecke f, Hauptspindel f, Mittelschnecke f, Zentralschnecke f, Zentralspindel f, Zentralwelle f
  main stabilizer Primärstabilisator m
  main switch Hauptschalter m
  main take-off (unit) Primärabzug m
  main task Hauptaufgabe f
mains (system) Netz n [elektrischer Strom, selten: Wasser]
  mains cable Netzkabel n
  mains-powered mit Netzstromversorgung f
  mains pressure Netzdruck m
  mains voltage Netzspannung f
  mains water Netzwasser n
  mains water pressure Wassernetzdruck m
  water mains Leitungswassernetz n, Wassernetz n
maintenance Instandhaltung f, Unterhaltung f, Wartung f
  maintenance costs Unterhaltskosten pl, Wartungskosten pl
  maintenance-free wartungsfrei, pflegefrei
  maintenance personel Wartungspersonal n
  maintenance requirements Wartungsansprüche mpl
  freedom from maintenance Wartungsfreiheit f
  requiring little maintenance wartungsarm
  requiring no maintenance wartungsfrei
make/to machen, tun, herstellen
make/to elastic elastifizieren
make Fabrikat n, Bauart f
  make of extruder Extruder-Fabrikat n
maker Hersteller(firma) m(f)
  mo(u)ld maker Formenbauer m, Werkzeugbauer m, Werkzeugmacher m
making 1. Zubereitung f; 2. Einschalt- [Elektro]
  making device [used for impressing identifying marks into extruded pipe] Signiergerät n
  making impact resistant Schlagfestmachen n
  making something easier Vereinfachung f
  making thixotropic Thixotropierung f
  making visible Sichtbarmachen n
  making water-repellent 1. Hydrophobierung f; 2. hydrophobierend
  mo(u)ld making Werkzeugbau m
  non-paste making nichtverpastbar
  paste making Pastenbereitung f
male mo(u)ld Patrize f, Stempel m
male tool Positivwerkzeug n
maleate Maleat n
  dialkyltin maleate Dialkylzinnmaleat n
  lead maleate Bleimaleat n
  propylene glycol maleate Prophylenglykolmaleat n
maleic acid Maleinsäure f
  maleic acid ester Maleinsäureester m
  maleic anhydride Maleinsäureanhydrid n

maleic resin Maleinatharz n
maleinate Maleinsäureester m
  di-2-ethylhexyl maleinate Di-2-ethylhexylmaleinat n
  di-n-butyltin maleinate Di-n-butylzinnmaleinat n
  dibutyltin maleinate Dibutylzinnmaleinat n
  dioctyl maleinate Di-2-ethylhexylmaleinat n
  tin maleinate Zinnmaleinat n
malfunction Fehlfunktion f, Fehlleistung f, Störung f, Betriebsstörung f
  malfunction alarm Störwertmeldung f, Störmeldungsanzeige f, Störmeldeeinrichtung f
  malfunction diagnostics Störungsdiagnose f
  malfunction indicator Störungsanzeige f, Störungsmeldegerät n, Störsignalisierung f, Fehlerursachen-Anzeige f
  malfunction monitoring (device) Störungsüberwachung f
  malfunction signal Störungsmeldung f, Störanzeige f
  in case of malfunction im Störfall m
  machine malfunction Betriebsstörung f, Maschinenstörung f
management Verwaltung f, Management n
  management consultancy Unternehmungsberatung f
  data management Datenverwaltung f
  diskette management Diskettenverwaltung f
  process management Prozeßbeherrschung f
managing directors Vorstand m
mandrel Dorn m, Pinole f, Pinolenkörper m, Wickeldorn m, Kern m, Werkzeugdorn m
  mandrel flex test Dornbiegeversuch m
  mandrel flexibility Dornbiegefestigkeit f
  mandrel holding plate Dornhalteplatte f, Dornhalterung f, Dornhalter m, Dornträger m
  mandrel support Tragring m, Dornhalter m
  mandrel support design Dornhalterkonstruktion f
  mandrel with a heart-shaped groove Herzkurvendorn m
  mandrel with a ring-shaped groove Ringrillendorn m
  blowing mandrel support Blasdornträger m
  bolt-type mandrel support Bolzendornhalterung f
  breaker plate-type mandrel support Lochtragring m, Loch(scheiben)dornhalter m, Siebdornhalter m, Siebkorbhalterung f, Siebkorbdornhalter m
  cooling mandrel Kühldorn m
  dipping mandrel Tauchdorn m
  incandescent mandrel test Glühdornprüfung f
  inflating mandrel Blaspinole f, Aufblasdorn m, Blasdorn m, Spritzdorn m
  parison stretching mandrel Schlauchspreizvorrichtung f, Spreizdorn(vorrichtung) m(f), Spreizdornanlage f, Spreizvorrichtung f
  rotating mandrel Dralldorn m
  spider-type mandrel support Stegtragring m, Stegdornhalter m, Dornsteghalter m, Dornsteghalterung f, Radialsteghalter m
  spiral mandrel Spiraldorn m
  spiral mandrel (blown film) die Spiraldorn(blas)-kopf m, Wendelverteilerkopf m, Schmelzewendelverteilerkopf m, Schmelzewendelverteilerwerkzeug n, Spiralverteiler-blaskopf m
  spiral mandrel distributor Wendelverteiler m, Schmelzewendelverteiler m
  threaded mandrel unscrewing device Gewindedorn-Ausdrehvorrichtung f
  twin spider-type mandrel support Doppelstegdornhalter(ung) m (f)
  winding mandrel Bobine f, Wickelkern m, Wickelkörper m
manganese violet Manganviolett n
manifold Rohrverzweigung f, Düsenkanal m, Sammelbohrung f, Schmelzeverteiler m, Verteilerrohr n, Verteiler(kanal) m, Verteilerbohrung f, Verteilungskanal m, Zuführ(ungs)kanal m
  manifold block Verteilerstück n, Verteilerblock m, Verteilerbalken m
  manifold system Verteiler(röhren)system n
  manifold-type die Verteilerwerkzeug n, Verteilerkanaldüse f
  central cooling water manifold Zentralkühlwasserverteilung f
  coathanger manifold Kleiderbügel(-Verteiler)kanal m
  cooling water manifold Kühlwasserverteiler m
  fishtail manifold Fischschwanzkanal m
  hot runner manifold Heißkanal(verteiler)platte f
  hot runner manifold block Heizblock m, Heißkanalverteilerbalken m, Heißkanal-(verteiler)block m, Querverteiler m
  twin manifold Doppelverteilerkanal m
Mannich base Mannichbase f
manometer Manometer n
  precision manometer Feinmeßmanometer n
manual manuell, mit der Hand, Hand-
  manual butt welding Handstumpfschweißen n
  manual drive Handgetriebe n
  manual operation Handbedienung f, Handbetrieb m
  manual setting Handeinstellung f
manually operated valve Handventil n
manufacture Fertigung f, Herstellung f, Produktion f
  computer aided manufacture [CAM] CAM-Rechentechnik f
  machine and equipment manufacture Maschinen- und Apparatebau m
manufacturer Produzent m, Hersteller m
  manufacturer's technical literature Firmenmerkblätter npl
  bottle manufacturer Hohlkörperhersteller m, Flaschenhersteller m
  extruder manufacturer Extruderbaufirma f
  machine manufacturer Maschinenhersteller m
manufacturing costs Fertigungskosten pl
manufacturing process Herstellungsverfahren n, Fertigungsverfahren n
marble Marmor m
marbling Marmorieren n

marbling Marmorieren n
margin Grenze f
  profit margin Verdienstspanne f
  safety margin Sicherheitsgrenze f, Sicherheitsreserve f, Sicherheitsspielraum m, Sicherheitsabstand m
marginal marginal
marine paint Schiffsfarbe f
mark Markierung f, Kennzeichen n
  mark left by the pressure transducer Druckaufnehmerabdruck m
  mark-sense card Markierungsbeleg m
  ejector mark Auswerfermarkierung f
  gate mark [if referring to a mo(u)lded part] Angußstelle f, Angußpunkt m, Anschnittstelle f, Anspritzstelle f, Angußmarkierung f, Anschnittmarkierung f
  sink mark Einfallstelle f
  surface mark Oberflächendefekt m
  trade mark Warenzeichen n
market Markt m
  market breakdown Marktaufteilung f, Marktgliederung f
  market conditions Marktverhältnisse npl
  market expansion Marktausweitung f
  market forecast Marktprognose f
  market-orientated marktorientiert
  market share Marktanteil m
  bank marks Oberflächenmarkierungen fpl
  common market countries EWG-Länder npl
  flow marks Fließmarkierungen fpl
  home market Binnenmarkt m
  longitudinal marks Linienmarkierungen fpl, Längsmarkierungen fpl
  sink marks Sinkmarkierungen fpl
  surface marks Oberflächenmarkierungen fpl
Martens heat distortion temperature Formbeständigkeit f (in der Wärme) nach Martens, Martens-Wärmeformbeständigkeit f
Martens temperature Martensgrad m, Martenszahl f, Martenswert m
mask Maske f
  face mask Atem(schutz)maske f, Filtermaske f, Gesichtsmaske f, Schutzmaske f, Staubmaske f
  input mask Eingabemaske f
mass Masse f
  mass flow rate Massedurchsatz m
  mass production Massenfertigung f, Massenproduktion f, Serienfertigung f, Serienproduktion f
  mass production conditions Serienbedingungen fpl
  mass production mo(u)ld Serienwerkzeug n
  mass spectrometry Massenspektrometrie f
  mass spectroscopy Massenspektroskopie f
  mass storage unit Massenspeicher m
master Muster(stück) n, Origianl n, Urform f, Standard m, Haupt-
  master file Stammdatei f
  master model Ausgangsmodell n, Bezugsmodell n, Urmodell n
  master pattern plate Urmodellplatte f
  master terminal Leitstandterminal n
masterbatch Vormischung f, Grundmischung f, Masterbatch m
  pigment masterbatch Pigment-Kunststoffkonzentrat n, Farbgranulat n
  polyethylene masterbatch PE-Farbkonzentrat n
  TiO_2-PE masterbatch TiO_2-PE-Konzentrat n
masticate/to [e.g. physical or chemical breakdown of natural rubber to reduce viscosity for further processing] kneten; zerkleinern, mastizieren
mat Matte f, Vlies n
  chopped strand (glass fiber) mat Glasseiden-Schnittmatte f, Glasfaservliesstoff m, Faserschnittmatte f, Kurzfaservlies n
  continuous strand (glass fiber) mat Glasseiden-Endlosmatte f, Endlosfasermatte f, Endlosmatte f
  glass fiber mat Glasfasermatte f, Textilglasmatte f, Textilglasvlies n
  glass mat Glasseidenmatte f, Glasvlies n, Glasmatte f
  glass mat reinforced glasmattenverstärkt
  glass mat reinforced polyester laminate GF-UP-Mattenlaminat n
  overlay mat Oberflächenvlies n
  surfacing mat Oberflächenmatte f, Vlies n [GFK]
matched metal mo(u)lding Warmpreßverfahren n [GFK]
matched metal press mo(u)lding Heißpreßverfahren n, Heißpressen n [GFK]
material Material n, Stoff m, Werkstoff m
  material accumulations Masseanhäufung f, Werkstoffanhäufungen fpl
  material asset Sachanlage f
  material being ground Mahlgut n
  material being transported Fördergut n
  material being weighed Wiegegut n
  material being welded Schweißgut n
  material being wound up Wickelgut n
  material constant Stoffkonstante f, Werkstoffkennwert m, Stoffkennwert m, Werkstoffkenngröße f
  material constants Stoffkenndaten pl
  material failure Werkstoffversagen n
  material feed Masseeinzug m
  material flow Massefluß m, Produktstrom m
  material properties Stoffeigenschaften fpl, Werkstoffeigenschaften fpl
  material property Stoffgröße f
  material-related werkstoffspezifisch
  material testing laboratory Werkstoffprüflabor n
  material to be ground Mahlgut n
  material to be transported Fördergut n
  material to be weighed Wiegegut n
  material to be welded Schweißgut n
  material transport blower Fördergebläse n
  material wear Werkstoffabtrag m
  materials testing Werkstoffprüfung f
  amount of material in the mo(u)ld Werkzeug-

füll(ungs)grad m
amount of material in the screw flight(s)
Schneckengangfüllung f
amount of material injected Einspritzmenge f
amount of raw material required for a day's production Rohstofftagesmenge f
base material Basismaterial n, Grundmaterial n, Grundwerkstoff m
choice of material Werkstoffauswahl f
coarse material Grobgut n
crazed material Fließzonenmaterial n
cushioning material Polstermaterial n
die material Düsenbaustoff m, Düsenwerkstoff m
electrical insulating material Elektroisolierstoff m
flat material [e.g. paper, film, fabric] Flächengebilde n
glass fiber material Textilglas n, Glasfaser f
glass fiber materials Textilglaserzeugnisse npl, Textilglasprodukte npl
glazing material Verglasungsmaterial n
group of materials Werkstoffgruppe f
heat insulating material Wärmedämmstoff m
incoming raw materials control Rohstoffeingangskontrolle f
insulating material Dämmaterial n, Dämmstoff m
jet of material Freistrahl m
lightweight material Leicht(bau)werkstoff m
low-temperature insulating material Kältedämmstoff m
model making material Modellwerkstoff m
mo(u)ld material Formwerkstoff m, Werkzeugwerkstoff m
nozzle material Düsenwerkstoff m, Düsenbaustoff m
packaging material Packstoff m, Verpackungsmaterial n, Verpackungsrohstoff m, Verpackungswerkstoff m
phenolic material Phenolharz-Werkstoff m
plastics material Kunststoffwerkstoff m
polymer material Polymerwerkstoff m
polymer raw material Kunststoffrohstoff m
raw material Rohstoff m, Rohmaterial n
raw material availability Rohstoffverfügbarkeit f
raw material costs Rohmaterialkosten pl
raw material data file Rohstoffdatei f
raw material feed unit Rohstoffzuführgerät n
raw material recovery plant Rohstoffrückgewinnungsanlage f
raw material savings Rohmaterialersparnis f
raw material shortage Rohstoffverknappung f
raw material source Rohstoffquelle f
raw material supplier Rohstofflieferant m
reinforcing material Verstärkungsstoff m, Verstärkungsmaterial n
rigid foam insulating material Hartschaumisolierstoff m
solid-borne sound insulating material Körperschallisoliermaterial n

solid material Feststoff m
sound insulating material Schalldämmstoff m
specific to the material werkstoffspezifisch
starting material Ausgangsstoff m
substitute material Substitutionswerkstoff m
supporting material Trägermaterial n
tarpaulin material Planenstoff m
thermoset material Duromer
upholstery material Polstermaterial n
virgin material Neumaterial n, Neuware f, Originalmaterial n
mathematical mathematisch
mathematical equation Rechenansatz m, mathematische Gleichung f
mathematical example Rechenbeispiel n
mathematical formula Berechnungsformel f, Rechenformel f
mathematical method Rechenmethode f
mathematical model Berechnungsmodell n, Rechenmodell n
mating surfaces Paßflächen fpl
matrix Matrix f
LED matrix Leuchtdiodenmatrix f
polymer matrix Polymermatrix f, Kunststoff-Matrix f
resin matrix Harzmatrix f
matter Materie f, Stoff m
foreign matter Fremd(bestand)teile npl, Fremdteile npl
volatile matter flüchtige Bestandteile mpl
maturing Reifung f
maturing process [e.g. of a PVC paste] Reifevorgang m
maximization Maximierung f
output maximization Durchsatzmaximierung f
maximum Maximum n
maximum allowable concentration [MAC] MAK-Wert m, maximale Arbeitsplatzkonzentration f
maximum continuous operating temperature Dauerbetriebsgrenztemperatur f
maximum daylight between platens größte lichte Weite f zwischen den Platten
maximum dimension Größtmaß n
maximum force Höchstkraft f
maximum memory capacity maximaler Speicherausbau m
maximum output Spitzenleistung f
maximum permissible höchstzulässig
maximum pressure Druckgrenze f, Druckmaximum n, Druckobergrenze f, Höchstdruck m, Druckspitze f
maximum requirements Spitzenbedarf m
maximum screw speed Drehzahlgrenze f
maximum shot weight maximales Teilgewicht n
maximum speed Drehzahlgrenze f
maximum temperature Höchsttemperatur f, Spitzentemperatur f
maximum tensile stress Zugspannungsmaximum n
maximum throughputs Höchstdurchsätze f
maximum torque Drehmomentgrenze f

maximum value Größtwert m, Höchstwert m, Spitzenwert m
maximum working temperature Einsatzgrenztemperatur f
damping maximum Dämpfungsmaximum n
MDI [diphenylmethane-di-isocyanate] Diphenylmethan-diisocyanat n
mean 1. Mittel-, Durchschnitts-; 2. Mittel(wert) n(m)
  mean temperature Mitteltemperatur f, Temperaturmittelwert m
  mean value Mittelwert m
  arithmetic mean arithmetischer Mittelwert m
meaningful aussagekräftig [Ergebnis]
means Mittel npl
  cooling by means of long, slender cooling channels [method of cooling long core inserts in injection mo(u)lding] Fingerkühlung f
  means of production Produktionsmittel n
measure/to (aus)messen
measure Maß n; Maßnahme f
  economic measures Rationalisierungsmaßnahmen fpl
  legislative measures Gesetzgebungsmaßnahmen fpl
  noise reduction measures Geräuschdämpfungsmaßnahmen fpl, Lärmminderungsmaßnahmen fpl
  pollution control measures Umweltschutzmaßnahmen fpl
  safety measures Schutzmaßnahmen fpl, Vorsichtsmaßnahmen fpl
  technical measures verfahrenstechnische Maßnahmen fpl
measured gemessen
  accurately measured out genau (gemessen)
measurement Messung f
  measurement of flow rate Durchflußmessung f
  measurement of pressure Druckmessung f
  comparative measurement Vergleichsmessung f
  precision measurement Feinmessung f
  stroke measurement Wegerfassung f
  temperature measurement Temperaturmessung f, Temperaturerfassung f
  wall thickness measurement Wanddickenmessung f
measuring Meß -
  measuring and control instruments MSR-Geräte npl
  device for measuring pipe wall thickness Rohrwanddicken-Meßgerät n
  electronic measuring and control system MSR-Elektronik f
  film ga(u)ge measuring instrument Foliendickenmeßgerät n
  method of measuring temperature Temperaturmeßmethode f
  pressure measuring point Druckmeßstelle f
  pressure measuring system Druckmeßsystem n
  stroke measuring system Wegmeßsystem n
  temperature measuring device Temperaturmessung f, Temperaturmeßvorrichtung f
  temperature measuring point Temperaturmeßstelle f
  web tension measuring device Bahnzugkraftmeßvorrichtung f
  web tension measuring unit Bahnzug(kraft)meßstation f
mechanical mechanisch, maschinell
  mechanical blowing (process) [method of making foam] Direktbegasungsverfahren n, Begasungsverfahren n, Begasen n, mechanische Verschäumung f
  mechanical blowing unit Direktbegasungsanlage f, Begasungsanlage f
  mechanical engineering Maschinenbau m
  mechanical loading mechanische Beanspruchung f
  mechanical properties mechanisches Niveau n, mechanische Werte mpl
  mechanical strength mechanische Festigkeit f
mechanically blown mechanisch getrieben
  mechanically blown foam mechanischer Schaum m, Schlagschaum m, Begasungsschaum m
  paste for making mechanically blown foam Schlagschaumpaste f
mechanics Mechanik f
  fracture mechanics Bruchmechanik f
mechanism Mechanismus m, Wirkungsweise f, Wirkungsmechanismus m
  adhesion mechanism Haftmechanismus m
  adjusting mechanism Einstelleinrichtung f, Verstelleinrichtung f, Verstellmöglichkeit f
  adjustment mechanism Einstelleinrichtung f, Verstelleinrichtung f, Verstellmöglichkeit f
  anti-drift control mechanism Antidrift-Regelung f
  automatic changing mechanism Wechselautomatik f
  automatic locking mechanism Zwangsverriegelung f
  automatic operating mechanism Bedienungsautomatik f
  automatic starting mechanism Anfahrautomatik f
  automatic switch-off mechanism Abschaltautomatik f
  automatic switch-on mechanism Einschaltautomatik f
  back pressure adjusting mechanism Staudruckeinstellung f
  back pressure relief mechanism Staudruckentlastung f
  clamping mechanism Formschluß m, Formschließsystem n, Schließeinrichtung f, Schließsystem n
  core puller control mechanism Kernzugsteuerung f
  core pulling mechanism Kernzugvorrichtung f
  counting mechanism Zählwerk n
  coupling mechanism Kopplungsmechanismus m
  crosslinking mechanism Vernetzungsmechanismus m
  curing mechanism Härtungsmechanismus m
  damping mechanism Dämpfungseinrichtung f
  degradation mechanism Zersetzungsmechanismus m
  delaying mechanism Verzögerungseinrichtung f
  die adjusting mechanism Düseneinstellung f, Düsenjustiereinrichtung f
  ejector control mechanism Auswerfersteuerung f
  ejector damping mechanism Auswerferdämpfung f

ejector mechanism Auswerfereinrichtung f, Auswerfersystem n
ejector plate safety mechanism Auswerferplattensicherung f
ejector release mechanism Auswerferfreistellung f
end-of-travel damping mechanism Endlagendämpfung f
failure mechanism Bruchmechanismus m, Versagensmechanismus m
feed mechanism Einzugswerk n, Speisevorrichtung f
film width control mechanism Folienbreitenregelung f
flash trimming mechanism Butzenabtrennung f
forced feed mechanism Zwangszuführeinrichtung f
free-radical chain growth mechanism Radikalkettenmechanismus m
gentle mo(u)ld clamping mechanism sanfter Werkzeugschluß m
high-speed mo(u)ld clamping mechanism Werkzeugschnellspannvorrichtung f
high-speed nozzle retraction mechanism Düsenschnellabhebung f
hot runner needle shut-off mechanism Heißkanal-Nadelverschlußsystem n
interval timing mechanism Pausenzeituhr f
lip adjusting mechanism Lippenverstellung f
locking mechanism Arretiervorrichtung f, Verriegelung(seinrichtung) f, Verriegelungssystem n, Maschinenzuhaltung f, Zuhalteinrichtung f, Zuhaltemechanismus m
mo(u)ld clamp delaying mechanism Formschließverzögerung f
mo(u)ld height adjusting mechanism Werkzeughöhenverstelleinrichtung f
mo(u)ld locking mechanism Formzuhaltung f
mo(u)ld safety mechanism Formschließsicherung f, Formschutz m, Schließhubsicherung f, Werkzeug-Sicherheitsbalken m, Werkzeugschutz m, Zufahrsicherung f, Werkzeugschließsicherung f
needle shut-off mechanism Nadelverschlußsystem n, Nadelventilverschluß m
nip adjusting mechanism Spaltverstellung f
nip setting mechanism Walzenspalteinstellung f
override mechanism Überfahreinrichtung f
parison gripping mechanism Schlauchgreifvorrichtung f
parison length control mechanism Vorformlingslängenregelung f
part delivery safety mechanism Ausfallsicherung f
piston drive mechanism Kolbentriebwerk n
plunger injection mechanism Kolbenspritzsystem n
quick-action coupling mechanism Schnellspannvorrichtung f
reaction mechanism Reaktionsmechanismus m
reel changing mechanism Rollenwechselsystem n
remote speed control mechanism Geschwindigkeitsfernsteuerung f
restrictor bar adjusting mechanism Staubalkenverstellung f
reversing rod mechanism Wendestangensystem n
roll adjusting mechanism Walzenverstellung f
roll bending mechanism Walzenbiegeeinrichtung f, Walzendurchbiegevorrichtung f, Walzengegenbiegeeinrichtung f
screw drive mechanism Schneckenantriebssystem n
screw retraction mechanism Schneckenrückholvorrichtung f
screw rotating mechanism Schneckenrotation f
screw stroke adjusting mechanism Schneckenhubeinstellung f, Schneckenwegeinstellung f
self-locking mechanism Selbstverriegelung f
setting mechanism Einstelleinrichtung f
shut-off mechanism Absperrmechanismus m, Verschlußmechanismus m
side ejector mechanism Seitenauswerfer m
speed setting mechanism Drehzahleinstellung f
spur gear speed reduction mechanism Stirnrad-Untersetzungsgetriebe n
stripper mechanism Abstreifvorrichtung f, Abstreifsystem n
stroke adjusting mechanism Hubeinstellung f
stroke sensing mechanism Weggebersystem n
temperature control mechanism Temperaturausgleichsystem n
temperature setting mechanism Temperatureinstellung f
tension control mechanism [for film as it comes off the calender] Spannungskontrolle f
three-plate clamping mechanism Dreiplatten-Schließsystem n
three-speed control mechanism Dreigang-Schaltgetriebe f
tilting mechanism Schwenkeinrichtung f
timing mechanism Zeitgebersystem n
toggle clamp mechanism Kniehebelverschluß m
toggle lock mechanism Kniehebelverriegelung f
toggle mechanism Kniehebelmechanismus m, Kniehebelsystem n
torque adjusting mechanism Drehmomenteinstellung f
unscrewing mechanism Ausdrehmechanik f, Ausschraubvorrichtung f
vertical adjusting mechanism Vertikalverstellung f
wall thickness control mechanism Wanddickenregulierungssystem n
web tension control mechanism Bahnspannungsregelung f
weighing mechanism Wägemechanismus m
width control mechanism Breitenregelung f, Breitenregulierung f
medium 1. Medium n; 2. mittel, Mittel-
medium-boiling mittelsiedend
medium-boiling solvent Mittelsieder m
medium-chain mittelkettig

medium-hard halbhart, mittelhart
medium-impact mittelzäh, halbschlagfest, mittelschlagfest, mittelschlagzäh, mittelstoßfest
medium molecular weight mittelmolekular
medium-oil mittelölig
medium-oil alkyd (resin) mittelfettes Alkydharz n, Mittelölalkydharz n
medium runs mittlere Serien fpl
medium vacuum Feinvakuum n
medium-viscosity mittelzäh, mittelviskos
medium voltage Mittelspannung f
medium-voltage range Mittelspannungsbereich m
medium-wave mittelwellig
attacking medium Angriffsflüssigkeit f, Angriffsmittel n
blowing medium Blasmedium n
circulating medium Umlaufmedium n
constant temperature medium Temperierflüssigkeit f, Temperiermittel n, Temperiermedium n
cooling medium Kühlmedium n
damping medium Dämpfungsflüssigkeit f, Dämpfungsmedium n
dispersion medium Dispersionsträger m
filter medium Filtermedium n, Filtermittel n
heat transfer medium Wärmeträger(mittel) m(n), Wärmeübertragungsmittel n, Wärmeübertragungsmedium n
heating-cooling medium Temperiermittel n, Temperierflüssigkeit f, Temperiermedium n
mo(u)ld cooling medium Werkzeugkühlmedium n
mo(u)ld heating medium Werkzeugheizmedium n
surrounding medium Umgebungsmedium n
temperature control medium Temperiermittel n, Temperiermedium n, Temperierflüssigkeit f
test medium Prüfmedium n, Prüfmittel n
meet/to zusammentreffen, entsprechen [Vorgabe]; einhalten [Toleranz]
meeting practical requirements praxisgerecht
meeting Versammlung f
general meeting [of shareholders etc.] Hauptversammlung f
melt/to (auf)schmelzen
melt Schmelze f; Plastifikat n
melt accumulation Schmelzedepot n
melt accumulator Schmelzebehälter m, Zylinderspeicher m, Speicher m, Schmelzespeicher m, Kopfspeicher m, Stauzylinder m, Massespeicher m, Plastikakkumulator m
melt accumulator machine Kopfspeichermaschine f
melt accumulator ram Schmelzespeicherkolben m
melt accumulator system Kopfspeichersystem n, Massespeichersystem n
melt back-flow Schmelzerückfluß m
melt back flushing Schmelzenrückspülung f
melt circulation Schmelzumlagerung f
melt cushion Massepolster n, Schmelzepolster n, Polster n

melt decompression Schmelzedekompression f
melt decompression system Schmelzedekompression f
melt degassing (unit) Schmelzeentgasung f
melt delivery Schmelzeaustritt m, Schmelzeaustrag m
melt delivery screw Schmelzeaustragsschnecke f
melt devolatilizing (system) Schmelzedekompression f
melt distribution Schmelzeverteilung f
melt dividing schmelzeteilend
melt elasticity Schmelzeelastizität f
melt exit temperature Masseaustrittstemperatur f
melt extensibility Dehnviskosität f, Schmelzedehnung f
melt (fed) extruder Heißschmelzextruder m, Schmelzeextruder m, Schmelzeaustragsextruder m, Ausstoßextruder m
melt feed (system) Schmelzezufluß m, Schmelzeeinspeisung f
melt filter Schmelzefilter n,m
melt filtration Schmelzefiltrierung f, Schmelzenfilterung f
melt filtration unit Rohmaterialfilter n,m, Schmelzenfiltereinrichtung f, Schmelzefilter n,m, Schmelzenfilterung f
melt flow index [MFI] Schmelzindex m, Fließindex m
melt flow (rate) Schmelzestrom m, Massefluß m, Schmelzefluß m
melt flow-way (system) Schmelzeleitsystem n, Massekanal m, Schmelzeleitung f, Schmelze(führungs)bohrung f, Schmelze(führungs)kanal m, Strömungskanal m
melt fracture Schmelzbruch m
melt front Fließfront f
melt homogeneity Schmelzehomogenität f
melt layer Schmelzeschicht f
melt metering Schmelzedosierung f
melt metering pump Schmelzedosierpumpe f
melt metering unit Schmelzedosierung f
melt particle Schmelzeteilchen n
melt pressure Massedruck m, Schmelzedruck m
melt pressure indicator Massedruckanzeige f
melt pressure profile Massedruckverlauf m
melt pressure sensor Massedruckaufnehmer m, Massedruckgeber m
melt quality Plastifikatgüte f, Schmelzequalität f
melt spinning Schmelzespinnen n
melt stability Schmelzestabilität f
melt strand Massestrang m, Schmelzstrang m
melt stream Produktstrom m, Schmelzestrom m
melt stream-dividing strömungsteilend
melt temperature Formmassetemperatur f, Kunststofftemperatur f, Schmelzetemperatur f, Spritzguttemperatur f, Massetemperatur f
melt temperature at the feed end [e.g. of an extruder] Einlaufmassetemperatur f, Eingangsmassetemperatur f

melt temperature distribution Massetemperaturverteilung f
melt temperature indicator Massetemperaturanzeige f
melt thermocouple Massethermoelement n, Massetemperaturfühler m, Schmelzetemperaturfühler m
melt throughput Schmelzestrom m, Massedurchsatz m
melt transport Schmelzeförderung f
melt viscosity Schmelzeviskosität f
melt vortex Schmelzewirbel m
amount of melt passing through the die Düsenfluß m
axial melt stream Axialstrom m
condition of the melt Schmelzezustand m
device which divides the melt stream Schmelzestromteiler m
exposed melt surface [i.e. during degassing] Entgasungsoberfläche f
final melt temperature Schmelzeendtemperatur f
hot melt adhesive Schmelzkleber m, Schmelzklebstoff m
hot melt adhesive applicator Schmelzklebstoff-Auftragsgerät n
hot melt contact adhesive Schmelzhaftklebstoff m
hot melt extruder Heißschmelzextruder m, Schmelze(austrags)extruder m
hot melt granulation Heißgranulierung f
hot melt granulator Heißgranuliervorrichtung f, Heißgranuliermaschine f
hot melt sealant Schmelzmasse f
large-area melt filter Großflächen-Schmelzefilter n,m
passage of the melt Schmelzeführung f
polymer melt Kunststoffschmelze f, Polymerisatschmelze f, Polymerschmelze f, Rohstoffschmelze f, Thermoplastschmelze f
polymer melt filtration Kunststoffiltration f
reverse melt flow Schmelzerückfluß m
separate melt streams Partialströme mpl, Teilströme mpl, Masseteilströme mpl, Schmelzeteilströme mpl
spiral melt stream Wendelstrom m
melted schmelzflüssig, geschmolzen
melting Schmelz-, schmelzend
melting characteristics Aufschmelzverhalten n
melting point Schmelzpunkt m, Schmelztemperatur f
melting range Schmelzbereich m, Schmelzintervall m
crystallite melting point Kristallitschmelztemperatur f
low-melting niedrigschmelzend
member Glied n, Element n, Mitglied n
member of the board of directors Aufsichtsratmitglied n
EEC member countries EG-Mitgliedsländer npl
memory Gedächtnis n, Speicher m, Erinnerung f
memory address Speicheradresse f
memory capacity Speicherkapazität f
memory content Speicherinhalt m

memory effect Erinnerungseffekt m
memory module Speicherbaustein m, Speichermodul m
alarm memory Alarmgedächtnis n
cycle memory Zyklusgedächtnis n
internal memory interner Speicher m
maximum memory capacity maximaler Speicherausbau m
peak memory Spitzenspeicher m
random access memory [RAM] RAM-Speicher m
read-only memory [ROM] Nur-Lesespeicher m, ROM-Speicher m
semi-conductor memory Halbleiterspeicher m
setpoint memory Sollwertspeicher m
write-read memory Schreib-Lese-Speicher m
mercaptide Merkaptid n [auch: Mercaptid n]
butyltin mercaptide Butylzinnmerkaptid n
dialkyltin mercaptide Dialkylzinnmerkaptid n
dibutyltin mercaptide Dibutylzinnmerkaptid n
dioctyltin mercaptide Dioctylzinnmerkaptid n
octyltin mercaptide Octylzinnmerkaptid n
organo-tin mercaptide Organozinnmerkaptid n
tin mercaptide Zinnmerkaptid n
mercapto -merkapto-
dialkyltin mercaptochloride Dialkylzinnmerkaptochlorid n
dibutyltin mercapto compound Dibutylzinnmerkaptoverbindung f
organo-tin mercaptocarboxylate Organozinnmerkaptocarbonsäureester m
mercury Quecksilber n
mercury column Hg-Säule f, Quecksilbersäule f
mesh Masche f
mesh width Maschenweite f
coarse-mesh grobmaschig
copper wire mesh Kupfergeflecht n
number of meshes Maschenzahl f
stainless steel wire mesh Edelstahlmaschengewebe n
wire mesh screen Drahtsiebboden m
metaborate Metaborat n
barium metaborate Bariummetaborat n
metal Metall n
metal guide plate Leitblech n
alkali metal Alkalimetall n
alkaline earth metal Erdalkalimetall n
containing heavy metals schwermetallhaltig
ferrous metal FE-Metall n, Eisenmetall n
heavy metal Schwermetall n
heavy metal complex Schwermetallkomplex m
heavy metal salt Schwermetallsalz n
light metal housing Leitmetallgehäuse n
matched metal mo(u)lding Warmpreßverfahren n, Heißpreßverfahren n, Heißpressen n
non-ferrous metal Buntmetall n, NE-Metall n, Nichteisenmetall n
non-ferrous metal alloy Buntmetall-Legierung f
sheet metal Stahlblech n

metallization Bedampfen n, Aufbringen n von Metallüberzügen
  high vacuum metallization Hochvakuumbedampfen n
meter Messer m, Meter m
  torque meter Drehmomentwaage f
metered amount Dosiermenge f
metering messend, Dosier-
  metering accuracy Dosiergenauigkeit f
  metering and mixing unit Dosier- und Mischmaschine f
  metering cylinder Dosierzylinder m
  metering equipment Dosieranlage f, Dosiereinrichtung f
  metering-mixing unit Dosiermischmaschine f
  metering performance Dosierleistung f
  metering pump Dosierpumpe f
  metering section Austragsbereich m, Ausstoßzone f, Ausbringungszone f, Austragszone f, Pumpzone f, Ausstoßteil m
  metering station Dosierstation f
  metering stroke Dosierhub m, Dosierweg m
  metering system Dosiersystem n
  metering unit Dosieraggregat n, Dosiereinheit f, Dosierwerk n, Dosierelement n, Dosiergerät n
  metering valve Dosierschieber m
  additional metering unit Zusatz-Dosieraggregat n
  automatic metering unit Dosierautomat m
  granule metering unit Granulatdosiergerät n
  high-pressure metering unit Hochdruckdosieranlage f
  inaccurate metering Dosierungenauigkeit f
  melt metering pump Schmelzedosierpumpe f
  melt metering unit Schmelzedosierung f
  precision metering unit Präzisionsdosiereinheit f
  twin screw metering section Zweischnecken-austragszone f
methanol solution methanolische Lösung f
method Methode f, Verfahren n
  method of analysis Analysenmethode f, Analysenverfahren n
  method of application Applikationsmethode f, Auftragsmethode f
  method of construction Bauweise f
  method of demo(u)lding Entformungsprinzip n
  method of determination Bestimmungsmethode f
  method of estimating Abschätzmethode f
  method of heating Aufheizmethode f
  method of installation Einbaumethode f
  method of joining Verbindungsverfahren n
  method of measuring temperature Temperaturmeßmethode f
  method of operation Fahrweise f, Funktionsweise f, Betriebsweise f
  method of polymerization Polymerisationstechnik f
  method of processing Verarbeitungsverfahren n
  method of separation Trennmethode f
  finite element method [FEM] FEM-Verfahren n
  mathematical method Rechenmethode f

  test method Prüfmethode f, Prüfmethodik f, Prüfverfahren n, Testmethode f
  time-lag method Zeitverzögerungsmethode f
  wet-in-wet method Naß-in-Naß-Verfahren n
methyl acrylate Acrylsäuremethylester m
methylated methyliert
  partly methylated partiellmethyliert
MFI [melt flow index] Schmelzindex m
mica Glimmer m
  mica-insulated heater bands Glimmerheizbänder npl
  mica platelet Glimmerplättchen n
  potash mica Muscovit-Glimmer m
  white mica Muscovit-Glimmer m
micaceous iron ore Eisenglimmer m
micell Mizelle f, Mizell n [Kolloidchemie]
micrograph Mikroaufnahme f, mikroskopische Aufnahme f
  electron micrograph elektronenmikroskopische Abbildung f, elektronenmikroskopische Aufnahme f
  scanning electron micrograph rasterelektronenmikroskopische Aufnahme f, REM-Aufnahme f
microscope Mikroskop n
  electron microscope Elektronenmikroskop n
  light microscope Auflichtmikroskop n
  optical microscope Durchlichtmikroskop n, Lichtmikroskop n
  polarizing microscope Polarisationsmikroskop n
  scanning electron microscope Rasterelektronenmikroskop n
microscopy Mikroskopie f
  optical microscopy Lichtmikroskopie f
  scanning electron microscopy Rasterelektronenmikroskopie f
microsection Dünnschliff m, mikroskopischer Schnitt m
microtome section Dünnschnitt m, Feinschliff m, Querschliff m
microtomy Mikrotomie f
  ultra-thin section microtomy Ultradünnschnittmikrotomie f
migration Wandern n
  migration behavio(u)r Wanderungsverhalten n
  migration resistance Wanderungsbeständigkeit f, Wanderungsfestigkeit f
  migration tendency Wanderungstendenz f
  acting as a barrier against plasticizer migration weichmachersperrend
  plasticizer migration Weichmachermigration f, Weichmacherwanderung f
  plasticizer migration resistance Weichmacherwanderungsbeständigkeit f
mill Walze f, Mühle f
  ball mill Kugelmühle f
  colloid mill Kolloidmühle f
  feed mill Speisewalzwerk n
  hammer mill Hammermühle f
  impact mill Prallmühle f
  laboratory two-roll mill Zweiwalzen-Labormischer m

roll mill Walzenreibstuhl m, Walzenstuhl m, Walzwerk n
toothed disk mill Zahnscheibenmühle f
triple-roll mill Dreiwalze f, Dreiwalzenmaschine f, Dreiwalzenstuhl m
two-roll mill Doppelwalze f, Zweiwalze f, Zweiwalzenmaschine f
milled glass fibers Mahlglasfasern fpl
milling Fräsen n, Walzprozeß m; Fräsrotor m
  milling cutter Walzenfräser m
  milling temperature Walztemperatur f
  milling test Walztest m
  milling time Walzzeit f
  milling-type cutter Fräsrotor m, Walzenfräser m
  copy milling machine Kopierfräsmaschine f
  long-term milling test Dauerwalztest m
mine Mine f, Bergbaubetrieb m
mineral powder Gesteinsmehl n
minicomputer Kleinrechner m
minimization Minimierung f
  downtime minimization Stillstandminimierung f
  minimizing of waste Abfallminimierung f
  minimizing the amount of flash produced Butzenminimierung f
minimum Minimum n
  minimum dimension Kleinstmaß n
  minimum value Kleinstwert m
  reducing waste to a minimum Ausschußminimierung f
mining Bergbau m
misalignment Fluchtungsfehler m
  mo(u)ld misalignment Werkzeugversatz m
miscible with water wassermischbar, wasserverträglich
misinterpretation Fehlinterpretation f
mistake Fehler m
  formulation mistake Rezepturfehler m
miter (joint) [GB: mitre] Gehrungsfuge f
mix/to (ein)mischen, einarbeiten
mix Mischung f, Ansatz m
  adhesive mix Klebstoffansatz m
  polyester mix Polyesteransatz m
  polyester resin mix UP-Reaktionsharzmasse f
  ready-to-use mix Fertigmischung f
  rendering mix Putz m
  resin-catalyst mix Harzansatz m, Reaktionsmasse f, Reaktionsharzmischung f
mixed-cell [foam] gemischtzellig
mixer Mischer m, Rührwerk n
  continuous mixer Durchlaufmischer m
  cooling mixer Kühlmischer m
  drum mixer Trommelmischer m
  fluid mixer Fluidmischer m
  general-purpose mixer Universalmischer m
  hand mixer Handrührgerät n
  heavy-duty high speed mixer Hochleistungsschnellmischer m
  heavy-duty internal mixer Hochleistungsinnenmischer m
  heavy-duty mixer Hochleistungsmischer m
  high-speed mixer Schnellmischer m
  internal mixer Stempelkneter m
  laboratory twin-screw mixer Labor-Zweiwellenkneter m
  paddle mixer Paddelmischer m, Schaufelrührer m
  planetary mixer Planetenmischer m, Planetenmischkneter m, Planetenrührwerk n
  ribbon mixer Bandmischer m
  slow-speed mixer Langsamläufer m
  tumble mixer Taumelmischer m
mixing Vermischung f, Abmischung f
  mixing blades Knetschaufeln fpl, Knetflügel mpl
  mixing chamber Knetraum m, Knetkammer f
  mixing flight [of a screw] Knetsteg m, Knetgang m
  mixing gap [of a mixing screw] Knetspalt m
  mixing ratio Abmischungsverhältnis n
  mixing screw Knetschnecke f, Knet(schnecken)welle f
  axial mixing Axialvermischung f
  faceted mixing section [of screw] Rautenmischteil n
  high pressure mixing unit Hochdruckmischanlage f
  hot mixing Heißmischung f
  knurled mixing section [of screw] Nockenmischteil n
  laboratory mixing rolls Labormischwalzwerk n
  longitudinal mixing Axialvermischung f, Längsmischung f
  metering and mixing unit Dosier- und Mischmaschine f
  paste mixing Verpastung f
  pigment mixing unit Einfärbegerät n
  static mixing element Statikmischelement n
  transverse mixing Quermischung f
  twin mixing head Doppelmischkopf m
mixture Gemisch n
  reaction mixture Reaktionsgemisch n
  solvent mixture Lösemittelgemisch n
mobility Beweglichkeit f
  molecule mobility Molekülbeweglichkeit f
mock Leno weave Scheindreherbindung f [Textil]
mode Modus m
  interactive mode Dialogbetrieb m, Dialogführung f, Dialogverkehr m
  mode of action Wirkungsmechanismus m, Wirkungsweise f
  mode of operation Arbeitsweise f, Betriebsweise f
model Modell n, Ausführung f
  model concept Modellvorstellung f
  model law Modellgesetz n
  model-making material Modellwerkstoff m
  model-making resin Modellharz n
  model quantity Modellgröße f
  model test Modelluntersuchung f
  basic model Grundtyp m
  copy-milling model Kopier(fräs)modell n
  duplicate model Duplikatmodell n, Modellduplikat n
  general-purpose model Universaltyp m

Urmodell n
mathematical model Berechnungsmodell n, Rechenmodell n
process model Prozeßmodell n
standard model Standardausführung f
subsequent model Nachfolgemodell n
table-top model Tischgerät n, Tischmodell n
working model Arbeitsmodell n
modelling Modellieren n
  modelling paste Urmodellpaste f
modem [acronym for MOdulator/DEModulator] Modem n [Signalumformer]
moderately flexible mittelflexibel
moderately reactive mittelreaktiv
modernization Modernisierung f
modification Modifizierung f
  design modification Konzeptionsänderung f
  impact modification Schlagzähmodifizierung f, Zähmodifizierung f
  program modifications Programm-Abwandlungen fpl
modified modifiziert, abgeändert
  impact modified schlagzähmodifiziert
  oil modified ölmodifiziert
  polymer modified polymermodifiziert
  synthetic resin modified kunststoffmodifiziert, kunststoffvergütet
modifier Modifizier(ungs)mittel n, Modifikator m, Modifier m
  impact modifier Schlagzähmodifikator m, Schlagzähmodifiziermittel n, Schlagzähmodifizierharz n, Schlagfestmacher m, Schagzähmodifier m, Schlagzähigkeitsverbesserer m, Schlagzähkomponente f, Schlagzähmacher m
modifying agent Modifizierungsmittel n, Modifikator m, Modifier m
modifying resin Modifizierharz n
modular baukastenartig, baukastenmäßig, bausteinartig, in Blockbauweise f, modulartig, Modul-
  modular character Modularität f
  modular construction (system) Modulbauweise f, Baukastenbauweise f, Blockaufbau m, Elementbauweise f
  modular drive Bausteingetriebe n
  modular electronic system Bausteinelektronik f
  modular injection mo(u)lding machine Baukasten-Spritzgießmaschine f
  modular machine Baukastenmaschine f
  modular machines Maschinenkonzepte npl in Bausatzform
  modular principle Baukastenprinzip n
  modular range Baukastenprogramm n, Baukastenreihe f
  modular screw Bausatzschnecke f, Schneckenbaukasten m, Baukastenschnecke f
  modular screw assembly Baukastenschneckensatz m
  modular system Bausteinsystem n
module Modul m, Baustein m, Baueinheit f, Bausatz m

  adaptor module Anpaßbaustein m
  analog(ue) module Analogbaugruppe f
  display module Anzeigebaugruppe f
  electronic module Elektronikbaustein m
  extra module Zusatzbaustein m, Zusatzeinheit f
  hardware module Hardwarebaustein m
  injection module Einspritzbaustein m
  input module Eingabebaustein m, Eingabebaugruppe f, Eingabestein m, Eingangsmodul m
  input-output module Ein-Ausgabestein m
  logic module Logikbaustein m
  memory module Speicherbaustein m, Speichermodul m
  output module Ausgabegruppe f, Ausgabestein m, Ausgangsmodul m
  program module Programmbaustein m
  programming module Programmierbaustein m
  semi-conductor module Halbleiterbaustein m
  separate module Einzelbaustein m
  slide-in module Einschub m
  slide-in temperature control module Temperatur-Regeleinschub m
  software module Software-Baustein m
  time module Zeitbaustein m, Zeitmodul m
  transmission module Übertragungsmodul m
modulus Modul m, Modulwert m, Spannungswert m
  modulus of elasticity E-Modul m, Elastizitätsmodul m
  modulus of elasticity in bending Biege-E-Modul m, Biegeelastizitätsmodul m
  modulus of elasticity in compression Druck-E-Modul m
  modulus of elasticity in tension Zug-E-Modul m
  apparent modulus of rigidity Torsionssteifheit f
  compressive modulus Druckmodul m, Kompressionsmodul m
  creep modulus Kriechmodul m
  creep modulus curve Kriechmodulkurve f
  damping modulus Dämpfungsmodul m
  elastic modulus E-Modul m, Elastizitätsmodul m
  flexural creep modulus Biege-Kriechmodul m
  flexural modulus Biegemodul m
  loss modulus Verlustmodul m
  relaxation modulus Relaxationsmodul m
  shear modulus Gleitmodul m, Schubmodul m
  storage modulus Speichermodul m
  tensile creep modulus Zug-Kriechmodul m
  tensile modulus Zugmodul m
  torsion modulus G-Modul m, Torsionsmodul m
moisture Feuchte f, Feuchtigkeit f
  moisture absorption Feuchtigkeitsaufnahme f
  moisture content Feuchtewert m, Feuchtigkeitsniveau n, Feuchtegehalt m, Feuchtigkeitsgehalt m
  moisture content of the atmosphere Luftfeuchtegehalt m
  moisture loss Feuchtigkeitsverlust m
  moisture-proof feuchtigkeitsundurchlässig

moisture-proof feuchtigkeitsundurchlässig
moisture resistance Feuchtigkeitsresistenz f, Feuchtigkeitsbeständigkeit f
moisture resistant feuchtigkeitsbeständig
affected by moisture feuchtigkeitsempfindlich
condensed moisture Kondenswasser n, Schwitzwasser n
condensed moisture atmosphere Schwitzwasserklima n
condensed moisture test Schwitzwasserprüfung f
effect of moisture Feuchteeinfluß m
final moisture content Endfeuchte f
impermeable to moisture feuchtigkeitsundurchlässig
initial moisture content Ausgangsfeuchte f
original moisture content Ausgangsfeuchte f
residual moisture content Restfeuchtigkeit f, Restfeuchte f, Restfeuchtigkeitsgehalt m
sensitivity to moisture Feuchteempfindlichkeit f
surface moisture Oberflächenfeuchte f
susceptibility to moisture Feuchteempfindlichkeit f
traces of moisture Feuchtigkeitsspuren fpl
unaffected by moisture feuchteunempfindlich
molar molar, Mol-
  molar fraction Molenbruch m
  molar heat Molwärme f
  molar ratio Molverhältnis n
molecular molekular
  molecular lattice structure Molekülgitter n
  molecular movement Molekularbewegung f
  molecular orientation Molekülorientierung f
  molecular structure Molekülstruktur f, Molekülaufbau m, Molekülgebilde n, Molekularaufbau m
  molecular weight molare Masse f, Molekülmasse f, Molekulargewicht n, Molekularmasse f, Molgewicht n, Molmasse f
  molecular weight distribution Molekulargewichtsverteilung f, Molmassenverteilung f
  molecular weight distribution curve Molekulargewichtsverteilungskurve f
  average molecular weight mittleres Molekulargewicht n, Molekulargewicht-Mittelwert m
  change in molecular weight Molekulargewichtsänderung f
  depending on the molecular weight molekulargewichtsabhängig
  limiting molecular weight Grenzmolekulargewicht n
  medium molecular weight mittelmolekulares Gewicht n
  number-average molecular weight Molekulargewicht-Zahlenmittel n
  ultra-high molecular weight ultrahochmolekulares Gewicht n
  weight-average molecular weight gewichtsmittleres Molekulargewicht n
molecule Molekül n
  molecule chain Molekülkette f
  molecule chain length Molekülkettenlänge f

molecule length Moleküllänge f
molecule mobility Molekülbeweglichkeit f
molecule segment Molekülsegment n
adjacent molecules Nachbarmoleküle npl
dipole molecule Dipolmolekül n
giant molecule Riesenmolekül n
monomer molecule Monomermolekül n
peroxide molecule Peroxidmolekül n
plasticizer molecule Weichmachermolekül n
polymer molecule Kunststoffmolekül n, Polymermolekül n
thread-like molecule Fadenmolekül n
molten tube [as it comes out of the extruder, before being expanded into a film bubble] Schmelzeschlauch m
molybdenum Molybdän n, Mo
  molybdenum disulphide Molybdändisulfid n
  molybdenum orange Molybdänorange n
  molybdenum red Molybdatrot n
  molybdenum trioxide Molybdäntrioxid n
moment 1. Moment n, Drehmoment n [Mechanik]; 2. Moment m, Augenblick m [Zeit], Zeitpunkt m
  moment of demo(u)lding Entformungszeitpunkt m
  moment of inertia Flächen-Trägheitsmoment n, Massenträgheitsmoment n, Trägheitsmoment n
  bending moment Biegemoment n
  dipole moment Dipolmoment n
  torsional moment Torsionsmoment n
monitor Monitor m, Überwachungsgerät n [s.a. monitoring]
  charge monitor Ladungsmonitor m
  contamination monitor Verschmutzungsüberwachung f [Gerät]
  cycle monitor Taktüberwachung f [Gerät]
  discharge monitor Entleerüberwachung f [Gerät]
  flow monitor Strömungswächter m
  level monitor Füllstandüberwachung f [Gerät]
  malfunction monitor Störungsüberwachung f [Gerät]
  oil level monitor Ölniveauwächter m, Ölstandüberwachung f [Gerät]
  oil temperature monitor Öltemperaturwächter m
  process monitor Prozeßüberwachung f [Gerät]
  temperature monitor Temperaturüberwachung f [Gerät]
  web monitor Bahnwächter m
  zero point monitor Nullpunktüberwachung f [Gerät]
monitoring Überwachung f
  monitoring devices Überwachungselemente npl, Überwachungsarmaturen fpl, Überwachungsorgane npl
  monitoring equipment Überwachungseinrichtung f
  monitoring function Überwachungsfunktion f
  monitoring instrument Überwachungsgerät n
  monitoring system Überwachungssystem n
  monitoring task Überwachungsaufgabe f
  ejector monitoring device Auswerfer-Überwachungseinheit f
  level monitoring device Niveauwächter m,

Niveauüberwachung f, Niveaukontrolle f [Gerät]
   level monitoring (unit) Füllstandüberwachung f
   malfunction monitoring (device) Störungsüberwachung f
   mo(u)ld filling monitoring (system) Formfüllüberwachung f
   mo(u)lding cycle monitoring (system) Zyklusüberwachung f
   oil level monitoring Ölstandüberwachung f [Gerät]
   process monitoring unit Prozeßüberwachung f
   production monitoring (department/system)
   Fertigungsüberwachung f
   web monitoring device Bahnwächter m
mono Mono-
   mono-alkyl tin chloride Monoalkylzinnchlorid n
   mono-alkyl tin compound Monoalkylzinnverbindung f
   mono-alkyl tin stabilizer Monoalkylzinnstabilisator m
   mono-alkyl tin thioglycolate Monoalkylzinnthioglykolsäureester m
   mono-layer Monolage f
   mono-organosilane Monoorganosilan n
   mono-organotin compound Monoorganozinn-
   Verbindung f
   mono-substituted monosubstituiert
   fatty acid mono-ester Fettsäuremonoester m
   glycerin mono-oleate Glycerinmonooleat n
monoacid Monosäure f
monocarboxylate Monocarbonsäureester m
monocarboxylic acid Monocarbonsäure f
   monocarboxylic acid ester Monocarbonsäureester m
monochrome visual display unit S/W-Videogerät n,
   Schwarz-Weiß-Sichtgerät n
monoethanolamine Monoethanolamin n
   dodecyl monoethanolamine Dodecylmonoethanolamin n
monofilament Elementarfaden m, Monofil(ament) n
   monofilament die Monofilwerkzeug n
   monofilament extruder Spinnextruder m
   monofilament extrusion line Monofilanlage f
   monofilament grade [of compound] Monofilament-
   Typ m
monofunctional monofunktionell
monohydric [if an alcohol] einwertig [Alkohol]
monomer Monomer n
   monomer concentration Monomerkonzentration f
   monomer droplet Monomertröpfchen n
   monomer molecule Monomermolekül n
   monomer radical Monomerradikal n
   monomer soluble monomerlöslich
   monomer unit Monomerbaustein m,
   Monomereinheit f
   original monomer content Ausgangsmonomergehalt m
   removal of residual monomer Restmonomerentfernung f
   residual monomer Restmonomer n

   residual monomer content Restmonomergehalt m
   starting monomer Ausgangsmonomer n
   styrene monomer Monostyrol n
   vinyl monomer Vinylmonomer n
   with a low monomer content monomerarm
monomeric monomer
   monomeric plasticizer Monomerweichmacher m
monomethylol urea Monomethylolharnstoff m
monovalent einbindig, einwertig
montan wax Montanwachs n
montanate Montansäureester m
   calcium montanate Calciummontanat n
montanic acid Montansäure f
   montanic acid ester Montansäureester m
Mooney plasticity Mooney-Plastizität f
Mooney viscosity Mooney-Viskosität f
morphological morphologisch
morphology Morphologie f
   two-phase morphology Zweiphasenmorphologie f
mortar Mörtel m
   mortar screed Mörtelbelag m
   epoxy-based mortar EP-Mörtel m
   polymer mortar Polymermörtel m
   repair mortar Reparaturmörtel m
   self-levelling mortar Verlaufsmörtel m
most suitable optimal
motor Motor m
   motor drive motorischer Antrieb m
   motor driven motorbetrieben, motorisch angetrieben
   motor power Motorleistung f
   motor safety switch Motorschutzschalter m
   motor speed Motordrehzahl f
   motor vehicle Kraftfahrzeug n
   axial piston motor Axialkolbenmotor m
   commutator motor Kommutatormotor m
   d.c. motor Gleichstrommotor m
   d.c. permanent magnet motor Gleichspannungs-
   Drehankermagnet-Motor m
   d.c. shunt motor Gleichstrom-Nebenschlußmotor m
   drive motor Antriebsmotor m
   effective motor power Motorwirkleistung f
   electric drive motor Elektro-Antriebsmotor m
   electric motor E-Motor m, Elektromotor m
   gear(ed) motor Getriebemotor m, Zahnradmotor m
   hydraulic motor Ölmotor m
   linear motor Linearmotor m
   main motor Hauptmotor m
   pump motor Pumpen(antriebs)motor m
   radial piston motor Radialkolbenmotor m
   screw drive motor Schneckenantriebsmotor m
   shunt motor Nebenschlußmotor m
   shunt-wound geared DC motor Nebenschluß-
   Gleichstrom-Getriebemotor m
   squirrel cage motor Kurzschlußankermotor m
   stepper (motor) Schrittmotor m
   synchronous motor Synchronmotor m
   three-phase geared motor Drehstromgetriebemotor m

three-phase motor Drehstrommotor m
torque motor Torquemotor m, Drehmomentantrieb m
total installed motor power gesamte installierte Motorenleistung f
vane-type motor Flügelzellenmotor m
vibrator motor Rüttelmotor m
mo(u)ld/to spritzen
mo(u)ld 1. Werkzeug n, Form f, Form(en)werkzeug n, Gießform f, Hohlform f, Hohlformwerkzeug n, Fertigungsmittel n; 2. Schimmel(pilz) m
mo(u)ld advance speed Formfahrgeschwindigkeit f
mo(u)ld assembly Formeinheit f
mo(u)ld attachment Formenanschluß m, Werkzeugbefestigung f, Werkzeugaufspannung f
mo(u)ld attachment holes Aufspannbohrungen fpl
mo(u)ld breathing Werkzeugatmung f
mo(u)ld carriage Werkzeugschlitten m, Werkzeugträger m
mo(u)ld carrier [of a carousel-type foam **mo(u)lding or injection mo(u)lding machine**] Formenträger m, Formträgerplatte f
mo(u)ld cavity Spritzgießgesenk n, Einarbeitung f, Spritzgießkavität f, Werkzeugkontur f
mo(u)ld cavity venting Formnestentlüftung f
mo(u)ld cent(e)ring device Werkzeugzentrierung f
mo(u)ld cent(e)ring hole Werkzeugzentrierbohrung f
mo(u)ld changing time Werkzeugwechselzeit f
mo(u)ld clamp delaying mechanism Formschließverzögerung f
mo(u)ld clamping control Formschließregelung f
mo(u)ld clamping frame Schließgestell n
mo(u)ld cleaning device Werkzeugreinigungsvorrichtung f
mo(u)ld closing movement Formschließbewegung f, Schließbewegung f, Werkzeugschließbewegung f
mo(u)ld closing speed Formenschließgeschwindigkeit f, Formzufahrgeschwindigkeit f, Werkzeugschließgeschwindigkeit f, Schließgeschwindigkeit f
mo(u)ld closing time Formschließzeit f, Schließzeit f
mo(u)ld components Formenbauteile npl, Werkzeugbauteile npl
mo(u)ld construction Werkzeugausführung f, Werkzeugaufbau m
mo(u)ld construction time Formenbauzeit f
mo(u)ld contents Forminhalt m
mo(u)ld cooling Werkzeugkühlung f
mo(u)ld cooling (channel) system Werkzeugkühlkanalanlage f
mo(u)ld cooling medium Werkzeugkühlmedium n
mo(u)ld cooling system Werkzeugkühlsystem n
mo(u)ld core Spritzgießkern m
mo(u)ld correction factor Formenkorrekturmaß n
mo(u)ld defects Werkzeugmängel mpl
mo(u)ld deflection Werkzeugdurchbiegung f
mo(u)ld-dependent werkzeuggebunden
mo(u)ld-dependent dimensions werkzeuggebundene Maße npl

mo(u)ld design Formenkonstruktion f, Werkzeugkonzept n, Werkzeuggestaltung f, Werkzeugauslegung f
mo(u)ld designer Formgestalter m, Werkzeugkonstrukteur m
mo(u)ld dimensions Formenmaße npl, Werkzeugmaße npl
mo(u)ld drawing Werkzeugzeichnung f
mo(u)ld filling monitoring system Formfüllüberwachung f
mo(u)ld filling operation Formfüllprozeß m, Werkzeugfüllvorgang m, Formenfüllung f, Formfüllvorgang m
mo(u)ld filling phase Formfüllphase f, Füllphase f
mo(u)ld filling pressure Formfülldruck m
mo(u)ld filling speed Formfüllgeschwindigkeit f, Werkzeugfüllgeschwindigkeit f, Füllgeschwindigkeit f
mo(u)ld filling study Füllstudie f
mo(u)ld filling time Füllzeit f, Spritzzeit f, Einspritzzeit f, Formfüllzeit f
mo(u)ld fixing details [diagram showing pattern of holes in platen for **mo(u)ld mounting**] Werkzeugaufspannzeichnung f, Bohrbild n, Lochbild n, Aufspannplan m
mo(u)ld fixing dimensions Formaufspannmaße npl, Aufspannmaße npl
mo(u)ld half Formhälfte f, Werkzeughälfte f
mo(u)ld heater Formheizgerät n
mo(u)ld heating Werkzeugbeheizung f
mo(u)ld heating medium Werkzeugheizmedium n
mo(u)ld heating system Werkzeugbeheizung f
mo(u)ld height Formaufspannhöhe f, Formeinbauraum m
mo(u)ld height adjusting mechanism Werkzeughöhenverstelleinrichtung f, Werkzeughöhenverstellung f
mo(u)ld height adjustment Werkzeughöhenverstellung f
mo(u)ld impression dimensions formgebende Werkzeugmaße npl
mo(u)ld-independent nicht werkzeuggebunden
mo(u)ld-independent dimensions nicht werkzeuggebundene Maße npl
mo(u)ld insert Formeinsatz m, Kontureinsatz m, Werkzeugeinsatz m
mo(u)ld life Werkzeugstandzeit f
mo(u)ld maker Formenbauer m, Werkzeugbauer m, Werkzeugmacher m
mo(u)ld making Werkzeugbau m
mo(u)ld material Werkzeugwerkstoff m, Formwerkstoff m
mo(u)ld misalignment Werkzeugversatz m
mo(u)ld mounting Werkzeugeinbau m
mo(u)ld mounting dimensions Werkzeugaufspannmaße npl, Werkzeugeinbaumaße npl, Einbaumaße npl, Werkzeuganschlußmaße npl
mo(u)ld movement Werkzeugbewegung f
mo(u)ld movement-related werkzeugbewe-

gungsabhängig
mo(u)ld opening and closing cycle Formenöffnungs- und Schließspiel n
mo(u)ld opening and closing speed Werkzeuggeschwindigkeit f
mo(u)ld opening and closing valve Formenöffnungs- und Schließventil n
mo(u)ld opening cylinder Formaufdrückzylinder m
mo(u)ld opening direction Formöffnungsrichtung f
mo(u)ld opening force Formauftreibdruck m, Öffnungskraft f, Werkzeugöffnungskraft f, Werkzeugauftriebskraft f, Aufreißkraft f, Auftreibkraft f
mo(u)ld opening movement Formöffnungsbewegung f, Öffnungsvorgang m, Öffnungsbewegung f, Werkzeugöffnungsvorgang m, Werkzeugöffnungsbewegung f
mo(u)ld opening speed Öffnungsgeschwindigkeit f, Werkzeugöffnungsgeschwindigkeit f, Formauffahrgeschwindigkeit f
mo(u)ld opening stroke Formöffnungsweg m, Formöffnungshub m, Öffnungsweg m, Öffnungsweite f, Öffnungshub m, Werkzeugöffnungsweg m, Werkzeugöffnungshub m
mo(u)ld opening stroke limiting device Öffnungswegbegrenzung f, Werkzeugöffnungshub-Begrenzung f
mo(u)ld plate Formplatte f
mo(u)ld plate assembly Plattenpaket n
mo(u)ld protection Werkzeugschonung f
mo(u)ld proving [to check mo(u)ld performance] Abmusterung f
mo(u)ld release agent Formtrennmittel n
mo(u)ld release effect Formtrennwirkung f
mo(u)ld return stroke Werkzeugrückhub m
mo(u)ld rigidity Werkzeugsteifigkeit f
mo(u)ld safety Werkzeugsicherheit f
mo(u)ld safety mechanism Formschutz m, Formschließsicherung f, Schließhubsicherung f, Werkzeugschutz m, Werkzeug-Sicherheitsbalken m, Zufahrsicherung f, Werkzeugschließsicherung f
mo(u)ld sealant Formversiegler m
mo(u)ld servicing and maintenance Werkzeugwartung- und Instandhaltung f
mo(u)ld setter Werkzeugeinrichter m
mo(u)ld setting record card Formeinrichterkarte f
mo(u)ld setting time Werkzeugeinrichtezeit f
mo(u)ld space Einbauhöhe f, Formeinbauraum m, Formeinbauhöhe f, Formaufspannhöhe f, Werkzeug(bau)höhe f, Werkzeugeinbaulänge f, Werkzeugeinbauraum m
mo(u)ld surface Formoberfläche f
mo(u)ld temperature Formtemperatur f, Werkzeugtemperatur f
mo(u)ld temperature control (system) Formentemperierung f, Werkzeugtemperaturregelung f, Werkzeugtemperierung f
mo(u)ld temperature distribution Werkzeugtemperaturverteilung f
mo(u)ld thermocouple Werkzeugtemperaturfühler mpl
mo(u)ld tolerances Werkzeugtoleranzen fpl
mo(u)ld unit Werkzeugrohling m
mo(u)ld venting Konturenlüftung f, Werkzeugbelüftung f, Werkzeugentlüftung f
mo(u)ld wall Formwandung f
mo(u)ld wear Werkzeugverschleiß m
mo(u)ld weight Werkzeuggewicht n
alumin(i)um mo(u)ld Aluminiumform f
amount of material in the mo(u)ld Werkzeugfüllungsgrad m
automatic mo(u)ld safety device Automatikwerkzeugschutz m
blow mo(u)ld cavity Blasgesenk n
blow(ing) mo(u)ld Blasform f, Blas(form)werkzeug n
bottom mo(u)ld half Werkzeugunterteil n
careful mo(u)ld clamping mechanism sanfter Werkzeugschluß m
cast alumin(i)um mo(u)ld Aluminiumguß-Werkzeug n
casting mo(u)ld Gießform f
center part of the mo(u)ld Werkzeugmittelteil n
cold runner mo(u)ld Kaltkanalform f, Kaltkanalwerkzeug n
compression mo(u)ld Preßwerkzeug n
diecasting mo(u)ld Druckgußform f, Druckgußwerkzeug n
double-daylight mo(u)ld Zweietagenwerkzeug n, Dreiplattenwerkzeug n
eight-cavity mo(u)ld Achtfachwerkzeug n
experimental mo(u)ld Versuchswerkzeug n, Prüfwerkzeug n
extremely small mo(u)ld Kleinstwerkzeug n
fixed mo(u)ld half Gesenkseite f, Einspritzseite f, Düsenseite f, feststehende Werkzeughälfte f
flash mo(u)ld Abquetschwerkzeug n, Überlaufwerkzeug n
foaming mo(u)ld Schäumform f
four-cavity mo(u)ld Vierfachform f, Vierfachwerkzeug n
gentle treatment of the mo(u)ld Werkzeugschonung f
heavy-duty mo(u)ld Hochleistungsform f, Hochleistungswerkzeug n
high-speed mo(u)ld changing system Werkzeugschnellwechselsystem n
high-speed mo(u)ld clamping mechanism Werkzeugschnellspannvorrichtung f
hot runner injection mo(u)ld Heißkanal-Spritzgießwerkzeug n
hot runner mo(u)ld Heißkanalwerkzeug n, Heißkanalform f
hot runner multi-daylight mo(u)ld Heißkanal-Etagenwerkzeug n
hot runner stack mo(u)ld Heißkanal-Etagenwerkzeug n
hot runner two-cavity mo(u)ld Heißkanal-

Doppelwerkzeug n
housing containing the mo(u)ld clamping mechanism Formschließgehäuse n
injection blow mo(u)ld Spritzblaswerkzeug n
injection mo(u)ld Spritzgießwerkzeug n, Spritz(gieß)form f
insulated runner mo(u)ld Isolierkanalform f, Isolierkanalwerkzeug n
laminating mo(u)ld Laminierform f
large mo(u)ld Großwerkzeug n
longitudinal mo(u)ld axis Werkzeuglängsachse f
low-pressure mo(u)ld safety device Niederdruckwerkzeugschutz m
male mo(u)ld Patrize f, Stempel m
moving mo(u)ld half Auswerferseite f, auswerferseitige [oder: schließseitige, bewegliche, spritzseitige] Werkzeughälfte f, Schließseite f
multi-cavity mo(u)ld Vielfachwerkzeug n
multi-daylight injection mo(u)ld Etagenspritzgießwerkzeug n, Stockwerk-Spritzgießwerkzeug n
multi-part mo(u)ld mehrteiliges Werkzeug n
neck mo(u)ld Halswerkzeug n
on the fixed mo(u)ld half spritzseitig, düsenseitig
on the moving mo(u)ld half auswerferseitig, schließseitig
on the stationary mo(u)ld half spritzseitig, düsenseitig
positive mo(u)ld Füll(raum)werkzeug n, Tauchkantenwerkzeug n
precision mo(u)ld Präzisionsform f, Präzisionswerkzeug n
production mo(u)ld Produktionswerkzeug n
prototype mo(u)ld Prototypwerkzeug n
record of mo(u)ld data Werkzeugdatenkatalog m
re-designing of the mo(u)ld Werkzeugumgestaltung f
reduced mo(u)ld clamping pressure [to prevent damage to the mo(u)ld] Werkzeugsicherungsdruck m
removal of the mo(u)ld Werkzeugausbau m
semi-positive mo(u)ld Abquetschwerkzeug n
separate mo(u)ld Einzelform f
silicone rubber mo(u)ld Silikonkautschukform f
single-cavity injection mo(u)ld Einfachspritzgießwerkzeug n
single-cavity mo(u)ld Ein-Kavitätenwerkzeug n, Einfachform f
single-daylight mo(u)ld Doppelplattenwerkzeug n,
six-cavity mo(u)ld Sechsfachwerkzeug n
sliding split mo(u)ld Schieberform f, Schieber(platten)werkzeug n
split mo(u)ld Backenwerkzeug n, zweiteiliges Werkzeug n
stack mo(u)ld Etagenwerkzeug n
standard hot runner mo(u)ld units Heißkanalnormalien fpl
standard mo(u)ld Serienwerkzeug n
standard mo(u)ld unit Stammwerkzeug n, Stammform f

standard mo(u)ld units Baukasten-Normen fpl, genormte Formenbauteile npl, Norm-Formwerkzeuge npl; Normwerkzeuge npl, Normen-Bauelemente npl, Werkzeugnormalien fpl
stationary mo(u)ld half Düsenseite f, Spritzseite f, Einspritzseite f
structural foam mo(u)ld TSG-Werkzeug n [TSG = Thermoplast-Schaum-Guß]
(thermoforming) mo(u)ld Formwerkzeug n, Tiefziehwerkzeug n
three-cavity mo(u)ld Dreifachform f, Dreifachwerkzeug n
three-part mo(u)ld Zweietagenwerkzeug n
three-plate compression mo(u)ld Dreiplattenpreßwerkzeug n
three-plate mo(u)ld Zweietagenwerkzeug n, Dreiplatten(-Abreiß)werkzeug n
toggle mo(u)ld clamping unit Kniehebel-Formschließaggregat n
top mo(u)ld half Werkzeugoberteil n
transfer mo(u)ld Spritzpreßform f
twelve-cavity injection mo(u)ld Zwölffach-Spritzwerkzeug n
two-cavity injection blow mo(u)ld Zweifachspritzblaswerkzeug n
tow-cavity injection mo(u)ld Zweifachspritzgießwerkzeug n
two-cavity mo(u)ld Zweifachform f, Doppelwerkzeug n
two-part mo(u)ld Doppelplattenwerkzeug n, Einetagenwerkzeug n, Zweiplattenwerkzeug n
two-part sliding split mo(u)ld Doppelschieberwerkzeug n
unscrewing mo(u)ld Schraubwerkzeug n
valve gated mo(u)ld Nadelventilwerkzeug n
mo(u)ldability Verpreßbarkeit f, Spritzbarkeit f, Formbarkeit f, Spritzfähigkeit f
mo(u)ldable formbar
  blow mo(u)ldable blasbar
  injection mo(u)ldable spritzgießfähig, spritzgießbar
mo(u)lded geformt, Form-
  mo(u)lded article Formartikel m
  mo(u)lded foam Formschaumstoff m
  mo(u)lded-in umspritzt, eingespritzt
  mo(u)lded-in stresses Spannungseinschlüsse mpl
  mo(u)lded-on angeformt, angegossen, angespritzt
  mo(u)lded part Formkörper m, Formteil n, Fertigteil n
  mo(u)lded-part cross section Formteilquerschnitt m
  mo(u)lded-part design Formteilgestaltung f, Formteilkonstruktion f
  mo(u)lded-part designer Formteilkonstrukteur m
  mo(u)lded-part dimensions Formteilmaße npl
  mo(u)lded-part drawing Formteilzeichnung f, Produktionsteilzeichnung f
  mo(u)lded-part properties Formteileigenschaften fpl
  mo(u)lded-part quality Formteilqualität f
  mo(u)lded-part shape Formteilgestalt f

mo(u)lded-part shrinkage Formteilschwindung f
mo(u)lded-part size Formteilgröße f
mo(u)lded-part surface Formteiloberfläche f
mo(u)lded-part tolerances Formteiltoleranzen fpl
mo(u)lded-part volume Formteilvolumen n
mo(u)lded-part wall thickness Formteilwanddicke f
mo(u)lded-part weight Formteilmasse f, Formteilgewicht n
mo(u)lded phenolic material Phenoplast-Preßstoff m
mo(u)lded thermoset resin Preßstoff m
blow mo(u)lded geblasen, blasgeformt
blow mo(u)lded part Blaskörper m, Blasformteil n, blasgeformter Hohlkörper m
compression mo(u)lded (form)gepreßt
extrusion blow mo(u)lded extrusionsgeblasen
foam mo(u)lded formgeschäumt
freshly mo(u)lded spritzfrisch
hot-cured mo(u)lded foam Heißformschaumstoff m
hot press mo(u)lded laminate Warmpreßlaminat n
in the mo(u)lded state im geformten Zustand m
injection blow mo(u)lded spritzgeblasen
injection mo(u)lded spritzgegossen
injection mo(u)lded consumer goods Konsumspritzteile npl
injection mo(u)lded fittings Spritzguß-Fittings npl
injection mo(u)lded part Spritzgießformteil n
injection mo(u)lded rubber article Gummi-Spritzgußteil n
precision mo(u)lded part Präzisionsteil n
press mo(u)lded using liquid resin naßgepreßt
sandwich mo(u)lded part Sandwich-Spritzgußteil n
transfer mo(u)lded spritzgepreßt
mo(u)lder Spritzer m
  automatic injection blow mo(u)lder Spritzblasautomat m
  blow mo(u)lder Blasmaschine f
  dip blow mo(u)lder Tauchblasmaschine f
  extrusion blow mo(u)lder Extrusionsblas-(form)maschine f
  injection mo(u)lder Spritzgießaggregat n, Spritzgießmaschine f, Spritzgießer m [Person]
  range of injection mo(u)lders Spritzgießmaschinenprogramm n, Spritzgießmaschinenbaureihe f
  stretch blow mo(u)lder Streckblas(form)maschine f
mo(u)lding 1. Formung f, Formling m, Formteilherstellung f, Formkörper m, Formteil n, Preßling m, Preßteil n; 2. Schimmeln n, Verschimmeln n
  mo(u)lding compound Formmasse f, Preßmasse f, Spritzgießmaterial n
  mo(u)lding compound producer Formmassehersteller m
  mo(u)lding conditions Formgebungsbedingungen fpl, Spritzbedingungen fpl, Verarbeitungsbedingungen fpl
  mo(u)lding cycle Arbeitstakt m, Arbeitszyklus m, Fertigungszyklus m, Formzyklus m, Preßzyklus m, (zyklischer) Prozeßablauf m, Spritztakt m, Spritzzyklus m
  mo(u)lding cycle monitoring (system) Zyklusüberwachung f
  mo(u)lding fault Spritzfehler m
  mo(u)lding jobs Spritzgießaufgaben fpl
  mo(u)lding operation Form(teil)bildungsprozeß m, Formbildungsvorgang m, Formteilbildungsvorgang m,
  mo(u)lding pressure Preßdruck m, Preßkraft f, Verarbeitungsdruck m
  mo(u)lding process Formgebungsverfahren n, Formgebungsprozeß m, Formteilherstellung f, Umformvorgang m, Formbildungsvorgang m
  mo(u)lding properties Formbarkeit f
  mo(u)lding shop Fromerei f, Presserei f, Spritzerei f
  mo(u)lding shrinkage Formschwund m, Formschwindung f, Formschrumpf m, Verarbeitungsschrumpf m
  mo(u)lding stresses Verarbeitungsspannungen fpl
  mo(u)lding technique Formtechnik f
  mo(u)lding temperature Verformungstemperatur f, Preßtemperatur f, Umformungstemperatur f, Urformtemperatur f
  mo(u)lding tolerance Fertigungstoleranz f
  mo(u)lding trials Musterabspritzungen fpl, Probe(ab)spritzungen fpl, Spritzversuche mpl
  accumulator head blow mo(u)lding (process) Staukopfverfahren n
  amino mo(u)lding compound Aminoplast-Preßmasse f
  automatic blow mo(u)lding machine Blas(form)automat m, Hohlkörperblasautomat m
  automatic carousel-type injection mo(u)lding machine Revolverspritzgießautomat m
  automatic compression mo(u)lding machine Preßautomat m
  automatic downstroke transfer mo(u)lding press Oberkolbenspritzpreßautomat m
  automatic fast cycling injection mo(u)lding machine Schnellspritzgießautomat m
  automatic-filling machine Füllautomat m
  automatic foam mo(u)lding unit Formteilschäumautomat m
  automatic high-capacity blow mo(u)lding machine Hochleistungsblasformautomat m
  automatic injection mo(u)lding machine Spritz(gieß)automat m
  automatic mo(u)lding machine Formteilautomat m
  automatic oriented PP stretch blow mo(u)lding machine OPP-Streckblasautomat m
  automatic rotary injection mo(u)lding machine Rotations-Spritzgußautomat m
  automatic rotary-table injection mo(u)lding machine Revolverspritzgießautomat m, Revolverspritzgießmaschine f
  automatic screw-plunger injection mo(u)lding machine Schneckenspritzgießautomat m
  automatic transfer mo(u)lding press Spritz(preß)-

automat m
automatic two-colo(u)r injection mo(u)lding machine Zweifarben-Spritzgußautomat m
blow mo(u)lding Blasformen n, Blasformung f, Blasverarbeitung f, Blasvorgang m, Blasen n, Blas(form)teil n, Blaskörper m
blow mo(u)lding and filling machine Blas(form)- und Füllmaschine f
blow mo(u)lding compound Blas(form)masse f
blow mo(u)lding grade [of mo(u)lding compound] Blasmarke f
blow mo(u)lding line Blas(form)anlage f, Blaslinie f
blow mo(u)lding machine Blas(form)maschine f
blow mo(u)lding of oriented polypropylene OPP-Verfahren n
blow mo(u)lding station Blas(form)station f
blow mo(u)lding system Blasformsystem n
blow mo(u)lding technology Blasformtechnik f
blow mo(u)lding unit Blasaggregat n
blow mo(u)ldings blasgeformte Teile npl
bulk mo(u)lding compound [BMC] Feuchtpreßmasse f
canister blow mo(u)lding machine Kanisterblasmaschine f
carousel-type injection mo(u)lding machine Drehtisch-Spritzgußmaschine f, Spritzgießrundtischanlage f, Revolverspritzgießmaschine f
coextrusion blow mo(u)lding Koextrusionsblasen n
coextrusion blow mo(u)lding line Koextrusions-Blasanlage f
coextrusion blow mo(u)lding machine Koextrusionsblasformmaschine f
coextrusion blow mo(u)lding plant Koextrusions-Blas(form)anlage f
coextrusion blow mo(u)lding trials Koextrusionsblasversuche mpl
cold press mo(u)lding Kaltpressen n
cold press mo(u)lding tool Kaltpreßwerkzeug n
compression blow mo(u)lding (process) Kompressionsblasen n, Kompressionsblas(form)verfahren n
compression mo(u)lding Formpressen n, Kompressionsformen n, Preßformen n, Pressen n, Preßverfahren n
contact mo(u)lding (process) Kontaktverfahren n
conventional injection mo(u)lding Kompaktspritzen n, Normalspritzguß m
conventional injection mo(u)lding machine Kompaktspritzgießmaschine f
diallyl phthalate mo(u)lding compound Diallylphthalatpreßmasse f
dip blow mo(u)lding Tauchblasen n, Tauchblasformen n
dip mo(u)lding Tauchen n, Tauchverfahren n
direction of mo(u)lding Spritzrichtung f
double-daylight mo(u)lding Zweietagenspritzen n
encapsulated by injection mo(u)lding [insert] umspritzt
epoxy mo(u)lding compound Epoxidpreßmasse f, Epoxidharz-Preßmasse f, Epoxidformmasse f
extrusion blow mo(u)lding Blasextrusion f, Extrusionsblasen n, Extrusionsblasformen n
extrusion blow mo(u)lding line Blasextrusionsanlage f
extrusion blow mo(u)lding machine Extrusionsblas(form)maschine f
extrusion blow mo(u)lding of expandable thermoplastics Thermoplast-Extrusions-Schaumblasen n
extrusion blow mo(u)lding plant Extrusionsblas(form)anlage f
extrusion stretch blow mo(u)lding Extrusions-Streckblasen n, Extrusions-Streckblasformen n
fast cycling injection mo(u)lding machine Hochleistungsspritzgießmaschine f, Schnelläufer-Spritzgießmaschine f
flow mo(u)lding Fließguß m
flow mo(u)lding process Fließgießverfahren n
foam mo(u)lding Form(teil)schäumen n, Formverschäumung f
foam mo(u)lding machine Schäummaschine f
foam mo(u)lding plant Schäumanlage f, Formschäumanlage f
foam mo(u)lding tool Schäumform f
four-daylight mo(u)lding Vier-Etagenspritzen n
high-capacity blow mo(u)lding line/plant Großblasformanlage f, Hochleistungsblasformanlage f
high-capacity blow mo(u)lding machine Großblasformmaschine f
high-capacity extrusion blow mo(u)lding plant Hochleistungsblasextrusionsanlage f
high-speed injection mo(u)lding Hochleistungsspritzgießen n, Leistungsspritzguß m
hot press mo(u)lding Heißpreßverfahren n, Heißpressen n, Warm(form)pressen n, Warmpreßverfahren n
hot-runner injection mo(u)lding Durchspritzverfahren n
hot-runner injection mo(u)lding system Spritzgieß-Heißkanalsystem n
independent of the mo(u)lding cycle zyklusunabhängig
injection blow mo(u)lding Spritzblasen n, Spritzblasformen n, Spritzgieß-Blasformen n
injection blow mo(u)lding machine Spritzblasmaschine f
injection-compression mo(u)lding Prägespritzen n, Spritzprägen n
injection mo(u)lding Spritzgießen n, Spritzgießfertigung f, Spritzguß m, Spritz(gieß)teil n, Spritzgießformteil n, Formling m
injection mo(u)lding [inserts] Umspritzen n
injection mo(u)lding compound Spritzteilmasse f,

Spritzgießmaterial n
injection mo(u)lding formulation Spritzgießrezeptur f
injection mo(u)lding grade [of mo(u)lding compound] Spritzgießmarke f
injection mo(u)lding line Spritzgießanlage f
injection mo(u)lding machine Spritzgießeinheit f, Spritzgießaggregat n
injection mo(u)lding of consumer goods Konsumentenspritzguß m
injection mo(u)lding of packaging containers Verpackungsspritzgießen n
injection mo(u)lding of plasticized PVC Weichspritzguß m, Weich-PVC-Spritzguß m
injection mo(u)lding of unplasticized PVC Hart-PVC-Spritzguß m, Hartspritzguß m
injection mo(u)lding process Spritzgießvorgang m
injection mo(u)lding resin Spritzgießharz n
injection mo(u)lding shop Spritzgießbetrieb m, Spritzgießfabrik f
injection mo(u)lding tool Spritz(gieß)form f, Spritzgießwerkzeug n
injection stretch blow mo(u)lding Spritzgieß-Streckblasen n, Spritzstreckblasen n
injection stretch blow mo(u)lding plant Spritzstreckblasanlage f
interruption of the mo(u)lding cycle Zyklusstörung f, Zyklusunterbrechung f
isostatic mo(u)lding isostatisches Pressen n
large mo(u)lding Großformteil n
liquid resin press mo(u)lding Naßpreßverfahren n
low-pressure injection mo(u)lding machine Niederdruckspritzgießmaschine f
low-pressure mo(u)lding compound Niederdruckformmasse f
machine for blow mo(u)lding drums Faßblasmaschine f
machine for blow mo(u)lding fuel oil storage tanks Heizöltank-Blasmaschine f
matched metal mo(u)lding Warmpreßverfahren n
matched metal press mo(u)lding Heißpressen n, Heißpreßverfahren n
modular injection mo(u)lding machine Baukasten-Spritzgießmaschine f
multi-component structural foam mo(u)lding machine TSG-Mehrkomponenten-Rundläufermaschine f [Thermoplast-Schaum-Guß]
multi-daylight injection mo(u)lding Etagenspritzen n
one-component structural foam mo(u)lding (process) Einkomponenten-Schaumspritzgießverfahren n
oriented polypropylene blow mo(u)lding machine OPP-Maschine f
outsert mo(u)lding Outsert-Technik f
phenolic mo(u)lding Phenoplast-Formteil n
phenolic mo(u)lding compound Phenolharz-Preßmasse f, Phenoplast-Preßmasse f, Phenoplast-Formmasse f
plastics injection mo(u)lding shop Kunststoffspritzerei f
plastics mo(u)lding Kunststoffteil n
plastics mo(u)lding compound Kunststofformmasse f
plunger injection mo(u)lding machine Kolbeninjektions-Spritzgießmaschine f, Kolben(spritzgieß)maschine f
polyester dough mo(u)lding compound Feuchtpolyester n
polyester mo(u)lding compound Polyesterharz-Preßmasse f, Polyesterpreßmasse f
polyolefin mo(u)lding compound Polyolefin-Formmasse f
polystyrene mo(u)lding compound Polystyrol-Formmasse f
precision injection mo(u)lding Präzisionsspritzgußteil n, Präzisionsspritzgießverarbeitung f, Präzisionsspritzguß m, Qualitätsspritzguß m
precision mo(u)lding Präzisionsteil n, Präzisionsteilfertigung f
press mo(u)lding Preßverfahren n, Verpressen n
projected mo(u)lding area projizierte Formteilfläche f, Formteilprojektionsfläche f
range of thermoset mo(u)lding compounds Duroplastformmassen-Sortiment n
reaction foam mo(u)lding (process) Reaktions-Schaumgießverfahren n
reaction injection mo(u)lding [RIM] Reaktions-(spritz)gießen n, Reaktionsguß m, Reaktionsgießverarbeitung f
reaction injection mo(u)lding machine [RIM machine] Reaktionsgießmaschine f
reciprocating injection mo(u)lding machine Schubschneckenspritzgießmaschine f
reject mo(u)ldings Abfallteile npl, Ausfallproduktion f, Ausschuß m, Ausschußstücke npl, Ausschußteile npl, Produktionsrückstände mpl
removal of blow mo(u)ldings Blaskörperentnahme f
rigid PVC mo(u)lding Hart-PVC-Formteil n
rotary blow mo(u)lding unit Karussell-Blasaggregat n
rotary table injection mo(u)lding machine Spritzgießrundtischanlage f, Drehtisch-Spritzgußmaschine f
rotational mo(u)lding Rotationsformen n, Rotationsgießen n, Rotationsschmelzen n
rotational mo(u)lding machine Rotationsgießmaschine f
rubber injection mo(u)lding machine Gummi-Spritzgießmaschine f
sandwich mo(u)lding Sandwich-Spritzgußteil n, Verbundspritztechnik f
sandwich mo(u)lding process Sandwich-Spritzgießverfahren n
screw injection mo(u)lding machine Schneckenspritzgießmaschine f
screw-plunger injection mo(u)lding machine Schneckenkolbenspritzgießmaschine f
semi-automatic transfer mo(u)lding press Spritz-

preß-Halbautomat m
sheet mo(u)lding compound Harzmatte f
short mo(u)ldings unvollständige Teile npl
silicone resin mo(u)lding compound Siliconharz-preßmasse f
single-screw injection mo(u)lding machine Einschnecken-Spritzgießmaschine f
single-stage blow mo(u)lding process Einstufen-Blasverfahren n
single-stage extrusion stretch blow mo(u)lding Einstufen-Extrusionsstreckblasen n
single-station automatic blow mo(u)lding machine Einstationenblasformautomat m
single-station blow mo(u)lding machine Einstationenblas(form)maschine f
single-station extrusion blow mo(u)lding machine Einstationenextrusionsblas(form)anlage f
slush mo(u)lding Schalengießverfahren n
small blow mo(u)lding Kleinhohlkörper m
special-purpose mo(u)lding compound Spezialformmasse f
sprueless injection mo(u)lding Durchspritzverfahren n, Vorkammerdurchspritzverfahren n, angußloses Spritzen n
standard injection mo(u)lding compound Standardspritzgußmarke f
stretch blow mo(u)lding Streckblas(form)en n
stretch blow mo(u)lding plant Streckblasanlage f
structural foam mo(u)lding machine Strukturschaummaschine f, Thermoplast-Schaumgießmaschine f, TSG-Maschine f, TSG-Spritzgießmaschine f
structural foam mo(u)lding (process) TSG-Verfahren n, TSG-Verarbeitung f, Thermoplast-Schaumguß m, Schaumspritzgießen n
structural foam mo(u)lding technology TSG-Technik f
structural foam mo(u)ldings TSG-Teile npl
thermoplastic foam mo(u)lding Thermoplast-Schaumteil n
thermoset injection mo(u)lding machine Duroplast-Spritzgießmaschine f
thermoset mo(u)lding Duroplast-Formteil n
thermoset mo(u)lding compound Duroplastmasse f, Duroplast-Formmasse f
three-layer blow mo(u)lding Dreischichthohlkörper m
transfer mo(u)lding Preßspritzen n, Spritzpressen n
transfer mo(u)lding plant Spritzpreßanlage f
transfer mo(u)lding press Spritzpresse f
tubular ram injection mo(u)lding (process) Ringkolbeninjektionsverfahren n
two-component injection mo(u)lding Zweikomponenten-Spritzgießen n
two-component low-pressure injection mo(u)lding machine Zweikomponenten-Niederdruck-Spritzgießmaschine f
two-layer blow mo(u)lding Zweischichthohl-körper m
two-stage extrusion blow mo(u)lding Zweistufen-Extrusionsblasformen n
two-stage injection blow mo(u)lding Zweistufen-spritzblasen n
two-station blow mo(u)lding machine Zweistationenblasmaschine f
unplasticized PVC mo(u)lding compound Hart-PVC-Formmasse f
urea mo(u)lding compound Harnstoff-Preßmasse f
vacuum bag mo(u)lding Vakuumfolienverfahren n
vacuum injection mo(u)lding (process) Vakuumeinspritzverfahren n
vented injection mo(u)lding machine Entgasungsmaschine f
mount/to montieren
mount Fassung f, Halterung f
  die mount Düsenhalterung f
mounting Montage f
  mounting flange Befestigungsflansch m
  flange mounting Flanschbefestigung f
  foot mounting Fußbefestigung f
  mo(u)ld mounting Werkzeugeinbau m (Der Werkzeugeinbau ist leicht zu bewerkstelligen / the mo(u)ld is easily mounted)
  mo(u)ld mounting dimensions Werkzeuganschlußmaße npl, Werkzeugaufspannmaße npl, Werkzeugeinbaumaße npl
move/to (sich) bewegen
  move in/to einfahren
  to move into position/to einfahren
  move out/to ausfahren
movement Bewegung f
  basic movements Grundbewegungen fpl
  Brownian movement Brownsche Bewegung f
  closing and opening movements Schließ- und Öffnungsbewegungen fpl
  closing movement Schließvorgang m
  direction of movement Bewegungsrichtung f
  downward movement Abwärtsbewegung f
  ejector forward movement Auswerfervorlauf m
  ejector movement Auswerferbewegung f
  ejector return movement Auswerferrücklauf m
  longitudinal movement Längsbewegung f
  machine movements Maschinenbewegungen fpl, Bewegungsabläufe mpl
  main direction of movement Hauptbewegungsrichtung f
  molecular movement Molekularbewegung f
  mo(u)ld closing movement Formschließbewegung f, Werkzeugschließbewegung f
  mo(u)ld movement Werkzeugbewegung f
  mo(u)ld movement-related werkzeugbewegungsabhängig
  mo(u)ld opening movement Formöffnungsbewegung f, Öffnungsvorgang m,

Werkzeugöffnungsbewegung f,
Werkzeugöffnungsvorgang m
nozzle forward movement Düsenanlegebewegung f
open-close movement Auf-Zu-Bewegung f
opening movement Öffnungsbewegung f, Öffnungsvorgang m
plunger movement Kolbenbewegung f
reciprocating movement Schubbewegung f
rotary movement Drehbewegung f, Rotationsbewegung f
screw forward movement Schneckenvorlaufbewegung f
sliding movement Schiebebewegung f
tumbling movement Taumelbewegung f
turning movement Schwenkbewegung f
up and down movements Auf- und Abbewegungen fpl
upward movement Augwärtsbewegung f
moving beweglich, (sich) bewegend
  moving carriage Arbeitsschlitten m, Schlitten m
  moving mo(u)ld half bewegliche Werkzeughälfte f, Kernseite f, schließseitige Werkzeughälfte f, Schließseite f, Auswerferseite f
  moving platen bewegliche Werkzeugaufspannplatte f, Schließplatte f, Werkzeugträgerschlitten m, Werkzeugschließplatte f, bewegliche Aufspannplatte f
  on the moving mo(u)ld half schließseitig, auswerferseitig
mucous membranes Schleimhäute fpl
mudguard Kotflügel m
multi viel-, mehr-, vielfach-, Mehrfach-
  multi-cavity mo(u)ld Vielfachwerkzeug n
  multi-component structural foam mo(u)lding (process) TSG-Mehrkomponentenverfahren n
  multi-daylight design Spritzgießwerkzeug n in Etagenbauweise
  multi-daylight injection mo(u)ld Stockwerk-Spritzgießwerkzeug n, Etagenspritzgießwerkzeug n
  multi-daylight injection mo(u)lding Etagenspritzen n
  multi-daylight mo(u)ld Etagenwerkzeug n
  multi-daylight press Etagenpresse f
  multi-daylight transfer mo(u)lding Etagenspritzen n
  multi-orifice die Vielfachwerkzeug n
  multi-part mo(u)ld mehrteiliges Werkzeug n
  multi-point gating mehrfaches Anbinden n
  multi-point pin gate Reihenpunktanschnitt m
  multi-screw extruder Vielschneckenmaschine f
  multi-stage die Stufenwerkzeug n
  multi-stage screw Stufenschnecke f
  multi-strand die Vielfachstrangdüse f
  hot runner multi-daylight mo(u)ld Heißkanal-Etagenwerkzeug n
  three-plate multi-cavity mo(u)ld Dreiplatten-Mehrfachwerkzeug n
multifunctional multifunktionell
multiple mehrfach
  multiple-gated mehrfach angespritzt
  multiple pin gate Reihenpunktanschnitt m
  hot runner multiple gating Heißkanal-Mehrfachanguß m
multiplexor Multiplexer m
muscovite Muscovit-Glimmer m
name Name m
  common name [of a chemical compound] Trivialname m
  trade name Handelsname m, Markenname m
naphthenate Naphthenat n
  cobalt naphthenate Kobaltnaphthenat n
naphthenic naphthenisch
narrow construction Schmalbauweise f
narrowing Verjüngung f
  narrowing cross-sections Querschnittsverengungen fpl
  narrowing flow channels Querschnittsverengungen fpl
  narrowing runners Querschnittsverengungen fpl
natural natürlich, Natur-, Eigen-, Roh-, naturfarben
  natural chalk Naturkreide f
  natural frequency Eigenfrequenz f
  natural resin Naturharz n
  natural rubber Naturgummi n,m, Naturkautschuk m
  natural weathering Naturbewitterung f
naturally weathered freibewittert
near nah
  near the gate angußnah, in Angußnähe f, im Angußbereich m
  near the sprue angußnah, in Angußnähe f, im Angußbereich m
  near the surface oberflächennah
neck Hals m
  neck and base flash Abquetschlinge mpl
  neck calibration Halskalibrierung f
  neck calibrating device Halskalibrierung f [Gerät]
  neck flash Halsabfälle mpl, Halsbutzen m, Halsüberstände mpl
  neck mo(u)ld Halswerkzeug n
  neck pinch-off Halsquetschkante f
  neck section Halspartie f
  bubble neck Folienhals m
  petrol tank neck Tankeinfüllstutzen m
necking [of film] Einsprung m, Seiteneinsprung m
need for foreign exchange Devisenbedarf m
needle Nadel f
  needle pyrometer Einstichpyrometer n
  needle shut-off mechanism Nadelventilverschluß m, Nadelverschlußsystem n, Schließnadel f
  needle valve Nadelventil n, Nadelverschluß m
  needle valve nozzle Nadelverschluß(spritz)düse f
  inflation needle Blasnadel f, Blasstift m
  shut-off needle Verschlußnadel f
negative negativ
negligible vernachlässigbar
neighbo(u)ring countries Nachbarländer npl
neopentyl diglycidyl ether Neopentyldiglycidylether m
neopentyl glycol Neopentylglykol n
neoprene rubber Neoprenkautschuk m

net increase Netto-Zugang m
net profit Reingewinn m
network Netz(werk) n
  network structure Netz(werk)struktur f
neutral neutral
neutralization Neutralisieren n, Neutralisierung f, Neutralisation f
neutralized neutralisiert
neutralizing agent Neutralisationsmittel n
neutron Neutron n
new neu
  new design Neukonstruktion f
  new paint film Neuanstrich m
newly developed neuentwickelt
Newtonian Newtonsch(e, er, es), reinviskos
  Newtonian flow Newtonsches Fließen n
  Newtonian liquid Newtonsche Flüssigkeit f
nickel titanium yellow Nickeltitangelb n
nickel-plated vernickelt
night shift Nachtschicht f
nip Spalt m, Quetschspalt m, Walzenspalt m; Abkneifen n
  nip adjusting gear Walzenanstellung f
  nip adjusting mechanism Spaltverstellung f
  nip pressure Spaltdruck m, Walzenspaltdruck m, Walzenspaltkraft f, Walzentrennkraft f, Spaltkraft f, Spaltlast f, Walzenlast f
  nip rolls Abquetschwalzenpaar n, Abquetschwalzen fpl, Abquetschvorrichtung f, Spaltwalzenpaar n, Quetschwalzenpaar n, Quetschwalzen fpl
  nip setting (mechanism) Walzenspalteinstellung f
  nip width Walzenabstand m, Spaltweite f, Spaltbreite f
  distance between die and nip rolls Abzughöhe f
  embossing nip Prägespalt m
  polishing nip Glättspalt m
  take-off nip Abzugswalzenspalt m
  total nip pressure Gesamtspaltlast f
nipple Nippel m
  connecting nipple Anschlußnippel m
nitrate Nitrat n
  cellulose nitrate Cellulosenitrat n
nitric acid Salpetersäure f
nitrided Nitrier-gehärtet, gasnitriert
  nitrided layer Nitrierschicht f
  nitrided liner [of an extruder barrel] nitriergehärtete Bohrung f
  nitrided steel Nitrierstahl m
nitriding Nitrierung f
  gas nitriding Gasnitrierung f
nitrile Nitril n
  nitrile group Nitrilgruppe f
  nitrile rubber Nitrilkautschuk m
  nitrile rubber latex Nitrillatex m
nitrocellulose Nitrocellulose f, Collodiumwolle f
  nitrocellulose lacquer Nitrocelluloselack m
nitrogen Stickstoff m
  nitrogen permeability Stickstoffdurchlässigkeit f
  atmosphere of nitrogen Stickstoffatmosphäre f
  internal cooling with liquid nitrogen Stickstoffinnenkühlung f
  liquid nitrogen cooling system Flüssig-Stickstoffkühlung f
no-load power Leerlaufleistung f
noise Lärm m, Geräusch n
  noise level Geräuschniveau n, Geräuschpegel m, Schallpegel m, Lärmemissionswert m
  noise (level) reduction Schallpegelreduzierung f, Geräuschreduzierung f, Lärmminderung f, Schalldämpfung f
  noise reduction measures Geräuschdämpfungsmaßnahmen fpl, Lärmminderungsmaßnahmen fpl
  overall noise level Gesamtschallpegel m
nomenclature Nomenklatur f
nominal nominal, Nenn-
  nominal capacity Nenninhalt m
  nominal diameter Nenndurchmesser m
  nominal dimensions Nennmaße npl
  nominal elongation Nenndehnung f
  nominal pressure Nenndruck m
  nominal size Nenngröße f
  nominal strain Nenndehnung f
  nominal torque Nenndrehmoment n
  nominal viscosity Nennviskosität f
  nominal volume Nennvolumen n
  nominal width Nennweite f
non nicht-
  non-absorbent nichtsaugend
  non-aqueous nichtwässrig
  non-blocking verblockfrei
  non-cash contribution Sacheinlage f
  non-cellular nichtzellular, nichtzellig
  non-combustible unbrennbar
  non-conductive nichtleitend
  non-conductor Nichtleiter m
  non-contact berührungslos, kontaktlos
  non-corrosive nichtkorrosiv
  non-crystalline nichtkristallin
  non-crystallizing kristallisationsstabil, kristallisationsbeständig, kristallisationsfrei, nichtkristallisierend
  non-destructive [test] nichtzerstörend, zerstörungsfrei
  non-directional nichtdirektional
  non-drying nichttrocknend
  non-dusting staubfrei
  non-elastic nichtelastisch
  non-equilibrium state Nicht-Gleichgewichtszustand m
  non-ferrous metal Buntmetall n, NE-Metall n, Nichteisenmetall n
  non-ferrous metal alloy Buntmetall-Legierung f
  non-fibrous nichtfasrig
  non-flammability Unbrennbarkeit f, Nichtbrennbarkeit f, Nichtentflammbarkeit f
  non-flam(mable) nichtbrennbar, nichtentflammbar, unbrennbar
  non-food packaging Nicht-Nahrungsmittel-Verpackung f

non-intermeshing [screws] nichtkämmend, tangierend, nichtineinandergreifend
non-ionic nichtionisch, nichtionogen
non-isothermal nichtisotherm
non-leaking leckagefrei
non-linear nichtlinear
non-linearity Nichtlinearität f
non-metallic nichtmetallisch
non-migrating wanderungsbeständig
non-Newtonian Nicht-Newtonsch(e, er, es), strukturviskos
non-Newtonian liquid Nicht-Newtonsche Flüssigkeit f
non-oriented orientierungsfrei
non-paste making nichtverpastbar
non-polar nichtpolar, unpolar
non-porous unporös
non-positive [connection of joint] kraftschlüssig
non-pressure pipeline Freispiegelleitung f
non-productive time Stillstandzeit f, Stehzeit f, Totzeit f, Nebenzeit f
non-radical nichtradikalisch
non-recurring einmalig
non-return valve Rückschlagventil n, Rücklaufsperre f, Rückströmsperre f, Rückstausperre f
non-rusting nichtrostend, rostfrei, rostbeständig, unverrostbar
non-sag standfest
non-sag properties Standfestigkeit f
non-shrink(ing) nichtschrumpfend, schrumpffrei
non-skid gleitsicher, rutschfest
non-skid coating Rutschfestbeschichtung f
non-skid properties Rutschfestigkeit f
non-solvating [plasticizer] nichtgelierend
non-solvent Nichtlösemittel n, Nichtlöser m
non-staining nichtverfärbend
non-stick nichthaftend, abhäsiv, antiadhäsiv, haftabweisend
non-stick coating Abhäsivbeschichtung f, Antihaftbelag m, Antihaftüberzug m
non-stick effect Antiklebewirkung f
non-stick properties Trenneigenschaften fpl, Trennfähigkeit f, Trenn(mittel)wirkung f
non-stressed unbelastet
non-sulphur schwefelfrei
non-tin stabilized nicht-zinnstabilisiert
non-toxic nichttoxisch, ungiftig, untoxisch
non-toxicity Ungiftigkeit f
non-transparent nichttransparent
non-twisting verdrillungsfrei
non-utilized unausgelastet
non-volatile nichtflüchtig
non-warping verzugsfrei
non-warping properties Verzugsfreiheit f
non-woven nichtgewebt
non-yellowing vergilbungsfrei, gilbungsstabil
normal normal
normal impact normal(schlag)zäh, normalschlagfest, normalstoßfest
normal position Grundstellung f
normally flammable normalentflammbar
normally reactive normalaktiv
not nicht, un- [s.a. non]
not aged ungealtert
not containing lubricant gleitmittelfrei
not resistant [material in presence of chemicals] unbeständig, nichtbeständig
not sensitive to oxidation oxidationsunempfindlich
notch Kerbe f
double-V notch Doppel-V-Kerbe f
original notch Anfangskerbe f
notched (an)gekerbt
novolak Novolak n
novolak resin Novolakharz n
cresol novolak Kresolnovolak n
phenol novolak Phenolnovolak n
nozzle Düse f
nozzle advance cylinder Düsenanpreßzylinder m
nozzle advance speed Düsenvorlaufgeschwindigkeit f
nozzle aperture Düsenbohrung f, Düsenöffnung f
nozzle approach speed Düsenanfahrgeschwindigkeit f
nozzle block Düsenblock m
nozzle carriage Düsenhalteplatte f
nozzle contact Düsenanlage f, Düsenanpressung f
nozzle contact pressure Düsenanpreßkraft f, Düsenanpreßdruck m, Düsenanlagedruck m, Düsenanlagekraft f, Düsenanpressung f
nozzle deformation Düsendeformation f
nozzle design Düsenkonstruktion f, Düsenbauart f
nozzle dimensions Düsenabmessungen fpl
nozzle displacement Düsenversatz m
nozzle forward movement Düsenanlagebewegung f
nozzle heater Düsenheizkörper m, Düsenheizung f
nozzle heater band Düsenheizband n
nozzle material Düsenbaustoff m, Düsenwerkstoff m
nozzle orifice Düsenmund m
nozzle point Düsenspitze f
nozzle retraction Düsenabhebung f, Düsenabhub m
nozzle retraction speed Düsenabhebegeschwindigkeit f, Abhebegeschwindigkeit f
nozzle retraction stroke Düsenabhebeweg m, Abhebehub m
nozzle seating Düsenkalotte f, Düsensitz m
nozzle speed Düsen(fahr)geschwindigkeit f
nozzle stroke Düsenhub m
ante-chamber nozzle Vorkammerdüse f
extended nozzle Tauchdüse f, verlängerte Düse f
free-flow nozzle offene Düse f
high-speed nozzle retraction (mechanism) Düsenschnellabhebung f
injection nozzle Einspritzdüse f, Spritzdüse f
long-reach nozzle Tauchdüse f, verlängerte Düse f
machine nozzle Maschinendüse f
near the nozzle in Düsennähe f
needle valve nozzle Nadelverschluß(spritz)düse f

pin gate nozzle Punktangußdüse f, Punktanschnittdüse f
screw-in nozzle Einschraubdüse f
shut-off nozzle Düsenverschluß m, Verschlußdüse f
sliding shut-off nozzle Schiebe(verschluß)-(spritz)düse f
spraying nozzle Spritzdüse f
thermally conductive nozzle Wärmeleitdüse f
three-channel nozzle Dreikanaldüse f
type of nozzle Düsenbauart f, Düsenform f, Düsenart f
welding nozzle Schweißdüse f
nucleating nukleierend
  nucleating agent Nukleierungs(hilfs)mittel n
  nucleating effect Nukleierungswirkung f
  nucleating process Nukleierungsvorgang m
nucleation Nukleierung f
nucleophilic nukleophil
number Nummer f, Zahl f
  number average Zahlenmittel n
  number-average molecular weight Molekulargewicht-Zahlenmittel n
  number of breaking stress cycles Bruchlastspielzahl f
  number of cavities Formnestzahl f
  number of (complete) vibrations Schwingspielzahl f
  number of dry cycles [usually per minute] Trockenlaufzahl f, Anzahl der Leerlaufzyklen mpl
  number of employees Beschäftigtenzahl f
  number of flights Schneckengangzahl f, Gangzahl f
  number of meshes Maschenzahl f
  number of (mo(u)lding) cycles Taktzahl(en) f(pl)
  number of rejects Ausschußzahlen fpl, Ausschußquote f
  number of revolutions Drehzahl f, Umdrehungszahl f, Tourenzahl f
  number of shots Schußzahl f
  number of starts Gängigkeit f
  number of stress cycles Lastwechselzahl f, Lastspielzahl f
  number of vibrations to failure Bruch-Schwingspielzahl f
  batch number Chargennummer f
  binary number Binärzahl f
  elasticity number Elastizitätszahl f
  Gardner colo(u)r standard number Gardner-Farbzahl f
  irrespective of the number of items stückzahlunabhängig
  job number Auftragsnummer f
  large numbers große Stückzahlen fpl
  limiting viscosity number Staudinger-Index m, Grenzviskosität f
  lines of numbers Zahlenzeilen fpl
  oxidation number Oxidationsstufe f
  Prandtl number Prandtl-Zahl f
  Reynolds number Reynoldzahl f
  small numbers kleine Stückzahlen fpl
  total number Gesamtzahl f
  total number of employees Gesamtbeschäftigtenzahl f, Gesamtbelegschaft f
  viscosity number Viskositätszahl f
  wave number Wellenzahl f
numeric numerisch
  numeric display Ziffernanzeige f
  numeric key Zahlentaste f
  numeric keypad Zifferntastatur f
  numerical equation Zahlenwertgleichung f
  numerical example Zahlenbeispiel n
  numerical value Zahlenwert m
nut Mutter f
  cap nut Überwurfmutter f
  check nut Kontermutter f
  lock nut Kontermutter f
nylon Nylon n [Produktname], Polyamid n
objective Zielsetzung f
obligation Verbindlichkeit f
observation period Beobachtungsdauer f
occlusions Einschlüsse mpl
octoate Octoat n
  barium octoate Bariumoctoat n
  cobalt octoate Kobaltoctoat n
octyl acrylate Octylacrylat n
octyl tin Octylzinn n
  octyl tin carboxylate Octylzinncarbocylat n
  octyl tin compound Octylzinnverbindung f
  octyl tin mercaptide Octylzinnmerkaptid n
  octyl tin stabilizer Octylzinnstabilisator m
odor [GB: odour] Geruch m
  odo(u)r permeability Aromadurchlässigkeit f
  freedom from odo(u)r Geruchslosigkeit f, Geruchsfreiheit f, Geruchsneutralität f
odo(u)rless geruchlos, geruchsfrei, geruchsneutral
odourproof aromadicht
off-center außermittig
off-line Offline-, selbständig (betrieben), rechnerunabhängig (arbeitend)
offer Angebot n
office Büro n
  office equipment Bürogeräte npl, Büromaschinen fpl
  office furniture Büromöbel npl
  Federal German Public Health Office Bundesgesundheitsamt n
offset 1. abgesetzt; 2. Absatz m
  offset yield stress Dehnspannung f
Ohm's Law Ohmsches Gesetz n
oil Öl n
  oil absorption value Ölzahl f
  oil bath Ölbad n
  oil circuit Ölkreislauf m
  oil compression Ölkompression f
  oil cooling (system) Ölkühlung f
  oil cooling unit Ölkühler m
  oil decompression Ölentspannung f
  oil diffusion pump Öldiffusionspumpe f
  oil-extended ölverstreckt

oil fatty acid Ölfettsäure f
oil filtration (unit) Ölfilterung f
oil flow regulator Ölstromregler m
oil heated ölbeheizt, öltemperiert
oil heating unit Öltemperiergerät n
oil-hydraulic system Ölhydraulik f
oil inlet Ölzufluß m
oil length Öllänge f
oil level Ölniveau n, Ölspiegel m, Ölstand m
oil level control (unit) Ölstandkontrolle f
oil level indicator Ölstandanzeiger m
oil level monitor Ölniveauwächter m, Ölstandüberwachung f
oil level monitoring Ölstandüberwachung f
oil modified ölmodifiziert
oil outlet Ölabfluß m
oil-producing countries Ölländer npl
oil requirement Öl(mengen)bedarf m
oil reservoir Ölbehälter m
oil residues Ölreste mpl
oil resistance Ölbeständigkeit f
oil resistant ölbeständig, ölfest
oil seal Öldichtung f [z.B. Simmerring]
oil tank ventilation Öltankbelüftung f
oil temperature control (system) Öltemperierung f
oil temperature monitor Öltemperaturwächter m
amount of oil needed Öl(mengen)bedarf m
castor oil Rizinusöl n
circulating oil heating system Ölumlaufheizung f
circulating oil lubricating system Ölumlaufschmierung f, Umlaufölung f
circulating oil temperature control (system) Ölumlauftemperierung f
coconut oil fatty acid Kokosölfettsäure f
cottonseed oil fatty acid Baumwollsamenölfettsäure f
counterflow oil cooler Gegenstromölkühler m
counterflow oil cooling system Gegenstromölkühler m
cutting oil Bohröl n, Schneidöl n
edible oil Speiseöl n
extremely fine-mesh oil filter Feinstölfilter n,m
hydraulic oil Drucköl n, Hydrauliköl n
hydraulic oil flow Hydraulikölstrom m
internal oil heating system Ölinnentemperierung f
linseed oil Leinöl n
linseed oil fatty acid Leinölfettsäure f
lubricating oil Schmieröl n
lubrication oil reservoir Schmieröltank m
pressurized oil lubricating (system) Drucköllung f
recirulating oil cooling system Umlaufölkühlung f
soluble oil Kühlemulsion f
soya bean oil Sojaöl n
soya bean oil fatty acid Sojaölfettsäure f
tall oil Tallöl n
tall oil fatty acid Tallölfettsäure f
transformer oil Trafoöl n, Transformatoröl n
weight without oil Gewicht n ohne Ölfüllung

old concrete Altbeton m
oleate Ölsäureester m
olefin Olefin n
olefinic olefinisch
oleic acid ester Ölsäureester m
oleum Oleum n
oligomer Oligomer m
oligomerous oligomer
on-line Online-, rechnerabhängig, während der Fertigung
 on-line operation On-line-Betrieb m
on-off switch Ein-Aus-Schalter m
one 1. Eins f; 2. ein(e)
 one-bit processor Ein-Bit-Prozessor m
 one-coat paint Einschichtlack m
 one-coat system Einschichtsystem n
 one-component compound Einkomponentenmasse f
 one-component structural foam mo(u)lding (process) Einkomponenten-Schaumspritzgießverfahren n
 one-component system Einkomponentensystem n
 one-dimensional eindimensional
 one-man operation Einmannbedienung f
 one-off production Einzel(an)fertigung f
 one-pack einkomponentig
 one-pack adhesive Einkomponentenklebstoff m, Einkomponentenkleber m
 one-pack silicone rubber Einkomponentensilikonkautschuk m
 one-pack system Einkomponentensystem n
 one-piece screw Einstückschnecke f
 one-section screw Einzonenschnecke f
 one-shift operation Einschichtenanlage f
 one-shift plant Einschichtenbetrieb m
 one-sided einseitig
opacity [of a pigment] Deckung f, Deckfähigkeit f, Deckvermögen n, Deckungsgrad m
opaque deckend, nichttransparent, opak, undurchsichtig, lichtundurchlässig
open offen, frei
 open and closed-loop controls Steuer- und Regeleinrichtung f, Steuerungen fpl und Regelungen fpl
 open-cell character Ofenporigkeit f
 open-cell [foam] offenporig, offenzellig
 open-close movement Auf-Zu-Bewegung f
 open-loop control circuit (offener) Steuerkreis m
 open-loop control instrument Steuergerät n
 open-loop pressure control circuit Drucksteuerkreis m
 open-loop process control Prozeßsteuerung f
 open position Öffnungsstellung f
opening Öffnen n, Öffnung f; Auffahren n [Form]
 opening force Öffnungskraft f
 opening movement Öffnungsbewegung f, Öffnungsvorgang m
 opening of the mo(u)ld Werkzeugöffnung f, Auffahren n
 opening speed Öffnungsgeschwindigkeit f

opening stroke Öffnungsweite f, Öffnungsweg m, Öffnungshub m
die opening Düsenausgang m
die opening force Aufreißkraft f, Auftreibkraft f, Düsenöffnungsdruck m, Werkzeugauftriebskraft f
discharge opening Entleerungsstutzen m
feed opening Einspeisebohrung f, Zuführungsbohrung f, Einfüllschacht m, Zulauföffnung f, Einfüllöffnung f, Einzugsöffnung f, Einlauföffnung f, Zuführöffnung f, Extrudereinfüllöffnung f, Zugabeöffnung f, Mahlguteinlauf m
mo(u)ld opening and closing cycle Formenöffnungs- und Schließspiel n
mo(u)ld opening and closing speed Werkzeuggeschwindigkeit f
mo(u)ld opening and closing valve Formenöffnungs- und Schließventil n
mo(u)ld opening cylinder Formaufdrückzylinder m
mo(u)ld opening direction Formöffnungsrichtung f
mo(u)ld opening force Formauftreibdruck m, Öffnungskraft f, Werkzeugöffnungskraft f, Werkzeugauftriebskraft f, Aufreißkraft f, Auftreibkraft f
mo(u)ld opening movement Formöffnungsbewegung f, Öffnungsvorgang m, Öffnungsbewegung f, Werkzeugöffnungsbewegung f, Werkzeugöffnungsvorgang m
mo(u)ld opening speed Öffnungsgeschwindigkeit f, Werkzeugöffnungsgeschwindigkeit f, Formauffahrgeschwindigkeit f
mo(u)ld opening stroke Formöffnungshub m, Formöffnungsweg m, Öffnungshub m, Öffnungsweite f, Öffnungsweg m, Werkzeugöffnungsweg m, Werkzeugöffnungshub m
mo(u)ld opening stroke limiting device Vorrichtung f zur Öffnungswegbegrenzung (Werkzeugöffnungshub-Begrenzung)
operate/to bedienen
operated betrieben, bedient
  foot operated switch Fußschalter m
  hand operated handbedient, handbetätigt, handbetrieben
  manually operated handbedient, handbetätigt, handbetrieben
  manually operated valve Handventil n
  pneumatically operated druckluftbetrieben
operating Arbeits-, Betriebs-
  operating conditions Arbeitsbedingungen fpl, Betriebszustände mpl, Betriebsbedingungen fpl
  operating costs Betriebsaufwand m, Betriebskosten pl, Unterhaltskosten pl
  operating curve Arbeitskennlinie f
  operating elements Funktionselemente npl
  operating in cycles periodisch arbeitend
  operating instruction Bedienbefehl m, Betriebsanweisung f, Betriebsanleitung f
  operating lever Betätigungshebel m

operating period Betriebszeit f, Betriebsdauer f
operating personnel Bedienungsmannschaft f, Bedienungspersonal n
operating platform Arbeitsbühne f
operating point Arbeitspunkt m
operating positition Arbeitsstellung f, Betriebsstellung f
operating pressure Betriebsdruck m
operating principle Funktionsprinzip n, Arbeitsprinzip n
operating radius Arbeitsradius m
operating range Leistungsfahrbreite f
operating screw speed Betriebsdrehzahl f, Arbeitsdrehzahl f [Extruder]
operating sequence Funktionsablauf m
operating sequence display Funktionsablaufanzeige f
operating speed Betriebsdrehzahl f, Arbeitsdrehzahl f
operating speed range Betriebsdrehzahlbereich m
operating switch Bedienungsschalter m
operating temperature Einsatztemperatur f, Betriebstemperatur f
operating temperature range Betriebstemperaturbereich m, Temperatureinsatzbereich m, Temperatur-Anwendungsbereich m, Einsatztemperaturbereich m
operating torque Betriebsmoment n
operating variables Betriebswerte mpl
operating voltage Betriebsspannung f
automatic operating mechanism Bedienungsautomatik f
machine operating program Maschinenprogramm n
operation Bedienung f, Betätigung f, Arbeitsgang m, Arbeitsweise f
operation with a pressure accumulator Druckspeicherbetrieb m
adiabatic operation adiabatische Arbeitsweise f
automatic operation Automatikbetrieb m
continuous operation Dauereinsatz m, permanenter Betrieb m
coupled operation [of several machines] Verbundbetrieb m
discontinuous operation intermittierender Betrieb m
ease of operation Bedienungsfreundlichkeit f, Bedienungskomfort m
faulty operation Fehlbedienung f
forming operation Umformvorgang m
hours of operation Betriebsstunden fpl
intermittent operation intermittierender Betrieb m
machine operation Maschinenbetrieb m, Maschinenlauf m
manual operation Handbedienung f, Handbetrieb m
method of operation Arbeitsweise f, Betriebsweise f, Fahrweise f, Funktionsweise f
mode of operation Arbeitsweise f
mo(u)ld filling operation Formfüllprozeß m, Formfüllung f, Werkzeugfüllvorgang m, Formfüllvorgang m
on-line operation On-line-Betrieb m

one-man operation Einmannbedienung f
one-shift operation Einschichtenbetrieb m
packaging operation Packvorgang m
principle of operation Arbeitsprinzip n, Funktionsprinzip n
program operation Programmablauf m
putting into operation [plant, machine etc.] Inbetriebnahme f
quiet in operation schallarm, geräuscharm, lärmarm, laufruhig
quietness in operation Laufruhe f
set-up operation Rüstvorgang m,
shift operation Schichtbetrieb m
simultaneous operation Simultanbetrieb m
three-shift operation Dreischichtbetrieb m
time of operation Betriebszeit f, Betriebsdauer f
type of operation [manual or automatic] Betriebsart f
very quiet in operation hohe Laufruhe f
welding operation Schweißvorgang m
operational betrieblich, Betriebs-, Funktions-, funktionsfähig
  operational software Betriebssoftware f
operator Bediener m, Bedienungsmann m, Bedien(ungs)person f, Betriebsmann m
  operator comfort Bedienungskomfort m
  operator panel Bedienungspult n, Bedientableau n, Bedienungsfront f, Bedientafel f, Bedienkonsole f, Bedienungsfeld n
  operator prompt Bedienerführung f
  operator's guard door Bedienungsschutztür f
  operator's side [of machine] Bedienungsseite f
  machine operator Maschinenbediener m, Maschinenführer m
  machine operators Maschinen(bedien)personal n, Bedienungspersonal n, Bedienungsmannschaft f
o-phthalic acid Orthophthalsäure f
opposing screw Gegenschnecke f
opposite gegenüber(liegend), entgegengesetzt
  rotating in opposite directions [twin screws] gegenläufig, gegeneinanderlaufend
  the side opposite the operator's side Bedienungsgegenseite f
optic optisch
  optic fiber cable Glasfaserkabel n
  optical cable Lichtwellenleiterkabel n
  optical fiber Lichtleitfaser f
  optical microscope Durchlichtmikroskop n, Lichtmikroskop n
  optical microscopy Lichtmikroskopie f
optimization Optimierung f
optimize/to optimieren
optimizing the design of single-screw extruders Optimierung f der verfahrenstechnischen Auslegung von Einschneckenextrudern
optimum 1. optimal; 2. Optimum n
option Möglichkeit f

  control options Regelmöglichkeiten fpl, Steuervarianten fpl
  injection options Einspritzmöglichkeiten fpl
  input options Eingabemöglichkeiten fpl
  software options Softwaremenü n
optional equipment Wahlausrüstung f
orange 1. orange; 2. Apfelsine f, Orange f
  orange peel effect Apfelsinenschalenstruktur f, Apfelsinenschaleneffekt m
  cadmium orange Cadmiumorange n
  molybdenum orange Molybdatorange n
orbital pump Kreiskolbenpumpe f
order Ordnung f, Reihenfolge f, Grad m; Befehl m; Auftrag m
  order (of magnitude) Größenordnung f
  first order reaction Reaktion f erster Ordnung
  in working order betriebsfähig
  incoming orders Auftragseingang m
  out of order betriebsunfähig
  second order reaction Reaktion f zweiter Ordnung
ordinary gewöhnlich, regelmäßig, ordentlich, marktüblich
  ordinary capital Stammkapital n
  ordinary chute Normalschurre f
organic organisch
  organic group Organogruppe f, Organorest m
organization Organisation f
  sales organization Vertriebsorganisation f
organo- organo-, Organo-
  organo-functional organofunktionell
  organo-lithium lithiumorganisch
  organo-metallic organometallisch
  organo-modified organomodifiziert
  organo-phosphite Organophosphit n
  organo-polysiloxane Organopolysiloxan n
  organo-silane Organosilan n
  organo-silicon siliciumorganisch
  organo-tin zinnorganisch
  organo-tin chloride Organozinnchlorid n
  organo-tin compound Organozinnverbindung f
  organo-tin mercaptide Organozinnmercaptid n
  organo-tin mercaptocarboxylate Organozinn-merkaptocarbonsäureester m
  organo-tin stabilizer Organozinnstabilisator m
  organo-titanate Organotitanat n
organosol Organosol n
orientation Orientierung f, Anordnung f, Ausrichtung f, Verlauf m
  orientation distribution Orientierungsverteilung f
  orientation effect Ausrichteffekt m
  orientation in machine direction Längsorientierung f
  crystallite orientation Kristallitorientierung f
  degree of orientation Orientierungsgrad m
  direction of orientation Orientierungsrichtung f
  fiber orientation Faseranordnung f, Faserausrichtung f, Faserorientierung f, Faserverlauf m
  filler particle orientation Füllstofforientierung f

flat film orientation (process) Flachfolienstreckverfahren n
glass fiber orientation Glasfaserorientierung f
increase in orientation Orientierungszuwachs m
initial orientation Anfangsorientierung f
molecular orientation Molekülorientierung f
residual orientation Restorientierung f
signs of orientation Orientierungserscheinungen fpl
state of orientation Orientierungszustand m
transverse orientation Querorientierung f
oriented (aus)gerichtet
  oriented polypropylene blow mo(u)lding machine OPP-Maschine f
  automatic oriented PP stretch blow mo(u)lding machine OPP-Streckblasautomat m
  biaxially oriented biaxial gestreckt, biaxial verstreckt
  blow mo(u)lding of oriented polypropylene OPP-Verfahren n
  longitudinally oriented [film] längsverstreckt, längsgerichtet
  uniaxially oriented monoaxial verstreckt
orifice Öffnung f; Düse f; Mundstück n; Meßblende f
  cooling ring orifice Kühlring-Lippenspalt m
  die orifice Lippenspalt m
  nozzle orifice Düsenmund m
origin [of materials] Provenienz f, Herkunft f
original 1. originalgetreu, detailgetreu, Ausgangs-; 2. Original n,
  original acrylonitrile content Ausgangsacrylnitrilgehalt m
  original concentration Ausgangskonzentration f
  original container Originalgebinde n
  original contents Originalfüllung f
  original design Ausgangsbauart f
  original dimensions Ausgangsmaße npl
  original equipment Erstausrüstung f
  original figure Ausgangswert m
  original hardness Ausgangshärte f
  original length Ausgangslänge f
  original measured length Ausgangsmeßlänge f
  original moisture content Ausgangsfeuchte f
  original monomer content Ausgangsmonomergehalt m
  original notch Anfangskerbe f
  original packaging Originalverpackung f
  original position Ausgangsstellung f, Ausgangsposition f
  original sample Ausgangsprobe f
  original state Ausgangszustand m
  original strength Ausgangsfestigkeit f
  original styrene content Ausgangsstyrolgehalt m
  original viscosity Ausgangsviskosität f
  original weight Ausgangsgewicht n
O-ring O-Ring m
ortho-phthalic acid Orthophthalsäure f
ortho-position Orthostellung f
orthotropic orthotrop

oscillating oszillierend
oscillograph Oszillograph m
  cathode ray oscillograph Kathodenstrahloszillograph m
  electron beam oscillograph Elektronenstrahloszillograph m
  light beam oscillograph Lichtstrahloszillograph m
osmometry Osmometrie f
out aus, außerhalb; nicht [allg. Negation]
  out-of-balance Unwucht f
  out-of-balance forces Unwuchtkräfte fpl
  out-of-order betriebsunfähig
  crystallize out/to auskristallisieren
  leach out/to auslaugen
  swung out/to ausgeschwenkt
  wear out/to verschleißen
  weigh out/to einwiegen
outdoor draußen, außen, Außen-, Freiluft-
  outdoor conditions Freiluftklima n
  outdoor insulator Freiluftisolator m
  outdoor use Freiluftbetrieb m
  outdoor weathering Außenlagerung f, Frei(luft)bewitterung f
  outdoor weathering performance Freibewitterungsverhalten n
  outdoor weathering resistance Außenbewitterungsbeständigkeit f
  outdoor weathering station Freiluftbewitterungsanlage f, Freiluftprüfstand m, Freibewitterungsstation f,
  outdoor weathering test Freiluftversuch m, Freibewitterungsversuch m, Außenbewitterungsversuch m
outer Außen-
  outer barrel surface Zylinderaußenfläche f
  outer chamber Außenkammer f
  outer cylinder surface Zylinderaußenfläche f
  outer fiber Außenfaser f
  outer fiber strain Randfaserdehnung f
  outer fiber zone Randfaserbereich m
  outer layer Randschicht f, Deckschicht f
  outer skin Oberflächenhaut f, Außenhaut f
  outer surface Außen(ober)fläche f
  outer zone Randzone f
outlet Ablauf m, Abfluß m, Entleerungsstutzen m, Entleerungsöffnung f, Austrittsöffnung f
  outlet channel Abströmkanal m
  outlet temperature Auslauftemperatur f, Austrittstemperatur f
  air outlet Luftaustritt m
  cooling water outlet Kühlwasserablauf m
  cooling water outlet temperature Kühlwasserablauftemperatur f
  oil outlet Ölabfluß m
  water outlet Wasserabfluß m, Wasserauslaß m, Wasserauslauf m, Wasseraustritt m
output Ausgabe f, Ausgang m, Ausstoß m, Leistung f, Durchsatzmenge f, Ausstoßmenge f, Ausstoßleistung f,

Durchsatzleistung f, Durchsatz m, Leistungsfähigkeit f, Förderrate f
output card Ausgangskarte f
output channel Ausgangskanal m
output current Ausgangsstrom m
output data Leistungsangaben fpl, Ausgabedaten npl
output-dependent leistungsabhängig
output facility Ausgabemöglichkeit f
output fluctuations Ausstoßschwankungen fpl
output impedance Ausgangsimpedanz f
output maximization Durchsatzmaximierung f
output module Ausgabebaugruppe f, Ausgabestein m, Ausgangsmodul m
output per hour Stundenleistung f
output rate Durchsatzmenge f, Durchsatzleistung f, Durchsatz m, Ausstoßgeschwindigkeit f, Ausstoßleistung f, Ausstoßmenge f
output rate curve Ausstoßkennlinie f
output record Ausgabeliste f
output side Ausgabeseite f
output signal Ausgangssignal n
output temperature Ausstoßtemperatur f
output unit Ausgabeeinheit f, Ausgabegerät n
output variable Ausgangsgröße f
output voltage Ausgangsspannung f
output volume Ausstoßvolumen n
analog output Analogausgabe f, Analogausgang m
average output Durchschnittsleistung f
control output Steuer(ungs)ausgang m
data output Datenausgabe f
depending on the output leistungsabhängig
digital output Digitalausgang m
extra output Zusatzausgang m
extruder output (rate) Extruderausstoß m, Extruderleistung f, Extrusionsleistung f
extruder output equation Extruderkennliniengleichung f
guaranteed output Garantieausstoß m, Leistungsgarantie f
hourly output Stundenleistung f
increased outputs Leistungserhöhung f, Leistungssteigerungen fpl
logic output Logikausgang m
machine output Maschinenleistung f
material output (rate) Massedurchsatz m
maximum output Spitzenleistung f
power output Leistungsabgabe f
range of outputs Leistungsfahrbreite f
reduced output Ausstoßverringerung f
reduction in output Durchsatzreduktion f, Ausstoßleistungsminderung f, Durchsatzeinbuße f
relay output Relaisausgang m
screw output Schneckenleistung f
signal output Signalausgang m
total output Gesamt(produktions)leistung f
outsert mo(u)lding Outsert-Technik f
outside außen, Außen-

outside air Außenluft f
outside control Fremdüberwachung f
outside diameter [of pipes] Außendurchmesser m
outside energy Fremdenergie f
outside pipe diameter Rohraußendurchmesser m
outside temperature Außentemperatur f
outside the instrument geräteextern
outstanding hervorragend
oven Ofen m, Heizschrank m
air circulating oven Luftumwälzofen m
circulating air drying oven Umluftwärmeschrank m
circulating air oven Umluftofen m
conditioning oven Temperofen m
curing oven Aushärteofen m, Härtungsofen m
drying oven Trockenofen m, Wärmeschrank m
fusion oven [for PVC pastes] Gelierofen m
high-speed drying oven Schnelltrockner m
hot air oven Heißluftofen m, Heizluftschrank m
laboratory oven Wärmeschrank m
preheating oven Vorheizofen m, Vorwärmofen m
stoving oven Einbrennofen m
tunnel oven Durchlaufofen m
vacuum oven Vakuumofen m
overall gesamt, total, Gesamt-, Total-
overall design Gesamtkonzept n
overall dimensions Gesamtabmessungen fpl
overall noise level Gesamtschallpegel m
overall temperature Gesamttemperaturniveau n
overbaking [paint-film] Überbrennen n
overcapacity Überkapazität f
overcured überhärtet
overcuring Überhärtung f, Übervulkanisation f
overfeed/to überfüllen, überfüttern, überladen
overflow Überlauf m
overflow valve Überströmventil n
overheads Gemeinkosten pl
production overheads Fertigungsgemeinkosten pl
overheating thermische Überlastung f, Überhitzung f
local overheating Punktüberheizung f, örtliche Überhitzung f
protection against overheating Überhitzungsschutz m, Übertemperaturschutz m
risk of overheating Überhitzungsgefahr f
thermostat to prevent overheating Überhitzungs-Sicherheitsthermostat n
overlap/to überlappen
overlap Überlappung f
overlap gate Überlappungsanschnitt m
overlapping überlappend
overlay mat Oberflächenvlies n
overload/to überladen
overload Überlast f
overload prevention device Überlast(ungs)sicherung f, Überlastschutz m
overload-proof überlastsicher
overload protection Überlastungsschutz m
overload signal Überlastanzeige f

overloaded überlastet
overloading Überbeanspruchung f, Überlastung f
  risk of overloading Überlastungsgefahr f
  safe against overloading überlastsicher
overpack/to [a mo(u)ld] überladen, überfüttern, überfüllen
overpaint/to überstreichen
overpainting tolerance Überstreichbarkeit f
override/to überfahren, übersteuern, eingreifen [in einen Regelvorgang]
  override mechansim Überfahreinrichtung f
oversea(s) überseeisch, Übersee-, außereuropäisch [aus europäischer Sicht]
oversize übergroß, überdimensioniert
overspray/to überspritzen
overspray tolerance Überspritzbarkeit f
overstoving Überbrennen n
  overstoving resistance Überbrennstabilität f
overview Übersicht f
  production overview Produktionsübersicht f
owing to the nature of the process aus verfahrenstechnischen Gründen mpl
own eigen
  own capital [of a company] Eigenkapital n
  on its own unverschnitten
ox-bow profile Ochsenjochprofil n
oxidation Oxidation f
  oxidation catalyst Oxidationskatalysator m
  oxidation number Oxidationsstufe f, Oxidationszahl f
  oxidation product Oxidationsprodukt n
  oxidation rate Oxidationsgeschwindigkeit f
  oxidation resistance Oxidationsbeständigkeit f, Oxidationsstabilität f
  oxidation sensitivity Oxidationsempfindlichkeit f, Oxidationsanfälligkeit f
  degree of oxidation Oxidationsgrad m
  not sensitive to oxidation oxidationsunempfindlich
  radiation-induced oxidation Strahlungsoxidation f
  rate of oxidation Oxidationsgeschwindigkeit f
  surface oxidation Oberflächenoxidation f
  susceptible to oxidation oxidationsanfällig, oxidationsempfindlich
  thermal oxidation Wärmeoxidation f
oxidative oxidativ
oxide Oxid n
  oxide film Oxidschicht f
  alkaline earth oxide Erdalkalioxid n
  alumin(i)um oxide Aluminiumoxid n
  brown iron oxide Eisenoxidbraun n
  calcium oxide Calciumoxid n
  ethylene oxide Ethylenoxid n
  ethylene oxide group Ethylenoxidgruppe f
  lithium oxide Lithiumoxid n
  magnesium oxide Magnesiumoxid n
  polyethylene oxide Polyethylenoxid n
  polyphenylene oxide [PPO] Polyphenylenoxid n
  polypropylene oxide Polypropylenoxid n
  propylene oxide Propylenoxid n
  red iron oxide Eisenoxidrot n
  zinc oxide Zinkoxid n
  zinc phosphate-iron oxide primer Zinkphosphat-Eisenoxid-Grundierung f
oxidizable oxidierbar
oxidized oxidiert
oxidizing agent Oxidationsmittel n
oxychloride Oxychlorid n
  zinc oxychloride Zinkoxychlorid n
oxygen Sauerstoff m
  oxygen atom Sauerstoffatom n
  oxygen-containing sauerstoffhaltig
  oxygen index Sauerstoffindex m
  oxygen permeability Sauerstoffdurchlässigkeit f
  oxygen sensitivity Sauerstoffempfindlichkeit f
  active oxygen content Aktivsauerstoffgehalt m
  amount of oxygen available Sauerstoffangebot n
  excess oxygen Sauerstoffüberschuß m
  lack of oxygen Sauerstoffmangel m
  under the exclusion of oxygen unter Sauerstoffausschluß m
oxyhydrogen flame Knallgasflamme f
oxymethylene formaldehyde Oximethylenformaldehyd n
oxymethylene group Oxymethylengruppe f
ozone Ozon n
  ozone resistance Ozonfestigkeit f
pack Packung f, Verpackung f
  blister pack Blisterpackung f
  bubble pack Bubblepackung f
  expanded polystyrene pack Polystyrolschaumstoff-Verpackung f
  food pack Lebensmittelverpackung f, Lebensmittelbehälter m
  screen pack Filterpaket n, Siebblock m, Siebeinsatz m, Siebgewebepackung f, Siebkorb m, Siebpaket n, Siebsatz m, Filtergewebepaket n, Maschengewebepaket n
  shrink wrapped pack Schrumpfverpackung f
  skin pack Skinpackung f
  stretch wrapped pack Streckfolien-Verpackung f, Streckverpackung f
  twin screen pack Doppelsiebkopf m
package Paket n, Packung f, Verpackung f
  package deal Paketlieferung f
  hydraulic monitoring package Hydraulik-Überwachunspaket n
  program package Programmpaket n
  software package Softwarepaket n
packaging Verpackungs-
  packaging container Verpackungsbehälter m, Verpackungshohlkörper m, Verpackungsgebinde n, Verpackungsteil n, Verpackungsmittel n
  packaging film Verpackungsfolie f
  packaging industry Emballageindustrie f, Verpackungswesen n
  packaging line Verpackungsstraße f

packaging material Packstoff m, Verpackungsmaterial n, Verpackungsrohstoff m, Verpackungswerkstoff m
packaging operation Packvorgang m
packaging plant Verpackungsanlage f
packaging sector Verpackungsgebiet n
packaging tapes Verpackungsbänder npl
blister packaging machine Blisterformmaschine f
blow-fill-seal packaging line Blasform-, Füll- und Verschließanlage f
food packaging film Lebensmittelverpackungsfolie f
injection mo(u)lding of packaging containers Verpackungsspritzgießen n
non-food packaging Nicht-Nahrungsmittel-Verpackung f
original packaging Originalverpackung f
skin packaging machine Skinformmaschine f
standard packaging Standardverpackung f
packing [i.e. a seal] Packung f, Dichtung f, Futter n; Einpacken n
  packing density Packungsdichte f, Rütteldichte f, Rüttelgewicht n
  packing phase Kompressionsphase f, Verdichtungsphase f, Kompressionszeit f
  packing pressure Verdichtungsdruck m
  packing profile Verdichtungsprofil n
  packing time Kompressionszeit f
  degree of packing Verdichtungsgrad m
  high temperature packing Hochtemperaturpackung f
pad Polster n, Einlage f, Unterlage f, Auflage f, Puffer m
  bridge bearing pad Brückenlager n
  pressure pad Druckkissen n, Druckplatte f
paddle Paddel n, Flügel m
  paddle mixer Paddelmischer m, Schaufelrührer m
page Seite f, Blatt n
  page printer Blattschreiber m
  display page Bildschirmseite f
paint Anstrichfarbe f, Anstrichmittel n, Anstrichstoff m, Lack m, Farbe f, Lackfarbe f [s.a. varnish/lacquer]
  paint additive Lackadditiv n
  paint auxiliary Lackhilfsmittel n
  paint film Anstrichfilm m, Lackierung f, Anstrich m, Lackfilm m, Lackschicht f
  paint film defect Lackierungsdefekt m
  paint film quality Beschichtungsqualität f
  paint film structure Anstrichaufbau m
  paint film surface Lackoberfläche f
  paint formulation Anstrichformulierung f
  paint industry Farbenindustrie f, Lackindustrie f
  paint residues Farbreste mpl, Lackreste mpl
  paint resin Lackharz n, Lackbindemittel n, Lackrohstoff m
  paint solids content Lackfeststoffgehalt m
  paint solvent Lacklösemittel n
  paint spraying booth Spritzkabine f
  paint spraying unit Farbspritzanlage f
  paint system Anstrich(mittel)system n, Lacksystem n
  alkyd paint Alkydharzlack m
  anti-corrosive paint Korrosionsschutzfarbe f, Korrosionsschutzlack m, Korrosionsschutzsystem n
  anti-fouling paint Antifoulingfarbe f
  automotive paint Fahrzeuglack m
  chlorinated rubber paint Chlorkautschukfarbe f, Chlorkautschuklack m
  coil coating paint Coil-Coating-Decklack m, Walzlack m
  dipping paint Tauchlack m
  emulsion paint Dispersionsanstrichmittel n, Dispersionsfarbe f, Dispersionslack m
  epoxy(-based) paint EP-Anstrichmittel n, Epoxidharzfarbe f
  exterior paint Fassadenfarbe f
  finishing paint Decklack m
  hammer finish paint Hammerschlaglack m
  high-solids paint lösemittelarmer Lack m
  household appliance paint Haushaltsgerätelack m
  marine paint Schiffsfarbe f
  new paint film Neuanstrich m
  one-coat paint Einschichtlack m
  powdered paint Pulverfarbe f
  road marking paint Straßenmarkierungsfarbe f
  solvent-based paint Lösungsmittellack m
  spraying paint Spritzlack m
  synthetic paint resin Lackkunstharz n
  synthetic resin-based paint Kunstharzlack m
  two-pack paint Zweikomponentenanstrich(mittel) m(n)
  underwater paint Unterwasseranstrich m
  water-based paint Wasserlack m
  zinc-rich paint Zinkstaubfarbe f, Zinkstaubformulierung f
paintability Lackierbarkeit f
paintable lackierfähig
painting Lackierung f, Anstrich m, Farbanstrich m [s.a. paint]
  painting trials Beschichtungsversuche mpl [Farbe]
  continuous painting line Durchlauflackieranlage f
  spray painting Spritzlackierung f
pair Paar n
  pair of rolls Walzenpaar n
  electron pair Elektronenpaar n
pallet Palette f
  cage pallet Gitterboxpalette f
palmitic acid Palmitinsäure f
panel Tafel f, Platte f, Tableau n
  cladding panel Fassadenplatte f, Verkleidungsplatte f
  control panel Bedienungstableau n, Frontplatte f, Fronttafel f, Schalttableau n, Schalttafel f, Bedientafel f, Bedientableau n, Bedienungsfront f, Bedienungsfeld n, Überwachungstafel f
  front panel Frontplatte f
  instrument front panel Gerätefrontplatte f
  instrument panel Armaturentafel f, Armaturenbrett n
  light signal indicator panel Leuchtanzeigetableau n
  operator panel Bedienungskonsole f, Bedientableau n,

Bedientafel f, Bedienungsfeld n, Bedienungsfront f, Bedienungspult n
  rib-reinforced sandwich panel Stegdoppelplatte f
  side panel [of a machine] Seitenschild n
  sound-insulating panel Schalldämmplatte f
paper Papier n
  paper chromatography Papierchromatographie f
  paper-based laminate Hartpapier(laminat) n, Papierschichtstoff m
  paper-like film papierähnliche Folie f, Papierfolie f
  paper-like polyethylene film [high molecular weight polyethylene film] HM-Folie f
  paper sack Papiersack m
  paper tape Lochstreifen m
  paper tape punch Lochstreifenstanzer m
  paper varnish Papierlack m
  paper web Papierbahn f
  backing paper Trägerpapier n
  decorative paper Dekorpapier n
  emery paper Schleifpapier n, Schmirgelpapier n
  epoxy paper laminate Epoxidharzhartpapier n
  glass paper Glaspapier n
  graph paper Diagrammpapier n
  phenolic paper laminate Phenolharz-Hartpapier n
  release paper Mitläuferpapier n, Trennpapier n
  varnished paper Lackpapier n
para-dichlorobenzene Paradichlorbenzol n
para-position Parastellung f
paraffin Paraffin n
  chlorinated paraffin Chlorparaffin n
paraffinic paraffinisch
parallel parallel
  parallel arrangement Parallelschaltung f
  parallel runners Parallelverteiler mpl
  parallel travel Parallelführung f
  arranged in parallel parallelgeschaltet
  axially parallel achsparallel
  plane parallel planparallel
parallelism Parallelität f
parameter Parameter m [s.a. factor]
  machine parameter Maschinenparameter m
  process parameter Prozeßgröße f
  processing parameter Verarbeitungsgröße f, Verarbeitungsparameter m, Verfahrensparameter m
  technically important parameters verfahrenstechnisch relevante Parameter mpl
  test parameter Versuchsparameter m
parcel shelf [in a car] Hutablage f
parent company Muttergesellschaft f
parison Schlauch m, Rohrstück n, Schlauchabschnitt m, Schlauchstück n, Schlauchvorformling m, schlauchförmiger Vorformling m, Vorformling m
  parison circumference Schlauchumfang m
  parison coextrusion die Koextrusionsschlauchkopf m
  parison curl Rollen fpl [des Schmelzeschlauches]
  parison cutter Schlauchabschneider m
  parison delivery rate Schlauchausstoß-geschwindigkeit f, Schlauchaustrittsgeschwindigkeit f
  parison diameter Schlauchdurchmesser m
  parison diameter control(ler) Schlauchdurchmesserregelung f
  parison die Extrusionsblaskopf m, Hohlkörperblasdüse f, Vorformlingswerkzeug n, Schlauch(extrusions)düse f, Schlauch(spritz)kopf m, Schlauchformeinheit f, Schlauchwerkzeug n
  parison gripper Schlauchgreifer m
  parison gripping mechanism Schlauchgreifvorrichtung f
  parison length Vorformlingslänge f
  parison length control (mechanism) Vorformlingslängenregelung f, Schlauchlängenregelung f
  parison length programming (device) Schlauchlängen-Programmierung f
  parison receiving station Schlauchübernahmestation f
  parison stretching mandrel Spreizdorn(anlage) m(f), Spreiz(dorn)vorrichtung f
  parison support Vorformlingsträger m
  parison take-off Schlauchabzug m
  parison wall Vorformlingswand f
  parison wall thickness control(ler) Schlauchdickenregelung f
  parison waste Blas(teil)butzen m, Butzen(abfall) m, Butzenmaterial n
  center-fed parison die Dornhalterschlauchkopf m
  side-fed parison die seitlich angestömter Pinolenkopf m
  spider-type parison die Stegdornhalterwerkzeug n, Stegdornhalter(blas)kopf m
  twin parison die Doppelschlauchkopf m, Zweifachschlauchkopf m, Zweifachblaskopf m
part Teil n; Formkörper m, Formteil n [s.a. parts]
  part cross-section Formteilquerschnitt m
  part delivery safety mechanism Ausfallsicherung f
  part design Formteilkonstruktion f, Formteilgestaltung f
  part designer Formteilkonstrukteur m
  part dimensions Formteilmaße npl
  part drawing Formteilzeichnung f, Produktionsteilzeichnung f
  part properties Formteileigenschaften fpl
  part quality Formteilqualität f, Teilqualität f
  part shape Formteilgestalt f
  part shrinkage Formteilschwindung f
  part size Formteilgröße f
  part surface Formteiloberfläche f
  part tolerances Formteiltoleranzen fpl
  part volume Formteilvolumen n
  part wall thickness Formteilwanddicke f
  part weight Formteilmasse f, Formteilgewicht n
  blow mo(u)lded part geblasener Hohlkörper m, blasgeformter Hohlkörper m, Blas(form)teil n, Blaskörper m
  blown part removal (device) Blaskörperentnahme f
  bottom part Unterteil n

center part of the mo(u)ld Werkzeugmittelteil n
extra part Zusatz(bau)teil n
injection mo(u)lded part Spritz(gieß)(form)teil n
mo(u)lded part Formteil n
precision injection mo(u)lded part Präzisions-(spritzguß)teil n
sandwich mo(u)lded part Sandwich(-Spritzguß)teil n
space for part ejection Ausfallöffnung f
standard part Serienteil n
thermoplastic foam mo(u)lding part Thermoplast-Schaumteil n
threaded part Gewindepartie f
top part Oberteil n
top part of the machine Maschinenoberteil n
partial partiell, teilweise, Teil-
   partial crystallinity Teilkristallinität f
   partial discharge Teilentladung f
   partial ester Partialester m
   partial gelation Vorgelierung f, Angelieren n
   partial hot runner Teilheißläufer m
   partial plastification Anplastifizieren n
   partial pressure Partialdruck m
   partial print-out Teilausdruck m
   partial vacuum Teilvakuum n
   partial vapo(u)r pressure Dampfteildruck m
   partial view Teilansicht f
   partial vulcanization Anvulkanisieren n
   partial water vapo(u)r pressure Wasserdampf-partialdruck m
partially crystalline teilkristallin
partially oxidized teiloxidiert
participation Beteiligung f
particle Partikel m, Teilchen n, Kornpartikel m, Pulverkorn n, Pulverteilchen n
   particle bridge Partikelbrücke f
   particle characteristics Kornbeschaffenheit f
   particle flow Teilchenfluß m
   particle interface Partikelgrenzfläche f
   particle porosity Kornporosität f
   particle shape Kornform f, Partikelform f, Teilchenform f
   particle size Korndurchmesser m, Kornfeinheit f, Korngröße f, Partikelgröße f, Teilchengröße f, Teilchendurchmesser m, Körnung f
   particle size distribution Kornzusammensetzung f, Korn(größen)verteilung f, Teilchen(größen)verteilung f, Partikel(größen)verteilung f
   particle size range Kornklasse f, Korngrößenbereich m, Kornspektrum n
   particle structure Kornstruktur f, Teilchenstruktur f
   particle surface Kornoberfläche f, Teilchenoberfläche f
   air between the particles [e.g. of resin] Zwischenkornluft f
   coarse particles Grobbestandteile mpl
   dust particles Staubteilchen npl
   elementary particle Elementarteilchen n
   extremely fine particle Feinstpartikel m
   filler particle Füllstoffkorn n, Füllstoffteilchen n
   filler particle orientation Füllstofforientierung f
   foreign particle Fremdpartikel m
   gel particle Gelpartikel m, Gelteilchen n
   gel particles Gele npl
   latex particle Latexteilchen n
   melt particle Schmelzeteilchen n
   pigment particle Pigmentkorn n, Pigmentteilchen n
   polymer particle Kunststoffteilchen n, Polymerteilchen n
   predominant particle size häufigste Korngröße f
   primary particle Primärkorn n, Primärteilchen n
   PVC particle PVC-Pulverkorn n, PVC-Teilchen n
   secondary particle Sekundärkorn n, Sekundärteilchen n
   solid particle Feststoffpartikel m
   space between the particles Teilchenzwischenräume mpl
particulate partikulär, aus (einzelnen) Teilchen bestehend
parting line Formtrennaht f, Teilungslinie f, Trennaht f
   gating at the (mo(u)ld) parting line Trennlageneinspritzung f
   mo(u)ld parting line Formtrennlinie f
parting surface Formteilung f, Formteilebene f, Werkzeugteilungsebene f, Werkzeugtrennfläche f, Formtrennfläche f, Trennfläche f, Teilungsfuge f, Formtrennebene f, Teilungsfläche f, Teilungsebene f
partition wall Trennwand f
partly teilweise
   partly automated teilautomatisiert
   partly branched teilverzweigt
   partly crosslinked teilvernetzt
   partly dissolved angelöst
   partly etherified partiellverethert
   partly evacuated teilevakuiert
   partly filled teilgefüllt
   partly full teilgefüllt
   partly fused angeschmolzen
   partly gelled angeliert, vorgeliert
   partly heated teilbeheizt
   partly methylated partiellmethyliert
   partly saponified teilverseift
   partly used angebrochen
parts Teile npl, Anteile mpl
   parts by volume Raumteile mpl
   parts by weight [p.b.w.] Gewichtsteile mpl, Massenteile mpl
   parts counter Stückzähler m
   parts of the extruder Extruderabschnitte mpl
   parts-removal robot Entnahmeautomat m, Entnahmeroboter m
   parts-removal station Entnahmestation f
   parts subject to wear Verschleißteile npl
   parts to be joined Fügeteile npl
   cost of spare parts Ersatzteilkosten pl
   list of spare parts Ersatzteilliste f

spare parts Ersatzteile npl
passage Passage f, Führung f
  passage of melt Schmelze(strom)führung f
  smooth passage through the machine ruhige Laufeigenschaften fpl
paste Paste f
  paste deaeration Pastenentlüftung f
  paste dipping (process) Pastentauchverfahren n
  paste extender resin Pastenverschnittharz n
  paste extrusion Pastenextrusion f
  paste for making mechanically blown foam Schlagschaumpaste f
  paste formulation Pastenrezeptur f, Pastenansatz m
  paste-like breiartig, pastös
  paste making Pasten(zu)bereitung f
  paste-making pastenbildend, verpastbar
  paste-making grade [of PVC] Pastentyp m
  paste-making polymer Pastenpolymerisat n
  paste-making PVC Pasten-PVC n, PVC-Pastentyp m
  paste mixing Verpastung f
  paste polymer Pastenpolymerisat n
  paste preparation Pastenherstellung f
  paste viscosity Pastenviskosität f
  anchor coating paste Grundierungspaste f
  benzoyl peroxide paste Benzoylperoxid-Paste f
  casting paste Gießpaste f
  catalyst paste Härterpaste f
  curing agent paste Vernetzerpaste f
  dipping paste Tauchpaste f
  expandable paste Schaumpaste f
  heat sink paste Wärmeleitpaste f
  impregnating paste Tränkpaste f
  laminating paste Laminierpaste f
  modelling paste Urmodellpaste f
  pigment paste Farb(pigment)paste f, Pigmentanreibung f, Pigmentpaste f
  PVC paste PVC-Paste f, Plastisol n
  PVC paste resin Pasten-PVC n, PVC-Pastentyp m, Pastenware f
  silicone paste Siliconpaste f
  spread coating paste Streichpaste f
  underbody sealant paste Unterbodenschutzpaste f
pastel shade Pastellton m
patent rights Schutzrechte npl, Patentrechte npl
paternoster-type conveyor Paternosteranlage f
path Weg m, Fließweg m, Strömungsweg m
pattern Muster n, Modell n, Form f
  pattern guides Modellführungen fpl
  pattern-making shop Modellwerkstatt f
  pattern plate Modellplatte f
  curing pattern Härtungsverlauf m
  flow pattern Fließfiguren fpl
  foundry pattern Gießereimodell n
  master pattern plate Urmodellplatte f
  weave pattern Webart f
pay-off creel Abspulgatter n
PBTP [polybutylene terephthalate] Polybutylenterephthalat n
p.b.w. [parts by weight] Massenteile npl
PC [polycarbonate] Polycarbonat n
PE [polyethylene] Polyethylen n
peak Spitze f
  peak memory Spitzenspeicher m
  peak pressure Druckmaximum n, Druckspitze f, Spitzendrücke mpl
  peak temperature Temperaturspitze f
  pressure peak Druckspitze f
  stress peak Belastungsspitze f, Spannungsspitze f
  temperature peak Temperaturspitze f
  viscosity peak Viskositätsberg m
  without pressure peaks druckspitzenlos
peel Schale f
  peel force Schälkraft f
  peel strength Abschälfestigkeit f, Schälfestigkeit f, Schälwert m, Schälwiderstand m
  peel stress Schälbeanspruchung f, Schälbelastung f
  peel test Schälversuch m
  orange peel effect Apfelsinenschaleneffekt m, Apfelsinenschalenstruktur f
  under peel stress schälbeansprucht
pellet Granulat n, Kügelchen n, Körnchen n, Krümmel m, Tablette f
  pellet cooling unit Granulatkühlvorrichtung f
  pellet pre-heating unit Granulatvorwärmer m
  pellet size Granulatgröße f
  high-performance pellet drier Hochleistungsgranulattrockner m
pelleting Pelletieren n, Tablettierung f
  pelleting machine Tablettenmaschine f
  pelleting press Tablettierpresse f
pelletization Granulierung f
  die-face pelletization Heißgranulierung f, Kopfgranulierung f, Heißabschlaggranulierung f
  strand and strip pelletization Kaltgranulierverfahren n
  strand pelletization Stranggranulierung f
  water-cooled die face pelletization Wasserringgranulierung f
pelletize/to zerkleinern, granulieren
pelletizer Granulator m, Zerkleinerer m, Granuliermaschine f, Schneidgranulator m, Zerkleinerungsmaschine f, Zerkleinerungsaggregat n, Zerkleinerungsanlage f
  pelletizer knife Granulier(messer)flügel m
  pelletizer range Mühlenbaureihe f
  die-face pelletizer Heißabschlaggranuliereinrichtung f, Heißabschlaggranulator m
  hot-cut pelletizer Heißabschlagvorrichtung f
  hot-cut water-cooled pelletizer Heißabschlag-Wassergranulierung f
  strand pelletizer Stranggranulator m, Strangschneider m
  strip pelletizer Brandgranulator m, Kaltgranuliermaschine f, Würfelschneider m,

Bandschneider m
twin screw pelletizer Zweiwellen-Granuliermaschine f
underwater pelletizer Unterwassergranulator m
underwater strand pelletizer Unterwasserstranggranulator m
water-cooled die face pelletizer Wasserringgranulierung f
pelletizing Granulier-
  pelletizing die Granulierplatte f, Granulierdüse f, Granulierkopf m, Granulierwerkzeug n, Granulierlochplatte f, Lochdüse f
  pelletizing line Granulieranlage f
  die face pelletizing system Direktabschlagsystem n
  hot cut pelletizing system Direktabschlagsystem n
  strand pelletizing line Stranggranulieranlage f
  strand pelletizing system Strangabschlagsystem n
  strip pelletizing line Bandgranulierstraße f
  strip pelletizing system Bandabschlagsystem n
  twin screw pelletizing line Doppelschnecken-Granulieranlage f
  underwater pelletizing system Unterwassergranuliersystem n
pellets Granulat n, Linsengranulat n, Granulatkörner npl
  cylindrical and diced pellets Kaltgranulat n
  cylindrical pellets Stranggranulat n, Zylindergranulat n
  hot-cut pellets Heißgranulat n
  plastics pellets Kunststoffgranulat n
pendulum Pendelhammer m, Schlaghammer m, Schlagpendel n
  pendulum hardness Pendelhärte f
  pendulum impact speed Pendelhammergeschwindigkeit f, Schlagpendelgeschwindigkeit f
  pendulum impact test Pendelschlagversuch m
  pendulum impact tester Pendelschlagwerk n, Pendelhammergerät n
  residual pendulum hardness Restpendelhärte f
  retained pendulum hardness Restpendelhärte f
  torsion pendulum (apparatus) Torsionspendel n, Torsionsschwingungsgerät n, Schwingungsgerät n
  torsion pendulum test Schwingungsversuch m, Torsionsschwingversuch m, Torsionsschwingungsmessung f, Torsionsversuch m
penetration Durchdringung f, Eindringen n
  penetration energy Durchstoßarbeit f
  penetration power Eindringvermögen n
  penetration test Durchstoßversuch m
  degree of penetration Eindringgrad n
  depth of penetration Eindringtiefe f
  fabric penetration Durchschlag m [PVC-Paste in Gewebe]
  resistance to root penetration Wurzelfestigkeit f
  resistant to root penetration wurzelfest
  speed of penetration Eindringgeschwindigkeit f
pentaerythritol Pentaerythrit(ol) n
  pentaerythritol ester Pentaerythritester m
pentane Pentan n

peptizer [promotes breakdown of rubber during mastication] Plastikator m
per per, durch, für, pro
  per-acid Persäure f
  per capita consumption Pro-Kopf-Verbrauch m
  per-ester Perester m
  alkyl per-ester Alkylperester m
  peracetic acid Peressigsäure f
peradipic acid Peradipinsäure f
perbenzoic acid Perbenzoesäure f
perbutyric acid Perbuttersäure f
percarbonate Percarbonat n
percent Prozent n
  percent by volume [v/v] Volumenprozent n
  percent by weight [w/w] Gewichts-Prozent n
percentage Prozentzahl f, Prozentanteil m
  percentage conversion prozentualer Umsatz m
  percentage light transmission Lichtdurchlässigkeitszahl f
  percentage turnover prozentualer Umsatz m
perchlorinated perchloriert
perchloroethylene Perchlorethylen n
perfect/to optimieren, vervollkommnen
perforated perforiert
  perforated die Lochdüse f
  perforated disk Ringlochplatte f
perforation Lochen n, Perforation f
  tear-off perforation Abreißperforation f
performance Leistungsfähigkeit f, Leistung f, Leistungsniveau n, Leistungsvermögen n; Verhalten n, Betriebsverhalten n
  performance characteristics Gebrauchseigenschaften fpl
  performance data [of a machine] Leistungsdaten npl, Leistungsangaben fpl
  performance features Leistungsmerkmale npl
  performance limit Leistungsgrenze f
  performance requirements Funktionsanforderungen fpl
  performance test Funktionsprüfung f
  assessment of performance Leistungsbewertung f
  decline in performance Leistungsabfall m
  extruder performance Extruderverhalten n
  extrusion performance Extrudierbarkeit f, Extrusionsverhalten n
  feed performance Einzugsverhalten n, Einziehverhalten n
  general performance Gesamtverhalten n
  high-performance leistungsstark
  long term performance Zeitstandverhalten n, Dauergebrauchseigenschaften fpl, Langzeitverhalten n
  low-temperature performance Tieftemperaturverhalten n
  lubricant performance Gleitmittelverhalten n
  machine performance Maschinenverhalten n, Maschinenleistung f
  metering performance Dosierleistung f

outdoor weathering performance Freibewitterungsverhalten n
plasticizing performance Plastifizierverhalten n, Plastifizierstrom m, Plastifizierleistung f, Plastifiziervermögen n
screw performance Schneckeneigenschaften fpl, Schneckenleistung f
performic acid Perameisensäure f
period Zeitdauer f, Zeitraum m, Zeitspanne f
 ag(e)ing period Alterungszeit f, Lagerungszeit f, Lager(ungs)dauer f
 amortization period Amortisationszeit f
 annealing period Temperzeit f
 conditioning period Temperzeit f
 half life period Halbwertzeit f
 heat ag(e)ing period Warmlagerungszeit f
 heating-up period Aufheizzeit f
 immersion period Lagerdauer f, Lagerungszeit f, Lagerungsdauer f [in Flüssigkeiten]
 observation period Beobachtungsdauer f
 operating period Betriebsdauer f, Betriebszeit f
 pre-heating period Vorwärmzeit f
 recovery period Erholungszeitspanne f
 reference period Bezugszeitraum m
 retardation period Retardationszeit f
 shut-down period Stehzeit f, Totzeit f, Stillstandzeit f, Nebenzeit f, Abstellzeit f, Leerlaufzeit f
 starting-up period Anfahrzeit f
 test period Prüfzeit f, Prüfzeitraum m, Untersuchungszeitraum m, Versuchszeit f
 transitional period Übergangszeit f
 weathering period Bewitterungszeit(raum) f(m), Bewitterungsdauer f
periodic periodisch
peripheral 1. peripher, Rand-, Umfangs-; 2. Peripherie f, Peripheriebaustein m, Peripheriebauteil n
 peripheral equipment Peripheriegerät n
 peripheral processor Peripherieprozessor m
 peripheral screw speed Schneckenumfangsgeschwindigkeit f
 peripheral unit Peripheriebauteil n, Peripheriebaustein m
 peripheral velocity Umfangsgeschwindigkeit f
periphery Peripherie f, Rand m
 roll periphery Ballenrand m, Walzenballenrand m
perketal Perketal
permanence Permanenz f, Dauerhaftigkeit f
permanent [e.g. dimensional change] bleibend, dauerhaft, Dauer-
 permanent contact Dauerkontakt m
 permanent deformation Dauerverformung f
 permanent flexibility Dauerelastizität f
 permanent joint unlösbare Verbindung f
 d.c. permanent magnet motor Gleichstrom-Drehankermagnet-Motor m
permanently antistatic dauerantistatisch
permanently flexible dauerplastisch, dauerelastisch

permeability Durchlässigkeit f, Permeabilität f
 permeability behavio(u)r Permeabilitätsverhalten n
 permeability coefficient Permeabilitätswert m, Permeabilitätskoeffizient m, Durchlässigkeitswert m
 permeability data Permeabilitätsdaten npl
 gas permeability Gasdurchlässigkeit f
 nitrogen permeability Stickstoffdurchlässigkeit f
 odo(u)r permeability Aromadurchlässigkeit f
 oxygen permeability Sauerstoffdurchlässigkeit f
 vapo(u)r permeability Dampfdurchlässigkeit f
 water vapo(u)r permeability Wasserdampfdurchlässigkeit f, Wasserdampfpermeabilität f
permeable durchlässig, permeabel
 permeable to X-rays röntgenstrahldurchlässig
permeation Permeation f
 gel permeation chromatography Gel-Permeationschromatographie f
permissible zulässig
permittivity Dielektrizitätskonstante f, DK
 relative permittivity relative Dielektrizitätskonstante f
peroctoate Peroctoat n
 butyl peroctoate Butylperoctoat n
peroxide Peroxid n
 peroxide concentration Peroxidkonzentration f
 peroxide crosslinkage peroxidische Vernetzung f
 peroxide-crosslinking peroxidisch-vernetzend
 peroxide decomposition Peroxidzerfall m, Peroxidzersetzung f
 peroxide-free peroxidfrei
 peroxide group Peroxidgruppe f
 peroxide molecule Peroxidmolekül n
 peroxide radical Peroxidradikal n
 peroxide suspension Peroxid-Suspension f
 acetylacetone peroxide Acetylacetonperoxid n
 amount of peroxide Peroxiddosierung f, Peroxidmenge f
 benzoyl peroxide Benzoylperoxid n
 benzoyl peroxide paste Benzoylperoxid-Paste f
 cyclohexanone peroxide Cyclohexanonperoxid n
 dialcyl peroxide Dialcylperoxid n, Dialkylperoxid n
 dialkyl peroxide group Dialkylperoxidgruppe f
 dicumyl peroxide Dicumylperoxid n
 dilauroyl peroxide Dilauroylperoxid n
 di-tertiary butyl peroxide [DTBB] Di-tert.-butylperoxid n
 hydrogen peroxide Wasserstoffsuperoxid n, Wasserstoffperoxid n
 lauroyl peroxide Lauroylperoxid n
 low-temperature peroxide Tieftemperaturperoxid n
 tertiary-butyl peroxide tert.-Butylperoxid n
 triphenylmethyl peroxide Triphenylmethylperoxid n
peroxy compound Peroxyverbindung f
peroxydicarbonate Peroxydicarbonat n
perpropionic acid Perpropionsäure f
personnel Personal n
 personnel expenditure Personalaufwand m
 maintenance personnel Wartungspersonal n

operating personnel Bedienungsmannschaft f, Bedienungspersonal n
persuccinic acid Perbernsteinsäure f
persulphate Persulfat n
PET(P) [polyethylene terephthalate] Polyethylenterephthalat n
petrochemical petrochemisch
petrol Benzin n
  petrol consumption Benzinverbrauch m
  petrol ga(u)ge Tankgeber m
  petrol tank Benzintank m
  petrol tank cap Tankverschluß m
  petrol tank float Tankschwimmer m
  petrol tank neck Tankeinfüllstutzen m
petroleum ether Petrolether m
petroleum industry Erdölindustrie f
pH (value) pH-Wert m
pharmaceutical industry Pharmaindustrie f
pharmaceutical products Pharmazeutika pl
phase Phase f
  phase angle Phasenwinkel m
  phase boundary Phasengrenze f, Phasengrenzfläche f
  phase distribution Phasenverteilung f
  phase equilibrium Phasengleichgewicht n
  phase interface control Phasenanschnittsteuerung f
  phase reversal Phasenumschlag m, Phasenumwandlung f
  phase reversal point Phasenumschlagpunkt m
  phase separation Phasentrennung f
  phase stability Phasenstabilität f
  cooling phase Abkühlphase f
  development phase Entwicklungsphase f
  disperse phase dispergierte Phase f, disperse Phase f
  dwell pressure phase Formgebungsphase f
  elastomer phase Elastomerphase f
  expansion phase Expansionsphase f
  gaseous phase Gasphase f
  hold(ing) pressure phase Nachdruckphase f, Formgebungsphase f
  injection phase Einspritzphase f, Formfüllphase f, Füllphase f
  liquid phase Flüssigkeitsphase f
  mo(u)ld filling phase Füllphase f, Formfüllphase f
  packing phase Kompressionsphase f, Kompressionszeit f, Verdichtungsphase f
  polymer phase Polymerphase f
  pressure build-up phase Druckaufbauphase f
  pressure decreasing phase Druckabfallphase f
  setting phase Erstarrungsphase f
  solid phase Festphase f, Feststoffbereich m
  solid phase condensation Festphasenkondensation f
  solid phase forming Schlagpressen n
  solid phase pyrolysis Festphasenpyrolyse f
  solvent phase Lösemittelphase f
  starting-up phase Anfahrphase f
phenol Phenol n
  phenol-formaldehyde condensate Phenol-Formaldehyd-Kondensat n
  phenol-formaldehyde resin Phenol-Formaldehydharz n
  phenol-modified phenolmodifiziert
  phenol novolak Phenolnovolak n
  phenol resol Phenolresol n
  alkyl phenol Alkylphenol n
  dimethyl phenol Dimethylphenol n
phenolic phenolisch
  phenolic engineering resin technisches Phenolharz n
  phenolic foam Phenolharzschaumstoff m, Phenolschaum m
  phenolic laminate Phenolharz-Schichtstoff m
  phenolic material Phenolharz-Werkstoff m
  phenolic mo(u)lding Phenoplast-Formteil n
  phenolic mo(u)lding compound Phenoplast-Formmasse f, Phenolharz-Preßmasse f, Phenoplast-Preßmasse f
  phenolic paper laminate Phenolharz-Hartpapier n
  phenolic resin Phenolharz n
  liquid phenolic resin Phenolflüssigharz n
  mo(u)lded phenolic material Phenoplast-Preßstoff m
  powdered phenolic resin Phenolpulverharz n
phenolics Phenoplaste mpl
phenolphthalein Phenolphthalein n
phenomena Erscheinungen fpl, Phänomene npl
  ag(e)ing phenomena Alterungserscheinungen fpl
phenoxy radical Phenoxy-Radikal n
phenyl- Phenyl-
  phenyl group Phenylgruppe f
  phenyl naphthylamine Phenylnaphthylamin n
  phenyl radical Phenylrest m
  phenyl ring Phenylring m
  phenyl silicone resin Phenylsiliconharz n
  phenyl urea Phenylharnstoff m
phenylene diamine Phenylendiamin n
phenylmethyl silicone Phenylmethylsilicon n
phenylmethyl silicone fluid Phenylmethylsiliconöl n
phenylmethyl silicone resin Phenylmethylsiliconharz n
phenylmethyl vinyl polysiloxane Phenylmethylvinylpolysiloxan n
phenylpropyl siloxane Phenylpropylsiloxan n
phosphate Phosphat n, Phosphorsäureester m
  phosphate plasticizer Phosphat-Weichmacher m
  aryl alkyl phosphate Arylalkylphosphat n
  dicresylphenyl phosphate Dikresylphenylphosphat n
  diphenylcresyl phosphate Diphenylkresylphosphat n
  diphenyl ethylhexyl phosphate Diphenyl-(2-ethylhexyl-)phosphat n
  diphenyloctyl phosphate Diphenyloctylphosphat n, Diphenyl-(2-ethylhexyl-)phosphat n
  trialkyl phosphate Trialkylphosphat n
  triaryl phosphate Triarylphosphat n
  tricresyl phosphate [TCP] Trikresylphosphat n
  trioctyl phosphate triphenyl phosphate [TPP] Triphenylphosphat n
  trixylenyl phosphate [TXP] Trixylenylphosphat n

zinc phosphate-iron oxide primer Zinkphosphat-
  Eisenoxid-Grundierung f
phosphated phosphatiert
phosphating Phosphatierung f
phosphite Phosphit n, Phosphorigsäureester m
  alkyl phosphite Alkylphosphit n
  aryl phosphite Arylphosphit n
  lead phosphite Bleiphosphit n
  triphenyl phosphite Triphenylphosphit n
phosphoric acid Phosphorsäure f
  phosphoric acid ester Phosphorsäureester m
phosphorus 1. Phosphor m; 2. Phospor-
phosphorus compound Phosphorverbindung f
phosphorus-containing phosphorhaltig
phosphorus-halogen compound
  Phosphor-Halogen-Verbindung f
  containing phosphorus phosphorhaltig
photo Photo n, Foto n
photocatalytic photokatalytisch
photochemical photochemisch
photodegradation Photoabbau m, Photodegradation f
photoelastic spannungsoptisch
photoelasticity Spannungsoptik f
photoelectric cell Photozelle f
photograph Photographie f, Fotografie f
  X-ray diffraction photograph Röntgenbeugungsbild n
  X-ray photograph Röntgenaufnahme f
photolytic photolytisch
photomicrograph lichtmikroskopische Aufnahme f
  fracture photomicrograph Bruchbild n
photo-oxidation Photooxidation f
photosensitivity Photoempfindlichkeit f
phthalate Phthalat n, Phthalsäureester m
  phthalate plasticizer Phthalatweichmacher m
  phthalate resin Phthalatharz n
  aryl alkyl phthalate Arylalkylphthalat n
  benzyl butyl phthalate [BBP] Benzylbutylphthalat n
  butyl benzyl phthalate [BBP] Butylbenzylphthalat n
  di-2-ethylhexyl phthalate [DOP] Di-2-ethylhexyl-
    phthalat n
  diallyl phthalate Diallylphthalat n,
    Phthalsäurediallylester m
  diallyl phthalate mo(u)lding compound
    Diallylphthalatpreßmasse f
  dibutyl phthalate [DBP] Dibutylphthalat n,
    Phthalsäuredibutylester m
  dicyclohexyl phthalate Dicyclohexylphthalat n
  diethyl phthalate [DEP] Diethylphthalat n
  diethylhexyl phthalate [DOP] Phthalsäurediethyl-
    hexylester m
  diglycidyl phthalate Phthalsäurediglycidylester m
  di-isodecyl phthalate [DIDP] Diisodecylphthalat n
  di-isononyl phthalate [DINP] Diisononylphthalat n
  di-isooctyl phthalate [DIOP] Diisooctylphthalat n
  di-isotridecyl phthalate Diisotridecylphthalat n
  dimethoxyethyl phthalate Dimethoxyethylphthalat n
  dimethylcyclohexyl phthalate Dimethylcyclohexyl-
    phthalat n
  dimethyl glycol phthalate Dimethylglykolphthalat n
  dimethyl phthalate [DMP] Dimethylphthalat n
  dioctyl phthalate [DOP] Di-2-ethylhexylphthalat n,
    Dioctylphthalat n
  glyceryl phthalate Glycerylphthalat n
  lead phthalate Bleiphthalat n
  linear phthalate Linearphthalat n
  polydiallyl phthalate Polydiallylphthalat n
phthalic acid Phthalsäure f
  phthalic acid ester Phthalsäureester m
  phthalic anhydride Phthalsäureanhydrid n
phthalocyanine blue Phthalocyaninblau n
phthalocyanine green Phthalocyaningrün n
physical physikalisch
  physical drying physikalische Trocknung f
  physical form as supplied Lieferform f
physiologically inert physiologisch indifferent
pickle/to [e.g. metal in acid prior to bonding or
  painting] beizen, ätzen
pickling bath Ätzbad n
pickling solution Ätzflüssigkeit f, Ätzlösung f
PID characteristics PID-Verhalten n
  [Proportional-Integral-Differential]
piece Stück n
  piece of furniture Möbelstück n
  cut-to-size piece of sheet Plattenzuschnitt m
  cut-to-size pieces [of sheet stock, prepregs etc.]
    Zuschnitte mpl
  distance piece Abstandhalter m, Abstandstück n,
    Distanzbolzen m, Distanzhalter m, Distanzstück n
  distance pieces Dichtbacken fpl,
    Begrenzungsbacken fpl, Abdichtungsbacken fpl
  dumbbell(-shaped) tensile test piece Schulter-
    zugstab m
  dumbbell(-shaped) test piece Schulterprobe f,
    Schulterstab m, Hantelstab m
  flexural impact test piece Schlagbiegestab m
  flexural test piece Biegestab m
  pure-resin test piece Reinharzstab m
  removal of stress from the test piece Prüfkörper-
    entlastung f
  shrinkage test piece Schwindungsstab m
  small standard test piece Normkleinstab m
  square-section test piece Vierkantstab m
  standard dumbbell-shaped test piece Norm-
    Schulterprobe f
  standard tensile test piece Normzugstab m
  standard test piece Norm-Probekörper m,
    Normstab m
  tensile test piece Zugprobe f, Zugprobekörper m,
    Zugstab m
  test piece Probe f, Probekörper m, Probestab m,
    Prüfkörper m, Prüfling m, Prüfstab m, Prüfstück n
  test piece dimensions Probenabmessungen fpl,
    Probendimensionen fpl
  test piece insertion device Probeneinlegevorrichtung f

test piece preparation Probenaufbereitung f
test piece support Probenträger m
thermal expansion piece Wärmedehnbolzen m
piezo-electric effect Piezoeffekt m
piezoelectrical piezoelektrisch
piezoresistive piezoresistiv
pigment/to (ein)färben
pigment Pigment n, Pigmentfarbstoff m; Füllstoff m [Gummi]
   pigment accumulations Pigmentnester npl
   pigment agglomerate Pigmentagglomerat n
   pigment agglomerates Farbnester npl
   pigment binding powder Pigmentbindevermögen n
   pigment compatibility Pigmentverträglichkeit f
   pigment concentrate Farbkonzentrat n
   pigment content Farbanteil m, Pigmentierungshöhe f
   pigment dispersion Farbverteilung f, Pigmentverteilung f
   pigment enrichment Pigmentanreicherung f
   pigment grinding Pigmentvermahlen n
   pigment masterbatch Pigment-Kunststoffkonzentrat n
   pigment mixing unit Einfärbegerät n
   pigment particle Pigmentteilchen n, Pigmentkorn n
   pigment paste Farb(pigment)paste f, Pigmentanreibung f, Pigmentpaste f, Farbbatch n
   pigment slurry Pigmentaufschlämmung f
   pigment stabilizer Pigmentstabilisator m
   pigment strength Farbkraft f
   pigment volume concentration [p.v.c.] Pigmentvolumenkonzentration f
   pigment wetting power Pigmentaufnahmevermögen n
   anti-corrosive pigment Korrosionsschutzpigment n, Rostschutzpigment n
   azo pigment Azopigment n
   cadmium pigment Cadmiumpigment n
   chrome pigment Chromatpigment n
   colo(u)red pigment Farbpigmente npl, Buntpigmente npl
   fluorescent pigment Fluoreszenz-Pigment n
   lead chromate pigment Bleichromatpigment n
   lead pigment Bleipigment n
   powdered pigment Farbpulver n
   rutile pigment Rutilpigment n
   total pigment content Gesamtpigmentierung f
   type of pigment Pigmentierungsart f
   white pigment Weißpigment n
   with a low pigment content niedrigpigmentiert, schwachpigmentiert
pigmentability Einfärbbarkeit f
pigmentation Pigmentierung f, Einfärbung f
pigmented pigmentiert, eingefärbt
   cadmium pigmented cadmiumpigmentiert
   easily pigmented leicht pigmentierbar
   slightly pigmented schwachpigmentiert
pilot Versuchs...
   pilot light Kontrollampe f
   pilot plant Pilotanlage f, Technikum n

   pilot plant machine Technikumaschine f
   pilot plant production Vorserienfertigung f
   pilot plant run Nullserie f
   pilot plant-scale 1. kleintechnisch; 2. Technikummaßstab m, Versuchsmaßstab m
   pilot plant-scale machine Pilotmaschine f
   pilot plant trial Technikumversuch m
pin Stift m, Bolzen m, Nadel f
   pin bearing Nadellager n
   pin gate Punktanguß m, Punktanschnitt(kanal) m, punktförmiger Anschnittkanal m
   pin gate nozzle Punktangußdüse f, Punktanschnittdüse f
   pin gate sprue Punktangußkegel m
   pin inserts eingelegte Stifte mpl
   pin-lined barrel Stiftzylinder m
   pin roller fibrillator Nadelwalzenfibrillator m
   pin roller process Nadelwalzenverfahren n
   ante-chamber pin gate Vorkammerpunktanguß m, Punktanguß m mit Vorkammer, Stangenpunktanguß m
   antechamber-type pin gate Vorkammerpunktanguß m, Punktanguß m mit Vorkammer, Stangenpunktanguß m
   blowing pin Blasstift m, Blasnadel f
   central ejector pin Zentralausdrückstift m
   code pin Lesestift m
   cooling pin Kühlstift m
   core pin Bohrungskern m, Bohrungsstift m
   ejector pin Ausstoßstift m, Auswerferstift m, Auswerferstempel m
   ejector plate return pin Rückdrückstift m
   fixing pin Fixierstift m
   four-point pin gate Vierfach-Punktanschnitt m
   guide pin Führungssäule f, Führungsholm m, Führungsstift m
   knock-out pin Ausstoßstift m, Auswerferstempel m, Auswerferstift m
   multi-point pin gate Reihenpunktanschnitt m
   multiple pin gate Reihenpunktanschnitt m
   sensor pin Fühlerstift m
   single pin gate Einzelpunktanschnitt m
   sprue ejector pin Angußausdrückstift m
   sprue puller pin Angußziehstift m
   sprue puller pin gate Abreißpunktanschnitt m
   square-section pin Vierkantstift m
   stop pin Anschlagbolzen m
   tunnel gate with pin-point feed Tunnelpunktanguß m, Tunnelanguß m mit Punktanschnitt
   valve pin Düsennadel f
   vent(ing) pin Entlüftungsstift m
pinch Quetschung f, Kniff m
   pinch-off Schweißkanten fpl, Quetschkanten fpl, Schneidkanten fpl, Abquetschstelle f
   pinch-off area Quetschzone f
   pinch-off bars Schweißkanten fpl, Schneidkanten fpl
   pinch-off temperature Abquetschtemperatur f

pinch-off weld Quetschnaht f, Abquetschstelle f, Schweißnaht f, Abquetschmarkierung f
   neck pinch-off Halsquetschkante f
   shoulder pinch-off Schulterquetschkante f
   take-off pinch rolls Abzugs-Abquetschwalzenpaar n
pinholing [of a paint film] Nadelstichbildung f
pinion Ritzel n, Zahnritzel n
pipe Rohr n
   pipe bend Rohrbogen m, Rohrkrümmer m
   pipe bore Rohrinnenfläche f
   pipe calibrating unit Rohrkalibrierung f [Gerät]
   pipe calibration Rohrkalibrierung f
   pipe circumference Rohrumfang m
   pipe (compound) formulation Rohrrezeptur f
   pipe cutting device Rohrtrennvorrichtung f
   pipe die Rohr(spritz)kopf m, Rohrdüsenkopf m, Rohr(extrusions)werkzeug n
   pipe die design Rohrwerkzeugkonstruktion f
   pipe die head Rohr(spritz)kopf m
   pipe extrusion compound Rohrwerkstoff m, Rohrgranulat n, Rohrware f
   pipe extrusion head Rohr(spritz)kopf m
   pipe extrusion line Rohr(fertigungs)straße f, Rohrlinie f, Rohrherstellungsanlage f
   pipe fitting Rohrfitting n
   pipe joint Rohrverbindung f
   pipe production (line) Rohrfabrikation f
   pipe quality Rohrgüte f
   pipe scrap Ausschußrohre npl, Rohrabfälle mpl
   pipe scrap granulator Rohrschneidmühle f
   pipe section Rohrabschnitt m, Rohrzuschnitt m
   pipe sheathing die Rohrbeschichtungswerkzeug n
   pipe take-off (unit) Rohrabzug m, Rohrabzugswerk n
   pipe welding instrument Rohrschweißgerät n
   pipe welding unit Rohrschweißvorrichtung f
   pipe winding plant Rohrwickelanlage f
   automatic pipe cutter Rohrtrennautomat m
   centrifugally cast pipe Schleuderrohr n
   connecting pipe Verbindungsrohr n
   corrugated pipe extrusion line Wellrohranlage f
   device for measuring pipe wall thickness Rohrwanddicken-Meßanlage f
   domestic waste pipe Hausabflußrohr n
   drinking water pipe Trinkwasserrohr n
   effluent pipe [industrial] Abwasserrohr n [in der Industrie]
   feed pipe Einlaufrohr n, Zuleitungsrohr n
   filament wound pipe Wickelrohr n
   gas pipe Gasrohr n
   glass cloth laminate pipe Glashartgeweberohr n
   heat exchanger pipe Wärmetauschrohr n
   heat pipe Wärmerohr n
   heated tool pipe welding line Heizelement-Rohrschweißanlage f
   hexagonal-section pipe Sechskantrohr n
   high-speed pipe extrusion line Hochleistungs-Rohrextrusionslinie f
   inside pipe diameter Rohrinnendurchmesser m
   irrigation pipe Bewässerungsrohr n
   laminate pipe Schichtpreßstoffrohr n
   large-bore pipe Großrohr n
   large-bore pipe calibration (unit) Großrohrkalibrierung f
   large-bore pipe die Großrohrwerkzeug n
   large-bore pipe extrusion line Großrohrstraße f
   long-term internal pressure test for pipes Rohrinnendruckversuch m [Langzeitversuch]
   outside pipe diameter Rohraußendurchmesser m
   plate-type pipe die Siebkorbrohrkopf m
   pressure pipe Druckrohr n
   round-section pipe Rundrohr n
   small-bore pipe Kleinrohr n
   waste pipe Abflußrohr n
   waste pipe [domestic] Abwasserrohr n [Haushalte]
   water supply pipe Wasserrohr n
pipeline Rohrleitung f
   pipeline system Rohrleitungssystem n
   non-pressure pipeline Freispiegelleitung f
   pressure pipeline Druckrohrleitung f
piston Kolben m
   piston drive mechanism Kolbentriebwerk n
   piston pump Kolbenpumpe f
   piston ring Kolbenring m
   piston rod Kolbenstange f
   annular piston pump Kreiskolbenpumpe f
   axial piston motor Axialkolbenmotor m
   radial piston motor Radialkolbenmotor m
   radial piston pump Radialkolbenpumpe f
   rotary piston pump Sperrschieberpumpe f
pitch Steigung f, Ganghöhe f [Schraube], (gleichmäßiger) Abstand m
   decreasing pitch screw steigungsdegressive Schnecke f
   increasing pitch screw Progressivspindel f
   with decreasing pitch [screw] steigungsdegressiv
   with increasing pitch [screw] steigungsprogressiv
pitched roof Steildach n, Schrägdach n
plain language Klartext m
   plain language display Klartextanzeige f
   plain language print-out Klartextausdruck m
   plain language signal Klartextmeldung f
   plain weave Leinwandbindung f
plan Plan m
   production plan Fertigungsprogramm n
plane parallel planparallel
plane polished planpoliert
planet(ary) screw Planetschnecke f, Planetenspindel f
   planetary gear Planetengetriebe n
   planetary gear extruder Walzenextruder m, Planetwalzenextruder m
   planetary-gear section Planetwalzenteil n
   planetary mixer Planetenmischer m, Planetenmischkneter m, Planetenrührwerk n
planing Hobeln n

planing tool Schälwerkzeug n
planning Planung f
  planning stage Projektphase f, Fertigungsplanung f
  amount of planning (necessary) Planungsaufwand m
  production planning (department) Fertigungsplanung f
  work planning Arbeitsvorbereitung f
plant Betriebsanlage f, Werk n, Fertigungsstätte f, Fabrikationsstätte f, Produktionsstätte f, Anlage f, Werksanlagen fpl
  plant design and lay-out Anlagenkonzept n
  plant scrap Produktionsabfälle mpl
  batch-type plant Batchanlage f
  blown film plant Folienblasbetrieb m
  bottle blowing plant Flaschenblasanlage f
  cable reclaiming plant Kabelaufbereitungsanlage f
  casting plant Gießanlage f
  centrifugal casting plant Schleuderanlage f
  chemical plant construction Chemieanlagenbau m
  coextrusion blow mo(u)lding plant Coextrusions-Blasformanlage f, Coextrusionsblasanlage f
  complete plant Komplettanlage f
  compounding plant Compoundieranlage f
  continuous plant Kontianlage f
  cooling plant Kühlanlage f
  discontinuous plant Batchanlage f
  electroplating plant Galvanikanlage f
  experimental plant Versuchsanlage f
  extrusion blow mo(u)lding plant Extrusionsblas(form)anlage f
  film conversion plant Folienweiterverarbeitungsanlage f
  film laminating plant Folienkaschieranlage f
  film scrap reclaim plant Folienregranulieranlage f, Folienschnitzel-Granulieranlage f
  film scrap re-processing plant Folienaufbereitungsanlage f
  film stretching plant Folienreckanlage f
  film tape stretching plant Bändchenreckanlage f
  foam mo(u)lding plant Formschäumanlage f
  foaming plant Schäumanlage f
  heat laminating plant Thermokaschieranlage f
  high-capacity blown film plant Hochleistungsblasfolienanlage f
  high-capacity extrusion blow mo(u)lding plant Hochleistungsblasextrusionsanlage f
  high-capacity plant Hochleistungsanlage f
  high-pressure spraying plant Hochdruckspritzanlage f
  high-speed cable sheathing plant Hochleistungs-Kabelummantelungsanlage f
  high-speed plant Hochgeschwindigkeitsanlage f
  HM-HDPE blown film plant HM-Anlage f
  impregnating plant Tränkanlage f
  injection stretch blow mo(u)lding plant Spritzstreckblasanlage f
  laminating plant Laminieranlage f
  one-shift plant Einschichtenbetrieb m
  packaging plant Verpackungsanlage f
  pilot plant Pilotanlage f, Technikum n
  pilot plant machine Technikummaschine f
  pilot plant production Vorserienfertigung f
  pilot plant run Nullserie f
  pilot plant-scale 1. kleintechnisch; 2. Technikummaßstab m, Versuchsmaßstab m
  pilot plant-scale machine Pilotmaschine f
  pilot plant trial Technikumversuch m
  pipe winding plant Rohrwickelanlage f
  plastics coating plant Kunststoffbeschichtungsanlage f
  polymerization plant Polymerisationsanlage f
  raw material recovery plant Rohstoffrückgewinnungsanlage f
  reclaim plant Regenerieranlage f
  refuse sorting plant Müllaufbereitungsanlage f
  scrap reprocessing plant Abfallaufbereitungsanlage f
  sheathing plant Ummantelungsanlage f
  spray lay-up plant Faserharzspritzanlage f
  spread coating plant Streichanlage f
  spreading plant Streichanlage f
  steam vulcanizing plant Dampfvulkanisieranlage f
  stretch blow mo(u)lding plant Streckblasanlage f
  thermoforming sheet plant Tiefziehfolienanlage f
  transfer mo(u)lding plant Spritzpreßanlage f
  waste gas purification plant Abgasreinigungsanlage f
plasma Plasma n
  cut with a plasma arc plasmageschnitten
plaster Putz m
plastic 1. plastisch; 2. Plastik n, Kunststoff m, Chemiewerkstoff m
  plastic sheathed kunststoffummantelt
  barrier plastic Barriere-Kunststoff m
  commodity plastic Standardkunststoff m
  engineering plastic technischer Kunststoff m
  fluorocarbon plastic Fluorkunststoff m, Fluorkunstharz n
  glass fiber reinforced plastic [GRP] Glasfaserkunststoff m
  special plastic Spezialkunststoff m
  special-purpose plastic Spezialkunststoff m
plasticate/to plastifizieren
plasticating Plastifizier-
  plasticating chamber Plastifizierkammer f
  plasticating shaft Plastifizierschaft m
  plasticating unit Aufschmelzeinheit f
plastication Plastifiziervorgang m, Plastifizierungsverfahren n, Plastifizierung f
plastication temperature Plastifizierungstemperatur f
plasticator Plastifikator m, Plastifiziereinheit f, Plastifiziereinrichtung f, Plastifiziermaschine f, Plastifizieraggregat n
  Diskpack plasticator [tradename of a machine made by the Farrel machinery group] Scheibenplastifizieraggregat n, Scheibenextruder m

heavy-duty plasticator Hochleistungsplastifizieraggregat n
ultra-high pressure plasticator Höchstdruckplastifiziereinheit f
plasticized weichmacherhaltig, weichgemacht, Weich-
  plasticized polyvinyl chloride Polyvinylchlorid n weich, PVC n weich
  plasticized PVC Polyvinylchlorid n weich, PVC n weich
  plasticized (PVC) compound Weich-PVC-Masse f, Weichmischung f, Weich-Granulat n
  extrusion of plasticized PVC Weich-Extrusion f [PVC]
  injection mo(u)lding of plasticized PVC Weich-PVC-Spritzguß m
  internally plasticized innerlich weichgemacht
  processing in plasticized form [PVC] Weichverarbeitung f
plasticizer Weichmacher m, Weichmachungsmittei n
  plasticizer absorption Weichmacherabsorption f, Weichmacheraufnahme f
  plasticizer action Weichmacherwirkung f
  plasticizer blend Weichmachermischung f
  plasticizer concentration Weichmacherkonzentration f
  plasticizer content Weichmachergehalt m, Weichmacheranteil m
  plasticizer effect Weichmachereffekt m
  plasticizer efficiency Weichmacherwirksamkeit f
  plasticizer evaporation Weichmacherverdampfung f
  plasticizer extraction Weichmacherextraktion f
  plasticizer loss Weichmacherverlust m
  plasticizer migration Weichmachermigration f, Weichmacherwanderung f
  plasticizer migration resistance Weichmacherwanderungsbeständigkeit f
  plasticizer molecule Weichmachermolekül n
  plasticizer range Weichmachersortiment n
  plasticizer resistant weichmacherfest
  plasticizer vapo(u)r extractor Weichmacherdampfabsaugung f
  plasticizer vapo(u)rs Weichmacherdämpfe mpl
  plasticizer volatility Weichmacherflüchtigkeit f
  acting as a barrier against plasticizer migration weichmachersperrend
  additional plasticizer Zusatzweichmacher m
  amount of plasticizer Weichmacheranteil m, Weichmachermenge f
  effect of plasticizer Weichmachereinfluß m
  epoxy plasticizer Epoxidweichmacher m
  ester plasticizer Esterweichmacher m
  extender plasticizer Verschnittkomponente f, Verschnittmittel n [Weichmacher]
  monomeric plasticizer Monomerweichmacher m
  phosphate plasticizer Phosphat-Weichmacher m
  phthalate plasticizer Phthalatweichmacher m
  polymeric plasticizer Polymerweichmacher m
  primary plasticizer Primärweichmacher m
  processing with plasticizer Weichverarbeitung f
  processing without plasticizer Hartverarbeitung f
  range of plasticizers Weichmachersortiment n
  rate of plasticizer loss Weichmacher-Verlustrate f
  required amount of plasticizer Weichmacherbedarf m
  secondary plasticizer Sekundärweichmacher m, Zweitweichmacher m
  sole plasticizer Alleinweichmacher m
  special-purpose plasticizer Spezialweichmacher m
  speciality plasticizer Spezialweichmacher m
  standard plasticizer Standardweichmacher m
  total amount of plasticizer Gesamtweichmachermenge f
  type of plasticizer Weichmacherart f, Weichmachertyp m
  with a high plasticizer content weichmacherreich
  with a low plasticizer content weichmacherarm
plasticity Plastizität f, Verformungsvermögen n, Verformbarkeit f
  Mooney plasticity Mooney-Plastizität f
plasticization Weichmachung f, Plastifiziervorgang m, Plastifizierungsverfahren n, Plastifizierung f [s.a. plasticizing]
  at the start of plasticization bei Plastifizierbeginn m
  degree of plasticization Weichmachungsgrad m, Plastifiziergrad m
  end of plasticization Plastifizierende n
  external plasticization äußere Weichmachung f
  high-pressure plasticization Hochdruckplastifizierung f
  internal plasticization innere Weichmachung f
  partial plasticization Anplastifizieren n
  plunger plasticization Kolbenplastifizierung f
  pressure plasticization Druckplastifizierung f
  screw plasticization Schneckenplastifizierung f
  shear plasticization Scherplastifizierung f
  ultra-high pressure plasticization Höchstdruckplastifizierung f
  vented plasticization Entgasungsplastifizierung f
plasticize/to plastifizieren
plasticizing weichmachend, plastifizierend [s.a. plasticization]
  plasticizing aid Plastifizierhilfe f
  plasticizing capacity Plastifizierleistung f, Plastifizierkapazität f, Plastifiziermenge f, Verflüssigungsleistung f, Weichmachervermögen n, Plastifiziervolumen n
  plasticizing chamber Plastifizierkammer f
  plasticizing cylinder Plastifizierzylinder m
  plasticizing element Plastifizierorgan n
  plasticizing grooves Plastifizierkanäle mpl
  plasticizing performance Plastifizierverhalten n, Plastifiziervermögen n
  plasticizing process Plastifizierungsverfahren n, Plastifiziervorgang m
  plasticizing screw Plastifizierspindel f,

Plastifizierschnecke f
plasticizing section Plastifizierzone f
plasticizing system Plastifiziersystem n
plasticizing time Dosierzeit f, Plastifizierzeit f
plasticizing unit Plastifiziereinheit f, Plastifiziereinrichtung f, Plastifizieraggregat n
accumulator head plasticizing unit Staukopfplastifiziergerät n
high-pressure plasticizing (unit) Hochdruckplastifizierung f, Hochdruckplastifiziereinheit f
plunger plasticizing cylinder Kolbenplastifizierzylinder m
plunger plasticizing (unit) Kolbenplastifizierung f
reciprocating-screw plasticizing unit Schneckenschub-Plastifizieraggregat n
required plasticizing rate Plastifizierstrombedarf m
screw plasticizing cylinder Schneckenplastifizierungszylinder m
screw plasticizing unit Schneckenplastifiziereinheit f, Schneckenplastifizieraggregat n
standard plasticizing unit Normalplastifiziereinheit f
three-section plasticizing unit Dreizonen-Plastifiziereinheit f
vented plasticizing (unit) Entgasungsplastifiziereinheit f

plastic(s) Kunststoff m, Plast m
  plastics additive Kunststoffadditiv n
  plastics coating plant Kunststoffbeschichtungsanlage f
  plastics compounder Kunststoffaufbereitungsmaschine f
  plastics extruder Kunststoffschneckenpresse f
  plastics film Kunststoff-Folie f
  plastics foam Schaumkunststoff m
  plastics fuel tank Kunststofftank m [Öltank]
  plastics industry Kunststoffbranche f
  plastics injection mo(u)lding shop Kunststoffspritzerei f
  plastics material Kunststoffwerkstoff m
  plastics mo(u)lding Kunststoffteil n
  plastics mo(u)lding compound Kunststofformmasse f
  plastics pellets Kunststoffgranulat n
  plastics processing Kunststoffverarbeitung f
  plastics processing machine Kunststoffmaschine f
  plastics raw material Kunststoffrohstoff m
  plastics scrap Kunststoffabfälle mpl, Kunststoffausschuß m
  plastics technology Kunststofftechnologie f
  automatic device for cutting plastics profiles into lengths Kunststoffprofil-Ablängeautomat m
  bonded plastics joint Kunststoffverklebung f
  bonding of plastics Kunststoffverklebung f
  bulk plastics Konsumkunststoffe mpl, Massenkunststoffe mpl
plastisized [PVC] compound Weichcompound m,n
plastisol Plastisol n
  plastisol coat(ing) Plastisolschicht f
  PVC plastisol PVC-Plastisol n

plate Platte f, Scheibe f [s.a. plates]
  plate electrode Plattenelektrode f
  plate heat exchanger Plattenwärmeaustauscher m
  plate-out Belagbildung f
  back(ing) plate Stützplatte f, Rückplatte f
  baffle plate Umlenkblech n, Prallplatte f, Schikane f
  base plate Grundplatte f
  breaker plate Siebträgerscheibe f, Brecherplatte f, Extruderlochplatte f, Lochring m, Lochscheibe f, Sieblochplatte f, Siebstützplatte f, Stützlochplatte f, Stützplatte f
  breaker plate-type die Siebkorbkopf m
  breaker plate-type mandrel support Loch(scheiben)-dornhalter m, Sieb(korb)(dorn)halter m, Lochtragring m
  breaker plate-type pipe die Siebkorbrohrkopf m
  calibrating plate assembly Scheibenpaket n
  cavity plate düsenseitige Formplatte f, Formnestblock m, Formnestplatte f, Gesenk n, Gesenkplatte f, Konturplatte f
  cone and plate viscometer Kegel-Platte-Viskosimeter m
  core plate Patrize f, schließseitige Formplatte f, Stempelplatte f
  die plate Düsenteller m, Werkzeughalteplatte f, Siebträger m, Düsenplatte f
  draw plate assembly Blendenpaket n
  draw plate calibrator Blendenkalibrator m
  ejector plate (assembly) Auswerferteller m, Auswerfer(halte)platte f, Plattenauswerfer m
  ejector plate return pin Rückdrückstift m
  ejector plate safety mechanism Auswerferplattensicherung f
  ejector retaining plate Auswerfergrundplatte f
  feed plate Anguß(verteiler)platte f, Verteilerplatte f
  filter plate Filterplatte f
  guide plate Leitblech n
  hot plate Heizelement n
  hot plate welding Spiegel(schweiß)verfahren n
  hot runner plate Verteilerplatte f
  insert plate Einsatzplatte f
  insulating plate Temperierschutzplatte f
  mandrel holding plate Dornhalteplatte f
  master pattern plate Urmodellplatte f
  mo(u)ld plate Formplatte f, Werkzeugplatte f
  mo(u)ld plate assembly Plattenpaket n
  pattern plate Modellplatte f
  sinter plate extruder Schmelztellerextruder m
  sizing plate assembly Blendenpaket n
  slide plate Schieberplatte f
  sprue puller plate Abreißplatte f
  stop plate Anschlagplatte f
  stripper plate Abschiebeplatte f, Abstreifplatte f
  vacuum suction plate Vakuumsaugteller m
  wear plate Verschleißplatte f
plated (metall)überzogen, (galvanisch) beschichtet; plattiert [Textil]

chrome-plated verchromt
chromium-plated verchromt
high-polish chromium-plated hochglanzverchromt
platelet Plättchen n
  platelet-like blättchenförmig, blättrig
  platelet structure Blättchenstruktur f
  mica platelet Glimmerplättchen n
platen Platte f, Formenträger m, Formaufspannplatte f, Aufspannplatte f, Maschinenaufspannplatte f, Trägerplatte f, Werkzeugaufspannplatte f, Werkzeugträger m, Werkzeuggrundplatte f, Werkzeugträgerplatte f
  platen area Werkzeugaufspannfläche f, Formaufspannfläche f, Aufspannfläche f
  platen deflection Plattendurchbiegung f
  platen dimensions Werkzeuganschlußmaße npl, Werkzeugeinbaumaße npl, Werkzeugaufspannmaße npl, Einbaumaße npl
  platen heaters Plattenheizung f
  platen heating system Plattenheizung f
  platen press Plattenpresse f
  platen pressure Plattendruck m
  platen size Plattengröße f
  platen temperature Plattentemperatur f
  fixed platen feststehende Werkzeugaufspannplatte f, Werkzeugdüsenplatte f, spritzseitige Aufspannplatte f, Düsenplatte f
  moving platen auswerferseitige/bewegliche Werkzeugaufspannplatte f, schließseitige Aufspannplatte f, Werkzeugschließplatte f, bewegliche Werkzeugträgerseite f
  press platen Preßblech n, Preßplatte f
  standard platen details Werkzeugaufspannzeichnung f, Lochbild n, Aufspannplan m, Bohrbild n
  stationary platen feststehende Werkzeugaufspannplatte f, Werkzeugdüsenplatte f, spritzseitige Aufspannplatte f, Düsenplatte f
plates Platten fpl
  capacitor plates Kondensatorplatten fpl
  check plates Abdichtbacken fpl, Begrenzungsbacken fpl, Dichtbacken fpl
  draw plates (used to calibrate very thin tubing) Ziehblenden fpl, Ziehscheiben fpl
  sizing plates (used to calibrate very thin tubing) Ziehblenden fpl, Ziehscheiben fpl
platform Plattform f, Bühne f
  machine platform Bühne f
  operating platform Arbeitsbühne f
  weighing platform Wägeplattform f
platinum Platin n
  platinum catalyst Platinkatalysator m
  platinum catalyzed platinkatalysiert
plausibility check Plausibilitätskontrolle f
play Spiel n, Betriebsspiel n
pleated plissiert, gefaltet
pleating Plissierung f, Plissieren n
plotter Plotter m, Diagrammschreiber m

plug Stopfen m, Stecker m
  plug flow Pfropfenströmung f
  plug-in (ein)steckbar
  plug-in card Steckkarte f
  plug-in connection Steckanschluß m
  plug-in coupling Steckkupplung f
  connecting plug Anschlußstecker m
  floating plug Schleppstopfen m
plunger Druckstempel m, Kolben m, Tauchkolben m
  plunger advance speed Kolbenvorlaufgeschwindigkeit f
  plunger injection cylinder Kolbenspritzzylinder m
  plunger injection mechanism Kolbenspritzsystem n
  plunger injection mo(u)lding machine Kolbeninjektions-Spritzgießmaschine f, Kolben(spritzgieß)maschine f
  plunger injection system Kolbeneinspritzsystem n
  plunger injection unit Kolbenspritzeinheit f
  plunger movement Kolbenbewegung f
  plunger plasticization Kolbenplastifizierung f
  plunger plasticizing cylinder Kolbenplastifizierzylinder m
  plunger plasticizing unit Kolbenplastifizierung f [Gerät]
  plunger stroke-dependent kolbenwegabhängig
  depending on the plunger stroke kolbenwegabhängig
  injection plunger Spritzkolben m, Einspritzkolben m
  transfer plunger Spritzkolben m, Einspritzkolben m
plywood Sperrholz n
p-methyl styrene Paramethylstyrol n
PMMA Polymethylmethacrylat n
pneumatic pneumatisch, Druckluft-
  pneumatic control valve Luftsteuerventil n
  pneumatic conveyor Druckluftfördergerät n
  pneumatic cylinder Pneumatikzylinder m
  pneumatic ejector Druckluftauswerfer m, Luftauswerfer m
  pneumatic suction system Saugpneumatik f
pneumatically operated druckluftbetrieben
processing conditions verfahrenstechnische Bedingungen fpl
pocket Tasche f
  pocket calculator Taschenrechner m
  pocket-size Taschenformat n
  feed pocket Einzugstasche f
point Punkt m, Stelle f, Spitze f
  point angle Spitzenwinkel m
  point of impact Auftreffstelle f, Auftreffpunkt m
  point of intersection Kreuzungspunkt m, Schnittpunkt m
  point of view Standpunkt m, Gesichtspunkt m
  boiling point Siedepunkt m, Siedetemperatur f
  branch point Verzweigungsstelle f
  change-over point Umschaltpunkt m, Umschaltschwelle f, Umschaltniveau n, Umschaltzeitpunkt m
  cloud point Trübungspunkt m

contact point Kontaktstelle f
crosslink point Vernetzungsstelle f
crystallite melting point Kristallitschmelztemperatur f
culminating point Kulminationspunkt m
dew point Taupunkt m, Taupunkttemperatur f
feed point Einspeisepunkt m, Einspeisestelle f, Einspritzpunkt m, Anschnittstelle f, Anschnittpunkt m, Angußpunkt m
fixed point Fixpunkt m
flash point Flammpunkt m
floating decimal point Fließkomma n
four point toggle Vierpunktkniehebel m
freezing point Erstarrungspunkt m, Gefrierpunkt m
from the energy point of view energetisch (betrachtet)
fusion point Geliertemperatur f
gate point Anspritzpunkt m, Anspritzstelle f
gate sealing point Siegelzeitpunkt m
gel point [of UP, EO resins] Geliertemperatur f, Gelierpunkt m
ignition point Entzündungspunkt m, Zündpunkt m, Zündtemperatur f
important from the processing point of view verfahrenstechnisch wesentlich
injection point Anspritzpunkt m, Anschnittpunkt m, Einspritzpunkt m
interesting from the processing point of view verfahrenstechnisch interessant
low-temperature brittleness point Kältesprödigkeitspunkt m
lubricating points Schmierstellen fpl
melting point Schmelzpunkt m, Schmelztemperatur f
nozzle point Düsenspitze f
operating point Arbeitspunkt m
phase reversal point Phasenumschlagpunkt m
pour point Pourpoint m
pressure measuring point Druckmeßstelle f
salient points Besonderheiten fpl
softening point Erweichungspunkt m, Erweichungstemperatur f, Plastifizierungstemperatur f
speed change-over point Geschwindigkeitsumschaltpunkt m
switching point Schaltpunkt m
temperature measuring point Temperaturmeßstelle f
transition point Umwandlungspunkt m
tunnel gate with pin point feed Tunnelanguß m mit Punktanschnitt
Vicat softening point Formbeständigkeit f in der Wärme mit Vicatnadel, Vicat-Erweichungspunkt m, Vicat-Erweichungstemperatur f, Vicat-Wärmeformbeständigkeit f, Vicatwert m, Vicatzahl f, Wärmeformbeständigkeit f nach Vicat
weakened points Sollbruchstellen fpl
with a low fusion point [PVC] niedriggelierend
yield point Fließgrenze f, Streckgrenze f
zero point monitor Nullpunktüberwachung f
polar polar

polarity Polarität f
polarizability Polarisierbarkeit f
polarizable polarisierbar
polarization Polarisierung f, Polarisation f
polarized polarisiert
polarizing microscope Polarisationsmikroskop n
pole-reversing polumschaltbar
policy Politik f, Strategie f
  company policy Firmenstrategie f
polish Politur f, Glanz m
  surface polish Oberflächenglanz m
polished poliert
  polished to a mirror finish spiegelhochglanzpoliert
  capable of being polished polierfähig
  highly polished feinstpoliert
  plane polished planpoliert
  with a highly polished surface oberflächenpoliert
polisher Poliergerät n
  abrasive belt polisher Bandschleifpoliermaschine f
  flame polisher Flammpoliergerät n
polishing Polieren n
  polishing and take-off unit Glätt- und Abziehmaschine f
  polishing calender Glättkalander m
  polishing nip Glättspalt m
  polishing roll Glättwalze f
  polishing stack Glätteinheit f, Glättwalzen fpl, Glättwerk n
  polishing wheel Polierscheibe f
  flame polishing Flammpolieren n
  four-point polishing stack Vierwalzen-Glättwerk n
  lambswool polishing wheel Lammfellscheibe f
  sheet polishing unit Plattenglättanlage f
  three-roll polishing stack Dreiwalzen-Glättwerk n, Dreiwalzen-Glättkalander m
  twin-roll polishing stack Zweiwalzenglättwerk n
polluted verschmutzt
pollution Verschmutzung f
  pollution control Umweltschutz m
  pollution control equipment Umweltschutzanlage f
  pollution control measures Umweltschutzmaßnahmen fpl
  pollution control regulations Umweltschutzverordnungen fpl
  pollution of the enviroment Umweltverschmutzung f
  atmospheric pollution Luftverunreinigung f, Luftverschmutzung f
poly-4-methylpentene-1 Poly-4-methylpenten-1 n
polyacetal [POM] Polyacetal n, Polyoxymethylen n
polyacrylamide Polyacrylamid n
polyacrylate Polyacrylat n, Polyacryl(säure)ester m
  polyacrylate rubber Acrylatkautschuk m
polyacrylic acid Polyacrylsäure f
polyacrylonitrile Polyacrylnitril n
polyaddition Polyaddition f
  polyaddition product Polyadditionsprodukt n
  polyaddition reaction Polyadditionsreaktion f

polyaddition resin Polyadditionsharz n
polyadduct Polyaddukt n
polyalkylene terephthalate Polyalkylenterephthalat n
polyamide [PA] Polyamid n
  polyamide imide Polyamidimid n
  polyamide resin Polyamidharz n
  cast polyamide Gußpolyamid n
polyamine Polyamin n
  polyamine hardener Polyaminhärter m
polyaminoamide Polyaminoamid n
  polyaminoamide hardener Polyaminoamidhärter m
polyaryl ether Polyarylether m
polyaryl sulphone Polyarylsulfon n
polyblend Polymer(isat)gemisch n, Polymerlegierung f, Polymermischung f, Kunststoff-Legierung f
polybutadiene Polybutadien n
polybutene Polybuten n
polybutyl titanate Polybutyltitanat n, Polytitansäurebutylester m
polybutylene terephthalate [PBTP] Polybutylen-terephthalat n
polycaprolactam Polycaprolactam n
polycarbonate [PC] Polycarbonat n
polycarboxylic acid Polycarbonsäure f
  polycarboxylic acid anhydride Polycarbon-säureanhydrid n
polychlorobutadiene Polychlorbutadien n
polychloroprene Polychloropren n
  polychloroprene rubber Polychloroprenkautschuk m
polycondensate Polykondensat n
polycondensation Polykondensation f
  polycondensation product Polykondensations-produkt n
  polycondensation resin Polykondensationsharz n
polycyclic polyzyklisch
polydialkyl siloxane chain Polydialkylsiloxankette f
polydiallyl phthalate Polydiallylphthalat n
polydimethyl siloxane Polydimethylsiloxan n
polyene Polyen n
polyepoxide Polyepoxid n
polyester Polyester m
  polyester adhesive [unsaturated polyester] UP-Harzkleber m
  polyester bulk mo(u)lding compound [polyester BMC] Feuchtpolyester m
  polyester concrete Polyesterbeton m
  polyester DMC Feuchtpolyester m
  polyester dough mo(u)lding compound Feuchtpolyester m
  polyester formulation Polyesterrezeptur f
  polyester glycol Polyesterglykol n
  polyester imide Polyesterimid n
  polyester mix Polyesteransatz m
  polyester mo(u)lding compound Polyesterharz-Formmasse f, Polyesterpreßmasse f
  polyester prepreg Polyesterharzmatte f
  polyester resin [UP resin] Polyesterharz n, UP-Harz n
  polyester resin mix UP-Reaktionsharzmasse f
  polyester sheet mo(u)lding compound Polyesterharzmatte f
  polyester surface coating resin Lackpolyester m
  polyester unit Polyesterbaustein m
  catalyzed polyester resin UP-Reaktionsharzmasse f
  cured polyester resin Polyesterharzformstoff m, UP-Harzformstoff m
  glass mat reinforced polyester laminate GF-UP-Mattenlaminat n
polyether Polyether m
  polyether amide Polyetheramid n
  polyether ester Polyetherester m
  polyether glycol Polyetherglykol n
  polyether imide Polyetherimid n
  polyether polyol Polyetherpolyol n
  polyether sulphone Polyethersulfon n
  polyether urethane Polyetherurethan n
polyethylene Polyethylen n, PE n
  polyethylene bag PE-Beutel m
  polyethylene chain Polyethylenkette f
  polyethylene film Polyethylenfolie f
  polyethylene foam Polyethylenschaumstoff m, Zellpolyethylen n
  polyethylene glycol Polyethylenglykol n
  polyethylene liner PE-Innensack m
  polyethylene masterbatch PE-Farbkonzentrat n
  polyethylene mo(u)lding compound Polyethylen-Formmasse f
  polyethylene oxide Polyethylenoxid n
  polyethylene terephthalate [PETP] Polyethylen-terephthalat n
  polyethylene wax Polyethylenwachs n
  blown polyethylene film Polyethylenschlauchfolie f
  cellular polyethylene Zellpolyethylen n
  chlorinated polyethylene Chlorpolyethylen n
  high density polyethylene [HDPE] Niederdruck-Polyethylen n, PE n hart, Polyethylen n hart, Polyethylen n hoher Dichte
  high-molecular weight polyethylene film HM-Folie f
  low-density polyethylene [LDPE] Hochdruck-Polyethylen n, PE n weich, Polyethylen n niedriger Dichte, Polyethylen n weich
polyfunctional polyfunktionell
polyglycol terephthalate Polyglycolterephthal-säureester m
polyhydroxybenzoate Polyhydroxybenzoat n
polyhydroxyl compound Polyhydroxyverbindung f
polyimide Polyimid n
polyisobutylene Polyisobutylen n
polyisocyanate Polyisocyanat n
polyisocyanurate Polyisocyanurat n
polyisoprene Polyisopren n
polymer Polymer(isat) n
  polymer alloy Kunststoff-Legierung f
  polymer blend Kunststoff-Legierung f, Polymerisatgemisch n, Polymergemisch n,

Polymerlegierung f, Polymermischung f
polymer chain Polymerkette f
polymer concentration Polymerkonzentration f
polymer concrete Kunstharzbeton m, Polymerbeton m
polymer damage Polymerenschädigung f
polymer degradation Kunststoffabbau m
polymer dispersion Kunststoffdispersion f, Polymerdispersion f
polymer material Polymerwerkstoff m
polymer matrix Polymermatrix f, Kunststoff-Matrix f
polymer melt Kunststoffschmelze f, Polymer(isat)schmelze f
polymer melt filtration Kunststoffiltration f
polymer modified polymermodifiziert
polymer molecule Kunststoffmolekül n, Polymermolekül n
polymer mortar Polymermörtel m
polymer particle Kunststoffteilchen n, Polymerteilchen n
polymer phase Polymerphase f
polymer radical Polymerradikal n
polymer residues Kunststoffrückstände mpl
polymer solution Kunststofflösung f, Polymer(isat)lösung f
polymer structure Polymerstruktur f
acrylic polymer Acrylpolymer n
addition polymer Additionspolymer n
bulk polymer Blockpolymerisat n, Massepolymerisat n, Schmelzpolymerisat n
condensation polymer Kondensationspolymer n
emulsion polymer Emulsions(homo)polymerisat n
extender polymer Fremdharz n
fluorocarbon polymer Fluorpolymer n
graft polymer Pfropfpolymer(isat) n
ladder polymer Leiterpolymer n
main polymer chain Polymerhauptkette f
paste-making polymer Pastenpolymerisat n
paste polymer Pastenpolymerisat n
pure polymer Reinpolymerisat n
sequential polymer Sequenzpolymer n, Sequenztyp m
solution polymer Lösungspolymerisat n
styrene polymer Styrol-Polymerisat n
suspension polymer Suspensionspolymerisat n, Suspensionstype f
vinyl chloride polymer Vinylchloridpolymerisat n
virgin polymer Kunststoffneuware f
when the polymer starts to melt bei Plastifizierbeginn m
polymeric polymer
   polymeric plasticizer Polymerweichmacher m
polymerize/to polymerisieren
polymerized polymerisiert
   graft polymerized pfropfpolymerisiert
   solution polymerized lösungspolymerisiert
polymerizable polymerisierbar, polymerisationsfähig
polymerization Polymerisation f
   polymerization aid Polymerisationshilfsmittel n

polymerization conversion Polymerisationsumsatz m
polymerization initiator Polymerisationsstarter m, Polymerisations-Initiator m
polymerization plant Polymerisationsanlage f
polymerization process Polymerisationsverfahren n, Polymerisationsverlauf m
polymerization rate Polymerisationsgeschwindigkeit f
polymerization-retarding polymerisationsverzögernd
polymerization temperature Polymerisationstemperatur f
polymerization termination Polymerisationsabbruch m
bead polymerization Perlpolymerisation f
bulk polymerization Blockpolymerisation f, Massepolymerisation f, Substanzpolymerisation f, Massepolymerisationsverfahren n
condensation polymerization Kondensationspolymerisation f
degree of polymerization Polymerisationsgrad m
emulsion polymerization Emulsionsverfahren n, Emulsionspolymerisation f
emulsion polymerization process Emulsionspolymerisationsverfahren n, E-Polymerisationsverfahren n
free-radical (chain growth) polymerization Radikal(ketten)polymerisation f
graft polymerization Pfropfpolymerisation f
heat of polymerization Polymerisationswärme f
high-pressure polymerization Hochdruckpolymerisation f
low-pressure polymerization Niederdruckpolymerisation f
method of polymerization Polymerisationstechnik f
precipitation polymerization Fällungspolymerisation f
rate of polymerization Polymerisationsgeschwindigkeit f
solution polymerization Lösungspolymerisation f
suspension polymerization Suspensionspolymerisation f
suspension polymerization process Suspensionspolymerisationsverfahren n, S-Polymerisationsverfahren n
polymerizing conditions Polymerisationsbedingungen fpl
polymethacrylate Polymethacrylat n, Polymethacrylsäureester m
polymethyl methacrylate [PMMA] Polymethylmethacrylat n
polymethyl styrene Polymethylstyrol n
polymethylene bridge Polymethylenbrücke f
polymethylene urea Polymethylenharnstoff m
polynomial theorem Polynomansatz m
polyol Polyol n
   polyol-cured polyolgehärtet
   polyol hardener Polyolhärter m
   polyether polyol Polyetherpolyol n
polyolefin Polyolefin n

polyolefin mo(u)lding compound
Polyolefin-Formmasse f
polyolefin rubber Polyolefinkautschuk m
polyoxymethylene [POM] Polyoxymethylen n
polyphenylene oxide Polyphenylenoxid n
polyphenylene sulphide Polyphenylensulfid n
polyphosphoric acid Polyphosphorsäure f
polypropylene [PP] Polypropylen n
  polypropylene glycol Polypropylenglykol n
  polypropylene oxide Polypropylenoxid n
  polypropylene wax Polypropylenwachs n
  blow mo(u)lding of oriented polypropylene OPP-Verfahren n
  oriented polypropylene blow mo(u)lding machine OPP-Maschine f
polysiloxane Polysiloxan n
  dimethyl polysiloxane Dimethylpolysiloxan n
  phenylmethyl vinyl polysiloxane Phenylmethyl-vinylpolysiloxan n
polystyrene Polystyrol n, PS n
  polystyrene granules Polystyrol-Granulat n
  polystyrene mo(u)lding compound Polystyrol-Formmasse f
  expanded polystyrene Polystyrolschaum(stoff) m, Schaumpolystyrol n, Polystyrol-Hartschaum m
  expanded polystyrene pack Polystyrolschaumstoff-Verpackung f
  expanded polystyrene production line Polystyrol-Schaumanlage f
  single-layer expanded polystyrene container Einschicht-PS-Schaumhohlkörper m
  standard polystyrene Normalpolystyrol n, Standardpolystyrol n
polysulphide resin Polysulfidharz n
polysulphide rubber Polysulfidkautschuk m, Thiokol n
polysulphone Polysulfon n
polyterephthalate Polyterephthalat n
polytetrafluoroethylene [PTFE] Polytetrafluorethylen n
polytetramethylene glycol Polytetramethylenglykol n
polythene [not used in the USA] Polyethylen n
polytrifluorochloroethylene Polytrifluorchlorethylen n
polytrifluoroethylene Polytrifluorethylen n
polyurea Polyharnstoff m
polyurethane [PU] Polyurethan n
  polyurethane foam Polyurethanschaum(stoff) m
  polyurethane integral foam Polyurethan-Integralschaum m, Polyurethan-Strukturschaumstoff m
  polyurethane resin Polyurethanharz n
  rigid polyurethane foam Hart-Urethanschaum m
polyvinyl Polyvinyl-
  polyvinyl acetal Polyvinylacetal n
  polyvinyl alcohol Polyvinylalkohol m
  polyvinyl butyral Polyvinylbutyral n
  polyvinyl carbazole Polyvinylcarbazol n
  polyvinyl chloride [PVC] Polyvinylchlorid n
  polyvinyl ether Polyvinylether m
  polyvinyl fluoride Polyvinylfluorid n
  polyvinyl formal Polyvinylformal n
  plasticized polyvinyl chloride [PVC] Polyvinylchlorid n weich
  rigid polyvinyl chloride Polyvinylchlorid n hart
polyvinylidene chloride Polyvinylidenchlorid n
polyvinylidene fluoride [PVDF] Polyvinylidenfluorid n
POM Polyacetal n, Polyoxymethylen n
poor schlecht, arm, mangelhaft, mager
  poor adhesion Haftungsmängel mpl
  poor-flow schwerfließend, hartfließend
  having poor solubility schlecht löslich
  with poor flow schwerfließend, hartfließend
porcelain Porzellan n
pore Pore f
  pore detector Lochsuchgerät n, Porensuchgerät n
porosity Blasigkeit f, Porosität f
  particle porosity Kornporosität f
porous porös
port 1. Hafen m; 2. Öffnung f, Eingang m
  inlet port [e.g. for cooling water] Einlaufstutzen m
  water inlet port Wasserzugabestutzen m
portable tragbar
position Position f, Stellung f
  position control circuit Lageregelkreis m
  position-controlled lagegeregelt
  position feedback Lagerückmeldung f
  position of the gate Anschnittlage f
  position sensor Stellungsgeber m, Wegaufnehmer m, Wegmeßgeber m
  position transducer Stellungsgeber m, Wegaufnehmer m, Wegmeßgeber m
  blowing position Blasposition f
  change in position Positionsänderung f
  closed position Schließstellung f
  ejector end position Auswerferendstellung f
  final position Endposition f
  in position [part of a machine] eingeschwenkt
  initial position Anfangsposition f
  key position Schlüsselstellung f
  move into position/to einfahren
  normal position Grundstellung f
  open position Öffnungsstellung f
  operating position Arbeitsstellung f
  original position Ausgangsposition f, Ausgangsstellung f
  screw position Schneckenstellung f
  start-up position Anfahrstellung f
  zero position Nullstellung f
positioning Positionierung f
  positioning of the gate Angußpositionierung f
  accurate positioning Positioniergenauigkeit f
positive positiv; formschlüssig [Verbindung]
  positive displacement circulating pump Umlaufverdrängerpumpe f
  positive displacement pump Verdrängerpumpe f
  positive displacement vacuum pump Verdränger-

Vakuumpumpe f
positive mo(u)ld Füll(raum)werkzeug n, Tauchkantenwerkzeug n
positive template Positivschablone f
rotary positive displacement pump Rotationsverdrängerpumpe f
possibility Möglichkeit f
  possibility of adjusting Verstellmöglichkeit f
  possibility of adjusting the die lips Lippenverstellmöglichkeit f
  application possibilities Einsatzmöglichkeiten fpl, Anwendungsmöglichkeiten fpl
  construction possibilities Konstruktionsmöglichkeiten fpl
  design possibilities Gestaltungsmöglichkeiten fpl, Konstruktionsmöglichkeiten fpl
  processing possibilities verfahrenstechnische Möglichkeiten fpl, Verarbeitungsmöglichkeiten fpl
  substitution possibilities Substitutionsmöglichkeiten fpl
possible linkages Verknüpfungsmöglichkeiten fpl
possible uses Einsatzmöglichkeiten fpl
post-chlorinated nachchloriert
post-condensation Nachkondensation f
post-cooling section Nachkühlstrecke f
post-crystallization Nachkristallisation f
post-cure Nachhärtung f, Nachvernetzung f, Temperung f
post-curing Nachhärtung f, Nachvernetzung f, Temperung f
post-curing conditions Nachhärtungsbedingungen fpl
post-curing temperature Nachhärtungstemperatur f
post-expansion [of foam] Nachblähen n
post-extrusion equipment Extrudernachfolge(maschine) f
post insulator Stützisolator m
post-polymerization Nachpolymerisation f
post-shrinkage Nachschwund m, Nachschwindung f
pot Topf m
  pot life Gebrauchsdauer f, Topfzeit f, Verarbeitungsperiode f, Verarbeitungsspielraum m, Verarbeitungszeit f [bei Mehrkomponentenlacken]
  having a pot life of... verarbeitungsfähig (bis)...
  required pot life Verarbeitungsbedarf m
potash mica Muscovit-Glimmer m
potassium Kalium n
  potassium dichromate Kaliumdichromat n
  potassium tetrafluoroborate Kaliumtetrafluorborat n
potentially explosive explosionsgefährlich
potentiometer Potentiometer n
potentiometric potentiometrisch
pour/to schütten, gießen
  pour point Fließpunkt m, Pourpoint m [Öl]
pourable fließfähig, gießbar, gießfähig, schüttbar
pouring Eingießen n, Schütten n
  pouring hole Eingußloch n
  fluidized pouring Wirbelschütten n
powder Pulver n, Puder m

  powder coating Pulverbeschichtung f, Pulverlack m
  mineral powder Gesteinsmehl n
  redispersible powder Dispersionspulver n
  slate powder Schiefermehl n
powdered 1. pulverförmig, Pulver-; 2. -pulver n
  powdered filler Pulverfüllstoff m
  powdered graphite Graphitpulver n
  powdered paint Pulverfarbe f
  powdered phenolic resin Phenolpulverharz n
  powdered pigment Farbpulver n
  powdered resin Pulverharz n
  finely powdered feinpulverig
power Kraft f, Leistung f, Energie f, Kapazität f
  power cable Starkstromkabel n
  power consumption Leistungsbedarf m
  power failure Spannungsausfall m, Stromausfall m
  power law Potenzansatz m, Potenz(fließ)gesetz n
  power law behavio(u)r/characteristics Potenzgesetzverhalten n
  power law fluid Potenzgesetz-Flüssigkeit f, Potenzgesetzstoff m
  power law index Potenzgesetzexponent m
  power loss Verlustleistung f
  power output Leistungsabgabe f
  power requirement(s) Kraftbedarf m, Leistungsbedarf m
  power reserve [e.g. of a timer or time switch] Gangreserve f
  power reserves Leistungsreserven fpl
  power saving leistungssparend
  power supply Stromversorgung f, Stromzufuhr f
  power supply cable Zuführungskabel n
  power switch Leistungsschalter m
  power tool Elektrowerkzeug n
  power transformer Leistungstransformator m
  power used aufgenommene Leistung f
  adhesive power Klebekraft f, Klebevermögen n
  binding power Bindekraft f
  breakdown of the power supply Stromausfall m
  continuous power consumption Dauerleistungsbedarf m
  drive power Antriebsleistung f
  drop in power Leistungsabfall m
  earning power Ertragskraft f
  effective motor power Motorwirkleistung f
  effective power aufgenommene Wirkleistung f, aufgenommene Leistung f, Nutzleistung f
  emergency power supply unit Notstromaggregat n
  gelling power Geliervermögen n
  hiding power [of a pigment] Deckkraft f, Deckfähigkeit f, Deckvermögen n, Deckungsgrad m
  motor power Motorleistung f
  no-load power Leerlaufleistung f
  penetrating power Eindringvermögen n
  pigment binding power Pigmentbindevermögen n
  pigment wetting power Pigmentaufnahmevermögen n
  pump drive power Pumpenantriebsleistung f

resolving power Auflösungsvermögen n
screw drive power Schneckenantriebsleistung f
semi-conductor power switch Halbleiterleistungsschalter m
solvating power Gelierkraft f, Gelierfähigkeit f,
Löseeigenschaften fpl, Lösefähigkeit f,
Lösevermögen n
spreading power Streichvermögen n, Ausgiebigkeit f
[Farbe]
swelling power Quellvermögen n
total drive power Gesamtantriebsleistung f
total installed motor power gesamte installierte
Motorenleistung f
total power Gesamtleistung f
total power consumption Gesamtleistungsbedarf m
powered [e.g. a machine] angetrieben
powered drum Antriebstrommel f
battery-powered mit Batteriestromversorgung f
mains-powered mit Netzstromversorgung f
powerful leistungsfähig, leistungsstark
PP Polypropylen n
PPO Polyphenylenoxid n
practical praktisch, nützlich, verwendbar, Praxis-
practical conditions Praxisbedingungen fpl
practical experience Praxiserfahrung f
practical importance Praxisrelevanz f
practical test Praxistest m, Praxisversuch m
meeting practical requirements praxisgerecht
Prandtl number Prandtl-Zahl f
pre-accelerated vorbeschleunigt
pre-blowing Vorblasen n
pre-blowing device Vorblaseinrichtung f
pre-condensate Vorkondensat n
pre-condensation Vorkondensation f
pre-condensed vorkondensiert
pre-cooked meal Fertiggericht n
pre-cooked food Fertiggericht n
pre-dispersed vordispergiert
pre-dispersion Vordispergierung f
pre-drying Vortrocknung f
pre-drying unit Vortrockengerät n
pre-expanded vorgeschäumt, vorexpandiert
pre-expanded beads Vorschaumperlen fpl
pre-expanding Vorblähen n, Vorschäumvorgang m,
Vorblasen n
pre-expansion Vorblähen n, Vorschäumvorgang m,
Vorblasen n
pre-fabricated vorgefertigt
pre-foamed vorexpandiert, vorgeschäumt
pre-foaming (process) Vorschäumprozeß m
pre-foaming unit Vorschäumer m
pre-forming Vorformeinheit f
pre-gelation temperature Vorgeliertemperatur f
pre-gelled vorgeliert
pre-gelling tunnel Vorgelierkanal m
pre-heat/to vortemperieren, vorwärmen
pre-heated vorgeheizt

pre-heating Vorheizung f, Vorwärmung f
pre-heating oven Vorwärmofen m
pre-heating period Vorwärmzeit f
pre-heating rolls Vorwärmwalzwerk n
pre-heating unit Vorwärmgerät n
pre-homogenize/to vorhomogenisieren
pre-mix/to vormischen
pre-mixed vorgemischt
pre-mixing unit Vormischer m
pre-plasticized vorplastifiziert
pre-plasticization Vorplastifizierung f
pre-plasticizing screw Vorplastifizierschnecke f
pre-plasticizing system Vorplastifizierungssystem n
pre-plasticizing unit Vorplastifizierungsaggregat n,
Vorplastifizierung f
pre-polymer Präpolymer n, Prepolymer n, Vorpolymer n
pre-production trial Nullserie f
pre-programmed vorprogrammiert
pre-selectable vorwählbar
pre-selection Vorwahl f
time pre-selection switch Zeitvorwahlschalter m
pre-selector switch Vorwahlschalter m
pre-set/to voreinstellen, vorgeben
pre-set processing conditions Sollverarbeitungsbedingungen fpl
pre-stabilization Vorstabilisierung f
pre-stabilized vorstabilisiert
pre-stressing Vorspannung f
tie bar pre-stressing Säulenvorspannung f
pre-stretched vorgereckt
pre-treated vorbehandelt
pre-treating unit [e.g. for film] Vorbehandlungsgerät n
pre-treatment Vorbehandlung f
precautions Maßnahmen fpl, Vorkehrungen fpl,
Sicherheitsmaßnahmen fpl
precautions against inhalation of dust Staubschutz-
Vorsichtsmaßnahmen fpl
fire precautions Brandschutzmaßnahmen fpl
safety precautions Sicherheitsmaßnahmen fpl,
Sicherheitsvorkehrungen fpl
preceding process Vorprozeß m
precipitate/to ausfällen
precipitated (aus)gefällt, präzipitiert
precipitated barium sulphate Blanc fixe n
precipitating agent Fäll(ungs)mittel n
precipitation Fällung f
precipitation conditions Fällungsparameter mpl
precipitation polymerization Fällungspolymerisation f
precisely repeatable/reproducible reproduziergenau
precision feinmechanisch, feinwerktechnisch,
Genauigkeit f, Präzision f
precision balance Feinwaage f, Genauigkeitswaage f
precision determination Feinmessung f
precision die Präzisionswerkzeug n
precision engineering Feinmechanik f,
Feinwerktechnik f

precision filament winding Präzisionswickeltechnik f
precision granulator Präzisionsschneidmühle f
precision grinding Präzisionsschleifen n
precision ground feingeschliffen
precision injection mo(u)lded part Präzisionsspritzgußteil n
precision injection mo(u)lding Qualitätsspritzguß m, Präzisionsspritzgußteil n, Präzisionsspritzguß m, Präzisionsspritzgießverarbeitung f
precision manometer Feinmanometer n
precision measurement Feinmessung f
precision metering unit Präzisionsdosiereinheit f
precision mo(u)ld Präzisionsform f, Präzisionswerkzeug n
precision mo(u)lded part Präzisionsteil n
precision mo(u)lding Präzisionsteilfertigung f, Präzisionsteil n
precision standard unit Präzisionsnormteil n
precise repeatability/reproducibility Repetitionsgenauigkeit f, Reproduziergenauigkeit f
prediction Vorhersagen n
predominant particle size häufigste Korngröße f
prefab(ricated) vorfabriziert, Fertig(teil)-
 prefabricated roof liner Fertighimmel m
preferred colo(u)r Vorzugsfarbe f
preform Vorpreßling m, Vorformling m, Rohling m
preheat/to vorheizen
preheating Vorheizen n
 preheating drum Vorheiztrommel f
 preheating oven Vorheizofen m
preliminary vorbereitend, einleitend, vorläufig, Vor-
 preliminary granulation Vorzerkleinern n
 preliminary granulator Vorzerkleinerungsmühle f, Vorzerkleinerer m
 preliminary size reduction Vorzerkleinern n
 preliminary size reduction unit Vorzerkleinerungsmühle f, Vorzerkleinerer m
 preliminary stage Vorstufe f
 preliminary test Vorprüfung f, Vorversuch m, Vortest m
 preliminary testing Vortesten n
premature vorzeitig, Früh-
 premature vulcanization Vorvernetzung f
 risk of premature vulcanization Scorchgefahr f
preparation Präparation f, Vorbereitung f, Vorbehandlung f
 preparation of surfaces to be bonded Klebflächenvorbehandlung f
 preparation of the substrate Untergrundvorbehandlung f
 data preparation Datenaufbereitung f
 paste preparation Pastenherstellung f
 program preparation Programmherstellung f
 surface preparation Oberflächenvorbehandlung f
 test piece preparation Probenaufbereitung f
preplasticizer Vorplastifiziergerät n
 screw preplasticizer Schneckenvorplastifiziergerät n

preplasticization Vorplastifizierung f
 screw preplasticization Schneckenvorplastifizierung f
 screw preplasticizing unit Schneckenvorplastifiziergerät n
prepreg Glasfaserprepreg n, harzvorimprägniertes Halbzeug n, vorimprägniertes Textilglas n, Harzmatte f
 chopped strand prepreg Kurzfaser-Prepreg n
 cut-to-size prepregs Harzmattenzuschnitte mpl
 polyester prepreg [polyester sheet mo(u)lding compound] Polyesterharzmatte f
presentation Darstellung f
 data presentation Datendarstellung f
preservative Konservierungsmittel n, Konservierungsstoff m, Lagerkonservierungsmittel n
press Presse f
 press-fit Preßsitz m
 press-mo(u)lded laminate Preßlaminat n
 press mo(u)lded using liquid resin naßgepreßt
 press mo(u)lding Preßverfahren n
 press platen Preßblech n, Preßplatte f
 press-side granulator Beistellgranulator m
 press table Pressentisch m
 press vulcanization Preßvulkanisation f
 automatic downstroke press Oberkolbenpreßautomat m
 automatic downstroke transfer mo(u)lding press Oberkolbenspritzpreßautomat m
 automatic transfer mo(u)lding press Spritzpreßautomat m
 bedding-down press Tuschierpresse f
 downstroke press Oberkolbenpresse f
 embossing press Prägepresse f
 filter press Filterpresse f
 folding press Abkantbank f
 hot press mo(u)lded laminate Warmpreßlaminat n
 hot press mo(u)lding Warm(form)pressen n, Warmpreßverfahren n
 hydraulic press ölhydraulische Presse f
 liquid resin press mo(u)lding Naßpreßverfahren n
 matched metal press mo(u)lding Heißpreßverfahren n, Heißpressen n
 multi-daylight press Etagenpresse f
 pelleting press Tablettierpresse f
 platen press Plattenpresse f
 semi-automatic transfer mo(u)lding press Spritzpreß-Halbautomat m
 toggle press Kniehebelpresse f
 transfer mo(u)lding press Spritzpresse f
 upstroke press Unterkolbenpresse f
 welding press Schweißpresse f
pressed sheet Preßplatte f
pressing pressend, Preß -
 by pressing a key auf Tastendruck m
presspahn Preßspan m
 electrical grade presspahn Elektropreßspan m
pressure Druck m; Spannung f [Dampf]
 pressure accumulator Druckspeicher m

pressure adjusting device Druckeinstellorgan n
pressure adjusting valve Druckeinstellventil n
pressure build-up Druckaufbau m, Druckaufbauphase f
pressure change-over Druckumschaltung f
pressure change-over time Druckumschaltzeit f
pressure control Druckführung f
pressure control circuit Druckregelkreis m, Drucksteuerkreis m
pressure control device Druckregelgerät n
pressure control valve Druck(regel)ventil n, Druckregulierventil n
pressure correction Druckkorrektur f
pressure curve Druckkennlinie f, Druckverlaufkurve f
pressure cut-out Druckabschneidung f
pressure decrease Druckabfall m
pressure decreasing phase Druckabfallphase f
pressure-dependent druckabhängig
pressure difference Druckdifferenz f, Differenzdruck m, Druckunterschied m
pressure distribution Druckverteilung f
pressure filter Druckfilter m,n
pressure flow Druckfluß m, Druckströmung f, Strauströmung f
pressure flow-drag flow ratio Drosselquotient m, Drosselkennzahl f
pressure fluctuations Druckpulsationen fpl
pressure ga(u)ge Druckanzeigegerät n, Druckmeßgeber m, Druckmeßeinrichtung f, Druckmeßgerät n, Druckmessung f [Gerät]
pressure gelation Druckgelierverfahren npl
pressure gradient Druckgefälle n, Druckgradient m
pressure hose Druckschlauch m
pressure-less drucklos
pressure limit Drucklimit n
pressure limitation Druckbegrenzung f
pressure line Druckleitung f
pressure loss Druckverlust m
pressure-lubricated zwangsgeschmiert
pressure measuring point Druckmeßstelle f
pressure measuring system Druckmeßsystem n
pressure pad Druckkissen n, Druckplatte f
pressure peak Druckspitze f
pressure per unit area Flächenpressung f
pressure pipe Druckrohr n
pressure pipeline Druckrohrleitung f
pressure plasticization Druckplastifizierung f
pressure profile Druckprofil n, Druckverlauf m, Druckführung f
pressure program Druckprogramm n
pressure propagation Druckfortpflanzung f
pressure ram Druckkolben m
pressure range Druckbereich m
pressure recorder Druckschreiber m
pressure recording device Druckschreiber m
pressure reduction Druckreduzierung f
pressure regulator Druckwaage f

pressure release Druckentlastung f
pressure release system Druckentspannungssystem n
pressure release valve Druckminderventil n, Druckreduzierventil n, Druckentlastungsventil n, Druckbegrenzungsventil n
pressure reserve Druckreserve f
pressure roll Andrückwalze f, Anpreßwalze f, Kontaktwalze f
pressure sensitive druckempfindlich
pressure sensor Drucksensor m, Druckaufnehmer m, Druckfühler m, Drucksonde f
pressure servo-valve Druck-Servoventil n
pressure surge Druckstoß m
pressure test for pipes Rohrinnendruckversuch m
pressure transducer Druck(meß)umformer m, Druck(meß)dose f, Druckgeber m
pressure transfer Druckübertragung f
pressure variations Druckschwankungen fpl, Druckschwingungen fpl
absolute pressure Absolutdruck m
actual pressure Druckistwert m
at constant pressure isobar
atmospheric pressure Atmosphärendruck m, Luftdruck m
back pressure Staudruck m, Extrusionsstaudruck m, Gegendruck m, Rückdruck m, Rückstau(druck) m
back pressure adjusting mechanism Staudruckeinstellung f [Gerät]
back pressure adjustment Staudruckeinstellung f
back pressure controller Staudruckregler m
back pressure forces Rückdruckkräfte fpl
back pressure program Staudruckprogramm n
back pressure reduction Staudruckabbau m
back pressure relief (mechanism) Staudruckentlastung f
barometric pressure Barometerdruck m
bearing pressure Lagerkraft f
blow-up pressure Aufblasdruck m
blowing air pressure Blasluftdruck m
blowing pressure Blasdruck m
buckling pressure Beuldruck m
bursting pressure Berstdruck m
capacity to build up pressure Druckaufbauvermögen n
cavity pressure Form(nest)(innen)druck m, Werkzeug(innen)druck m
cavity pressure-dependent forminnendruckabhängig, werkzeuginnendruckabhängig
cavity pressure-independent forminnendruckunabhängig, werkzeuginnendruckunabhängig
cavity pressure profile Werkzeug(innen)druckverlauf m
cavity pressure transducer Werkzeuginnendruck(-Meßwert)aufnehmer m
change-over pressure Umschaltdruck m
change-over to hold(ing) pressure Nachdruck-

umschaltung f
clamping pressure Einspanndruck m, Schließdruck m, Niederhaltedruck m
contact pressure Anpreßdruck m, Anpreßkraft f, Berührungsdruck m, Kontaktdruck m
conveying pressure Förderdruck m
counter pressure Gegendruck m
decrease in pressure Druckabfall m
deformation through internal pressure Innendruckverformung f
delaminating pressure Spaltlast f
demo(u)lding pressure Entformungsdruck m
depending on the cavity pressure werkzeuginnendruckabhängig
depending on the pressure druckabhängig
die back pressure Werkzeug(gegen)druck m, Werkzeugrückdruck m
die head pressure Düseneintrittdruck m
die pressure Düsendruck m
difference in pressure Druckdifferenz f, Druckunterschiede mpl
differential pressure Differenzdruck m
differential pressure transducer Differenzdruckumformer m
downward pressure on prices Preisdruck m
drop in pressure Druckeinbruch m
dwell pressure Formgebungsdruck m
dwell pressure phase Formgebungsphase f
embossing pressure Prägedruck m
excess pressure Überdruck m
excess water pressure Wasserüberdruck m
extrudate delivery pressure Ausstoßdruck m [Extrusion]
extrusion pressure Extrusionsdruck m, Spritzdruck m
foaming pressure Blähdruck m
follow-up pressure Nachdruckniveau n, Nachdruck m
gas pressure Gasdruck m
high-pressure plasticator Hochdruckplastifiziergerät n
high-pressure plasticizing unit Hochdruckplastifiziergerät n
hold(ing) pressure Haltedruck m, Formgebungsdruck m, Druckstufe II f
hold(ing) pressure fluctuations Nachdruckpulsationen fpl
hold(ing) pressure phase Nachdruckphase f, Druckstufe II f, Formgebungsphase f
hold(ing) pressure profile Nachdruckverlauf m
hold(ing) pressure program Nachdruckprogramm n
hold(ing) pressure stages Nachdruckstufen fpl
hold(ing) pressure time Druckhaltezeit f, Nachdruckzeit f, Nachdruckdauer f
hydraulic pressure (hydraulischer) Systemdruck m
hydraulic system pressure ga(u)ge Hydrauliköl-Druckanzeige f
increase in pressure Druckanstieg m
independence of pressure Druckunabhängigkeit f
independent of cavity pressure innendruckunabhängig
inflating pressure Aufblasdruck m
initial injection pressure Anfangsspritzdruck m
injection pressure Druckstufe I f, Einspritzdruck m, Fülldruck m, Spritzdruck m
injection pressure regulator Spritzdruckregler m
injection pressure stage Spritzdruckstufe f
injection pressure time Spritzdruckzeit f
internal air pressure Innenluftdruck m
internal air pressure calibrating unit Überdruckkalibrator m, Überdruckkalibrierung f, Stützluftkalibrierung f
internal gas pressure Gasinnendruck m
joining pressure Fügedruck m
laminating pressure Laminierdruck m
limiting pressure Grenzdruck m
load pressure Lastdruck m
long-term failure test under internal hydrostatic pressure [test used for plastics pipes] Zeitstand-Innendruckversuch m, Innendruck-Zeitstanduntersuchung f
long-term resistance to internal hydrostatic pressure [of plastics pipes] Innendruck-Zeitstandverhalten n, Innendruck-Zeitstandwert m
loss of pressure Druckabfall m, Druckverlust m
low-pressure injection mo(u)lding machine Niederdruckspritzgießmaschine f
low-pressure mo(u)ld safety device Niederdruckwerkzeugschutz m
low-pressure mo(u)lding compound Niederdruckformmasse f
low-pressure polymerization Niederdruckpolymerisation f
low-pressure process Niederdruckverfahren n
low-pressure reactor Niederdruckreaktor m
low-pressure scanning (device) Niederdruckabtastung f
low-pressure separator Niederdruckabscheider m
mains pressure Netzdruck m
mains water pressure Wassernetzdruck m
mark left by the pressure transducer Druckaufnehmerabdruck m
maximum pressure Druck(ober)grenze f, Druckmaximum n, Druckspitze f
measurement of pressure Druckmessung f
melt pressure Massedruck m, Schmelzedruck m
melt pressure indicator Massedruckanzeige f
melt pressure profile Massedruckverlauf m
melt pressure sensor Massedruckaufnehmer m, Massedruckgeber m
mo(u)ld filling pressure Formfülldruck m
mo(u)lding pressure Preßdruck m, Preßkraft f, Verarbeitungsdruck m
nip pressure Walzenspaltkraft f, Walzenspaltdruck m, Walzentrennkraft f, Walzenlast f, Spaltlast f, Spaltkraft f, Spaltdruck m

nominal pressure Nenndruck m
nozzle contact pressure Düsenanlagekraft f, Düsenanlagedruck m, Düsenanpressung f, Düsenanpreßdruck m, Düsenanpreßkraft f
operating pressure Betriebsdruck m
packing pressure Verdichtungsdruck m
partial pressure Partialdruck m
partial vapo(u)r pressure Dampfteildruck m, Dampfpartialdruck m
partial water vapo(u)r pressure Wasserdampfpartialdruck m, Wasserdampfteildruck m
peak pressure Druckmaximum n, Druckspitze f, Spitzendrücke mpl
platen pressure Plattendruck m
quartz pressure transducer Quarzkristall-Druckaufnehmer m
ram pressure Stößeldruck m
recording of pressure Druckaufzeichnung f
reduced mo(u)ld clamping pressure Werkzeugsicherungsdruck m
reduction in pressure Druckminderung f
reference pressure Bezugsdruck m
reference pressure curve Solldruckkurve f
reference pressure range Referenzdruckbereich m
relative pressure Relativdruck m
required pressure Drucksollwert m, Solldruck m
residual pressure Restdruck m
resistance to internal pressure Innendruckfestigkeit f
rubber covered pressure roll Anpreßgummiwalze f
saturation pressure Sättigungsdruck m
saturation vapo(u)r pressure Sättigungsdampfdruck m, Wasserdampfsättigungsdruck m
screw back pressure Schneckenrückdruck m, Schneckenrückdruckkraft f, Schneckenstaudruck m, Schneckengegendruck m
set pressure Drucksollwert m
sound pressure Schalldruck m
sound pressure intensity Schalldruckstärke f
sudden pressure increase Druckstoß m
threshold pressure Druckschwellwert m
total nip pressure Gesamtspaltlast f
ultra-high pressure plasticator Höchstdruckplastifiziergerät m
under pressure druckbeansprucht, druckbelastet
vapo(u)r pressure Dampfdruck m
welding pressure Schweiß(preß)druck m
without pressure peaks druckspitzenlos
working pressure Betriebsdruck m
pressurized Druck-
 pressurized oil lubricating (system) Druckölung f
 pressurized vessel Druckbehälter m, Druckgefäß n
 pressurized water Druckwasser n
 pressurized water unit Druckwassergerät n
 space pressurized during closing movement [of toggle clamp unit] Schließraum m, Öffnungsraum m
pressurizing valve Vorspannventil n

pretreating Vorbehandlung f
 corona pretreating unit Corona-Vorbehandlungsanlage f
 film pretreating instrument Folienvorbehandlungsgerät n
pretreatment Vorbehandlung f
 corona pretreatment Coronavorbehandlung f, Korona-Vorbehandlung f
prevent/to vorbeugen, sichern, schützen
 device to prevent screw retraction Schneckenrückdrehsicherung f
 thermostat to prevent overheating Überhitzungs-Sicherheitsthermostat m
prevention Schutz m, Sicherung f
 accident prevention Unfallschutz m, Unfallverhütung f
 accident prevention regulations Unfall-Verhütungsvorschriften fpl, UVV
 overload prevention device Überlastschutz m, Überlast(ungs)sicherung f
previous history [i.e. the thermal or mechanical history of a test piece or mo(u)lded part] Vorgeschichte f, Fließgeschichte f
previous year Vorjahr n
price Preis m, Preisniveau n
 price increase Preiserhöhung f, Preisanhebungen fpl, Preissteigerungen fpl
 price per unit volume Volumenpreis m
 price per unit weight Gewichtspreis m
 price reduction Preisnachlaß m, Preisreduzierung f, Verbilligung f
 downward pressure on prices Preisdruck m
 guide price Richtpreis m
 selling price Verkaufspreis m
 volume price advantage Volumenpreisvorteil m
primary primär, Primär-
 primary particle Primärteilchen n, Primärkorn n
 primary plasticizer Primärweichmacher m
 primary radical Primärradikal n
 primary valency bond Hauptvalenzbindung f
primed vorbereitet [allg.]; grundiert
primer Grundfarbe f, Voranstrich m, Grundierung f
 primer coated grundiert, primerlackiert
 primer coating Grundlackierung f, Haftbrücke f, Primerschicht f, Haftgrundierung f
 anti-corrosive primer Korrosionsschutzgrundierung f, Rostschutzgrundierung f
 application of primer Primerlackierung f
 coated with primer grundiert, primerlackiert
 coating with primer Grundierung f, Primerlackierung f
 coil coating primer Coil-Coating-Grundierung f
 low viscosity primer Einlaßgrundierung f
 zinc chromate primer Zinkchromatgrundierung f
 zinc phosphate-iron oxide primer Zinkphosphat-Eisenoxid-Grundierung f
 zinc-rich primer Zinkstaub-Primer m,

Zinkstaubgrundierung f
principal characteristics Hauptmerkmale npl
principle Prinzip n
  basic principle Grundprinzip n
  construction principles Konstruktionsprinzipien npl
  design principles Konstruktionsprinzipien npl
  forced circulation principle Zwangslaufprinzip n
  modular principle Baukastenprinzip n
  operating principle Funktionsprinzip n, Arbeitsprinzip n
  principle of operation Funktionsprinzip n, Arbeitsprinzip n
  single-screw principle Einschneckenprinzip n
  twin screw principle Doppelschneckenprinzip n
print-out/to ausdrucken, protokollieren
print-out Ausdruck m, Protokollierung f
  computer print-out Computer-Ausdruck m
  partial print-out Teilausdruck m
  plain language print-out Klartextausdruck m
  program print-out Programmausdruck m
printability Bedruckbarkeit f
printable bedruckbar
printed bedruckt, gedruckt
  printed board Leiterplatte f
  printed circuit gedruckte Schaltung f
  printed circuit board Leiterplatte f, Elektrolaminat n, Elektroschichtpreßstoff m
  printed image Druckbild n, Druckfilm m
  printed in bold type fettgedruckt
  printed out ausgedruckt
  capable of being printed out protokollierbar
printer Drucker m
  digital printer Digitaldrucker m
  hard copy printer Protokolldrucker m
  high-speed printer Schnelldrucker m
  page printer Blattschreiber m
printing 1. Bedrucken n, Drucken n, Druck m; 2. Druck-
  printing industry Druckindustrie f
  printing ink Druckfarbe f
  printing ink binder Druckfarbenbindemittel n
  colo(u)r printing unit Farbdruckvorrichtung f
  flexographic printing Flexodruck m
  flexographic printing ink Flexodruckfarbe f
  gravure printing ink Tiefdruckfarbe f
  intaglio printing Tiefdruck m
  letterpress printing Buchdruck m
  rotary screen printing Rotationssiebdruck m
  screen printing (process) Siebdruckverfahren n
printout Protokoll n
prism-shaped prismenförmig
prismatic prismenförmig, prismatisch
probe Fühler m, Geber m, Sensor m, Sonde f
problem Problem n, Schwierigkeit f
  problem-oriented problemorientiert
  adhesion problems Haftschwierigkeiten fpl, Haftungsprobleme npl
  compatibility problems Verträglichkeitsschwierigkeiten fpl
  demo(u)lding problems Entformungsschwierigkeiten fpl
  dispersing problems Dispergierprobleme npl, Verteilungsprobleme npl
  ecological problems Umwelt(schutz)probleme npl
  environmental problems Umweltprobleme npl
  feed problems Einspeisungsschwierigkeiten fpl, Einzugsschwierigkeiten fpl
  flow problems Verlaufschwierigkeiten fpl, Verlaufstörungen fpl
  processing problems verfahrenstechnische Schwierigkeiten fpl
  shelf life problems Lagerstabilitätsschwierigkeiten fpl
  staff problems Personalschwierigkeiten fpl
  starting-up problems Anfahrschwierigkeiten fpl
  technical problems technische Schwierigkeiten fpl
  transport problems Förderprobleme npl, fördertechnische Schwierigkeiten fpl
  wetting problems Benetzungsschwierigkeiten fpl
procedure Verfahren(sweise) n (f)
  experimental procedure Versuchsdurchführung f
process Vorgang m, Arbeitsablauf m, Arbeitsprozeß m, Prozeß(ablauf) m, Verfahrensablauf m
  process analysis Prozeßanalyse f, Verfahrensanalyse f
  process automation Prozeßautomatisierung f
  process computer Prozeßrechner m
  process conditions Prozeßbedingungen fpl
  process control Prozeßführung f, Prozeßsteuerung f, Prozeßregelung f, Prozeßkontrolle f, Arbeitsablaufsteuerung f, Ablaufsteuerung f
  process control equipment Prozeßsteueranlage f
  process control system Prozeßleitsystem n, Prozeßführungseinrichtung f, Prozeßführungssystem n
  process control (unit) Arbeitsablaufsteuerung f, Prozeßkontrolle f, Prozeßführungsinstrument n
  process controlled prozeßgeführt
  process data Prozeßdaten npl
  process data control unit Prozeßdatenüberwachung f
  process data gathering Prozeßdatenerfassung f
  process data monitor(ing) Prozeßdatenüberwachung f
  process management Prozeßbeherrschung f
  process model Prozeßmodell n
  process monitoring (unit) Prozeßüberwachung f
  process parameter Prozeßgröße f
  process segment Prozeßabschnitt m
  process sequence Arbeitsablauf m
  process stabilizer Verarbeitungsstabilisator m
  process stages Prozeßschritte mpl, Verfahrensschritte mpl
  process variable Prozeßgröße f, Prozeßvariable f, Verfahrensvariable f
  adsorption process Adsorptionsvorgang m
  ag(e)ing process Alterungsprozeß m, Alterungsvorgang m
  blown film process Schlauchfolienkonzept n
  breakdown process Versagensprozeß m

combustion process Verbrennungsprozeß m
continuous process Fließprozeß m
conveying process Fördervorgang m
cooling process Abkühlvorgang m
course of the process Prozeßablauf m
crosslinking process Vernetzungsprozeß m
deformation process Verformungsvorgang m
degradation process Abbauprozeß m
diffusion process Diffusionsvorgang m
discontinuous process Stückprozeß m
dissolving process Lösungsvorgang m
downstream process Nachfolgeprozeß m
drying process Trocknungsablauf m
easy to process verarbeitungsfreundlich
energy conversion process Energieumwandlungsprozeß m
extrusion process Extrusionsverlauf m
factors influencing the process Verfahrenseinflußgrößen fpl
flow processes Strömungsvorgänge mpl
foaming process Aufschäumvorgang m
fracture process Bruchvorgang m, Schädigungsprozeß m
high-pressure process Hochdruckverfahren n
high-temperature process Hochtemperaturverfahren n, HT-Verfahren n
injection mo(u)lding process Spritzgießvorgang m
laminating process Laminierverfahren n
low-pressure process Niederdruckverfahren n
low-temperature process LT-Verfahren n
manufacturing process Fertigungsverfahren n
maturing process [e.g. of a PVC paste] Reifevorgang m
mo(u)lding process Form(teil)bildungsprozeß m, Form(teil)bildungsvorgang m, Formgebungsprozeß m, Formgebungsverfahren n, Formteilherstellung f
nucleating process Nukleierungsvorgang m
owing to the nature of the process aus verfahrenstechnischen Gründen mpl
pin roller process [for making fibrillated film] Nadelwalzenverfahren n
plasticating process Plastifiziervorgang m, Plastifizier(ungs)verfahren n
plasticizing process Plastifiziervorgang m, Plastifizier(ungs)verfahren n
polymerization process Polymerisationsverlauf m
preceding process Vorprozeß m
production process Fertigungsverfahren n, Herstellungsverfahren n, Produktionsprozeß m
recovery process Erholungsprozeß m
relaxation process Relaxationsprozeß m, Relaxationsvorgang m
retardation process Retardationsvorgang m
separation process Trennungsprozeß m
single-stage blow mo(u)lding process Einstufen-Blasverfahren n
single-stage process Einstufenverfahren n

subsequent process Nachfolgeprozeß m
swelling process Quell(ungs)vorgang m
thermoforming process Umformvorgang m
two-component process Zweikomponentenverfahren n
two-stage process Zweistufenverfahren n
processability Verarbeitbarkeit f
processed verarbeitet
   capable of being processed verarbeitungsfähig
processing Verarbeitung f, Fertigung f, Arbeitsweise f; Abarbeitung f [von Daten]
   processing advantages Verarbeitungsvorteile mpl
   processing aid Herstellungshilfsmittel n, Verarbeitungshilfsmittel n
   processing characteristics Verarbeitungsmerkmale npl, Verarbeitungsverhalten n
   processing conditions Fahrbedingungen fpl, Verarbeitungsbedingungen fpl, verfahrenstechnische Bedingungen fpl
   processing cycle Verarbeitungszyklus m
   processing differences Verarbeitungsunterschiede mpl
   processing difficulties Verarbeitungsschwierigkeiten fpl
   processing disadvantages verfahrenstechnische Nachteile mpl
   processing equipment Verarbeitungsgeräte npl, Verarbeitungsmaschine f
   processing guidelines Verarbeitungshinweise mpl, Verarbeitungsrichtlinien fpl
   processing in plasticized form [PVC] Weichverarbeitung f
   processing in unplasticized form [PVC] Hartverarbeitung f
   processing latitude Verarbeitungsbreite f, Verarbeitungsspielraum m
   processing of thermoplastics Thermoplastverarbeitung f
   processing of thermosets Duroplastverarbeitung f
   processing parameter Verarbeitungsgröße f, Verfahrensparameter m, Verarbeitungsparameter m
   processing possibilities verfahrenstechnische Möglichkeiten fpl, Verarbeitungsmöglichkeiten fpl
   processing problems verfahrenstechnische Schwierigkeiten fpl
   processing program Verarbeitungsprogramm n
   processing requirements verfahrenstechnische Anforderungen fpl, verfahrenstechnische Bedürfnisse npl
   processing sequence verfahrenstechnische Ablauf m
   processing stages Verarbeitungsschritte mpl, Verarbeitungsphasen fpl
   processing technique Verarbeitungsverfahren n
   processing temperature Verarbeitungstemperatur f
   processing temperature range Verarbeitungstemperaturbereich m
   processing unit verfahrenstechnische Einheit f
   processing with plasticizer Weichverarbeitung f
   processing without plasticizer Hartverarbeitung f
   approximate processing conditions Verarbeitungs-

richtwerte mpl
automatic processing equipment Verarbeitungsautomat m
batch processing Stapelverarbeitung f
central processing unit [CPU] zentrale Prozeßeinheit f, zentrales Rechenwerk n, Zentraleinheit f
data processing Datenverarbeitung f
data processing facilities Datenverarbeitungsmöglichkeiten fpl
data processing unit Datenverarbeitungsanlage f
dry blend processing Pulververarbeitung f
effect of processing Verarbeitungseinfluß m
high-temperature processing Hochtemperaturverarbeitung f
important from the processing point of view verfahrenstechnisch wesentlich
interesting from the processing point of view verfahrenstechnisch interessant
method of processing Verarbeitungsverfahren n
plastics processing Kunststoffverarbeitung f
plastics processing machine Kunststoffmaschine f
pre-set processing conditions Sollverarbeitungsbedingungen fpl
ready for processing verarbeitungsfertig
real time processing Echtzeitverarbeitung f
stability during processing Verarbeitungsstabilität f
processor Prozessor m, Verarbeiter m
   bit processor Bitprozessor m
   central data processor BDE-Zentrale f
   central processor Zentralprozessor m
   food processor Küchengerät n, Küchenmaschine f
   one-bit processor Ein-Bit-Prozessor m
   peripheral processor Peripherieprozessor m
   word processor Textprozessor m, Texteditor m
producer Hersteller m, Produzent m
   mo(u)lding compound producer Formmassehersteller m
   raw material producer Rohstoffhersteller m
producing a lot of wear verschleißintensiv
product Produkt n, Erzeugnis n [s.a. products]
   product code Typenbezeichnung f
   product constant Produktkennzahl f
   product data sheet Typenmerkblatt n
   product-dependent produktabhängig
   product design Produktgestaltung f
   product group Produktgruppe f
   product quality Fertigungsgüte f
   product range Verkaufsprogramm n, Herstellungsprogramm n, Produkt(ions)palette f, Produktsortiment n, Produkt(ions)programm n
   product recovery Produktrückgewinnung f
   product specification Produktnorm f
   product uniformity Produktkonstanz f
   product variations Produktschwankungen fpl
   combustion product Verbrennungsprodukt n
   commercial product Handelsprodukt n

   condensation product Kondensationsprodukt n
   decomposition product Spaltprodukt n, Zersetzungsprodukt n, Zerfallsprodukt n
   degradation product Abbauprodukt n
   depending on the product produktabhängig
   development product Entwicklungsprodukt n
   end product Enderzeugnis n, Endprodukt n, Fertigteil n, Fertigerzeugnis n, Fertigartikel m, Fertigprodukt n
   end product properties Endprodukteigenschaften fpl, Fertigprodukteigenschaften fpl, Produkteigenschaften fpl
   experimental product Versuchsprodukt n
   extruded product Extrusionsartikel m
   flexible PVC product Weich-PVC-Artikel m
   gross national product Bruttosozialprodukt n
   key product Schlüsselprodukt n
   oxidation product Oxidationsprodukt n
   polyaddition product Polyadditionsprodukt n
   polycondensation product Polykondensationsprodukt n
   pyrolysis product Pyrolyseprodukt n
   reaction product Reaktionsprodukt n, Umsetzungsprodukt n
   rival product Konkurrenzfabrikat n
   secondary product Folgeprodukt n
   starting product Ausgangsprodukt n
production Produktion f, Fertigung f
   production area [of a factory] Produktionsbereich m
   production batch Fertigungscharge f
   production breakdown Produktionsstörung f, Produktionsunterbrechung f
   production calender Betriebskalander m
   production capacity Produktionskapazität f, Produktionsleistung f
   production compounder Produktionskneter m
   production conditions Fertigungsparameter mpl, Herstellungsparameter mpl, Herstellungsbedingungen fpl, Produktionsverhältnisse npl, Produktionsbedingungen fpl
   production control Fabrikationskontrolle f, Fabrikationsprüfung f, Fabrikationsüberwachung f, Fertigungskontrolle f, Produktionskontrolle f, Produktionsüberwachung f, Fertigungssteuerung f
   production control system Fertigungssteuerung f
   production costs Fertigungskosten pl, Herstellungskosten pl, Produktionskosten pl
   production cycle Fertigungsfluß m, Produktionszyklus m, Produktionsablauf m, Fertigungszyklus m
   production data Betriebsgrößen fpl, Betriebsdaten npl, Produktionsdaten npl, Produktionswerte mpl
   production data acquisition Produktionsdatenerfassung f, Betriebsdatenerfassung f, BDE f
   production data collecting system BDE-System n
   production data collecting unit Betriebsdaten-Erfassungegerät n, Produktionsdaten-Erfassungsgerät n

production data collection Betriebsdatenerfassung f, Produktionsdatenerfassung f
production data record Betriebsdatenprotokoll n, Produktionsdatenprotokoll n
production data terminal Betriebsdatenerfassungsstation f, BDE-Terminal n
production date Herstelldatum n, Produktionsdatum n
production die Produktionswerkzeug n
production downtime Produktionsausfall m
production equipment Fertigungseinrichtungen fpl, Fertigungsmittel n
production extruder Fertigungsextruder m, Produktionsextruder m
production facility Fabrikationsstätte f
production figures Produktionszahlen fpl
production guidelines Herstellungshinweise mpl
production hours Produktionsstunden fpl
production increase Produktionszuwachs m
production line Anlage f, Fertigungsanlage f, Fertigungslinie f, Fertigungsstraße f
production logger Produktionsschreiber m
production conditions Fertigungsparameter mpl
production process Produktionsprozess m
production monitoring (department) Fertigungsüberwachung f [Abteilung]
production monitoring system Fertigungsüberwachung f
production mo(u)ld Produktionswerkzeug n
production of hard copy Protokollherstellung f
production of large containers Großhohlkörperfertigung f
production of reject mo(u)ldings Ausfallproduktion f
production of rejects Ausfallproduktion f
production of rigid film Hartfolienherstellung f
production of single units Einzel(an)fertigung f
production overheads Fertigungsgemeinkosten pl
production overview Produktionsübersicht f
production plan Fertigungsprogramm n
production planning (department) Fertigungsplanung f
production process Herstellungsverfahren n
production rate Fertigungsgeschwindigkeit f
production records Fertigungsunterlagen fpl
production requirements Produktionserfordernisse npl
production runs Auflagenhöhen fpl
production scale Produktionsmaßstab m
production-scale machine Produktionsmaschine f
production-scale trial Großversuch m
production schedule Fertigungsprogramm n, Produktionsprogramm n, Herstellungsprogramm n, Fabrikationsprogramm n
production shed Fertigungshalle f, Produktionshalle f, Fabrikationshalle f
production speed Fertigungsgeschwindigkeit f, Produktionsgeschwindigkeit f
production stage Produktionsabschnitt m

production status Produktionszustand m
production tasks Produktionsaufgaben fpl
production time Fertigungszeit f, Herstellungszeit f
production tolerance Herstellungstoleranz f, Fertigungstoleranz f
production trial Fertigungsversuch m
production unit Fertigungseinheit f, Produktionseinheit f
production variable Fertigungsparameter m, Herstellungsparameter m, Arbeitswert m
annual world production Weltjahresproduktion f
assembly line production Fließbandfertigung f
break in production Produktionsstörung f, Produktionsunterbrechung f, Verarbeitungsunterbrechung f
constant production conditions Produktionskonstanz f
continuous production Dauerproduktion f
expanded polystyrene production line Polystyrol-Schaumanlage f
film production machine Folienproduktionsmaschine f
film tape production line Folienbändchenanlage f
high-speed production line Hochgeschwindigkeitsanlage f
increase in production Produktionserhöhung f, Produktionszuwachs m
large-bore pipe production line Großrohrstraße f
large-container production Großhohlkörperfertigung f
large-scale production Großproduktion f, Großserienfertigung f
large-scale production unit Großproduktionsmaschine f
mass production Massenfertigung f, Massenproduktion f, Serienfertigung f, Serienproduktion f
mass production conditions Serienbedingungen fpl
mass production mo(u)ld Serienwerkzeug n
means of production Produktionsmittel npl
one-off production Einzelanfertigung f
pilot plant production Vorserienfertigung f
pipe production Rohrfabrikation f
pipe production line Rohrherstellungsanlage f, Rohrstraße f, Rohrfertigungsstraße f
prototype production Prototypenfertigung f
small-scale production Kleinstserienfertigung f
start of production Produktionsbeginn m
total production Gesamtproduktion f
productive time [of a machine] Produktivzeit f
productivity Arbeitsproduktivität f, Ausnutzungsgrad m, Nutzungsgrad m, Produktivität f
productivity increase Produktivitätssteigerung f
machine productivity Maschinennutzungsgrad m
products Produkte npl [s.a. product]
coated products Streichartikel mpl
curing agent decomposition products Vernetzerspaltprodukte npl

dairy products Molkereiprodukte npl
description of products Typenbeschreibung f
finished products Fertigwaren fpl
pharmaceutical products Pharmazeutika pl
range of products Angebotspalette f, Fabrikationsprogramm n, Produktenreihe f, Produkt(ions)palette f, Produktprogramm n, Produktsortiment n, Verkaufssortiment n, Verkaufsprogramm n, Herstellungprogramm n
semi-finished products Halbfabrikate npl, Halbfertigerzeugnisse npl, Halbzeug n
profile 1. Profilstrang m, Profil n; 2. Profil-
profile calibrating unit Profilkalibrierung f [Gerät]
profile calibration Profilkalibrierung f
profile calibrator Profilkalibrierung f [Gerät]
profile die Profildüse f, Profil(spritz)kopf m, Profilwerkzeug n
profile extrusion downstream equipment Profilnachfolge f
profile extrusion line Profil(extrusions)straße f, Profilanlage f
profile scrap Ausschußprofile npl
profile scrap granulator Profilschneidmühle f
profile section Profilabschnitt m
profile take-off (unit) Profilabzug m
automatic profile welding unit Profilschweißautomat m
casement profile Flügelprofil n
cavity pressure profile Werkzeug(innen)druckverlauf m
channel profile Gangprofil n
clamping force profile Schließkraftverlauf m
flow profile Strömungsprofil n
handrail profile Handlaufprofil n
heavy-duty window profile Fensterhochleistungsprofil n
hold(ing) pressure profile Nachdruckverlauf m
injection speed profile Einspritzgeschwindigkeitsverlauf m
melt pressure profile Massedruckverlauf m
ox-bow profile Ochsenjochprofil n
packing profile Verdichtungsprofil n
pressure profile Druckprofil n, Druckverlauf m, Druckführung f
rectangular profile die Rechteckprofildüse f
residence time profile Verweilzeitspektrum n
roller blind profile Rolladenprofil n
round profile die Kreisprofildüse f, Rundstrangdüse f
runner profile Verteiler(kanal)querschnitt m
screw channel profile Schneckengangprofil n, Schneckenkanalprofil n
screw profile Schneckenprofil n
self-sealing profile Dichtprofil n
self-wiping profile Dichtprofil n
speed profile Drehzahlprogrammablauf m
stair edging profile Treppenkantenprofil n
temperature profile Temperaturverlauf m, Temperaturprofil n
velocity profile Geschwindigkeitsprofil n
window profile Fensterprofil n
window profile compound Fensterformulierung f
window sill profile Fensterbankprofil n
profit Gewinn m, Profit m; Ergebnis n
profit and loss account Gewinn- und Verlustrechnung f
profit margin Verdienstspanne f
net profit Reingewinn m
profitability Rentabilität f
profitability calculation Wirtschaftlichkeitsberechnung f
profits Ertrag m
drop in profits Gewinnrückgang m
program Programm n
program control Programmsteuerung f, Programmregelung f
program control unit Programmregler m, Programmsteuerung f
program-controlled programmgesteuert
program controller Programmregler m, Programmsteuerung f
program development Programmentwicklung f
program implementation Programmimplementierung f
program instruction Programmbefehl m
program key Programmtaster m
program language Programmsprache f
program maintenance Programmpflege f
program modifications Programm-Abwandlungen fpl
program module Programmbaustein m
program operation Programmablauf m
program package Programmpaket n
program preparation Programmerstellung f
program print-out Programmausdruck m
program selection Programmwahl f
program selection key Programmwahltaste f
program selection switch Programmwahlschalter m
program step Programmschritt m
program storage Programmspeicherung f
program store Programmspeicher m, Programmspeicherung f
application program Anwendungsprogramm n, Benutzerprogramm n
assembly program Assembler m
back pressure program Stauchdruckprogrammverlauf m
central program unit Programmzentraleinheit f
computer program Computerprogramm n, Berechnungsprogramm n, EDV-Programm n, EDV-Rechenprogramm n, Rechenprogramm n, Rechnerprogramm n
diagnostics program Diagnoseprogramm n
emergency program Notprogramm n
error diagnostics program Fehlerdiagnoseprogramm n

fault location program Fehlersuchprogramm n
hold(ing) pressure program Nachdruckprogramm n
injection program Einspritzprogramm n
machine control program Maschinensteuerprogramm n
machine operating program Maschinenprogramm n
machine setting program Maschineneinstellprogramm n
pressure program Druckprogramm n
processing program Verarbeitungsprogramm n
ready-made program fremdbezogenes Programm n, käufliches Programm n
service program Dienstprogramm n
speed program Drehzahlprogramm n, Geschwindigkeitsprogramm n
standardizing program Eichprogramm n
stored program control speicherprogrammierte Steuerung f
temperature program Temperaturprogramm n
test program Prüfprogramm n, Versuchsprogramm n
programmability Programmierbarkeit f
programmable programmierbar
  electrically erasable programmable read-only memory [EEPROM] EEPROM n, Eeprom-Speicher m
  freely programmable freiprogrammierbar
programmer Programmgeber m
programming Programmierung f
  programming instrument Programmiergerät n
  programming language Programmiersprache f
  programming module Programmierbaustein m
  programming unit Programmiereinheit f
  parison length programming [device] Schlauchlängen-Programmierung f
  wall thickness programming [device] Wanddicken-Progarmmiergerät n, Wanddickenprogrammierung f
progressive fortschreitend, progressiv
projected projiziert
  projected area Projektionsfläche f, projizierte Fläche f
  projected mo(u)lding area projizierte Formteilfläche f, Formteilprojektionsfläche f, projizierte Spritz(lings)fläche f
  projected runner surface area projizierte Verteilerkanalfläche f
projecting ausladend
prolonged immersion Dauerlagerung f, Langzeitlagerung f [in Flüssigkeit]
prolonged immersion in water Langzeitwasserlagerung f
proof 1. Nachweis m, Beweis m, Prüfung f; 2. beständig, dicht, undurchlässig
  moisture-proof feuchtigkeitsundurchlässig
  rot-proof verrottungsbeständig
  scratch-proof kratzbeständig, kratzfest
  vapo(u)r-proof dampfdicht
proofing Dichtmachen n, Imprägnieren n
  fire proofing Flammfestausrüsten n, Flammschutz m

  flame proofing Flammfestausrüsten n, Flammschutz m
propagation Ausbreitung f, Wachstum n, Fortschritt m
  crack propagation Bruchausbreitung f, Bruchfortschritt m, Rißausbreitung f, Rißfortpflanzung f
  crack propagation force Rißausbreitungskraft f
  crack propagation rate Riß(fortpflanzungs)geschwindigkeit f
  crack propagation resistance Rißausbreitungswiderstand m
  crack propagation test Rißfortpflanzungsversuch m
  craze propagation Craze-Wachstum n
  fatigue crack propagation Ermüdungsrißausbreitung f
  pressure propagation Druckfortpflanzung f
  tear propagation Weiterreißen n
  tear propagation force Weiterreißkraft f
  tear propagation resistance Weiterreißwiderstand m, Weiterreißfestigkeit f
propane Propan n
properties kennzeichnende Eigenschaften fpl, Eigenschaftswerte mpl, Eigenschaftsmerkmale npl, Eigenschaftsniveau n, Kenndaten npl [s.a. property]
  adhesive properties Hafteigenschaften fpl, Klebeeigenschaften fpl
  ag(e)ing properties Alterungskennwerte mpl, Alterungseigenschaften fpl
  anti-blocking properties Antiblockeigenschaften fpl
  anti-corrosive properties Korrosionsschutzeigenschaften fpl
  anti-friction properties Gleitfähigkeitsverhalten n, Gleitfähigkeit f, Gleiteigenschaften fpl
  barrier properties Barriereeigenschaften fpl, Barrierewerte mpl, Sperreigenschaften fpl
  basic properties Grundeigenschaften fpl
  change in properties Eigenschaftsänderung f
  cold bending properties Kaltbiegeeigenschaften fpl
  combination of properties Eigenschaftskombination f
  comparison of properties Eigenschaftsvergleich m
  deterioration of properties Werteabfall m
  drying properties Trocknungsverhalten n
  electrical insulating properties Elektroisolierverhalten n
  electrical properties elektrische Werte mpl
  end product properties Endprodukteigenschaften fpl, Fertigprodukteigenschaften fpl, Produkteigenschaften fpl
  end properties Endeigenschaften fpl
  film properties Filmeigenschaften fpl
  flame retardant properties Flammschutzeigenschaften fpl
  flow properties Verlaufseigenschaften fpl
  foaming properties Verschäumungseigenschaften fpl
  free-flowing properties Rieselfähigkeit f, Rieselverhalten n
  functional properties Gebrauchseigenschaften fpl
  general properties Allgemeineigenschaften fpl, Eigenschaftsbild n
  handling properties Handlingeigenschaften fpl

hard wearing properties Verschleißarmut f
heat ag(e)ing properties Wärmealterungswerte mpl
heat stabilizing properties Wärmestabilisiervermögen n
high-temperature properties Hochtemperatureigenschaften fpl
hydrophobic properties Hydrophobie f
long-term properties Langzeiteigenschaften fpl
low-temperature properties Tieftemperatureigenschaften fpl
low-warpage properties Verzugsarmut f
lubricating properties Gleiteigenschaften fpl, Schmiereigenschaften fpl
machining properties Bearbeitbarkeit f
material properties Stoffeigenschaften fpl, Werkstoffeigenschaften fpl
mechanical properties mechanisches Niveau n, mechanische Werte mpl
mo(u)lded-part properties Formteileigenschaften fpl
mo(u)lding properties Formbarkeit f
non-sag properties Standfestigkeit f
non-skid properties Rutschfestigkeit f
non-stick properties Trenneigenschaften fpl, Trennfähigkeit f
non-warping properties Verzugsfreiheit f
part properties Formteileigenschaften fpl
range of properties Eigenschaftsspektrum n
raw material properties Rohstoffeigenschaften fpl
release properties Entformbarkeit f
self-lubricating properties Selbstschmierfähigkeit f
short-term properties Kurzzeiteigenschaften fpl
shrinkage properties Schwindungseigenschaften fpl
sound-absorbent properties Schallschluckeigenschaften fpl
thermal insulating properties Wärmedämmeigenschaften fpl, Wärmeisoliereigenschaften fpl
thermal properties thermische Eigenschaften fpl, thermische Werte mpl
thermoforming properties Tiefzieheigenschaften fpl, Tiefziehfähigkeit f, Tiefziehverhalten n, Warmformeigenschaften fpl
vulcanizate properties Vulkanisateigenschaften fpl
weathering properties Bewitterungsverhalten n
property Eigenschaft f, Vermögen n [s.a. properties]; Zähigkeit f
property-related eigenschaftsbezogen
property variations Eigenschaftsschwankungen fpl
barrier property Sperrschichteigenschaft f
material property Stoffgröße f
propionate Propionat n
propionic acid Propionsäure f
proportional proportional
proportional control valve Proportionalstellventil n
proportional hydraulic system Proportionalhydraulik f
proportional hydraulics Proportionalhydraulik f
proportional limit Proportionalitätsgrenze f
proportional solenoid Proportionalmagnet m
proportional valve Proportionalventil n
proportionality constant Proportionalitätskonstante f
propylene Propylen n
propylene chain Propylenkette f
propylene glycol Propylenglykol n
propylene glycol ether Propylenglykolether m
propylene glycol maleate Propylenglykolmaleat n
propylene glycol methyl ether Propylenglykolmethylehter m
propylene oxide Propylenoxid n
protected [e.g. data in a memory] geschützt
protected against wear verschleißgeschütz
protecting against corrosion korrosionsschützend
protection Absicherung f, Schutz m
protection against corrosion Korrosionsschutz m
protection against flying stones Steinschlagschutzwirkung f
protection against overheating Überhitzungsschutz m, Übertemperaturschutz m
protection against wear Verschleißschutz m
protection of the environment Umweltschutz m
data protection Datensicherung f
environmental protection Umweltschutz m
fire protection Brandschutz m
mo(u)ld protection Werkzeugschonung f
overload protection Überlastungsschutz m
surface protection Oberflächenschutz m
protective Schutz-
protective clothing Schutz(be)kleidung f
protective coating Schutzüberzug m, Schutzanstrich m, Schutzbeschichtung f, Schutzschicht f
protective colloid Schutzkolloid n
protective gloves Schutzhandschuhe mpl
protective sleeve Schutzhülse f
proton Proton n
prototype Prototyp m
prototype die Prototypwerkzeug n
prototype mo(u)ld Prototypwerkzeug n
prototype production Prototypenfertigung f
proven bewiesen, erprobt, bewährt
provisional solution Übergangslösung f
provisional specification Vornorm f
PS [polystyrene] Polystyrol n
pseudoplastic nicht-newtonsch, pseudoplastisch, strukturviskos
PTFE [polytetrafluoroethylene] Polytetrafluorethylen n
p-toluenesulphonic acid Paratoluolsulfonsäure f
PU [polyurethane] Polyurethan n
publication Druckschrift f, Publikation f
publicity Propaganda f, Reklame f, Werbung f
word-of-mouth publicity Mund-zu-Mund-Propaganda f
puller Zieher m, Reißer m, Zugartikel m
sprue puller Angußauszieher m, Angußausreißer m
sprue puller bush Angußziehbuchse f
sprue puller gate Abreißanschnitt m

sprue puller pin Angußziehstift m
sprue puller pin gate Abreißpunktanschnitt m
sprue puller plate Abreißplatte f
   core puller control mechanism Kernzugsteuerung f
pulsate/to pulsieren; ungleichmäßig fördern [Pumpe]
pulsating pulsierend
pulsation Pulsation f, Pulsieren n; ungleichmäßige Förderung f
pulse-like impulsartig
pultruded stranggezogen
pultrusion Strangziehen n, Düsenziehverfahren n, Profilziehverfahren n, Ziehverfahren n
pulverization Feinmahlen n, Feinvermahlen n
pulverize/to pulverisieren
pulverizer Feinmahlaggregat n
pulverizing unit Feinmahlaggregat n
pump/to pumpen
pump Pumpe f
   pump capacity Pumpenleistung f, Pumpenfördermenge f
   pump circuit Pumpenkreis m
   pump displacement chamber Pumpenverdrängerraum m
   pump drive power Pumpenantriebsleistung f
   pump motor Pumpen(antriebs)motor m
   pump speed Pumpendrehzahl f
   adsorption pump Adsorptionspumpe f
   amount conveyed by the pump Pumpenleistung f, Pumpenfördermenge f
   annular piston pump Kreiskolbenpumpe f
   axial piston pump Axialkolbenpumpe f
   booster pump Druckerhöhungspumpe f
   centrifugal pump Zentrifugalpumpe f
   circulating pump Umwälzpumpe f
   constant delivery pump Konstantpumpe f
   diffusion pump Diffusionspumpe f
   extraction pump Austragspumpe f
   feed pump Dosierpumpe f, Speisepumpe f
   fixed displacement pump Konstantpumpe f
   fuel pump Benzinpumpe f, Treibmittelpumpe f
   gear pump Getriebepumpe f, Zahnradpumpe f
   generated-rotor pump Wälzkolbenpumpe f
   gerotor pump Wälzkolbenpumpe f
   heat pump Wärmepumpe f
   high-pressure pump Hochdruckpumpe f
   high-pressure pump unit Hochdruckpumpenaggregat n
   internal gear pump Innenzahnradpumpe f
   liquid-ring pump Wasserringpumpe f
   liquid-ring vacuum pump Wasserringvakuumpumpe f
   melt metering pump Schmelzedosierpumpe f
   metering pump Dosierpumpe f
   oil diffusion pump Öldiffusionspumpe f
   orbital pump Kreiskolbenpumpe f
   piston pump Kolbenpumpe f
   positive displacement circulating pump Umlaufverdrängerpumpe f
   positive displacement pump Verdrängerpumpe f
   radial piston pump Radialkolbenpumpe f
   reciprocating piston pump Hubkolbenpumpe f
   rotary piston pump Sperrschieberpumpe f
   rotary positive displacement pump Rotationsverdrängerpumpe f
   rotary pump Drehkolbenpumpe f, Kreiselpumpe f
   turbo-vacuum pump Turbovakuumpumpe f
   vacuum pump Unterdruckpumpe f, Vakuumpumpe f
   vane-type pump Flügelzellen(regel)pumpe f
   variable delivery pump Regelpumpe f, Verstellpumpe f
   variable displacement pump Regelpumpe f, Verstellpumpe f
pumpable pumpbar
pumping capacity [of screw] Förderkapazität f
pumping space Schöpfraum m
punch Stempel m
   paper tape punch Lochstreifenstanzer m
punchability Stanzbarkeit f, Stanzfähigkeit f
punched gestanzt, Loch-
   punched card Lochkarte f
   punched card control (system) Lochkartensteuerung f
   punched card reader Lochkartenleser m
   punched holes Stanzlöcher npl
   punched tape Lochstreifen m
   punched tape reader Lochstreifenleser m
   controlled by punched cards lochkartengesteuert
punching Stanzen n
   punching characteristics Stanzfähigkeit f, Stanzbarkeit f
   punching tool Stanzwerkzeug n
puncture resistance Durchstichfestigkeit f, Durchstoßfestigkeit f, Durchdrückzähigkeit f
purchasing power Kaufkraft f
pure rein, pur, Rein-
   pure polymer Reinpolymerisat n
   pure resin Reinharz n
   pure-resin test piece Reinharzstab m
purge/to (aus)spülen, reinigen
purging [emptying the machine by operating it without adding further material] Leerfahren n
   purging air Spülluft f
   purging compound Reinigungscompound m,n, Reinigungsgranulat n
purgings Produktionsabfälle mpl
   machine purgings Anfahrbrocken mpl, Anfahrfladen mpl
purification Reinigung f
   waste gas purification plant Abgasreinigungsanlage f
purity Reinheit f, Reinheitsgrad m
purpose Vorsatz m, Absicht f, Zweck m, Zielsetzung f
   general-purpose universell einsetzbar, Allzweck-, Universal-
   general-purpose extruder Allzweckextruder m, Universalextruder m
   general-purpose extruder head Universalkopf m

general-purpose grade Universaltyp m
general-purpose granulator Universal-Zerkleinerer m
general-purpose machine Universalmaschine f
general-purpose model Universaltyp m
general-purpose rubber Allzweckkautschuk m
general-purpose screw Allzweckschnecke f, Standardschnecke f
special-purpose design Einzweckausführung f
special-purpose extruder Einzweckextruder m, Sonderextruder m
special-purpose machine Einzweckausführung f, Sondermaschine f
special-purpose screw Sonderschnecke f, Spezialschnecke f
push button Druckknopf m, Drucktaste f, Taste f
  push-button console Druckknopf-Station f
  push-button control (unit) Druckknopfsteuerung f, Drucktastensteuerung f
  push-button controlled drucktastengesteuert
  push-button switch Druckknopfschalter m
putting into operation [plant, machine etc.] Inbetriebnahme f
putty knife Ziehklinge f, Spachtel m
p.v.c. [pigment volume concentration] Pigmentvolumenkonzentration f
PVC PVC n, Polyvinylchlorid n
  PVC cable compound Kable-PVC n
  PVC foam Schaum-PVC n
  PVC particle PVC-Pulverkorn n, PVC-Teilchen n
  PVC paste PVC-Paste f, PVC-Plastisol n
  PVC paste resin Pastenware f, Pasten-PVC n, PVC-Pastentyp m
  PVC plastisol PVC-Plastisol n
  PVC resin Roh-PVC n
  PVC screw PVC-Schnecke f
  bulk PVC Masse-PVC n
  emulsion PVC Emulsions-PVC n
  extender PVC Extender-PVC n
  extrusion of plasticized PVC PVC-Weich-Extrusion f
  flexible PVC film PVC-Weichfolie f, Weich-PVC-Folie f
  flexible PVC product Weich-PVC-Artikel m
  injection mo(u)lding of plasticized PVC Weich-PVC-Spritzguß m
  paste-making PVC Pasten-PVC n, Pastenware f, PVC-Pastentyp m
  plasticized PVC Polyvinylchlorid n weich
  plasticized PVC compound Weich-PVC-Masse f
  rigid PVC Hart-PVC n, Polyvinylchlorid n hart, PVC n hart
  rigid PVC film Hart-PVC-Folie f, PVC-Hartfolie f
  rigid PVC mo(u)lding Hart-PVC-Formteil n
  rigid PVC sheet Hart-PVC-Platte f
  suspension PVC Suspensions-PVC n, Suspensionstype f
  unplasticized PVC compound Hart-PVC-Formmasse f, Hart-PVC-Compound m,n
  unplasticized PVC mo(u)lding compound Hart-PVC-Formmasse f, Hart-PVC-Compound m,n
PVF [polyvinylidene fluoride] Polyvinylidenfluorid n
pyknometer Pyknometer n
pyrogenic pyrogen
pyrogram Pyrogramm n
pyrolysis Pyrolyse f
  pyrolysis product Pyrolyseprodukt n
  laser pyrolysis Laserpyrolyse f
  solid phase pyrolysis Festphasenpyrolyse f
pyrolyze/to pyrolysieren
pyromellitic acid Pyromellitsäure f
pyromellitic anhydride Pyromellitsäureanhydrid m
pyrometer Pyrometer n
  needle pyrometer Einstichpyrometer n
  radiation pyrometer Strahlungspyrometer n
quadrihydric [if an alcohol] vierwertig
quadrivalent vierbindig, vierwertig
quadruple screw Vierfachschnecke f
qualitative qualitativ
quality Qualität f, Gütegrad m, Qualitätsniveau n
  quality assessment Qualitätsbeurteilung f
  quality assurance Gütesicherung f, Qualitätssicherung f
  quality control Güteüberwachung f, Qualitätskontrolle f, Qualitätsüberwachung f
  quality criteria Qualitätskriterien npl
  quality-enhancing qualitätssteigernd
  quality factor Gütefaktor m
  quality guidelines Qualitätsrichtlinien fpl, Güterichtlinien fpl
  quality requirements Güteanforderungen fpl, Qualitätsanforderungen fpl
  quality variations Qualitätsschwankungen fpl
  coating quality Beschichtungsqualität f
  constant quality Qualitätskonstanz f
  cutting quality Schnittgüte f
  extrusion quality Extrusionsqualität f
  film quality Filmqualität f, Folienbeschaffenheit f
  foam quality Schaumqualität f
  high-quality qualitativ hochwertig
  improvement in quality Qualitätsverbesserungen fpl
  incoming goods quality Wareneingangsparameter m
  loss of quality Qualitätseinbuße f
  melt quality Plastifikatgüte f, Schmelzequalität f
  mo(u)lded-part quality Formteilqualität f
  of constant quality qualitätskonstant
  paint film quality Beschichtungsqualität f
  part quality Formteilqualität f, Teilqualität f
  pipe quality Rohrgüte f
  product quality Fertigungsgüte f, Produktqualität f
  variations in quality Qualitätsschwankungen fpl
  weld quality Nahtgüte f, Schweißnahtgüte f, Schweißnahtqualität f
quantification Quantifizierung f
quantitative quantitativ
quantity Quantität f, Menge f, (zahlenmäßige) Größe f

experimental quantities Versuchsmengen fpl
in commercial quantities in Handelsmengen fpl
model quantity Modellgröße f
reference quantity Bezugsgröße f, Referenzgröße f
target quantity Zielgröße f
quartz Quarz n
  quartz force transducer Quarzkristall-Kraftaufnehmer m
  quartz pressure transducer Quarzkristall-Druckaufnehmer m
  quartz transducer Quarzkristallaufnehmer m, Quarzkristall-Meßwertaufnehmer m
quasi-static quasistatisch
quaternary quartär, quarternär
quench Abschreck-, Kühl-, Lösch-
  quench bath Kühlbad n
  quench tank Kühltrog m, Kühlwanne f
quenched schockgekühlt
quick schnell
  quick-acting clamp Schnellverschluß m
  quick-action coupling mechanism Schnellspannvorrichtung f, Schnellspannsystem n
  quick-action screen changer Sieb-Schnellwechseleinrichtung f
  quick-change filter unit Schnellwechselfilter n,m
quiet ruhig, leise
  quiet in operation lärmarm, laufruhig
  quiet-running mit geräuscharmem Lauf m, laufruhig
  very quiet-running mit hoher Laufruhe f
quietness in operation Laufruhe f
quinone Chinon n
rack Baugruppenträger m, Zahnstange f
  rack-and-pinion drive Zahnstangenantrieb m
  gear rack Zahnstange f
radial radial
  radial clearance Radialspiel n, radiales Spiel n, Schmierspalt m
  radial piston motor Radialkolbenmotor m
  radial piston pump Radialkolbenpumpe f
  radial screw clearance Schneckenscherspalt m
  radial system of runners Verteilerkreuz n, Verteilerstern m, Verteilerspinne f, Sternverteiler m, Anschnittstern m, Angußspinne f, Angußstern m
radiant heaters Strahlerheizung f, Strahlungsheizung f
radiation Abstrahlung f, Strahlung f
  radiation crosslinkage Strahlenvernetzung f, Strahlungsvernetzung f
  radiation-crosslinked strahlenvernetzt
  radiation curing Strahlungshärtung f
  radiation dosage Strahlungsdosis f, Bestrahlungsdosis f
  radiation heat loss Strahlungs(wärme)verlust m
  radiation-induced strahlungsinduziert, strahleninduziert
  radiation-induced oxidation Strahlungsoxidation f
  radiation intensity Bestrahlungsintensität f, Strahlungsintensität f
  radiation pyrometer Strahlungspyrometer n
  radiation resistance Strahlungsbeständigkeit f, Strahlenbeständigkeit f
  radiation-resistant strahlungsbeständig
  radiation sensitivity Bestrahlungsempfindlichkeit f
  radiation source Strahlungsquelle f
  radiation stabilizer Strahlenschutzmittel n
  heat loss due to radiation Strahlungs(wärme)verlust m
  heat radiation Wärmestrahlung f
  solar radiation Sonnenstrahlung f
  UV radiation UV-Strahlung f
radiator Kühler m, Heizkörper m, Radiator m
  radiator grill(e) Kühlergrill m, Kühlerschutzgitter n
  radiator surround Kühlerzarge f
  radiator tank Kühlwasserkasten m, Kühlwasserausgleichsbehälter m, Wasserkasten m
radical 1. radikalisch; 2. Radikal n, Rest m
  radical attack Radikalangriff m
  radical availability Radikalangebot n
  radical concentration Radikalkonzentration f
  radical-donating radikalspendend
  radical donor Radikalspender m
  radical formation Radikalbildung f
  radical former Radikalbildner m
  radical interception Radikaleinfang m
  radical interceptor Radikalfänger m
  radical yield Radikalausbeute f
  acid radical Säureradikal n, Säurerest m
  alkoxy radical Alkoxyradikal n
  alkyl radical Alkylrest m
  alkylperoxy radical Alkylperoxidradikal n
  amino radical Aminorest m
  ester alcohol radical Esteralkoholrest m
  glycidyl radical Glycidylrest m
  halogen radical Halogenradikal n
  hydrocarbon radical Kohlenwasserstoffrest m
  initiator radical Startradikal n
  monomer radical Monomerradikal n
  organic radical Organorest m
  peroxide radical Peroxidradikal n, Peroxyradikal n
  phenoxy radical Phenoxy-Radikal n
  phenyl radical Phenylrest m
  polymer radical Polymerradikal n
  primary radical Primärradikal n
  secondary radical Sekundärradikal n
  thioglycolate radical Thioglykolsäurerest m
  triphenylmethyl radical Triphenylmethylradikal n
  vinyl radical Vinylrest m
radius Radius m, Halbmesser m
  radius of curvature Krümmungsradius m
  radius of gyration Trägheitshalbmesser m
  film bubble radius Folienschlauchhalbmesser m
  operating radius Arbeitsradius m
rail Stange f, Schiene f
  rail tanker Eisenbahnkesselwagen m

rail transport Eisenbahntransport m
curtain rail Vorhangschiene f, Gardinenstange f
guide rail Führungsschiene f
rain Regen m
   driving rain Schlagregen m
   resistance to driving rain Schlagregensicherheit f
rainwater Regenwasser n
rake angle Spanwinkel m
ram Ramme f, Kolben m, Stößel m, Druckstempel m
   ram accumulator Kolbenspeicher m, Kolbenakku(mulator) m
   ram extruder Kolbenextruder m, Kolbenmaschine f, Kolbenstrangpresse f
   ram feeder Kolbenstopfaggregat n
   ram injection unit Kolbeninjektor m
   ram pressure Stößeldruck m
   ram stroke Stößelhub m
   clamp ram Schließkolben m
   clamping cylinder ram Schließkolben m
   clamping unit ram Schließkolben m
   feed ram Dosierkolben m
   melt accumulator ram Schmelzespeicherkolben m
   pressure ram Druckkolben m
   stuffing ram Stopfkolben m
   tubular ram Ringkolben m
   tubular ram accumulator Ringkolbenspeicher m
   tubular ram accumulator head Ringkolbenspeicherkopf m
   tubular ram injection Ringkolbeninjektion f
   tubular ram injection mo(u)lding (process) Ringkolbeninjektionsverfahren n
random zufällig, wirr, Zufalls-
   random access memory [RAM] RAM-Speicher m
   random sample Stichprobe f
   random test Stichprobenprüfung f
randomly distributed glass fibers wirre Glasfaserverteilung f
range Bereich m, Reichweite f; Bauserie f, Baureihe f
   range of accessories Zubehörprogramm n
   range of adhesives Klebstoffsortiment n
   range of applications Anwendungsbreite f, Anwendungsbereich m, Anwendungsspektrum n
   range of automatic blow mo(u)lding machines Blasformautomaten-Baureihe f
   range of concentration Konzentrationsbereich m
   range of equipment Ausrüstungsumfang m
   range of extruders Extruderbaureihe f
   range of grades Typenprogramm n, Typenübersicht f, Typensortiment n, Typenpalette f
   range of injection mo(u)lders Spritzgießmaschinenbaureihe f
   range of injection mo(u)lding machines Spritzgießmaschinenprogramm n
   range of machines Anlagenspektrum n, Anlagenprogramm n, Maschinenreihe f, Maschinenprogramm n
   range of outputs Leistungsfahrbreite f
   range of plasticizers Weichmachersortiment n
   range of products Angebotspalette f, Fabrikationsprogramm n, Produktenreihe f, Produktprogramm n, Produkt[ions]palette f, Produktsortiment n, Verkaufssortiment n, Verkaufsprogramm n, Fertigungsprogramm n, Herstellungsprogramm n
   range of properties Eigenschaftsspektrum n
   range of screws Schneckensortiment n
   range of standard units Standardgeräteserie f
   range of thermoset mo(u)lding compounds Duroplastformmassen-Sortiment n
   range of uses Einsatzbreite f, Einsatzspektrum n
   range selector Bereichswähler m
   basic range [of machines, equipment etc.] Grundreihe f
   boiling range Siedeintervall n
   decomposition range Zersetzungsbereich m
   extruder range Extruderbaureihe f
   frequency range Frequenzbereich m
   glass transition range Einfrierbereich m
   glass transition temperature range Glasumwandlungstemperaturbereich m
   granulator range Mühlenbaureihe f
   high-voltage range Hochspannungsbereich m
   low-voltage range Niederspannungsbereich m
   medium-voltage range Mittelspannungsbereich m
   melting range Schmelzbereich m, Schmelzintervall n
   modular range [of machines] Baukastenreihe f, Baukastenprogramm n
   operating range Leistungsfahrbreite f
   operating speed range Betriebsdrehzahlbereich m
   operating temperature range Betriebstemperaturbereich m, Einsatztemperaturbereich m, Temperatur-Anwendungsbereich m
   particle size range Korngrößenbereich m, Kornklasse f, Kornspektrum n, Korngrößenverteilung f
   pelletizer range Mühlenbaureihe f
   plasticizer range Weichmachersortiment n
   processing temperature range Verarbeitungstemperaturbereich m
   product range Verkaufsprogramm n, Produktsortiment n, Produktionspalette f, Produktprogramm n, Produktpalette f
   reference pressure range Referenzdruckbereich m
   scatter range Streubereich m, Streubreite f
   screw range Schneckensortiment n
   screw speed range Drehzahl-Regelbereich m, Drehzahlbereich m, Schneckendrehzahlbereich m, Drehzahlbereich m der Schnecke
   setting range Verstellbereich m
   softening range Erweichungsbereich m, Erweichungsgebiet n
   solubility range Löslichkeitsbereich m
   spectral range Spektralbereich m
   speed range Drehzahlbereich m

standard colo(u)r range Standardfarben-Palette f
standard range Normprogramm n, Serienprogramm n, Standardsortiment n
stoving temperature range Einbrennbereich m
temperature range Temperatur(einsatz)bereich m, Temperaturintervall m, Temperaturspanne f
tolerance range Toleranzfeld n
temperature tolerance range Temperaturtoleranzfeld n
transition range Umwandlungsbereich m
ultrasonic range Ultraschallbereich m
weighing range Wägebereich m, Wiegebereich m
raschel machine Raschelmaschine f
raschel-knit fabric Raschelgewirke n
rate Rate f, Geschwindigkeit f, Leistung f
  rate of compression Stauchungsgeschwindigkeit f
  rate of conversion Umsatzrate f
  rate of crystal growth Kristallwachstumsgeschwindigkeit f
  rate of crystallization Kristallisationsgeschwindigkeit f
  rate of decomposition Zerfallgeschwindigkeit f
  rate of deformation Verformungsgeschwindigkeit f, Deformationsgeschwindigkeit f
  rate of diffusion Diffusionsgeschwindigkeit f, Permeationsrate f
  rate of elongation Dehn(ungs)geschwindigkeit f
  rate of extrusion Extrusionsgeschwindigkeit f, Spritzgeschwindigkeit f
  rate of flame spread Flammenausbreitungsgeschwindigkeit f
  rate of growth Wachstumsgeschwindigkeit f
  rate of increase Steigerungsrate f
  rate of injection Einspritzgeschwindigkeit f, Einspritzstrom m, Einspritz-Volumenstrom m, Einspritzrate f, Spritzgeschwindigkeit f
  rate of investment Investitionsrate f
  rate of oxidation Oxidationsgeschwindigkeit f
  rate of plasticizer loss Weichmacher-Verlustrate f
  rate of poymerization Polymerisationsgeschwindigkeit f
  rate of shear Schergeschwindigkeit f
  rate of speed increase Drehzahlsteigerungsrate f
  rate of spherulite growth Sphärolithwachstumsgeschwindigkeit f
  air circulation rate Luftzirkulationsrate f
  burning rate Brenngeschwindigkeit f
  conversion rate Umsatzrate f
  conveying rate Förderleistung f
  cooling air delivery rate Kühlluftaustrittsgeschwindigkeit f
  cooling rate Abkühlgeschwindigkeit f
  cooling water flow rate Kühlwasserdurchflußmenge f
  crack growth rate Rißausbreitungsgeschwindigkeit f, Rißerweiterungsgeschwindigkeit f
  crack propagation rate Rißfortpflanzungsgeschwindigkeit f, Riß(wachstums)geschwindigkeit f, Rißwachstumsrate f
  creep rate Kriechgeschwindigkeit f, Kriechrate f
  curing rate Aushärtungsgeschwindigkeit f, Vernetzungsrate f
  decomposition rate Zerfallgeschwindigkeit f, Zersetzungsgeschwindigkeit f
  delivery rate Austrittsgeschwindigkeit f
  diffusion rate Diffusionsgeschwindigkeit f, Permeationsrate f
  discharge rate Entladungsgeschwindigkeit f
  emission rate Abgaberate f
  energy input rate Energieanlieferungsrate f
  energy release rate Energiefreisetzungsrate f
  evaporation rate Verdampfungsrate f
  extraction rate Extraktionsgeschwindigkeit f
  feed rate Einzugsgeschwindigkeit f
  flow rate Durchflußgeschwindigkeit f, Durchflußleistung f, Durchflußstrom m, Fließgeschwindigkeit f, Durchfluß m, Durchlaufmenge f, Durchflußmenge f, Fördermenge f, Strömungsgeschwindigkeit f
  flow rate transducer Strömungsgeschwindigkeitsaufnehmer m
  fusion rate Geliergeschwindigkeit f
  growth rate Wachstumsgeschwindigkeit f, Steigerungsrate f, Zuwachsrate f
  heating-up rate Aufheizgeschwindigkeit f, Aufheizrate f
  injection rate Einspritzmenge f
  machine hour rate Maschinenstundensatz m
  mass flow rate Massedurchsatz m
  measurement of flow rate Durchflußmessung f
  melt flow rate Schmelzestrom m
  output rate Ausstoß m, Ausstoßgeschwindigkeit f, Ausstoßleistung f, Ausstoßmenge f, Durchsatz m, Durchsatzleistung f, Massedurchsatz m
  output rate curve Ausstoßkennlinie f
  oxidation rate Oxidationsgeschwindigkeit f
  parison delivery rate Schlauchausstoßgeschwindigkeit f
  polymerization rate Polymerisationsgeschwindigkeit f
  production rate Fertigungsgeschwindigkeit f
  reaction rate Reaktionsgeschwindigkeit f
  required plasticizing rate Plastifizierstrombedarf m
  saponification rate Verseifungsgeschwindigkeit f, Verseifungsrate f
  scrap rate Ausschußquote f, Ausschußzahlen fpl
  shear rate Schergeschwindigkeit f
  stressing rate Beanspruchungsgeschwindigkeit f, Belastungsgeschwindigkeit f
  stretch rate Reckgeschwindigkeit f
  throughput (rate) Förderleistung f, Förderrate f, Förderungsgeschwindigkeit f
  volume flow rate Volumendurchsatz m
  volumetric flow rate volumetrischer Durchsatz m, Volumendurchsatz m
  vulcanizing rate Vulkanisationsgeschwindigkeit f
rating Schätzung f; Nennleistung f

drive rating [of motor] Antriebsleistung f
flow rating [of pump] Schluckvolumen n
power rating Anschlußleistung f
ratio Verhältnis n
   blow-up ratio Aufblasverhältnis n
   channel depth ratio Kanaltiefenverhältnis n
   compression ratio Kompressionsverhältnis n, Schneckenkompression f, Verdichtungsverhältnis n, Volumenkompressionsverhältnis n
   concentration ratio Konzentrationsverhältnis n
   cost-benefit ratio Preis-Wirkungs-Relation f
   cost-performance ratio Aufwand-Nutzen-Relation f, Preis-Durchsatz-Verhältnis n, Preis-Leistungs-Verhältnis n
   cost-service life ratio Preis-Lebensdauer-Relation f
   draw-down ratio Unterzugsverhältnis n
   flight depth ratio Gangtiefenverhältnis n
   flow length-wall thickness ratio Fließweg-Wanddickenverhältnis n
   friction (ratio) Friktion f
   haul-off ratio Abzugsverhältnis n
   mixing ratio Abmischverhältnis n
   molar ratio Molverhältnis n
   pressure flow-drag flow ratio Drosselkennzahl f, Drosselquotient m
   reduction ratio Untersetzungsverhältnis n
   slenderness ratio Schlankheitsgrad m
   stretch ratio Reckverhältnis n, Streckverhältnis n, Verstreckungsgrad m, Verstreckungsverhältnis n
   surface area-weight ratio Oberflächengewichtsverhältnis n
   viscosity ratio Viskositätsverhältnis n
rationalization Rationalisierung f
rationalize/to rationalisieren
raw rauh, roh, Roh-
   raw material Rohstoff m
   raw material availability Rohstoffverfügbarkeit f
   raw material costs Rohmaterialkosten pl
   raw material data file Rohstoffdatei f
   raw material feed unit Rohstoffzuführgerät n
   raw material producer Rohstoffhersteller m
   raw material recovery plant Rohstoffrückgewinnungsanlage f
   raw material requirements Rohstoffbedarf m
   raw material savings Rohmaterialersparnis f
   raw material shortage Rohstoffverknappung f
   raw material source Rohstoffquelle f
   raw material supplier Rohstofflieferant m
   amount of raw material required for a day's production Rohstofftagesmenge f
   incoming raw material control Rohstoffeingangskontrolle f
   plastics raw material Kunststoffrohstoff m
rays Strahlen mpl
   gamma rays Gammastrahlen mpl
reactant Reaktand m, Reaktionspartner m
reaction Reaktion f, Umsetzung f

reaction conditions Reaktionsparameter mpl, Reaktionsbedingungen fpl
reaction enthalpy Reaktionsenthalpie f
reaction environment Reaktionsmilieu n
reaction foam mo(u)lding (process) Reaktions-Schaumgießverfahren n
reaction-inhibiting reaktionshemmend
reaction injection mo(u)lding [RIM] Reaktions(spritz)gießen n, Reaktionsguß m, Reaktionsgießverarbeitung f
reaction injection mo(u)lding machine [RIM machine] Reaktionsgießmaschine f
reaction mechanism Reaktionsmechanismus m
reaction mixture Reaktionsgemisch n
reaction product Reaktionsprodukt n, Umsetzungsprodukt n
reaction rate Reaktionsgeschwindigkeit f
reaction stages Reaktionsschritte mpl
reaction temperature Reaktionstemperatur f
reaction time Reaktionsdauer f
reaction vessel Reaktionsgefäß n
addition reaction Additionsreaktion f
branching reaction Verzweigungsreaktion f
chain transfer reaction Kettenübertragungsreaktion f
chemical reaction chemische Umsetzung f, chemischer Umsatz m, chemische Reaktion f
condensation reaction Kondensationsreaktion f
coupling reaction Kupplungsreaktion f
crosslinking reaction Verknüpfungsreaktion f
curing reaction Aushärtungsreaktion f, Härtungsablauf m, Härtungsreaktion f, Vernetzungsreaktion f
decomposition reaction Zerfallsreaktion f, Zersetzungsreaktion f
degradation reaction Abbaureaktion f
dimerization reaction Dimerisierungsvorgang m
disproportioning reaction Disproportionierungsreaktion f
equilibrium reaction Gleichgewichtsreaktion f
first order reaction Reaktion f erster Ordnung
general reaction Reaktionsschema n
heat of reaction Reaktionswärme f
initial reaction Startreaktion f
intermediate reaction Zwischenreaktion f
polyaddition reaction Polyadditionsreaktion f
second order reaction Reaktion f zweiter Ordnung
secondary reaction Folgereaktion f, Sekundärreaktion f
side reaction Nebenreaktion f
speed of reaction Reaktionsgeschwindigkeit f
substitution reaction Substitutionsreaktion f
termination reaction Abbruchreaktion f
trimerization reaction Trimerisierungsvorgang m
vulcanizing reaction Vulkanisationsreaktion f
reactive reaktiv, reaktionsfähig, reaktionsfreudig
reactive thinner Reaktivverdünner m
less reactive reaktionsträger
moderately reactive mittelreaktiv

normally reactive normalaktiv
reactivity Reaktionsbereitschaft f, Reaktionsfähigkeit f, Reaktivität f
   reduction of reactivity Reaktivitätsminderung f
reactor Reaktor m
   high-pressure reactor Hochdruckreaktor m
   low-pressure reactor Niederdruckreaktor m
   top of the reactor Reaktorkopf m
read/to lesen
read-only memory [ROM] Nur-Lesespeicher m, ROM-Speicher m
   erasable and programmable read-only memory [EPROM] Eprom-Speicher m
reader Leseeinrichtung f
   magnetic card reader Magnetkartenleser m
   punched card reader Lochkartenleser m
   punched tape reader Lochstreifenleser m
readily leicht, gern
   readily soluble gut löslich
   readily volatile leichtflüchtig
readiness Bereitwilligkeit f, Schnelligkeit f
   readiness to gel Gelierfreudigkeit f
   readiness to invest Investitionsbereitschaft f
readjust/to nachstellen, nachjustieren
ready bereit, fertig
   ready for injection einspritzfertig
   ready for installation einbaufertig
   ready for processing verarbeitungsfertig
   ready for retrieval abrufbereit
   ready to be connected anschlußfertig
   ready-to-use verarbeitungsfertig, anwendungsfertig, gebrauchsfertig
   ready-to-use mix Fertigmischung f
real wirklich, echt, real
   real time Echtzeit f, Realzeit f
   real-time computer Realzeitrechner m
   real time control Echtzeitsteuerung f
   real time processing Echtzeitverarbeitung f
rear 1. Rück-; 2. Hinterseite f, Heck n
   rear apron Heckschürze f
   rear axle Hinterachse f
   rear bumper Heckstoßfänger m
   rear face of flight passive Stegflanke f, hintere Flanke f, passive Gewindeflanke f, passive Schneckenflanke f, hintere Stegflanke f, passive Flanke f
   rear hatch Heckklappe f
   rear light Rückleuchte f
   rear light housing Rückleuchtengehäuse n
   rear spoiler Heckspoiler m
   rear window Heckscheibe f
reasonable [e.g. costs] tragbar, vernünftig
reasonably priced preiswürdig, preisgünstig, kostengünstig
reasons Gründe mpl
   economic reasons Wirtschaflichkeitsgründe mpl
   for financial reasons aus Kostengründen mpl
   for reasons of safety aus Sicherheitsgründen mpl

rebound Rückprall m
   rebound resilience Rückstellelastizität f, Rückprallelastizität f, Rücksprunghärte f
rebuild/to [a machine] umbauen
rebuilding Umbau m
   machine rebuilding costs Maschinenumbaukosten pl
receive/to empfangen, erhalten, bekommen
   parison receiving station Schlauchübernahmestation f
recent neu, heute vorkommend
   recent development Neuentwicklung f
recess Aussparung f
   spare wheel recess Reserveradmulde f
   swallowtail recess Schwalbenschwanzausnehmung f
recession Rezession f
   economic recession Konjunkturschwäche f
reciprocal reziprok
reciprocating reversierend, Schub-
   reciprocating barrel accumulator Zylinderschubspeicher m
   reciprocating extruder reversierender Extruder m
   reciprocating movement Schubbewegung f
   reciprocating piston pump Hubkolbenpumpe f
   reciprocating screw Schneckenschub m, Schubschnecke f
   reciprocating-screw accumulator Schneckenschubspeicher m, Schneckenkolbenspeicher m
   reciprocating-screw cylinder Schneckenschubzylinder m
   reciprocating-screw extruder Schubschneckenextruder m
   reciprocating-screw injection mo(u)lding machine Schubschneckenspritzgießmaschine f
   reciprocating-screw machine Schubschneckenmaschine f, Schneckenschubmaschine f
   reciprocating-screw plasticizing unit Schneckenschub-Plastifizieraggregat n
   reciprocating-screw unit Schubschneckeneinheit f, Schneckenschubaggregat n
   single reciprocating screw Einfach-Schubschnecke f
recirculation 1. Rezirkulieren n; 2. Umlauf-
   recirculation cooling (system) Umlaufkühlung f
   recirculation oil cooling (system) Ölumlaufkühlung f
   cascade-type melt recirculation extruder Kaskadenumlaufextruder m
reclaim/to wiedergewinnen, wiederaufbereiten, aufarbeiten, regenerieren
reclaim Regenerat n, Rückführmaterial n, Rückware f
   reclaim extruder Regranulierextruder m
   reclaim plant Regenerieranlage f
   addition of reclaim Regeneratzusatz m
   edge trim reclaim(ing) (unit) Folienrandstreifenaufbereitung f
   film scrap reclaim plant Foliengranulieranlage f
reclamation Rückgewinnung f, Aufbereitung f, Regenerierung f
re-clamping Wiederaufspannen n
re-combination Rekombination f

recommendation Empfehlung f
  standard recommendation Standardempfehlung f
recommended empfohlen
recondition/to aufarbeiten
record/to protokollieren, aufzeichnen
record Aufzeichnung f, Protokoll n, Akte f
  machine setting record Einstellprotokoll n
  mo(u)ld setting record card Formeinrichtekarte f
  output record Ausgabeliste f
  production data record Betriebsdatenprotokoll n, Produktionsdatenprotokoll n
  record of mo(u)ld data Werkzeugdatenkatalog m
  set-up time record Rüstzeitaufnahme f
  shrinkage record Schwindungskatalog m
  status record Zustandsprotokoll n
recorder Registrator m, Aufnahmegerät n, Schreiber m
  cassette recorder Kassettenrecorder m, Bandkassettengerät n, Magnetband-Kassettengerät n
  chart recorder Linienschreiber m
  four-channel chart recorder Vierkanal-Linienschreiber m
  light spot line recorder Lichtpunktlinienschreiber m
  pressure recorder Druckschreiber m
  temperature recorder Temperaturschreiber m
recording Registrierung f, Protokollierung f
  recording device Registriereinrichtung f
  recording instrument Registriergerät n, Aufzeichnungsgerät n
  recording of pressure Druckaufzeichnung f
  recording of temperature Temperaturaufzeichnung f
  pressure recording device Druckschreiber m
records Unterlagen fpl
  production records Fertigungsunterlagen fpl
recovery Rückgewinnung f, Regenerierung f, Rückfederung f, Rückstellung f, Rückschrumpf m, Rückverformung f, Rückdeformation f
  recovery behavio(u)r Erholungsverhalten n
  recovery period Erholungszeit(spanne) f
  recovery process Erholungsprozeß m
  cooling water recovery unit Kühlwasserrückgewinnungsanlage f
  economic recovery Konjunkturerholung f
  elastic recovery elastische Rückfederung f, elastische Rückdeformation f
  energy recovery Energierückgewinnung f
  heat recovery Wärmerückgewinnung f
  product recovery Produktrückgewinnung f
  raw material recovery plant Rohstoffrückgewinnungsanlage f
  solvent recovery unit Lösemittelrückgewinnung f
recrystallize/to umkristallisieren
recrystallizing temperature Rekristallisationstemperatur f
rectangular rechteckig
  rectangular flow channel Rechteckkanal m
  rectangular profile die Rechteckprofildüse f
  rectangular runner Rechteckkanal m
  rectangular strip Rechteckstreifen m
  rectangular test piece Flachstab m
  rectangular test specimen Flachstab m
rectifier Gleichrichter m
recycled air Regenerierluft f
recycling Verwertung f, Rückführung f, Wiederverarbeiten n, Recycling n
  edge trim recycling system Randstreifenrückführsystem n
  edge trim recycling (unit) Randstreifenrückspeiseeinrichtung f, Randstreifenrückführanlage f, Randstreifenrezirkulierung f, Randstreifenrückführung f, Randstreifenrückspeisung f
  scrap recycling Abfallverwertung f
  waste heat recycling Abwärmenutzung f
red rot
  red iron oxide Eisenoxidrot n
  cadmium red Cadmiumrot n
  molybdenum red Molybdatrot n
re-designed neukonzipiert
re-designing Überarbeitung f, Umkonstruktion f, Neuauslegung f
  re-desgning of the mo(u)ld Werkzeugumgestaltung f
redispersibility Redispergierbarkeit f
redispersible redispergierbar
  redispersible powder Dispersionspulver n
reduced herabgesetzt, reduziert
  reduced mo(u)ld clamping pressure Werkzeugsicherungsdruck m
  reduced output(s) Ausstoßverringerung f, Leistungsminderung f
  reduced viscosity reduzierte Viskosität f
reducing Reduktions-, Reduzier-
  reducing agent Reduktionsmittel n
  reducing gear Reduziergetriebe n
  reducing waste to a minimum Ausschußminimierung f
  V-belt speed reducing drive Keilriemenuntersetzung f
reduction Reduktion f, Reduzierung f, Verringerung f, Einbruch m
  reduction gear Zahnraduntersetzungsgetriebe n
  reduction in gloss Glanz(grad)verlust m, Glanzminderung f
  reduction in output Durchsatzreduktion f, Ausstoßleistungsminderung f
  reduction in pressure Druckabfall m, Druckminderung f
  reduction of reactivity Reaktivitätsminderung f
  reduction ratio Untersetzungsverhältnis n
  back pressure reduction Staudruckabbau m
  cost reduction Kostensenkung f
  cycle time reduction Zykluszeitreduzierung f, Zykluszeitverkürzung f
  data reduction Datenreduktion f
  gear reduction [unit] Getriebeuntersetzung f, Zahnraduntersetzung f
  impact strength reduction Schlagzähigkeits-

(ver)minderung f
noise level reduction Schallpegelreduzierung f
noise reduction Schalldämpfung f, Geräuschreduzierung f, Lärmminderung f
noise reduction measures Geräuschdämpfungsmaßnahmen fpl, Lärmminderungsmaßnahmen fpl
preliminary size reduction Vorzerkleinern n
preliminary size reduction unit Vorzerkleinerungsmühle f, Vorzerkleinerer m
pressure reduction Druckreduzierung f
price reduction Preisnachlaß m, Preisreduzierung f, Verbilligung f
size reduction unit Zerkleinerer m, Zerkleinerungsmühle f, Zerkleinerungsmaschine f, Zerkleinerungsanlage f, Zerkleinerungsaggregat n
speed reduction Drehzahlerniedrigung f, Untersetzung f
speed reduction gear Untersetzungsgetriebe n
spur gear speed reduction mechanism Stirnrad-Untersetzungsgetriebe n
viscosity reduction Viskositätserniedrigung f, Viskositätsabsenkung f, Viskositätsabfall m, Viskositätsminderung f
volume reduction Volumenverkleinerung f
reel Rolle f, Spule f, Folienwickel m, Bobine f, Wickel m
reel changing mechanism Rollenwechselsystem n
reel diameter Wickeldurchmesser m
reel tension Wickelzug m, Wickelspannung f
reel trimmer Rollenschneidmaschine f
reel winder Rollenwickelmaschine f
reeling Roll-, Aufroll-, Abspul-
reeling speed Aufspulgeschwindigkeit f
reeling tension Aufwickelspannung f
reeling unit Spulwerk n
reference Bezugnahme f, Referenz f
reference curve Bezugskurve f
reference period Bezugszeitraum m
reference pressure Bezugsdruck m
reference pressure curve Solldruckkurve f
reference pressure range Referenzdruckbereich m
reference quantity Bezugsgröße f, Referenzgröße f
reference sample Vergleichsmuster n, Vergleichsprobe f
reference solution Vergleichslösung f
reference temperature Bezugstemperatur f, Referenztemperatur f
reflect/to reflektieren
reflecting surface Abstrahlfläche f
reflection Reflexion f
reflector Reflektor m
headlamp reflector Scheinwerferreflektor m
reflux Rückfluß m
reflux condenser Rückflußkühler m
under reflux unter Rückfluß m
reformation Rückbildung f
refraction Brechung f [Strahlen]
double refraction Doppelbrechung f

refractive index Brechungsindex m, Brechungszahl f, Lichtbrechungsindex m
refractometer Refraktometer n
refrigeration machine Kältemaschine f
refrigerator Kühlschrank m
refrigerator liner Kühlschrankinnenverkleidung f, Kühlschrank-Innengehäuse n
refuse Müll m
refuse bag Müllsack m
refuse sorting plant Müllaufbereitungsanlage f
domestic refuse Hausmüll m
domestic refuse collection Hausmüllsammlung f
regenerated cellulose Regeneratcellulose f
register/to erfassen
register Register n, Verzeichnis n
register ring Zentrierring m
data register Datenregister n
instruction register Befehlsregister n
secondary register Sekundär-Register n
regrind Mahlgut n, Regenarat n, Regranulat n, Rückführmaterial n, Rückware f [aufgearbeiteter Abfall]
regular regelmäßig
regulations Regelungen fpl, Vorschriften fpl
accident prevention regulations Unfall-Verhütungsvorschriften fpl, UVV fpl
food regulations lebensmittelrechtliche Anforderungen fpl, lebensmittelrechtliche Bestimmungen fpl
pollution control regulations Umweltschutzvorschriften fpl, Umweltschutzbedingungen fpl
safety regulations Sicherheitsregeln fpl, Sicherheitsvorschriften fpl
transport regulations Beförderungsvorschriften fpl
regulator Regulator m, Regler m
cell regulator Zellregulierungsmittel n
foam regulator Schaumregulator m
injection pressure regulator Spritzdruckregler m
oil flow regulator Ölstromregler m
pressure regulator Druckwaage f, Kraftwaage f
speed regulator Drehzahlregler m, Geschwindigkeitsregler m
viscosity regulator Viskositätsregler m
voltage regulator Spannungsgeber m
re-heat wiederaufwärmen
re-heating Wiedererwärmung f
re-heating section Wiederaufwärmstrecke f
re-heating time Wiederaufheizzeit f
time required for re-heating Wiederaufheizzeit f
reinforced armiert, verstärkt
reinforced concrete construction Stahlbetonbau m
reinforced with continuous glass strands langglasverstärkt
carbon fiber reinforced kohle(nstof)faserverstärkt
chopped strand reinforced kurzglasfaserverstärkt
fabric reinforced textilverstärkt

fiber reinforced faserverstärkt
glass fiber reinforced glasfaserhaltig, glasfaserverstärkt, textilglasverstärkt
glass fiber reinforced plastic [GRP] Glasfaserkunststoff m, GFK
glass mat reinforced glasmattenverstärkt
glass mat reinforced polyester laminate GF-UP-Mattenlaminat n
synthetic fiber reinforced chemiefaserarmiert
reinforcement Armierung f
  carbon fiber reinforcement Kohle(n)faserfüllung f, Kohlenstoffaserverstärkung f
  glass cloth reinforcement Glasgewebeverstärkung f, Glasfaserverstärkung f, Glasseidenverstärkung f, Textilglasverstärkung f
  glass reinforcement Glasverstärkung f
  woven roving reinforcement Rovinggewebeverstärkung f
reinforcing Verstärkungs-, Armierungs-
  reinforcing agent Verstärkungsmittel n
  reinforcing effect Verstärkungseffekt m, Verstärkungswirkung f
  reinforcing fabric Armierungsgewebe n, Verstärkungsgewebe n
  reinforcing fibers Verstärkungsfasern fpl
  reinforcing filler Verstärkerstoff m, Verstärkungsfüllstoff m, Verstärkungsadditiv n
  reinforcing material Verstärkungsstoff m, Verstärkungsmaterial n
  reinforcing ribs Versteifungsrippen fpl
reject/to wegwerfen, verwerfen, ablehnen
  reject articles Ausschußware f
  reject mo(u)ldings Abfallteile npl, Ausschuß m, Ausschußstücke npl, Ausschußteile npl, Produktionsrückstände mpl
rejects Ausschuß m
  number of rejects Ausschußzahlen fpl, Ausschußquote f
  production of rejects Ausfallproduktion f
related [materials, e.g. PE and PP] artverwandt
relating to control steuerungstechnisch
relation(ship) Beziehung f
relative relativ
  relative determination Relativmessung f
  relative humidity relative Luftfeuchte f
  relative permittivity Dielektrizitätszahl f, dielektrischer Konstant m, relative Dielektrizitätskonstante f
  relative pressure Relativdruck m
  relative tear strength relative Reißfestigkeit f
  relative velocity Relativgeschwindigkeit f
  relative viscosity relative Viskosität f
relax/to relaxieren
  ability to relax Relaxationsvermögen n
relaxation Relaxation f, Entspannungsvorgang m, Entspannung f
  relaxation behavio(u)r Relaxationsverhalten n

relaxation curve Relaxationskurve f
relaxation modulus Relaxationsmodul m
relaxation process Relaxationsvorgang m, Relaxationsprozeß m
relaxation test Relaxationsversuch m
relaxation zone [part of screw] Beruhigungszone f
stress relaxation Spannungsabbau m, Spannungsrelaxation f
stress relaxation speed Spannungsabbaugeschwindigkeit f
stress relaxation test Spannungsrelaxationsversuch m
relay Schütz n, Relais n
  relay control Relais-Steuerung f, Schützensteuerung f
  relay-controlled schützengesteuert
  relay output Relaisausgang m
  time relay Zeitrelais n
release/to freigeben, (her)auslösen
release Trennung f, Freisetzung f
  release agent Entformungs(hilfs)mittel n, Trennmittel n
  release agent residues Trennmittelreste mpl
  release coating Trennbeschichtung f, Trennfilm m
  release film Mitläuferfolie f, Trennfolie f
  release paper Mitläuferpapier n, Trennpapier n
  release properties Entformbarkeit f
  application of release agent Trennmittelbehandlung f
  ejector release (mechanism) Auswerferfreistellung f
  energy release Energiefreisetzung f
  energy release rate Energiefreisetzungsrate f
  mo(u)ld release agent Formtrennmittel n
  mo(u)ld release effect Formtrennwirkung f
  pressure release Druckentlastung f
  pressure release system Druckentspannungssystem n
  pressure release valve Druckreduzierventil n, Druckbegrenzungsventil n, Druckentlastungsventil n, Druckminderventil n
  silicone release agent Silicontrennmittel n
  silicone release resin Silicontrennharz n
reliability [in operation] Betriebssicherheit f, Funktionssicherheit f
reliable funktionssicher, betriebssicher
relief Entspannung f
  relief valve Dekompressionsventil n, Schnüffelventil n
  back pressure relief (mechanism) Staudruckentlastung f
re-machining [e.g. a worn machine part] Nachbearbeitung f
re-melting Wiederaufschmelzen n
remote fern, entfernt, Fern-
  remote control Fernsteuerung f, Fernkontrolle f
  remote control adjusting Ferneinstellung f
  remote control unit Fernkontrolle f, Fernsteuerung f [Gerät]
  remote speed control (mechanism) Geschwindigkeitsfernsteuerung f
  remote-controlled ferngesteuert, fernsteuerbar
re-mounting [e.g. a mo(u)ld] Wiederaufspannen n

removable demontierbar, herausnehmbar
removal Entfernen n, Beseitigen n, Abtransport m
  removal of blow mo(u)ldings Blaskörperentnahme f
  removal of heat Wärmeabfuhr f, Wärmeentzug m
  removal of hydrogen Wasserstoff-Abstraktion f
  removal of residual gases Restentgasung f
  removal of residual monomer Restmonomerentfernung f
  removal of screw Schneckenausbau m
  removal of stress from the test piece Prüfkörperentlastung f
  removal of the die Werkzeugausbau m
  removal of the mo(u)ld Werkzeugausbau m
  removal of waste Abfallentfernung f
  removal of water Entwässerung f
  blown part removal (device) Blaskörperentnahme f
  dust removal Entstauben n
  dust removal system Entstaubungssystem n
  edge trim removal unit Randstreifenabsauganlage f
  flash removal Entbutzen n, Butzenbeseitigung f, Abfallentfernung f
  load removal device Entlastungsvorrichtung f
  scrap removal system Abfallentfernungssystem n
  scrap removal unit Abfallentfernvorrichtung f
  stress removal Entlastung f
  stress removal time Entlastungszeit f
remove/to entfernen, ausbauen
rendering (coat) Putz m
  rendering mix Putz m
renewal Erneuerung f, Neubildung f
  surface renewal Oberflächenneubildung f
renovation Sanierung f, Renovierung f
re-painting Neuanstrich m
repair/to reparieren
repair(s) Reparatur(arbeiten) f(pl)
  repair costs Reparaturaufwand m, Reparaturkosten pl
  repair mortar Reparaturmörtel m
  easy to repair reparaturfreundlich
repeat test Wiederholungsprüfung f, Wiederholungsmessung f
  repeat test series Wiederholungsmeßreihen fpl
repeatability Repetierbarkeit f, Reproduzierfähigkeit f, Wiederholbarkeit f
repeatable reproduzierbar, wiederholbar
repeated wiederholt, Dauer-, Wechsel-, wechselnd
  repeated stress Wechselbeanspruchung f
  repeated tensile stress Zugschwellast f, Zugschwellbelastung f
repellency Abweisung f
  imparting water repellency Hydrophobierung f
  water repellency Wasserabweisung f
repellent abstoßend, abweisend
  dirt repellent schmutzabweisend
  making water repellent hydrophobierend
  water repellent wasserabstoßend, wasserabweisend, hydrophob
repelletize/to regranulieren

re-pelletizing Umgranulieren n
repelletizing regranulierend, Regranulier-
  scrap repelletizing line Regranulieranlage f
replacement Ersetzen n, Austausch m
  replacement value Wiederbeschaffungswert m
replacing 1. ersetzend, austauschend; 2. -wechsel m
  replacing the die Düsenwechsel m
  replacing the screen Siebwechsel m
report Bericht m
  annual report (jährlicher) Geschäftsbericht m
  status report Zustandsbericht m
  test report Prüfbericht m
reprocess/to wiederaufbereiten, aufarbeiten
reprocessing Aufbereitung f, Aufbereitungsverfahren n, Wiederverarbeiten n
  reprocessing line Aufbereitungsanlage f, Aufbereitungsstraße f
  reprocessing system Aufbereitungssystem n
  reprocessing unit Aufbereitungsmaschine f, Aufbereitungsaggregat n
  film scrap re-processing plant Folienaufbereitungsanlage f
  scrap reprocessing Abfallaufbereitung f
  scrap reprocessing plant Abfallaufbereitungsanlage f
  twin reprocessing unit Aufbereitungs-Doppelanlage f
reproducibility Reproduzierbarkeit f, Reproduzierfähigkeit f
reproducible wiederholbar, reproduzierbar
  precisely reproducible reproduziergenau
reproduction Reproduktion f
  accurate reproduction of detail Abbildegenauigkeit f
re-programming Neuprogrammierung f, Umprogrammierung f
require/to erfordern, verlangen
  requiring a lot of energy energieaufwendig
  requiring a lot of space platzaufwendig, platzintensiv
  requiring little maintenance wartungsarm
  requiring little space platzsparend
  requiring no maintenance wartungsfrei
required 1. erforderlich, Soll-; 2. -bedarf m
  required amount of plasticizer Weichmacherbedarf m
  required diameter Solldurchmesser m
  required dimension Sollmaß n
  required plasticizing rate Plastifizierstrombedarf m
  required pot life Verarbeitungsbedarf m [Lagerfähigkeit]
  required pressure Drucksollwert m, Solldruck m
  required screw speed Extruderdrehzahl-Sollwert m
  required temperature Solltemperatur f, Temperatursollwert m
  required value Sollgröße f, Sollwert m
  required wall thickness Sollwanddicke f, Wanddickensollwert m
  required weight Sollgewicht n
  amount of energy required Energieaufwand m
  amount of raw material required for a day's

requirement
    production Rohstofftagesmenge f
  as required wählbar
  space required Platzbedarf m
  time required Zeitaufwand m, Zeitbedarf m
  time required for blowing Blaszeit f
  time required for cleaning Reinigungszeit f
  time required for re-heating Wiederaufheizzeit f
  time required to evaporate Ablüftungszeit f [Lösungsmittel]
requirement Forderung f, Bedarf m, Anforderung f
  customers' requirements Kundenbedürfnisse npl
  energy requirements Energiebedarf m
  floor space requirements [to install a machine] Grundflächenbedarf m, Stellflächenbedarf m
  heat requirements Wärmebedarf m
  hydraulic oil requirements Hydraulikölbedarf m
  increased energy requirements Energiemehraufwand m
  list of requirements Anforderungskatalog m, Lastenheft n
  maintenance requirements Wartungsansprüche mpl
  maximum requirements Spitzenbedarf m
  meeting practical requirements praxisgerecht
  oil requirements Ölbedarf m, Ölmengenbedarf m
  performance requirements Funktionsanforderungen fpl
  power requirements Leistungsbedarf m, Kraftbedarf m
  processing requirements verfahrenstechnische Bedürfnisse npl, verfahrenstechnische Anforderungen fpl
  production requirements Produktionserfordernisse npl
  quality requirements Güteanforderungen fpl, Qualitätsanforderungen fpl
  raw material requirements Rohstoffbedarf m
  safety requirements Sicherheitsbedürfnis n, Sicherheitsanforderungen fpl
  space requirements Raumbedarfsplan m
  styling requirements Stylinganforderungen fpl
  tolerance requirements Toleranzanforderungen fpl
re-sealable wiederverschließbar
research Forschung f, Untersuchung f, Nachforschung f
  research costs Forschungsaufwand m
  research facilities Forschungseinrichtungen fpl
  research work Forschungsarbeit(en) f(pl)
reserve Reserve f
  reserve locking force Zuhaltekraftreserve f
  energy reserve Leistungsreserve f
  power reserve Gangreserve f
  pressure reserve Druckreserve f
  speed reserve Drehzahlreserve f
reserves Rücklagen fpl, Rückstellungen fpl
  stability reserves Stabilitätsreserven fpl
reservoir Reservoir n, Sammelbecken n, Staubecken n
  hydraulic oil reservoir Hydrauliköltank m
  lubricating oil reservoir Schmieröltank m
  oil reservoir Ölbehälter m

re-set wiedereinstellen
re-setting Wiedereinrichten n
residence Wohnen n, Verweilen n
  residence time Aufenthaltszeit f, Verweilzeit f
  residence time profile Verweilzeitspektrum n
  residual residence time Verweilzeitschwanz m
residual übrigbleibend, Rest-
  residual acid content Restsäuregehalt m
  residual acrylonitrile content Acrylnitrilrestgehalt m
  residual amount Restmenge f
  residual elongation bleibende Dehnung f
  residual elongation at break Restreißdehnung f
  residual epoxide group content Rest-Epoxidgruppengehalt m
  residual ethylene content Restethylengehalt m
  residual moisture content Restfeuchtigkeitsgehalt m, Restfeuchte f, Restfeuchtigkeit f
  residual monomer Restmonomer m
  residual monomer content Restmonomergehalt m
  residual orientation Restorientierung f
  residual pressure Restdruck m
  residual residence time Verweilzeitschwanz m
  residual solvent Lösungsmittelrest m
  residual solvent content Lösungsmittelrestgehalt m
  residual stability Reststabilität f
  residual stresses Restspannungen fpl
  residual styrene Styrolreste mpl
  residual styrene content Styrolrestgehalt m, Reststyrolanteil m, Reststyrolgehalt m
  residual tack Restklebrigkeit f
  residual ultimate strength Restreißkraft f
  residual vinyl chloride (monomer) content VC-Restmonomergehalt m
  residual water Restwasser n
  residual welding stresses Schweißrestspannungen fpl
  distribution of residual welding stresses Schweißrestspannungsverteilung f
  removal of residual gases Restentgasung f
  removal of residual monomer Restmonomerentfernung f
residue Rest m, Rückstand m
  carbon residues Kohlenstoffreste mpl, Kohlenstoffrückstände mpl
  dry residue Trockenrückstand m
  emulsifier residues Emulgatorreste mpl
  filtration residue Filterrückstand m
  grease residues Fettreste mpl
  oil residues Ölreste mpl
  paint residues Farbreste mpl, Lackreste mpl
  polymer residues Kunststoffrückstände mpl
  release agent residues Trennmittelreste mpl
  sieving residue Siebrückstand m
  stoving residue Einbrennrückstand m
  tar residues Teerreste mpl
  wax residues Wachsreste mpl
resilience Feder(ungs)eigenschaften fpl, Rückfedervermögen n, Rückstellfähigkeit f,

Rückstellkraft f, Rückstellvermögen n
impact resilience Stoßelastizität f
rebound resilience Rückprallelastizität f, Rücksprunghärte f, Rückstellelastitzität f
resilient rückstellfähig
tough and resilient zähelastisch
resin Harz n
  resin-catalyst mix Harzansatz m, Reaktionsharzmischung f, Reaktions(harz)masse f
  resin content Harzanteil m, Harzgehalt m
  resin impregnated harzgetränkt, harzimprägniert
  resin matrix Harzmatrix f
  resin pick-up Harzaufnahme f
  resin-rich harzreich
  resin solution Harzlösung f
  acetal resin Acetalharz n
  acrylic resin Acryl(säure)harz n
  addition of resin Harzzusatz m
  adhesive resin Klebeharz n, Klebstoffharz n, Leimharz n
  alkyd resin Alkydharz n
  alkyl silicone resin Alkylsilikonharz n
  allyl resin Allylharz n
  amino resin Amino(plast)harz n
  amount of resin applied Harzauftragsmenge f
  base resin Basisharz n, Grundharz n
  benzoguanamine resin Benzoguanaminharz n
  bisphenol resin Bisphenolharz n
  blending resin Fremdharz n, Verschnittharz n
  brake lining resin Bremsbelagharz n
  cast resin Gießharzformstoff m
  casting resin Gießharz n
  casting resin technology Gießharztechnik f
  catalyzed epoxy resin EP-Reaktionsharzmasse f
  catalyzed polyester resin UP-Reaktionsharzmasse f
  catalyzed resin Reaktionsharzmischung f, Reaktions(harz)masse f
  condensation resin Kondensationsharz n
  copolymer resin Copolymerharz n
  coumarone-indene resin Cumaron-Indenharz n
  cresol resin Kresolharz n
  cured casting resin Gießharzformstoff m
  cured epoxy resin Epoxidharz-Formstoff m
  cured polyester resin Polyesterharzformstoff m, UP-Harzformstoff m
  cured resin Harzformstoff m, Preßstoff m
  cyclohexanone resin Cyclohexanonharz n
  dry resin content Trockenharzgehalt m
  electrical insulating resin Elektroisolierharz n
  encapsulating resin Vergußharz n
  epoxide resin EP-Reaktionsharz n, Epoxidharz n
  epoxy casting resin Epoxidgießharz n, Epoxygießharz n
  epoxy resin EP-Harz n
  epoxy resin solution Epoxidharzlösung f
  extender resin Fremdharz n, Verdünnerharz n, Verschnittharz n
  filament winding resin Wickelharz n
  foundry resin Gießereibindemittel n, Gießereiharz n
  furane resin Furanharz n
  gel coat resin Deckschichtharz n, Oberflächenharz n [GFK]
  grinding wheel resin Schleifscheibenharz n
  hard resin Hartharz n
  hydrocarbon resin Kohlenwasserstoffharz n
  impregnating resin Tränkharz n
  injection mo(u)lding resin Spritzgießharz n
  laminating resin Laminierharz n
  liquid phenolic resin Phenolflüssigharz n
  liquid resin Flüssigharz n
  liquid resin blend Flüssigharzkombination f
  liquid resin press mo(u)lding Naßpreßverfahren n
  maleic resin Maleinsäureharz n
  model-making resin Modelharz n
  modifying resin Modifizierharz n
  mo(u)lded thermoset resin Preßstoff m
  natural resin Naturharz n
  novolak resin Novolakharz n
  paint resin Lackrohstoff m, Lackharz n, Lackbindemittel n
  paste extender resin Pastenverschnittharz n
  phenol-formaldehyde resin Phenol-Formaldehydharz n
  phenolic engineering resin technisches Phenolharz n
  phenolic resin Phenolharz n
  phenyl silicone resin Phenylsiliconharz n
  phenylmethyl silicone resin Phenylmethylsiliconharz n
  phthalate resin Phthalatharz n
  polyaddition resin Polyadditionsharz n
  polyamide resin Polyamidharz n
  polycondensation resin Polykondensationsharz n
  polyester resin Polyester-Kunstoff m, Polyesterharz n
  polyester resin mix UP-Reaktionsharzmasse f
  polyester surface coating resin Lackpolyester m
  polysulphide resin Polysulfidharz n
  polyurethane resin Polyurethanharz n
  powdered phenolic resin Phenolpulverharz n
  powdered resin Pulverharz n
  press mo(u)lded using liquid resin naßgepreßt
  pure resin Reinharz n
  PVC paste resin PVC-Pastentyp m, Pasten-PVC n
  PVC resin Roh-PVC n
  resol resin Resolharz n
  resorcinol(-formaldehyde) resin Resorcinharz n
  silicone casting resin Silicongießharz n
  silicone release resin Silicontrennharz n
  silicone resin Siliconharz n
  silicone resin mo(u)lding compound Siliconharzpreßmasse f
  soft resin Weichharz n
  solid resin Festharz n
  solid resin blend Festharzkombination f
  solid silicone resin Siliconfestharz n

special-purpose resin Spezialharz n
speciality resin Spezialharz n
surface coating resin Lackbindemittel n, Lackharz n, Lackrohstoff m
synthetic paint resin Lackkunstharz n
synthetic resin Kunstharz n
synthetic resin-based paint Kunstharzlack m
synthetic resin bound kunstharzgebunden
synthetic resin casting Gießharzkörper m, Kunststoffgießharzkörper m
synthetic resin dispersion Kunstharzdispersion f
synthetic resin insulator Kunstharzisolator m
synthetic resin laminate Kunstharzlaminat n
synthetic resin latex Kunstharzlatex n
synthetic resin modified kunstharzvergütet, kunstharzmodifiziert
terpene resin Terpenharz n
thiourea-formaldehyde resin Thioharnstoffharz n
tooling resin Werkzeugharz n
typical of epoxy resins epoxidtypisch
unsaturated polyester resin [UP resin] UP-Harz n, UP-Reaktionsharz n
UP resin UP-Harz n, UP-Reaktionsharz n
urea(-formaldehyde) resin Harnstoff-Formaldehydharz n, Harnstoffharz n
vinyl ester resin Vinylesterharz n
vinyl resin Vinylharz n
with a high resin content harzreich
with a low resin content harzarm
resistance Widerstand(sfähigkeit) m(f), Beständigkeit f, Resistenz f
resistance band heater Widerstandsheizband n
resistance heater Widerstandsheizelement n, Widerstandsheizung f, Widerstandsheizkörper m
resistance thermocouple Widerstands-(temperatur)fühler m
resistance thermometer Widerstandsthermometer n
resistance to chalking Kreidungsresistenz f, Kreidungsbeständigkeit f
resistance to changing climatic conditions Klimawechselbeständigkeit f
resistance to deformation through impact Schlagunverformbarkeit f
resistance to driving rain Schlagregensicherheit f
resistance to flying stones Steinschlagfestigkeit f
resistance to high and low temperatures Temperaturbeständigkeit f
resistance to internal pressure Innendruckfestigkeit f
resistance to root penetration Wurzelfestigkeit f
resistance to sterilizing temperatures Sterilisierfestigkeit f
resistance to swelling Schwellfestigkeit f
resistance to torsional stress Torsionsbelastbarkeit f
resistance to yellowing Gilbungsresistenz f
abrasion resistance Abrasionsfestigkeit f, Abrasionswiderstand m, Abriebbeständigkeit f, Abriebfestigkeit f, Abriebwiderstand m

accelerated weathering resistance Kurzzeitbewitterungsverhalten n
acid resistance Säurebeständigkeit f
ag(e)ing resistance Alterungsbeständigkeit f
alkali resistance Alkalibeständigkeit f, Laugenbeständigkeit f
arc resistance Bogenwiderstand m, Lichtbogenfestigkeit f
bitumen resistance Bitumenbeständigkeit f
blocking resistance Blockfestigkeit f
buckling resistance Beulsteifigkeit f, Knickfestigkeit f
changes in flow resistance Fließwiderstandsänderungen fpl
chemical resistance Chemikalien-Tauglichkeit f, Chemikalienbeständigkeit f, Chemikalienresistenz f, chemische Beständigkeit f
chemical resistance table Beständigkeitstabelle f [gegenüber Chemikalien]
cold crack resistance Kältebruchfestigkeit f
compression resistance Druckbeanspruchbarkeit f
corona resistance Koronabeständigkeit f, Koronafestigkeit f
corrosion resistance Korrosionsbeständigkeit f
crack propagation resistance Rißausbreitungswiderstand m
crack resistance Rißbildungsresistenz f, Rißwiderstand m
creep resistance Kriechwiderstand m
current-limiting resistance Strombegrenzungswiderstand m
deformation resistance Deformationswiderstand m, Verformungsstabilität f
detergent resistance Detergentienfestigkeit f
die resistance Düsenwiderstand m, Spritzkopfwiderstand m, Werkzeugwiderstand m
diffusion resistance Diffusionswiderstand m
erosion resistance Erosionsbeständigkeit f, Erosionsfestigkeit f
extraction resistance Auslaugebeständigkeit f, Extraktionsbeständigkeit f
fire resistance Brandsicherheit f, Feuerwiderstandsfähigkeit f, Flammfestigkeit f
flash-over resistance Überschlagfestigkeit f
flow resistance Durchflußwiderstand m, Fließwiderstand m, Strömungswiderstand m
flow resistance changes Fließwiderstandsänderung f
frictional resistance Reibungswiderstand m
gelation resistance Gelierungsbeständigkeit f
grease resistance Fettbeständigkeit f
heat ag(e)ing resistance Wärmealterungsbeständigkeit f
heat resistance Temperatur(stand)festigkeit f, Temperaturbeständigkeit f, Temperaturstabilität f, Wärme(form)beständigkeit f, thermische Beständigkeit f, Warmfestigkeit f, Wärme(stand)festigkeit f
heat transfer resistance Wärmedurchgangs-

widerstand m
high-temperature resistance Hochtemperaturbeständigkeit f
hot air resistance Heißluftbeständigkeit f
impact resistance Schlagbeständigkeit f
incandescence resistance Glutbeständigkeit f
indentation resistance Eindruckwiderstand m
input resistance Eingangswiderstand m
insulation resistance Widerstand m zwischen Stöpseln
light resistance Lichtbeständigkeit f, Lichteigenschaften fpl
limited resistance bedingt beständig
long-term chemical resistance Chemikalien-Zeitstandverhalten n
long-term heat resistance Dauerwärmebelastbarkeit f, Dauerwärmestabilität f, Dauerwärmebeständigkeit f, Dauertemperaturbeständigkeit f, Dauertemperaturbelastungsbereich m [Hitze]
long-term temperature resistance Dauertemperaturbelastungsbereich m [Hitze und Kälte]
long-term tracking resistance Kriechstromzeitbeständigkeit f
long-term weathering resistance Dauerwitterungsstabilität f
low-temperature resistance Tieftemperaturbeständigkeit f, Tieftemperaturfestigkeit f
low-temperature shock resistance Kälteschockfestigkeit f
migration resistance Wanderungsbeständigkeit f, Wanderungsfestigkeit f
moisture resistance Feuchtigkeitsresistenz f
oil resistance Ölbeständigkeit f
outdoor weathering resistance Außenwitterungs-Beständigkeit f
overstoving resistance Überbrennstabilität f
oxidation resistance Oxidationsbeständigkeit f, Oxidationsstabilität f
ozone resistance Ozonfestigkeit f
plasticizer migration resistance Weichmacherwanderungsbeständigkeit f
puncture resistance Durchdrückzähigkeit f, Durchstichfestigkeit f, Durchstoßfestigkeit f
radiation resistance Strahlenbeständigkeit f
root resistance Wurzelfestigkeit f
rot resistance Verrottungsbeständigkeit f, Verrottungsfestigkeit f
saponification resistance Verseifungsbeständigkeit f
scorch resistance Scorchsicherheit f
scratch resistance Kratzfestigkeit f, Oberflächenkratzfestigkeit f, Ritzfestigkeit f
screw-in resistance thermometer Einschraub-Widerstandsthermometer n
scrub resistance Scheuerbeständigkeit f, Scheuerfestigkeit f, Scheuerresistenz f
shock resistance Erschütterungsfestigkeit f, Schockfestigkeit f, Schockzähigkeit f
skinning resistance Hautbildungsresistenz f

solder bath resistance Lötbadfestigkeit f
solvent resistance Lösemittelbeständigkeit f, Lösungsmittelresistenz f
stain resistance Fleckenunempfindlichkeit f
stress cracking resistance Spannungskorrosionsbeständigkeit f, Spannungsrißbeständigkeit f, Spannungsrißkorrosionsbeständigkeit f
surface resistance Oberflächenwiderstand m
swelling resistance Quellbeständigkeit f
tear propagation resistance Weiterreißfestigkeit f, Weiterreißwiderstand m
tear resistance Einreißfestigkeit f, Einreißwiderstand m
tear resistance test Einreißversuch m
thermal shock resistance Wärmeschockverhalten n
tracking resistance Kriechstromfestigkeit f
tracking resistance test Kriechstromprüfung f
vapo(u)r diffusion resistance Dampfdiffusionswiderstand m
volume resistance Volumenwiderstand m
warp resistance Verformungswiderstand m
washing resistance Waschbeständigkeit f
water resistance Naßfestigkeit f, Wasserbeständigkeit f, Wasserfestigkeit f
water vapo(u)r resistance Wasserdampfstabilität f
wear resistance Verschleißfestigkeit f, Verschleißwiderstandsfähigkeit f, Verschleißwiderstand m
weathering resistance Bewitterungsbeständigkeit f, Bewitterungsstabilität f, Freiluftbeständigkeit f, Wetterbeständigkeit f, Wetterfestigkeit f, Witterungsbeständigkeit f, Witterungsstabilität f
wet scrub resistance Naßabriebfestigkeit f
with enhanced impact resistance erhöht schlagzäh
resistant widerstandsfähig, resistent, beständig
resistant to boiling water kochfest
resistant to compression druckfest
resistant to deformation through impact schlagunverformbar
resistant to dry sliding friction trockengleitverschleißarm
resistant to high and low temperatures temperaturbeständig
resistant to pressure druckfest
resistant to root penetration wurzelfest
resistant to solvents lösungsmittelfest
resistant to sterilizing temperatures sterilisationsbeständig
resistant to stress cracking spannungsrißbeständig
resistant to tropical conditions tropenbeständig, tropenfest
resistant to yellowing gilbungsstabil, vergilbungsbeständig
abrasion resistant abrasionsfest, abriebfest
acid resistant säurebeständig
alkali resistant alkalibeständig
arc resistant lichtbogenfest
bitumen resistant bitumenbeständig

break resistant bruchsicher
chemical resistant chemikalienbeständig, chemikalienfest, chemikalienresistent
corrosion resistant korrosionsfest, korrosionsbeständig
extraction resistant extraktionsbeständig
extremely wear resistant hochverschleißfest
fire resistant feuerbeständig
flame resistant brandgeschützt, brandschutzausgerüstet, feuerbeständig, flammfest, flammsicher, flammwidrig
fuel oil resistant heizölbeständig
fuel resistant benzinfest
grease resistant fettbeständig
heat resistant wärme(form)beständig, wärmestabil, wärmestandfest
high-temperature resistant hochhitzebeständig, hochtemperaturbeständig, hochwärme(form)beständig
highly abrasion resistant hochabriebfest
hot water resistant heißwasserbeständig
impact resistant schlagzäh, schlagfest, stoßfest, stoßsicher
light resistant lichtbeständig
low-temperature impact resistant kaltschlagzäh
making impact resistant Schlagfestmachen n
moisture resistant feuchtigkeitsbeständig
not resistant [material in presence of chemicals] unbeständig, nichtbeständig
oil resistant ölbeständig, ölfest
plasticizer resistant weichmacherfest
radiation resistant strahlungsbeständig
root resistant wurzelfest
saponification resistant verseifungsbeständig
scorch resistant scorchsicher
shock resistant stoßfest, schockresistent, schockfest
solvent resistant lösungsmittelfest
tear resistant reißfest
torsion resistant verwindungssteif
tracking resistant kriechstromfest
vibration resistant erschütterungsfest
water resistant wasserbeständig, wasserfest
wear resistant verschleißarm, verschleißbeständig, verschleißfest, verschleißfrei, verschleißwiderstandsfähig
wear resistant bushing Verschleißbüchse f
wear resistant coating Panzerung f, Panzerschicht f, Oberflächenpanzerung f, Verschleißschicht f
wear resistant layer Verschleißschicht f
weather resistant (be)witterungsstabil, wetterbeständig, witterungsbeständig, witterungfest
resistivity spezifischer Widerstand m
  surface resistivity spezifischer Oberflächenwiderstand m
  volume resistivity spezifischer Durchgangswiderstand m
resol Resol n
  resol resin Resolharz n
  cresol resol Kresolresol n

phenol resol Phenolresol n
xylenol resol Xylenolresol n
resolution Auflösung f
resolving power Auflösungsvermögen n
resonance Resonanz f
  resonance frequency Resonanzfrequenz f
  resonance spectroscopy Resonanzspektroskopie f
resorcinol Resorcin n
  resorcinol(-formaldehyde) resin Resorcinharz n
response 1. Antwort f, Reaktion f; 2. Ansprech-
  response sensitivity Ansprechempfindlichkeit f
  response threshold Ansprechschwelle f
  response time Ansprechzeit f, Reaktionszeit f
  frequency response Frequenzgang m
  speed of response Ansprechgeschwindigkeit f
re-starting [a machine] Wiederanfahren n, Wiederinbetriebnahme f
restoration Restaurierung f, Sanierung f
restricted eingeschränkt
  restricted flow zone Dammzone f, Drosselstelle f, Stauzone f, Drosselfeld n
restriction Beschränkung f, Einschränkung f
  flow restriction bush Staubüchse f
  flow restriction device Drosselorgan n, Drosselkörper m, Drossel(vorrichtung) f
  flow restriction effect Drosselwirkung f
restrictor 1. Drossel f; 2. Stau-
  restrictor bar Stauleiste f, Staubalken m
  restrictor bar adjusting mechanism Staubalkenverstellung f [Gerät]
  restrictor bar adjustment Staubalkenverstellung f
  restrictor ring Stauring m
  flow restrictor Drosselorgan n, Drossel(vorrichtung) f, Drosselkörper m
  flow restrictor gap Drosselspalt m
  flow restrictor grid Drosselgitter n
results Ergebnisse npl, Resultate npl
  accelerated test results Kurzzeitwerte mpl
  short-term test results Kurzzeitwerte mpl
  test results Prüfergebnisse npl, Prüfresultate npl, Testergebnisse npl, Versuchsergebnisse npl, Versuchswerte mpl
  weathering results Bewitterungsergebnisse npl
retain/to festhalten
retained restlich, Rest-
  retained elongation Restdehnung f
  retained elongation at break Restbruchdehnung f
  retained flexural strength Restbiegefestigkeit f
  retained pendulum hardness Restpendelhärte f
  retained tear strength Restreißfestigkeit f
  retained tensile strength Rest(zug)festigkeit f, Restreißkraft f
  retained ultimate strength Restreißkraft f
retainer Halterung f
  cold slug retainer Pfropfenhalterung f
retaining Halte-
  retaining bolt Haltebolzen m

ejector retaining plate Auswerfergrundplatte f
retard/to retardieren, verzögern, hemmen
retardancy Verzögerung f, Hemmung f
   flame retardancy Flammhemmung f
retardant 1. hemmend; 2. Flammhemmittel n
   fire retardant feuerhemmend
   flame retardant 1. feuerhemmend, flammhemmend ausgerüstet, schwer brennbar; 2. Flammhemmer m, Flammschutz-Additiv n, Flammschutzkomponente f, Flammschutzmittel-Zusatz m, Flammschutzmittel n, Flammschutzsystem n
   flame retardant effect Flammschutzwirkung f, Flammschutzeffekt m
   flame retardant properties Flammschutzeigenschaften fpl, Schwerentflammbarkeit f
retardation Retardation f, Verzögerung f
   retardation behavio(u)r Retardationsverhalten n
   retardation period Retardationszeit f
   retardation process Retardationsvorgang m
   retardation test Retardationsversuch m
retention Beibehaltung f, Zurückhaltung f, Retention f
   data retention Datenhaltung f
   gloss retention Glanzbeständigkeit f, Glanzerhalt m
   shape retention Formänderungsfestigkeit f, Formbeständigkeit f
   solvent retention Lösungsmittelretention f
retractable zurückfahrbar, zurückziehbar
retracted zurückgezogen
retraction Zurücknahme f, Rücklauf m, Rückzug m
   retraction force Rückzugkraft f
   retraction speed Rücklaufgeschwindigkeit f, Rückzuggeschwindigkeit f
   device to prevent screw retraction Schneckenrückdrehsicherung f
   ejector retraction force Auswerferrückzugkraft f
   high-speed nozzle retraction mechanism Düsenschnellabhebung f
   nozzle retraction Düsenabhebung f, Düsenabhub m
   nozzle retraction speed Abhebegeschwindigkeit f, Düsenabhebgeschwindigkeit f
   nozzle retraction stroke Düsenabhebeweg m, Abhebehub m
   screw retraction Schneckenrückzug m, Schneckenrückdrehung f, Schneckenrückholung f, Schneckenrücklauf m
   screw retraction force Schneckenrückzugkraft f
   screw retraction mechanism Schneckenrückholvorrichtung f
retrieval Aufruf m, Abruf m [Daten]
   data retrieval Datenabfrage f
   ready for retrieval abrufbereit
retrieve/to aufrufen
return Rückkehr f, Wiederkehr f, Rücklauf m
   return temperature Rücklauftemperatur f
   ejector plate return pin Rückdrückstift m
   ejector return movement Auswerferrücklauf m
   ejector return speed Auswerferrücklaufgeschwindigkeit f
   mo(u)ld return stroke Werkzeugrückhub m
   screw return speed Schneckenrücklaufgeschwindigkeit f
   screw return stroke Schneckenrückhub m
   screw return time Schneckenrücklaufzeit f
returning air Rückluft f
returning water Rücklaufwasser n
re-usable wiederverwendbar
re-use Wiederverwendung f, Wiederverwertung f
reversal Umkehrung f, Wende f; Umsteuerung f [Motor]
   phase reversal Phasenumschlag m, Phasenumwandlung f
   phase reversal point Phasenumschlagpunkt m
reverse Gegenteil n, Rückseite f
   reverse flow Rückströmung f, Rückfluß m
   reverse melt flow Schmelzerückfluß m
   reverse roll coater Umkehrwalzenbeschichter m
   reverse roll coating Umkehrbeschichtung f
   reverse wind-up (unit) Wendewickler m
reversible reversibel, umkehrbar
reversing reversierend
   reversing rod mechanism Wendestangensystem n
   reversing rods Wendestangen fpl
   reversing system Reversier-System n
reversion Reversion f, Umkehrung f
   reversion stabilized reversionsstabilisiert
revolutions Umdrehungen fpl
   number of revolutions Tourenzahl f, Umdrehungszahl f, Drehzahl f
   screw revolutions Schneckenumdrehungen fpl
re-winding unit Umbäumstuhl m
Reynolds number Reynoldssche Zahl f
rheological rheologisch
rheology Rheologie f
rheometer Rheometer n
   extrusion rheometer Extrusionsviskosimeter n
   torque rheometer Drehmoment-Rheometer n
rheopectic rheopex
rheopexy Rheopexie f
rheostat Drehwiderstand m, Regelwiderstand m
rhombic rautenförmig
rib Rippe f
   rib-reinforced sandwich panel Stegdoppelplatte f
   reinforcing ribs Versteifungsrippen fpl
   supporting ribs Stützrippen fpl
ribbed torpedo Rippentorpedo m
ribbon mixer Bandmischer m
right rechts; richtig
   right-hand thread Rechtsgewinde n
   of the right size formatgerecht
rigid (biege)steif, starr
   rigid composite Hartverbund m
   rigid film Hartfolie f
   rigid foam Hartschaum(stoff) m
   rigid foam core Hartschaumkern m
   rigid foam insulating material Hartschaum-

isolierstoff m
rigid foam system Hartschaumsystem n
rigid polyurethane foam Hart-Urethanschaum m
rigid polyvinyl chloride Polyvinylchlorid n hart, Hart-PVC n
rigid PVC film Hart-PVC-Folie f, PVC-Hartfolie f
rigid PVC mo(u)lding Hart-PVC-Formteil n
rigid PVC sheet Hart-PVC-Platte f
fabrication of rigid film Hartfolienherstellung f
production of rigid film Hartfolienherstellung f
rigidity Starrheit f, Steifheit f, Steifigkeit f, Biegesteifigkeit f, Standfestigkeit f
  apparent modulus of rigidity Torsionssteifheit f
  mo(u)ld rigidity Werkzeugsteifigkeit f
  shear rigidity Schubsteifigkeit f
  torsional rigidity Torsionssteifheit f
rim Rand m, Bord m, Kante f, Felge f
  car wheel rim Felge f, Radfelge f
RIM [reaction injection mo(u)lding] Reaktionsgießverarbeitung f, Reaktionsguß m, Reaktions(spritz)gießen n
  RIM machine Reaktionsgießmaschine f
ring Ring m
  ring-closing ringschließend
  ring closure Ringschluß m
  ring formation Ringbildung f
  ring gate ringförmiger Bandanschnitt m, Ringanschnitt m
  ring heater Ringheizelement n
  ring-shaped groove Ringnut n
  ring system Ringsystem n
  ring-type distributor Ringverteiler m
  ring-type spiral distributor Ring-Wendelverteiler m
  adaptor ring Zwischenring m
  air ring Luftkühlring m
  benzene ring Benzolring m
  caprolactam ring Caprolactamring m
  clamping ring Spannring m
  cooling ring lips Kühlringlippen fpl
  cooling ring orifice Kühlring-Lippenspalt m, Lippenspalt m, Kühlluftring m
  die ring Düsenmundstück n, Düsenring m, Extrudermundstück n, Zentrierring m
  ejector ring Auswerferring m
  guide ring Führungsring m
  insert ring Einsatzring m
  mandrel with a ring-shaped groove Ringrillendorn m
  outer die ring Extrusionsmundstück n, Mundstück n, Mundstückring m
  phenyl ring Phenylring m
  piston ring Kolbenring m
  register ring Zentrierring m
  restrictor ring Stauring m
  sealing ring Dicht(ungs)ring m
  slip ring Schleifring m
  spacer ring Distanzring m
  stripper ring Abstreifring m
  supporting ring Stützring m
  thrust ring Druckring m
  water cooling ring Wasserkontaktkühlung f
rinse/to (ab)spülen, ausspülen
risk Risiko n, Gefahr f
  risk of contamination Verunreinigungsgefahr f, Verunreinigungsrisiko n
  risk of corrosion Korrosionsgefahr f
  risk of damage Beschädigungsgefahr f
  risk of overheating Überhitzungsgefahr f
  risk of premature vulcanization Scorchgefahr f
  risk of scorching Scorchgefahr f
  fire risk Brandgefahr f, Feuerrisiko n
  health risk Gesundheitsrisiko n
  safety risk Sicherheitsrisiko n
rival product Konkurrenzfabrikat n
riveting Nieten n
road Straße f
  road building Straßenbau m
  road marking paint Straßenmarkierungsfarbe f
  road salt Auftausalz n, Streusalz n
  road tanker Silofahrzeug n, Straßensilofahrzeug n, Straßentankwagen m, Tankwagen m
  road transport Straßentransport m
robot Roboter m, Handhabungsautomat m, Handhabungsgerät n, Handlingautomat m
  automatic workpiece robot Werkstück-Handhabungsautomat m
  demo(u)lding robot Entnahmeroboter m, Entnahmeautomat m
  filament winding robot Wickelroboter m
  insert-placing robot Einlegeroboter m, Einlegeautomat m
  parts-removal robot Entnahmeroboter m, Entnahmeautomat m
robotic system Handhabungssystem n, Robotersystem n
rocker arm Schwinge f
Rockwell hardness Rockwellhärte f
rod Stange f, Stab m
  connecting rod Pleuel m, Pleuelstange f
  ejector rod Auswerferkolben m, Auswerferstange f, Auswerferstößel m, Stangenauswerfer m
  fishing rod Angelrute f
  guide rod Führungsstange f
  heating rod Stabheizkörper m
  hexagonal-section solid rod Sechskant-Vollstab m
  piston rod Kolbenstange f
  reversing rod mechanism Wendestangensystem n
  reversing rod system Wendestangensystem n
  reversing rods Wendestangen fpl
  round-section rod Rund(voll)stab m
  solid rod Vollstab m
  square-section rod Vierkant(-Voll)stab m
  tension rod Zuganker m
  welding rod Schweißschnur f, Schweißstab m, Schweiß(zusatz)draht m, Schweißzusatzstoff m, Zusatzdraht m, Zusatzstoff m, Zusatzwerkstoff m

roll Walze f, Rolle f [s.a. roller]
   roll adjusting mechanism Walzenanstellung f, Walzenverstellung f
   roll adjustment Walzenanstellung f, Walzenverstellung f
   roll bearing Walzenlager n, Walzenlagerung f
   roll bending Gegenbiegen n, Walzengegenbiegung f
   roll bending mechanism Walzenbiegeeinrichtung f, Walzendurchbiegevorrichtung f, Walzengegenbiegeeinrichtung f
   roll configuration Walzenanordnung f
   roll covering Walzenbezug m, Walzenmantel m
   roll deflection Walzendurchbiegung f, Walzenverbiegung f
   roll diameter Walzendurchmesser m
   roll face Ballen m, Walzenballen m
   roll face center Walzenballenmitte f
   roll face width Ballenbreite f, Ballenlänge f, Walzenballenlänge f, Walzenballenbreite f
   roll grinding Walzenschliff m
   roll journal Walzenzapfen m
   roll mill Walzen(reib)stuhl m, Walzwerk n
   roll periphery Ballenrand m, Walzenballenrand m
   roll surface Walzenmantel m
   roll width Walzenbreite f
   applicator roll Auftragwalze f
   backing roll Gegendruckwalze f, Pressurwalze f
   bottom roll Unterwalze f
   calender take-off roll Kalanderabzugswalze f
   casting roll Gießwalze f
   casting roll unit Gießwalzeneinheit f
   center roll Mittelwalze f
   chill-cast roll Kokillen-Hartgußwalze f
   chill roll Kühlwalze f
   chill roll casting Chillroll-Verfahren n, Kühlwalzenverfahren n, Extrusionsgießverfahren n
   chill roll casting line Chillroll-Folienanlage f, Chillroll-Flachfolien-Extrusionsanlage f, Chillroll-Anlage f, Chillroll-Filmgießanlage f
   chill roll unit Chillroll-Walzengruppe f
   constant temperature rolls Temperierwalzen fpl
   cooling roll Kühlwalze f
   cooling rolls Temperierwalzen fpl [Kühlung]
   cross-axis roll adjustment Walzenschrägverstellung f, Walzenschräg(ein)stellung f
   delivery rolls Auslaufwalzen fpl
   discharge rolls Auslaufwalzen fpl
   distance between die and rolls Abzugshöhe f
   drilled roll Bohrungswalze f
   eccentric rolls unrund laufende Walzen fpl
   feed roll Dosierwalze f, Einlaufwalze f, Einzugswalze f, Speisewalze f, Zuführ(ungs)walze f
   film guide rolls Folienführungswalzen fpl
   fixed roll Festwalze f
   glue applicator roll Beleimungswalze f
   godet roll stretch unit Galettenstreckwerk n
   grinding rolls Mahlwalzwerk n
   guide roll Führungsrolle f, Führungswalze f, Leitwalze f
   guide rolls Rollenführung f
   heating rolls Temperierwalzen fpl [Heizung]
   highly polished roll Hochglanzwalze f
   laboratory mixing rolls Labormischwalzwerk n
   nip rolls Abquetschwalzen fpl, Abquetschwalzenpaar n, Quetschwalzen fpl, Quetschwalzenpaar n
   pair of rolls Walzenpaar n
   pinch rolls Abquetschwalzen fpl, Abquetschwalzenpaar n, Spaltwalzenpaar n
   polishing roll Polierwalze f
   pre-heating rolls Vorwärmwalzwerk n
   pressure roll Andrückwalze f, Anpreßwalze f, Kontaktwalze f
   reverse roll coater Umkehrwalzenbeschichter m
   reverse roll coating Umkehrbeschichtung f
   rubber covered pressure roll Anpreßgummiwalze f
   rubber covered roll Gummiwalze f
   spreader roll Breithalter m, Breitstreckwalze f
   spreader roll unit Breithaltevorrichtung f, Breitstreckwerk n
   stretching roll Reckwalze f, Streckwalze f
   stretching roll section Rollenreckstrecke f
   stretching roll unit Rollenreckwerk n
   take-off pinch rolls Abzugs-Abquetschwalzenpaar n
   take-off roll Abreißwalze f, Abzugswalze f
   take-off rolls Abzugsrollen fpl, Abzugswalzenstation f, Walzenabzug m, Abzugswalzenpaar n
   tension roll Spannrolle f, Zugwalze f
   top roll Oberwalze f
   transverse stretching roll Breitstreckwalze f
   triple roll mill Dreiwalze f, Dreiwalzenmaschine f, Dreiwalzenstuhl m
   unwinding roll Abwickelwalze f
roller Rolle f, Walze f [s.a. roll]
   roller application Rollen n, Walzenauftrag m
   roller bearing Rollenlager n, Zylinderrollenlager n
   roller blind profile Rolladenprofil n
   roller blind slats Rolladenstäbe mpl
   roller blinds Rolladen mpl
   roller die extruder Doppelwalzenextruder m
   roller-type rotor Walzenrotor m
   adjustable roller Stellwalze f
   compensating roller Tänzerwalze f
   deflecting roller Umlenkwalze f
   embossing roller Prägewalze f
   floating roller Pendelwalze f
   hot-melt roller application [of an adhesive] Walzenschmelzverfahren n
   pin roller fibrillator Nadelwalzenfibrillator m
   pin roller process Nadelwalzenverfahren n
   self-aligning roller bearing Pendelrollenlager n
   triple roller Dreiwalze f
roof Dach n
   roof gutter Dachrinne f

roof insulation Dachdämmung f, Dachisolierung f
roof spoiler Dachspoiler m
roof window Dachfenster n
car roof liner Auto(dach)himmel m
flat roof Flachdach n
pitched roof Steildach n
prefabricated roof liner Fertighimmel m
single-shell roof Warmdach n
sun roof Sonnendach n
roofing felt Bitumenpappe f, Teerpappe f
roofing sheet Dachbahn f, Dachfolie f
room Raum m
   room temperature Raumtemperatur f, Zimmertemperatur f
   room temperature ag(e)ing Raumtemperaturlagerung f
   room temperature curing Normalklimahärtung f, Raumtemperaturhärtung f, Normaltemperatur-Aushärtung f
   room temperature vulcanization Kaltvulkanisation f
   room temperature vulcanizing kaltvulkanisierend
   main control room Hauptsteuerwarte f
   wet-process room Naßraum m
root Wurzel f, Fuß m
   root diameter Kerndurchmesser m [Gewinde]
   root resistance Wurzelfestigkeit f
   root resistant wurzelfest
   root surface Grund m [Schnecke]
   resistance to root penetration Wurzelfestigkeit f
   resistant to root penetration wurzelfest
   screw root Schneckenkern m
   screw root surface Ganggrund m, Schneckenkanalgrund m, Schneckenkanaloberfläche f, Schneckengrund m
   screw root temperature control system Schneckenkerntemperierung f
   with constantly increasing root diameter [screw] kernprogressiv
rosin Kolophonium n
rot Fäulnis f, Fäule f, Verrottung f
   rot proof verrottungsbeständig, fäulnisfest, verrottungsfest, unverrottbar
   rot resistance Verrottungsbeständigkeit f, Verrottungsfestigkeit f
rotary rotierend, drehend, Rotations-, Dreh- [s.a. rotating]
   rotary extruder Drehextruder m
   rotary feed unit Rotationsdosiereinrichtung f
   rotary knife Kreismesser n
   rotary movement Drehbewegung f, Rotationsbewegung f
   rotary (piston) pump Drehkolbenpumpe f, Kreiselpumpe f
   rotary screen printing Rotationssiebdruck m
   rotary slitter Rundlaufspaltmaschine f
   rotary table Revolverteller m, Rundtisch m, Drehteller m, Drehtisch m
   rotary-table arrangement Rundtischanordnung f

   rotary table design Drehtischbauweise f
   rotary table injection mo(u)lding machine Drehtisch-Spritzgußmaschine f, Spritzgießrundtischanlage f, Revolverspritzgießmaschine f
   rotary table machine Drehtischmaschine f, Revolvermaschine f, Rundläufermaschine f, Rundläuferanlage f, Rundtischanlage f, Rundläufer m, Rundtischmaschine f
   rotary table system Drehtischsystem n
   rotary-table unit Revolvereinheit f
   rotary take-off unit Entnahmedrehtisch m
   rotary transverse cutter Rotationsquerschneider m
   rotary-type blow mo(u)lding unit Karussell-Blasaggregat n
   rotary-type cutter Karussellspaltmaschine f
   automatic rotary-table injection mo(u)lding machine Rotations-Spritzgußautomat m, Revolverspritzgießautomat m
   automatic rotary(-table) machine Rundläuferautomat m
   16-station rotary table machine Sechszehn-Stationen-Drehtischmaschine f
rotatable drehbar
rotating drehend, rotierend, Dreh-, Rotations- [s.a. rotary]
   rotating in opposite directions gegenläufig, gegeneinanderlaufend
   rotating in the same direction gleichsinnig, gleichläufig, gleichlaufend
   rotating mandrel Dralldorn m
   screw rotating mechanism Schneckenrotation f
   time during which the screw is rotating Schneckendrehzeit f
rotation Rotation f, Umdrehung f
   rotation mo(u)lding Rotationsschmelzen n, Rotationsgießen n, Rotationsformen n
   rotation mo(u)lding machine Rotationsgießmaschine f
   angle of rotation Drehwinkel m
   axis of rotation Rotationsachse f
   direction of rotation Drehrichtung f, Drehsinn m
   direction of screw rotation Schneckendrehrichtung f, Schneckendrehsinn m
   screw rotation Schneckendrehbewegung f, Schneckendrehung f, Schneckenrotation f
rotational Rotations-
   rotational sintering Rotationssintern n
   rotational speed Umlaufgeschwindigkeit f
   rotational viscometer Rotationsviskosimeter n
rotomo(u)lder Rotationsgießmaschine f
   carousel-type rotomo(u)lder kreisförmige Rotationsschmelzanlage f
rotomo(u)lding Rotationsformen n, Rotationsgießen n, Rotationsschmelzen n
rotor Rotor m, Laufrad n, Drehflügel m
   rotor bearing Rotorlagerung f
   rotor blade Laufradschaufel f
   rotor knife Rotormesser n, Rotorscheibe f, Kreismesser n

rotor knife block Rotormesserbalken m
rotor knife cutting circle Rotormesserkreis m
rotor shaft Rotorwelle f
rotor speed Rotordrehzahl f
cam-type rotor Nockenrotor m
roller-type rotor Walzenrotor m
separator rotor Trennwalze f
sigma-type rotor Sigmarotor m
rough-cut vorzerkleinert
roughen/to aufrauhen
roughened angerauht, aufgerauht
roughly shredded vorzerkleinert
roughness Rauhigkeit f
   roughness height Rauhtiefe f
   surface roughness Oberflächenrauhigkeit f
round rund, mit kreisförmigem Querschnitt
   round-bottom flask Rundkolben m
   round profile die Kreisprofildüse f
   round-section die Rund(strang)düse f, Rundstrangdüsenkopf m, Extrusionsrunddüse f
   round-section pipe Rundrohr n
   round-section profile die Rundstrangdüse f
   round-section rod Rund(voll)stab m
   round-section runner Rundlochkanal m
   long-neck round-bottom flask Langhals-Rundkolben m
roundtable Rundtisch m
routine Routine f
   routine check Routinekontrolle f
   routine determination Routinemessung f
   routine test Routineprüfung f, Routineuntersuchung f
   interface routine Anschlußprogramm n
roving Glasfaserroving n, Glasseidenroving n, Textilglasroving n
   roving cloth Glasrovinggewebe n
   roving for filament winding Wickelroving n
   roving laminate Rovinglaminat n
   roving strand Rovingstrang m
   chopping roving Schneidroving n
   woven roving Glasseidenrovinggewebe n, Rovinggewebe n
   woven roving reinforcement Rovinggewebeverstärkung f
row Reihe f, Zeile f
   rows of digital switches Digitalschalterreihen fpl
rub down/to anschleifen, anschmirgeln, schmirgeln, abreiben
   rub fastness Reibechtheit f
rubber Kautschuk m, Gummi n,m
   rubber component Kautschukkomponente f
   rubber compound Kautschukmischung f
   rubber-containing kautschukhaltig
   rubber content Kautschukanteil m
   rubber covered gummiert, kautschuküberzogen
   rubber covered (back-up) roll Gummiwalze f
   rubber covered pressure roll Anpreßgummiwalze f
   rubber-elastic kautschukelastisch
   rubber elasticity Kautschukelastizität f
   rubber-glass transition Zäh-Sprödübergang m
   rubber gloves Gummihandschuhe mpl
   rubber injection mo(u)lding machine Gummi-Spritzgießmaschine f
   rubber latex Kautschuklatex m
   rubber-like gummiähnlich, gummiartig
   rubber-like elasticity Entropieelastizität f, Kautschukelastizität f, Gummielastizität f
   rubber modified kautschukmodifiziert, kautschukhaltig
   rubber particle Kautschukteilchen n
   ABS rubber [ABS rubber] Acrylnitril-Butadien-Styrol-Kautschuk m
   acrylonitrile-butadiene rubber Acrylnitril-Butadien-Kautschuk m
   acrylonitrile-butadiene-styrene rubber [ABS rubber] Acrylnitril-Butadien-Styrol-Kautschuk m
   butadiene-acrylonitrile rubber Butadien-Acrylnitril-Kautschuk m
   butadiene rubber Butadienkautschuk m
   butadiene-styrene rubber Butadien-Styrol-Kautschuk m
   butyl rubber Butylkautschuk m
   chlorinated butyl rubber Chlorbutylkautschuk m
   chlorinated rubber Chlorkautschuk m
   chlorinated rubber coating Chlorkautschukanstrich m
   chlorinated rubber paint Chlorkautschukfarbe f, Chlorkautschuklack m
   chloroprene rubber Chloroprenkautschuk m
   crude rubber Rohkautschuk m
   epichlorhydrin rubber Epichlorhydrinkautschuk m
   ethylene-propylene-diene rubber [EPDM] Ethylen-Propylen-Dien-Kautschuk m
   ethylene-propylene rubber Ethylen-Propylen-Kautschuk m, Ethylen-Propylen-Elastomer m
   ethylene-vinyl acetate rubber Ethylen-Vinylacetat-Kautschuk m
   fluorocarbon rubber Fluorcarbonkautschuk m
   fluorosilicone rubber Fluorsilikonkautschuk m
   foam rubber Moosgummi n,m, Zellgummi n,m
   general-purpose rubber Allzweckkautschuk m
   hard rubber Hartgummi n,m
   injection mo(u)lded rubber article Gummi-Spritzgußteil n
   isobutylene-isoprene rubber Isobutylen-Isopren-Kautschuk m
   liquid rubber Flüssigkautschuk m
   liquid silicone rubber Flüssigsilikonkautschuk m
   natural rubber Naturgummi n,m, Naturkautschuk m
   neoprene rubber Neoprenkautschuk m
   nitrile rubber Nitrilkautschuk m
   nitrile rubber latex Nitrillatex m
   one-pack silicone rubber Einkomponentensilikonkautschuk m
   polyacrylate rubber Acrylatkautschuk m
   polychloroprene rubber Polychloroprenkautschuk m
   polyolefin rubber Polyolefinkautschuk m

polysulphide rubber Polysulfidkautschuk m, Thiokol m
silicone rubber Silicongummi n,m, Silikonkautschuk m
silicone rubber compound Silikonkautschukmischung f
silicone rubber mo(u)ld Silikonkautschukform f
solid rubber Vollgummi n,m
special-purpose rubber Spezialkautschuk m, Sonderkautschuk m
speciality rubber Sonderkautschuk m
styrene-butadiene rubber [SBR] Styrol-Butadien-Kautschuk m
synthetic rubber Synthesegummi n,m, Synthesekautschuk m
synthetic rubber coated [e.g. rolls] kunstkautschukbeschichtet
two-pack liquid silicone rubber Zweikomponentenflüssigsilikonkautschuk m
rubbery gummiartig, gummiähnlich, gummielastisch, entropieelastisch
 rubbery-elastic entropieelastisch, gummielastisch
rugged robust
ruggedly constructed robust
ruggedness Robustheit f
rule Regel f
 rule of thumb Faustregel f
run Lauf m, Gang m; Serie f
 extremely short runs Kleinstserien fpl
 long runs große Serien fpl
 medium runs mittlere Serien fpl
 pilot plant run Nullserie f
 production runs Auflagenhöhen fpl
 short runs Kleinserien fpl, kleine Produktionszahlen fpl, kleine Serien fpl
 tendency to run [paint] Ablaufneigung f
 trial run Nullserie f
runner Angußverteiler m, Angußweg m, Anschnittweg m, Anspritzkanal m, Fließkanal m, Fließquerschnitt m, Flußkanal m, Fließweg m, Kanal m, Massekanal m, Querschnitt m, Renner m, Verteiler(kanal) m, Verteilerrohr n, Angußtunnel m, Zuführ(ungs)kanal m, Anschnittkanal m, Anströmkanal m, Schmelze(führungs)kanal m, Schmelze(führungs)bohrung f, Strömungsweg m, Strömungskanal m
 runner cross-section Verteilerkanalquerschnitt m
 runner design Fließkanalgestaltung f
 runner profile Verteiler(kanal)querschnitt m
 runner shape Verteiler(kanal)querschnitt m
 runner system Verteiler(röhren)system n, Spinne f, Verteilerspinne f, Sternverteiler m, Angußverteilersystem m
 runners arranged side by side Reihenverteiler m
 circular runner Ringkanal m
 double runner Doppelverteilerkanal m
 examples of hot runner systems Heißkanalbeispiele npl
 hot runner Heißkanal m
 hot runner feed system Heißkanalanguß m, Heißkanal-Angußsystem n
 hot runner injection mo(u)lding system Spritzgieß-Heißkanalsystem n
 hot runner manifold block Heißkanalblock m, Heizblock m, Heißkanalverteilerbalken m, Querverteiler m, Heißkanalverteilerblock m
 hot runner stack mo(u)ld Heißkanal-Etagenwerkzeug n
 hot runner unit Querverteiler m
 hot runner unit construction Heißkanalblock-Ausführung f
 insulated runner feed system Isolierkanal-Angußplatte f
 insulated runner unit Isolierkanalanguß m, Isolierkanal-Angußplatte f
 main runner Anströmkanal m, Angußtunnel m, Anschnittkanal m, Hauptverteilersteg m, Hauptverteilerkanal m
 parallel runners Parallelverteiler m
 partial hot runner Teilheißläufer m
 projected runner surface area projizierte Verteilerkanalfläche f
 radial system of runners Spinne f, Angußstern m, Angußspinne f, Anschnittstern m
 rectangular runner Rechteckkanal m
 round-section runner Rundlochkanal m
 secondary runner Nebenkanal m
 standard hot runner mo(u)ld units Heißkanalnormalien pl
 standard hot runner unit Norm-Heißkanalblock m, Heißkanal-Normblock m
 triple runner Dreifachverteilerkanal m
 types of hot runner Heißkanalbauarten fpl
running costs Betriebskosten pl
rupture Bruch m
 creep rupture curve Zeitbruchkurve f, Zeitbruchlinie f
 creep rupture strength [of pipes] Zeitstandfestigkeit f
rust Rost m
 rust formation Rostbildung f
rusting Rostbildung f
rutile Rutil n
 rutile pigment Rutilpigment n
sack Sack m
 heavy-duty sack Schwergutsack m
 heavy-duty sack film Schwergutsackfolie f
 paper sack Papiersack m
safe sicher, unfallsicher, gefahrlos
 safe against overloading überlastsicher
 safe to handle handhabungssicher
 safe working stress [of pipes] zulässige Wandbeanspruchung f, zulässige Beanspruchung f, Sigmawert m
 ecologically safe ökologisch unbedenklich,

umweltfreundlich
safety Sicherheit f
  safety cover Schutzverdeck n
  safety data Sicherheitskennzahlen fpl
  safety device Schutzeinrichtung f, Schutzvorrichtung f, Sicherheitseinrichtung f
  safety factor Sicherheitszahl f, Sicherheitsbeiwert m, Sicherheitskoeffizient m, Sicherheitsfaktor m
  safety glass Sicherheitsglas n, Verbundsicherheitsglas n
  safety guard Schutzgitter n
  safety guidelines Sicherheitsrichtlinien fpl
  safety helmet Schutzhelm m
  safety hood Schutzabdeckung f, Schutzhaube f
  safety in use Arbeitssicherheit f
  safety instructions Sicherheitshinweise mpl, Sicherheitsratschläge mpl
  safety interlock system Sicherheitsverriegelung f
  safety lock Sicherheitsverschluß m
  safety margin Sicherheitsgrenze f, Sicherheitsreserve f, Sicherheitsspielraum m
  safety measures Schutzmaßnahmen fpl, Vorsichtsmaßnahmen fpl, Sicherheitsmaßnahmen fpl, Sicherheitsvorkehrungen fpl
  safety precautions Schutzmaßnahmen fpl, Vorsichtsmaßnahmen fpl, Sicherheitsmaßnahmen fpl, Sicherheitsvorkehrungen fpl
  safety regulations Sicherheitsvorschriften fpl, Sicherheitsregeln fpl
  safety requirements Sicherheits(an)forderungen fpl, Sicherheitsbedürfnisse npl
  safety risk Sicherheitsrisiko n
  safety shield Schutzverkleidung f
  safety switch Sicherheitsschalter m
  safety valve Sicherheitsventil n
  automatic mo(u)ld safety device Automatikwerkzeugschutz m
  ejector plate safety mechanism Auswerferplattensicherung f
  excess-pressure safety device Überdrucksicherung f
  for reasons of safety aus Sicherheitsgründen mpl
  low-pressure mo(u)ld safety device Niederdruckwerkzeugschutz m
  motor safety switch Motorschutzschalter m
  mo(u)ld safety mechanism Formschließsicherung f, Formschutz m, Schließhubsicherung f, Werkzeug-Sicherheitsbalken m, Werkzeugschutz m, Zufahrsicherung f, Werkzeugschließsicherung f
  part delivery safety mechanism Ausfallsicherung f
  sliding safety guard Schiebe(schutz)gitter n
  works safety guidelines Arbeitsschutz-Richtlinien fpl
sag Auslängen n, Aushängen n, Durchhang m, Durchhängen n
salary Gehalt n
sale Verkauf m, Vertrieb m
sales Absatz m, Verkauf m
  sales figures Absatzzahlen fpl

sales organization Vertriebsorganisation f
sales proceeds Umsatzerlös m
  drop in sales Absatzrückgang m
  improvement in sales Absatzverbesserung f
  increasing sales Absatzausweitung f
salicylate Salicylsäureester m
  lead salicylate Bleisalicylat n
salicylic acid Salicylsäure f
  salicylic acid ester Salicylsäureester m
salient points Besonderheiten fpl
salt Salz n
  salt solution Salzlösung f
  salt spray test Salzsprühtest m
  heavy metal salt Schwermetallsalz n
  road salt Auftausalz n, Streusalz n
sample Probe f
  sample for inspection Anschauungsmuster n
  customer's sample Kundenvorlage f
  original sample Ausgangsprobe f
  random sample Stichprobe f
  reference sample Vergleichsmuster n, Vergleichsprobe f
SAN Styrol-Acrylnitril-Copolymerisat n
sand Sand m
  sand-lime brick Kalksandstein m
  silica sand Quarzsand m
sand (down)/to schmirgeln, anschleifen, anschmirgeln
sandblasted sandgestrahlt
sandblasting Sandstrahlen n
sandwich Sandwich n, Schichtwerkstoff m
  sandwich construction Depotverfahren n, Sandwich-Verbundweise f
  sandwich mo(u)lded part Sandwich-Spritzgußteil n
  sandwich mo(u)lding (process) Sandwich-Spritzgießverfahren n, Sandwichverfahren n, Sandwichschäumverfahren n, Verbundspritztechnik f
  rib-reinforced sandwich panel Stegdoppelplatte f
sanitary sanitär
  sanitary equipment Sanitäreinrichtungen fpl, Sanitärartikel mpl, Sanitärteile npl
sanitaryware Sanitärkeramik f
saponifiable verseifbar
saponification Verseifung f
  saponification rate Verseifungsrate f, Verseifungsgeschwindigkeit f
  saponification resistance Verseifungsbeständigkeit f
  saponification resistant verseifungsbeständig
  saponification value Verseifungszahl f
saponified verseift
  partly saponified teilverseift
satin weave Atlasbindung f [Textil]
saturated gesättigt, satt-
  saturated steam Sattdampf m
  saturated vapo(u)r Sattdampf m
  saturated with water vapo(u)r wasserdampfgesättigt
saturation Absättigung f, Sättigung f

saturation concentration Sättigungskonzentration f
saturation pressure Sättigungsdruck m
saturation vapo(u)r pressure Wasserdampfsättigungsdruck m, Sättigungsdampfdruck m
degree of saturation Sättigungsgrad m
state of saturation Sättigungszustand m
saving 1. (ein)sparend; 2. Einsparung f
  cost saving Kosteneinsparung f, Kostenersparnis f
  energy saving 1. energiesparend; 2. Energieeinsparung f
  power saving leistungssparend, stromsparend
  raw material savings Rohmaterialersparnis f
  space saving platzsparend, raumsparend
  stabilizer saving Stabilisatoreinsparung f
  time saving zeitoptimal
  weight savings Gewichtseinsparungen fpl
saw Säge f
  saw blade Sägeblatt n
  band saw Bandsäge f
  circular saw Kreissäge f
  floating saw Pendelsäge f
SBR Styrol-Butadien-Kautschuk m, Styrol-Butadienpfropfpolymerisat n
scale Skala f, Maßeinteilung f, Waagschale f, Waage f
  scale division Skalenteilung f
  colo(u)r scale Farbskala f
  differential scales Differentialwaage f
  gray scale Graumaßstab m
  Hazen colo(u)r scale Hazen-Farbskala f
  iodine colo(u)r scale Jodfarbskala f
  laboratory scale Labormaßstab m
  pilot plant-scale 1. kleintechnisch; 2. Technikummaßstab m, Versuchsmaßstab m
  production scale Produktionsmaßstab m
  time scale Zeitmaßstab m
scanning Abtastung f
  scanning electron micrograph rasterelektronenmikroskopische Aufnahme f, REM-Aufnahme f
  scanning electron microscope Rasterelektronenmikroskop n
  scanning electron microscopy Rasterelektronenmikroskopie f
  scanning head Abtastkopf m
  differential scanning calorimetry DSC-Methode f
  laser scanning analyzer Laser-Abtastgerät n
  low-pressure scanning (device) Niederdruckabtastung f
scatter range Streubereich m, Streubreite f
scattering [of test results] Streuung f
  light scattering Lichtstreuung f
schedule (zeitliches) Programm n, Zeitplan m
  curing schedule Härteprogramm n
  production schedule Herstellungsprogramm n, Produktionsprogramm n, Fabrikationsprogramm n
scheduling department Terminbüro n
schematic drawing Prinzipskizze f, Schema n, Schemaskizze f

schematically in schematischer Darstellung f, schematisch
scheme Schema n
scleroscope Skleroskop n
scorch/to versengen, verbrennen, scorchen
  scorch resistance Scorchsicherheit f
  scorch resistant scorchsicher
  scorch temperature Scorchtemperatur f
  scorch time Sicherheitszeit f vor Einsetzen des Scorchens
scorching [premature curing] Vorvernetzung f
  risk of scorching Scorchgefahr f
  susceptible to scorching scorchanfällig
scrap Ausschuß m, Abfall m
  scrap material Produktionsrückstände mpl
  scrap rate Ausschußzahlen fpl, Ausschußquote f
  scrap recycling Abfallverwertung f
  scrap removal system Abfallentfernungssystem n
  scrap removal unit Afallentfernvorrichtung f
  scrap repelletizing line Regranulieranlage f
  scrap reprocessing Abfallaufbereitung f
  scrap reprocessing plant Abfallaufbereitungsanlage f
  extrusion scrap Extrusionsabfälle mpl
  film scrap Folienabfall m
  film scrap reclaim plant Folienschnitzel-Granulieranlage f, Folienregranulieranlage f
  film scrap re-processing plant Folienaufbereitungsanlage f
  pipe scrap Ausschußrohre npl, Rohrabfälle mpl
  plant scrap Produktionsabfälle mpl
  plastics scrap Kunststoffabfälle mpl, Kunststoffausschuß m
  profile scrap Ausschußprofile npl
  profile scrap granulator Profilschneidmühle f
  web scrap Folienbahnabfall m, Stanzgitterabfall m, Stanzgitter n
  web scrap granulator Stanzgittermühle f
scraper Auskratzer m, Kratzer m, Schaber m
scratch Kratzer m, Schramme f
  scratch hardness Ritzhärte f
  scratch proof kratzbeständig, kratzfest
  scratch resistance Kratzfestigkeit f, Oberflächenkratzfestigkeit f, Ritzfestigkeit f
scratched verkratzt, zerkratzt
  easily scratched kratzempfindlich
screed Estrich m, Aufbeton m
  floating screed schwimmender Estrich m
  floor(ing) screed Estrich m, Fußbodenausgleichmasse f, Bodenspachtelmasse f, Bodenausgleichmasse f
  industrial floor screed Industriebodenbelag m
  mortar screed Mörtelbelag m
  self-levelling screed Fließbelag m, Verlaufsbelag m
screen 1. Filtersieb n, Sieb n, Filtergewebe n; 2. Bildschirm m, Anzeigebildschirm m
  screen area Filterfläche f
  screen cassette Siebkassette f

screen changer Siebwechselkassette f, Siebwechsler m, Siebwechseleinrichtung f, Siebwechselgerät n, Siebscheibenwechsler m
screen changer body Siebwechslerkörper m
screen fabric Siebgewebe n
screen guard Sicherheitsgitter n
screen pack Filterpaket n, Siebblock m, Siebeinsatz m, Siebgewebepackung f, Siebkorb m, Siebpaket n, Siebsatz m, Filtergewebepaket n, Maschengewebepaket n
screen printing process Siebdruckverfahren n
screen support Siebträger m
colo(u)r screen Farbbildschirm m
computer with integral screen Bildschirmcomputer m
effective screen area freie Siebfläche f
filter screen Gewebesieb n, Siebgewebe n, Filtergewebe n
fine-mesh screen Feinsieb n, feinmaschiges Sieb n
high-speed screen changer Sieb-Schnellwechseleinrichtung f, Schnellsiebwechsel-Einrichtung f
quick-action screen changer Sieb-Schnellwechseleinrichtung f
replacing the screen Siebwechsel m
rotary screen printing Rotationssiebdruck m
stainless steel filter screen Edelstahlmaschengewebe n
twin screen pack Doppelsiebkopf m
vibrating screen Schwingsieb n
video screen Anzeigebildschirm m
wire gauze screen Drahtgewebefilter n,m
wire mesh screen Drahtsiebboden m
screening Filterung f, Filtration f, Filtrieren n, Rastern n
screening effect Abschirmeffekt m
screw Schnecke f, Schraube f
  screw advance Schneckenvorschub m, Schneckenvorlauf m
  screw advance speed Schneckenvorlaufgeschwindigkeit f
  screw assembly Schneckeneinheit f, Schneckensatz m, Schneckenausrüstung f, Schneckenbaukasten m
  screw assembly components Schneckensatzelemente npl
  screw axis Schneckenachse f, Wellenachse f
  screw back pressure Schneckenrückdruck m, Schneckenstaudruck m, Schneckenrückdruckkraft f, Schneckengegendruck m
  screw built up from modules Bausatzschnecke f, Baukastenschnecke f
  screw bushing Schneckenbuchsen fpl
  screw cap Schraubverschluß m
  screw center height Extrudierhöhe f, Extruderhöhe f, Extrusionshöhe f
  screw channel Schnecken(gang)kanal m, Schmelzekanal m
  screw channel volume Schneckenkanalvolumen n
  screw characteristic Schneckenkennlinie f, Schneckenkennzahl f, Schneckeneigenschaften fpl
  screw circumference Schneckenumfang m
  screw clamp Schraubzwinge f
  screw clearance Scherspalt m, Schneckenspiel n, Schneckenscherspalt m
  screw compounder Knetscheiben-Schneckenpresse f
  screw configuration Schneckenkonfiguration f, Schneckenbauform f, Schneckengestaltung f, Schneckengeometrie f
  screw conveyor Schneckenförderer m, Schneckenfördergerät n
  screw cooling (system) Schneckenkühlung f
  screw coupling Verschraubung f, Schraubenanschluß m
  screw depth Schneckentiefe f
  screw design Schneckenentwurf m, Schneckenkonzeption f, Schneckenkonzept n, Schneckenkonstruktion f, Schneckenauslegung f, Schneckenausführung f, Schneckenausbildung f
  screw devolatilizer Schneckenverdampfer m
  screw diameter Schneckendurchmesser m
  screw drive Schneckengetriebe n
  screw drive mechanism Schneckenantriebssystem n
  screw drive motor Schneckenantriebsmotor m, Schneckenantriebsleistung f
  screw drive shaft Schneckenantriebswelle f
  screw equipped with a shear section Scherteilschnecke f, Scherzonenschnecke f
  screw feeder Schneckendosierer m, Scheckendosiervorrichtung f, Schneckendosiereinheit f, Schneckendosiergerät n, Schneckenspeiseeinrichtung f
  screw flight Schneckenelement n, Schneckenwendel m
  screw flight contents Schneckengangfüllung f
  screw flight flank Schneckenflügelflanke f
  screw flights Schneckengewinde n
  screw flights and kneader disks Schnecken- und Knetelemente npl
  screw forward movement Schneckenvorlaufbewegung f
  screw fracture Schneckenbruch m
  screw head Schraubenkopf m
  screw-in nozzle Einschraubdüse f
  screw-in resistance thermometer Einschraub-Widerstandsthermometer n
  screw injection cylinder Schneckenspritzzylinder m
  screw injection mo(u)lding machine Schneckenspritzgießmaschine f
  screw injection unit Schneckeneinspritzaggregat n, Schneckenspritzeinheit f
  screw intermeshing Schneckeneingriff m
  screw length Schnecken(bau)länge f, Arbeitslänge f, Extruderlänge f
  screw output Schneckenleistung f
  screw performance Schneckeneigenschaften fpl, Schneckenleistung f
  screw plasticization Schneckenplastifizierung f
  screw plasticizing cylinder Schneckenplastifizierungszylinder m

screw plasticizing unit Schneckenplastifiziereinheit f, Schneckenplastifizieraggregat n, Schneckenplastifizierung f
screw-plunger Schneckenkolben m
screw-plunger injection Schneckenkolbeninjektion f, Schneckenkolbeneinspritzung f
screw-plunger injection mo(u)lding machine Schneckenkolbenspritzgießmaschine f
screw-plunger injection system Schneckenkolbeneinspritzsystem n
screw-plunger injection unit Schneckenkolbeninjektion f, Schneckenkolbeneinspritzaggregat n
screw-plunger machine Schneckenkolbenmaschine f, Schneckenkolbeneinheit f
screw position Schneckenstellung f
screw preplasticizer Schneckenvorplastifiziergerät n
screw preplasticizing (unit) Schneckenvorplastifiziergerät n, Schneckenvorplastifizierung f
screw profile Schneckenprofil n
screw range Schneckensortiment n
screw retraction Schneckenrückholung f, Schneckenrückzug m, Schneckenrücklauf m, Schneckenrückdrehung f
screw retraction force Schneckenrückzugkraft f
screw retraction mechanism Schneckenrückholvorrichtung f
screw return speed Schneckenrücklaufgeschwindigkeit f
screw return stroke Schneckenrückhub m
screw return time Schneckenrücklaufzeit f
screw revolution Schneckenumdrehung f
screw root Schneckenkern m
screw root temperature control (system) Schneckenkerntemperierung f
screw rotation Schneckendrehung f, Schneckendrehbewegung f, Schneckenrotation f
screw section Schneckenabschnitt m, Schneckenstufe f, Schneckenteil m, Schneckenzone f
screw shank Schnecken(wellen)schaft m
screw size Schneckengröße f
screw speed Schneckenumdrehungszahl f, Schnecken(umlauf)geschwindigkeit f, Schnecken(wellen)drehzahl f, Extruderdrehzahl f, Wellendrehzahl f, Umdrehungszahl f
screw speed range Schneckendrehzahlbereich m, Umdrehungszahlbereich m der Schnecke
screw stroke Schneckenhub m, Schneckenweg m
screw stroke adjusting mechanism Schneckenhubeinstellung f, Schneckenwegeinstellung f
screw stroke adjustment Schneckenhubeinstellung f, Schneckenwegeinstellung f
screw stroke transducer Schneckenwegaufnehmer m
screw support Schneckenabstützung f, Schneckenlagerung f
screw surface (area) Schneckenoberfläche f
screw system Schneckensystem n
screw temperature control (system) Schneckentemperierung f
screw threads Schneckenwindungen fpl
screw tip Schneckenende n, Schneckenspitze f
screw torque Schneckendrehmoment m
screw turn Schneckenumgang m
screw unit Schneckenaggregat n, Schneckeneinheit f
screw venting system Schneckendekompressionseinrichtung f
screw wear Schneckenabnutzung f, Schneckenverschleiß m
screw wearing surface Schneckenlauffläche f
adjusting screw Einstellschraube f, Stellschraube f, Verstellschraube f
amount of material in the screw flight(s) Schneckengangfüllung f
ancillary screw Nebenschnecke f, Seitenschnecke f, Zusatzschnecke f
automatic screw-plunger injection mo(u)lding machine Schneckenspritzgießautomat m, Schneckenkolbenspritzsystem n
auxiliary twin screw extruder Doppelschnecken-Seitenextruder m
basic screw Ausgangsschnecke f, Grundschnecke f
cent(e)ring screw Zentrierschraube f
changing the screw Schneckenaustausch m, Schneckenwechsel m
closely intermeshing twin screws engkämmende Doppelschnecken fpl
compounding screw unit Knetschneckeneinheit f, Knetschneckenwelle f
compression screw Kompressionsschnecke f, Verdichtungsschnecke f
constant taper screw kernprogressive Schnecke f, Kernprogressivschnecke f
conventional screw Normalschnecke f
conveying screw kompressionslose Förderschnecke f
co-rotating twin screw Gleichdrall-Doppelschnecke f, Gleichdrallschnecke f
co-rotating twin screw extruder Gleichdrall-Schneckenextruder m, Gleichdrallmaschine f
co-rotating twin screw system Gleichdrallsystem m
counter-rotating twin screw extruder Gegendrallmaschine f, Gegendrall-Schneckenextruder m
counter-rotating twin screw (system) Gegendrallsystem n, Gegendrall-Doppelschnecke f
decreasing pitch screw steigungsdegressive Schnecke f
delivery screw Auspreßschnecke f
depending on the screw stroke schneckwegabhängig
device for preventing screw fracture Schneckenbruchsicherung f
device to prevent screw retraction Schneckenrückdrehsicherung f
direct screw injection system Schneckendirekteinspritzung f
direction of screw rotation Schneckendrehrichtung f,

Schneckendrehsinn m
discharge screw Auspreßschnecke f, Ausstoßschnecke f, Austragsschnecke f
double-conical screw Doppelkonusschnecke f
dry blend screw Pulverschnecke f
effective screw length Schneckenarbeitslänge f
experimental screw Versuchsschnecke f
external screw diameter Schneckenaußendurchmesser m
extruder screw Extruderschnecke f, Extrusionsschnecke f
feed screw Aufgabeschnecke f, Beschickungsschnecke f, Einspeiseschnecke f, Einzugsschnecke f, Füllschnecke f, Speiseschnecke f, Dosierschnecke f, Zuführschnecke f
feed screw unit Speiseschneckeneinheit f
feed twin screws Zuführschneckenpaar n
fixing screw Befestigungsschraube f
flattened screw tip Schneckenballen m
flighted screw tip Schneckenförderspitze f, Förderspitze f
four-section screw Vierzonenschnecke f
fully-flighted screw durchgeschnittene Schnecke f
general-purpose screw Allzweckschnecke f, Standardschnecke f, Universalschnecke f
grooved torpedo screw Nutentorpedoschnecke f
heavy-duty screw Hochleistungsschnecke f
hexagonal screw Sechskantschraube f
high-speed single screw extruder Hochleistungseinschneckenextruder m
high-speed twin screw extruder Hochleistungsdoppelschneckenextruder m
homogenizing screw Knetwelle f
increasing pitch screw Progressivspindel f
initial screw speed Eingangsdrehzahl f
injection screw Spritzgießmaschinenschnecke f
intermeshing screws with a self-sealing Dichtprofilschnecken fpl
internal screw cooling system Schneckeninnenkühlung f
internal screw diameter Schneckeninnendurchmesser m
laboratory twin screw extruder Doppelschnecken-Laborextruder m
limiting screw speed Grenzdrehzahl f, Drehzahlgrenze f [Schnecke]
locking screw Feststellschraube f
long-compression zone screw Langkompressionsschnecke f
low-compression screw Niedrigkompressionsschnecke f
main screw Hauptschnecke f, Hauptspindel f, Mittelschnecke f, Zentralschnecke f, Zentralspindel f, Zentralwelle f
maximum screw speed Grenzdrehzahl f, Drehzahlgrenze f [Schnecke]
melt delivery screw Schmelzeaustragsschnecke f

mixing screw Knetschnecke f
modular screw Bausatzschnecke f, Baukastenschnecke f
modular screw assembly Baukastenschneckensatz m
multi-stage screw Stufenschnecke f
one-piece screw Einstückschnecke f
one-section screw Einzonenschnecke f
operating screw speed Arbeitsdrehzahl f, Betriebsdrehzahl f
opposing screw Gegenschnecke f
peripheral screw speed Schneckenumfangsgeschwindigkeit f
planet(ary) screw Planetschnecke f, Planetenspindel f, Planet m
plasticating screw Plastifizierschnecke f, Plastifizierspindel f
plasticizing screw Plastifizierschnecke f, Plastifizierspindel f
pre-plasticizing screw Vorplastifizierschnecke f
PVC screw PVC-Schnecke f
quadruple screw Vierfachschnecke f
radial screw clearance radiales Spiel n, Stauspalt m, Spalthöhe f
range of screws Schneckensortiment n
reciprocating screw Schubschnecke f, Schneckenschub m
removal of screw Schneckenausbau m
required screw speed Extruderdrehzahl-Sollwert m
shallow-flighted screw flache Schnecke f
shearing-mixing screw Scher-Mischschnecke f
short-compression zone screw Kurzkompressionsschnecke f
single reciprocating screw Einfach-Schubschnecke f
single screw Einzelschnecke f, Einschnecke f, Einfachschnecke f, Einspindelschnecke f, Einwellenschnecke f
single screw compounding extruder Plastifizier-Einschneckenextruder m
space in front of the screw Stauraum m, Schneckenstauraum m, Schneckenvorraum m
special-purpose screw Sonderschnecke f, Spezialschnecke f
stuffing screw Stopfschnecke f, Einpreßschnecke f
thermoplastic screw Thermoplast-Schnecke f
thermoset screw Duroplastschnecke f
three-section screw Dreizonenschnecke f
thrust screw Druckschraube f
time during which the screw is rotating Schneckendrehzeit f
total screw torque Schneckengesamtdrehmoment n
transport screw kompressionslose Förderschnecke f, kompressionslose Schnecke f
triple screw Dreifachschnecke f
triple screw extruder Dreischneckenextruder m
twin feed screw Doppel-Dosierschnecke f, Doppel-Einlaufschnecke f
twin screw Doppelschnecke f, Zweischnecke f,

Zweiwellenschnecke f, zweiwellige Schnecke f, Schneckenpaar n
twin screw assembly Schneckenpaar n
twin screw barrel Doppelschneckenzylinder m
twin screw compounder Doppelschneckenmaschine f, Zweischneckenmaschine f, Zweiwellenmaschine f, Zweiwellenkneter m, zweiwellige Schneckenmaschine f, zweiwelliger Schneckenkneter m
twin screw compounding extruder Zweiwellen-Knetscheiben-Schneckenpresse f
twin screw design Doppelschneckenausführung f
twin screw extruder Doppelschneckenextruder m, Zweischneckenextruder m, Zweiwellenextruder m, zweiwellige Schneckenmaschine f, Zweischneckenmaschine f, Doppelschneckenmaschine f, Zweiwellenmaschine f
twin screw extrusion Doppelschneckenextrusion f
twin screw feed section Zweischneckeneinzugszone f
twin screw metering section Zweischneckenaustragszone f
twin screw pelletizer Zweiwellen-Granuliermaschine f
twin screw pelletizing line Doppelschnecken-Granulieranlage f
twin screw principle Doppelschneckenprinzip n
twin screw system Zweiwellensystem n
two-section screw Zweizonenschnecke f
two-stage twin screw extruder Zweistufen-Doppelschneckenextruder m
type of screw Schnecken(bau)art f
type of screw drive Schneckenantriebsart f, Schneckenantrieb m
unflighted screw tip glatte Spitze f, glatte Schneckenspitze f
variable-design screw Aufsteckschnecke f, Schaftschnecke f
variable-geometry screw Schaftschnecke f
vented screw Entgasungsschnecke f, Vakuumschnecke f, Dekompressionsschnecke f
vented twin screw extruder Doppelschnecken-Entgasungsextruder m
vented twin screws Entgasungsschneckenpaar n
vertical twin screw extruder Vertikaldoppelschneckenextruder m
worn screw Schrottschnecke f
scrub/to schrubben, scheuern
  scrub resistance Scheuerfestigkeit f, Scheuerresistenz f, Scheuerbeständigkeit f
scrubber Wäscher m
  air scrubber Abluftwäscher m
seal/to abdichten
seal Abdichtung f, Dichtung f
  seal assembly Dichtungssatz m
  door seal Türabdichtung f
  shaft seal Wellenabdichtung f
sealant Dichtstoff m, Dichtungsmasse f
  hot melt sealant Schmelzmasse f
  mo(u)ld sealant Formversiegler m

structural sealant Bautendichtungsmasse f, Konstruktionsdichtungsmasse f
underbody sealant Autounterbodenschutzmasse f, Unterbodenschutz m, Unterbodenschutzmasse f
underbody sealant paste Unterbodenschutzpaste f
sealing Siegel-, Dicht-
  sealing bar Siegelbacke f [Hitze]
  sealing edges Dichtkanten fpl
  sealing face Dichtfläche f
  sealing instrument Siegelwerkzeug n
  sealing machine Verschließanlage f
  sealing ring Simmerring m
  sealing tape Dichtungsband n
  sealing temperature Siegeltemperatur f
  automatic heat sealing machine Heißsiegelautomat m
  gate sealing Angußversiegeln n
  gate sealing point Siegelzeitpunkt m
  thermal sealing Wärmekontaktschweißen n
seam Nahtverbindung f, Schweißnaht f
seamless nahtlos
seat Sitz(fläche) m(f)
  seat cover Sitzbezug m
  seat upholstery Sitzpolster n
  bearing seat Lagerstelle f
  valve seat Ventilsitz m
seating 1. sitzend, Sitz-; 2. -sitz m
  seating valve Sitzventil n
  nozzle seating Düsenkalotte f, Düsensitz m
seawater Seewasser n
sebacate Sebacat n, Sebacinsäureester m
  di-2-ethylhexyl sebacate [DOS] Di-2-ethylhexyl-sebacat n
  dibutyl sebacate [DBS] Dibutylsebacat n
  diethyl sebacate [DES] Diethylsebacat n
  dioctyl sebacate [DOS] Di-2-ethylhexylsebacat n, Dioctylsebacat n
sebacic acid Sebacinsäure f
  sebacic acid ester Sebacinsäureester m
sebacic anhydride Sebacinsäureanhydrid n
second 1. Sekunde f; 2. zweite(r, s)
  second order reaction Reaktion f zweiter Ordnung
  fraction of a second Sekundenbruchteil m
  within seconds in Sekundenschnelle f
secondary sekundär, untergeordnet
  secondary cooling (system) Sekundärkühlung f
  secondary linkage Sekundärbindung f
  secondary particle Sekundärteilchen npl, Sekundärkorn n
  secondary plasticizer Sekundärweichmacher m, Zweitweichmacher m
  secondary product Folgeprodukt n
  secondary radical Sekundärradikal n
  secondary reaction Folgereaktion f, Sekundärreaktion f
  secondary register Sekundär-Register n
  secondary runner Nebenkanal m
  secondary structure Sekundärstruktur f

secondary take-off (unit) Sekundärabzug m
secondary valency forces Nebenvalenzkräfte fpl
section Strecke f, Zone f, Bereich m, Abschnitt m, Schnitt m, Teil m
  section thickness [if the part has no true walls, e.g. a gear wheel] Wanddicke f
  barrel cooling sections Zylinderkühlzonen fpl, Zylinderschüsse mpl
  barrel section Gehäuseabschnitt m, Gehäuseschuß m, Zylinderelement n
  barrel sections Gehäuseteile npl, Schneckengehäuseschüsse mpl
  compounding section Aufbereitungsteil m
  compression section Kompressionsbereich m, Komprimierzone f
  conditioning section Temperstrecke f
  conveying section [of screw] Transportteil m
  cooling section Abkühlstrecke f, Kühlabschnitt m, Kühlpartie f, Kühlstrecke f, Kühlzone f, Kühlkanal m
  decompression section Dekompressionszone f
  devolatilizing section Dekompressionszone f, Entgasungszone f, Entgasungsbereich m, Entgasungsschuß m, Entspannungszone f, Zylinderentgasungszone f
  die section Düsenteil m
  discharge section Auslaufgehäuse n, Austragsteil n
  drying section Trockenstrecke f
  faceted mixing section [of screw] Rautenmischteil m
  faceted smear section [of screw] Flächenscherteil n
  feed section Einfüllabschnitt m, Eingangszone f, Eingangsteil m, Einlaufteil m, Einlaufstück n, Einzugszone f, Einzugszonenabschnitt m, Einzugs(zonen)teil n, Einzugszonenbereich m, Förderlänge f, Förderzone f, Füllzone f, Trichterzone f, Zylindereinzug m, Zylindereinzugszone f, Zylindereinzugsteil n, Einzugsbereich m
  feed section flight depth Einzugsgangtiefe f
  filament wound section Wickelschuß m
  grooved feed section Nuteneinzugszone f
  grooved shear section Nutenscherteil n
  heating section Heizstrecke f, Temperierstrecke f
  homogenizing section [of screw] Aufschmelzzone f, Aufschmelzbereich m
  incoming goods section Wareneingangsbereich m
  kneading section Knetgehäuse n
  knurled mixing section [of screw] Igelkopf m, Nockenmischteil n
  longitudinal section Längsschnitt m
  metering section Ausbringungszone f, Ausstoßteil m, Ausstoßzone f, Austragsbereich m, Austragszone f, Austragsteil n, Pumpzone f
  microtome section Dünnschnitt m, Querschliff m
  neck section Halspartie f
  pipe section Rohrabschnitt m, Rohrzuschnitt m
  planetary-gear section Planetwalzenteil n
  plasticating section (of screw) Schmelzone f
  post-cooling section Nachkühlstrecke f
  profile section Profilabschnitt m
  re-heating section Wiederaufwärmstrecke f
  round section die Extrusionsrunddüse f, Rundstrangdüse f
  screw equipped with a shear section Scherzonenschnecke f
  screw section Schneckenabschnitt m, Schneckenstufe f, Schneckenteil n, Schneckenzone f
  shear section Scherzone f, Scherteil m
  spray cooling section Sprühkühlstrecke f
  stretching roll section Rollenreckstrecke f
  take-off section Abzugsstrecke f
  transition section Aufschmelzbereich m, Aufschmelzzone f, Plastifizierbereich m, Übergangsbereich m, Verdichtungszone f, Plastifizierzone f, Kompressionsbereich m, Umwandlungszone f, Übergangszone f
  twin screw metering section Zweischneckenaustragszone f
  ultra-thin (microtome) section Ultradünnschnitt m
  ultra-thin section microtomy Ultradünnschnittmikrotomie f
  vacuum calibration section Vakuumkalibrierstrecke f
  vent section Entgasungsschuß m, Entgasungsteil n
  vulcanizing section Vulkanisationszone f
  water cooled feed section Naßbüchse f
sectional drawing Schnittdarstellung f, Schnittbild n
sector Bereich m, Gebiet n, Sektor m
  sector of the economy Wirtschaftszweig m
  food sector Lebensmittelbereich m
  packaging sector Verpackungsgebiet n
sedimentation Sedimentation f
segment Segment n, Abschnitt m
  barrel segment Zylindersegment n
  molecule segment Molekülsegment n
  process segment Prozeßabschnitt m
seize (up)/to (fest)fressen
selectable (an)wählbar, auswählbar, selektierbar
selection Auswahl f, Selektion f
  selection criterion [plural: criteria] Auswahlkriterium n
  program selection key Programmwahltaste f, Programmwahlschalter m
selective selektiv, gezielt
selectivity Selektivität f
selector 1. Wähler m, Auswähler m, Selektor m; 2. Wahl-
  selector key Wahltaster m
  selector switch Wahlschalter m
  function selector switch Betriebsartenwahlschalter m
  range selector Bereichswähler m
  setpoint selector key Sollwertauswahltaste f
self selbst
  self-accelerating selbstbeschleunigend
  self-adhesive selbstklebend
  self-adhesive letter Klebefolienbuchstabe m
  self-adjusting selbsteinstellend, selbstnachstellend, selbstregulierend

self-aligning roller bearing Pendelrollenlager n
self-cent(e)ring selbstzentrierend
self-cleaning (effect) Selbstreinigung f
self-colo(u)red durchgefärbt
self-configuring selbstkonfigurierend
self-contained [machine unit] geschlossen
self-crosslinking selbstvernetzend
self-curing selbsthärtend
self-diagnosis Selbstdiagnose f
self-extinguishing selbst(ver)löschend
self-feeding selbstdosierend
self-ignition Selbstentzündung f
self-ignition temperature Selbstentzündungstemperatur f
self-insulating selbstisolierend
self-levelling selbstverlaufend [Ausgleichsmasse]
self-levelling mortar Verlaufsmörtel m
self-levelling screed Fließbelag m, Verlaufsbelag m
self-locking selbstverriegelnd, selbstsichernd
self-locking mechanism Selbstverriegelung f
self-lubricating selbstschmierend
self-lubricating properties Selbstschmierfähigkeit f
self-regulating selbstregulierend
self-sealing selbstdichtend
self-sufficient autark
self-supporting selbsttragend
self-test Selbsttest m
semi halb
  semi-automatic(ally) halbautomatisch, teilautomatisch
  semi-automatic machine Halbautomat m
  semi-automatic system Halbautomatik f
  semi-automatic transfer mo(u)lding press Spritzpreß-Halbautomat m
  semi-circle Halbkreis m
  semi-circular halbkreisförmig
  semi-conductor Halbleiter m
  semi-conductor memory Halbleiterspeicher m
  semi-conductor module Halbleiterbaustein m
  semi-conductor power switch Halbleiterleistungsschalter m
  semi-conductor switch Halbleiterschalter m
  semi-finished products Halbfertigerzeugnisse npl, Halbfabrikate npl, Halbzeug n
  semi-flexible halbflexibel
  semi-logarithmic halblogarithmisch
  semi-microscale Halbmikromaßstab m
  semi-permeability Semipermeabilität f
  semi-quantitative halb-quantitativ
  semi-rigid halbsteif, halbhart, mittelhart
  semi-solid halbfest
semicarbazide Semicarbazid n
sensing fühlend
  flow sensing element Durchflußfühler m
  stroke sensing mechanism Weggebersystem n
sensitive empfindlich, sensitiv, feinfühlig
  heat sensitive wärmeempfindlich, temperaturempfindlich
  light sensitive lichtempfindlich
  pressure sensitive druckempfindlich
  shear sensitive scherempfindlich
sensitivity Empfindlichkeit f
  sensitivity shift Empfindlichkeitsänderung f
  sensitivity threshold Empfindlichkeitsschwelle f
  sensitivity to atmospheric humidity Luftfeuchteempfindlichkeit f
  sensitivity to impact Stoßempfindlichkeit f
  sensitivity to moisture Feuchteempfindlichkeit f
  sensitivity to water Wasserempfindlichkeit f
  input sensitivity Eingangsempfindlichkeit f
  light sensitivity Lichtempfindlichkeit f
  oxidation sensitivity Oxidationsanfälligkeit f, Oxidationsempfindlichkeit f
  oxygen sensitivity Sauerstoffempfindlichkeit f
  radiation sensitivity Bestrahlungsempfindlichkeit f
  response sensitivity Ansprechempfindlichkeit f
  shear sensitivity Scherempfindlichkeit f
  transducer sensitivity Aufnehmerempfindlichkeit f
sensitization Sensibilisierung f
sensitizer Sensibilisator m
sensor Sonde f, Sensor m, Fühler m, Geber m, Aufnehmer m
  sensor pin Fühlerstift m
  sensor tip Fühlerspitze f
  compensating arm sensor Tänzerarmfühler m
  control sensor Reguliersonde f
  flow sensor Durchflußfühler m
  melt pressure sensor Massedruckgeber m, Massedruckaufnehmer m
  temperature sensor Temperatur(meß)fühler m, Temperatur(meß)geber m, Temperatursensor m, Temperatursonde f, Thermodraht m, Thermoelement(fühler) n (m)
  torque sensor Drehmomentaufnehmer m, Drehmomentsensor m
separate/to trennen
separate einzel, freistehend, getrennt, separat
  separate additives Einzeladditive npl
  separate control circuit Einzelregelkreis m
  separate cores Einzelkerne npl
  separate determinations Einzelmessungen fpl
  separate drive Einzelantrieb m
  separate feed unit Fremddosierung f
  separate functions Einzelfunktionen fpl
  separate heating Fremdbeheizung f
  separate melt streams Partialströme mpl, Teilströme mpl, Masseteilströme mpl, Schmelzeteilströme mpl
  separate module Einzelbaustein m
  separate mo(u)ld Einzelform f
  consisting of separate units in offener Bauweise f, in offener Elementbauweise f
separately controlled fremdgesteuert
separating column [in chromatography] Trennsäule f
separation Abtrennung f, Auftrennung f, Abspaltung f,

Ausscheidung f, Trennung f, Entmischung f
separation effect Abscheidewirkung f
separation energy Trennungsenergie f
separation of water Wasserabspaltung f, Wasserentmischung f
separation process Trennungsprozeß m
degree of separation Abscheidegrad m
method of separation Trennmethode f
phase separation Phasentrennung f
separator Separator m, Abscheider m, Trennwalze f
battery separator Batterieseparator m
centrifugal separator Zentrifugalabscheider m
contaminant separator Fremdkörperabscheider m
cyclone separator Zyklonabscheider m
dust separator Staubabscheider m
low-pressure separator Niederdruckabscheider m
sprue separator Anguß-Separator m
sequence Ablauf m, Abfolge f, Sequenz f
sequence control Folgeregelung f, Ablaufsteuerung f, Schrittfolgesteuerung f
cycle sequence Zyklusablauf m
operating sequence Funktionsablauf m
operating sequence display Funktionsablaufanzeige f
process sequence Arbeitsablauf m
processing sequence verfahrenstechnischer Ablauf m
speed sequence Geschwindigkeitsfolge f
switch sequence Schaltfolge f
sequential seriell, Sequenz-
sequential polymer Sequenzpolymer n, Sequenztyp m
serial aufeinanderfolgend, sequentiell, seriell
series Bauserie f
series arrangement Reihenschaltung f, Serienschaltung f
arranged in series hintereinandergeschaltet, in Reihenanordnung f
in series in Reihe f
repeat test series Wiederholungsmeßreihen fpl
test series Versuchsreihe f, Versuchsserie f
service Dienst m, Service m, Dienstleistung f
service conditions Beanspruchungsbedingungen fpl, Beanspruchungsverhältnisse npl
service engineer Wartungstechniker m
service program Dienstprogramm n
after-sales service Kundendienst m
continuous service temperature Dauerbetriebstemperatur f
easy to service servicefreundlich, wartungsfreundlich
long-term service temperature Dauergebrauchstemperatur f
serviceability Funktionsfähigkeit f, Funktionstüchtigkeit f, Gebrauchstüchtigkeit f, Gebrauchswert m, Gebrauchstauglichkeit f
serviceable funktionstüchtig
services offered [by a company] Leistungsangebot n
servicing Unterhaltung f, Wartung f
servicing instructions Wartungsanweisungen fpl
ease of servicing Servicefreundlichkeit f

mo(u)ld servicing and maintenance Werkzeugwartung und Instandhaltung f
servo Servo-
servo-control Vorsteuerung f, Servoregelung f
servo-control unit Vorsteuereinheit f
servo-controlled vorgesteuert
servohydraulic servohydraulisch
servo-impulse Verstellimpuls m
servo-motor Servomotor m, Stellmotor m
servo-valve Servoventil n, Vorsteuerventil n
directional servo-valve Wege-Servoventil n
pressure servo-valve Druck-Servoventil n
set/to vorgeben, abbinden, erstarren, einfrieren
set pressure Drucksollwert m
sum set aside for depreciation Abschreibungsvolumen n
set Satz m, Reihe f, Folge f, Serie f, Kollektion f
set of tools Werkzeugsatz m
compression set Druckverformungsrest m, Verformungsrest m
tension set Zugverformungsrest m
setpoint Sollwert m
setpoint adjustment Sollwerteinstellung f
setpoint curve Soll(wert)kurve f
setpoint deviations Sollwertabweichungen fpl
setpoint display Sollwertanzeige f
setpoint input Sollwertvorgabe f, Sollwerteingabe f
setpoint memory Sollwertspeicher m
setpoint selector key Sollwertauswahltaste f
setter Einsteller m
machine setter Einrichter m, Maschineneinsteller m
mo(u)ld setter Werkzeugeinrichter m
setting Vorgabe f, Einstellung f, Einrichten n [Maschine]; Abbinden n
setting accuracy Einstellfeinheit f
setting characteristics Abbindeverhalten n
setting error Einstellfehler m
setting mechanism Einstelleinrichtung f
setting phase Erstarrungsphase f
setting range Verstellbereich m
setting record Einstellprotokoll n
setting speed Abbindegeschwindigkeit f, Erstarrungsgeschwindigkeit f
setting time Abbindezeit f, Abkühldauer f, Abkühlzeit f, Erstarrungszeit f, Erstarrungsdauer f, Kühlzeit f, Einrichtezeit f
setting-up Rüstvorgang m
setting-up instructions Einrichtblätter npl
digital setting Digitaleinstellung f
extruder settings Extrudereinstellwerte mpl
heat setting Wärmestabilisierung f, Thermofixierung f
hot setting warmabbindend
machine setting Maschineneinstellparameter m, Maschineneinstellgröße f
machine setting program Maschineneinstellprogramm n

machine settings Maschineneinstellungen fpl, Maschineneinstellwerte mpl, Maschineneinstelldaten npl
machine settings store Einstelldatenabspeicherung f
manual setting Handeinstellung f
mo(u)ld setting record card Formeinrichtekarte f
mo(u)ld setting time Werkzeugeinrichtezeit f
nip setting (mechanism) Walzenspalteinstellung f
setpoint setting Sollwertvorgabe f
speed setting (mechanism) Drehzahleinstellung f
standard setting Standardeinstellung f
temperature setting (mechanism) Temperatureinstellung f
wrong settings Fehleinstellungen fpl
settle out/to sedimentieren
tendency to settle (out) Sedimentationsneigung f, Absetzneigung f
settling out Sedimentation f, Absetzen n
set-up Aufbau m, Anordnung f
set-up operation Rüstvorgang m
set-up time Rüstzeit f
set-up time record Rüstzeitaufnahme f
experimental set-up Versuchsaufbau m
test set-up Prüfanordnung f, Versuchseinrichtung f, Versuchsanordnung f
sewage Abwasser n
sewage pipe Kanal(isolations)rohr n
shade Farbton m
pastel shade Pastellton m
shaded schraffiert, schattiert
shaft Welle f, Spindel f, Schacht m
shaft end Wellenstummel m
shaft seal Wellenabdichtung f
cardan shaft Kardanwelle f, Gelenkwelle f
central shaft Zentralwelle f
drive shaft Antriebswelle f, Getriebewelle f
hollow drive shaft Getriebehohlwelle f
impeller shaft Rührwelle f
main drive shaft Hauptantriebswelle f
plasticating shaft Plastifizierschaft m
rotor shaft Rotorwelle f
screw drive shaft Schneckenantriebswelle f
ventilation shaft Lüftungskanal m
wind-up shaft Wickelwelle f
shallow flach, seicht
shallow-flighted seichtgeschnitten, flachgängig, flachgeschnitten
shallow-flighted screw flache Schnecke f
shank Schaft m
screw shank Schnecken(wellen)schaft m
shape Gestalt f, Form f
shape retention Formänderungsfestigkeit f, Formbeständigkeit f
mo(u)lded-part shape Formteilgestalt f
of complex shape formkompliziert
part shape Formteilgestalt f
particle shape Kornform f, Partikelform f, Teilchenform f
runner shape Verteilerkanalquerschnitt m
shaped fassoniert, geformt
bar-shaped stabförmig
funnel-shaped trichterförmig
suitably shaped formgerecht
shaping Formgebung f, spanlose Verarbeitung f, spanlose Bearbeitung f
share Aktie f, Anteil m
share capital Aktienkapital n
export share Exportquote f
market share Marktanteil m
sharp-edged scharfkantig
shear 1. Scherung f, innere Reibung f; 2. Scher-,
shear deformation Scherdeformation f
shear effect Scherwirkung f
shear element Scherelement n
shear energy Scherenergie f
shear force Scherkraft f, Schubkraft f
shear-intensive scherintensiv
shear modulus Gleitmodul m, Schubmodul m
shear plasticization Scherplastifizierung f
shear rate Deformationsgeschwindigkeit f, Geschwindigkeitsgefälle n, Geschwindigkeitsgradient m, Schergefälle n, Schergeschwindigkeit f
shear rigidity Schubsteifigkeit f
shear section Scherzone f, Scherteil m
shear sensitive scherempfindlich
shear sensitivity Scherempfindlichkeit f
shear strain Scherdehnung f
shear strength Scher(stand)festigkeit f, Schubfestigkeit f
shear stress Scherbeanspruchung f, Scherspannung f, Schubspannung f
shear stress at failure Abscherspannung f
shear test Scherversuch m
excessive shear Überscherung f
grooved shear section Nutenscherteil m
introduction of shear forces Scherkrafteinleitung f
limiting shear stress Grenzschubspannung f
longitudinal shear Längsscherung f
rate of shear Schergeschwindigkeit f
screw equipped with a shear section Scherteilschnecke f, Scherzonenschnecke f
stiffness in shear Schubsteifigkeit f
tensile shear strength Zugscherfestigkeit f
tensile shear stress Zugscherbelastung f
tensile shear test Zugscherversuch m
transverse shear Querscherung f
shearing-mixing screw Scher-Mischschnecke f
sheath(e)/to [cables, wires etc.] umspritzen, ummanteln
sheathed umspritzt, ummantelt
plastic sheathed kunststoffummantelt
sheathing Ummantelung f
sheathing plant [e.g. for pipes] Ummantelungsanlage f
cable sheathing die Ummantelungswerkzeug n

cable sheathing extruder Ummantelungsextruder m
cable sheathing line Ummantelungsanlage f
high-speed cable sheathing plant Hochleistungs-Kabelummantelungsanlage f
pipe sheathing die Rohrbeschichtungswerkzeug n
shed Halle f
  production shed Fertigungshalle f, Produktionshalle f, Fabrikationshalle f
sheet Platte f, Tafel f; Blech n
  sheet calendering Plattenziehen n
  sheet die Plattenkopf m, Platten-Breitschlitzdüse f, Plattendüse f, Plattenwerkzeug n
  sheet extrusion Breitschlitzplattenextrusion f, Plattenextrusion f
  sheet extrusion head Plattenkopf m
  sheet extrusion line Plattenanlage f, Plattenstraße f, Platten-Extrusionsanlage f, Plattenextrusionslinie f
  sheet-like plattenartig
  sheet metal Blech n, Feinblech n
  sheet mo(u)lding compound [SMC] Harzmatte f
  sheet polishing unit Plattenglättanlage f
  sheet slippage [during thermoforming] Folienschlupf m
  sheet take-off (unit) Plattenabzug m
  sheet thermoforming line Plattenformmaschine f
  sheet web Plattenbahn f
  acrylic (glazing) sheet Acrylplatte f
  alumin(i)um sheet Aluminiumblech n
  asbestos sheet Asbestplatte f
  barrier sheet(ing) Sperrfolie f
  cast sheet Gießplatte f
  composite sheet Verbundplatte f
  cut-to-size piece of sheet Plattenzuschnitt m
  extruded sheet Breitschlitz(düsen)platte f
  fabric-based laminate sheet Hartgewebetafel f
  glass cloth laminate sheet Glashartgewebeplatte f
  GRP sheet GF-UP-Platte f, GFK-Platte f
  insulating sheet Dämmplatte f
  iron sheet Eisenblech n
  laminate sheet Schichtpreßstoffplatte f, Schichtpreßstofftafel f
  milled sheet Fell n
  polyester sheet mo(u)lding compound Polyesterharzmatte f
  pressed sheet Preßplatte f
  product data sheet Typenmerkblatt n
  PVC sheet PVC-Platte f
  rigid PVC sheet Hart-PVC-Platte f
  roofing sheet Dachbahn f, Dachfolie f
  thermoforming sheet Tiefziehfolie f
  thermoforming sheet extrusion line Tiefziehfolien-Extrusionsanlage f
  thermoforming sheet plant Tiefziehfolienanlage f
  two-layer sheet extrusion Zweischicht-Tafelherstellung f
  wide sheet extrusion line Platten-Großanlage f
sheeted-out compound Fell n, Walzfell n
sheeting Folie f [Kunststoff], Blech [Metall]
  sheeting calender Folienziehkalander m
  sheeting-out [e.g. PVC, rubber etc. on a mill] Ausziehen n
sheets plattenförmiges Halbzeug n, Plattenmaterial n, Plattenware f
shelf Regal n, Gestell n
  shelf life Lagerfähigkeit f, Lagerstabilität f, Lagerbeständigkeit f, Lagerzeit f, Haltbarkeit f
  shelf life problems Lagerstabilitätsschwierigkeiten fpl
  having a long shelf life lagerstabil
  guaranteed shelf life Lagerfähigkeitsgarantie f
shell Schale f, Muschel f
  shell-type construction Schalenbauweise f
  electron shell Elektronenschale f
shield Schild n
  engine sound shield Geräuschkapsel f, Schallverkleidung f
  safety shield Schutzverkleidung f
shift 1. Verschiebung f, Verstellung f, Umstellung f; 2. Gußfehler m; 3. Schicht f
  shift log Schichtprotokoll n
  shift operation Schichtbetrieb m
  day shift Tagesschicht f
  night shift Nachtschicht f
  sensitivity shift Empfindlichkeitsänderung f
  unmanned shift Geisterschicht f
  zero shift Nullpunktverschiebung f
ship Schiff n
shipbuilding industry Schiffsbauindustrie f
shock Schock m, Stoß m
  shock-absorbent stoßabsorbierend, stoßdämmend
  shock absorber Stoßdämpfer m
  shock resistance Erschütterungsfestigkeit f, Schockfestigkeit f, Schockzähigkeit f
  shock resistant stoßfest, erschütterungsfest, schockfest, schockresistent
  shock-sensitive erschütterungsempfindlich
  low-temperature shock resistance Kälteschockfestigkeit f
  thermal shock resistance Wärmeschockverhalten n
shockproof stoßfest, schockfest, schockresistent
shop Halle f, Laden m, Geschäft n, Betrieb m
  extrusion shop Extruderhalle f, Extrusionshalle f, Spritzerei f
  hardening shop Härterei f
  injecton mo(u)lding shop Spritzgießbetrieb m, Spritzgießfabrik f
  mo(u)lding shop Presserei f, Spritzerei f
  pattern-making shop Modellwerkstatt f
  plastics injection mo(u)lding shop Kunststoffspritzerei f
shopfitting Ladenbau m
Shore hardness Shore-Härte f
  Shore-A hardness Shore-Härte A f
  Shore-D hardness Shore-Härte D f
  change in Shore hardness Shoreänderung f

intermediate Shore hardness Zwischenshorehärte f
with a high Shore hardness hochshorig
short kurz, von geringer Baulänge f, von kurzer Baulänge f
  short-chain kurzkettig
  short-chain branching Kurzkettenverzweigung f
  short circuit Kurzschluß m
  short-compression zone screw Kurzkompressionsschnecke f
  short mo(u)ldings unvollständige Teile npl
  short-oil kurzölig
  short-oil alkyd (resin) Kurzölalkydharz n
  short runs kleine Produktionszahlen fpl, kleine Serien fpl, Kleinserien fpl
  short-screw extruder Kurzschneckenextruder m
  short single-screw extruder Einschnecken-Kurzextruder m
  short stroke Kurzhub m
  short-stroke kurzhubig
  short-term kurzzeitig, kurzfristig
  short-term behavio(u)r Kurzzeitverhalten n
  short-term breaking stress Kurzzeitbruchlast f
  short-term dielectric strength Kurzzeit-Durchschlagfestigkeit f
  short-term elongation Kurzzeitlängung f
  short-term immersion Kurzzeitlagerung f [in Flüssigkeit]
  short-term properties Kurzzeiteigenschaften fpl
  short-term stress Kurzzeitbeanspruchung f
  short-term tensile stress Kurzzeitzugbeanspruchung f
  short-term test Kurzzeit-Prüfung f, Kurzzeitversuch m
  short-term test results Kurzzeitwerte mpl
  short-wave kurzwellig
  extremely short runs Kleinstserien fpl
  for a short time kurzfristig
shortage Knappheit f, Verknappung f
  energy shortage Energieknappheit f
  raw material shortage Rohstoffverknappung f
shot Schuß m, Spritze f
  shot capacity mögliches Schußvolumen n, Schußleistung f, Spritzkapazität f
  shot volume Dosiervolumen n, Schußvolumen n, Einspritzvolumen n, Spritzvolumen n
  shot weight maximales Teilgewicht n, Einspritzgewicht n, Schußmasse f, Schußgewicht n, Schußgröße f, Spritz(teil)gewicht n
  lead shot Bleischrot n
shoulder Schulter f
  shoulder flash Schulterbutzen m
  shoulder pinch-off Schulterquetschkante f
shredder Zerkleinerer m, Zerhacker m, Granulator m, Schnitzelmühle f, Zerkleinerungsmaschine f, Zerkleinerungsaggregat n
  edge trim shredder Randstreifenzerhacker m
shrink/to schrumpfen, einlaufen
  shrink wrapped pack Schrumpfverpackung f
  shrink wrapping Schrumpfverpackung f
  shrink wrapping film Schrumpfolie f
  tendency to shrink Schrumpfneigung f
shrinkage Schwund m, Schrumpfung f, Schwindung f
  shrinkage compensation Schrumpfungskompensation f
  shrinkage crack Schwundriß m
  shrinkage difference Schwindungsunterschied m
  shrinkage differences Schwindungsdifferenzen fpl
  shrinkage on drying Trocknungsschrumpf m
  shrinkage on solidification Erstarrungsschrumpf m
  shrinkage properties Schwindungseigenschaften fpl
  shrinkage record Schwindungskatalog m
  shrinkage-reducing schrumpfmindernd
  shrinkage test piece Schwindungsstab m
  shrinkage value Schrumpf(ungs)wert m
  anisotropic shrinkage Schwindungsanisotropie f
  curing shrinkage Härtungsschrumpf m, Härtungsschwund m
  degree of shrinkage Schwindmaß n
  longitudinal shrinkage Längenschrumpf m, Längenschwund m, Längskontraktion f, Längsschrumpf(ung) m(f), Längsschwindung f
  mo(u)lded-part shrinkage Formteilschwindung f
  mo(u)lding shrinkage Formschrumpf m, Formschwindung f, Formschwund m, Verarbeitungsschrumpf m, Verarbeitungsschwindung f
  part shrinkage Formteilschwindung f
  total shrinkage Gesamtschrumpfung f
  transverse shrinkage Querkontraktion f, Querschrumpf m, Querschwindung f
  volume shrinkage Volumenkontraktion f, Volumenschrumpfung f, Volumenschwindung f
shunt Nebenschluß m
  shunt motor Nebenschlußmotor m
  shunt-wound geared DC motor Nebenschluß-Gleichstrom-Getriebemotor m
  d.c. shunt motor Gleichstrom-Nebenschlußmotor m
shut geschlossen
  shut-down period Leerlaufzeit f, Stillstandzeit f, Totzeit f, Nebenzeit f, Stehzeit f
  shut-off mechanism Absperrmechanismus m, Verschlußmechanismus m
  shut-off needle Verschlußnadel f
  shut-off nozzle Düsenverschluß m, Verschlußdüse f
  shut-off slide valve Absperrschieber m
  shut-off valve Verschlußventil n, Sperrventil n
  hot runner needle shut-off mechanism Heißkanal-Nadelverschlußsystem n
  needle shut-off mechanism Schließnadel f, Nadelverschlußsystem n, Nadelventilverschluß m
  sliding shut-off nozzle Schiebeverschluß(spritz)düse f, Schiebedüse f
shuttle Schützen m [Textil]
side Seite f
  side chain Seitenkette f
  side effect Nebeneffekt m
  side ejector mechanism Seitenauswerfer m

side exposed to view Sichtseite f
side-fed seitlich eingespeist, seitlich angespeist, radial angeströmt, quer angeströmt, seitlich angeströmt
side-fed blown film die stegloser Folienblaskopf m, seitlich eingespeister Folienblaskopf m, Umlenkblaskopf m, Pinolenblaskopf m
side-fed die Krümmerkopf m, Pinolen(schlauch)kopf m, Pinolenwerkzeug n, Pinolenspritzkopf m
side-fed parison die seitlich angeströmter Pinolenkopf m
side feed seitliches Anspritzen n
side gate seitlicher Anschnitt m
side-gated seitlich angeschnitten, seitlich angespritzt
side group Seitengruppe f
side-gussetted blown film Seitenfaltenschlauchfolie f
side-gusseting device Seitenfalteneinrichtung f
side panel [of a machine] Seitenschild n
side panels [of a car] Seitenverkleidungen fpl
side reaction Nebenreaktion f
side spoiler Seitenspoiler m
side wall Seitenwand f
side window Seitenscheibe f
double sided doppelseitig
feed side Beschickungsseite f
injection side spritzseitig
input side Eingabeseite f
on all sides allseitig
on both sides doppelseitig, beidseitig
on one side einseitig
on the ejector side of the mo(u)ld auswerferseitig
on the feed side angußseitig
operator's side [of machine] Bedienungsseite f
output side Ausgabeseite f
runners arranged side by side Reihenverteiler mpl
the side opposite the operator's side Bedienungsgegenseite f
visible side Sichtseite f
sidewall Seitenwand f
  tyre sidewall compound Reifenseitenwandmischung f
sieve/to (ab)sieben
sieve Sieb n [für Feststoffe]
  sieve analysis Siebanalyse f, Siebkontrolle f
  test sieve Prüfsieb n
sieving residue Siebrückstand m
sift/to sichten; sieben
sight Sicht f
  sight glass Einblickfenster n, Schauglas n, Sichtscheibe f, Sichtglas n, Sichtfenster n
sigma value Sigmawert m
sigma-type rotor Sigmarotor m
signal Signal n
  signal flowchart Signalflußplan m
  signal generator Signalgeber m
  signal input Signaleingang m
  signal output Signalausgang m
  signal status Signalzustand m
  signal status display Signalzustandsanzeige f

signal transducer Signalumwandler m
alarm signal Alarmgeber m, Alarmanzeige f, Störungsmeldung f, Alarmmeldung f
analog(ue) signal Analogsignal n
audible warning signal akustische Störanzeige f
charge signal Ladungssignal n
clock signal generator Taktgenerator m
control signal Kontrollsignal n, Steuersignal n
d.c. signal Gleichspannungssignal n
digital signal Digitalsignal n
flashing light signal Blinksignal n
flashing light warning signal Blink-Störungslampe f, Störungsblinkanzeige f
input signal Eingangssignal n
light signal Leuchtanzeige f
light signal indicator panel Leuchtanzeigetableau n
limiting value signal Grenzwertmeldung f
logic signal Logiksignal n
malfunction signal Störanzeige f
output signal Ausgangssignal n
overload signal Überlastanzeige f
plain language signal Klartextmeldung f
status signal Zustandsmeldung f
transducer signal Aufnehmersignal n
visual warning signal optische Störanzeige f
warning signal Störwertmeldung f, Alarmanzeige f, Alarmgeber m, Störmeldeeinrichtung f, Störmeldung f, Störmeldungsanzeige f, Warnsignal n
signalling Signalisierung f
signs Zeichen npl, Erscheinungen fpl
  signs of charring Verbrennungserscheinungen fpl
  signs of corrosion Korrosionserscheinungen fpl
  signs of decomposition Zersetzungserscheinungen fpl
  signs of gelling Gelierungserscheinungen fpl
  signs of orientation Orientierungserscheinungen fpl
  signs of wear Verschleißerscheinungen fpl
silane Silan n
  tetramethyl silane Tetramethylsilan n
  vinyl silane Vinylsilan n
silanization Silanisierung f
silanized silanisiert
silanol group Silanolgruppe f
silanol-functional silanolfunktionell
silent [e.g. machine, pump etc.] geräuscharm, lärmarm
silica Kieselsäure f, Siliciumdioxid n
  silica flour Quarzmehl n
  silica gel Kieselgel n, Silikagel n
  silica sand Quarzsand m
  silica skeleton Kieselsäuregerüst n
  synthetic silica Quarzgut n
  synthetic silica flour Quarzgutmehl n
silicate 1. silikatisch; 2. Silikat n
  silicate filler Silikatfüllstoff m
  alumin(i)um silicate Aluminiumsilikat n
  ethyl silicate Ethylsilikat n
  magnesium silicate Magnesiumsilikat n
  zirconium silicate Zirkonsilikat n

silicon Silicium n
  silicon carbide Siliciumkarbid n
  silicon chip Silicium-Einkristall-Plättchen n, Siliciumchip m
  silicon-oxygen linkage Silicium-Sauerstoffbindung f
  silicon tetrachloride Siliciumtetrachlorid n
silicone Silicon n [auch: Silikon]
  silicone casting resin Silicongießharz n
  silicone content Siliconanteil m, Silicongehalt m
  silicone elastomer Siliconelastomer n
  silicone fluid Siliconöl n
  silicone-free siliconfrei
  silicone-modified siliconmodifiziert
  silicone paste Siliconpaste f
  silicone release agent Silicontrennmittel n
  silicone release resin Silicontrennharz n
  silicone resin Siliconharz n
  silicone resin mo(u)lding compound Siliconharzpreßmasse f
  silicone rubber compound Siliconkautschukmischung f
  silicone rubber mo(u)ld Siliconkautschukform f
  silicone-treated siliconisiert
  alkyl silicone resin Alkylsiliconharz n
  liquid silicone rubber Flüssigsiliconkautschuk m
  one-pack silicone rubber Einkomponentensiliconkautschuk m
  phenyl silicone resin Phenylsiliconharz n
  phenylmethyl silicone Phenylmethylsilicon n
  phenylmethyl silicone fluid Phenylmethylsiliconöl n
  phenylmethyl silicone resin Phenylmethylsiliconharz n
  solid silicone resin Siliconfestharz n
  two-pack liquid silicone rubber Zweikomponentenflüssigsiliconkautschuk m
siliconized siliconisiert
silk-finish seidenglänzend
silo Silo m, Vorratsbehälter m [für Feststoffe]
  silo containing a day's supply of mo(u)lding compound Tagesbehälter m, Tagessilo m
  silo installation Siloanlage f
  bulk storage silo Lagersilo m
  degassing silo Entgasungssilo m
  large-capacity silo Großsilo m
  suitable for storing in silos silierbar
  suitable for transferring to silos silierbar
  transfer to silos Silieren n
siloxane Siloxan n
  siloxane group Siloxangruppe f
  siloxane linkage Siloxanbindung f
  siloxane unit Siloxanbaustein m
  dihydroxypolydimethyl siloxane Dihydroxypolydimethylsiloxan n
  phenylpropyl siloxane Phenylpropylsiloxan n
  polydialkyl siloxane chain Polydialkylsiloxankette f
  polydimethyl siloxane Polydimethylsiloxan n

silyl ether Silylether m
silylation Silylierung f
similar [materials, e.g. two grades of PE] artgleich
simple einfach
  technically simple verfahrenstechnisch einfach
simplification Vereinfachung f
simply constructed in einfacher Bauart f
simulated simuliert
simultaneous simultan, gleichzeitig
  simultaneous operation Simultanbetrieb m
  simultaneous stretching (process) Simultanreckverfahren n, Simultanstrecken n
sine curve Sinuskurve f
single einzeln, Einzel-, Ein-
  single bond Einfachbindung f
  single-cavity injection mo(u)ld Einfachspritzgießwerkzeug n
  single-cavity mo(u)ld Einfachform f, Einfachwerkzeug n, Ein-Kavitätenwerkzeug n
  single-circuit cooling system Einkreis-Kühlsystem n
  single-circuit system Einkreissystem n
  single-daylight mo(u)ld Doppelplattenwerkzeug n, Einetagenwerkzeug n, Zweiplattenwerkzeug n
  single-die extruder head Einfach(extrusions)kopf m, Einfachwerkzeug n
  single-die extrusion line Einkopfanlage f
  single-layer einschichtig
  single-layer die Einschichtdüse f, Einschichtwerkzeug n
  single-layer expanded polystyrene container Einschicht-PS-Schaumhohlkörper m
  single-layer extrusion Einschichtextrusion f
  single-layer film Einschichtfolie f
  single-parameter control circuit Einzelparameter-Regelkreis m
  single-parison die Einfach-Schlauchkopf m
  single-phase einphasig
  single-phase system Einphasensystem n
  single pin gate Einzelpunktanschnitt m
  single reciprocating screw Einfach-Schubschnecke f
  single screw Einspindelschnecke f, Einwellenschnecke f, Einzelschnecke f, Einschnecke, Einfachschnecke f
  single-screw einwellig
  single-screw arrangement Einschneckenanordnung f
  single-screw barrel Einschneckenzylinder m
  single-screw compounder Einschneckenmaschine f, Einwellenmaschine f, Einschnecken-Plastifizieraggregat n, einwelliger Schneckenkneter m, Einschneckenkneter m
  single-screw compounding extruder Einschnecken-Plastifizierextruder m
  single-screw devolatilizer Einwellenmaschine f
  single-screw extruder Einschneckenpresse f, Einwellenextruder m, Einschneckenaggregat n, Einschnecken(extrusions)anlage f, Einschneckenextruder m, Einwellenmaschine f,

Einschneckenmaschine f
single-screw extrusion Einschneckenextrusion f
single-screw extrusion line Einschnecken-(extrusions)anlage f
single-screw injection mo(u)lding machine Einschnecken-Spritzgießmaschine f
single-screw machine Einschneckenmaschine f, Einwellenmaschine f
single-screw principle Einschneckenprinzip n
single-screw system Einschneckensystem n, Einwellensystem n
single-shell einschalig
single-shell roof Warmdach n
single-speed clamp(ing) unit Eingeschwindigkeits-schließeinheit f
single-stage einstufig
single-stage blow mo(u)lding process Einstufen-Blasverfahren n
single-stage extrusion stretch blow mo(u)lding Einstufen-Extrusionsstreckblasen n
single-stage process Einstufenverfahren n
single-start [screw] eingängig
single-station automatic blow mo(u)lding machine Einstationenblasformautomat m
single-station blow mo(u)lding machine Einstationenblasformmaschine f
single-station extrusion blow mo(u)lding machine Einstationenextrusionsblasformanlage f
single-station machine Einstationenmaschine f, Einstationenanlage f
single-station winder Einstellen-Aufwicklung f
from a single supplier aus einer Hand f
high-speed single-screw extruder Hochleistungs-einschneckenextruder m, Einschnecken-Hochleistungsmaschine f
high-speed single-screw extrusion line Einschnecken-Hochleistungsextrusionsanlage f
laboratory single-screw extruder Labor-Einwellen-extruder m
production of single units Einzel(an)fertigung f
short single-screw extruder Einschnecken-Kurz-extruder m
vented single-screw extruder Einschnecken-Entgasungsextruder m
sink Spülbecken n, Vertiefung f
  sink mark Einfallstelle f [Preßfehler]
  heat sink paste Wärmeleitpaste f
sinter plate extruder Schmelztellerextruder m
sintering Versintern n
  free sintering Freisintern n
  rotational sintering Rotationssintern n
sinusoidal sinusförmig
situation Situation f, Lage f
  economic situation Konjunktur(lage) f
  employment situation Beschäftigungslage f
six-cavity mo(u)ld Sechsfachwerkzeug n
six-runner arrangement Sechsfachverteilerkanal m

size/to dimensionieren, kalibrieren
size 1. Größe f, Umfang m; 2. Schlichte f, Schlichtemittel n, Schmälze f
  size deposits Schlichteablagerungen fpl
  size reduction unit Zerkleinerer m, Zerkleinerungsmühle f, Zerkleinerungsmaschine f, Zerkleinerungsaggregat n
  batch size [e.g. of a two pack adhesive after mixing] Ansatzgröße f, Chargenmenge f
  cell size distribution Porengrößenverteilung f, Zellgrößenverteilung f
  cut to size/to formatschneiden
  gate size Angußgröße f, Anschnittgröße f
  mo(u)lded-part size Formteilgröße f
  nominal size Nenngröße f
  part size Formteilgröße f
  particle size Korndurchmesser m, Kornfeinheit f, Korngröße f, Körnung f, Partikelgröße f, Teilchengröße f
  particle size distribution Korn(größen)verteilung f, Kornzusammensetzung f, Partikelgrößenverteilung f, Teilchen(größen)verteilung f
  particle size range Korngrößenbereich m, Kornklasse f, Kornspektrum n
  pellet size Granulatgröße f
  platen size Plattengröße f
  predominant particle size häufigste Korngröße f
  preliminary size reduction Vorzerkleinern n
  preliminary size reduction unit Vorzerkleinerer m, Vorzerkleinerungsmühle f
  textile size Textilschlichte f
sizing 1. Dimensionierung f, Kalibrieren; 2. Schlichten n [Textil]
  sizing plate assembly Blendenpaket n, Kalibrierblendenpaket n
  sizing plates Ziehblenden fpl, Ziehscheiben fpl
  sizing unit Kalibrator m, Kalibriereinrichtung f
skidproof gleitsicher, rutschfest
skin Haut f
  skin contact Hautkontakt m
  skin pack Skinpackung f
  skin packaging machine Skinformmaschine f
  outer skin Oberflächenhaut f, Außenhaut f
  surface skin Oberflächenhaut f, Außenhaut f
skinning Hautbildung f
  skinning resistance Hautbildungsresistenz f
skirting board Fußbodenleiste f, Sockelleiste f
slabstock foam Blockschaumstoff m, Blockware f, Schaumstoffblock(material) m(n)
slabstock foaming Blockschäumung f
slate powder Schiefermehl n
sleeve Hülse f
  sleeve ejection system Hülsenausdrücksystem n
  sleeve ejector Hülsenauswerfer m
  compressed air calibrating sleeve Druckluft-kalibrierhülse f
  protective sleeve Schutzhülse f

253

slender cooling channels Fingerkühlung f
slenderness ratio Schlankheitsgrad m
slide Schieber m
    slide bar Schieber m
    slide-in einschiebbar
    slide-in module Einschub m
    slide-in temperature control module Temperatur-Regeleinschub m
    slide-in tray Einschiebehorde f
    slide plate [of a screen changer] Schieberplatte f
    slide valve Schieber m
    shut-off slide valve Absperrschieber m
sliderule Rechenschieber m
sliding Gleit-, Schiebe-
    sliding bearing Gleitlager n
    sliding friction Bewegungsreibung f, Gleitreibung f
    sliding guard door Schiebetür f, Schiebe(schutz)gitter n
    sliding insert Schiebeeinsatz m
    sliding movement Schiebebewegung f
    sliding safety guard Schiebe(schutz)gitter n
    sliding shut-off nozzle Schiebeverschluß(spritz)düse f
    sliding split mo(u)ld Schieberform f, Schieber(platten)werkzeug n
    sliding table Schiebetisch m
    sliding table machine Schiebetischmaschine f
    coefficient of sliding friction Gleitreibungszahl f, Gleitreibungskoeffizient m
    resistant to dry sliding friction trockengleitverschleißarm
    two-part sliding split mo(u)ld Doppelschieberwerkzeug n
slightly schwach
    slightly pigmented schwachpigmentiert
slip/to gleiten, schlüpfen
slip Gleiten n, Rutschen n, Gleitverhalten n
    slip agent Slipmittel n
    slip behavio(u)r Gleitverhalten n
    slip ring Schleifring m
    having good surface slip gleitfähig
    having very good surface slip hochgleitfähig
    surface slip Gleitfähigkeit f, Gleitfähigkeitsverhalten n
    surface slip characteristics Gleiteigenschaften fpl
    with very good surface slip hochgleitfähig
slippage Gleiten n
    sheet slippage [during thermoforming] Folienschlupf m
    wall slippage Wandgleiten n
slit Schlitz m, Spalt m
    slit die Schlitzdüse f, Breitschlitzwerkzeug n, Breitschlitzdüse f
    slit die extrusion Breitschlitzextrusion f, Breitschlitz(düsen)verfahren n
    slit die extrusion line Breitschlitzextrusionsanlage f
    slit die film extrusion Breitschlitzfolienverfahren n, Breitschlitzfolienextrusion f
    slit die film extrusion line Folienextrusionsanlage f, Breitschlitzfolienanlage f
    annular slit Ringspalt(e) m(f), Ringspaltöffnung f
    barrel vented through longitudinal slits Längsschlitz-Entgasungsgehäuse n
    suction slit Saugschlitz m
    two-layer slit die Zweischicht-Breitschlitzwerkzeug n
    ventilation slit Lüftungsschlitz m
    venting slit Entlüftungsspalt m, Entlüftungsschlitz m
slitter Schlitzvorrichtung f
    slitter device Schneidvorrichtung f, Schneidaggregat n, Schlitzvorrichtung f
    automatic film slitter Folienschneidautomat m
    film slitter (unit) Folienschneidaggregat n
    high-speed slitter Hochgeschwindigkeitsschneider m
    rotary slitter Rundlaufspaltmaschine f
slope Gefälle n, Neigung f, Steilheit f, Steigung f
    angle of slope Neigungswinkel m
slot die Schlitzdüse f, Breitschlitzdüse f, Breitschlitzwerkzeug n
slow im Schleichgang m, langsam
    slow-speed langsamlaufend, niedertourig
    slow-speed machine Langsamläufer m
    slow-speed mixer Langsamläufer m
slug Pfropfen m
    cold slug kalter Pfropfen m
    cold slug retainer Pfropfenhalterung f
slurry Schlämme f
    corundum slurry Korundschlämme f
    made into a slurry aufgeschlämmt
    pigment slurry Pigmentaufschlämmung f
slush mo(u)lding Schalengießverfahren n
small klein
    small-area kleinflächig
    small-bore pipe Kleinrohr n
    small blow mo(u)lding Kleinhohlkörper m
    small numbers kleine Stückzahlen fpl
    small-scale kleintechnisch
    small-scale production Kleinserienfertigung f
    small-scale unit Kleingerät n
    small standard specimen Normkleinstab m
    small standard test piece Normkleinstab m
    extremely small amount Kleinstmenge f
    extremely small mo(u)ld Kleinstwerkzeug n
    too small unterdimensioniert
smear Schmiere f
    smear device [used to obliterate spider lines in the melt flow] Stegverwischungseinrichtung f, Verwischungsgewinde n
    smear head Scherkopf m, Schertorpedo m, Schmierkopf m, Torpedoscherteil n
    faceted smear [section of screw] Flächenscherteil n
    grooved smear head Nutenscherteil n
smearing [e.g. when polishing] Schmiereffekt m, Schmierwirkung f
smell Geruch m
    inherent smell Eigengeruch m
smoke Rauch m, Rauchgas n, Qualm m

smoke density Rauch(gas)dichte f, Rauchgaswert m
smoke-suppressant rauchunterdrückend
smooth glatt, sanft
  smooth-bore [pipe] glattwandig
  smooth passage through the machine ruhige Laufeigenschaften fpl
smoothness Glätte f
  surface smoothness Oberflächenglätte f
snap-in joint Schnappverbindung f
snap-on assembly Steckmontage f
soap Seife f
  soap solution Seifenlösung f
  barium soap Bariumseife f
  calcium soap Calciumseife f
  fatty acid soap Fettsäureseife f
  lead soap Bleiseife f
  zinc soap Zinkseife f
socket Hülse f, Muffe f, Tülle f; Steckdose f
  socket construction Muffenkonstruktion f
  socket fusion welding Muffenschweißung f
  socket joint Muffenverbidung f
  socket-type construction Muffenkonstruktion f
  lamp socket Lampensockel m, Leuchtenfassung f
  welded socket joint Muffenschweißverbindung f
soda Soda n
  soda solution Natronlauge f
  caustic soda solution Natriumhydroxidlösung f
sodium Natrium n
  sodium benzoate Natriumbenzoat n
  sodium dichromate Natriumdichromat n
  sodium hydroxide solution Natriumhydroxidlösung f
  sodium tetrafluoroborate Natriumtetrafluorborat n
soft weich
  soft resin Weichharz n
  soft rubber Weichgummi m,n
softener Weichmacher m
  fabric softener Textilweichmacher m
  water softener Wasserenthärtungsmittel n
softening [solvents etc. acting on certain materials] 1. Weichmachen n; 2. weichmachend
  softening point Erweichungspunkt m, Erweichungstemperatur f, Plastifizierungstemperatur f
  softening range Erweichungsgebiet n, Erweichungsbereich m
  Vicat softening point Formbeständigkeit f in der Wärme mit Vicatnadel, Vicat-Erweichungspunkt m, Vicat-Erweichungstemperatur f, Vicat-Wärmeformsteständigkeit f, Vicatwert m, Vicatzahl f
softness Weichheitsgrad m
software Software f
  software house Softwarehaus n
  software menu [list of software options] Softwaremenü n
  software module Software-Baustein m
  software options Softwaremenü n
  software package Softwarepaket n, fremde Software f
  application software Anwendersoftware f
  operational software Betriebssoftware f
  packaged software fremde Software f
  ready-made software fremde Software f
  system software Systemsoftware f
  user software Benutzersoftware f
soiled verschmutzt
solar irradiation Sonnenbestrahlung f
solar radiation Sonnenstrahlung f
solder Lot n, Lötmittel n, Lötnaht f
  solder bath Lötbad n
  solder bath immersion Lötbadlagerung f
  solder bath resistance Lötbadfestigkeit f
sole plasticizer Alleinweichmacher m
solenoid Magnet m
  solenoid valve Magnetventil n
  solenoid valve driver Magnetventiltreiber m
  control solenoid Regelmagnet m
  proportional solenoid Proportionalmagnet m
  wet-armature solenoid Naßankermagnet m
solid 1. Festkörper m, Feststoff m; 2. nichtzellig, nichtzellular, fest
  solid-borne sound Körperschall m
  solid-borne sound insulating material Körperschallisoliermaterial n
  solid casting Vollguß m
  solid core [of material] Feststoffkern m
  solid flow Blockströmung f
  solid material Feststoff m
  solid particle Feststoffpartikel m
  solid phase Festphase f, Feststoffbereich m
  solid phase condensation Festphasenkondensation f
  solid phase forming Schlagpressen n
  solid phase pyrolysis Festphasenpyrolyse f
  solid resin Festharz n
  solid resin blend Festharzkombination f
  solid rod Vollstab m
  solid rubber Vollgummi n,m
  solid silicone resin Siliconfestharz n
  solid(-)state 1. Festzustand m; 2. kontaktlos, berührungslos
  solid-state controls kontaktlose Steuerung f
  hexagonal-section solid rod Sechskant-Vollstab m
  square-section solid rod Vierkant-Vollstab m
solidification Erstarrung f
  shrinkage on solidification Erstarrungsschrumpf m
solidify/to erstarren, verfestigen
solids Feststoffe mpl
  solids accumulations Feststoffansammlungen fpl
  solids content Festkörperanteil m, Festkörpergehalt m, Feststoffanteil m, Feststoffgehalt m
  solids conveying (system) Feststoff-Förderung f
  dry solids content Trockengehalt m
  paint solids content Lackfeststoffgehalt m
solubility Löslichkeit f
  solubility characteristics Löseverhalten n, Löslichkeitsverhalten n, Löslichkeitseigenschaften fpl
  solubility in ethanol Ethanollöslichkeit f

solubility range Löslichkeitsbereich m
having poor solubility schlecht löslich
water solubility Wasserlöslichkeit f
soluble löslich
  soluble in acids säurelöslich
  soluble oil Kühlemulsion f
  alcohol soluble alkohollöslich
  completely soluble klarlöslich
  largely soluble weitgehend löslich
  monomer soluble monomerlöslich
  readily soluble gut löslich
  sparingly soluble begrenzt löslich, beschränkt löslich
  water soluble wasserlöslich
solution Lösung f
  solution aid Lösungsvermittler m
  solution polymer Lösungspolymerisat n
  solution polymerized lösungspolymerisiert
  solution polymerization Lösungspolymerisation f
  solution viscosity Lösungsviskosität f
  alkali solution Lauge f
  caustic potash solution Kalilauge f
  caustic soda solution Natriumhydroxidlösung f, Natronlauge f
  compromise solution Kompromißlösung f
  detergent solution Detergentienlösung f, Waschmittellauge f
  electroplating solution Galvanisierungslösung f
  epoxy resin solution Epoxidharzlösung f
  ethanol solution ethanolische Lösung f
  methanol solution methanolische Lösung f
  pickling solution Ätzflüssigkeit f, Ätzlösung f
  polymer solution Kunststofflösung f, Polymer(isat)lösung f
  provisional solution Übergangslösung f
  reference solution Vergleichslösung f
  resin solution Harzlösung f
  salt solution Salzlösung f
  soap solution Seifenlösung f
  test solution Prüflösung f
  wetting agent solution Netzmittellösung f
solvating gelierend, solvatisierend
  solvating power [of plasticizer for PVC] Gelierfähigkeit f, Gelierkraft f, Lösungsvermögen n, Löseeigenschaften fpl, Lösefähigkeit f
  solvating temperature Gelierpunkt m, Geliertemperatur f
solvation Solvation f
  solvation temperature Lösetemperatur f
  degree of solvation Geliergrad m
solvent Löser m, Lösemittel n, Lösungsmittel n
  solvent activation Lösemittelaktivierung f
  solvent-based adhesive Lösemittelkleber m, Lösemittelklebstoff m
  solvent-based paint Lösungsmittellack m
  solvent-containing lösungsmittelhaltig
  solvent content Lösemittelgehalt m, Lösungsmittelanteil m
  solvent evaporation Lösungsmittelabgabe f, Lösemittelverdunstung f, Lösungsmittelabdunstung f
  solvent-free lösemittelfrei
  solvent loss Lösungsmittelverlust m
  solvent mixture Lösemittelgemisch n
  solvent phase Lösemittelphase f
  solvent recovery (unit) Lösungsmittelrückgewinnung f
  solvent resistance Lösemittelbeständigkeit f, Lösungsmittelresistenz f
  solvent resistant lösungsmittelfest
  solvent retention Lösungsmittelretention f
  solvent vapo(u)rs Lösungsmitteldämpfe mpl
  solvent weld Quellschweißnaht f
  solvent welding Quellschweißen n
  affected by solvents lösungsmittelempfindlich
  aliphatic solvent Aliphat n
  aromatic solvent Aromat m
  grease solvent Fettlösungsmittel n
  low-boiling solvent Niedrigsieder m
  medium-boiling solvent Mittelsieder m
  paint solvent Lacklösemittel n
  residual solvent Lösungsmittelrest m
  residual solvent content Lösungsmittelrestgehalt m
  resistant to solvents lösungsmittelfest
  susceptibility to solvent attack Lösungsmittelempfindlichkeit f
  type of solvent Lösungsmitteltyp m
  welding solvent Quellschweißmittel n
sonic converter Schallwandler m
soot Ruß m
sorption Sorption f
sorting Sortier-
  refuse sorting plant Müllsortieranlage f
sound Schall m
  sound-absorbent schallschluckend
  sound-absorbent ceiling tile Schallschluckdeckenplatte f
  sound-absorbent properties Schallschluckeigenschaften fpl
  sound absorption Schallabsorption f
  sound barrier Schallverkleidung f
  sound emission value Lärmemissionswert m
  sound-insulating schalldämmend
  sound-insulating material Schalldämmstoff m
  sound-insulating panel Schalldämmplatte f
  sound insulation Schalldämmung f, Schallschutz m, Lärmdämmung f
  sound pressure Schalldruck m
  sound pressure intensity Schalldruckstärke f
  sound pressure level Schalldruckstärke f, Schalldruckpegel m
  sound propagation Schallausbreitung f
  air-borne sound Luftschall m
  engine sound shield Geräuschkapsel f, Schallverkleidung f
  solid-borne sound Körperschall m

solid-borne sound insulating material Körperschall-isoliermaterial n
soundproof schalldämmend
   soundproof chute Schalldämmschurre f
   soundproof hood Schallschutzhaube f
soundproofed schallgedämpft, schallisoliert, lärmgedämpft
soundproofing Schalldämmung f, Schallschutz m
   soundproofing elements Schallschutzelemente npl
source Quelle f
   source of constant voltage Konstantspannungsquelle f
   source of energy Energiequelle f
   source of error Fehlerquelle f
   source of ignition Entzündungsquelle f, Zündquelle f
   source of information Informationsquelle f
   a.c. voltage source Wechselstromquelle f
   d.c. source Gleichstromquelle f
   energy source Energiequelle f
   from one source aus einer Hand f
   heat source Wärmequelle f
   light source Lichtquelle f
   radiation source Strahlungsquelle f
   raw material source Rohstoffquelle f
   voltage source Spannungsquelle f
soya bean [USA: soybean] Sojabohne f
   soya bean oil Soja(bohnen)öl n
   soya bean oil fatty acid Sojaöl-fettsäure f
space Platz m, Raum m
   space between the particles Teilchenzwischenräume mpl
   space for part ejection Ausfallöffnung f
   space in front of the screw Schneckenvorraum m, Schneckenstauraum m
   space pressurized during closing movement [of toggle clamp unit] Schließraum m, Öffnungsraum m
   space required Platzbedarf m
   space requirements Raumbedarf(splan) m
   space saving platzsparend, raumsparend
   space travel Raumfahrt f
   annular space Ringraum m
   floor space Aufstellfläche f, Bodenfläche f
   floor space requirements [to install a machine] Grundflächenbedarf m, Stellflächenbedarf m
   mo(u)ld space Werkzeugeinbaulänge f, Werkzeughöhe f, Werkzeugeinbauraum m, Werkzeug(ein)bauhöhe f
   pumping space Schöpfraum m
   requiring a lot of space platzaufwendig, platzintensiv
   requiring little space platzsparend
   storage space Lagerplatz m
spacecraft Raumfahrzeug n
spacer Distanzhalter m
   spacer disk Distanzscheibe f
   spacer ring Distanzring m
   threaded spacer Distanzbolzen m
spare parts Ersatzteile npl

list of spare parts Ersatzteilliste f
spare wheel recess Reserveradmulde f
spares Ersatzteile npl
sparingly begrenzt, beschränkt
   sparingly soluble begrenzt löslich, beschränkt löslich, schwerlöslich
spark Funke m
   spark eroded funkenerodiert
   spark erosion Funkenerosion f
   spark-over Überschlag m
sparkle Brillanz f
spatial räumlich
special besonders, speziell, Sonder-
   special equipment Sonderausrüstung f
   special features Besonderheiten fpl
   special formulation Sondereinstellung f, Spezialeinstellung f
   special-purpose design Einzweckausführung f
   special-purpose extruder Sonderextruder m, Einzweckextruder m
   special-purpose grade Sonderqualität f
   special-purpose machine Sondermaschine f, Einzweckmaschine f
   special-purpose mo(u)lding compound Spezialformmasse f
   special-purpose plastic Spezialkunststoff m
   special-purpose plasticizer Spezialweichmacher m
   special-purpose resin Spezialharz n
   special-purpose rubber Spezialkautschuk m, Sonderkautschuk m
   special-purpose screw Sonderschnecke f, Spezialschnecke f
speciality Spezialität f, Besonderheit f
   speciality plastic Spezialkunststoff m
   speciality resin Spezialharz n
   speciality rubber Sonderkautschuk m, Spezialkautschuk m
specific spezifisch, gezielt
   specific gravity spezifisches Gewicht n
   specific heat spezifische Wärme f
   specific to the material werkstoffspezifisch
   specific viscosity spezifische Viskosität f
   specific volume spezifisches Volumen n
specification Spezifikation f; Lastenheft n, Pflichtenheft n
   product specification Produktnorm f
   provisional specification Vornorm f
   standard specification Norm f, Normvorschrift f
   test specification Prüfbestimmung f, Prüfvorschrift f
specified vorgegeben
specimen Probe f, Muster n
   dumbbell test piece specimen Schulterprobe f, Schulterstab m
   flexural specimen Biegestab m
   notched impact test specimen Kerbschlagprobe f
   rectangular test specimen Flachstab m
   small standard specimen Normkleinstab m
   tensile specimen Zerreißprobe f, Zugprobekörper m,

Zugprobe f, Zugstab m
test specimen Versuchsprobe f
spectacle frame Brillengestell n
spectral range Spektralbereich m
spectrometric spektrometrisch
spectrometry Spektrometrie f
  infra-red differential spectrometry Infrarot-Differenz-spektrometrie f
  mass spectrometry Massenspektrometrie f
spectroscopy Spektroskopie f
  absorption spectroscopy Absorptionsspektroskopie f
  mass spectroscopy Massenspektroskopie f
  nuclear resonance spectroscopy Kernresonanz-spektroskopie f
  resonance spectroscopy Resonanzspektroskopie f
  UV spectroscopy UV-Spektroskopie f
  X-ray spectroscopy Röntgenspektroskopie f
speed Geschwindigkeit f, Schnelligkeit f, Fahrgeschwindigkeit f, Drehzahl f
  speed change-over point Geschwindigkeitsumschaltpunkt m
  speed changing device Geschwindigkeitsumschaltung f
  speed control Drehzahlregulierung f, Drehzahlregelung f
  speed control device Drehzahlregler m, Drehzahlregulierung f
  speed control system Fahrgeschwindigkeitsüberwachung f
  speed counter Drehzahlmesser m
  speed increase Drehzahlerhöhung f, Geschwindigkeitsanzeige f
  speed indicator Drehzahlanzeige f, Geschwindigkeitsanzeige f
  speed of advance Vorschubgeschwindigkeit f, Vorlaufgeschwindigkeit f
  speed of impact Schlaggeschwindigkeit f, Stoßgeschwindigkeit f
  speed of penetration Eindringgeschwindigkeit f
  speed of reaction Reaktionsgeschwindigkeit f
  speed of response Ansprechgeschwindigkeit f
  speed profile Drehzahlprogrammablauf m
  speed program Drehzahlprogramm n, Geschwindigkeitsprogramm n
  speed reduction Drehzahlerniedrigung f, Untersetzung f
  speed reduction gear Untersetzungsgetriebe n
  speed regulator Drehzahlregler m, Geschwindigkeitsregler m
  speed reserve Drehzahlreserve f
  speed sequence Geschwindigkeitsfolge f
  speed setting (mechanism) Drehzahleinstellung f
  speed stages Drehzahlstufen fpl
  speed transducer Geschwindigkeitsfühler m, Geschwindigkeitsaufnehmer m
  speed variations Drehzahlschwankungen fpl
  adjustment speed Anstellgeschwindigkeit f

  change in speed Drehzahländerung f
  closing speed Schließgeschwindigkeit f
  constant speed Drehzahlkonstanz f
  crack speed Rißgeschwindigkeit f
  curing speed Härtungsgeschwindigkeit f
  cutting speed Schnittgeschwindigkeit f
  ejector forward speed Auswerfervorlaufgeschwindigkeit f
  ejector return speed Auswerferrücklaufgeschwindigkeit f
  ejector speed Auswerfergeschwindigkeit f
  extrusion speed Extrudiergeschwindigkeit f, Extrusionsgeschwindigkeit f, Spritzgeschwindigkeit f
  filling speed Füllgeschwindigkeit f
  forward speed Einfahrgeschwindigkeit f
  gelation speed Geliergeschwindigkeit f
  heavy-duty high speed mixer Hochleistungsschnellmischer m
  high-speed clamping cylinder Schnellschließzylinder m
  high-speed coupling Schnellkupplung f, Schnellverbindung f
  high-speed cylinder Eilgangzylinder m
  high-speed drier Schnelltrockner m
  high-speed drying oven Schnelltrockner m
  high-speed (injection) cylinder Schnellfahrzylinder m
  high-speed mixer Schnellmischer m, Schnellrührer m
  high-speed mo(u)ld changing system Werkzeugschnellwechselsystem n
  high-speed mo(u)ld clamping mechanism Werkzeugschnellspannvorrichtung f
  high-speed nozzle retraction mechanism Düsenschnellabhebung f
  high-speed production line Hochgeschwindigkeitsanlage f
  high-speed screen changer Sieb-Schnellwechseleinrichtung f, Schnellsiebwechsel-Einrichtung f
  high-speed single-screw extruder Einschnecken-Hochleistungsmaschine f
  high-speed single-screw extrusion line Einschnecken-Hochleistungsextrusionsanlage f
  high-speed slitter [especially for film] Hochgeschwindigkeitsschneider m
  high-speed steel Schnellarbeitsstahl m
  high-speed stirrer Dissolver m, Schnellrührer m
  high-speed welding Schnellschweißen n
  impact speed Aufprallgeschwindigkeit f, Stoßgeschwindigkeit f, Schlaggeschwindigkeit f
  initial screw speed Eingangsdrehzahl f der Schnecke
  initial speed Eingangsdrehzahl f, Startdrehzahl f
  injection speed Spritzgeschwindigkeit f
  injection speed profile Einspritzgeschwindigkeitsverlauf m
  limiting speed Grenzdrehzahl f
  machine speed Fahrgeschwindigkeit f
  maximum speed Drehzahlgrenze f
  motor speed Motordrehzahl f

mo(u)ld closing speed Formzufahrgeschwindigkeit f, Schließgeschwindigkeit f, Werkzeugschließgeschwindigkeit f
mo(u)ld filling speed Formfüllgeschwindigkeit f, Werkzeugfüllgeschwindigkeit f
mo(u)ld opening and closing speed Werkzeuggeschwindigkeit f
mo(u)ld opening speed Öffnungsgeschwindigkeit f, Werkzeugöffnungsgeschwindigkeit f, Formauffahrgeschwindigkeit f
nozzle advance speed Düsenvorlaufgeschwindigkeit f
nozzle approach speed Düsenanfahrgeschwindigkeit f
nozzle retraction speed Düsenabhebegeschwindigkeit f
opening speed Öffnungsgeschwindigkeit f
operating speed Arbeitsdrehzahl f, Betriebsdrehzahl f, Arbeitsgeschwindigkeit f
operating speed range Betriebsdrehzahlbereich m
parison delivery speed Schlauchaustrittsgeschwindigkeit f
pendulum impact speed Pendelhammergeschwindigkeit f, Schlagpendelgeschwindigkeit f
peripheral screw speed Schneckenumfangsgeschwindigkeit f
plunger advance speed Kolbenvorlaufgeschwindigkeit f
production speed Produktionsdrehzahl f, Produktionsgeschwindigkeit f
pump speed Pumpendrehzahl f
rate of speed increase Drehzahlsteigerungsrate f
reeling speed Aufspulgeschwindigkeit f
remote speed control (mechanism) Geschwindigkeitsfernsteuerung f
required screw speed Extruderdrehzahl-Sollwert m
retraction speed Rücklaufgeschwindigkeit f, Rückzuggeschwindigkeit f
rotational speed Umlaufgeschwindigkeit f
rotor speed Rotordrehzahl f
screw advance speed Schneckenvorlaufgeschwindigkeit f
screw operating speed Arbeitsdrehzahl f, Betriebsdrehzahl f [Schnecke]
screw return speed Schneckenrücklaufgeschwindigkeit f
screw speed Schneckengeschwindigkeit f, Schneckenumdrehungszahl f, Schneckenumlaufgeschwindigkeit f, Schnecken(wellen)drehzahl f, Extruderdrehzahl f, Wellendrehzahl f
screw speed range Schneckendrehzahlbereich m, Drehzahlbereich m der Schnecke
setting speed Abbindegeschwindigkeit f, Erstarrungsgeschwindigkeit f
slow-speed langsamlaufend, niedertourig
slow-speed machine Langsamläufer m
slow-speed mixer Langsamläufer m
spur gear speed reduction mechanism Stirnrad-Untersetzungsgetriebe n

stress relaxation speed Spannungsabbaugeschwindigkeit f
take-off speed Abzugsgeschwindigkeit f
testing speed Prüfgeschwindigkeit f
transmission speed Übertragungsrate f, Übertragungsgeschwindigkeit f
unwinding speed Abwickelgeschwindigkeit f
welding speed Schweißgeschwindigkeit f
wind-up speed Wickelgeschwindigkeit f
working speed Arbeitsgeschwindigkeit f
spherical kugelähnlich, kugelförmig, kugelig, sphärisch
spherical electrode Kugelelektrode f
spherical tank Kugeltank m
spherulite Sphärolit m
spherulite structure Sphärolithgefüge n
rate of spherulite growth Sphärolithwachstumsgeschwindigkeit f
spider Stegtragring m [s.a. spider-type mandrel support]
spider leg Steg m, Dornsteghalter m, Dornsteghalterung f, Stegdornhalter m, Radialsteghalter m, Dornhaltersteg m, Dornträgersteg m, Haltesteg m
spider lines Stegdornmarkierungen fpl, Dornhaltermarkierungen fpl, Dornstegmarkierungen fpl, Fließschatten m, Steg(halter)markierungen fpl, Strömungsschatten m, Fließmarkierungen fpl, Längsmarkierungen fpl
spider-type die Stegdornhalterwerkzeug n, Steg(dorn)halterkopf m
spider-type mandrel support Stegdornhalter m, Dornsteghalter m, Dornsteghalterung f, Radialsteghalter m, Stegtragring m
spider-type parison die Stegdornhalter(blas)kopf m, Stegdornhalterwerkzeug n
spider with staggered legs Versetztstegdornhalter m, Dornhalter m mit versetzten Stegen
twin spider-type mandrel support Doppelstegdornhalter(ung) m(f)
spin welding Rotations(reib)schweißen n
spindle Spindel f
ejector spindle Austriebsspindel f
threaded spindle Gewindespindel f
trapezoidal spindle Trapezspindel f
spinneret Spinndüse f, Spinnkopf m
spinneret extrusion Extrusionsspinnprozeß m
spinneret extrusion line Extrusionsspinnanlage f
spinneret extrusion unit Extrusionsspinneinheit f
spinning Spinnen n
melt spinning Schmelzespinnen n
spiral spiralförmig, wendelförmig
spiral channel Wendelkanal m
spiral coil vaporizer Schlangenrohrverdampfer m
spiral flow Wendelströmung f
spiral flow length Spirallänge f
spiral flow test Spiraltest m
spiral grooves Wendelnuten fpl
spiral mandrel Spiraldorn m

spiral mandrel blown film die Wendelverteilerkopf m, Schmelzewendelverteilerwerkzeug n, Schmelzewendelverteilerkopf m, Spiraldornkopf m, Wendelverteilerwerkzeug n, Spiraldornblaskopf m
spiral mandrel die Spiraldorn m, Spiralverteilerblaskopf m
spiral mandrel (melt) distributor Schmelzewendelverteiler m, Wendelverteiler m
spiral melt stream Wendelstrom m
spiral spring Schraubenfeder f
ring-type spiral distributor Ring-Wendelverteiler m
star-type spiral distributor Stern-Wendelverteiler m
spirally grooved spiralförmig genutet, spiralgenutet
spirit Spiritus m
  white spirit Testbenzin n
split/to spleißen
split [mo(u)ld] zweiteilig
  split mo(u)ld Backenwerkzeug n, zweiteiliges Werkzeug n
  sliding split mo(u)ld Schieberform f, Schieber(platten)werkzeug n
  tendency to split Spleißneigung f
  two-part sliding split mo(u)ld Doppelschieberwerkzeug n
spoiler Spoiler m
  front spoiler Frontspoiler m
  rear spoiler Heckspoiler m
  roof spoiler Dachspoiler m
  side spoiler Seitenspoiler m
spot Punkt m
  spot welding Punkt(ver)schweißen n
  dead spots tote Ecken fpl, fließtote Räume mpl, tote Stellen fpl, Toträume mpl, tote Zonen fpl
  light spot line recorder Lichtpunktlinienschreiber m
  ultrasonic spot welding Ultraschallpunktschweißen n
  ultrasonic spot welding instrument Ultraschall-Punktschweißgerät n
  weak spot Schwachstelle f
spotting aid Tuschierlehre f
spray Spray n, zerstäubte Flüssigkeit f, Sprühregen m
  spray cooling Sprühkühlung f
  spray cooling section Sprühkühlstrecke f
  spray cooling unit Sprühkühlung f
  spray dried sprühgetrocknet
  spray drying Sprühtrocknen n
  spray lay-up Faser(harz)spritzen n
  spray lay-up equipment Faser(harz)spritzanlage f
  spray lay-up laminate Spritzlaminat n, Faserspritzlaminat n
  spray lay-up plant Faser(harz)spritzanlage f
  spray painting Spritzlackierung f
  salt spray test Salzsprühtest m
  water spray Beregnung f
sprayability Spritzbarkeit f, Spritzfähigkeit f
sprayable spritzbar
spraygun Spritzpistole f, Sprühpistole f
  airless spraygun Airless-Gerät n

  two-component spraygun Zweikomponentenspritzpistole f
spraying Spritz-, Sprüh-
  spraying consistency Spritzviskosität f
  spraying nozzle Spritzdüse f
  spraying paint Spritzlack m
  spraying varnish Spritzlack m
  airless spraying luftloses Sprühverfahren n, Airless-Spritzverfahren n
  electrostatic spraying Elektrostatikspritzen n
  flame spraying Flammspritzen n
  flock spraying Beflocken n
  hand spraying Handspritzen n
  high-pressure spraying plant Hochdruckspritzanlage f
  paint spraying booth Spritzkabine f
  paint spraying unit Farbspritzanlage f, Spritzaggregat n
  water spraying device Beregnungsvorrichtung f
spread Ausbreitung f, Verbreitung f
  spread coating Streichbeschichten n
  spread coating compound Streichmasse f
  spread coating machine Streichmaschine f
  spread coating paste Streichpaste f
  spread coating plant Streichanlage f
  spread coating process Streichverfahren n
  direction of flame spread Flammenausbreitungsrichtung f
  flame spread Brandausbreitung f, Brandweiterleitung f, Feuerweiterleitung f, Flammenausbreitung f
  rate of flame spread Flammenausbreitungsgeschwindigkeit f
spreadable streichbar
spreader roll Breithalter m, Breitstreckwalze f
spreader roll unit Breithaltevorrichtung f, Breitstreckwerk n
spreading Ausbreiten, Verteilen n, Streichen n
  spreading knife Streichmesser n
  spreading plant Streichanlage f
  spreading power Streichvermögen n
spring Feder f
  spring actuated federbetätigt
  spring assembly Federpaket m
  spring housing Federgehäuse n, Federhaus n
  spring-loaded federbelastet, gefedert
  spring steel blade Federstahlmesser n
  compression spring Druckfeder f
  disk spring Tellerfeder f
  helical spring Schraubenfeder f
  leaf spring Blattfeder f
  spiral spring Schraubenfeder f
sprocket chain Zahnkette f
  sprocket chain drive Zahnkettenantrieb m
  sprocket wheel Zahnkettenrad n
sprue Angußkegel m, Formteileanguß m, Formteilanschnitt m, Düsenzapfen m,

Stangenangußkegel m, Stangenanguß m, Angußbutzen m, Angußstange f, Angußzäpfchen n, Angußzapfen m, Anschnittkegel m, Anspritzkegel m, Anspritzling m
sprue and runners Verteilerspinne f, Verteilerstern m, Verteilerkreuz n, Angußstern m, Anschnittstern m, Angußspinne f
sprue bush Angußbuchse f, Anschnittbuchse f
sprue disposal unit Angußabführung f
sprue ejector cylinder Anguß-Ausstoßzylinder m
sprue ejector pin Angußausdrückstift m, Angußauswerfer m, Angußauswerfvorrichtung f
sprue gate Stangenanguß m, Kegelanguß m, Kegelanschnitt m, Stangenanschnitt m
sprue insert Angußeinsatz m
sprue location Angußlage f
sprue puller Angußauszieher m, Angußausreißer m
sprue puller bush Angußziehbuchse f
sprue puller gate Abreißanschnitt m
sprue puller pin Angußziehstift m
sprue puller pin gate Abreißpunktanschnitt m
sprue puller plate Abreißplatte f
sprue separator Anguß-Separator m
sprue waste Angußverlust m, Angußabfall m
ante-chamber sprue bush Vorkammerangußbuchse f
away from the sprue angußfern, in Anschnittferne f
disposal of sprues Angußabführung f
near the sprue angußnah, in Anschnittnähe f
pin gate sprue Punktangußkegel m
position of the sprue Angußlage f
sprueless angußfrei, angußlos
  sprueless injection mo(u)lding angußloses Spritzen n, Durchspritzverfahren n, Vorkammerdurchspritzverfahren n
spur gear Stirn(zahn)rad n
  spur gear speed reduction mechanism Stirnrad-Untersetzungsgetriebe n
square quaderförmig
  square meter Quadratmeter (m^2) m
  square-section knife Vierkantmesser n
  square-section pin Vierkantstift m
  square-section rod Vierkantstab m
  square-section solid rod Vierkant-Vollstab m
  square-section test piece Vierkantstab m
  weight per square meter [weight/m^2] Quadratmetergewicht n
squirrel cage motor Kurzschlußankermotor m
stability Stabilität f, Beständigkeit f, Standfestigkeit f
  stability characteristics Beständigkeitseigenschaften fpl
  stability during processing Verarbeitungsstabilität f
  stability in use Gebrauchsstabilität f
  stability reserves Stabilitätsreserven fpl
  bubble stability Blasenstabilität f
  colo(u)r stability Farbbeständigkeit f
  dimensional stability Dimensionsbeständigkeit f, Dimensionsstabilität f, Formstabilität f, Formbeständigkeit f, Formänderungsfestigkeit f, Maßbeständigkeit f
  film bubble stability Schlauchblasenstabilität f
  light stability Lichtstabilität f
  long-term stability Langzeitstabilität f
  long-term thermal stability Dauertemperaturbeständigkeit f, Dauerwärmebeständigkeit f
  melt stability Schmelzestabilität f
  phase stability Phasenstabilität f
  residual stability Reststabilität f
  storage stability Lagerbeständigkeit f
  thermal stability thermische Stabilität f, Temperaturstabilität f, Thermostabilität f, Wärmestabilität f
stabilization Stabilisierung f
  heat stabilization Thermostabilisierung f, Wärmestabilisierung f
  light stabilization Lichtschutz m, Lichtstabilisierung f
stabilized stabilisatorhaltig, stabilisiert
  heat stabilized wärmestabilisiert
  high-temperature stabilized hochwärmestabilisiert
  lead stabilized bleistabilisiert
  light stabilized lichtstabilisiert
  non-tin stabilized nicht-zinnstabilisiert
  reversion stabilized reversionsstabilisiert
  tin stabilized zinnstabilisiert
stabilizer [e.g. antioxidant protecting rubber from polymerization] Stabilisator m
  stabilizer blend Stabilisatorkombination f, Stabilisatormischung f
  stabilizer efficiency Stabilisatorwirksamkeit f
  stabilizer-lubricant blend Stabilisator-Gleitmittel-Kombination f
  stabilizer saving Stabilisatoreinsparung f
  stabilizer system Stabilisierungssystem n
  all-round stabilizer tolerance universelle Stabilisierbarkeit f
  amount of stabilizer Stabilisatormenge f
  barium-cadmium stabilizer Barium-Cadmium-Stabilisator m
  butyl-tin stabilizer Butylzinnstabilisator m
  calcium-zinc stabilizer Calcium-Zink-Stabilisator m
  containing stabilizer stabilisatorhaltig
  dialkyltin stabilizer Dialkylzinnstabilisator m
  dibutyltin stabilizer Dibutylzinnstabilisator m
  foam stabilizer Schaumstabilisator m, Zellstabilisator m
  free from stabilizer stabilisatorfrei
  heat stabilizer Thermostabilisator m, Wärmestabilisator m
  lead stabilizer Bleistabilisator m
  light stabilizer Lichtschutzmittel n, Lichtschutzzusatz m
  long-term stabilizer Langzeitstabilisator m
  main stabilizer Primärstabilisator m
  mono-alkyl tin stabilizer Monoalkylzinnstabilisator m
  octyl tin stabilizer Octylzinnstabilisator m

organo-tin stabilizer Organozinnstabilisator m
pigment stabilizer Pigmentstabilisator m
process stabilizer Verarbeitungsstabilisator m
radiation stabilizer Strahlenschutzmittel n
tin stabilizer Zinnstabilisator m
UV stabilizer UV-Stabilisator m
stabilizing stabilisierend
  stabilizing effect Stabilisierwirkung f
  heat stabilizing thermostabilisierend
  heat stabilizing properties Wärmestabilisierungsvermögen n
  light stabilizing lichtstabilisierend
  light stabilizing effect Lichtstabilisierwirkung f, Lichtstabilisatorwirkung f, Lichtschutzeffekt m
  light stabilizing system Lichtschutzsystem n
stable stabil, beständig
  dimensionally stable formbeständig, formstabil, dimensionsstabil
  dimensionally stable at elevated temperatures warmformbeständig
  thermally stable temperaturstabil, thermostabil
stack Stapel m
  stack mo(u)ld Etagenwerkzeug n
  four-point polishing stack Vierwalzen-Glättwerk n
  hot runner stack mo(u)ld Heißkanal-Etagenwerkzeug n
  polishing stack Glätteinheit f, Glättwalzen fpl, Glättwerk n
  three-roll polishing stack Dreiwalzen-Glättkalander m, Dreiwalzen-Glättwerk n
  twin-roll polishing stack Zweiwalzen-Glättwerk n
stackability Stapelbarkeit f, Stapelfähigkeit f
stacker Stapelmaschine f, Ablegeeinrichtung f
  automatic stacker Stapelautomat m
stacking 1. Stapeln n; 2. Stapel-
  stacking container Stapelbehälter m
  stacking device Stapeleinrichtung f, Stapelvorrichtung f
  stacking table Ablagetisch m
  stacking unit Ablage f, Stapelanlage f
staff Personal n
  staff problems Personalschwierigkeiten fpl
stage Stufe f, Etappe f, Abschnitt m
  curing stage Härtungsstufe f
  degassing stages Entgasungsstufen fpl
  design stage Konstruktionsphase f, Konzeptphase f
  experimental stage Versuchsstadium n
  hold(ing) pressure stages Nachdruckstufen fpl
  in the early stages im Frühstadium n
  in three stages dreifach gestaffelt
  injection pressure stage Spritzdruckstufe f
  intermediate stage Zwischenstadium n, Zwischenstufe f
  loading stages Belastungsschritte mpl
  planning stage Projektphase f
  preliminary stage Vorstufe f
  process stages Prozeßschritte mpl, Verfahrensschritte mpl
  processing stages Verarbeitungsphasen fpl, Verarbeitungsschritte mpl
  production stage Produktionsabschnitt m
  reaction stages Reaktionsschritte mpl
  speed stages Drehzahlstufen fpl
staggered versetzt
  spider with staggered legs Versetztstegdornhalter m, Dornhalter m mit versetzten Stegen
stagnation zone Ruhezone f, Stagnationsstelle f, Stagnationszone f
stain resistance Fleckenunempfindlichkeit f
staining [e.g. blue or purple coloration of rubber surface due to oxidation] Verfärbung f
stainless steel Edelstahl m, nichtrostender Stahl m, rostfreier Stahl m
  stainless steel filter screen Edelstahlmaschengewebe n
  stainless steel wire mesh Edelstahlmaschengewebe n
stair edging profile Treppenkantenprofil n
stair step dicer Bandgranulator mit Stufenschnitt m, Stufenschnitt-Granulator m
stamping Stanzen n, Prägen n
standard standardisiert, normgerecht, serienmäßig, marktüblich, normalisiert, gängig
  standard component Normteil n
  standard feed chute Serienaufgabeschurre f
  standard formulation Standardrezeptur f, Standardeinstellung f
  standard grade Standardqualität f, Standardtype f
  standard hot runner mo(u)ld units Heißkanal-Normblock m, Standard-Heißkanalblock m, Heißkanalnormalien pl
  standard injection mo(u)lding compound Standardspritzgußmarke f
  standard length Fixlänge f, Standardlänge f
  standard model Standardausführung f
  standard mo(u)ld Serienwerkzeug n
  standard mo(u)ld unit Stammwerkzeug n, Stammform f
  standard mo(u)ld units Baukastennormalien pl, normalisierte Bauteile npl, Formwerkzeug-Normalien pl, Formenbauteile npl, Normalien-Bauelemente npl, Werkzeugnormalien pl, Normalien pl
  standard packaging Standardverpackung f
  standard part Serienteil n
  standard plasticizer Standardweichmacher m
  standard plasticizing unit Normalplastifiziereinheit f
  standard platen details Bohrbild n, Lochbild n, Werkzeugaufspannzeichnung f
  standard polystyrene Normalpolystyrol n, Standardpolystyrol n
  standard range Normprogramm n, Serienprogramm n, Standardsortiment n
  standard recommendation Standardempfehlung f
  standard setting Standardeinstellung f
  standard specification Normvorschrift f
  standard tensile test piece Normzugstab m

standard test piece Normstab m, Norm-Probekörper m
standard thickness Normdicke f
standard unit Normbaugruppe f, Normbaustein m, Seriengerät n
standard units Normalien-Bauelemente npl, Normalien pl
standard vehicle Serienfahrzeug n
Gardner colo(u)r standard number Gardner-Farbzahl f
precision standard unit Präzisionsnormteil n
range of standard units Standardgeräteserie f
small standard specimen Normkleinstab m
small standard test piece Normkleinstab m
standardization Standardisierung f, Normung f
standardized genormt, normalisiert, normiert, standardisiert
staple Stapel m
staple fiber Stapelfaser f
staple glass fiber Glasstapelfaser f
woven staple fiber Glasstapelfasergewebe n
star Stern m
star-delta start Sterndreieckanlauf m
star-shaped (runner) arrangement Sternanordnung f
star-type distributor Sternverteiler m
starch Stärke f [Polysaccharid]
start Start m, Beginn m, Anfang m
start button Startknopf m, Starttaste f
start of production Produktionsbeginn m
start-up/to [a machine] anfahren
start-up behavio(u)r Anfahrverhalten n
start-up position Anfahrstellung f
start-up waste Anfahrausschuß m
at the start of plasticization bei Plastifizierbeginn m
extruder start-up waste Extruderanfahrfladen m
false start Fehlstart m
number of starts Gängigkeit f
star-delta start Sterndreieckanlauf m
starting Anfangs-, Ausgangs-
starting conditions Ausgangsbedingungen fpl
starting formulation Rahmenrezeptur f, Richtrezeptur f, Schemarezeptur f
starting material Ausgangsstoff m
starting monomer Ausgangsmonomer m
starting product Ausgangsprodukt n
starting temperature Vorlauftemperatur f
starting-up Inbetriebsetzung f, Anfahren n
starting-up period Anfahrzeit f
starting-up phase Anfahrphase f
starting-up problems Anfahrschwierigkeiten fpl
automatic starting mechanism Anfahrautomatik f
starve-feed/to unterfüttern
starve feeding Unterdosierung f
state Zustand m
state of aggregation Aggregatzustand m
state of deformation Verformungszustand m
state of equilibrium Gleichgewichtszustand m
state of internal stress Eigenspannungszustand m
state of orientation Orientierungszustand m
state of saturation Sättigungszustand m
state of stress Belastungszustand m, Spannungszustand m
state of the art Stand m der Technik
crystalline state Kristallisationszustand m
dry state Trockenzustand m
equation of state Zustandsgleichung f
glassy state Glaszustand m
in the mo(u)lded state im geformten Zustand m
in the unmo(u)lded state im ungeformten Zustand m
non-equilibrium state Nicht-Gleichgewichtszustand m
original state Ausgangszustand m
solid state Festzustand m, berührungslos, kontaktlos
wet state Naßzustand m
static statisch
static friction Haftreibung f
static mixing element Statikmischelement n
static stress ruhende Belastung f, ruhende Beanspruchung f
coefficient of static friction Haftreibungszahl f, Haftreibungskoeffizient m, Startreibung f
station Standort f, Stelle f; Station f
blow mo(u)lding station Blasformstation f
blowing station Blasstation f
cooling station Kühlstation f
data station Datenstation f
deflashing station Entbutzstation f
feed station Dosierstation f
filling station Füllstation f
flame treatment station Beflammstation f
heating-up station Aufheizstation f
injection station Spritzstation f
input station Eingabestation f
loading station Beladestation f
mo(u)lding station Formstation f
nuclear power station Kernkraftwerk n
outdoor weathering station Freibewitterungsstation f, Freiluftbewitterungsanlage f
parison receiving station Schlauchübernahmestation f
parts-removal station Entnahmestation f
take-off station Abzugsstation f
trimming and take-off station Trimm- und Entladestation f
unloading station Entladestation f
unwinding station Abrollstation f
weathering station Bewitterungsstation f
wind-up station Wickelstation f, Wickelstelle f, Aufwickelstelle f
stationary stationär, ortsfest
stationary knife Festmesser n, Gegenmesser n
stationary mo(u)ld half Düsenseite f, Einspritzseite f
stationary platen düsenseitige Formaufspannplatte f, Düsenplatte f
stationary knife Statormesser n
on the stationary mo(u)ld half düsenseitig

263

statistic statistisch
   statistic certainty statistische Sicherheit f
status Zustand m
   status display Zustandsanzeige f
   status record Zustandsprotokoll n
   status report Zustandsbericht m
   status signal Zustandsmeldung f
   control status Schaltzustand m
   job status Auftragszustand m
   machine status Maschinenzustand m
   production status Produktionsstatus m
   signal status Signalzustand m
   signal status display Signalzustandsanzeige f
steam Wasserdampf m, Dampf m
   steam boiler Dampfkessel m
   steam chamber Dampfkammer f
   steam chest Dampfkasten m
   steam generating unit Dampferzeuger m
   steam heated dampfbeheizt
   steam treatment Bedampfen f
   steam vulcanization Dampfvulkanisation f
   steam vulcanizing plant Dampfvulkanisieranlage f
   generation of steam Dampferzeugung f
   saturated steam Sattdampf m
   superheated steam Heißdampf m
stearate Stearat n
   calcium stearat Calciumstearat n
   calcium-zinc stearate Ca-Zn-Stearat n
   lead stearate Bleistearat n
   magnesium stearat Magnesiumstearat n
stearic acid Stearinsäure f
steel Stahl m
   steel ball Stahlkugel f
   steel calibrator Stahlkalibrierung f [Gerät]
   steel diaphragm Stahlmembrane f
   steel insert Stahleinsatz m
   steel plate Stahlblech n
   steel wool Schleifwolle f, Schmirgelwolle f, Stahlwatte f, Stahlwolle f
   bonded steel joint Stahlverklebung f
   bonding of steel Stahl(ver)klebung f
   case-hardened steel Einsatzstahl m
   cast steel Stahlguß m
   hardened steel Hartstahl m
   hardened steel (outer) layer Hartstahlmantel m
   high-speed steel Schnellarbeitsstahl m
   hot worked steel Warmarbeitsstahl m
   nitrided steel Nitrierstahl m
   nitrided steel barrel Nitrierstahlgehäuse n
   spring steel blade Federstahlmesser n
   stainless filter screen Edelstahlmaschengewebe n
   stainless steel Edelstahl m, nichtrostender Stahl m
   stainless steel wire mesh Edelstahlmaschengewebe n
   tool steel Werkzeugstahl m
   welded steel construction Stahlschweißkonstruktion f
steering wheel Lenkrad n
step Stufe f, Schritt m
   program step Programmschritt m
   stair step dicer Bandgranulator mit Stufenschnitt m
stepless stufenlos
steplessly variable stufenlos einstellbar
stepper motor Schrittmotor m
stereo chemical stereochemisch
stereo equipment Stereoausrüstung f
stereo-rubber Stereokautschuk m
stereoisomer Stereoisomer n
stereospecific stereospezifisch
steric sterisch
sterically hindered sterisch gehindert
sterilizability Sterilisierbarkeit f
sterilizable sterilisierbar
sterilizing Sterilisier-, sterilisierend
   resistance to sterilizing temperatures Sterilisierfestigkeit f
   resistant to sterilizing temperatures sterilisationsbeständig
stick-slip effect Haft-Gleit-Effekt m
sticky klebrig, klebend
stiffened verstrammt, versteift, verstärkt
stiffening [in consistency] Verstrammung f, Versteifen n
   stiffening effect Versteifungswirkung f
stiffness Starrheit f, Steifheit f, Steifigkeit f
   stiffness in bend Biegesteifigkeit f
   stiffness in shear Schubsteifigkeit f
   stiffness in torsion Torsionssteifigkeit f
   stiffness on demo(u)lding Entformungssteifigkeit f, Formsteifigkeit f
stir/to (ein)rühren
stirrer Rührgerät n, Rührwerk n
   high-speed stirrer Dissolver m, Schnellrührer m
stock Vorrat m
   capital stock Grundkapital n
stoichiometric stöchiometrisch
stone Gestein n, Stein m
   protection against flying stones Steinschlagschutzwirkung f
   resistance to flying stones Steinschlagfestigkeit f
stop/to arretieren, anhalten
stop Anschlag m
   stop bars Anschlagleisten fpl
   stop pin Anschlagbolzen m
   stop plate Anschlagplatte f
   emergency stop button Notausschalter m
   fixed stop Festanschlag m
stopcock Absperrhahn m
stoppage Stillstand m
   machine stoppage Maschinenstillstand m
stopper Stopfen m; Ausgleichsmasse f, Spachtel m, Spachtelmasse f
   epoxy-based stopper EP-Spachtel m
   glass stopper Glasstopfen m
   ground-glass stopper Schliffstopfen m
stopping of the machine Maschinenstillstand m, Maschinenstopp m

storage Lagerung f, Aufbewahrung f, Speicherung f
  storage conditions Lagerungsbedingungen fpl
  storage heater Wärmespeicherofen m
  storage modulus Speichermodul m
  storage of machine settings Einstelldatenabspeicherung f
  storage space Lagerplatz m
  storage stability Lagerbeständigkeit f, Lagerstabilität f
  storage tank Lagerbehälter m, Lagertank m, Vorratsbehälter m
  storage temperature Lager(ungs)temperatur f
  bulk storage Schüttgutlagerung f
  bulk storage silo Schüttgutlagersilo n
  bulk storage unit Massenspeicher m
  data storage Daten(ab)speicherung f
  machine for blow mo(u)lding fuel oil storage tanks Heizöltank-Blasmaschine f, Heizöltankmaschine f
  mass storage unit Massenspeicher m
  program storage Programmspeicherung f
  temporary storage Zwischenlagerung f
  working storage Arbeitsspeicher m
store/to speichern, lagern
  store/to temporarily zwischenlagern, zwischenspeichern
store Vorrat m
  data store Datenvorrat m
  external store Externspeicher m
  machine settings store Einstelldatenabspeicherung f
  program store Programmspeicher m, Programmspeicherung f
stored (ab)gespeichert
  stored program speicherprogrammierbar, speicherprogrammiert
  stored-program control PC-Steuerung f
stovable einbrennbar
stoved eingebrannt
  stoved finish Einbrennlackierung f
stoving Einbrennen n
  stoving alkyd Einbrennalkyd n
  stoving conditions Einbrennbedingungen fpl
  stoving enamel Einbrennlack m
  stoving oven Einbrennofen m
  stoving residue Einbrennrückstand m
  stoving temperature Einbrenntemperatur f
  stoving temperature range Einbrennbereich m
  stoving test Einbrennversuch m
  stoving time Einbrennzeit f, Einbrenndauer f
straight gerade, direkt
  straight-chain geradkettig, gradlinig
  straight-through die Geradeaus(spritz)kopf m, Längsspritzkopf m, Geradeaus(extrusions)werkzeug n, Längsspritzwerkzeug n
strain Spannung f, Dehnung f
  strain ga(u)ge Dehn(ungs)meßstreifen m
  strain transducer Dehnungsaufnehmer m
  creep strain Kriechdehnung f
  limiting strain Grenzdehnung f
  nominal strain Nenndehnung f
  outer fiber strain Randfaserdehnung f
  shear strain Scherdehnung f
strand Strang m, Faserbündel n
  strand and strip pelletization Kaltgranulierverfahren n
  strand die Strangdüse f, Strangwerkzeug n, Vollstrangdüse f
  strand die head Strangdüsenkopf m
  strand pelletization Stranggranulierung f
  strand pelletized kaltgranuliert
  strand pelletizer Kaltgranuliermaschine f, Kaltgranulator m, Strangschneider m, Stranggranulator m
  strand pelletizing line Stranggranulieranlage f
  strand pelletizing system Strangabschlagsystem n
  chopped strand mat Kurzfaservlies n, Glasfaservliesstoff m, Faserschnittmatte f, Glasseiden-Schnittmatte f
  chopped strand prepreg Kurzfaser-Prepreg n
  chopped strand reinforced kurzglas(faser)verstärkt
  chopped strands geschnittene Glasfasern fpl, geschnittenes Textilglas n, Glaskurzfasern fpl, Kurzglasfasern fpl, Schnittglasfasern fpl
  continuous glass strands Endlosglasfasern fpl
  continuous strand composite Endlosfaserverbund m
  continuous strand mat Endlos(faser)matte f, Glasseiden-Endlosmatte f
  continuous strand-reinforced langfaserverstärkt
  fiber strand Faserbündel n, Faserstrang m
  glass strand Glasfaserstrang m, Glasseidenstrang m
  melt strand Massestrang m, Schmelzestrang m
  reinforced with continuous glass strands langglasverstärkt
  roving strand Rovingstrang m
  underwater strand pelletizer Unterwasserstranggranulator m
streaking Schlierenbildung f
streaks Schlieren fpl
  colo(u)red streaks Farbschlieren fpl
  free from streaks schlierenfrei
  surface streaks Oberflächenschlieren fpl
stream Strom m
  air stream Luftstrom m, Luftströmung f
  axial (melt) stream Axialstrom m
  cooling air stream Kühlluftströmung f
  device which divides the melt stream Schmelzestromteiler m
  external cooling air stream Außenkühlluftstrom m
  internal air stream Innenluftstrom m
  internal cooling air stream Innenkühlluftstrom m
  melt stream Produktstrom m, Schmelzestrom m
  melt stream-dividing strömungsteilend
  separate melt streams Partialströme mpl, Teilströme mpl, Masseteilströme mpl, Schmelzeteilströme mpl
  spiral melt stream Wendelstrom m

strength Festigkeit f, Stärke f, Zähigkeit f
 strength of the substrate Untergrundfestigkeit f
 adhesive strength Adhäsionsfestigkeit f, Adhäsivfestigkeit f, Haftfestigkeit f
 bond strength Bindefestigkeit f, Klebfestigkeit f, Klebwert m, Trennfestigkeit f
 breakdown field strength Durchschlagfeldstärke f
 breaking strength Bruchfestigkeit f, Berstfestigkeit f,
 bursting strength [particularly of roofing sheet] Schlitzdruckfestigkeit f
 change in tensile strength Zugfestigkeitsänderung f
 cohesive strength Kohäsionsfestigkeit f, Kohäsivfestigkeit f
 compressive strength Druckfestigkeit f, Stauchfestigkeit f, Stauchhärte f
 compressive strength test machine Druckfestigkeitsprüfmaschine f
 corner strength [of a PVC window frame] Eckfestigkeit f
 creep (rupture) strength [of pipes] Zeitstandfestigkeit f
 creep strength curve Zeitstandfestigkeitskurve f, Zeitstandfestigkeitslinie f
 cross-breaking strength Biegefestigkeit f
 decrease in strength Festigkeitsabfall m, Festigkeitsminderung f
 development of compressive strength Druckfestigkeitsentwicklung f
 dielectric strength Durchschlagfestigkeit f
 dimensional strength Formbeständigkeit f, Gestaltfestigkeit f
 drop impact strength Fallbruchfestigkeit f
 dry strength Trockenfestigkeit f
 electric field strength elektrische Feldstärke f
 fatigue strength Schwingfestigkeit f
 field strength Feldstärke f
 final strength Endfestigkeit f
 financial strength Finanzkraft f
 flexural creep strength Zeitstandbiegefestigkeit f
 flexural fatigue strength Zeit-Biegewechselfestigkeit f, Biegeschwingfestigkeit f, Biegewechselfestigkeit f
 flexural impact strength Schlagbiegefestigkeit f, Schlagbiegezähigkeit f
 flexural strength Biegefestigkeit f
 impact strength Schlagfestigkeit f, Stoßzähigkeit f, Stoßfestigkeit f
 impact strength reduction Schlagzähigkeitsminderung f
 initial strength Anfangsfestigkeit f
 Izod notched impact strength Izod-Kerbschlagfähigkeit f
 long-term dielectric strength Dauerdurchschlagfestigkeit f, Langzeit-Durchschlagfestigkeit f
 long-term flexural strength Dauerbiegefestigkeit f
 loss of strength Festigkeitseinbuße f
 low-temperature impact strength Tieftemperatur-Schlagzähigkeit f, Kälteschlagbeständigkeit f, Kälteschlagwert m, Kälteschlagzähigkeit f
 mechanical strength mechanische Festigkeit f
 notched impact strength Kerbschlagfestigkeit f, Kerbschlagwert m
 notched tensile impact strength Kerbschlagzugzähigkeit f
 original strength Ausgangsfestigkeit f
 peel strength Abschälfestigkeit f, Schälfestigkeit f, Schälwert m, Schälwiderstand m
 pigment strength Farbkraft f
 relative tear strength relative Reißfestigkeit f
 residual tear strength Restreißfestigkeit f
 residual tensile strength Restzugfestigkeit f
 residual ultimate strength Restreißkraft f
 retained flexural strength Restbiegefestigkeit f
 retained tear strength Restreißfestigkeit f
 retained tensile strength Restzugfestigkeit f, Restreißkraft f
 shear strength Schub(scher)festigkeit f
 short-term dielectric strength Kurzzeit-Durchschlagfestigkeit f
 tear strength Zerreißfestigkeit f, Reißfestigkeit f
 tensile creep strength Zeitstandzugfestigkeit f
 tensile impact strength Schlagzugzähigkeit f
 tensile shear strength Zugscherfestigkeit f
 tensile strength Zugfestigkeit f
 tensile strength in bending Biegezugfestigkeit f
 transverse strength Querbelastbarkeit f, Querfestigkeit f
 transverse tensile strength Querzugfestigkeit f
 ultimate tensile strength Reißfestigkeit f, Zerreißfestigkeit f
 weld strength Bindenahtfestigkeit f
 wet strength Naßfestigkeit f
 wet strength agent Naßfestmittel n
 wet strength value Naßfestigkeitswert m
stress Spannung f, Beanspruchung f, Stress m
 stress amplitude Belastungsamplitude f, Lastamplitude f
 stress analysis Spannungsanalyse f
 stress application device Belastungseinrichtung f
 stress concentration Spannungskonzentration f
 stress crack Spannungsriß m
 stress cracking Spannungsrißkorrosion f
 stress cracking behavio(u)r Spannungskorrosionsverhalten n
 stress cracking resistance Spannungsriß(korrosions)beständigkeit f
 stress cycle Belastungszyklus m, Lastspiel n, Lastwechsel m, Lastzyklus m
 stress cycle frequency Lastspielfrequenz f
 stress direction Beanspruchungsrichtung f
 stress distribution Spannungsverteilung f
 stress duration Laststandzeit f, Beanspruchungsdauer f, Belastungsdauer f, Belastungszeit f

stress-elongation diagram Spannungs-Dehnungs-Diagramm n
stress equilibrium Spannungsgleichgewicht n
stress-free spannungsfrei
stress intensity Spannungsintensität f
stress intensity amplitude Spannungsintensitätsamplitude f
stress intensity factor Spannungsintensitätsfaktor m
stress level Belastungshöhe f, Beanspruchungsgrad m
stress peak Belastungsspitze f, Spannungsspitze f
stress-related spannungsbezogen
stress relaxation Spannungsrelaxation f, Spannungsabbau m
stress relaxation speed Spannungsabbaugeschwindigkeit f
stress relaxation test Spannungsrelaxationsversuch m
stress removal Entlastung f
stress removal time Entlastungszeit f
stress-strain behavio(u)r Spannungs-Dehnungs-Verhalten n, Spannungs-Verformungs-Verhalten n
stress-strain diagram Kraft-Längenänderungs-Diagramm n, Kraft-Dehnungs-Diagramm n, Kraft-Verformungsdiagramm n, Spannungs-Dehnungs-Diagramm n
stress-whitened zone Weißbruchzone f
stress-whitening Weißbruchbildung f, Weißbruch m
stress-whitening effect Weißbrucheffekt m
amount of stress applied Beanspruchungshöhe f, Beanspruchungsgrad m
application of stress Lastaufbringung f
bending stress Biegebelastung f, Biegespannung f
bending stress at break Bruchbiegespannung f
breaking stress Bruchlast f, Reißkraft f
bursting stress Berstspannung f
compressive stress Druckbelastung f, Drucklast f, Druckbeanspruchung f, Druckspannung f, Stauchbelastung f
cooling stresses Abkühlspannungen fpl
creep stress Kriechbeanspruchung f, Kriechspannung f, Zeit(dehn)spannung f, Zeitstandbeanspruchung f
direction of greatest stress Hauptspannungsrichtung f
direction of tensile stress Zugbeanspruchungsrichtung f
distribution of residual welding stresses Schweißrestspannungsverteilung f
dynamic stress zügige Beanspruchung f, zügige Belastung f
elongation at yield stress Dehnung f bei Streckspannung
environmental stress cracking umgebungsbeeinflußte Spannungsrißbildung f
external stress Fremdspannung f
fatigue stress Ermüdungsbelastung f, Schwellbeanspruchung f
flexural fatigue stress Biegewechselbeanspruchung f
flexural impact stress Schlagbiegebeanspruchung f

flexural stress Biegebeanspruchung f
frozen-in stresses Orientierungsspannungen fpl, eingefrorene Spannungen fpl
hoop stress [of pipes] Rohrwandbeanspruchung f, Vergleichsspannung f, (zulässige) Wandbeanspruchung f, zulässige Beanspruchung f
internal compressive stresses Druckeigenspannungen fpl
internal cooling stresses Abkühleigenspannungen fpl
impact stress schlagartige Beanspruchung f, Schlagbeanspruchung f, Stoßbeanspruchung f, Stoßbelastung f
internal stress Eigenspannung f
internal tensile stress Zugeigenspannung f
internal stresses Eigenspannungsfeld n
limiting shear stress Grenzschubspannung f
limiting stress Grenzspannung f
long-term stress Dauerbelastung f, Langzeitbeanspruchung f, Langzeitbelastung f, Zeitstandbeanspruchung f
long-term torsional bending stress Dauertorsionsbiegebeanspruchung f
longitudinal stress Längsspannung f
maximum tensile stress Zugspannungsmaximum n
mo(u)lded-in stresses Spannungseinschlüsse fpl
mo(u)lding stresses Verarbeitungsspannungen fpl
number of breaking stress cycles Bruchlastspielzahl f
number of stress cycles Lastspielzahl f, Lastwechselzahl f
offset yield stress Dehnspannung f
peel stress Schälbeanspruchung f, Schälbelastung f
removal of stress from the test piece Prüfkörperentlastung f
repeated stress Wechselbeanspruchung f
repeated tensile stress Zugschwellast f, Zugschwellbelastung f
residual welding stresses Schweißrestspannungen fpl
resistance to torsional stress Torsionsbelastbarkeit f
resistant to stress cracking spannungsrißbeständig
safe working stress [of pipes] zulässige Beanspruchung f, Sigmawert m
shear stress Scherbeanspruchung f, Scherspannung f, Schubspannung f
shear stress at failure Abscherspannung f
short-term breaking stress Kurzzeitbruchlast f
short-term stress Kurzzeitbeanspruchung f
short-term tensile stress Kurzzeitzugbeanspruchung f
state of internal stress Eigenspannungszustand m
state of stress Belastungszustand m, Spannungszustand m
static stress ruhende Beanspruchung f, ruhende Belastung f
sudden stress schlagartige Beanspruchung f
susceptibility to stress cracking Spannungsrißanfälligkeit f
susceptible to stress cracking spannungsrißempfindlich

tangential stress Tangentialspannung f
tensile breaking stress Zugbruchlast f
tensile creep stress Zeitstand-Zugbeanspruchung f
tensile shear stress Zugscherbelastung f
tensile stress Zugbeanspruchung f, Zugbelastung f, Zuglast f, Zugspannung f
tensile stress at break Bruchspannung f
tensile stress at yield Zugspannung f bei Streckgrenze, Streckspannung f, Fließspannung f
tensile stress-elongation curve Zugspannungs-Dehnungs-Kurve f, Zug-Dehnungs-Diagramm n
time under stress Laststandzeit f
torsional stress Torsionsbeanspruchung f, Torsionsbelastung f
transverse stress Querspannung f
transverse tensile stress Querzugspannung f
type of stress Beanspruchungsart f, Belastungsart f
ultimate tensile stress Bruchspannung f, Reißkraft f
under bending stress biegebeansprucht
under compressive stress druckbeansprucht, druckbelastet
under flexural stress biegebelastet
under impact stress schlagbeansprucht
under peel stress schälbeansprucht
under stress belastet
under tensile stress zugbeansprucht, zugbelastet
vibration stress Schwingungsbeanspruchung f
yield stress Fließspannung f, Reißspannung f, Streckspannung f
stressed beansprucht, belastet
stressing rate Beanspruchungsgeschwindigkeit f, Belastungsgeschwindigkeit f
stretch Strecken n, Dehnung f
  stretch blow mo(u)lder Streckblasmaschine f
  stretch blow mo(u)lding Streckblasformen n, Streckblasen n
  stretch blow mo(u)lding machine Streckblasformmaschine f
  stretch blow mo(u)lding plant Streckblasanlage f
  stretch forming tool Streckziehwerkzeug n
  stretch ratio Reckverhältnis n, Streckverhältnis n, Verstreckungsverhältnis n, Verstreckungsgrad m
  stretch wrapped pack Streckfolien-Verpackung f, Streckverpackung f
  stretch wrapping Streckfolien-Verpackung f, Streckverpackung f
  stretch wrapping film Dehnfolie f, Streckfolie f
  automatic oriented PP stretch blow mo(u)lding machine OPP-Streckblasautomat m
  extrusion stretch blow mo(u)lding Extrusions-Streckblasen n, Extrusions-Streckblasformen n
  godet roll stretch unit Galettenstreckwerk n
  injection stretch blow mo(u)lding Spritzstreckblasen n, Spritzgieß-Streckblasen n
  injection stretch blow mo(u)lding machine Spritzgieß-Streckblasmaschine f, Spritzstreckblasmaschine f
  injection stretch blow mo(u)lding plant Spritzstreckblasanlage f
  single-stage extrusion stretch blow mo(u)lding Einstufen-Extrusionsstreckblasen n
  two-stage injection stretch blow mo(u)lding (process) Zweistufen-Spritzstreckblasverfahren n
stretched verstreckt, gestreckt
  transversely stretched querverstreckt
stretching Recken n, Strecken n, Verstrecken n
  biaxial stretching biaxiales Recken n
  biaxial stretching unit Biaxial-Reckanlage f
  blown film stretching process Schlauchstreckverfahren n
  film stretching device Folienreckvorrichtung f
  film stretching plant Folienreckanlage f
  film tape stretching plant Bändchenreckanlage f
  film tape stretching unit Folienbandstreckwerk n
  longitudinal stretching Längsverstreckung f
  longitudinal stretching unit Längsstreckmaschine f, Längsstreckwerk n
  parison stretching mandrel Schlauchspreizvorrichtung f, Spreizdorn m, Spreiz(dorn)vorrichtung f, Spreizdornanlage f
  simultaneous stretching Simultanstrecken n
  simultaneous stretching (process) Simultanreckverfahren n
  stretching flow Dehnströmung f
  stretching rate Reckgeschwindigkeit f
  stretching roll Reckwalze f, Streckwalze f
  stretching roll section Rollenreckstrecke f
  stretching roll unit Rollenreckwerk n
  stretching temperature Recktemperatur f
  stretching unit Reckanlage f, Streckeinrichtung f, Streckwerk n
  transverse stretching Querverstreckung f
  transverse stretching machine Breitreckmaschine f
  transverse stretching roll Breitstreckwalze f
  two-stage stretching (process) [for film] Zweistufenreckprozeß m
  uniaxial stretching unit Monoaxial-Reckanlage f
strike-through Durchschlag m
stringing Fadenziehen n
strip Streifen m, Band n
  strip pelletizer Würfelschneider m, Bandgranulator m, Kaltgranuliermaschine f, Kaltgranulator m, Bandschneider m
  strip pelletizing line Bandgranulierstraße f
  strip pelletizing system Bandabschlagsystem n
  strand and strip pelletization Kaltgranulierverfahren n
  rectangular strip Rechteckstreifen m
  test strip Probestreifen m, Prüfstreifen m
stripper Abstreifer m
  stripper bush Abstreifhülse f
  stripper mechanism Abstreifsystem n, Abstreifvorrichtung f
  stripper plate Abstreifplatte f, Abschiebeplatte f
  stripper ring Abstreifring m

stripping force Abschiebekraft f
strips Walzfell n, Fell n
  strips for feeding to the calender Fütterstreifen m
stroke Hub m, Verschiebeweg m
  stroke adjusting mechanism Hubeinstellung f
  stroke control Wegesteuerung f
  stroke counter Hubzähler m, Hubzählwerk n
  stroke-dependent wegabhängig
  stroke-independent wegunabhängig
  stroke limit (switch) Hubbegrenzung f
  stroke limitation Hubbegrenzung f
  stroke measurement Wegerfassung f
  stroke measuring system Wegmeßsystem n
  stroke sensing mechanism Weggebersystem n
  stroke setting Hubeinstellung f
  stroke transducer Weg(meß)geber m, Wegaufnehmer m
  clamping stroke Formschließhub m
  demo(u)lding stroke Entformungsweg m
  depending on the screw stroke schneckenwegabhängig
  eccentric stroke Exzenterhub m
  ejector stroke Auswerferweg m, Ausstoßweg m, Auswerferhub m
  ejector stroke limiting device Auswerferhubbegrenzung f
  injection stroke Einspritzhub m, Einspritzweg m, Spritzhub m
  lifting stroke Hebehub m
  metering stroke Dosierhub m, Dosierweg m
  mo(u)ld clamping stroke Formschließhub m
  mo(u)ld opening stroke Formöffnungshub m, Formöffnungsweg m, Öffnungshub m, Öffnungsweg m, Öffnungsweite f, Werkzeugöffnungsweg m, Werkzeugöffnungshub m
  mo(u)ld opening stroke limiting device Werkzeugöffnungshub-Begrenzung f
  mo(u)ld return stroke Werkzeugrückhub m
  nozzle retraction stroke Düsenabhebeweg m
  nozzle stroke Düsenhub m
  opening stroke Öffnungsweite f, Öffnungsweg m, Öffnungshub m
  opening stroke limiting device Öffnungswegbegrenzung f
  piston stroke Kolbenhub m
  plunger stroke Kolbenhub m
  plunger stroke-dependent kolbenwegabhängig
  ram stroke Kolbenhub m, Stößelhub m
  screw return stroke Schneckenrückhub m
  screw stroke Schneckenhub m, Schneckenweg m
  screw stroke adjusting mechanism Schneckenhubeinstellung f, Schneckenwegeinstellung f
  screw stroke adjustment Schneckenhubeinstellung f, Schneckenwegeinstellung f
  screw stroke transducer Schneckenwegaufnehmer m
  short stroke Kurzhub m
  toggle stroke Kniehebelhub m

strong [i.e. able to support heavy loads] tragfähig
structural strukturell
  structural adhesive Baukleber m, Konstruktionsklebstoff m
  structural changes Gefügeänderungen fpl, Strukturveränderungen fpl
  structural damage Gefügeschädigung f
  structural defect Gefügestörung f, Strukturfehler m
  structural details Konstruktionseinzelheiten fpl
  structural differences Strukturunterschiede mpl
  structural element Strukturelement n
  structural features Konstruktionsmerkmale npl
  structural foam Strukturschaum(stoff) m
  structural foam mo(u)ld TSG-Werkzeug n [s.a. structural foam mo(u)lding]
  structural foam mo(u)lding Schaumspritzgießen n, TSG-Verarbeitung f, Thermoplast-Schaumguß-Verfahren n
  structural foam mo(u)lding machine TSG-Maschine f, Strukturschaummaschine f, TSG-Spritzgießmaschine f, Thermoplast-Schaumgießmaschine f
  structural foam mo(u)lding technology TSG-Technik f
  structural foam mo(u)ldings TSG-Teile npl
  structural foam profiles strukturgeschäumte Profile npl
  structural formula Strukturformel f
  structural sealant Bautendichtungsmasse f, Konstruktionsdichtungsmasse f
  structural unit Struktureinheit f, Baugruppe f
  multi-component structural foam mo(u)lding machine TSG-Mehrkomponentenmaschine f
  multi-component structural foam mo(u)lding (process) TSG-Mehrkomponentenverfahren n
  one-component structural foam mo(u)lding (process) Einkomponenten-Schaumspritzgießverfahren n
  polyurethane structural foam Polyurethan-Strukturschaumstoff m
structure Struktur f, Stoffgefüge n, Gefüge n
  allophanate structure Allophanatstruktur f
  alumin(i)um structure Aluminiumkonstruktion f
  biuret structure Biuretstruktur f
  cell structure Porenbild n, Porenstruktur f, Zellengefüge n, Zell(en)struktur f
  cost structure Kostenstruktur f
  craze zone structure Fließzonenstruktur f
  crosslink structure Vernetzungsstruktur f
  cyclic structure Ringstruktur f
  foam structure Schaumstoffgefüge n, Schaum(stoff)struktur f
  honeycomb structure Wabenstruktur f
  ionic lattice structure Ionengitter n
  lamellar structure Lamellenstruktur f, Laminataufbau m
  lattice structure Gitterstruktur f

molecular lattice structure Molekülgitter n
molecular structure Molekülaufbau m, Molekülstruktur f, Molekularaufbau m, Molekülgebilde n
network structure Netz(werk)struktur f
paint film structure Anstrichaufbau m
particle structure Kornstruktur f, Teilchenstruktur f
platelet structure Blättchenstruktur f
polymer structure Polymerstruktur f
ring structure Ringstruktur f
secondary structure Sekundärstruktur f
spherulite structure Sphärolithgefüge n
three-dimensional structure Raumstruktur f
two-phase structure Zweiphasenstruktur f
study Studie f
  feasibility study Machbarkeitsstudie f
  mo(u)ld filling study Füllstudie f
stuffer Stopfvorrichtung f, Stopfwerk n, Stopfaggregat n
stuffing Stopf-, stopfend
  stuffing device Stopfvorrichtung f
  stuffing ram Stopfkolben m
  stuffing screw Einpreßschnecke f, Stopfschnecke f
  stuffing unit Stopfwerk n, Stopfaggregat n
sturdiness Robustheit f
sturdy robust
styling requirements Stylinganforderungen fpl
styrene Styrol n
  styrene-acrylonitrile copolymer [SAN] Styrol-Acrylnitril-Copolymerisat n
  styrene-butadiene copolymer Styrol-Butadienpfropfpolymerisat n
  styrene-butadiene rubber [SBR] Styrol-Butadien-Kautschuk m
  styrene copolymer Styrolcopolymerisat n
  styrene-free styrolfrei
  styrene homopolymer Styrol-Homopolymerisat n
  styrene monomer Monostyrol n
  styrene polymer Styrol-Polymerisat n
  styrene vapo(u)rs Styroldämpfe mpl
  addition of styrene Styrolzussatz m, Styrolzugabe f
  original styrene content Ausgangsstyrolgehalt m
  p-methyl styrene Parametylstyrol n
  polymethyl styrene Polymethylstyrol n
  residual styrene Styrolreste mpl
  residual styrene content Reststyrolanteil m, Reststyrolgehalt m, Styrolrestgehalt m
sub Unter-, Sub-
  sub-contractor Unterlieferant m, Subunternehmer m
  sub-group Untergruppe f
  sub-program Unterprogramm n
  sub-station Unterstation f
  sub-structure Unterstruktur f, Unterbau m
subject(ed) (to) ausgesetzt, unterworfen
  areas subject to wear Verschleißstellen fpl
  parts subject to wear Verschleißteile npl
  subject to high temperatures hitzebeansprucht
  subject to wear verschleißbeansprucht

subject to hydrolysis hydrolyseanfällig
sublimation Sublimation f
submicroscopic submikroskopisch
subroutine Subroutine f
subsequent nachfolgend
  subsequent model Nachfolgemodell n
  subsequent process Nachfolgeprozeß m
subsidiary zusätzlich, Zusatz-, Neben-, Hilfs-
  subsidiary extruder Beistellextruder m, Beispritzextruder m
substance Stoff m, Substanz f
  substance under test Prüfsubstanz f, Versuchsmaterial n
  active substance Aktivsubstanz f, Wirkstoff m
  active substance concentration Wirkstoffkonzentration f
  active substance content Wirkstoffgehalt m
  foreign substance Fremdstoff m
  harmful substance Schadstoff m
  test substance Prüfsubstanz f, Versuchsmaterial n
substituent Substituent m
substitute material Substitutionswerkstoff m
substituted substituiert
substitution reaction Substitutionsreaktion f
substrate Substrat n, Trägermaterial n, Untergrund m
  preparation of the substrate Untergrundvorbehandlung f
  strength of the substrate Untergrundfestigkeit f
succinate Bernsteinsäureester m
  dimethyl succinate Bernsteinsäuredimethylester m
succinic acid Bernsteinsäure f
  succinic acid ester Bernsteinsäureester m
succinic anhydride Bernsteinsäureanhydrid n
  dodecyl succinic anhydride Dodecylbernsteinsäureanhydrid n
suction 1. Ansaugen n, Saugen n; 2. Saug-
  suction blower Sauggebläse n
  suction conveyor Saugfördergerät n
  suction filter Ansaugfilter n,m, Saugfilter n,m
  suction hole Saugloch n
  suction hose Saugschlauch m
  suction line Ansaugleitung f
  suction slit Saugschlitz m
  suction valve Saugventil n
  pneumatic suction system Saugpneumatik f
  vacuum suction holes Vakuumsauglöcher npl
  vacuum suction plate Vakuumsaugteller m
sudden(ly) plötzlich
  sudden cooling Schockkühlung f
  sudden increase in concentration Konzentrationssprung m
  sudden pressure increase Druckstoß m
  sudden stress schlagartige Beanspruchung f
suggested empfohlen
  suggested formulation Rezepturvorschlag m
suitability test Eignungsprüfung f
suitable geeignet, rationell, zweckentsprechend

suitable for food contact applications lebensmittelecht
suitable for storing in silos silierbar
suitable for thermoforming tiefziehfähig, thermoformbar
suitable for transferring to silos silierbar
most suitable optimal
suitably shaped formgerecht
sulphate [USA: sulfate] Sulfat n
  sulphate ion Sulfation n
  alkaline earth sulphate Erdalkalisulfat n
  barium sulphate Bariumsulfat n
  dimethyl sulphate Dimethylsulfat n
  lead sulphate Bleisulfat n
  precipitated barium sulphate Blanc fixe n
sulphated ash [USA: sulfated] Sulfatasche f
  sulphated ash content Sulfataschegehalt m
sulphenamide accelerator Sulfenamidbeschleuniger m
sulphide [USA: sulfide] Sulfid n
  alkyl sulphide Alkylsulfid n
  polyphenylene sulphide Polyphenylensulfid n
  zinc sulphide Zinksulfid n
sulphite [USA: sulfite] Sulphit n
  alkali sulphite Alkalisulphit n
sulphonate [USA: sulfonate] Sulfonat n
  alkyl sulphonate Alkylsulfonat n, Alkylsulfonsäureester m
sulphone [USA: sulfone] Sulfon-
  sulphone group Sulfongruppe f
  polyaryl sulphone Polyarylsulfon n
  polyether sulphone Polyethersulfon n
sulphonic acid [USA: sulfonic] Sulfonsäure f
  sulphonic acid ester Sulfonsäureester m
  alkyl sulphonic acid ester Alkylsulfonsäureester m
  toluene sulphonic acid Toluolsulfonsäure f
sulphoxide [USA: sulfoxide] Sulfoxid n
  dimethyl sulphoxide Dimethylsulfoxid n
sulphur [USA: sulfur] Schwefel m
  sulphur compound Schwefelverbindung f
  sulphur-containing schwefelhaltig
  sulphur donor [on organic polysulphide replacing some elemental sulphur in rubber compounds and promoting curing] Schwefellieferant m
  sulphur-free schwefelfrei
  containing sulphur schwefelhaltig
sulphuric acid [USA: sulfuric] Schwefelsäure f
  fuming sulphuric acid Oleum n, rauchende Schwefelsäure f
sun roof Sonnendach n
sun visor Sonnenblende f
superfinishing Kurz(hub)honen n, Schwingschleifen n, Ziehschleifen n
superheated steam Heißdampf m
superimposition Überlagerung f
supervisory computer Supervisorrechner m
supple biegsam
supplementary controls Zusatzsteuerungen fpl

supplementary unit Ergänzungseinheit f, Ergänzungsaggregat n
suppleness Biegsamkeit f
supplied geliefert
  as supplied im Anlieferungszustand m, in Lieferform f
  form in which supplied Lieferform f
  physical form as supplied Lieferform f
supplier Lieferant m, Zulieferer m
  change of supplier Lieferantenwechsel m
  choice of supplier Lieferantenwahl f
  from a single supplier aus einer Hand f
  list of suppliers Lieferantennachweis m, Lieferantenverzeichnis n
  machine supplier Maschinenlieferant m
  raw material supplier Rohstofflieferant m
supply Zufuhr f, Versorgung f
  supply of fresh air Frischluftzufuhr f
  supply units Versorgungseinheiten fpl
  supply voltage Speisespannung f
  air supply Luftzufuhr f
  air supply elements Luftzuführungselemente npl
  air supply line Luftzuführungsleitung f
  blowing air supply Blasluftzufuhr f
  breakdown of the power supply Stromausfall m
  compressed air supply system Preßluftnetz n
  cooling water supply Kühlwasserzulauf m, Kühlwasserzufuhr f
  drinking water supply (system) Trinkwasserversorgung f
  emergency power supply unit Notstromaggregat n
  energy supply (system) Energieversorgung f
  fresh air supply Frischluftzufuhr f
  gas supply Gasversorgung f
  power supply Stromversorgung f, Stromzufuhr f
  power supply cable Zuführungskabel n [Strom]
  silo containing a day's supply of mo(u)lding compound Tagesbehälter m
  vacuum supply Vakuumversorgung f
  water supply Wasserversorgung f, Wasserzufluß m, Wasserzufuhr f, Wasserzulauf m
  water supply pipe Wasserrohr n
  water supply unit Wasserversorgung f
  welding gas supply Schweißgasversorgung f
support Auflager n, Träger m, Halterung f
  blowing mandrel support Blasdornträger m
  bolt-type mandrel support Bolzendornhalterung f
  breaker plate-type mandrel support Lochscheibendornhalter m, Lochtragring m, Sieb(korb)dornhalter m, Siebkorbhalterung f
  die with breaker plate-type mandrel support Lochdornhalterkopf m
  distance between supports Widerlagerabstand m, Stützweite f
  filter element support Filterelementhalteplatte f
  machine support Maschinenlagerung f
  mandrel support Tragring m, Dornhalter m, Dornträger m

mandrel support design Dornhalterkonstruktion f
mandrel support [system] Dornhalterung f
parison support Vorformlingsträger m
screen support Siebträger m
screw support Schneckenabstützung f, Schneckenlagerung f
spider-type mandrel support Radialsteghalter m, Stegtragring m, Stegdornhalter m, Dornsteghalterung f, Dornsteghalter m
test piece support Probenträger m
three-point support [e.g. of a test specimen] Dreipunktauflage f
torpedo support Torpedohalterung f
twin spider-type mandrel support Doppelstegdornhalter(ung) m (f)
supporting tragen, Trag-, Träger-, Stütz-
   supporting frame Tragegestell n, Unterbau m
   supporting material Trägermaterial n
   supporting ribs Stützrippen fpl
   supporting surface Auflagefläche f, Abstützfläche f
supramolecular supramolekular
surface Oberfläche f
   surface active grenzflächenaktiv, oberflächenaktiv
   surface area-weight ratio Oberflächengewichtsverhältnis n
   surface attack Oberflächenangriff m
   surface changes Oberflächenveränderung f
   surface coating resin Lackbindemittel n, Lackharz n, Lackrohstoff m
   surface coating system Beschichtungssystem n, Lacksystem n
   surface conductivity Oberflächenleitfähigkeit f
   surface contaminants Oberflächenverunreinigungen fpl
   surface contours Oberflächenkonturen fpl
   surface corrosion Oberflächenkorrosion f
   surface damage Oberflächenschädigung f
   surface defect Oberflächendefekt m, Oberflächenstörung f
   surface deposit Oberflächenbelegung f, Oberflächenbelag m
   surface dirt Oberflächenverunreinigung f
   surface-drive winder Oberflächenwickler m
   surface embrittlement Oberflächenversprödung f
   surface energy Oberflächenenergie f
   surface erosion Oberflächenerosion f
   surface filter Flächenfilter n,m
   surface filtration Siebfiltration f
   surface finish Oberflächenqualität f, Oberflächenbeschaffenheit f, Oberflächengüte f
   surface finishing Veredeln n
   surface friction Oberflächenreibung f
   surface gloss [of a plastics article] Oberflächenglanz m
   surface hardened oberflächengehärtet
   surface hardening Oberflächenhärtung f
   surface hardness Oberflächenhärte f
   surface layer Oberflächenschicht f
   surface marks Oberflächenmarkierungen fpl
   surface moisture Oberflächenfeuchte f
   surface oxidation Oberflächenoxidation f
   surface polish Oberflächenglanz m
   surface preparation Oberflächenvorbehandlung f
   surface protection Oberflächenschutz m
   surface renewal Oberflächenneubildung f
   surface resistance Oberflächenwiderstand m
   surface resistivity spezifischer Oberflächenwiderstand m
   surface roughness Oberflächenrauhigkeit f
   surface skin Oberflächenhaut f
   surface slip Gleitfähigkeitsverhalten n, Gleitfähigkeit f
   surface slip characteristics Gleiteigenschaften fpl
   surface smoothness Oberflächenglätte f
   surface streaks Oberflächenschlieren fpl
   surface tack Oberflächenklebrigkeit f
   surface temperature Oberflächentemperatur f
   surface tension Oberflächenspannung f
   surface texture Oberflächenstruktur f
   surface thermocouple Oberflächentemperaturfühler m
   surface to be bonded Klebfläche f
   surface treated oberflächenbehandelt, randschichtbehandelt
   surface treatment Oberflächenbehandlung f, Randschichtbehandlung f, Oberflächenveredelung f
   surface wear Oberflächenabtrag m
   surfaces to be bonded Fügeteiloberflächen fpl
   surfaces to be joined Fügeflächen fpl
adherent surface Haftgrund m, Klebfläche f
catalyzed surface coating system Reaktionslacksystem n
cavity surface Form(nest)oberfläche f, Werkzeughohlraumoberfläche f
cavity surface temperature Formoberflächentemperatur f, Formnestoberflächentemperatur f
circular surface Kreisfläche f
contact surface Anlagefläche f, Kontaktfläche f
contact surfaces Berührungsflächen fpl
crack surface Rißfläche f
cut surface Schnittfläche f
exposed melt surface Entgasungsoberfläche f
exterior surface Außenoberfläche f
film surface Beschichtungsoberfläche f, Filmoberfläche f
film surface characteristics Filmoberflächeneigenschaften fpl
flush with the surface oberflächenbündig
fracture surface Bruchfläche f
gating surface Anschnittebene f
gel coat surface Deckschichtoberfläche f [GFK]
having good surface slip gleitfähig
having the same surface area flächengleich
having very good surface slip hochgleitfähig
heat reflecting surface Wärmeabstrahlfläche f
impact surface Stoßfläche f

interfacial surface tension Grenzflächenspannung f
laminate surface Laminatoberfläche f
mating surfaces Paßflächen fpl
mo(u)ld parting surface Formteilebene f, Formteilung f
mo(u)ld surface Formoberfläche f, Werkzeugoberfläche f
mo(u)ld surface temperature Werkzeugoberflächentemperatur f
mo(u)lded-part surface Formteiloberfläche f
near the surface oberflächennah
outer barrel surface Zylinderaußenfläche f
outer cylinder surface Zylinderaußenfläche f
outer surface Außen(ober)fläche f
paint film surface Lackoberfläche f
part surface Formteiloberfläche f
particle surface Kornoberfläche f, Teilchenoberfläche f
parting surface Formteilebene f
polyester surface coating resin Lackpolyester m
preparation of surfaces to be bonded Klebflächenvorbehandlung f
projected runner surface area projizierte Verteilerkanalfläche f
roll surface Walzenmantel m
root surface Ganggrund m, Grund m
screw root surface Schneckengrund m, Schneckenkanaloberfläche f, Schneckenkanalgrund m
screw surface (area) Schneckenoberfläche f
screw wearing surface Schneckenlauffläche f
supporting surface Abstützfläche f, Auflagefläche f
visible surface Sichtfläche f
wearing surface Lauffläche f
with a polished surface oberflächenpoliert
with very good surface slip hochgleitfähig
surfacer Ausgleichsmasse f, Spachtelmasse f
cement surfacer Zementspachtelmasse f
wall surfacer Wandspachtelmasse f
surfacing mat Oberflächenmatte f, Vlies n [GFK]
surfboard Surfbrett n
surge/to schwingen, schwanken; pumpen [Verdichter]
surge Druckstoß m, Stoß m
surge-free pulsationsfrei
surge frequency Pulsationsfrequenz f
pressure surge Druckstoß m
surging 1. pulsierend, Pulsieren n;
2. Pumperscheinungen fpl
surplus for the year Jahresüberschuß m
surround Eingrenzung f, Umrandung f, Zarge f
radiator surround Kühlerzarge f
surrounding atmosphere Umgebungsluft f, Umgebungsatmosphäre f
surrounding medium Umgebungsmedium n
survey Überblick m
general survey Übersicht f
susceptibility Anfälligkeit f, Empfindlichkeit f
susceptibility to corrosion Korrosionsanfälligkeit f

susceptibility to cracking Rißanfälligkeit f
susceptibility to solvent attack Lösungsmittelempfindlichkeit f
susceptibility to stress cracking Spannungsrißanfälligkeit f
ag(e)ing susceptibility Alterungsanfälligkeit f
susceptible empfindlich, empfänglich, anfällig
susceptible to ag(e)ing alterungsempfindlich
susceptible to corrosion korrosionsanfällig
susceptible to oxidation oxidationsanfällig, oxidationsempfindlich
susceptible to scorching scorchanfällig
susceptible to stress cracking spannungsrißempfindlich
susceptible to wear verschleißanfällig
suspend/to suspendieren
suspending agent Suspensionshilfsmittel n
suspension Suspension f, Aufhängung f
suspension copolymer Suspensionscopolymerisat n
suspension graft copolymer Suspensionspfropfcopolymerisat n
suspension homopolymer Suspensionshomopolymerisat n
suspension polymer Suspensionspolymerisat n, Suspensionstype f
suspension polymerization (process) Suspensionspolymerisationsverfahren n, Suspensionsverfahren n, Suspensionspolymerisation f, S-Polymerisationsverfahren n
suspension PVC Suspensions-PVC n, Suspensionstype f
extruder suspension (unit) Extruderaufhängung f
peroxide suspension Peroxid-Suspension f
swallowtail recess Schwalbenschwanzaussparung f
Sward hardness Schenkelhärte f
Sward hardness test Schenkelhärteprüfung f
swarf Späne mpl, Fließspan m
swell Schwellung f
die swell Düsenquellung f, Quellfaktor m, Schwellverhalten n, Strangaufweitung f, Strangaufweitungsverhältnis n
swelling Anquellung f, Quelldehnung f, Quellung f
swelling agent Quellmittel n
swelling behavio(u)r Quellverhalten n
swelling characteristics Schwellverhalten n
swelling effect Quellwirkung f
swelling power Quellvermögen n
swelling process Quellungsvorgang m
swelling resistance Quellbeständigkeit f
swelling test Quellversuch m
resistance to swelling Schwellfestigkeit f
swept volume Hubvolumen n
swing/to schwingen
swing-back wegschwenkbar, schwenkbar, klappbar, abschwenkbar
swing-hinged wegschwenkbar, schwenkbar, klappbar, abschwenkbar

swingdoor Pendeltür f
switch/to schalten
    switch off/to abschalten, ausschalten
    switch on/to einschalten
switch Schalter m
    switch box Schaltkasten m
    switch circuit Schaltkreis m
    switch limit Schaltschwelle f
    switch sequence Schaltfolge f
    automatic switch-off mechanism Abschaltautomatik f
    automatic switch-on mechanism Einschaltautomatik f
    decade switch Dekadenschalter m
    digital switch Digitalschalter m
    drive switch circuit Antriebsschaltkreis m
    foot operated switch Fußschalter m
    function selector switch Betriebsartenwahlschalter m
    high-voltage switch Hochspannungsschalter m
    key-operated switch Schlüsselschalter m
    limit switch Endschalter m, Grenztaster m, Grenzwertschalter m
    main switch Hauptschalter m
    motor safety switch Motorschutzschalter m
    on-off switch Ein-Aus-Schalter m
    operating switch Bedienungsschalter m
    power switch Leistungsschalter m
    pre-selector switch Vorwahlschalter m
    program selection switch Programmwahlschalter m
    push-button switch Druckknopfschalter m
    rows of digital switches Digitalschalterreihen fpl
    safety switch Sicherheitsschalter m
    selector switch Wahlschalter m
    semi-conductor power switch Halbleiterleistungsschalter m
    semi-conductor switch Halbleiterschalter m
    stroke limit (switch) Hubbegrenzung f
    time pre-selection switch Zeitvorwahlschalter m
    time switch Schaltuhr f, Zeitschalter m, Zeitschaltuhr f, Zeitschaltwerk n
    tumbler switch Kippschalter m
    weekly time switch Wochenschaltuhr f
switchgear Schalteinrichtung f, Schaltgerät n
switching Schalt-, schaltend
    switching-off Ausschaltung f
    switching point Schaltpunkt m
    switching time Schaltzeit f
    total switching time Gesamtschaltzeit f
swivel-mounted klappbar, (weg)schwenkbar
swivel-out ausschwenkbar
swivel-type [machine unit] schwenkbar, klappbar
symbolic symbolisch
    symbolic key Symboltaste f
symmetric symmetrisch
synchronous motor Synchronmotor m
syncronous timer Synchronschaltuhr f
syndiotactic syndiotaktisch
syneresis Synärese f
synergism Synergismus m

synergistic synergistisch
synthesis Synthese f
synthesize/to synthetisieren
synthetic künstlich, synthetisch
    synthetic fabric Chemiefasergewebe n, Chemiefaser f, Synthesefaser f
    synthetic fiber reinforced chemiefaserarmiert
    synthetic paint resin Lackkunstharz
    synthetic resin Kunstharz n
    synthetic resin-based paint Kunstharzlack m
    synthetic resin bound kunstharzgebunden
    synthetic resin casting Gießharzkörper m
    synthetic resin dispersion Kunststoffdispersion f
    synthetic resin insulator Kunststoffisolator m
    synthetic resin laminate Kunststofflaminat n
    synthetic resin latex Kunststofflatex n
    synthetic resin modified kunststoffmodifiziert, kunststoffvergütet
    synthetic rubber Kunstkautschuk m, Synthesekautschuk m, Synthesegummi n,m
    synthetic rubber coated kunstkautschukbeschichtet
    synthetic silica Quarzgut n
    synthetic silica flour Quarzgutmehl n
    completely synthetic vollsynthetisch
    fully synthetic vollsynthetisch
system System n
    system software Systemsoftware f
    ability of the system to expand Systemerweiterungsfähigkeit f
    accelerator system Beschleunigungssystem n
    adaptor system Adaptersystem n
    air circulating system Luftumwälzungssystem n
    air cooling system Kühlluftsystem n
    ante-chamber feed system Vorkammer-Angießtechnik f, Vorkammeranguß m
    automatic speed control system Drehzahl-Steuerautomatik f
    automatic winding system Wickelautomatik f
    back venting system Rückwärtsentgasung f
    basic system Grundsystem n
    blow mo(u)lding system Blasformsystem n
    blown film cooling system Schlauchfolienkühlung f
    braking system Bremssystem n
    calender heating system Kalanderbeheizung f
    cam system Nockensystem n
    catalyzed surface coating system Reaktionslacksystem n
    central grease lubricating system Fettzentralschmierung f
    central hydraulic system Zentralhydraulikanlage f
    central lubricating system Zentralschmierung f
    circulating hot water heating system Heißwasser-Umlaufheizung f
    circulating oil heating system Ölumlaufheizung f
    circulating oil lubricating system Ölumlaufschmierung f
    closed-circuit cooling system geschlossener

Kühlkreislauf m, Rückkühlung f
closed-loop control system Regel(ungs)system n
co-rotating twin screw system Gleichdrallsystem n
coating system Beschichtungssystem n
cold runner feed system Kaltkanalangußsystem n
composite control system Regelverbund m
composite system Verbundsystem n
compounding system Aufbereitungssystem n
continuous flow cooling system Durchflußkühlung f
control electronic system Elektroniksteuerung f
control system Steuer- und Regeleinrichtung f, Steuersystem n
cooling system Kühlsystem n, Kühlung f
counterflow oil cooling system Gegenstromölkühlanlage f
coupling system Kopplungssystem n
curing system Härtungssystem n, Vernetzungssystem n
data collecting system Datensammelsystem n
degassing system Entlüftungssystem n, Schmelzentgasung f
demo(u)lding system Entformungssystem n
devolatilizing system Schmelzentgasung f, Entlüftungssystem n
diagnostics system Diagnosesystem n
die face pelletizing system Direktabschlagssystem n
digital hydraulic system Digitalhydraulik f
direkt screw injection system Schneckendirekteinspritzung f
double toggle clamping system Doppelkniehebel-Schließsystem n
downstream venting system Stromabwärtsentgasung f
drinking water supply system Trinkwasserversorgung f
dual-circuit hydraulic system Zweikreis-Hydrauliksystem n
dual-circuit system Zweikreissystem n
dust removal system Entstaubungssystem n
electric control system Elektrosteuerung f
electronic measuring and control system MSR-Elektronik f
electronic system Elektronik f
emergency cooling system Notkühlsystem n
emulsifier system Emulgatorsystem n, Emulgiersystem n
energy supply system Energieversorgung f
epoxy coating system Epoxidharzanstrichsystem n
equilibrium system Gleichgewichtssystem n
external air cooling system Außenluftkühlung f, Außenluftkühlsystem n
extruder system Extrudersystem n
feed system Dosiersystem n, Einspeisevorrichtung f, Einspeisung f
feedback control system Regelungssystem n
feedback system Rückführsystem n
film bubble cooling system Folienkühlung f,
Schlauchkühlvorrichtung f
film winding system Folienwickelsystem n
fluid heating system Flüssigkeitsheizung f
foam system Schaumsystem n
forced circulation system Zwangsumlaufsystem n
forced conveying system Zwangsförderung f
forced feed system Zwangsbeschickung f, Zwangsfütterung f
forward venting system Vorwärtsentgasung f
fully automatic system Vollautomatik f
gating system Angußsystem n
gear tooth system Verzahnungssystem n
guard door interlock system Schutzgittersicherung f, Schutztürsicherung f
guide pillar system Holmführung f
heart-shaped groove-type of feed system Herzkurveneinspeisungssystem n
heating-cooling channel system Temperierkanalsystem n
heating system Temperiersystem n [zum Erhitzen]
high-build system Dickschichtsystem n
high-speed calibrating system Hochgeschwindigkeitskalibriersystem n
high-speed mo(u)ld change system Werkzeugschnellwechselsystem n
horizontal feed system Horizontalbeschickung f
hot cut pelletizing system Direktabschlagssystem n
hot runner feed system Heißkanalanguß m, Heißkanal-Angußsystem n, Heißkanalverteileranguß m
hot runner injection mo(u)lding system Spritzguß-Heißkanalsystem n
hot runner system Heißkanalsystem n, Heißkanal-Verteilersystem n, Heißkanal-Rohrsystem n
hydraulic pull-back system Vorspanneinrichtung f, Vorspannung f, Walzenvorspanneinrichtung f, Walzenvorspannsystem n
hydraulic system Hydraulik f
insulated runner feed system Isolierkanalangußplatte f, Isolierkanalanguß m
insulated runner system Isolierkanalanguß m, Isolierkanalsystem n, Isolierverteiler-Angießsystem n
internal air cooling system Innenluftkühlung f, Innenluftkühlsystem n
internal bubble cooling system Blaseninnenkühlung f, Folieninnenkühlvorrichtung f, Folieninnenkühlung f, Blaseninnenkühlsystem n
internal cooling system Innenkühlsystem n, Innenkühlung f
internal screw cooling system Schneckeninnenkühlung f
light stabilizing system Lichtschutzsystem n
liquid nitrogen cooling system Flüssig-Stickstoffkühlung f
lubricating system Schmiersystem n
main cooling system Primärkühlung f
main electronics system Primärelektronik f
manifold system Verteiler(röhren)system n

melt accumulator system Kopfspeichersystem n, Massespeichersystem n
melt decompression system Schmelzdekompression f
melt devolatilizing system Schmelzdekompression f
melt feed system Schmelzeinspeisung f, Schmelzezufluß m
melt flow-way system Schmelzeleitsystem n
metering system Dosiersystem n
modular constructing system Baukastenbauweise f
modular electronic system Bausteinelektronik f
modular system Bausteinsystem n
monitoring system Überwachungssystem n
mo(u)ld cooling channel system Werkzeugkühlkanalanlage f
mo(u)ld cooling system Werkzeugkühlsystem n, Werkzeugkühlung f
mo(u)ld filling monitoring system Formfüllüberwachung f
mo(u)ld heating system Werkzeugbeheizung f
mo(u)ld temperature control system Formentemperierung f, Werkzeugtemperaturregelung f, Werkzeugtemperierung f
mo(u)lding cycle monitoring system Zyklusüberwachung f
oil cooling system Ölkühlung f
oil hydraulic system Ölhydraulik f
oil temperature control system Öltemperierung f
one-coat system Einschichtsystem n
one-component system Einkomponentensystem n
one-pack system Einkomponentensystem n
paint system Anstrichmittelsystem n, Lacksystem n
pipeline system Rohrleitungssystem n
plasticizing system Plastifiziersystem n
platen heating system Plattenheizung f
plunger injection system Kolbeneinspritzsystem n
pneumatic suction system Saugpneumatik f
pre-plasticizing system Vorplastifizier(ungs)system n
pressure measuring system Druckmeßsystem n
pressure release system Druckentspannungssystem n
pressurized oil lubricating system Drucköung f
process control system Prozeßführungseinrichtung f, Prozeßführungssystem n, Prozeßleitsystem n
production control system Fertigungssteuerung f
production data collecting system BDE-System n
production monitoring system Fertigungsüberwachung f
proportional hydraulic system Proportionalhydraulik f
punched card control system Lochkartensystem n
quick-action mounting system Schnellspannsystem n
radial system of runners Sternverteiler m, Spinne f, Angußstern m, Anschnittstern m, Angußspinne f
reciprocating screw injection system Schneckenkolbeneinspritzsystem n
recirculating oil cooling system Umlaufölkühlung f
recirculation cooling system Umlaufkühlung f
reprocessing system Aufbereitungssystem n

reversing system Reversiersystem n
rigid foam system Hartschaumsystem n
ring system Ringsystem n
robotic system Handhabungssystem n, Robotersystem n
rotary table system Drehtischsystem n
runner system Verteilerröhrensystem n, Verteilersystem n
safety interlock system Sicherheitsverriegelung f
scrap removal system Abfallentfernungssystem n
screw cooling system Schneckenkühlung f
screw decompressing system Schneckendekompressionseinrichtung f
screw-plunger injection system Schneckenkolbeneinspritzsystem n
screw system Schneckensystem n
screw temperature control system Schneckentemperierung f
screw venting system Schneckendekompressionseinrichtung f
secondary cooling system Sekundärkühlung f
semi-automatic system Halbautomatiksystem n
single-circuit cooling system Einkreis-Kühlsystem n
single-circuit system Einkreissystem n
single-phase system Einphasensystem n
single-screw system Einschneckensystem n, Einwellensystem n
sleeve ejection system Hülsenausdrücksystem n
solids conveying system Feststoff-Förderung f
speed control system Drehzahlregelung f, Drehzahlregulierung f, Fahrgeschwindigkeitsregelung f
stabilizer system Stabilisierungssystem n
strand pelletizing system Strangabschlagsystem n
strip pelletizing system Bandabschlagsystem n
stroke measuring system Wegmeßsystem n
surface coating system Lacksystem n
temperature control system Temperaturregulierung f, Temperatursteuerung f, Temperiersystem n
thickness control system Wanddickenkontrolle f
three-phase system Dreiphasensystem n
toggle clamp system Kniehebelschließsystem n
twin screw system Zweiwellensystem n, Zweischneckensystem n
two-pack system Zweikomponentensystem n
two-phase system Zweiphasensystem n
type of gate system Angußart f
underwater pelletizing system Unterwassergranuliersystem n
unit construction system Baukastenbauweise f
vacuum calibration system Vakuumkalibriersystem n
valve gating system Nadelventilangußsystem n
vented hopper system Trichterentgasung f
venting system Entgasungseinrichtung f, Entgasungssystem n
venturi system Venturi-System n
water-hydraulic system Wasserhydraulik f
weighing system Wägesystem n

Taber abrasion Taber-Abrieb m
table 1. Tisch m; 2. Tabelle f
  table-top model Tischmodell n
  table-top unit Tischgerät n
  table-top version Tischversion f
  chemical resistance table Beständigkeitstabelle f
  comparative table Vergleichstabelle f
  decision table Entscheidungstabelle f
  general table Übersichtstabelle f
  rotary table Drehteller m, Drehtisch m
  rotary table design Drehtischbauweise f
  rotary table injection mo(u)lding machine Drehtisch-Spritzgußmaschine f, Spritzgießrundtischanlage f
  rotary table machine Drehtischmaschine f
  rotary table system Drehtischsystem n
  sliding table Schiebetisch m
  sliding table machine Schiebetischmaschine f
  stacking table Ablagetisch m
  tilting table Schwenktisch m
  16-station rotary table machine Sechzehn-Stationen-Drehtischmaschine f
tabular tabellarisch
tack Klebrigkeit f
  tack-free klebfrei
  free from tack klebfrei
  inherent tack Eigenklebrigkeit f
  surface tack Oberflächenklebrigkeit f
tackifier Klebrigmacher m
tackifying klebrigmachend
  tackifying agent Klebrigmacher m
tackiness Klebrigkeit f
tacky klebend, klebrig
tacticity Taktizität f
tail flash Bodenabfall m
tailgate Heckklappe f
take apart/to auseinandernehmen, zerlegen
take away/to abführen
take-off Abzug m, Abnahme f, Entnahme f
  take-off direction Abzugsrichtung f
  take-off nip Abzugswalzenspalt m
  take-off pinch rolls Abzugs-Abquetschwalzenpaar n
  take-off roll Abreißwalze f, Abzugswalze f
  take-off rolls Abzugsrollen fpl, Walzenabzug m, Abzugswalzenpaar n
  take-off section Abzugsstrecke f
  take-off speed Abzugsgeschwindigkeit f
  take-off station Abzugsstation f
  take-off tension Abzugskraft f
  take-off tower Abzugsturm m
  take-off unit Abzugsaggregat n, Abzugseinrichtung f, Abzugsvorrichtung f, Abzugswerk n, Abzugswalzenstation f
  caterpillar take-off Abzugsraupe f
  combined take-off and wind-up unit Abzugswicklerkombination f, Abzugswerk-Wickler-Kombination f
  cooling take-off unit Kühlabzugsanlage f
  film take-off (unit) Folienabziehwerk n, Folienabzug m
  parison take-off Schlauchabzug m
  polishing and take-off unit Glätt- und Abziehmaschine f
  rotary take-off unit Entnahmedrehtisch m
  secondary take-off (unit) Sekundärabzug m
  sheet take-off (unit) Plattenabzug m
  trimming and take-off station Trimm- und Entnahmestation f
  twin-belt take-off (unit/system) Doppelbandabzug m
  type of take-off Abzugstyp m
talc Talkum n
  talc-filled talkumverstärkt
tall oil Tallöl n
  tall oil fatty acid Tallölfettsäure f
tamping Einstampfen n, Feststampfen n
tandem arrangement Hintereinanderschaltung f, Tandemanordnung f
  arranged in tandem hintereinandergeschaltet
tangential tangential
  tangential chute Tangentialschurre f
  tangential direction Tangentialrichtung f
  tangential-drive winder Tangentialwickler m
  tangential flow Tangentialströmung f
  tangential force Tangentialkraft f
  tangential stress Tangentialspannung f
tank Gefäß n, Behälter m, Tank m
  tank lining Tankauskleidung f
  dipping tank Tauchwanne f
  expansion tank Ausdehnungsgefäß n, Ausgleichbehälter m
  feed tank Vorlaufbehälter m
  flat-bottom tank Flachbodentank m
  fuel oil storage tank Heizöl-Lagertank m
  fuel oil tank Heizöltank m
  fuel tank Brennstofftank m, Benzintank m, Kraftfahrzeugtank m, Treibstofftank m
  fuel tank cap Tankdeckel m
  fuel tank neck Tankeinfüllstutzen m
  oil tank ventilation Öltankbelüftung f
  petrol tank Benzintank m, Kraftstoffbehälter m, Kraftstofftank m
  petrol tank cap Tankverschluß m
  petrol tank float Tankschwimmer m
  plastics fuel tank Kunststofftank m
  quench tank Kühltrog m, Kühlwanne f
  radiator tank Kühlwasserausgleichsbehälter m, Kühlwasserkasten m, Wasserkasten m
  spherical tank Kugeltank m
  storage tank Lagerbehälter m, Lagertank m, Vorratsbehälter m
  vacuum calibrating tank Vakuumkalibrierbecken n
tanker Tanker m
  rail tanker Eisenbahnkesselwagen m
  road tanker Silofahrzeug n, Straßensilofahrzeug n,

Straßentankwagen m, Tankwagen m
tap Hahn m, Anzapfung f, Abzweig m
  tap water Leitungswasser n
  three-way tap Dreiwegehahn m
tape Band n
  adhesive tape Klebeband n, Klebestreifen m
  film tape Bändchen n, Folienband n, Folienbändchen n
  film tape extruder Folienbandextruder m
  film tape production line Folienbändchenanlage f
  film tape stretching plant Bändchenreckanlage f
  film tape stretching unit Folienbandstreckwerk n, Folienbandreckanlage f
  glass cloth tape Glasgewebeband n
  magnetic tape Magnetband n
  magnetic tape cassette Magnetbandkassette f
  magnetic tape unit Magnetbandgerät n
  packaging tape Verpackungsband n
  punched tape Lochstreifen m
  punched tape reader Lochstreifenleser m
  sealing tape Dichtungsband n
  woven glass tape Glasgewebeband n
tar Teer m
  tar fraction Teerfraktion f
  tar residues Teerreste mpl
  coal tar Steinkohlenteer m
tare weight Taragewicht n
target quantity Zielgröße f
tarpaulin Plane f, Abdeckplane f
  tarpaulin material Planenstoff m
task Aufgabe f
  control task Steuerungsaufgabe f
  extrusion task Extrusionsaufgabe f
  main task Hauptaufgabe f
  monitoring task Überwachungsaufgabe f
taste Geschmack m
  freedom from taste Geschmacksfreiheit f, Geschmacksneutralität f
tasteless geschmacklos, geschmacksneutral
TCP [tricresyl phosphate] Trikresylphosphat n
tear 1. Riß m; 2. Träne f, Tropfen [Tauchlackierung]
  tear initiation force Anreißkraft f
  tear propagation Weiterreißen n
  tear propagation force Weiterreißkraft f
  tear propagation resistance Weiterreißfestigkeit f, Weiterreißwiderstand m
  tear resistance Einreißwiderstand m, Einreißfestigkeit f
  tear resistance test Einreißversuch m
  tear strength Zerreißfestigkeit f
  trouser tear test Schenkel-Weiterreißversuch m
tear-off perforation Abreißperforation f
technical technisch [s.a. technically]
  technical advantages verfahrenstechnische Vorteile mpl
  technical assistance verfahrenstechnische Hilfestellung f
  technical conditions technische Gegebenheiten fpl

  technical considerations verfahrenstechnische Überlegungen fpl
  technical demands verfahrenstechnische Forderungen fpl
  technical design verfahrenstechnische Auslegung f
  technical development(s) verfahrenstechnische Entwicklung f
  technical features verfahrenstechnische Merkmale npl
  technical layout verfahrenstechnische Auslegung f
  technical limitations verfahrenstechnische Grenzen fpl
  technical literature Fachliteratur f
  technical measures verfahrenstechnische Maßnahmen fpl
  technical problems verfahrenstechnische Schwierigkeiten fpl
  manufacturers' technical literature Firmenmerkblätter npl
technically technisch [s.a. technical]
  technically feasible verfahrenstechnisch machbar
  technically important verfahrenstechnisch wesentlich
  technically important parameters verfahrenstechnisch relevante Parameter mpl
  technically impossible verfahrenstechnisch nicht möglich
  technically interesting verfahrenstechnisch interessant
  technically simple verfahrenstechnisch einfach
technique Technik f, Methode f, Verfahren n
  analytical technique Analysenmethode f, Analysenverfahren n
  cold forming techniques Kaltformmethoden fpl
  mo(u)lding technique Formtechnik f
  processing technique Verarbeitungsverfahren n
  welding techniques Schweißtechnik f
  working technique Arbeitstechnik f
technology Technologie f, Technik f
  blow mo(u)lding technology Blasformtechnik f
  casting resin technology Gießharztechnik f
  plastics technology Kunststofftechnologie f
  structural foam mo(u)lding technology TSG-Technik f
telecommunication cable Fernsprechkabel n, Fernmeldekabel n
telecommunication (engineering) Nachrichtentechnik f, Kommunikationstechnik f, Fernmeldetechnik f
temperature Temperatur f
  temperature adjustment Temperaturausgleich m
  temperature at the delivery end Ausstoßtemperatur f
  temperature build-up Temperatureinschwingen n
  temperature control Temperaturregelung f, Temperaturkontrolle f, Temperaturüberwachung f, Temperaturregulierung f, Temperaturausgleichssystem n
  temperature control circuit Temperaturregelkreis m, Temperierkreis(lauf) m
  temperature control instrument Temperaturregelgerät n, Temperaturregler m
  temperature control medium Temperiermittel n, Temperierflüssigkeit f, Temperiermedium n

temperature control system Temperatursteuerung f, Temperiersystem n, Temperaturregulierung f
temperature control unit Temperatursteuereinheit f, Temperaturüberwachung f, Temperiergerät n
temperature-controlled with water wassertemperiert
temperature controller Thermowächter m, Temperaturregelgerät n, Temperaturregler m
temperature decrease Temperaturabfall m, Temperaturabsenkung f
temperature dependence Temperaturabhängigkeit f
temperature-dependent temperaturabhängig
temperature difference Temperaturdifferenz f, Temperaturgefälle n
temperature differential Temperaturdifferenz f
temperature display Temperaturanzeige f
temperature distribution Temperaturverteilung f
temperature equilibrium Temperaturgleichgewicht n
temperature extremes Extremtemperaturen fpl
temperature fluctuations Temperaturschwankungen fpl
temperature gradient Temperaturgradient m
temperature increase Temperaturzunahme f, Temperaturerhöhung f
temperature-independent temperaturunabhängig
temperature indicator Temperaturanzeigegerät n
temperature level Temperaturniveau n
temperature limit Temperaturgrenze f, Temperaturgrenzwert m
temperature measurement Temperaturerfassung f, Temperaturmessung f
temperature measuring device Temperaturmeßgerät n, Temperaturmeßvorrichtung f
temperature measuring point Temperaturmeßstelle f
temperature peak Temperaturspitze f
temperature probe Temperatursonde f, Thermodraht m
temperature profile Temperaturverlauf m, Temperaturprofil n
temperature program Temperaturprogramm n
temperature range Temperaturintervall m, Temperaturbereich m, Temperaturspanne f
temperature recorder Temperaturschreiber m
temperature sensor Temperatursensor m, Temperatur(meß)fühler m, Thermoelement(fühler) n(m), Temperatur(meß)geber m, Thermoelement n
temperature setting (mechanism) Temperatureinstellung f
temperature tolerance range Temperaturtoleranzfeld n
temperature uniformity thermische Homogenität f
ag(e)ing temperature Alterungstemperatur f, Lagerungstemperatur f
alternating temperature test Temperaturwechselprüfung f
ambient temperature Umgebungstemperatur f
annealing temperature Temper-Temperatur f
approximate temperature Temperaturrichtwert m

barrel temperature Zylindertemperatur f
barrel temperature control Zylindertemperierung f
barrel wall temperature Zylinderwandtemperatur f
blocking temperature Blockpunkt m
blow-up temperature Aufblastemperatur f
brittleness temperature Kältesprödigkeitspunkt m
cascade temperature control unit Kaskadentemperaturregelung f
cavity surface temperature Form(nest)oberflächentemperatur f, Werkzeugoberflächentemperatur f
cavity wall temperature Werkzeugwandtemperatur f, Nestwandtemperatur f
change in temperature Temperaturwechsel m
changes in temperature Temperaturverlauf m
circulating oil temperature control (system) Ölumlauftemperierung f
circulating water temperature control unit Wasserumlauftemperiergerät n
cold crack temperature Kältebruchtemperatur f
conductor temperature Leitertemperatur f
conditioning temperature Temper-Temperatur f
constant temperature Temperaturkonstanz f
constant temperature chamber Temperierkammer f
constant temperature medium Temperierflüssigkeit f, Temperiermittel n, Temperiermedium n
constant temperature rolls Temperierwalzen fpl
constant temperature zone Temperierzone f
continuous service temperature Dauerbetriebstemperatur f
continuous temperature Dauertemperatur f
continuous working temperature Dauergebrauchstemperatur f
cooling water inlet temperature Kühlwasserzulauftemperatur f
cooling water outlet temperature Kühlwasserablauftemperatur f
core wall temperature Kernwandtemperatur f
crystallization temperature Kristallisationstemperatur f
cure temperature Aushärtungstemperatur f, Härtetemperatur f, Vernetzungstemperatur f, Härtungstemperatur f
curing temperature Aushärtungstemperatur f, Härtetemperatur f, Vernetzungstemperatur f, Härtungstemperatur f
cylinder temperature Zylindertemperatur f
cylinder temperature control system Zylindertemperierung f
cylinder wall temperature Zylinderwandtemperatur f
decomposition temperature Zersetzungspunkt m, Zersetzungstemperatur f
decrease in temperature Temperaturabsenkung f
deflection temperature Formbeständigkeitstemperatur f
demo(u)lding temperature Entformungstemperatur f
depending on the temperature temperaturabhängig
die head temperature Kopftemperatur f

dimensionally stable at elevated temperatures warmformbeständig
direct digital temperature control DDC-Temperaturregelung f
direct digital temperature control circuit DDC-Temperaturregelkreis m
discharge temperature Ausstoßtemperatur f
drop in temperature Temperaturabfall m
drying temperature Trocknungstemperatur f
elevated temperature erhöhte Temperatur f
even temperature Temperaturgleichmäßigkeit f
excessive temperature Übertemperatur f
exposure to alternating temperatures Temperaturwechselbeanspruchung f
exposure to high temperatures Wärmebeanspruchung f
feed temperature Einlauftemperatur f
filtration temperature Filtertemperatur f
final melt temperature Schmelzeendtemperatur f
final temperature Endtemperatur f
flash ignition temperature Fremdentzündungstemperatur f
foaming temperature Schäumtemperatur f
fusion temperature [of PVC paste] Gelierpunkt m
gelation temperature [of PVC paste] Vorgeliertemperatur f
glass transition temperature Einfrier(ungs)temperatur f, Glaspunkt m, Glastemperatur f, Glasübergangstemperatur f, Glasumwandlungstemperatur f, Glasumwandlungspunkt m, Tg-Wert m
glass transition temperature range Glasumwandlungstemperaturbereich m
heat ag(e)ing temperature Warmlagerungstemperatur f
heat distortion temperature Formbeständigkeitstemperatur f
high-temperature calendering (process) HT-Verfahren n, Hochtemperaturverfahren n
immersion temperature Lagerungstemperatur f [in Flüssigkeit]
independent of the temperature temperaturunabhängig
inflation temperature Aufblastemperatur f
initial temperature Anfangstemperatur f
injection temperature Einspritztemperatur f
inlet temperature Eingangstemperatur f, Eintrittstemperatur f, Einlauftemperatur f
kick-off temperature [of a catalyst] Anspringtemperatur f
limiting temperature Grenztemperatur f, Temperaturgrenzwert m
long-term service temperature Dauergebrauchstemperatur f
long-term temperature limit Dauertemperaturgrenze f
long-term temperature resistance Dauertemperaturbelastungsbereich m

low-temperature applications Tieftemperaturanwendungen fpl
low-temperature calendering (process) LT-Verfahren n, Niedertemperaturverfahren n
low-temperature flexibility Tieftemperaturflexibilität f
low-temperature impact strength Tieftemperatur-Schlagzähigkeit f
low-temperature initiator Tieftemperaturinitiator m
low-temperature limit Temperatur-Untergrenze f
low-temperature lubricating grease Tieftemperaturschmierfett n
low-temperature performance Tieftemperaturverhalten n
low-temperature peroxide Tieftemperaturperoxid n
low-temperature process LT-Verfahren n
low-temperature properties Tieftemperatureigenschaften fpl
low-temperature resistance Tieftemperaturfestigkeit f
low-temperature thermostat Kryostat m
low temperatures Tieftemperaturen fpl
Martens heat distortion temperature Formbeständigkeit f (in der Wärme) nach Martens
Martens temperature Martensgrad m, Martenswert m, Martenszahl f
material temperature Massetemperatur f
maximum continuous operating temperature Dauerbetriebsgrenztemperatur f
maximum temperature Spitzentemperatur f, Temperaturspitze f
maximum working temperature Einsatzgrenztemperatur f
mean temperature Mitteltemperatur f, Temperaturmittelwert m
melt exit temperature Masseaustrittstemperatur f
melt temperature Formmassetemperatur f, Schmelzetemperatur f, Spritzguttemperatur f, Massetemperatur f
melt temperature at the feed end [e.g. of an extruder] Einlaufmassetemperatur f, Eingangsmassetemperatur f
melt temperature distribution Massetemperaturverteilung f
melt temperature indicator Massetemperaturanzeige f
method of measuring temperature Temperaturmeßmethode f
milling temperature Walztemperatur f
mo(u)ld temperature Formtemperatur f, Werkzeugtemperatur f
mo(u)ld temperature control (system) Werkzeugtemperierung f, Werkzeugtemperaturregelung f
mo(u)ld temperature distribution Werkzeugtemperaturverteilung f
mo(u)lding temperature Umform(ungs)temperatur f, Verformungstemperatur f
oil temperature control Öltemperierung f
oil temperature monitor Öltemperaturwächter m

operating temperature Betriebstemperatur f, Einsatztemperatur f
operating temperature range Betriebstemperaturbereich m, Einsatztemperaturbereich m, Temperatur-Anwendungsbereich m, Temperatureinsatzbereich m
outlet temperature Ausgangstemperatur f, Auslauftemperatur f, Austrittstemperatur f, Ausstoßtemperatur f
outside temperature Außentemperatur f
overall temperature Gesamttemperaturniveau n
pinch-off temperature Abquetschtemperatur f
plastication temperature Plastifizierungstemperatur f
platen temperature Plattentemperatur f
polymerization temperature Polymerisationstemperatur f
post-curing temperature Nachhärtungstemperatur f
pre-gelation temperature Vorgeliertemperatur f
processing temperature Verarbeitungstemperatur f
processing temperature range Verarbeitungstemperaturbereich m
recording of temperature Temperaturaufzeichnung f
reference temperature Bezugstemperatur f
required temperature Solltemperatur f, Temperatursollwert m
resistance to sterilizing temperature Sterilisierfestigkeit f
resistance to high and low temperatures Temperaturbeständikeit f
resistant to high and low temperatures temperaturbeständig
resistant to sterilizing temperature sterilisationsbeständig
rise in temperature Temperaturanstieg m
room temperature Zimmertemperatur f, Raumtemperatur f
scorch temperature Scorchtemperatur f
screw root temperature control (system) Schneckenkerntemperierung f
screw temperature control (system) Schneckentemperierung f
sealing temperature Siegeltemperatur f
self-ignition temperature Selbstentzündungstemperatur f
slide-in temperature control module Temperatur-Regeleinschub m
solvation temperature [of plasticizer for PVC] Lösetemperatur f, Gelierpunkt m, Geliertemperatur f
starting temperature Ausgangstemperatur f
storage temperature Lagerungstemperatur f
stoving temperature Einbrenntemperatur f
stoving temperature range Einbrennbereich m
subject(ed) to high temperature hitzebeansprucht
surface temperature Oberflächentemperatur f
test temperature Untersuchungstemperatur f, Versuchstemperatur f
thermoforming temperature Warmformtemperatur f, Umformungstemperatur f

uniform temperature Temperaturhomogenität f
upper temperature limit Temperatur-Obergrenze f
vulcanizing temperature Vulkanisationstemperatur f
water-fed temperature control unit Wassertemperiergerät n
working temperature Arbeitstemperatur f, Betriebstemperatur f
temporal zeitlich
temporary zeitweilig, vorläufig, vorübergehend
temporary joint lösbare Verbindung f
temporary storage Zwischenlagerung f
tendency Neigung f, Tendenz f
tendency to become worn Verschleißneigung f
tendency to block Blockneigung f
tendency to char Verkohlungsneigung f
tendency to creep Kriechneigung f
tendency to crystallize Kristallisationsneigung f
tendency to discolo(u)r Verfärbungsneigung f
tendency to form lumps Verklumpungsneigung f
tendency to gel Gelierneigung f
tendency to run (paint) Ablaufneigung f
tendency to settle (out) Sedimentationsneigung f, Absetzneigung f
tendency to shrink Schrumpfneigung f
tendency to split Spleißneigung f
tendency to stick [e.g. compound to metal rolls] Klebneigung f
tendency to warp Verzugsneigung f
blocking tendency Blockneigung f
migration tendency Wanderungstendenz f
tensile dehnbar, streckbar, zugbelastbar
tensile behavio(u)r Zugverhalten n
tensile breaking stress Zugbruchlast f
tensile craze Zugspannungs-Craze f
tensile creep modulus Zug-Kriechmodul m
tensile creep strength Zeitstandzugfestigkeit f
tensile creep stress Zeitstand-Zugbeanspruchung f
tensile creep test Zeitstandzugversuch m
tensile deformation Zugspannungsverformung f
tensile energy Zugarbeit f
tensile fatigue strength Zugschwingungsversuch m
tensile force Zugkraft f
tensile force transducer Zugkraftaufnehmer m, Zugdose f
tensile impact strength Schlagzugzähigkeit f
tensile impact test Schlagzugversuch m
tensile modulus Zugmodul m
tensile shear strength Zugscherfestigkeit f
tensile shear stress Zugscherbelastung f
tensile shear test Zugscherversuch m
tensile specimen Zugstab m, Zugprobekörper m, Zugprobe f
tensile strength Zugfestigkeit f
tensile strength in bending Biegezugfestigkeit f
tensile stress Zugbeanspruchung f, Zugbelastung f, Zuglast f, Zugspannung f
tensile stress at break Bruchspannung f

tensile stress at yield Zugspannung f bei Streckgrenze, Streckspannung f, Fließspannung f
tensile stress-elongation curve Zug-Dehnungs-Diagramm, Zugspannungs-Dehnungs-Kurve f
tensile test Zug(dehnungs)versuch m
tensile test piece Zugprobekörper m, Zugprobe f, Zugstab m
tensile test specimen Zerreißprobe f [Material]
tensile testing machine Zerreiß(prüf)maschine f, Zugprüfmaschine f
   accelerated tensile test Kurzzeit-Zugversuch m
   change in tensile strength Zugfestigkeitsänderung f
   direction of tensile stress Zugbeanspruchungsrichtung f
   dumbbell-shaped tensile test piece Schulterzugstab m
   internal tensile stress Zugeigenspannung f
   long-term tensile stress Zeitstand-Zugbeanspruchung f
   maximum tensile stress Zugspannungsmaximum n
   repeated tensile stress Zugschwellast f, Zugschwellbelastung f
   short-term tensile stress Kurzzeitzugbeanspruchung f
   standard tensile test piece Normzugstab m
   ultimate tensile strength Zerreißfestigkeit f
   ultimate tensile stress Bruchspannung f
   under tensile stress zugbeansprucht, zugbelastet
tension Spannung f
   tension control Spannungskontrolle f
   tension control mechanism [for film as it comes off the calender] Spannungskontrolle f, Bahnspannungsregelung f
   tension rod Zuganker m
   tension roll Spannrolle f, Zugwalze f
   tension set Zugverformungsrest m
   interfacial surface tension Grenzflächenspannung f
   modulus of elasticity in tension Zug-E-Modul m
   reel tension Wickelspannung f, Wickelzug m, Waren(bahn)spannung f, Bahnspannung f
   reeling tension Aufwickelspannung f
   surface tension Oberflächenspannung f
   take-off tension Abzugskraft f
   web tension Bahnzugkraft f, Bandzug m, Folienbahnspannung f
   web tension control (unit) Zugspannungsregelung f, Bahnspannungskontrolle f, Bahnspannungsregelung f
   web tension measuring device Bahnzugkraftmeßvorrichtung f
   web tension measuring unit Bahnzug(kraft)-meßstation f
   winding tension Wickelspannung f, Wickelzug m
tensioning spannend, Spann-, Zug-
   tensioning unit Zugwerk n
   web tensioning device Folienbahnspannung f, Waren(bahn)spannung f, Bahnspannung f
tenter Spannrahmen m
   tenter frame Kluppenrahmen m, Spannrahmen m
terephthalate Theraphthalat n
   di-2-ethylhexyl terephthalate [DOTP] Di-2-ethylhexylterephthalat n
   dimethyl terephthalate Dimethylterephthalat n
   dioctyl terephthalate [DOTP] Di-2-ethylhexylterephthalat n
   polyalkylene terephthalate Polyalkylenterephthalat n
   polybutylene terephthalate [PBTP] Polybutylenterephthalat n
   polyethylene terephthalate [PETP] Polyethylenterephthalat n
   polyglycol terephthalate Polyglycolterephthalsäureester m
terephthalic acid Terephthalsäure f
term Frist f, Termin m
   long term performance Dauergebrauchseigenschaften fpl
terminal 1. endständig; 2. Terminal n
   terminal carboxyl group Carboxyl-Endgruppe f
   terminal hydroxyl group Hydroxyl-Endgruppe f
   area terminal Bereichsterminal n
   display terminal Bildschirmterminal n
   intelligent terminal intelligentes Terminal n
   master terminal Leitstandterminal n
   production data terminal BDE-Terminal n, Betriebsdatenerfassungsstation f
   video terminal Videoterminal n
termination Abschluß m, Beendigung f
   termination reaction Abbruchreaktion f
   polymerization termination Polymerisationsabbruch m
ternary ternär
terpene hydrocarbon Terpenkohlenwasserstoff m
terpene resin Terpenharz n
terpolymer Terpolymer n, Terpolymerisat n
tertiary tertiär
   tertiary-butyl hydroperoxide tert.-Butylhydroperoxid n
   tertiary-butyl perbenzoate tert.-Butylperbenzoat n
test Test m, Prüfung f, Versuch m
   test bed Testgelände n, Versuchsfeld n
   test conditions Versuchsbedingungen fpl
   test duration Versuchsdauer f
   test formulation Testformulierung f, Testrezeptur f, Versuchsrezeptur f
   test method Testmethode f
   test parameter Versuchsparameter m
   test period Untersuchungszeitraum m
   test piece Stab m, Testkörper m
   test program Versuchsprogramm n
   test result Testergebnis n, Versuchsergebnis n
   test series Versuchsreihe f, Versuchsserie f
   test set-up Versuchseinrichtung f, Versuchsanordnung f
   test specimen Versuchsprobe f
   test substance Versuchsmaterial n
   test temperature Untersuchungstemperatur f, Versuchstemperatur f

abrasion test Abriebprüfung f
accelerated bursting test Kurzzeit-Berstversuch m
accelerated tensile test Kurzzeit-Zugversuch m
accelerated test Kurzprüfung f, Kurzzeit-Prüfung f, Kurzzeitversuch m, Schnelltest m, Zeitraffertest m, Zeitrafferversuch m
accelerated test results Kurzzeitwerte mpl
accelerated weathering test Kurzzeitbewitterungsversuch m, Schnellbewitterung f
additional tests Zusatzsprüfungen fpl
ag(e)ing test Alterungsprüfung f, Alterungsuntersuchung f, Lagerungsversuch m
alternating temperature test Temperaturwechselprüfung f
ball indentation test Kugeldruckprüfung f, Kugeleindruckverfahren n
blank test Blindversuch m
boiling test Kochversuch m
breakdown test Durchschlagversuch m
bursting test Berstversuch m
comparative test Vergleichstest m, Vergleichsversuch m
compatibility test Verträglichkeitsprüfung f, Verträglichkeitsuntersuchung f
compression test Druckprüfung f, Druckversuch m
compressive creep test Zeitstanddruckversuch m
compressive strength test machine Druckfestigkeitsprüfmaschine f
condensed moisture test Schwitzwasserprüfung f
constant test atmosphere Konstantklima n
control test Überwachungsprüfung f
crash test Auffahrversuch m
creep test Kriechversuch m, Zeitstandprüfung f, Zeitstandversuch m, Langzeitprüfung f, Langzeittest m, Langzeitversuch m, Langzeituntersuchung f
creep test machine Zeitstandanlage f
delamination test Spaltversuch m
destructive test Zerrüttungsprüfung f, Zerrüttungsuntersuchung f
dumbbell(-shaped) tensile test piece Schulterzugstab m
dumbbell(-shaped) test piece Schulterprobe f, Schulterstab m
dumbbell(-shaped) test specimen Hantelstab m
ethyl acetate test Essigestertest m
falling ball test Kugelfallversuch m
falling dart test Falldorntest m
falling weight test Bolzenfallversuch m, Fallbolzenversuch m
fatigue test Dauerfestigkeitsversuch m, Dauerschwingversuch m, Zeitwechselfestigkeitsversuch m
flammability test Brandversuch m, Brennbarkeitstest m
flexural creep test Biegekriechversuch m, Zeitstandbiegeversuch m
flexural fatigue test Biegeschwingversuch m, Wechselbiegeversuch m
flexural impact test Schlagbiegeversuch m
flexural impact test piece Schlagbiegestab m
flexural test Biegeversuch m
flexural test piece Biegestab m
folding endurance test Dauer-Knickversuch m
four-point bending test Vierpunkt-Biegeversuch m
heat ag(e)ing test Wärmealterungsversuch m, Warmlagerungsversuch m
high-voltage test bed Hochspannungsprüffeld n
ignition test Zündversuch m
immersion test Einlagerungsversuch m, Tauchversuch m
impact test Schlagprüfung f, Schlagversuch m, Stoßversuch m
indentation test Tiefungsversuch m, Eindrückversuch m
laboratory test Laborversuch m
large scale burning test Brandgroßversuch m
light test Belichtungsprüfung f
loading test Belastungsprüfung, Belastungsversuch m
long-term failure test under internal hydrostatic pressure Innendruck-Zeitstanduntersuchung f, Zeitstand-Innendruckversuch m
long-term milling test Dauerwalzentest m
long-term test Langzeitprüfung f, Langzeittest m, Langzeituntersuchung f, Langzeitversuch m
long-term test voltage Dauerprüfspannung f
mandrel flex test Dornbiegeversuch m
milling test Walztest m
model test Modelluntersuchung f
natural weathering test Freibewitterungsprüfung f
notched impact test specimen Kerbschlagprobe f
outdoor weathering test Außenbewitterungsversuch m, Freibewitterungsprüfung f, Freibewitterungsversuch m
peel test Schälversuch m
pendulum impact test Pendelschlagversuch m
penetration test Durchstoßversuch m
preliminary test Vorprüfung f, Vorversuch m, Vortest m
random test Stichprobenprüfung f
rectangular test piece Flachstab m
rectangular test specimen Flachstab m
repeat test Wiederholungsmessung f, Wiederholungsprüfung f
repeat test series Wiederholungsmeßreihen fpl
shear test Scherversuch m
short-term test Kurzzeit-Prüfung f, Kurzzeitversuch m
short-term test results Kurzzeitwerte mpl
shrinkage test piece Schwindungsstab m
small standard test piece Normkleinstab m
spiral flow test Spiraltest m
square-section test piece Vierkantstab m
standard dumbbell-shaped test specimen Norm-Schulterprobe f
standard tensile test piece Normzugstab m

standard test piece Normstab m
standard test specimen Norm-Probekörper m
stoving test Einbrennversuch m
stress relaxation test Spannungsrelaxationsversuch m
suitability test Eignungsprüfung f
Sward hardness test Schenkelhärteprüfung f
tear resistance test Einreißversuch m
tensile creep test Zeitstandzugversuch m
tensile impact test Schlagzugversuch m
tensile shear test Zugscherversuch m
tensile test Zug(dehnungs)versuch m
tensile test piece Zugprobe f, Zugprobekörper m, Zugstab m
three-point bending test Dreipunkt-Biegeversuch m
torsion pendulum test Torsionsschwingungsmessung f, Torsions(schwing)versuch m
tracking resistance test Kriechstromprüfung f
trouser tear test Schenkel-Weiterreißversuch m
universal test apparatus Universalprüfmaschine f
universal test (method) Universalprüfmethode f
weathering test Bewitterungsprüfung f, Bewitterungsversuch m
tester Prüfgerät n
  abrasion tester Abriebprüfmaschine f
  automatic gel time tester Gelierzeitautomat m
  Charpy impact tester Charpygerät n
  Dynstat impact tester Dynstatgerät n
  falling weight tester Fallbolzenprüfgerät n
  gel time tester Gelzeitprüfgerät n
  hardness tester Härteprüfer m, Härteprüfgerät n
  impact tester Schlagwerk n
  leak tester Leckprüfgerät n
  pendulum impact tester Pendelhammergerät n, Pendelschlagwerk n
  universal tester Universalprüfmaschine f
testing testend, prüfend, Prüf-
  bottle testing machine Flaschenprüfstand m
  material testing laboratory Werkstoffprüflabor n
  materials testing Werkstoffprüfung f
  preliminary testing Vortesten n
  tensile testing machine Zerreiß(prüf)maschine f, Zugprüfmaschine f
tetra-alkyl tin compound Tetraalkylzinnverbindung f
tetra-substituted tetrasubstituiert
tetrabasic tetrabasisch
tetrabutyl titanate Tetrabutyltitanat n
tetracarboxylic acid Tetracarbonsäure f
tetrachloride Tetrachlorid n
  carbon tetrachloride Tetrachlorkohlenstoff m
  silicon tetrachloride Siliciumtetrachlorid n
tetrachloroethylene Perchlorethylen n
tetrachlorophthalic acid Tetrachlorphthalsäure f
tetrachlorophthalic anhydride Tetrachlorphthalsäureanhydrid n
tetrafluoroborate Tetrafluorborat n
  sodium tetrafluoroborate Natriumtetrafluorborat n
tetrafluoroethylene Tetrafluorethylen n

tetrafunctional tetrafunktionell
tetrahydrofurane Tetrahydrofuran n
tetrahydrophthalic acid Tetrahydrophthalsäure f
tetramer Tetramer n
tetrameric tetramer
tetramethyl silane Tetramethylsilan n
tetramine Tetramin n
  triethylene tetramine Triethylentetramin n
tetraphenyl ethane Tetraphenylethan n
text Text m
  diagnostic text Diagnosetext m
textile size Textilschlichte f
texture Gefüge n, Struktur f
  surface texture Oberflächenstruktur f
textured strukturiert, genarbt, gemasert
TGA [thermogravimetric analysis] thermogravimetrische Analyse f
thaw (out)/to auftauen
thermal thermisch, Wärme-
  thermal conductivity Wärmeleitfähigkeit f
  thermal conductivity coefficient Wärmeleitzahl f
  thermal conductivity equation Wärmeleitungsgleichung f
  thermal diffusivity Temperaturleitfähigkeit f
  thermal efficiency thermische Leistung f
  thermal endurance Wärmebelastbarkeit f
  thermal energy Wärmeenergie f
  thermal equilibrium thermisches Gleichgewicht n
  thermal expansion Wärme(aus)dehnung f
  thermal expansion piece Wärmedehnbolzen m
  thermal history Temperaturgeschichte f
  thermal insulating properties Wärmedämmeigenschaften fpl
  thermal insulation Wärmeisolierung f, Wärmedämmung f, Wärmeisolation f
  thermal isolation thermische Trennung f, Wärmetrennung f
  thermal load(ing) thermische Beanspruchung f, Wärmebeanspruchung f
  thermal oxidation Wärmeoxidation f
  thermal properties thermische Eigenschaften fpl, thermische Werte mpl
  thermal sealing Wärmekontaktschweißen n
  thermal shock resistance Wärmeschockverhalten n
  thermal stability Hitzestabilität f, thermische Stabilität f, Thermostabilität f, Wärmestabilität f, thermische Beständigkeit f, Temperaturstabilität f
  coefficient of thermal expansion thermischer Ausdehnungskoeffizient m, Wärmedehn(ungs)zahl f
  compensation for thermal expansion Wärmeausdehnungsausgleich m
  differential thermal analyzer Differentialthermoanalysengerät n
  long-term thermal stability Dauertemperaturbeständigkeit f, Dauerwärmestabilität f
thermally conductive nozzle Wärmeleitdüse f
thermally conductive torpedo Wärmeleittorpedo m

thermally insulated thermoisoliert, wärmegedämmt
thermally stable temperaturstabil, thermostabil
thermo-oxidative thermooxidativ
thermoanalysis Thermoanalyse f
   differential thermoanalysis [DTA] Differential-thermoanalyse f
thermocouple Temperaturaufnehmer m, Wärmefühler m, Thermoelementfühler m, Thermodraht m, Thermoelement n, Temperatur(meß)fühler m, Temperatursonde f, Temperatursensor m, Temperatur(meß)geber m, Temperaturmeßgerät n
   thermocouple well Thermofühlerbohrung f, Thermoelementbohrung f
   thermocouple wire Thermopaar n
   if the thermocouple breaks bei Thermoelementbruch m
   iron-constantan thermocouple Eisen-Konstantan-Thermoelement n, Fe-Ko-Thermoelement n
   jacketed thermocouple Mantelthermoelement n
   melt thermocouple Massetemperaturfühler m, Massethermoelement n
   mo(u)ld thermocouple Werkzeugtemperaturfühler m
   resistance thermocouple Widerstands(temperatur)fühler m
   surface thermocouple Oberflächentemperaturfühler m
thermodiffusion Thermodiffusion f
thermodynamic thermodynamisch
thermodynamics Thermodynamik f
thermoform/to thermo(ver)formen
thermoformability Warmformbarkeit f, Tiefziehfähigkeit f, Verformbarkeit f, Verformungsvermögen n
thermoformable warmformbar, thermoformbar, tiefziehfähig
thermoformed thermogeformt, warmgeformt, tiefgezogen
thermoforming spanloses Umformen n, Thermoformen n, thermoplastisches Umformen n, thermoplastische Verformung f, Warmformen n
   thermoforming draw Ziehen n
   thermoforming equipment Warmformanlage f
   thermoforming force Formungskraft f
   thermoforming machine Tiefziehmaschine f, Warmformmaschine f
   thermoforming mo(u)ld Tiefziehwerkzeug n
   thermoforming process Tiefziehvorgang m, Warmformverfahren n
   thermoforming properties Warmformeigenschaften fpl, Tiefziehfähigkeit f, Tiefzieheigenschaften fpl, Tiefziehverhalten n
   thermoforming sheet Tiefziehfolie f
   thermoforming sheet extrusion line Tiefziehfolien-Extrusionsanlage f
   thermoforming sheet plant Tiefziehfolienanlage f
   thermoforming station Warmformstation f
   thermoforming temperature Warmformtemperatur f
   automatic thermoforming machine Thermoform-Automat m, Tiefziehautomat m, Warmformautomat m
   laboratory thermoforming machine Laborziehgerät n
   sheet thermoforming line Plattenformmaschine f
   suitable for thermoforming thermoformbar, tiefziehfähig
   two layer thermoforming film extrusion line Zweischicht-Tiefziehfolienanlage f
thermogravimetric thermogravimetrisch
   thermogravimetric analysis [TGA] thermogravimetrische Analyse f
thermomechanical thermomechanisch
   thermomechanical analysis [TMA] thermomechanische Analyse f
thermometer Thermometer n, Temperaturmeßgerät n
   contact thermometer Berührungsthermometer n, Kontaktthermometer n
   resistance thermometer Widerstandsthermometer n
   screw-in resistance thermometer Einschraub-Widerstandsthermometer n
thermoplastic 1. thermoplastisch, warm verformbar; 2. Thermoplast m
   thermoplastic foam mo(u)lding part Thermoplast-Schaumteil n
   thermoplastic material Thermoplast m
   thermoplastic screw Thermoplast-Schnecke f
   engineering thermoplastics technische Thermoplaste mpl
   extrusion blow mo(u)lding of expandable thermoplastics Thermoplast-Extrusions-Schaumblasen fpl
   processing of thermoplastics Thermoplastverarbeitung f
thermoplasticity Thermoplastizität f
thermoset 1. duromer, duroplastisch; 2. Thermodur m, Duroplast n [s.a. thermosetting]
   thermoset injection mo(u)lding machine Duroplast-Spritzgießmaschine f
   thermoset injection mo(u)lding process Duroplast-Spritzgießverfahren n
   thermoset material Duroplast m, Duromer m
   thermoset mo(u)lding Duroplast-Formteil n
   thermoset mo(u)lding compound Duroplast-Formmasse f, Duroplastmasse f
   thermoset screw Duroplastschnecke f
   processing of thermosets Duroplastverarbeitung f
   range of thermoset mo(u)lding compounds Duroplastformmassen-Sortiment n
thermosetting thermohärtend, hitzehärtend [s.a. thermoset]
thermostat Thermostat m
   thermostat to prevent overheating Überhitzungs-Sicherheitsthermostat m
   bimetallic thermostat Bimetallthermostat m
   high-temperature thermostat Hochtemperaturthermostat m
   low-temperature thermostat Kältethermostat m, Kryostat m
thermostated thermostatisiert

thermostatically controllable thermostatisierbar
thermostatically controlled thermostatisch geregelt, thermostatisiert
thick dick
    thick-bed adhesive [for fixing tiles] Dickbettkleber m
    thick-section dickwandig
    thick-walled [e.g. a container] starkwandig, dickwandig
thicken/to eindicken
thickened verdickt, eingedickt
thickener Verdickungsmittel n, Eindick(ungs)mittel n
thickening Andickung f, Verdickung f
    thickening agent Eindick(ungs)mittel n, Verdickungsmittel n
    thickening effect Eindickungsaktivität f
thickness Dicke f, Stärke f
    thickness calibration (unit) Dickenkalibrierung f
    thickness ga(u)ge Dickenmeßgerät n, Dickenmeßeinrichtung f, Wanddickenmessung f
    thickness tolerance Dickentoleranz f
    thickness variations Dickenschwankungen fpl, Dickenabweichungen fpl
    adhesive film thickness Klebschichtstärke f
    coating thickness Beschichtungsdicke f, Schichtdicke f, Schichtstärke f
    depending on (the) wall thickness wanddickenabhängig
    differences in wall thickness Wanddickenunterschiede mpl
    dry film thickness Trockenfilmdicke f, Trockenschichtstärke f
    flow length-wall thickness ratio Fließweg-Wanddickenverhältnis n
    mo(u)lded-part wall thickness Formteilwanddicke f
    parison wall thickness control Schlauchdickenregelung f
    parison wall thickness controller Schlauchdickenregler m
    part wall thickness Formteilwanddicke f
    required wall thickness Sollwanddicke f, Wanddickensollwert m
    standard thickness Normdicke f
    total film thickness Gesamtschichtdicke f
    wall thickness Wanddicke f
    wall thickness control Wanddickenregulierung f, Wanddickensteuerung f
    wall thickness control mechanism Wanddickenregulierungssystem n
    wall thickness control unit Wanddickenregelgerät n
    wall thickness deviations Wanddickenstreuungen fpl
    wall thickness differences Wanddickenunterschiede mpl
    wall thickness distribution Wanddickenverteilung f
    wall thickness ga(u)ge Wanddickenmeßgerät n
    wall thickness measurement Wanddickenmessung f
    wall thickness programming Wanddickenprogrammierung f
    wall thickness programming device Wanddicken-Programmiergerät n
    wall thickness variations Wanddickenschwankungen fpl, Wanddickenabweichungen fpl
    wet coating thickness Naßschichtdicke f
    wet film thickness Naßfilmdicke f
thin dünn(schichtig)
    thin-bed adhesive [for fixing tiles] Dünnbettkleber m
    thin film Dünnfolie f
    thin film vaporizer Dünnschichtverdampfer m
    thin-layer chromatography Dünnschichtchromatographie f
    thin-section [e.g. mo(u)ldings] dünnwandig
thinner [specifically for paints] Verdünnungsmittel n, Verdünner m
thio-compound Thioverbindung f
thioacetate Thioacetat n
    dibutyltin thioacetate Dibutylzinnthioacetat n
thiodicarboxylate Thiodicarbonsäureester m
thioether Thioether m
thioglycolate Thioglykolat n, Thioglykolsäureester m
    di-n-butyltin thioglycolate Di-n-butylzinnthioglykolat n
    dialkyltin thioglycolate Dialkylzinnthioglycolsäureester m
    dibutyltin thioglycolate Dibutylzinnthioglykolat n
    dioctyltin thioglycolate Dioctylzinnthioglykolat n
    mono-alkyl tin thioglycolate Monoalkylzinnthioglykolsäureester m
    thioglycolate radical Thioglykolsäurerest m
thiokol Thiokol m
thiourea Thioharnstoff m
    thiourea(-formaldehyde) resin Thioharnstoffharz n
    diphenyl thiourea Diphenylthioharnstoff m, Thioharnstoff m
thixotropic thixotrop
    thixotropic agent Thixotropiermittel n
    thixotropic effect Thixotropiewirkung f
    making thixotropic Thixotropierung f
    imparting thixotropy Thixotropierung f
thixotropy Thixotropie f
thread 1. Gewinde n; 2. Faden m
    thread contours Stegkontur f [Schnecke]
    thread-cutting gewindeschneidend
    thread-forming gewindeformend
    thread insert Gewindeeinsatz m
    thread-like fadenartig
    thread-like molecule Fadenmolekül n
    external thread Außengewinde n
    left-hand thread Linksgewinde n
    screw threads Schneckenwindungen fpl
threaded gewunden, Gewinde-
    threaded bush Gewindebuchse f
    threaded connection Gewindeanschluß m
    threaded core Gewindekern m, Schraubkern m
    threaded hole Gewindebohrung f, Gewindeloch n
    threaded mandrel unscrewing device Gewinde-

dorn-Ausdrehvorrichtung f
threaded part Gewindepartie f
threaded spacer Distanzbolzen mpl
threaded spindle Gewindespindel f
three drei
  three-cavity injection mo(u)ld dreifaches Spritzgießwerkzeug n
  three-cavity mo(u)ld Dreifachwerkzeug n
  three-channel nozzle Dreikanaldüse f
  three-dimensional dreidimensional
  three-impression mo(u)ld Dreifachform f
  three-layer blown film die Dreischicht-Folienblaskopf m
  three-layer (coextrusion) die Dreischichtdüse f
  three-layer blow mo(u)lding Dreischichthohlkörper m
  three-layer coextrusion Dreischicht-Coextrusion f
  three-part mo(u)ld Zweietagenwerkzeug n, Dreiplattenwerkzeug n
  three-phase dreiphasig
  three-phase current Drehstrom m
  three-phase geared motor Drehstrommotor m
  three-phase system Dreiphasensystem n
  three-plate clamping mechanism Dreiplatten-Schließsystem n
  three-plate clamp(ing) unit Dreiplatten-Schließeinheit f
  three-plate compression mo(u)ld Dreiplattenpreßwerkzeug n
  three-plate mo(u)ld Zweietagenwerkzeug n, Dreiplattenwerkzeug n, Dreiplatten-Abreißwerkzeug n
  three-plate multi-cavity mo(u)ld Dreiplatten-Mehrfachwerkzeug n
  three-point bending Dreipunktbiegung f
  three-point bending test Dreipunkt-Biegeversuch m
  three-point controller Dreipunktregler m
  three-point support [e.g. of a test specimen] Dreipunktauflage f
  three-roll calender Dreiwalzenkalander
  three-roll offset calender Dreiwalzenkalander m in Schrägform
  three-roll polishing stack Dreiwalzen-Glättwerk n, Dreiwalzen-Glättkalander m
  three-roll vertical calender Dreiwalzen-I-Kalander m
  three-section plasticizing unit Dreizonen-Plastifiziereinheit f
  three-section screw Dreizonenschnecke f
  three-shift operation Dreischichtbetrieb m
  three-speed control mechanism Dreigang-Schaltgetriebe n
  three-start dreigängig
  three-station design Dreiplatzanordnung f
  three-way flow control (unit) Dreiwegestromregelung f
  three-way flow control valve Dreiwege-Mengen-regelventil n
  three-way tap Dreiwegehahn m
  three-way valve Dreiwegeventil n
  in three stages dreifach gestaffelt
threshold Schwelle f
  threshold pressure Druckschwellwert m
  response threshold Ansprechschwelle f
  sensitivity threshold Empfindlichkeitsschwelle f
throat Kehle f; Einschnürung f, Halsstück n
  feed throat Aufgabeschacht m, Aufgabeöffnung f, Beschickungsöffnung f, Einzug m, Einlauf m
throttle Drossel f
  throttle-check valve Drossel-Rückschlagventil n
  throttle control Drosselsteuerung f
  throttle valve Drosselklappe f, Drosselschieber m, Drosselventil n
throughput Durchfluß m, Durchsatz m, Massedurchsatz m, Förderrate f, Durchflußmenge f, Durchlaufmenge f, Förderstrom m
  throughput per hour Stundendurchsatz m
  throughput rate Förderrate f, Förder(ungs)geschwindigkeit f, Förderleistung f
  air throughput Luftdurchsatz m
  cooling water throughput Kühlwasser-durchflußmenge f
  hourly throughput Stundendurchsatz m
  material throughput Fördermenge f
  melt throughput Massedurchsatz m, Massestrom m, Schmelzestrom m
  volume throughput Volumendurchsatz m, volumetrischer Durchsatz m
thrust Schub m, Stoß m, Druck m
  thrust bearing Axiallagerung f, Drucklagerung f, Drucklager n
  thrust bearing unit Axiallagergruppe f
  thrust ring Druckring m, Stützring m
  thrust screw Druckschraube f
  forward thrust Vortriebskraft f
thumb Daumen m
  rule of thumb Faustregel f, Faustformel f
thyristor control unit Thyristorregler m
thyristor controlled thyristorgesteuert, thyristorgespeist
tie Verbindung f, Zugstab m
  tie bar Holm m
  tie coat Haftgrundierung f
  distance between tie bars Weite f zwischen den lichte Holmen, Holmabstand m, lichte Weite f zwischen den Säulen, lichter Abstand m zwischen den Säulen, lichter Holmabstand m, Lichtweite f
  with two tie bars zweiholmig
tight dicht
  tight fitting spielfrei, genau passend
  vacuum tight vakuumdicht
tighten/to (nach)spannen, straffen
tightening torque Anzugsdrehmoment n
tightly compressed hochverdichtet
tile Fliese f, Kachel f, Dachziegel m, Platte f
  tile adhesive Fliesenkleber m
  carpet tile Teppichauslegeware f
  vinyl-asbestos tile Vinyl-Asbest-Platte f

tilt/to klappen
tilting klappend, schwenkend, Kipp-, Schwenk-
  tilting angle Schwenkwinkel m
  tilting arrangement Kippvorrichtung f
  tilting device Schwenkeinrichtung f
  tilting mechanism Schwenkeinrichtung f
  tilting table Schwenktisch m
  tilting unit Schwenkaggregat n
time Zeit f, Uhrzeit f
  time consuming zeitaufwendig, zeitintensiv, zeitraubend
  time controller Zeitregler m
  time dependence Zeitabhängigkeit f
  time-dependent zeitabhängig
  time during which the screw is rotating Schneckendrehzeit f
  time factor Zeitfaktor m
  time-lag method Zeitverzögerungsmethode f
  time module Zeitbaustein m, Zeitmodul m
  time of day Tageszeit f, Uhrzeit f
  time of exposure [e.g. to chemicals] Einwirkungszeit f, Expositionszeit f, Expositionsdauer f
  time of exposure to heat thermische Belastungszeit f
  time of immersion Einlagerungszeit f, Lagerungszeit f, Lager(ungs)dauer f [in Flüssigkeit]
  time of operation Betriebsdauer f, Betriebszeit f
  time of the year Jahreszeit f
  time pre-selection switch Zeitvorwahlschalter m
  time-related zeitbezogen
  time relay Zeitrelais n
  time required Zeitaufwand m, Zeitbedarf m
  time required for blowing Blaszeit f
  time required for re-heating Wiederaufheizzeit f
  time required (for the solvent) to evaporate Ablüftungszeit f
  time scale Zeitmaßstab m
  time switch Schaltuhr f, Zeitschaltwerk n
  time-to-failure [expressed in hours] Zeitstandwert m, Zeitstandniveau n
  time under stress Laststandzeit f
  access time Zugriffzeit f
  automatic gel time tester Gelierzeitautomat m
  baking time Einbrennzeit f
  change-over time Umschaltzeit f
  cooling time Abkühldauer f, Abkühlzeit f, Erstarrungsdauer f, Erstarrungszeit f, Kühlzeit f
  cream time [in PU foaming] Startzeit f
  cure time Aushärtungszeit f, Härtungszeit f, Härtezeit f
  curing time Aushärtungszeit f, Härtungszeit f, Härtezeit f
  cycle time Taktzeit f, Zyklusdauer f, Zykluszeit f
  cycle time reduction Zykluszeitverkürzung f, Zykluszeitreduzierung f
  delay time Verzögerungszeit f
  dry cycle time Taktzeit f (im Trockenlauf), Trockenlauf-Zykluszeit f, Trockentaktzeit f, Trockenlaufzeit f
  drying time Trockenzeit f, Trocknungszeit f
  dust dry time Staubtrockenzeit f
  dwell time Aufenthaltzeit f, Verweilzeit f
  factor unfluencing cycle time Zykluszeitfaktor m
  flow time [in viscosity determinations] Durchflußzeit f, Auslaufzeit f
  for a short time kurzfristig
  fusion time Gelierzeit f, Gelierdauer f
  gate opening time Versiegelungszeit f
  gel time [of UP, EP resins] Gelierzeit f, Gelierdauer f
  gel time tester Gelzeitprüfgerät n
  heating-up time Erwärmungszeit f
  hold(ing) pressure time Nachdruckdauer f, Nachdruckzeit f, Druckhaltezeit f
  impregnating time Tränkungszeit f
  independent of time zeitunabhängig
  injection pressure time Spritzdruckzeit f
  injection time Formfüllzeit f, Werkzeugfüllzeit f
  loading time Beanspruchungsdauer f, Belastungsdauer f, Belastungszeit f
  milling time Walzzeit f
  mo(u)ld changing time Werkzeugwechselzeit f
  mo(u)ld closing time Formschließzeit f, Schließzeit f
  mo(u)ld construction time Formenbauzeit f
  mo(u)ld filling time Werkzeugfüllzeit f, Formfüllzeit f, Spritzzeit f, Einspritzzeit f, Füllzeit f
  mo(u)ld setting time Werkzeugeinrichtezeit f
  non-productive time Nebenzeit f, Stehzeit f, Totzeit f, Stillstandzeit f,
  packing time Kompressionszeit f
  passage of time Zeitverlauf m
  plasticizing time Plastifizierzeit f
  pressure change-over time Druckumschaltzeit f
  production time Fertigungszeit f
  re-heating time Wiederaufheizzeit f
  real time Echtzeit f
  real time control Echtzeitsteuerung f
  real time processing Echtzeitverarbeitung f
  residence time Aufenthaltszeit f, Verweilzeit f
  residence time profile Verweilzeitspektrum n
  residual residence time Verweilzeitschwanz m
  response time Ansprechzeit f
  screw return time Schneckenrücklaufzeit f
  setting time Abbindezeit f, Abkühldauer f, Abkühlzeit f, Erstarrungsdauer f, Erstarrungszeit f, Kühlzeit f; Einrichtezeit f [Maschine]
  stoving time Einbrenndauer f
  stress removal time Entlastungszeit f
  switching time Schaltzeit f
  total cycle time Gesamtzykluszeit f, Zyklusgesamtzeit f
  total switching time Gesamtschaltzeit f
  unit of time Zeiteinheit f
  variation with time [e.g. of a parameter] zeitlicher Verlauf m
  vulcanizing time Vulkanisationszeit f

weekly time switch Wochenschaltuhr f
timed zeitlich abgestimmt
  correctly timed zeitoptimal
timer Schaltuhr f, Zeitgeber m, Zeituhr f, Zeitmesser m
  digital timer Digital-Zeitmeßgerät n, Digitalzeituhr f
  synchronous timer Synchronschaltuhr f
timing Wahl f des Zeitpunkts, Timing n, Zeitsteuerung f
  timing device Zeitgeber m, Zeitmesser m, Zeituhr f
  timing mechanism Zeitgebersystem n
  interval timing mechanism Pausenzeituhr f
tin Zinn n
  tin carboxylate Zinncarboxylat n
  tin maleinate Zinnmaleinat n
  tin mercaptide Zinnmerkaptid n
  tin stabilized zinnstabilisiert
  tin stabilizer Zinnstabilisator m
  alkyl tin carboxylate Alkylzinncarboxylat n
  alkyl tin compound Alkylzinn-Verbindung f
  mono-alkyl tin chloride Monoalkylzinnchlorid n
  mono-alkyl tin compound Monoalkylzinnverbindung f
  mono-alkyl tin stabilizer Monoalkylzinnstabilisator m
  mono-alkyl tin thioglycolate Monoalkylzinnthioglykolsäureester m
  octyl tin carboxylate Octylzinncarboxylat n
  octyl tin compound Octylzinnverbindung f
  octyl tin mercaptide Octylzinnmercaptid n
  octyl tin stabilizer Octylzinnstabilisator m
  tetra-alkyl tin compound Tetraalkylzinnverbindung f
  tetra-aryl tin compound Tetraarylzinnverbindung f
  trialkyl tin compound Trialkylzinnverbindung f
tip Spitze f
  flattened screw tip Schneckenballen m
  flighted screw tip Schneckenförderspitze f, Förderspitze f
  screw tip Schneckenende n, Schneckenspitze f
  sensor tip Fühlerspitze f
  torpedo tip Torpedobug m, Torpedospitze f
  unflighted screw tip glatte Schneckenspitze f
titanate Titanat n
  butyl titanate Butyltitanat n
  polybutyl titanate Polybutyltitanat n
  tetrabutyl titanate Tetrabutyltitanat n
titanium Titan n
  titanium catalyst Titankatalysator m
  titanium dioxide Titandioxid n
  titanium white Titanweiß n
  nickel titanium yellow Nickeltitangelb n
titrate/to titrieren
titration Titration f
  direct titration Direkttitration f
thermomechanical analysis [TMA] thermomechanische Analyse f
toggle Kniehebel m
  toggle clamp machine Kniehebelmaschine f
  toggle clamp mechanism Kniehebelverschluß m
  toggle clamp system Kniehebelschließsystem n

toggle gear Kniehebelgetriebe n
toggle lever Einfachkniehebel m
toggle lock mechanism Kniehebelverriegelung f
toggle mo(u)ld clamping unit Kniehebel-Formschließaggregat n
toggle press Kniehebelpresse f
toggle stroke Kniehebelhub m
double toggle Doppelkniehebel m
double toggle clamp unit Doppelkniehebel-Schließeinheit f
double toggle clamping system Doppelkniehebel-Schließsystem n
double toggle machine Doppelkniehebelmaschine f
double toggle system Doppelkniehebelsystem n
five-point double toggle Fünfpunkt-Doppelkniehebel m
five-point toggle Fünfpunkt-Kniehebel m
four-point toggle Vierpunktkniehebel m
tolerance Toleranz f
  tolerance details Toleranzangaben fpl
  tolerance limit Toleranzgrenze f
  tolerance limits Toleranzbänder npl
  tolerance range Toleranzfeld n
  tolerance requirements Toleranzanforderungen fpl
  all-round stabilizing tolerance universelle Stabilisierbarkeit f
  dimensional tolerance Formattoleranz f
  ga(u)ge tolerance [of film] Dickentoleranz f
  machining tolerance Bearbeitungstoleranz f
  mo(u)ld tolerance Werkzeugtoleranz f
  mo(u)lded-part tolerance Formteiltoleranz f
  mo(u)lding tolerance Fertigungstoleranz f
  overpainting tolerance Überstreichbarkeit f
  overspray tolerance Überspritzbarkeit f
  part tolerance Formteiltoleranz f
  production tolerance Fertigungstoleranz f
  temperature tolerance range Temperaturtoleranzfeld n
  thickness tolerance Dickentoleranz f
toluene Toluol n
  toluene sulphonic acid Toluolsulfonsäure f
  vinyl toluene Vinyltoluol n
tool Werkzeug n
  tool box Werkzeugkasten m
  tool handle Werkzeuggriff m
  tool steel Werkzeugstahl m
  carbide tipped tool Hartmetallwerkzeug n
  cold press mo(u)lding tool Kaltpreßwerkzeug n
  cutting tool Schnittwerkzeug n
  drop hammer tool Schlaghammerwerkzeug n
  female tool Negativform f, Negativwerkzeug n
  foam mo(u)lding tool Schäumform f
  injection mo(u)lding tool Spritz(gieß)form f, Spritz(gieß)werkzeug n
  machine tool Werkzeugmaschine f
  planing tool Schälwerkzeug n
  power tool Elektrowerkzeug n

punching tool Stanzwerkzeug n
set of tools Werkzeugsatz m
stretch forming tool Streckziehwerkzeug n
turning tool Drehstahl m, Drehmeißel m
vacuum forming tool Vakuumtiefziehform f
welding tool Schweißwerkzeug n
tooling Werkzeug-
   tooling costs Werkzeug(herstellungs)kosten pl
   tooling resin Werkzeugharz n
   hot runner tooling Heißkanal-Formenbau m
   initial tooling costs Erstwerkzeugkosten pl
toolmaker Werkzeugmacher m, Werkzeugbauer m, Formenbauer m
   toolmaker blue Tuschierfarbe f
toolmaking department Werkzeugbau m [Abteilung]
tooth Zahn m
toothed disk mill Zahnscheibenmühle f
top Spitze f, oberstes Ende n
   top coat Schluß(an)strich m, Deckschicht f
   top coat(ing) Deckanstrich m, Decklackierung f, Deckstrich m
   top (mo(u)ld) force Oberwerkzeug n
   top mo(u)ld half Werkzeugoberteil n
   top part Oberteil n
   top part of the machine Maschinenoberteil n
   top roll Oberwalze f
torpedo Torpedo(körper) m, Verdrängerkörper m, Verdrängertorpedo m, Schmierkopf m, Scherkopf m, Schertorpedo m, Torpedoscherteil n
   torpedo support Torpedohalterung f
   torpedo tip Torpedobug m, Torpedospitze f
   grooved torpedo screw Nutentorpedo(schnecke) m(f)
   thermally conductive torpedo Wärmeleittorpedo m
   wire covering torpedo Drahtummantelungspinole f
torque Drehmoment n
   torque adjusting mechanism Drehmomenteinstellung f
   torque adjustment Drehmomenteinstellung f
   torque amplifier Drehmomentverstärker m
   torque dividing drive Drehmomentverteilergetriebe n
   torque increase Drehmomentanstieg m
   torque meter Drehmomentwaage f
   torque motor Torquemotor m
   torque rheometer Drehmoment-Rheometer m
   torque sensor Drehmomentaufnehmer m, Drehmomentsensor m
   torque transducer Drehmomentaufnehmer m, Drehmomentsensor m
   drive torque Antriebsdrehmoment n
   insufficient torque Drehmoment-Defizit n
   maximum torque Drehmomentgrenze f
   nominal torque Nenndrehmoment n
   operating torque Betriebsdrehmoment n
   screw torque Schneckendrehmoment n
   tightening torque Anzugsdrehmoment n
   total screw torque Schneckengesamtdrehmoment n
   with constant torque drehmomentkonstant

torsion Verdrehung f, Verdrillung f, Torsion f
   torsion bar Drehstab m
   torsion modulus G-Modul m, Torsionsmodul m
   torsion pendulum Torsionspendel n, Torsionsschwingungsgerät n
   torsion pendulum test Torsionsschwingversuch m
   torsion resistant verwindungssteif
   alternating torsion Wechseltorsion f
   angle of torsion Torsionswinkel m
   stiffness in torsion Torsionssteifheit f
torsional Torsions-
   torsional moment Torsionsmoment n
   torsional oscillation Torsionsschwingung f
   torsional rigidity Torsionssteifheit f
   torsional vibration Torsionsschwingung f
   long-term torsional bending stress Dauertorsionsbiegebeanspruchung f
   resistance to torsional stress Torsionsbelastbarkeit f
total total, gesamt, ganz, Gesamt-
   total amount of plasticizer Gesamtweichmachermenge f
   total connected load installierte Gesamtleistung f
   total consumption Gesamtverbrauch m
   total cost Gesamtkosten pl
   total cycle time Gesamtzykluszeit f, Zyklusgesamtzeit f
   total deformation Gesamtdeformation f, Gesamtverformung f
   total drive power Gesamtantriebsleistung f
   total electricity cost Gesamtstromkosten pl
   total elongation Gesamtdehnung f
   total energy input Gesamtenergieaufnahme f
   total energy used Gesamtenergieaufnahme f
   total film thickness Gesamtschichtdicke f
   total heating capacity Gesamtheizleistung f
   total installed load installierte elektrische Gesamtleistung f, Gesamtanschlußwert m, gesamtelektrischer Anschlußwert m
   total installed motor power gesamte installierte Motorenleistung f
   total investments Gesamtinvestitionen fpl
   total lead content Gesamtbleigehalt m
   total length Gesamtlänge f
   total nip pressure Gesamtspaltlast f
   total number Gesamtzahl f
   total number of employees Gesamtbeschäftigtenzahl f, Gesamtbelegschaft f
   total output Gesamtleistung f, Gesamtproduktionsleistung f
   total pigment content Gesamtpigmentierung f
   total power Gesamtleistung f
   total power consumption Gesamtleistungsbedarf m
   total production Gesamtproduktion f
   total screw torque Schneckengesamtdrehmoment n
   total shrinkage Gesamtschrumpfung f
   total switching time Gesamtschaltzeit f
   balance sheet total Bilanzsumme f

totally enclosed (machine) geschlossene Bauart f, Kompaktbauweise f
touch Berührung f, Tasten n
  touch of a button Tastendruck m
  touch-sensitive keyboard Folientastatur f
tough zäh
  tough and resilient zähelastisch
  tough fracture Zähbruch m
  hard and tough zähhart
toughened schlagzähmodifiziert, schlagzäh, schlagfest
toughening agent Schlagfestmacher m, Schlagzähmodifizierharz n, Schlagzähmodifikator m, Schlagzähmodifier m, Schlagzähkomponente f, Schlagzähmodifiziermittel n, Schlagzähigkeitsverbesserer m, Schlagzähmacher m
toughness Zähigkeit f
  fracture toughness Bruchzähigkeit f
tower Turm m
  cooling tower Kühlturm m
  take-off tower Abzugsturm m
toxic toxisch, giftig
toxicity Giftigkeit f, Toxizität f
toxicological toxikologisch
  toxicological data toxikologische Daten npl
  toxicological properties toxikologische Eigenschaften fpl
toxicology Toxikologie f
toys Spielwaren fpl, Spielzeugartikel mpl
TPP [triphenyl phosphate] Triphenylphosphat n
trace analysis Spurenanalyse f
traces of moisture Feuchtigkeitsspuren fpl
track Spur f
tracking Kriechwegbildung f, Kriechspurenbildung f
  tracking resistance Kriechstromfestigkeit f
  tracking resistance test Kriechstromprüfung f
  tracking resistant kriechstromfest
  long-term tracking resistance Kriechstromzeitbeständigkeit f
trade Handel m
  trade mark Warenzeichen n
  trade name Handelsname m, Markenname m
  foreign trade Außenhandel m
trading partner Handelspartner m
trailing edge of flight passive Gewindeflanke f, hintere Flanke f, hintere Schneckenflanke f, passive Flanke f, hintere Stegflanke f, passive Schneckenflanke f, passive Stegflanke f
train Bahn f, Zug m
  gear train Getriebezug m
trainee Anlernling m, Auszubildender m, Azubi m
transducer Fühler m, Geber m, Sensor m, Sonde f, Aufnehmer m
  transducer bridge Aufnehmer-Meßbrücke f
  transducer cable Aufnehmerkabel n
  transducer element Aufnehmerelement n
  transducer housing Aufnehmergehäuse n
  transducer sensitivity Aufnehmerempfindlichkeit f
  transducer signal Aufnehmersignal n
  acceleration transducer Beschleunigungsaufnehmer m
  actual value transducer Istwertgeber m
  cavity pressure transducer Werkzuginnendruckaufnehmer m, Werkzeuginnendruck-Meßwertaufnehmer m
  clamping force transducer Schließkraftmeßplatte f
  differential pressure transducer Differenzdruckumformer m
  flow rate transducer Strömungsgeschwindigkeitsaufnehmer m
  force transducer Kraftaufnehmer m, Kraftmeßdose f, Kraftmeßplatte f
  mark left by the pressure transducer Druckaufnehmerabdruck m
  position transducer Wegaufnehmer m, Weg(meß)geber m
  pressure transducer Druckaufnehmer m, Druckfühler m, Drucksensor m, Drucksonde f, Druckgeber m, Druck(meß)dose f, Druck(meß)umformer m
  screw stroke transducer Schneckenwegaufnehmer m
  signal transducer Signalumwandler m
  speed transducer Geschwindigkeitsfühler m, Geschwindigkeitsaufnehmer m
  strain transducer Dehnungsaufnehmer m
  stroke transducer Wegaufnehmer m, Weg(meß)geber m
  tensile force transducer Zugdose f, Zugkraftaufnehmer m
transfer/to übertragen
transfer Übertragung f, Überweisung f, Transfer m
  transfer chamber Füllraum m
  transfer mo(u)ld Spritzpreßform f
  transfer mo(u)lded spritzgepreßt
  transfer mo(u)lding Spritzpressen n
  transfer mo(u)lding plant Spritzpreßanlage f
  transfer mo(u)lding press Spritzpresse f
  transfer plunger Spritzkolben m, Einspritzkolben m
  transfer to silos Silieren n
  automatic downstroke transfer mo(u)lding press Oberkolbenspritzpreßautomat m
  automatic transfer mo(u)lding press Spritzpreßautomat m
  data transfer Datentransfer m, Datenübertragung f
  energy transfer Energieübertragung f
  heat transfer Wärmeübergang m, Wärmeübertragung f
  heat transfer coefficient Wärmeübergangszahl f, Wärmedurchgangskoeffizient m
  heat transfer data Wärmeübertragungskennwerte mpl
  heat transfer fluid Wärmeträgerflüssigkeit f
  heat transfer medium Wärmeträgermittel n, Wärmeübertragungsmittel n, Wärmeübertragungmedium n
  heat transfer oil Wärmeträgeröl n

heat transfer resistance Wärmedurchgangswiderstand m
multi-daylight transfer mo(u)lding Etagenspritzen n
pressure transfer Druckübertragung f
semi-automatic transfer mo(u)lding press Spritzpreß-Halbautomat m
transformation Umwandlung f, Transformation f
heat of transformation Umwandlungswärme f
transformer Umwandler m, Transformator m, Trafo m
transformer construction Trafobau m
transformer oil Trafoöl n
current transformer Stromwandler m
power transformer Leistungstransformator m
voltage transformer Spannungswandler m
transistor Transistor m
input transistor Eingangstransistor m
transistorized transistorisiert
transit Durchgang m, Durchlauf m
transit container Transportbehälter m
transition Übergang m
transition point Umwandlungspunkt m
transition range Umwandlungsbereich m
transition section Aufschmelzbereich m, Plastifizierzone f, Homogenisierzone f, Übergangszone f, Umwandlungszone f, Kompressionsbereich m, Aufschmelzzone f, Plastifizierbereich m, Übergangsbereich m
transition section [of screw] Schmelzzone f
transition zone [of a curve] Übergangszone f
brittle-tough transition Spröd-Zäh-Übergang m
glass transition [i.e. the transition of a polymer from the glassy to the viscoelastic state] Glasübergang m, Glasumwandlung f
glass transition range Einfrierbereich m
glass transition temperature Einfrier(ungs)temperatur f, Glas(umwandlungs)punkt m, Glasübergangstemperatur f, Glasumwandlungstemperatur f, Tg-Wert m
glass transition temperature range Glasumwandlungstemperaturbereich m
glass transition zone Glasübergangsgebiet n, Glasübergangsbereich m, Glasumwandlungsbereich m
rubber-glass transition Zäh-Spröd-Übergang m
transitional period Übergangszeit f
translation Übersetzung f
translucency Transluzenz f
translucent durchscheinend, lichtdurchlässig, transluzent, transparent
transmission Transmission f, Übertragung f, Übersetzung f, Getriebe n
transmission heat loss Transmissionswärmeverlust m
transmission module Übertragungsmodul m
transmission of force Kraftübertragung f
transmission speed Übertragungsrate f, Übertragungsgeschwindigkeit f
gear transmission Getriebe n
light transmission Lichtdurchgang m, Lichtdurchlässigkeit f
percentage light transmission Lichtdurchlässigkeitszahl f
transmitter Sender m, Meßwandler m
transparency Durchsichtigkeit f, Transparenz f
transparent durchsichtig, klarsichtig, transparent
transparent cover Klarsichtdeckel m
transparent film Klarsichtfolie f
transparent varnish Transparentlack m
transport/to fördern, transportieren
transport Transport m
transport cost(s) Transportkosten pl
transport direction Förderrichtung f
transport industry Transportwesen n
transport problems Transportprobleme npl
transport regulations Beförderungsvorschriften fpl
transport screw kompressionslose Förderschnecke f
film transport Folientransport m
material transport blower Fördergebläse n
melt transport Schmelzeförderung f
rail transport Eisenbahntransport m
road transport Straßentransport m
uneven transport Förder-Ungleichmäßigkeiten fpl
transportable transportabel
transported transportiert, gefördert, befördert
material to be transported Fördergut n
transverse transvers, quer (verlaufend)
transverse flow Transversalströmung f
transverse stretching machine Breitreckmaschine f
transverse stretching roll Breitstreckenwalze f
transverse stretching unit Breitstreckwerk n
trapezoidal trapezförmig
trapezoidal spindle Trapezspindel f
travel Lauf m, Bewegung f; Reise f
air travel Luftfahrt f
parallel travel Parallelführung f, Parallellauf m
tray Schale f, Mulde f, Trog m
drip tray Auffangwanne f
slide-in tray Einschiebehorde f
treat/to behandeln
treated behandelt
surface treated oberflächenbehandelt
treatment Behandlung f
careful treatment of the mo(u)ld Werkzeugschonung f
corona treatment Coronabehandlung f
flame treatment Flammstrahlen n, Abflammen n, Beflammen n
flame treatment station Beflammstation f
gentle treatment Schonbehandlung f, vorsichtiges Behandeln n
heat treatment Temperaturbehandlung f, Warmbehandlung f, Wärmebehandlung f
steam treatment Bedampfen n
surface treatment Oberflächenbehandlung f, Oberflächenveredelung f
treeing Bäumchenbildung f

water treeing [an electrical phenomenon] Bäumchenbildung f, Wasserbäumchenbildung f
trees Bäume mpl
  water trees [an electrical phenomenon] Wasserbäumchen npl
trend Tendenz f, Trend m
  development trend Entwicklungstendenz f
  upward trend Aufwärtstrend m
triacetate Triacetat n
  cellulose triacetate Cellulosetriacetat n
trial Erprobung f, Versuch m
  trial run Nullserie f, Probelauf m
  coating trials Beschichtungsversuche mpl
  coextrusion blow mo(u)lding trials Koextrusionsblasversuche mpl
  extrusion trial Extrusionsversuch m
  field trial Feldversuch m
  large-scale trial Großversuch m
  mo(u)lding trials Musterabspritzungen fpl, Spritzversuche mpl
  painting trials Beschichtungsversuche fpl
  pilot plant trial Technikumversuch m
  pre-production trial Nullserie f
  production-scale trial Großversuch m
  production trial Fertigungsversuch m
trialkyl phosphate Trialkylphosphat n
  trialkyl tin compound Trialkylzinnverbindung f
triallyl cyanurate Triallylcyanurat n
triaryl phosphate Triarylphosphat n
triazole Triazol n
tribasic dreibasisch, tribasisch
tribo-chemical tribochemisch
tribo-rheological triborheologisch
tribological tribologisch
tribology Tribologie f
tricarboxylic acid Tricarbonsäure f
trichloroethylene Tri n, Trichlorethylen n
tricresyl phosphate [TCP] Trikresylphosphat n
triethylene tetramine Triethylentetramin n
trifluoride Trifluorid n
trifluorochloroethylene Trifluorchlorethylen n
trifunctional trifunktionell
triglycidyl isocyanurate Triglycidylisocyanurat n
trihydric [if an alcohol] dreiwertig
trim Randbeschnitt m; Verzierung f, Deckleiste f
  door trim Türinnenverkleidung f
  edge trim Folienrandstreifen m
  edge trim reclaiming Folienrandstreifenaufbereitung f
  edge trim wind-up (unit) Abfallaufwicklung f
trimellitate Trimellitat n, Trimellitsäureester m
  trimellitic acid Trimellitsäure f
  trimellitic acid ester Trimellitsäureester m
trimer Trimer n, Trimerisat n
trimeric trimer
trimerization Trimerisation f
trimerized trimerisiert
  trimerization reaction Trimerisierungsvorgang m

trimethacrylate Trimethacrylat n
trimethyl hexamethylenediamine Trimethylhexamethylendiamin n
trimethylol propane Trimethylolpropan n
  trimethylol propane trimethacrylate Trimethylolpropantrimethacrylat n
trimmer Trimmer m, Beschneidemaschine f
  edge trimmer Kantenbeschneider m, Seitenschneider m
  flash trimmer Butzenabtrennvorrichtung f, Butzentrenner m
trimming 1. Beschneider m, Entgraten n; 2. Trimm-
  trimming and take-off station Trimm- und Entnahmestation f
  trimming unit Trimmstation f
  edge trimming device Seitenschneidvorrichtung f
  edge trimming (unit) Kantenbeschnitt m
  flash trimming (mechanism) Butzenabtrennung f
  width of film after trimming Folienfertigbreite f
trioctyl phosphate Trioctylphosphat n
triol Triol n
triorgano-tin compound Triorganozinnverbindung f
trioxide Trioxid n
  antimony trioxide Antimontrioxid n
  molybdenum trioxide Molybdäntrioxid n
triphenyl phosphate [TPP] Triphenylphosphat n
triphenyl phosphite Triphenylphosphit n
triphenylmethyl peroxide Triphenylmethylperoxid n
triphenylmethyl radical Triphenylmethylradikal n
triple dreifach
  triple-die (extruder) head Dreifachkopf m
  triple-flighted [screw] dreigängig
  triple-manifold die Dreikanalwerkzeug n
  triple roll mill Dreiwalze f, Dreiwalzenstuhl m, Dreiwalzenmaschine f
  triple roller Dreiwalze f, Dreiwalzenstuhl m, Dreiwalzenmaschine f
  triple runner Dreifachverteilerkanal m
  triple screw Dreifachschnecke f
  triple screw extruder Dreischneckenextruder m
trivalent dreibindig, dreiwertig
trixylenyl phosphate [TXP] Trixylenylphosphat n
tropical tropisch
  tropical conditions Tropen(klima)bedingungen fpl, Tropenklima n
  resistant to tropical conditions tropenfest, tropenbeständig
trouble Störung f
  trouble-free störungsfrei, störungssicher
  trouble-shooting Störungssuche f
  trouble-shooting chart Fehlersuchliste f
  almost trouble-free störungsarm
  liable to give trouble störanfällig, störempfindlich
trough Trog m
  cooling trough Kühlwanne f, Kühltrog m
trouser tear test Schenkel-Weiterreißversuch m
trowel Ziehklinge f, Spachtel m

true Leno weave Dreherbindung f [Textil]
tub Bottich m, Kübel m
  bath tub Badewanne f
tube Rohr n; Tube f; Schlauch m
  molten tube Schmelzeschlauch m
tubular film Blasfolie f, Schlauchfolie f
tumble/to taumeln
  tumble drier Taumeltrockner m
  tumble mixer Taumelmischer m
tumbler switch Kippschalter m
tumbling movement Taumelbewegung f
tunnel Tunnel m, (begehbarer) Kanal m
  tunnel gate Abscheranschnitt m, Tunnelanschnitt m, Tunnelanguß m
  tunnel gate with pin-point feed Tunnelanguß m mit Punktanschnitt, Tunnelpunktanguß m
  tunnel oven Durchlaufofen m
  cooling tunnel Kühltunnel m
  drying tunnel Trockenkanal m, Tunnelofen m
  fusion tunnel [for PVC coated fabrics] Tunnelofen m, Gelierkanal m
  gelling tunnel Gelierkanal m
  heating tunnel Heizkanal m, Temperierkanal m
  heating tunnel layout Temperierkanalauslegung f
  hot air tunnel Heißluft(düsen)kanal m
  infra-red heating tunnel Infrarottunnel m
  pre-gelling tunnel Vorgelierkanal m
  vulcanizing tunnel Vulkanisationstunnel m
turbid trübe
turbidity Trübung f
turbo Turbo-
  turbo-compressor Turboverdichter m
  turbo-drier Turbotrockner m
  turbo-vacuum pump Turbovakuumpumpe f
turbulence Turbulenz f
turbulent turbulent, Wirbel-
  turbulent flow turbulente Strömung f
turn/to drehen, wenden
  screw turn Schneckenumgang m
turning 1. Drehen n; 2. Dreh-
  turning movement Schwenkbewegung f
  turning off Ausschaltung f
  turning tool Drehstahl m, Drehmeißel m
turnkey [meaning that a plant is handed over to the client ready for operation] schlüsselfertig
turnover Umsatz m, Umschlag m
  turnover increase Umsatzausweitung f, Umsatzplus n, Umsatzzuwachs m
  drop in turnover Umsatzeinbruch m, Umsatzeinbuße f, Umsatzrückgang m
  foreign turnover Auslandsumsatz m
  percentage turnover prozentualer Umsatz m
  volume turnover Umsatzvolumen n
  world turnover Weltumsatz m
twelve-cavity injection mo(u)ld Zwölffach-Spritzwerkzeug n
twill Twill m, Köper m

twill weave Köperbindung f
  cross twill weave Kreuzköperbindung f
twin 1. Zwilling m; 2. Doppel-, Zwillings-, Tandem-
  twin barrel Doppelzylinder m
  twin-belt take-off (unit/system) Doppelbandabzug m
  twin compounding unit Aufbereitungs-Doppelanlage f
  twin die (extruder) head Doppelwerkzeug n, Zweifach(extrusions)kopf m, Doppel(spritz)kopf m, Zweifachwerkzeug n
  twin-die film blowing head Doppelblaskopf m
  twin-die film blowing line Doppelkopf-Blasfolienanlage f
  twin drive Doppellaufwerk n
  twin embossing unit Doppelprägewerk n
  twin feed screw Doppel-Dosierschnecke f, Doppel-Einlaufschnecke f
  twin floppy disk drive Doppel-Floppy-Disk(etten)-Laufwerk n
  twin manifold Doppelverteilerkanal m
  twin-manifold die Zweiverteilerwerkzeug n
  twin mixing head Doppelmischkopf m
  twin-orifice die Doppelstrangwerkzeug n
  twin parison die Zweifachschlauchkopf m, Zweifachblaskopf m, Doppelschlauchkopf m
  twin-roll polishing stack Zweiwalzenglättwerk n
  twin reprocessing unit Aufbereitungs-Doppelanlage f
  twin-screen pack Doppelsiebkopf m
  twin-screw Doppelschnecke f, Zwei(wellen)schnecke f, zweiwellige Schnecke f
  twin-screw assembly Schneckenpaar n
  twin-screw barrel Doppelschneckenzylinder m
  twin-screw compounder Doppelschneckencompounder m
  twin-screw compounding extruder Zweiwellen-Knetscheiben-Schneckenpresse f
  twin-screw design Doppelschneckenausführung f
  twin-screw extruder Doppelschneckenextruder m, Zweischneckenextruder m, Zweiwellenextruder m, zweiwellige Schneckenmaschine f, Zweischneckenmaschine f, Doppelschneckenmaschine f, Zweiwellenmaschine f, Doppelschneckenpresse f
  twin-screw extrusion Doppelschneckenextrusion f
  twin-screw feed section Zweischneckeneinzugszone f
  twin-screw metering section Zweischneckenaustragszone f
  twin-screw pelletizer Zweiwellen-Granuliermaschine f
  twin-screw pelletizing line Doppelschnecken-Granulieranlage f
  twin-screw plasticator Doppelschneckenkneter m
  twin screw principle Doppelschneckenprinzip n
  twin screw system Zweiwellensystem n, Zweischneckensystem n
  twin screws Schneckenpaar n
  twin spider-type mandrel support Doppelstegdornhalter m, Doppel-

stegdornhalterung f
twin-station wind-up (unit) Zweistellen-Aufwicklung f
twin vacuum hopper Doppelvakuumtrichter m, Vakuum-Doppeltrichter m
twin vacuum hopper assembly Vakuum-Doppeltrichteranlage f
twin wind-up (unit) Doppelaufwickler m
twin winder Doppelaufwickler m, Doppelwickler m
  auxiliary twin screw extruder Doppelschnecken-Seitenextruder m
  co-rotating twin screw Gleichdrallschnecke f, Gleichdrall-Doppelschnecke f
  co-rotating twin-screw compounder Gleichdralldoppelschneckenkneter m
  co-rotating twin screw extruder Gleichdrallmaschine f, Gleichdrall-Schneckenextruder m
  co-rotating twin screw system Gleichdrallsystem n
  co-rotating twin screws Gleichdrallsystem n
  counter-rotating twin screw Gegendrall-Doppelschnecke f, Gegendrallschnecke f
  counter-rotating twin screw extruder Gegendrallmaschine f, Gegendrall-Schneckenextruder m
  counter-rotating twin screw system Gegendrallsystem n
  feed twin screws Zuführschneckenpaar n
  high-speed twin screw extruder Hochleistungs-doppelschneckenextruder m
  laboratory twin screw extruder Doppelschnecken-Laborextruder m
  laboratory twin-screw mixer Labor-Zweiwellenkneter m
  two-stage twin screw extruder Zweistufen-Doppelschneckenextruder m
  vented twin screw extruder Doppelschnecken-Entgasungsextruder m
  vented twin screws Entgasungsschneckenpaar n
  vertical twin screw extruder Vertikaldoppelschneckenextruder m
twist drill Spiralbohrer m
twisting Torsion f, Verdrallung f, Verwindung f
  without twisting verdrillungsfrei
two zwei
  two-cavity injection blow mo(u)ld Zweifachspritzblaswerkzeug n, Zweifachspritzgießwerkzeug n
  two-cavity mo(u)ld Zweifachwerkzeug n, Doppelwerkzeug n, Zweikavitätenwerkzeug n, Zweifachform f
  two-compartment hopper Zweikammertrichter m
  two-component zweikomponentig
  two-component injection mo(u)lding Zweikomponenten-Spritzgießen n
  two-component low-pressure injection mo(u)lding machine Zweikomponenten-Niederdruck-Spritzgießmaschine f
  two-component process Zweikomponentenverfahren n
  two-component spraygun Zweikomponentenspritzpistole f
  two-dimensional zweidimensional
  two-directional zweidirektional, bidirektional
  two-layer doppellagig
  two-layer blow mo(u)lding Zweischichthohlkörper m
  two-layer blown film die Zweischichtfolienblaskopf m
  two-layer coextrusion Zweischicht-Koextrusion f
  two-layer coextrusion die Zweischichtdüse f
  two-layer composite Zweischichtverbund m
  two-layer film blowing line Zweischicht-Folienblasanlage f
  two-layer sheet extrusion Zweischicht-Tafelherstellung f
  two-layer slit die Zweischicht-Breitschlitzwerkzeug n
  two-layer thermoforming film extrusion line Zweischicht-Tiefziehfolienanlage f
  two-pack adhesive Zweikomponentenklebstoff m
  two-pack compound Zweikomponentenmasse f
  two-pack liquid silicone rubber Zweikomponentenflüssigsilikonkautschuk m
  two-pack paint Zweikomponentenanstrich(mittel) m(n)
  two-pack system Zweikomponentensystem n
  two-part zweiteilig
  two-part casting compound Zweikomponenten-Gießharzmischung f
  two-part mo(u)ld Zweiplattenwerkzeug n, Doppelplattenwerkzeug n, Einetagenwerkzeug n
  two-part sliding split mo(u)ld Doppelschieberwerkzeug n
  two-phase zweiphasig
  two-phase morphology Zweiphasenmorphologie f
  two-phase structure Zweiphasenstruktur f
  two-phase system Zweiphasensystem n
  two-plate mo(u)ld Doppelplattenwerkzeug n, Zweiplattenwerkzeug n, Einetagenwerkzeug n
  two-point controller Zweipunktregler m
  two-point gating Zweifachanspritzung f
  two-roll calender Zweiwalzenkalander m
  two-roll mill Doppelwalze f, Zweiwalze f, Zweiwalzenmaschine f
  two-roll vertical calender Zweiwalzen-I-Kalander m
  two-section screw Zweizonenschnecke f
  two-sided beidseitig
  two-speed zweitourig
  two-stage zweistufig
  two-stage blow mo(u)lding (process) Zweistufenblasverfahren n
  two-stage ejector Zweistufenauswerfer m
  two-stage extruder Zweistufenextruder m
  two-stage extrusion blow mo(u)lding Zweistufen-Extrusionsblasformen n, Zweistufenspritzblasen n
  two-stage injection stretch blow mo(u)lding (process) Zweistufen-Spritzstreckverfahren n
  two-stage injection unit Zweistufen-Einspritzaggregat n

two-stage process Zweistufenverfahren n
two-stage stretching (process) [for film] Zweistufenreckprozeß m
two-stage twin screw extruder Zweistufen-Doppelschneckenextruder m
two-start doppelgängig, zweigängig
two-station blow mo(u)lding machine Zweistationenblasmaschine f
two-station machine Zweistationenmaschine f
two-tie bar clamp unit Zweiholmenschließeinheit f
two-tie bar design Zweiholmenausführung f
two-way flow control unit Zweiwegestromregelung f
two-way valve Zweiwegventil n
automatic two-colo(u)r injection mo(u)lding machine Zweifarben-Spritzgußautomat m
hot runner two-cavity mo(u)ld Heißkanal-Doppelwerkzeug n
laboratory two-roll mill Zweiwalzen-Labormischer m
with two tie bars zweiholmig
TXP [trixylenyl phosphate] Trixylenylphosphat n
type Ausführung f, Typ m, Type f
    type of accumulator Speicherbauart f
    type of blowing agent Treibmittelart f
    type of construction Konstruktionsform f
    type of curing agent Vernetzerart f
    type of die Düsenform f
    type of design Konstruktionsform f
    type of emulsifier Emulgatorart f
    type of extruder Extrudertyp m
    type of filler Füllstoffart f, Füllstoffsorte f
    type of gate Angußart f
    type of heating Heizungsart f
    types of hot runner Heißkanalbauarten fpl
    type of loading Belastungsart f
    type of nozzle Düsenform f, Düsenbauart f, Düsenarten fpl
    type of operation [e.g. manual or automatic] Betriebsart f
    type of pigment Pigmentierungsart f
    type of plasticizer Weichmacherart f, Weichmachertyp m
    type of screw Schneckenart f, Schneckenbauart f
    type of screw drive Schneckenantriebsart f
    type of solvent Lösungsmitteltyp m
    type of stress Beanspruchungsart f
    type of take-off Abzugstyp m
    type of wind-up Wickler-Typ m
    ante-chamber type pin gate Vorkammer f mit Punktanguß
    heart-shaped groove type of feed system Herzkurveneinspeisung s
    milling type cutter Fräsrotor m
    printed in bold type fettgedruckt
typical typisch, kennzeichnend, charakteristisch
    typical formulation Formulierungsbeispiel n
    typical of epoxy resins epoxidtypisch
typification Typisierung f
typified typisiert
UHF [ultra-high frequency] Ultrahochfrequenz f, UHF
ultimate letzt, endlich, Höchst-, Grenz
    ultimate tensile strength Zerreißfestigkeit f, Reißfestigkeit f
    ultimate tensile stress Reißkraft f, Bruchspannung f
    residual ultimate strength Restreißkraft f
ultra ultra, Ultra-
    ultra-high frequency [UHF] Ultrahochfrequenz f, UHF f
    ultra-high molecular weight ultrahochmolekular
    ultra-high pressure plasticization Höchstdruckplastifizierung f
    ultra-high vacuum Ultrahochvakuum n
    ultra-thin (microtome) section Ultradünnschnitt m
    ultra-thin section microtomy Ultradünnschnittmikrotomie f
    ultra-violet [UV] ultraviolett, UV-
ultracentrifuge Ultrazentrifuge f
ultrasonic Ultraschall-
    ultrasonic energy Ultraschallenergie f
    ultrasonic flowmeter Ultraschalldurchflußmeßgerät n
    ultrasonic generator Ultraschall-Generator m
    ultrasonic range Ultraschallbereich m
    ultrasonic spot welding Ultraschallpunktschweißen n
    ultrasonic spot welding instrument Ultraschall-Punktschweißgerät n
    ultrasonic vibrations Ultraschallschwingungen fpl
    ultrasonic waves Ultraschallwellen fpl
    ultrasonic welding Ultraschallschweißen n, Ultraschallfügen n
    ultrasonic welding machine Ultraschallschweißmaschine f, Ultraschallschweißanlage f
    ultrasound Ultraschall m
un-crosslinked unvernetzt
un-cured nichtvernetzt
unaffected (by) unempfindlich, beständig
    unaffected by disturbing influences störgrößenunabhängig
    unaffected by heat hitzeunempfindlich, wärmeunempfindlich
    unaffected by moisture feuchtigkeitsunempfindlich, feuchteunempfindlich
    unaffected by water wasserunempfindlich
unaged ungealtert
unblended unverschnitten
unbranched unverzweigt
uncatalyzed unkatalysiert
uncharged ungeladen
uncoated nichtbeschichtet, unbeschichtet
uncoated [e.g. filler particles] unbehandelt
uncondensed unkondensiert
unconditioned ungetempert
uncontrollable unkontrollierbar
uncouple/to abkuppeln
uncured ungehärtet, unvulkanisiert
undecomposed [e.g. peroxide] unzersetzt

under unter
  under bending stress biegebeansprucht, biegebelastet
  under compressive stress druckbelastet, druckbeansprucht
  under impact stress schlagbeansprucht
  under load unter Last f
  under pressure druckbeansprucht, druckbelastet
  under tensile stress zugbeansprucht, zugbelastet
underbody Unterboden m
  underbody sealant Unterbodenschutzmasse f, Unterbodenschutz m
  underbody sealant paste Unterbodenschutzpaste f
undercoat Zwischenanstrich m
undercooling Unterkühlung f
undercured unterhärtet
undercuring Unterhärtung f, Untervernetzung f
undercut hinterschnitten, Hinterschneidung f, Unterschneidung f
underfeed/to unterfüttern
underfeeding Unterdosierung f
underfloor (central) heating Fußbodenheizung f
underground [e.g. pipes] 1. erdverlegt, unterirdisch; 2. Untergrund m
  for underground installation zur Erdverlegung f
underwater unter Wasser n, Unterwasser-
  underwater paint Unterwasseranstrich m
  underwater pelletizer Unterwassergranulator m, Unterwassergranuliervorrichtung f, Unterwassergranulieranlage f
  underwater pelletizing system Unterwassergranuliersystem n
  underwater strand pelletizer Unterwasserstranggranulator m
undiluted unverdünnt
undried ungetrocknet
uneconomic unwirtschaftlich
unequivocal zweifelsfrei
uneven ungleichmäßig
  uneven temperature thermische Inhomogenitäten fpl
  uneven transport Förder-Ungleichmäßigkeiten fpl
unfilled füllstofffrei, ungefüllt
unflighted screw tip glatte Schneckenspitze f
unfused [PVC paste] ungeliert
ungelled [UP, EP resins] ungeliert
uniaxial einachsig, einaxial, monoaxial
  uniaxial stretching unit Monoaxial-Reckanlage f
uniaxially oriented monoaxial verstreckt
unidirectional unidirektional
  unidirectional prepreg UD-Prepreg n
uniform colo(u)r Farbhomogenität f
uniform delivery [of extrudate] Ausstoßgleichmäßigkeit f
uniformity Konstanz f, Uniformität f
  uniformity of rotation Umdrehungskonstanz f
  product uniformity Produktkonstanz f
unique einmalig
unit Einheit f, Anlage f, Baueinheit f, Baustein m, Baukastenelement n, Aggregat n
unit area Flächeneinheit f
unit construction Elementbauweise f
unit construction system Baukastenbauweise f
unit of time Zeiteinheit f
accumulator head plasticizing unit Staukopfplastifiziergerät n
additional metering unit Zusatz-Dosieraggregat n
air circulation unit Luftumwälzvorrichtung f
air conditioning unit Klimagerät n
air cooling unit Luftkühlaggregat n, Luftkühlgerät n, Luftkühlung f
ancillary unit Nebenaggregat n, Zusatzgerät n
automatic deflashing unit Entgratautomat m
automatic feed unit Vorschubautomat m, Zuführungsautomat m
automatic filter unit Filterautomatik f
automatic foam moulding unit Formteilschäumautomat m
automatic hot gas welding unit Warmgas-Schweißautomat m
automatic metering unit Dosierautomat m
automatic profile welding unit Profilschweißautomat m
back of the unit Geräterückseite f
bagging unit Absackanlage f
barrel cooling unit Zylinderkühlaggregat n
basic unit Grundbaustein m, Grundgerät n
beaming unit Bäumanlage f [Textil]
biaxial stretching unit Biaxial-Reckanlage f
blow moulding unit Blasaggregat n
bottle gripping unit Behältergreifstation f
calibrating unit Kalibrierung f [Gerät]
carousel-type blow moulding unit Karussell-Blasaggregat n
carousel unit Revolvereinheit f
cascade control unit Kaskadenregelung f
cascade temperature control unit Kaskadentemperaturregelung f
casting roll unit Gießwalzeneinheit f
central lubricating unit Zentralschmieranlage f
central processing unit [CPU] Zentraleinheit f
central unit Zentraleinheit f
chill roll unit Chillroll-Walzengruppe f
circulating water temperature control unit Wasserumlauftemperiergerät n
circulation cooling unit Umwälzkühlung f
clamp(ing) unit Schließeinheit f, Schließhälfte f, Werkzeugschließeinheit f, Formschließeinheit f, Formschließaggregat n, Schließseite f
clamping unit ram Schließkolben m
closed-circuit cooling unit Rückkühlaggregat n, Rückkühlwerk n
coating unit Beschichtungseinheit f
colo(u)r printing unit Farbdruckvorrichtung f
colo(u)r visual display unit Farbsichtgerät n
combined take-off and wind-up unit Abzugswerk-

Wickler-Kombination f, Abzugswicklerkombination f
compact unit Kompaktgerät n
compounding screw unit Knetschneckeneinheit f
compounding unit Aufbereitungsmaschine f, Aufbereitungsaggregat n, Plastifizierteil n, Compoundiermaschine f
compressed air calibrator unit Druckluftkalibrierung f
compressed sizing unit Druckluftkalibrierung f
consisting of separate units in offener Bauweise f, in offener Elementbauweise f
constructed from units in Segmentbauweise f
consumer unit Verbraucher m
contact cooling unit Kontaktkühlung f
contact heating unit Kontaktheizung f
continuous control unit Stetigregler m
control unit Regeleinrichung f, Regelinstrument n, Steueraggregat n, Steueranlage f, Steuerblock m, Steuereinheit f, Steuerung f
conversion unit Umbausatz m
conveying unit Förderaggregat n, Transportanlage f
cooling take-off unit Kühlabzugsanlage f
cooling unit Kühlgerät n
cooling water recovery unit Kühlwasserrückgewinnungsanlage f
corona pretreating unit Corona-Vorbehandlungsanlage f
data acquisition unit Datenerfassungsanlage f, Datenerfassungsstation f
data display unit Datensichtgerät n, Datenanzeige f
data processing unit Datenverarbeitungsanlage f
deflashing unit Entgratungsstation f
degassing unit Entgasungsanlage f
destatisizing unit Entstatisierungseinrichtung f
devolatilizing unit Entgasungseinheit f, Entgasungsaggregat n
digital control unit Digitalsteuerung f
digital display unit digitale Anzeige f
double toggle clamp unit Doppelkniehebel-Schließeinheit f
downstream unit Folgeaggregat n, Folgeeinrichtung f, Folgegerät n, Nachfolgeaggregat n, Nachfolgevorrichtung f, Nachfolgeeinheit f, Nachfolge f
drive unit Antriebseinheit f
drive unit design Antriebsauslegung f
drum wind-up unit Trommelwickler m
drying unit Trockengerät n, Trockner m, Trocknungsanlage f, Trocknungsgerät n
dual-circuit unit Zweikreisgerät n
edge trim reclaim unit Folienrandstreifenaufbereitung f
edge trim recycling unit Randstreifenrückführanlage f, Randstreifenrückspeiseeinrichtung f, Randstreifenrückführsystem n
edge trim removal unit Randstreifenabsauganlage f
edge trimming unit Randstreifen-Beschneidstation f, Randstreifen-Schneidvorrichtung f

embossing unit Prägeeinrichtung f, Prägewerk n, Prägevorrichtung f
emergency power supply unit Notstromaggregat n
extra unit Zusatzeinheit f
extrusion welding unit Extrusionsschweißgerät n
feed screw unit Speiseschneckeneinheit f
feed unit Beschickungsanlage f, Beschickung f, Speiseeinrichtung f, Zuführmaschine f, Dosieraggregat n, Dosiereinheit f, Dosierelement n, Dosiergerät n, Dosierwerk n, Speisegerät n, Speisungsanlage f, Vorschubeinheit f, Vorschubgerät n, Zuführvorrichtung f
film blowing unit Blasaggregat n
film ga(u)ge equalizing unit Folienverlegeeinheit f
film inspection unit Folienbeobachtungsstand m
film slitting unit Folienschneidaggregat n
film tape stretching unit Folienbandreckanlage f, Folienbandstreckwerk n
filtration unit Filterapparatur f, Filtergerät n
fine filtration unit Feinfilter-Siebeinrichtung f
flame treating unit Beflammstation f
forced feed unit Zwangsdosiereinrichtung f
four-column clamp unit Vierholmschließeinheit f
gear reduction unit Getriebeuntersetzung f
godet roll stretch unit Galettenstreckwerk n
grading unit [for powders] Klassiereinheit f
granule metering unit Granulatdosiergerät n
grinding unit Mahlaggregat n
gripper unit Greifereinheit f
grit blasting unit Strahlanlage f
Hazen colo(u)r units Hazenfarbzahl f
heating unit Temperiergerät n
high-pressure metering unit Hochdruckdosieranlage f
high-pressure mixing unit Hochdruckmischanlage f
high-pressure plasticizing unit Hochdruckplastifiziereinheit f, Hochdruckplastifizierung f
high-pressure pump unit Hochdruckpumpenaggregat n
high-speed injection unit Hochleistungsspritzeinheit f
hot runner unit Heißkanal(verteiler)block m, Heizblock m, Querverteiler f, Heißkanalverteilerbalken m, Heißkanalelement n
hot runner unit construction Heißkanalblock-Ausführung f
hot runner unit design Heißkanalblock-Ausführung f
hot water heating unit Heißwasser-Heizanlage f
hydraulic injection unit Einspritzhydraulik f
hydraulic unit Hydraulikanlage f
individual units Einzelgeräte npl
injection control unit Einspritzsteuereinheit f
injection unit Einspritzeinheit f, Einspritzseite f, Einspritzaggregat n, Spritzaggregat n, Spritzeinheit f, Spritzseite f
input-output unit Ein-Ausgabeeinheit f
input unit Eingabeeinheit f
internal air pressure calibration unit Überdruckkalibrator m, Überdruckkalibrierung f,

Stützluftkalibrierung f
internal air pressure sizing unit Überdruckkalibrator m, Überdruckkalibrierung f, Stützluftkalibrierung f
laminating unit Doubliereinrichtung f, Laminator m
large-bore pipe calibration unit Großrohrkalibrierung f
large-container production unit Großhohlkörperfertigung f
large-scale production unit Großproduktionsmaschine f
level monitoring unit Füllstandüberwachung f
longitudinal stretching unit Längsstreckmaschine f, Längsstreckwerk n
machine conversion unit Maschinenumbausatz m
machine unit Anlagenaggregat n, Anlagenelement n
magnetic tape unit Magnetbandgerät n
mass storage unit Massenspeicher m
mechanical blowing unit Begasungsanlage f, Direktbegasungsanlage f
melt degassing unit Schmelzeentgasung f
melt devolatilization unit Schmelzeentgasung f
melt filtration unit Rohmaterialfilter n,m, Schmelzefilter n,m, Schmelzenfiltereinrichtung f, Schmelzenfilterung f
melt metering unit Schmelzedosierung f
metering and mixing unit Dosiermischmaschine f, Dosier- und Mischmaschine f
metering-mixing unit Dosiermischmaschine f, Dosier- und Mischmaschine f
metering unit Dosierwerk n, Dosieraggregat n, Dosiergerät n, Dosierelement n, Dosiereinheit f
monomer unit Monomerbaustein m, Monomereinheit f
mo(u)ld clamping unit Formschließeinheit f, Formschließaggregat n
mo(u)ld unit Werkzeugrohling m
oil cooling unit Ölkühler m
oil filtration unit Ölfilterung f
oil heating unit Öltemperiergerät n
oil level control unit Ölstandkontrolle f
output unit Ausgabegerät n, Ausgabeeinheit f, Stückzähler m
paint spraying unit Farbspritzanlage f
pellet cooling unit Granulatkühlvorrichtung f
pellet pre-heating unit Granulatvorwärmer m
pigment mixing unit Einfärbegerät n
pipe sizing unit Rohrkalibrierung f
pipe welding unit Rohrschweißvorrichtung f
plasticating unit Aufschmelzeinheit f, Plastifizierung f, Plastifiziereinheit f, Plastifizieraggregat n, Plastifiziereinrichtung f
plunger injection unit Kolbenspritzeinheit f
plunger plasticizing unit Kolbenplastifizierung f
polishing and take-off unit Glätt- und Abziehmaschine f
polyester unit Polyesterbaustein m

pre-drying unit Vortrockengerät n
pre-foaming unit Vorschäumer m
pre-heating unit Vorwärmgerät n
pre-mixing unit Vormischer m
pre-plasticizing unit Vorplastifizierungsaggregat n, Vorplastifizierung f
pre-treating unit [e.g. for film] Vorbehandlungsgerät n
precision metering unit Präzisionsdosiereinheit f
precision standard unit Präzisionsnormteil n
preliminary size reduction unit Vorzerkleinerer m, Vorzerkleinerungsmühle f
pressure per unit area Flächenpressung f
pressurized water unit Druckwassergerät n
price per unit volume Volumenpreis m
process control unit Arbeitsablaufsteuerung f, Prozeßführungsinstrument n, Prozeßkontrolle f
process data control unit Prozeßdatenüberwachung f
process monitoring unit Prozeßüberwachung f
processing unit Verarbeitungseinheit f, Verarbeitungsaggregat n, verfahrenstechnische Einheit f
production data acquisition unit Produktionsdatenerfassung f
production data collecting unit Produktionsdaten-Erfassungsgerät n
production of single units Einzelfertigung f
production unit Fertigungseinheit f, Produktionseinheit f
profile calibrating unit Profilkalibrierung f
pulverizing unit Feinmahlaggregat n
pump unit Pumpenaggregat n
push-button control Druckknopfsteuerung f, Drucktastensteuerung f
quick-change filter unit Schnellwechselfilter n,m
ram injector unit Kolbeninjektor m
range of standard units Standardgeräteserie f
raw material feed unit Rohstoffzuführgerät n
re-winding unit Umbäumstuhl m
reciprocating-screw plasticizing unit Schneckenschub-Plastifizieraggregat n
reciprocating-screw unit Schneckenschubaggregat n, Schubschneckeneinheit f
reeling unit Spulwerk n
remote control unit Fernsteuerung f, Fernkontrolle f
reprocessing unit Aufbereitungsaggregat n, Aufbereitungsmaschine f
rotary feed unit Rotationsdosiereinrichtung f
rotary-table unit Revolvereinheit f
rotary take-off unit Entnahmedrehtisch m
scrap removal unit Abfallentfernvorrichtung f
screw injection unit Schneckeneinspritzaggregat n, Schneckenspritzeinheit f
screw plasticizing unit Schneckenplastifiziereinheit f, Schneckenplastifizierung f, Schneckenplastifizieraggregat n
screw-plunger injection unit Schneckenkolbeneinspritzaggregat n, Schneckenkolbeneinspritzung f,

Schneckenkolbeninjektion f
screw-plunger unit Schneckenkolbeneinheit f
screw preplasticizing unit Schneckenvorplastifiziergerät f, Schneckenvorplastifizierung f
screw unit Schneckenaggregat n, Schneckeneinheit f
separate control unit Einzelregler m
separate feed unit Fremddosierung f
separate metering unit Fremddosierung f
sequence control unit Ablaufsteuerung f
servo-control unit Vorsteuereinheit f
sheet polishing unit Plattenglättanlage f
siloxane unit Siloxanbaustein m
single-speed clamp(ing) unit Eingeschwindigkeitsschließeinheit f
size reduction unit Zerkleinerungsmühle f, Zerkleinerungsmaschine f, Zerkleinerungsanlage f, Zerkleinerungsaggregat n, Zerkleinerer m
sizing unit Kalibrator m, Kalibriereinrichtung f
small-scale unit Kleingerät n
solvent recovery unit Lösemittelrückgewinnung f, Lösungsmittelrückgewinnung f
spinneret extrusion unit Extrusionsspinneinheit f
spray cooling unit Sprühkühlung f
spraying unit Spritzaggregat n
spreader roll unit Breithaltevorrichtung f, Breitstreckwerk n
sprue disposal unit Angußabführung f
stacking unit Ablage f, Ablegeeinrichtung f, Stapelanlage f, Stapeleinrichtung f
standard hot runner mo(u)ld units Heißkanalnormalien pl
standard hot runner unit Norm-Heißkanalblock m, Heißkanal-Normblock m
standard mo(u)ld unit Stammform f, Stammwerkzeug n
standard mo(u)ld units Baukastennormalien pl, normalisierte Bauteile npl, Formwerkzeug-Normalien pl, Normalien-Bauelemente npl, Normalien pl, Werkzeugnormalien pl
standard plasticizing unit Normalplastifiziereinheit f
standard unit Normbaugruppe f, Normbaustein m, Normteil n, Seriengerät n
standard units Normalien-Bauelemente npl, Normalien pl
steam generating unit Dampferzeuger m
stretching roll unit Rollenreckwerk n
stretching unit Reckanlage f, Streckeinrichtung f, Streckwerk n
structural unit Struktureinheit f, Baugruppe f
stuffing unit Stopfwerk n, Stopfaggregat n
supplementary unit Ergänzungsaggregat n, Ergänzungseinheit f, Zusatzaggregat n
supply units Versorgungseinheiten fpl
take-off unit Abzugsaggregat n, Abzugsvorrichtung f, Abzugswerk n
temperature control unit Temperatursteuereinheit f, Temperaturüberwachung f, Temperiergerät n

tensioning unit Zugwerk n
thickness calibration unit Dickenkalibrierung f
three-plate clamp(ing) unit Dreiplatten-Schließeinheit f
three-section plasticizing unit Dreizonen-Plastifiziereinheit f
three-way flow control unit Dreiwegestromregelung f
thrust bearing unit Axiallagergruppe f
thyristor control unit Thyristorregler m
tilting unit Schwenkaggregat n
toggle mo(u)ld clamping unit Kniehebel-Formschließaggregat n
transverse stretching unit Breitstreckwerk n
trimming unit Trimmstation f
twin compounding unit Aufbereitungs-Doppelanlage f
twin embossing unit Doppelprägewerk n
twin reprocessing unit Aufbereitungs-Doppelanlage f
two-stage injection unit Zweistufen-Einspritzaggregat n
two-tie bar clamp unit Zweiholmenschließeinheit f
two-way flow control unit Zweiwegestromregelung f
uniaxial stretching unit Monoaxial-Reckanlage f
unwind(ing) unit Abrolleinrichtung f, Abrollvorrichtung f, Abwickelgerät n, Abwickler m, Abwicklung f
vacuum calibrating unit Vakuumkalibrierung f
vacuum sizing unit Vakuumtank-Kalibrierung f
vented hopper degassing unit Vakuumtrichter-Entgasungsanlage f
vented plasticizing unit Entgasungsplastifizierung f, Entgasungsplastifiziereinheit f
vented unit Entgasungsaggregat n, Entgasungseinheit f
vibratory feed unit Vibratordosierung f
video display unit Sichtgerät n
visual inspection unit Sichtprüfstelle f, Sichtprüfstrecke f
volume control unit Volumenstromregler m
wall thickness control unit Wanddickenregelgerät n, Wanddickenregulierung f, Wanddickensteuerung f
water circulating unit Wasserumwälzeinheit f, Wasserumwälzung f
water cooling unit Wasserkühlung f
water-fed temperature control unit Wassertemperiergerät n
water supply unit Wasserversorgung f
waterbath cooling unit Wasserbadkühlung f
waterbath vacuum calibrating unit Vakuum-Kühltank-Kalibrierung f
web depositing unit Bandablegeeinheit f, Bandablegung f
web tension control unit Bahnspannungskontrolle f, Zugspannungsregelung f
web tension measuring unit Bahnzug-(kraft)meßstation f
weight per unit area Flächengewicht n

wind-up unit Aufwickelanlage f, Aufwicklung f
univalent einwertig
universal Universal-, universell einsetzbar
  universal joint Kreuzgelenk n, Kreuzgelenkkupplung f
  universal test apparatus Universalprüfmaschine f
  universal test (method) Universalprüfmethode f
  universal tester Universalprüfmaschine f
unlimited unbegrenzt
unloading station Entladestation f
unlocked entriegelt
unlubricated gleitmittelfrei
unmanned shift Geisterschicht f
unmodified nichtmodifiziert, unmodifiziert
unmo(u)lded ungeformt
  in the unmo(u)lded state im ungeformten Zustand m
unnotched ungekerbt
unoxidized nichtoxidiert
unpaired ungepaart
unpigmented pigmentfrei, ungefärbt, unpigmentiert
unplasticized nicht weichgemacht, unplastifiziert, weichmacherfrei
  unplasticized compound Hartcompound n,m, Hart-Granulat n
  unplasticized PVC compound Hart-PVC-Compound n,m
  unplasticized PVC mo(u)lding compound Hart-PVC-Formmasse f
  extrusion of unplasticized PVC Hart-PVC-Extrusion f
  injection mo(u)lding of unplasticized PVC Hart-PVC-Spritzguß m
  processing in unplasticized form [PVC] Hartverarbeitung f
unpolymerized nichtpolymerisiert
unprinted unbedruckt
unsupported trägerlos
untreated un(vor)behandelt
unvulcanized unvulkanisiert
unweathered nichtbewittert, unbewittert
unwind unit Abwickelgerät n, Abwickler m, Abwicklung f
unwinding roll Abwickelwalze f
unwinding speed Abwickelgeschwindigkeit f
unwinding station Abrollstation f
unwinding unit Abrolleinrichtung f, Abrollvorrichtung f
up and down movements Auf- und Abbewegungen fpl
UP resin [unsaturated polyester resin] UP-Harz n, UP-Reaktionsharz n
updating Aktualisierung f
upholstered furniture Polstermöbel npl
upholstery material Polstermaterial n
upper Ober-, oberhalb
  upper limit oberer Grenzwert m, Obergrenze f
  upper temperature limit Temperatur-Obergrenze f
upright (in construction) hohe Bauweise f
upstream stromaufwärts, vorgeschaltet
upstroke press Unterkolbenpresse f

upswing Aufschwung m
uPVC Hart-PVC n, Polyvinylchlorid n hart, PVC n hart
  uPVC sheet Hart-PVC-Platte f
upward aufwärts
  upward movement Aufwärtsbewegung f
  upward trend Aufwärtstrend m
urea Harnstoff m
  urea bridge Harnstoffbrücke f
  urea derivative Harnstoffderivat n
  urea-formaldehyde condensate Harnstoff-Formaldehyd-Kondensat n
  urea-formaldehyde foam Harnstoffharzschaum m
  urea-formaldehyde resin Harnstoffharz n, Harnstoff-Formaldehydharz n
  urea laminate Harnstoffharz-Schichtstoff m
  urea linkage Harnstoffbindung f
  urea linkage content Harnstoffbindungsanteil m
  urea mo(u)lding compound Harnstoff-Preßmasse f
  urea resin Harnstoffharz n
  dimethylol urea Dimethylolharnstoff m
  diphenyl urea Diphenylharnstoff m
  monomethylol urea Monomethylolharnstoff m
  phenyl urea Phenylharnstoff m
  polymethylene urea Polymethylenharnstoff m
urethane Urethan n
  urethane group Urethangruppe f
  urethane linkage Urethanbindung f
  urethane linkage content Urethanbindungsanteil m
  polyether urethane Polyetherurethan n
usable brauchbar, verarbeitungsfähig
use Gebrauch m, Benutzung f, Verwendung f, Einsatz m
  conditions of use Beanspruchungsbedingungen fpl, Beanspruchungsverhältnisse npl, Einsatzbedingungen fpl
  easy to use benutzerfreundlich, verarbeitungsfreundlich
  end use Einsatzzweck m, Endanwendung f
  everyday use Alltagsbetrieb m
  exterior use Außeneinsatz m
  in continuous use im Dauerbetrieb m
  main field of use Hauptanwendungsgebiet n
  outdoor use Außeneinsatz m, Außenverwendung f, Freiluftbetrieb m
  outside use Außeneinsatz m, Außenverwendung f, Freiluftbetrieb m
  possible uses Einsatzmöglichkeiten fpl
  range of uses Einsatzbreite f, Einsatzspektrum n
  ready to use verarbeitungsfertig
  safety in use Arbeitssicherheit f
  stability in use Gebrauchsstabilität f
used 1. gebraucht, Alt-; 2. Einsatz- [wenn nachgestellt]
  amount used Einsatzmenge f
  energy used aufgenommene Energie f, Energieaufnahme f
  partly used [contents of a drum etc.] angebrochen
  power used aufgenommene Leistung f
  total energy used Gesamtenergieaufnahme f

usefulness Gebrauchswert m
user Benutzer m, Anwender m
  user-friendly benutzerfreundlich, anwenderfreundlich
  first-time user Erstanwender m
  major user Großverbraucher m
UV [ultra-violet] ultraviolett, UV
  UV absorber UV-Absorber m
  UV absorption UV-Absorption f
  UV content UV-Anteil m
  UV radiation UV-Strahlung f
  UV spectroscopy UV-Spektroskopie f
  UV stabilizer UV-Stabilisator m
V-belt Keilriemen m
  V-belt drive Keilriemenantrieb m
  V-belt pulley Keilriemenscheibe f
  V-belt speed reducing drive Keilriemenuntersetzung f
vacuum Unterdruck m, Vakuum n
  vacuum bag mo(u)lding Vakuumfolienverfahren n
  vacuum cabinet Vakuumschrank m
  vacuum calibrating tank Vakuumkalibrierbecken n
  vacuum calibration Vakuumkalibrieren n, Vakuumkalibrierung f
  vacuum calibration section Vakuumkalibrierstrecke f
  vacuum calibration system Vakuumkalibriersystem n
  vacuum connection Vakuumanschluß m
  vacuum conveyor Vakuumfördergerät n
  vacuum drying cabinet Vakuumtrockenschrank m
  vacuum forming Vakuumformung f, Vakuumtiefziehen n
  vacuum forming machine Vakuumformmaschine f, Vakuumtiefziehmaschine f
  vacuum forming process Vakuumziehverfahren n
  vacuum forming tool Vakuumtiefziehform f
  vacuum hopper Entgasungstrichter m, Vakuumtrichter m, Vakuumspeisetrichter m
  vacuum impregnation (process) Vakuumtränkverfahren n
  vacuum injection mo(u)lding (process) Vakuumeinspritzverfahren n
  vacuum oven Vakuumofen m
  vacuum pump Unterdruckpumpe f, Vakuumpumpe f
  vacuum sizing Vakuumtank-Kalibrierung f
  vacuum suction holes Vakuumsauglöcher npl
  vacuum suction plate Vakuumsaugteller m
  vacuum supply Vakuumversorgung f
  vacuum tight vakuumdicht
  automatic vacuum forming machine Vakuumformautomat m
  liquid-ring vacuum pump Wasserringvakuumpumpe f
  low vacuum Grobvakuum n
  medium vacuum Feinvakuum n
  partial vacuum Teilvakuum n
  positive displacement vacuum pump Verdränger-Vakuumpumpe f
  twin vacuum hopper Doppelvakuumtrichter m
  twin vacuum hopper assembly Vakuum-Doppeltrichteranlage f

  ultra-high vacuum Ultrahochvakuum n
  waterbath vacuum calibrating unit Vakuum-Kühltank-Kalibrierung f
valency Valenz f, Wertigkeit f
  primary valency bond Hauptvalenzbindung f
  secondary valency forces Nebenvalenzkräfte fpl
validity Gültigkeit f
value Wert m
  acid value Säurezahl f
  amine value Aminzahl f
  approximate value Anhaltswert m
  calculated value Rechenwert m
  epoxide value Epoxidgehalt m, Epoxidwert m
  epoxy value Epoxidgehalt m, Epoxidwert m
  fixed value Festwert m
  hydroxyl value OH-Zahl f
  informative value Aussagekraft f, Aussagewert m
  initial value Anschaffungswert m
  instantaneous value Momentanwert m
  iodine colo(u)r value Jodfarbzahl f
  known value Erfahrungswert m
  limiting value Grenzwert m
  limiting value control Grenzwertüberwachung f
  limiting value signal Grenzwertmeldung f
  maximum value Größtwert m, Höchstwert m
  mean value Mittelwert m
  minimum value Kleinstwert m
  numerical value Zahlenwert m
  oil absorption value Ölzahl f
  peak value Spitzenwert m
  replacement value Wiederbeschaffungswert m
  required value Sollgröße f, Sollwert m
  saponification value Verseifungszahl f
  sigma value Sigmawert m
  sound emission value Lärmemissionswert m
  zero value Nullwert m
valve Ventil n
  valve ejector Ventilauswerfer m
  valve gated mo(u)ld Nadelventilwerkzeug n
  valve gating system Nadelventilangußsystem n
  valve pin Düsennadel f
  valve seat Ventilsitz m
  aerosol valve Aerosolventil n
  analog(ue) valve Analogventil n
  ball valve Kugelventil n
  braking valve Bremsventil n
  check valve Sperrventil n, Rückschlagventil n
  control valve Regulierventil n, Schaltventil n, Steuerventil n
  cooling water control valve Kühlwasserregulierventil n
  digital valve Digitalventil n
  directional control valve Wegeventil n
  discharge valve Entladeschieber m
  ejector valve Auswerferventil n
  flow control valve Durchflußsteuerventil n, Strom(regel)ventil n

high-pressure valve Hochdruckventil n
hot runner needle valve Heißkanal-Nadelventil n
manually operated valve Handventil n
metering valve Dosierschieber m
mo(u)ld opening and closing valve Formenöffnungs- und Schließventil n
needle valve Nadelventil n, Nadelverschluß m
needle valve nozzle Nadelverschluß(spritz)düse f
non-return valve Rücklaufsperre f, Rückstausperre f, Rückstromsperre f
overflow valve Überströmventil n
pneumatic control valve Luftsteuerventil n
pressure adjusting valve Druckeinstellventil n
pressure control valve Druck(regel)ventil n, Druckregulierventil n
pressure release valve Druckbegrenzungsventil n
pressure relief valve Druckentlastungsventil n, Druckminderventil n, Druckreduzierventil n
pressurizing valve Vorspannventil n
proportional control valve Proportionalstellventil n
proportional valve Proportionalventil n
relief valve Dekompressionsventil n, Schnüffelventil n
safety valve Sicherheitsventil n
seating valve Sitzventil n
shut-off slide valve Absperrschieber m
shut-off valve Verschlußventil n, Sperrventil n
slide valve Schieber m
solenoid valve Magnetventil n
solenoid valve driver Magnetventiltreiber m
suction valve Saugventil n
three-way flow control valve Dreiwege-Mengenregelventil n
three-way valve Dreiwegeventil n
throttle-check valve Drossel-Rückschlagventil n
throttel valve Drosselklappe f, Drosselschieber m, Drosselventil n
two-way valve Zweiwegventil n
van der Waals' forces van-der-Waalssche Kräfte fpl
vanadium accelerator Vanadiunbeschleuniger m
vane-type motor Flügelzellenmotor m
vane-type pump Flügelzellenpumpe f
variable-delivery vane-type pump Flügelzellen-regelpumpe f
vaporization Verdampfung f
　flash vaporization Entspannungsverdampfung f, Flashverdampfung f
vaporizer Verdampfer m
　falling film vaporizer Fallfilmverdampfer m
　spiral coil vaporizer Schlangenrohrverdampfer m
　thin film vaporizer Dünnschichtverdampfer m
　tubular vaporizer Röhrenverdampfer m
vapor [GB: vapour] Dampf m
　vapo(u)r barrier Dampfsperre f
　vapo(u)r barrier sheeting Dampfsperrbahn f
　vapo(u)r diffusion resistance Dampfdiffusions-widerstand m
　vapo(u)r impermeability Dampfundurchlässigkeit f, Dampfdichtheit f
　vapo(u)r permeability Dampfdurchlässigkeit f
　vapo(u)r pressure Dampfdruck m
　vapo(u)r proof gasdicht, dampfdicht
　exposure to water vapo(u)r Wasserdampflagerung f
　in vapo(u)r form dampfförmig
　partial vapo(u)r pressure Dampfteildruck m
　partial water vapo(u)r pressure Wasserdampfpartialdruck m, Wasserdampfteildruck m
　plasticizer vapo(u)r extractor Weichmacherdampfabsaugung f
　plasticizer vapo(u)rs Weichmacherdämpfe fpl
　saturated vapo(u)r Sattdampf m
　saturated with water vapo(u)r wasserdampfgesättigt
　saturation vapo(u)r pressure Sättigungsdampfdruck m, Wasserdampfsättigungsdruck m
　solvent vapo(u)rs Lösungsmitteldämpfe mpl
　styrene vapo(u)rs Styroldämpfe fpl
　vulcaniziation vapo(u)rs Vulkanizationsdämpfe fpl
　water vapo(u)r Naßdampf m, Wasserdampf m
　water vapo(u)r diffusion Wasserdampfdiffusion f
　water vapo(u)r permeability Wasserdampfpermeabilität f, Wasserdampfdurchlässigkeit f
　water vapo(u)r resistance Wasserdampfstabilität f
variable 1. veränderlich, variabel, einstellbar; 2. Variable f
variable delivery pump Regelpumpe f
variable-delivery vane-type pump Flügelzellen-regelpumpe f
variable-design screw Einstückschnecke f, Schaftschnecke f, Aufsteckschnecke f
variable displacement pump Verstellpumpe f
variable speed 1. regelbare Drehzahl f; 2. drehzahlveränderlich, drehzahlvariabel
　adjustable variable Stellgröße f, Stellwert m
　controlled variable Regelgröße f
　dependent variable Zielgröße f
　independent variable Einflußgröße f
　infinitely variable stufenlos einstellbar, stufenlos veränderbar
　input variable Eingabewert m, Eingangsgröße f
　machine variable Maschinenstellgröße f, Maschinenstellwert m
　output variable Ausgangsgröße f
　process variable Prozeßgröße f
　production variable Fertigungsparameter m
　steplessly variable stufenlos regelbar, stufenlos einstellbar
variations Schwankungen fpl, Unterschiede mpl, Abweichungen fpl
variations in colo(u)r Farbschwankungen fpl
variations in quality Qualitätsschwankungen fpl
　batch variations Chargenunterschiede mpl
　colo(u)r variations Farbschwankungen fpl
　film ga(u)ge variations Foliendickenabweichungen fpl, Foliendickenunterschiede mpl
　ga(u)ge variations [of film] Dickenabweichungen fpl,

Dickenschwankungen fpl
  pressure variations Druckschwankungen fpl, Druckschwingungen fpl
  product variations Produktschwankungen fpl
  quality variations Qualitätsschwankungen fpl
  thickness variations Dickenabweichungen fpl, Dickenschwankungen fpl
  wall thickness variations Wanddickenschwankungen fpl, Wanddickenstreuungen fpl, Wanddickenabweichungen fpl
varnish Firnis m, Lack m [meist farblos]
  dipping varnish Tauchlack m
  electrical insulating varnish Elektroisolierlack m
  paper varnish Papierlack m
  spraying varnish Spritzlack m
  transparent varnish Transparentlack m
varnished lackiert, Lack-
  varnished fabric Lackgewebe n
  varnished glass cloth Lackglasgewebe n
  varnished paper Lackpapier n
VDU [visual display unit] Bildschirm m
  colo(u)r VDU Farbsichtgerät n
vehicle Fahrzeug n
  vehicle construction Fahrzeugbau m
  commercial vehicle Nutzfahrzeug n
  motor vehicle Kraftfahrzeug n
  standard vehicle Serienfahrzeug n
velocity Geschwindigkeit f
  velocity constant Geschwindigkeitskonstante f
  velocity curve Geschwindigkeitskennlinie f
  velocity distribution Geschwindigkeitsverteilung f
  velocity gradient Geschwindigkeitsgefälle n, Geschwindigkeitsgradient m
  velocity profile Geschwindigkeitsprofil n
  angular velocity Winkelgeschwindigkeit f
  flow velocity Fließgeschwindigkeit f
  peripheral velocity Umfangsgeschwindigkeit f
  relative velocity Relativgeschwindigkeit f
vent/to entlüften
vent Entgasungsbohrung f, Entgasungsdom m, Entgasungskamin m, Entgasungsöffnung f, Entgasungsschacht m, Entgasungsstutzen m, Entgasungszapfen m, Kamin m, Entlüftungsbohrung f, Absaugöffnung f
  vent groove Entlüftungsspalt m, Entlüftungsschlitz m, Entlüftungsnute f
  vent section Entgasungsteil n, Entgasungsschuß m
  vent zone Dekompressionszone f, Entgasungsbereich m, Entgasungszone f, Entspannungszone f, Zylinderentgasungszone f, Vakuumzone f
  vent(ing) pin Entlüftungsstift m
vented entlüftet, Entgasungs-
  vented barrel Entgasungszylinder m
  vented barrel extruder Zylinderentgasungsextruder m
  vented cylinder Entgasungszylinder m
  vented extruder Vakuumextruder m, Entgasungsextruder m, Entgasungsmaschine f

  vented hopper Entgasungstrichter m
  vented hopper degassing unit Vakuumtrichter-Entgasungsanlage f
  vented hopper system Trichterentgasung f
  vented plasticizing (unit) Entgasungsplastifizierung f, Entgasungsplastifiziereinheit f
  vented screw Entgasungsschnecke f, Vakuumschnecke f, Dekompressionsschnecke f
  vented section [of screw] Vakuumteil n
  vented single-screw extruder Einschnecken-Entgasungsextruder m
  vented twin screw extruder Doppelschnecken-Entgasungsextruder m
  vented twin screws Entgasungsschneckenpaar n
  vented unit Entgasungsaggregat n, Entgasungseinheit f
  barrel vented through longitudinal slits Längsschlitz-Entgasungsgehäuse n
  with a vented barrel zylinderentgast
ventilation Lüftung f, Entlüftung f, Belüftung f
  ventilation shaft Lüftungskanal m
  ventilation slit Lüftungsschlitz m
  forced ventilation Zwangs(durch)lüftung f
  oil tank ventilation Öltankbelüftung f
venting Lüften n, Lüftung f, Entlüftung f
  venting channel Entlüftungskanal m
  venting device Entlüftungsvorrichtung f
  venting effect Entgasungswirkung f
  venting efficiency Entgasungsqualität f
  venting lamella Entlüftungslamelle f
  venting slit Entlüftungsschlitz m, Entlüftungsspalt m
  venting system Entgasungsvorrichtung f, Entgasungssystem n, Entlüftungssystem n
  back venting system Rückwärtsentgasung f
  downstream venting Stromabwärtsentgasung f
  forced venting Zwansentlüftung f
  forward venting Vorwärtsentgasung f
  forward venting system Vorwärtsentgasung f
  mo(u)ld cavity venting Formnestentlüftung f
  mo(u)ld venting Konturenentlüftung f, Werkzeugbelüftung f, Werkzeugentlüftung f
  screw venting system Schneckendekompressionseinrichtung f
venturi system Venturi-System n
vernier Nonius m, Noniusskala f
versatility Vielseitigkeit f
version Version f
  table-top version Tischversion f
vertical vertikal, senkrecht
  vertical adjusting mechanism Vertikalverstellung f
  vertical adjustment Höhenverstellung f, Vertikalverstellung f
  vertical construction Vertikalbauweise f
  vertical die head Senkrechtspritzkopf m
  vertical extruder Senkrechtextruder m, Vertikalextruder m
  vertical extruder head Senkrechtspritzkopf m
  vertical feeder Vertikalspeiseapparat m

vertical flash face Tauchkante f, Werkzeugtauchkante f
vertical granulator Vertikalschneidmühle f
vertical twin screw extruder Vertikaldoppelschneckenextruder m
- **four-roll vertical calender** Vierwalzen-I-Kalander m
- **three-roll vertical calender** Dreiwalzen-I-Kalander m
- **two-roll vertical calender** Zweiwalzen-I-Kalander m

vertically adjustable höhenverstellbar
vessel Gefäß n, Behälter m
- **pressurized vessel** Druckbehälter m, Druckgefäß n
- **reaction vessel** Reaktionsgefäß n

vibrating schwingend
- **vibrating chute** Vibrationsrinne f
- **vibrating screen** Schwingsieb n
- **vibrating table** Rütteltisch m

vibrator motor Rüttelmotor m
vibratory Vibrations-, Vibrator-
- **vibratory feed hopper** Vibrationseinfülltrichter m
- **vibratory feed unit** Vibratordosierung f
- **vibratory feeding** Vibratordosierung f

Vicat indentor Vicatnadel f
Vicat softening point Formbeständigkeit f in der Wärme mit Vicatnadel, Vicatzahl f, Vicatwert m, Vicat-Erweichungspunkt m, Vicat-Wärmeformbeständigkeit f, Vicat-Erweichungstemperatur f, Wärmeformbeständigkeit f nach Vicat
vicious circle Teufelkreis m
Vickers hardness Vickershärte f
video Video-
- **video disk** Videoplatte f
- **video display unit** Sichtgerät n
- **video terminal** Videoterminal n

view Sicht f, Blick m
- **view from above** Draufsicht f
- **view from below** Untersicht f
- **from the energy point of view** energetisch betrachtet
- **interesting from the processing point of view** verfahrenstechnisch interessant, verfahrenstechnisch wesentlich
- **partial view** Teilansicht f
- **side exposed to view** Sichtseite f

vinyl Vinyl n
- **vinyl acetate** Essigsäurevinylester m, Vinylacetat n
- **vinyl alcohol** Vinylalkohol m
- **vinyl-asbestos tile** Vinyl-Asbest-Platte f
- **vinyl carbazole** Vinylcarbazol n
- **vinyl chloride** Vinylchlorid n
- **vinyl chloride copolymer** VC-Copolymerisat n
- **vinyl chloride homopolymer** Vinylchlorid-Homopolymer n
- **vinyl chloride polymer** Vinylchloridpolymerisat n
- **vinyl chloride-vinyl acetate copolymer** Vinylchlorid-Vinylacetat-Copolymerisat n, Polyvinylchloridacetat n
- **vinyl ester** Vinylester m
- **vinyl ester resin** Vinylesterharz n
- **vinyl ether** Vinylether m
- **vinyl group** Vinylgruppe f
- **vinyl laurate** Vinyllaurat n
- **vinyl monomer** Vinylmonomer n
- **vinyl radical** Vinylrest m
- **vinyl resin** Vinylharz n
- **vinyl silane** Vinylsilan n
- **vinyl toluene** Vinyltoluol n
- **phenylmethyl vinyl polysiloxane** Phenylmethylvinylpolysiloxan n
- **residual vinyl chloride (monomer) content** VC-Restmonomergehalt m

vinylidene Vinyliden-
- **vinylidene chloride** Vinylidenchlorid n
- **vinylidene fluoride** Vinylidenfluorid n
- **vinylidene fluoride-tetrafluoroethylene copolymer** Vinylidenfluorid-Tetrafluorethylen-Copolymerisat n

virgin unberührt, rein, unbehandelt
- **virgin compound** Neugranulat n
- **virgin material** Neumaterial n, Neuware f, Originalmaterial n
- **virgin polymer** Kunststoffneuware f

viscoelastic viskoelastisch
visco-elasticity Viskoelastizität f
viscometer Viskosimeter n
- **Brookfield viscometer** Brookfield-Viskosimeter n
- **cone and plate viscometer** Kegel-Platte-Viskosimeter n
- **falling sphere viscometer** Kugelfallviskosimeter n
- **rotational viscometer** Rotationsviskosimeter n

viscometric viskometrisch
viscometry Viskometrie f
viscosity Viskosität f, Zähigkeit f
- **viscosity behavio(u)r** Viskositätsverhalten n
- **viscosity changes** Viskositätsänderungen fpl
- **viscosity-dependent** viskositätsabhängig
- **viscosity depressant** Viskositätserniedriger m
- **viscosity differences** Viskositätsunterschiede mpl
- **viscosity equation** Viskositätsansatz m
- **viscosity fluctuations** Viskositäts-Inhomogenitäten fpl
- **viscosity increase** Viskositätserhöhung f, Viskositätsanstieg m
- **viscosity limit** Viskositätsgrenze f
- **viscosity number** Viskositätszahl f
- **viscosity peak** Viskositätsberg m
- **viscosity ratio** Viskositätsverhältnis n
- **viscosity-reducing** viskositätserniedrigend, viskositätssenkend
- **viscosity reduction** Viskositätserniedrigung f, Viskositätsabfall m, Viskositätsabsenkung f, Viskositätsminderung f
- **viscosity regulator** Viskositätsregler m
- **apparent viscosity** scheinbare Viskosität f
- **average viscosity** Viskositätsmittel n
- **changes in viscosity** Viskositätsänderungen fpl
- **correct viscosity** Verarbeitungsviskosität f
- **drop in viscosity** Viskositätsabfall m, Viskositätsabsenkung f, Viskositätserniedrigung f,

Viskositätsminderung f
dynamic viscosity dynamische Viskosität f
having a constant viscosity viskositätsstabil
high viscosity Zähflüssigkeit f
intrinsic viscosity Staudinger-Index m
kinematic viscosity kinematische Viskosität f, kinematische Zähigkeit f
limiting viscosity Viskositätsgrenze f
limiting viscosity number Grenzviskosität f, Staudinger-Index m
low-viscosity primer Einlaßgrund m, Einlaßgrundierung f
medium-viscosity mittelzäh
melt viscosity Schmelzeviskosität f
nominal viscosity Nennviskosität f
paste viscosity Pastenviskosität f
reduced viscosity reduzierte Viskosität f
relative viscosity relative Viskosität f
rise in viscosity Viskositätsanstieg m, Viskositätserhöhung f
specific viscosity spezifische Viskosität f
viscous dickflüssig, viskos, zäh, zähflüssig
  viscous flow viskoses Fließen n
visible sichtbar
  visible side Sichtseite f
  visible surface Sichtfläche f
  making visible Sichtbarmachen n
visual visuell, optisch
  visual control Sichtkontrolle f
  visual display Bildschirmanzeige f
  visual display unit Datensichtgerät n
  visual inspection Sichtprüfung f
  visual inspection unit Sichtprüfstrecke f, Sichtprüfstelle f
  visual warning signal optische Störanzeige f
  colo(u)r visual display unit Farbsichtgerät n
  monochrome visual display unit S/W-Sichtgerät n
void Hohlraum m, Lunker m, Blase f
  void free blasenfrei, lunkerfrei
volatile flüchtig, verdampfbar
  volatile content Flüchtegehalt m
  volatile matter flüchtige Bestandteile npl
  readily volatile leichtflüchtig
volatility Flüchtigkeit f, Verdampfbarkeit f
  low volatility Schwerflüchtigkeit f
  plasticizer volatility Weichmacherflüchtigkeit f
voltage (elektrische) Spannung f
  a.c. voltage Wechselspannung f
  a.c. voltage source Wechselspannungserzeuger m
  breakdown voltage Durchschlagspannung f
  connected voltage Anschlußspannung f
  control voltage Steuerspannung f
  d.c. voltage Gleichspannung f
  exposure to high voltage Hochspannungsbelastung f
  input voltage Eingangsspannung f
  long-term test voltage Dauerprüfspannung f
  low voltage Niederspannung f
  low-voltage range Niederspannungsbereich m
  mains voltage Netzspannung f
  medium-voltage range Mittelspannungsbereich m
  operating voltage Betriebsspannung f
volume Volumen n
  volume change Volumenänderung f
  volume concentration Volumenkonzentration f
  volume content Volumenanteil m, Volumengehalt m
  volume control unit Volumensteuerung f
  volume decrease Volumenabnahme f, Volumen(ver)minderung f
  volume expansion Volumen(aus)dehnung f, Volumendilatation f
  volume flow rate Volumendurchsatz m, Volumenstrom m
  volume fraction Volumenbruch m
  volume increase Volumenvergrößerung f, Volumenzunahme f, Volumenerhöhung f
  volume of business Geschäftsvolumen n
  volume price advantage Volumenpreisvorteil m
  volume reduction Volumenverkleinerung f
  volume resistance Volumenwiderstand m
  volume resistivity spezifischer Durchgangswiderstand m
  volume shrinkage Volumenkontraktion f, Volumenschrumpfung f, Volumenschwindung f
  volume throughput Volumendurchsatz m, volumetrischer Durchsatz m
  volume turnover Umsatzvolumen n
  cavity volume Formhohlraumvolumen n, Werkzeughohlraumvolumen n
  change in volume Volumen(ver)änderung f
  compacted bulk volume Stampfvolumen n
  effective volume Nutzvolumen n
  having the same volume volumengleich
  increase in volume Volumenerhöhung f
  initial volume Anfangsvolumen n
  mo(u)ld cavity volume Formhohlraumvolumen n
  mo(u)lded-part volume Formteilvolumen n
  nominal volume Nennvolumen n
  output volume Ausstoßvolumen n
  parts by volume Raumteile mpl
  pigment volume concentration [p.v.c.] Pigmentvolumenkonzentration f
  price per unit volume Volumenpreis m
  screw channel volume Schneckenkanalvolumen n
  specific volume spezifisches Volumen n
volumetric volumetrisch
  volumetric extrusion rate Volumendurchsatz m
  volumetric feeder Volumendosierung f, Volumendosieraggregat n
  volumetric feeding Volumendosierung f
  volumetric flow rate volumetrischer Durchsatz m, Volumendurchsatz m, Volumenstrom m
vortex Wirbel m, Strudel m
  melt vortex Schmelzwirbel m
vulcanisate Vulkanisat n

vulcanisate properties Vulkanisateigenschaften fpl
vulcanizable vulkanisierbar
vulcanization Vulkanisation f
  vulcanization accelerator Vulkanisationsbeschleuniger m
  vulcanization vapo(u)rs Vulkanisationsdämpfe mpl
  complete vulcanization Ausvulkanisieren n, Durchvulkanisation f
  high-temperature vulcanization Hitzevulkanisation f
  partial vulcanization Anvulkanisieren n
  premature vulcanization Vorvernetzung f
  press vulcanization Preßvulkanisation f
  room temperature vulcanization Kaltvulkanisation f
  steam vulcanization Dampfvulkanisation f
vulcanized vulkanisiert
  vulcanized fiber Vulkanfiber n
  fully vulcanized durchvulkanisiert
vulcanizing Vulkanisier-, vulkanisierend, Vulkanisations-
  vulcanizing agent [catalyzes forming of crosslinks among polymer chains] Vulkanisationsmittel n, Vulkanisiermittel n, Vulkanisationskatalysator m
  vulcanizing characteristics Vulkanisierverhalten n
  vulcanizing conditions Vulkanisationsbedingungen fpl
  vulcanizing rate Vulkanisationsgeschwindigkeit f
  vulcanizing reaction Vulkanisationsreaktion f
  vulcanizing section Vulkanisationszone f
  vulcanizing temperature Vulkanisationstemperatur f
  vulcanizing time Vulkanisationszeit f
  vulcanizing tunnel Vulkanisationstunnel m
  high temperature-vulcanizing heißvulkanisierend
  room temperature-vulcanizing kaltvulkanisierend
  steam vulcanizing plant Dampfvulkanisieranlage f
wage Lohn m
  wages and salaries Personalkosten pl, Personalaufwand m
  wage-intensive lohnintensiv
wall Wand f
  wall-adhering [e.g. polymer melt] wandhaftend
  wall-slipping [e.g. polymer melt] wandgleitend
  wall thickness control Wanddickenregulierung f, Wanddickenkontrolle f, Wanddickensteuerung f
  wall thickness control mechanism Wanddickenregulier(ungs)system n
  wall thickness control unit Wanddickenregelgerät n, Wanddickenregler m
  wall thickness differences Wanddickenunterschiede mpl
  wall thickness distribution Wanddickenverteilung f
  wall thickness ga(u)ge Wanddickenmeßgerät n
  wall thickness measurement Wanddickenmessung f
  wall thickness programming device Wanddickenprogrammierung f, Wanddicken-Programmiergerät n
  wall thickness variations Wanddickenstreuungen fpl, Wanddickenabweichungen fpl, Wanddickenschwankungen fpl
  barrel wall Gehäusewand f, Zylinderwand f
  cavity wall Formnestwand(ung) f
  cooling channel walls Kühlkanalwandungen fpl
  cylinder wall Zylinderwand f
  differences in wall thickness Wanddickenunterschiede mpl
  parison wall thickness control Schlauchdickenregelung f
  parison wall thickness controller Schlauchdickenregler m
  part wall thickness Formteilwanddicke f
wallpaper adhesive Tapetenkleister m
warehouse Lagergebäude n, Lagerhalle f
warehousing Lagerhaltung f
warning Warnung f
  warning device Warneinrichtung f
  warning light Alarmlampe f, Warnlampe f, Warnleuchte f
  warning signal Störmeldung f, Warnsignal n
  audible warning signal akustische Störanzeige f
  flashing light warning signal Störungsblinkanzeige f
  visual warning signal optische Störanzeige f
warp/to verziehen
  tendency to warp Verzugsneigung f
warp Verziehen n, Werwerfen n; Kette f [Textil]
  warp beam Kettbaum m
  warp-beam film Kettbaumfolie f
  warp direction Kettrichtung f
  warp knitting Kettwirktechnik f
  warp resistance Verformungswiderstand m
  warp wire Kettdraht m
warpage Verzug m, Verziehen n
warping Verwerfung f, Verzugserscheinungen fpl
wash basin Waschbecken n
washing waschend, Wasch-
  washing machine Waschmaschine f
  washing resistance Waschbeständigkeit f
waste 1. Abfall m, Abfallmaterial n, Ausschuß m; 2. Alt-, Ab-
  waste air Fortluft f, Abluft f
  waste air exhaust fan Abluftventilator m
  waste air scrubber Abluftwäscher m
  waste gas purification plant Abgasreinigungsanlage f
  waste heat Abwärme f
  waste heat recycling Abwärmenutzung f
  waste pipe Abflußrohr n, Abwasserrohr n
  domestic waste pipe Hausabflußrohr n
  edge trim waste Randstreifenabfall m
  extruder start-up waste Extruderanfahrladen m
  in-plant waste Produktionsabfälle f
  minimizing of waste Abfallminimierung f
  parison waste Abfallbutzen m, Blas(teil)butzen m, Butzen(abfall) m, Butzenmaterial n
  reducing waste to a minimum Ausschußminimierung f
  removal of waste Abfallentfernung f
  sprue wastes Angußverlust m, Angußabfall m
  start-up waste Anfahrausschuß m
water Wasser n

water absorption Wasseraufnahme f
water-based paint Wasserlack m
water circulating unit Wasserumwälzeinheit f, Wasserumwälzung f
water circulation Wasserumwälzung f
water-cooled wassergekühlt
water-cooled die face pelletization Wasserringgranulierung f
water-cooled die face pelletizer Wasserringgranulierung f
water-cooled feed section/zone Naßbüchse f
water cooling ring Wasserkontaktkühlung f
water cooling (unit) Wasserkühlung f
water dispersible wasserdispergierbar
water flow-way Wasserfließweg m, Wasserführungsrohr n
water hammer Druckstoß m
water impermeability Wasserdichtheit f
water inlet Wasseranschluß m, Wassereintritt m, Wasserzulauf m, Wasserzufuhr f, Wasserzufluß m
water inlet port Wasserzugabestutzen m
water insoluble wasserunlöslich
water jet cutting Wasserstrahlschneiden n
water level Wasserstand m
water mains Leitungswassernetz n
water miscible wassermischbar
water of crystallisation Kristallwasser n, Hydratwasser n, chemisch gebundenes Wasser n, Hydrat(ions)wasser n
water outlet Wasserabfluß m, Wasserauslaß m, Wasserauslauf m, Wasseraustritt m
water repellency Wasserabweisung f, Hydrophobie f
water repellent Hydrophobierungsmittel n, wasserabstoßend, wasserabweisend
water resistance Naßfestigkeit f, Wasserbeständigkeit f, Wasserfestigkeit f
water vapo(u)r permeability Wasserdampfpermeabilität f, Wasserdampfdurchlässigkeit f
water vapo(u)r resistance Wasserdampfstabilität f
affected by water wasserempfindlich
boiling water Kochwasser n
central cooling water manifold Zentralkühlwasserverteilung f
circulating hot water heating (system) Heißwasser-Umlaufheizung f
circulating water Umlaufwasser n
circulating water temperature control unit Wasserumlauftemperiergerät n
containing water wasserhaltig
cooling water Kühlwasser n
cooling water circuit Kühlwasserkreislauf m
cooling water connection Kühlwasseranschluß m
cooling water consumer Kühlwasserverbraucher m
cooling water control valve Kühlwasserregulierventil n
cooling water flow rate Kühlwasserdurchflußmenge f
cooling water inlet Kühlwasserzulauf m
cooling water inlet temperature Kühlwasserzulauftemperatur f
cooling water line Kühlwasserleitung f
cooling water manifold Kühlwasserverteiler m
cooling water outlet Kühlwasserablauf m
cooling water outlet temperature Kühlwasserablauftemperatur f
cooling water recovery unit Kühlwasserrückgewinnungsanlage f
cooling water supply Kühlwasserzufuhr f
cooling water throughput Kühlwasserdurchflußmenge f
drinking water pipe Trinkwasserrohr n
drinking water supply (system) Trinkwasserversorgung f
elimination of water [e.g. during a chemical reaction] Wassereliminierung f
excess water pressure Wasserüberdruck m
exposure to water Wasserbelastung f
exposure to water vapo(u)r Wasserdampflagerung f
free from water wasserfrei
ground water Grundwasser n
pressurized water unit Druckwassergerät n
prolonged immersion in water Langzeitwasserlagerung f
removal of water Entwässerung f
residual water Restwasser n
resistant to boiling water kochfest
returning water Rücklaufwasser n
saturated with water vapo(u)r wasserdampfgesättigt
sensitivity to water Wasserempfindlichkeit f
separation of water Wasserabspaltung f, Wasserentmischung f
tap water Leitungswasser n
unaffected by water wasserunempfindlich
waste water [industrial] Abwasser n
weakened points Sollbruchstellen fpl
wear Abnutzung f, Verschleiß m
  wear characteristics Verschleißverhalten n
  wear-intensive verschleißintensiv
  wear plate Verschleißplatte f
  wear resistance Verschleißbeständigkeit f, Verschleißwiderstand m, Verschleißwiderstandsfähigkeit f, Verschleißfestigkeit f
  wear resistant verschleißfest, verschleißbeständig, verschleißwiderstandsfähig
  wear resistant bushing Verschleißbüchse f
  wear resistant coating Panzerschicht f, Panzerung f, Oberflächenpanzerung f, Verschleißschicht f
  wear resistant liner [of extruder barrel] Innenpanzerung f
  abrasive wear Abrasionsverschleiß m
  amound of wear Verschleißgrad m
  areas subject to wear Verschleißstellen fpl
  barrel wear Zylinderverschleiß m
  causing wear verschleißverursachend
  components subject to wear Verschleißteile npl

cylinder wear Zylinderverschleiß m
damage due to wear Verschleißschäden mpl
depth of wear Verschleißtiefe f
due to wear verschleißbedingt
machine wear Maschinenverschleiß m
material wear Werkstoffabtrag m
mo(u)ld wear Werkzeugverschleiß m
producing a lot of wear verschleißintensiv
protected against wear verschleißgeschützt
protection against wear Verschleißschutz m
screw wear Schneckenabnutzung f, Schneckenverschleiß m
signs of wear Verschleißerscheinungen fpl
subject to wear verschleißbeansprucht
surface wear Oberflächenabtrag m
susceptible to wear verschleißanfällig
wear out/to verschleißen
wearing Verschleiß-, Abnutzungs-
　wearing surface [e.g. of e screw] Lauffläche f
　extremely hard wearing hochverschleißfest, hochverschleißwiderstandsfähig
　hard wearing verschleißarm
　hard wearing properties Verschleißarmut f
　screw wearing surface Schneckenlauffläche f
weather Wetter n, Witterung f
　weather resistant (be)witterungsstabil, wetterbeständig, witterungsbeständig, witterungsfest
weathered bewittert, verwittert
　naturally weathered freibewittert
weathering Bewitterung f, Verwitterung f
　weathering behavio(u)r Witterungsverhalten n
　weathering characteristics Bewitterungseigenschaften fpl
　weathering conditions Bewitterungsbedingungen fpl, Witterungsbedingungen fpl
　weathering fastness [of pigments] Wetterechtheit f
　weathering influences Witterungseinflüsse mpl
　weathering instrument Bewitterungsgerät n
　weathering period Bewitterungszeitraum m, Bewitterungszeit f, Bewitterungsdauer f
　weathering properties Bewitterungsverhalten n
　weathering resistance Bewitterungsstabilität f, Bewitterungsbeständigkeit f, Freiluftbeständigkeit f, Wetterfestigkeit f, Wetterbeständigkeit f, Witterungsbeständigkeit f, Witterungsstabilität f
　weathering results Bewitterungsergebnisse npl
　weathering station Bewitterungsstation f
　weathering test Bewitterungsprüfung f
　weathering test results Bewitterungsergebnisse npl
　weathering trial Bewitterungsversuch m
　accelerated weathering Kurzbewitterung f
　accelerated weathering instrument Kurzbewitterungsgerät n, Schnellbewitterungsgerät n
　accelerated weathering resistance Kurzzeitbewitterungsverhalten n
　accelerated weathering test Kurzzeitbewitterungsversuch m, Schnellbewitterung f
　long-term weathering resistance Dauerbewitterungsstabilität f
　natural weathering Naturbewitterung f
　natural weathering test Freibewitterungsprüfung f, Freibewitterungsversuch m, Freiluftversuch m
　outdoor weathering Frei(luft)bewitterung f, Außenlagerung f
　outdoor weathering performance Freibewitterungsverhalten n
　outdoor weathering resistance Außenbewitterungs-Beständigkeit f
　outdoor weathering station Freibewitterungsstation f, Freibewitterungsstand m, Freiluftbewitterungsanlage f, Freiluftprüfstand m
　outdoor weathering test Freibewitterungsversuch m, Freibewitterungsprüfung f
weave/to weben
weave Webart f, Bindung f
　weave pattern Webart f
　cross twill weave Kreuzköperbindung f
　mock Leno weave Scheindreherbindung f
　plain weave Leinwandbindung f
　satin weave Atlasbindung f
　true Leno weave Drehbindung f
　twill weave Köperbindung f
web Schenkel m, Steg m; endlose Bahn f, Warenbahn n, Fell n
　web depositing (unit) Bandablegeeinheit f, Bandablegung f
　web guide Bahnführung f, Bahnsteuereinrichtung f, Bandführung f, Warenbahnführung f, Warenbahnsteuerung f
　web monitor(ing device) Bahnwächter m
　web scrap Folienbahnabfall m, Stanzgitter(abfall) n(m)
　web scrap granulator Stanzgittermühle f
　web speed Bahngeschwindigkeit f, Bandgeschwindigkeit f
　web tension control (mechanism) Bahnspannungsregelung f, Zugspannungsregelung f, Zugspannungskontrolle f
　web tension control unit Zugspannungsregelung f, Bahnspannungskontrolle f
　web tension (device) Bahnzugkraft f, Bandzug m, Folienbahnspannung f, Warenbahnspannung f, Bahnspannung f, Warenspannung f
　web tension measuring device Bahnzugkraftmeßvorrichtung f
　web tension measuring unit Bahnzugkraftmeßstation f, Bandzugmeßstation f
　film web Folienbahn f
　film web guide Folienbahnführung f, Folienführung f
　in web form bahnenförmig
　laminate web Laminatband n
　paper web Papierbahn f
　sheet web Plattenbahn f
wedge Zwickel m
　wedge angle Keilwinkel m

wedge-shaped keilförmig
weekly time switch Wochenschaltuhr f
weft of textile Schuß m
   weft direction Schußrichtung f
   weft wire Schußdraht m
weigh/to wägen
   weigh feeder Dosierwaage f, Gewichtsdosierung f, Waagedosierung f
   weigh feeding Waagedosierung f, Gewichtsdosierung f
   material being weighed Wiegegut n
weigh out/to einwiegen
   amount weighed out (material or sample) Einwaage f
weigher Waage f
   check weigher Ausfallwaage f
weighing 1. Verwiegung f; 2. Wäge-
   weighing compartment Wägeraum m
   weighing container Wägebehälter m
   weighing data Wägedaten npl
   weighing equipment Verwiegeanlage f, Wägeeinrichtungen fpl
   weighing error Wägefehler m
   weighing mechanism Wägemechanismus m
   weighing platform Wägeplattform f
   weighing range Wägebereich m, Wiegebereich m
weight Gewicht n
   weight average Gewichtsmittel n
   weight-average molecular weight gewichtsmittleres Molekulargewicht n
   weight content Gewichtsanteil m
   weight feeder Gewichtsdosiereinrichtung f
   weight increase Gewichtszunahme f
   weight loss Gewichtsabnahme f, Gewichtsverlust m
   weight loss determination Gewichtsabnahmebestimmung f
   weight per square meter [weight/m^2] Quadratmetergewicht n, Gewicht n pro m^2
   weight per unit area Flächengewicht n
   weight savings Gewichtseinsparungen fpl
   weight without oil Gewicht n ohne Ölfüllung
   approximate weight Cirka-Gewicht n
   atomic weight Atomgewicht n
   average molecular weight Molekulargewicht-Mittelwert m
   change in molecular weight Molekulargewichtsänderung f
   change in weight Gewichts(ver)änderung f
   coating weight Beschichtungsgewicht n
   constant weight Gewichtskonstanz f
   depending on the molecular weight molekulargewichtsabhängig
   dropping weight Fallhammer m
   dry weight Trockengewicht n
   equivalent weight Äquivalentmasse f
   falling weight Fallbolzen m, Fallgewicht n
   falling weight test Bolzenfallversuch m, Fallbolzenversuch m
   falling weight tester Fallbolzenprüfgerät n
   high-molecular weight höhermolekular
   high-molecular weight polyethylene film HM-Folie f
   increase in weight Gewichtszunahme f
   limiting molecular weight Grenzmolekulargewicht n
   loss of weight Gewichtsabnahme f, Gewichtsverlust m
   low-molecular weight niedermolekular, niedrigmolekular
   maximum shot weight maximales Teilgewicht n
   medium-molecular weight mittelmolekular
   molecular weight molare Masse f, Molekülmasse f, Molekulargewicht n, Molekularmasse f, Molgewicht n, Molmasse f
   molecular weight distribution Molekulargewichtsverteilung f, Molmassenverteilung f
   molecular weight distribution curve Molekulargewichtsverteilungskurve f
   mo(u)ld weight Werkzeuggewicht n
   mo(u)lded-part weight Formteilgewicht n, Formteilmasse f
   number-average molecular weight Molekulargewicht-Zahlenmittel n
   original weight Ausgangsgewicht n
   part by weight [p.b.w.] Massenteile npl, Gewichtsteile npl
   part weight Formteilmasse f, Formteilgewicht n
   percent by weight Gewichtsprozent n
   price per unit weight Gewichtspreis m
   required weight Sollgewicht n
   shot weight Einspritzgewicht n, Schußgewicht n, Schußgröße f, Schußmasse f, Spritz(teil)gewicht n
   tare weight Taragewicht n
   ultra-high molecular weight ultrahochmolekular
weld/to (ver)schweißen
weld Schweißnaht f
   weld line Bindenaht f, Fließnaht f, Fließlinie f, Strömungslinie f
   weld quality Nahtgüte f
   pinch-off welds Abquetschmarkierungen fpl
   solvent weld Quellschweißnaht f
welded geschweißt, verschweißt, Schweiß-
   welded joint Schweißfuge f, Schweißverbindung f
   welded socket joint Muffenschweißverbindung f
   welded steel construction Stahlschweißkonstruktion f
   butt welded stumpfgeschweißt
   butt welded joint Stumpfschweißverbindung f
   lap welded joint Überlappschweißnaht f
   material being welded Schweißgut n
welding Schweißen n, Schweißprozeß m
   welding bars Schweißbacken fpl
   welding conditions Schweißbedingungen fpl
   welding factor Schweißfaktor m
   welding fixture Schweißaufnahme f
   welding gas supply Schweißgasversorgung f
   welding instrument Schweißgerät n
   welding jig Schweißlehre f
   welding nozzle Schweißdüse f

welding operation Schweißvorgang m
welding press Schweißpresse f
welding pressure Schweiß(preß)druck m
welding rod Schweißstab m
welding solvent Quellschweißmittel n
welding speed Schweißgeschwindigkeit f
welding techniques Schweißtechnik f
welding tool Schweißwerkzeug n
automatic heat impulse welding instrument Wärmeimpulsschweißautomat m
automatic hot gas welding unit Warmgas-Schweißautomat m
automatic profile welding unit Profilschweißautomat m
butt welding Stumpfschweißen n
distribution of residual welding stresses Schweißrestspannungsverteilung f
extrusion welding Extrusionsschweißen n
extrusion welding unit Extrusionsschweißgerät n
friction welding Reibschweißen n
friction welding instrument Reibschweißgerät n
friction welding machine Reibschweißmaschine f
hand welding instrument Handschweißgerät n
heat impulse welding Wärmeimpulsschweißen n, Wärmeimpulssiegelung f
heat impulse welding machine Wärmeimpulsschweißmaschine f
heated tool pipe welding line Heizelement-Rohrschweißanlage f
heated tool welding Heizelementschweißen n, Heizkeilschweißen n
high-frequency welding machine Hochfrequenzschweißmaschine f
high-frequency welding (process) Hochfrequenzschweißverfahren n, Hochfrequenzschweißen n
high-speed hot air welding Warmgasschnellschweißen n
high-speed welding Schnellschweißen n
hot air welding Heißluftschweißen n
hot gas welding Warmgasschweißen n
hot plate butt welding Heizelementstumpfschweißen n
hot plate welding (process) Spiegel(schweiß)verfahren n
hot wire welding Trennahtschweißen n
lap welding Überlappschweißen n
long-term welding factor Langzeitschweißfaktor m
machine butt welding Maschinenstumpfschweißung f
manual butt welding Handstumpfschweißen n
pipe welding instrument Rohrschweißvorrichtung f, Rohrschweißgerät n
pipe welding unit Rohrschweißvorrichtung f, Rohrschweißgerät n
residual welding stresses Schweißrestspannungen fpl
socket fusion welding Muffenschweißung f
solvent welding Quellschweißen n
spin welding Rotations(reib)schweißen n

spot welding Punktverschweißen n
ultrasonic spot welding Ultraschallpunktschweißen n
ultrasonic spot welding instrument Ultraschall-Punktschweißgerät n
ultrasonic welding Ultraschallfügen n, Ultraschallschweißen n
ultrasonic welding machine Ultraschallschweißanlage f, Ultraschallschweißmaschine f
vibration welding Vibrationsschweißen n
well Bohrloch n, Schacht m, Angußvorkammer f, Vorkammer f, Vorkammerbohrung f, Vorkammerraum m
hot well contents Vorkammerkegel m
thermocouple well Thermoelementbohrung f, Thermofühlerbohrung f
wet naß
wet-armature solenoid Naßankermagnet m
wet film Naßfilm m
wet film thickness Naßfilmdicke f, Naßschichtdicke f
wet film viscosity Naßfilmviskosität f
wet grinding Naßschliff m
wet-in-wet method Naß-in-Naß-Verfahren n
wet-process room Naßraum m
wet scrub resistance Naßabriebfestigkeit f
wet state Naßzustand m
wet strength agent Naßfestmittel n
wet strength (value) Naßfestigkeit f, Naßfestigkeitswert m
wettability Benetzbarkeit f
wetting Benetzen n
wetting agent Benetzungsmittel n, Netzmittel n
wetting agent concentration Netzmittelkonzentration f
wetting agent solution Netzmittellösung f
wetting characteristics Benetzungvermögen n
wetting problems Benetzungsschwierigkeiten fpl
pigment wetting power Pigmentaufnahmevermögen n
Wheatstone bridge Wheatstone-Meßbrücke f
wheel Rad n
wheel arch Radhaus n, Radkasten m
abrasive wheel Schleifkörper m
buffing wheel Schwabbelscheibe f
car wheel rim Felge f, Radfelge f
driving wheel Antriebsrad n
gear wheel Zahnrad n
gear wheel blank Zahnradrohling m
grinding wheel Schleifscheibe f
grinding wheel resin Schleifscheibenharz n
lambswool polishing wheel Lammfellscheibe f
polishing wheel Polierscheibe f
spare wheel recess Reserveradmulde f
sprocket wheel Zahnkettenrad n
steering wheel Lenkrad n
white weiß
white goods Weißgeräte npl
white lead Bleiweiß n
white pigment Weißpigment n

white spirit Testbenzin n
titanium white Titanweiß n
zinc white Zinkweiß n
whiteness Weiß(heits)grad m
whiting Schlämmkreide f
whitish weißlich
wide weit, breit
  wide-mesh weitmaschig
  wide-ranging breitgefächert
  wide sheet extrusion line Platten-Großanlage f
widespread breitgefächert
width Breite f, Weite f
  width control (mechanism) Breitenregelung f, Breitensteuerung f, Breitenregulierung f
  width of film after trimming Folienfertigbreite f
  width variations Breitenschwankungen fpl
  channel width Gangbreite f
  constant width Breitenkonstanz f
  die gap width Düsenspaltbreite f, Düsenspaltweite f
  die width Düsenbreite f, Werkzeugbreite f
  effective width [e.g. of a sheet extrusion die] Verarbeitungsbreite f, Arbeitsbreite f
  film width control (mechanism) Folienbreitenregelung f
  flight land width Stegbreite f
  layflat width flachgelegte Breite f, Flachlegebreite f, Flachliegebreite f, flachgelegte Folienbreite f, Schlauchliegebreite f
  mesh width Maschenweite f
  nip width Walzenabstand m, Walzenspaltweite f, Spaltbreite f, Spaltweite f
  nominal width Nennweite f
  roll face width Ballenbreite f, Ballenlänge f, Walzenballenlänge f, Walzenballenbreite f
  roll width Walzenbreite f
Winchester disk Winchesterplatte f
wind-up (unit) Aufwickeleinrichtung f, Aufrollung f, Aufwicklung f, Aufwickelanlage f, Wickelanlage f, Wickeleinheit f, Wickelmaschine f, Wickler m, Wickelwerk n
  wind-up design Wickelkonzeption f
  wind-up drum Wickeltrommel f
  wind-up shaft Wickelwelle f
  wind-up station Aufwickelstelle f, Wickelstation f, Wickelstelle f
  combined take-off and wind-up unit Abzugswerk-Wickler-Kombination f, Abzugswicklerkombination f
  drum wind-up unit Trommelwickler m
  edge trim wind-up (unit) Randstreifenaufwicklung f
  film wind-up (unit) Folienaufwicklung f, Folienaufwickler m
  range of wind-up equipment Aufwickelmaschinenbaureihe f
  reverse wind-up (unit) Wendewickler m
  trim wind-up (unit) Abfallaufwicklung f
  twin-station wind-up (unit) Zweistellen-Aufwicklung f

  twin wind-up Doppelaufwickler m
  type of wind-up Wickler-Typ m
winder Wickelanlage f, Wickeleinheit f, Wickelmaschine f, Wickelwerk n, Wickler m
  automatic winder Wickelautomat m
  center-drive winder Direktwickler m, Achswickler m
  central winder Zentrumswickler m
  center winder Zentralwickler m
  drum winder Trommelwickler m
  film winder Folien(auf)wickler m, Folienaufwicklung f, Folienwickelmaschine f
  heavy-ga(u)ge winder Schwerfolienwickler m
  range of (film) winders Aufwickelmaschinenbaureihe f
  reel winder Rollenwickelmaschine f
  single-station winder Einstellen-Aufwicklung f
  surface-drive winder Oberflächenwickler m
  tangential-drive winder Tangentialwickler m
  twin winder Doppel(auf)wickler m
winding Wickeln n, Wickelverfahren n
  winding angle Wickelwinkel m
  winding mandrel Bobine f, Wickelkern m, Wickelkörper m
  winding speed Wickelgeschwindigkeit f
  winding tension Wickelspannung f, Wickelzug m
  automatic winding system Wickelautomatik f
  filament winding Faserwickelverfahren n
  filament winding resin Wickelharz n
  filament winding robot Wickelroboter m
  film winding mechanism Folienwickelsystem n
  film winding system Folienwickelsystem f
  GRP filament winding machine GFK-Wickelmaschine f
  pipe winding plant Rohrwickelanlage f
  precision filament winding Präzisionswickeltechnik f
  rovings for filament winding Wickelrovings pl
window Fenster n
  window frame Fensterrahmen m
  window frame assembly Rahmenkonfektionierung f
  window profile Fensterprofil n
  window profile compound Fensterformulierung f
  window sill Fensterbank f
  window sill profile Fensterbankprofil n
  casement window Flügelfenster n
  heavy-duty window profile Fensterhochleistungsprofil n
  rear window Heckscheibe f
  roof window Dachfenster n
  side window Seitenscheibe f
windscreen Frontscheibe f
  windscreen wash bottle Scheibenwaschbehälter m
wire Draht m
  wire brush Drahtbürste f
  wire cloth Drahtgewebe n
  wire covering compound Adermischung f
  wire covering die Drahtummantelungskopf m, Drahtummantelungs-Düsenkopf m

wire covering line Drahtummantelungsanlage f, Drahtisolierlinie f
wire covering torpedo Drahtummantelungspinole f
wire enamel Drahtlack m
wire gauze Drahtgewebe n
wire gauze filter Drahtgewebefilter n,m
wire gauze screen Drahtgewebefilter n,m
wire insulation Drahtisolierung f
wire mesh screen Drahtsiebboden m
copper wire mesh Kupfergeflecht n
hot wire welding Trennahtschweißen n
incandescent wire cutter Glühbandabschneider m, Glühdrahtschneiden n
incandescent wire test Glühdrahtprüfung f
stainless steel wire mesh Edelstahlmaschengewebe n
steel wire brush Stahldrahtbürste f
thermocouple wire Thermopaar n
weft wire Schußdraht m
wired verdrahtet
wiring Verdrahtung f, Verkabelung f
WLF curve [the initials stand for the names of the three people who first proposed the equation - Williams, Landel and Ferry] WLF-Kurve f
WLF equation WLF-Gleichung f
WLF figure WLF-Wert m
WLF value WLF-Wert m
Wöhler curve Wöhlerlinie f
wollastonite Wollastonit n
wood Holz n
  wood-based material Holzwerkstoff m
  wood flour Holzmehl n
  wood lacquer Harzfirnis m, Holzschutzlasur f
  wood oil Holzöl n
  wood oil alkyd (resin) Holzölalkydharz n
  compressed wood Preßholz n
  densified wood Preßholz n
  with a woodgrain finish holzgemasert
woodworking tools Holzbearbeitungswerkzeuge npl
wool Wolle f
  steel wool Schleifwolle f, Schmirgelwolle f, Stahlwatte f, Stahlwolle f
word Wort n
  word length Wortlänge f
  word-of-mouth publicity Mund-zu-Mund-Propaganda f
  word-of-mouth recommendation Mund-zu-Mund-Propaganda f
  word processor Wortprozessor m
work Arbeit(en) f(pl) [s.a. works]
  work place Arbeitsplatz m
  work planning Arbeitsvorbereitung f
  development work Entwicklungsarbeit(en) f(pl)
  hot worked steel Warmarbeitsstahl m
  research work Forschungsarbeit(en) f(pl)
work up/to aufarbeiten
worker Arbeiter m
  blue-collar worker gewerbliche Mitarbeiter mpl

working arbeitend, Arbeits-
  working model Arbeitsmodell n
  working-over Überarbeitung f
  working pressure Betriebsdruck m
  working speed Arbeitsgeschwindigkeit f
  working storage Arbeitsspeicher m
  working technique Arbeitstechnik f
  working temperature Arbeitstemperatur f, Betriebstemperatur f, Gebrauchstemperatur f
  continuous working temperature Dauergebrauchstemperatur f
  in working order betriebsfähig
  maximum working temperature Einsatzgrenztemperatur f
  safe working stress [of pipes] zulässige Wandbeanspruchung f, Sigmawert m, zulässige Beanspruchung f
workpiece Werkstück n
  automatic workpiece robot Werkstück-Handhabungsautomat m
workroom Arbeitsraum m
works Fertigungsbetrieb m
  works control Betriebskontrolle f
  works safety guidelines Arbeitsschutz-Richtlinien fpl
workshop Werkstätte f, Arbeitsraum m
  workshop hygiene Arbeitshygiene f
world Welt f
  world consumption Weltverbrauch m
  world turnover Weltumsatz m
  world-wide weltweit
  annual world capacity Weltjahreskapazität f
  annual world production Weltjahresproduktion f
worm gear Schneckengetriebe n, Schneckenvorgelege n
worn verschlissen, abgenutzt
  worn screw Schrottschnecke f
  tendency to become worn Verschleißneigung f
worse schlechter
  made worse verschlechtert
wound gewickelt
  filament wound pipe Wickelrohr n
  filament wound section Wickelschuß m
  material being wound up Wickelgut n
woven gewebt
  woven fabric Gewebe n
  woven glass tape Glasgewebeband n
  woven roving Glasseidenrovinggewebe n, Rovinggewebe n
  woven roving reinforcement Rovinggewebeverstärkung f
  woven staple fiber Glasstapelfasergewebe n
wrap/to wickeln, einhüllen
wrapped umwickelt
  shrink wrapped pack Schrumpfverpackung f
  stretch wrapped pack Streckfolien-Verpackung f, Streckverpackung f
wrapping Umhüllung f, Verpackung f
  wrapping film Einschlagfolie f

shrink wrapping Schrumpfverpackung f
shrink wrapping film Schrumpffolie f
stretch wrapping Streckverpackung f, Streckfolien-Verpackung f
stretch wrapping film Dehnfolie f, Streckfolie f
wrench Schraubenschlüssel m
  torque wrench Drehmomentschlüssel m
wrinkle-free faltenlos, faltenfrei, knitterfrei
write-read memory Schreib-Lese-Speicher m
write head Schreibkkopf m
wrong falsch
  wrong conclusions Fehlschlüsse mpl
  wrong settings Fehleinstellungen fpl
  liable to go wrong störanfällig, störempfindlich
wrongly designed falsch ausgelegt
X-ray Röntgenstrahl m, Röntgen-
  X-ray analysis Röntgenanalyse f
  X-ray diffraction photograph Röntgenbeugungsbild n
  X-ray photograph Röntgenaufnahme f
  X-ray spectroscopy Röntgenspektroskopie f
  impermeable to X-rays röntgenstrahlundurchlässig
  permeable to X-rays röntgenstrahldurchlässig
xenon Xenon n
  xenon arc Xenonbogen m
  xenon lamp Xenonbogenstrahler m
xylene Xylol n
xylenol Xylenol n
  xylenol resol Xylenolresol n
yarn Garn n
  film yarn Bändchen n, Folienbändchen n, Folienband n
year Jahr n
  financial year Geschäftsjahr n
  in the early years in den Anfangsjahren npl
  previous year Vorjahr n
  surplus for the year Jahresüberschuß m
  time of the year Jahreszeit f
yearly average Jahresdurchschnitt m
yellow/to vergilben
yellow gelb
  yellow tinge Gelbstich m
  become yellow/to vergilben
  cadmium yellow Cadmiumgelb n
  chrome yellow Chromgelb n
  nickel titanium yellow Nickeltitangelb n
  zinc yellow Zinkgelb n
yellowing Gelbtönung f, Gilbung f, Vergilbung f
  resistance to yellowing Gilbungsresistenz f
  resistant to yellowing gilbungsstabil, vergilbungsbeständig
yellowish gelblich
yellowness index Vergilbungszahl f, Vergilbungsgrad m, YI-Wert m
yield Ausbeute f, Ergiebigkeit f
  yield point Fließgrenze f, Streckgrenze f
  yield stress Fließspannung f, Reißspannung f, Streckspannung f
  elongation at yield Dehnung f bei Streckgrenze, Zugdehnung f bei Streckgrenze, Zugspannung f bei Streckgrenze
  elongation at yield stress Dehnung f bei Streckspannung f
  gas yield Gasausbeute f
  offset yield stress Dehnspannung f
  radical yield Radikalausbeute f
  tensile stress at yield Fließspannung f, Streckspannung f
zero Null f, Nullpunkt m
  zero adjustment Nullabgleich m, Nullpunkteinstellung f
  zero correction Nullpunktkorrektur f
  zero point monitor Nullpunktüberwachung f
  zero position Nullstellung f
  zero-potential potentialfrei
  zero setting Nullabgleichung f, Nullpunkteinstellung f
  zero shift Nullpunktverschiebung f
  zero value Nullwert m
Ziegler catalyst Ziegler-Katalysator m
zinc Zink n
  zinc borate Zinkborat n
  zinc casting alloy Zink-Gießlegierung f
  zinc chloride Zinkchlorid n
  zinc chromate Zinkchromat n
  zinc chromate primer Zinkchromatgrundierung f
  zinc dust Zinkstaub m
  zinc oxide Zinkoxid n
  zinc oxychloride Zinkoxychlorid n
  zinc phosphate-iron oxide primer Zinkphosphat-Eisenoxid-Grundierung f
  zinc-rich paint Zinkstaubformulierung f, Zinkstaubfarbe f
  zinc-rich primer Zinkstaubgrundierung f, Zinkstaub-Primer n
  zinc soap Zinkseife f
  zinc sulphide Zinksulfid n
  zinc white Zinkweiß n
  zinc yellow Zinkgelb n
zirconium Zirkonium n
  zirconium acetylacetonate Zirkonacetylacetonat n
  zirconium silicate Zirkonsilikat n
zone Zone f, Gebiet n
  barrel heating zones Zylinderheizzonen fpl
  boundary zone Grenzzone f
  bubble expansion zone Aufweitungszone f, Schlauchbildungszone f
  constant temperature zone Temperierzone f
  contact zone Berührungszone f
  cooling zone Kühlzone f, Temperierzone f
  craze zone Craze-Zone f, Crazefeld n, Fließzone f, Trübungszone f
  craze zone structure Fließzonenstruktur f
  crumple zone [in a car] Knautschzone f, Verformungsbereich m
  glass transition zone Glasübergangsbereich m, Glasübergangsgebiet n, Glasumwandlungsbereich m

heating zone Heizzone f, Temperierzone f
homogenizing zone Ausgleichszone f
intermeshing zone [between twin screws] Zwickel(bereich) m, Eingriffsbereich m
isolated glassy zones Glasinseln fpl
long-compression zone screw Langkompressionsschnecke f
outer fiber zone Randfaserbereich m
outer zone Randzone f
relaxation zone [part of screw] Beruhigungszone f
restricted flow zone Dammzone f, Drosselstelle f, Drosselfeld n, Stauzone f
short-compression zone screw Kurzkompressionsschnecke f
stagnation zone Ruhezone f, Stagnationsstelle f, Stagnationszone f
stress-whitened zone Weißbruchzone f
transition zone [of a curve] Übergangszone f
vent zone Vakuumzone f, Zylinderentgasungszone f, Entgasungszone f, Entspannungszone f, Entgasungsbereich m, Dekompressionszone f